HANDBOOK OF
SURFACE
AND
INTERFACE
ANALYSIS

Methods for
Problem-Solving

SECOND EDITION

HANDBOOK OF
SURFACE
AND
INTERFACE
ANALYSIS

Methods for
Problem-Solving

SECOND EDITION

Edited by

John C. Rivière
Sverre Myhra

CRC Press
Taylor & Francis Group
Boca Raton   London   New York

CRC Press is an imprint of the
Taylor & Francis Group, an **informa** business

CRC Press
Taylor & Francis Group
6000 Broken Sound Parkway NW, Suite 300
Boca Raton, FL 33487-2742

First issued in paperback 2017

© 2009 by Taylor & Francis Group, LLC
CRC Press is an imprint of Taylor & Francis Group, an Informa business

No claim to original U.S. Government works

ISBN 13: 978-1-138-11363-3 (pbk)
ISBN 13: 978-0-8493-7558-3 (hbk)

Library of Congress Cataloging-in-Publication Data

Handbook of surface and interface analysis methods for problem-solving / editors, John C. Rivière, Sverre Myhra. -- 2nd ed.
    p. cm.
Includes bibliographical references and index.
ISBN 978-0-8493-7558-3 (hardcover : alk. paper)
    1. Surfaces (Physics)--Analysis. 2. Interfaces (Physical sciences)--Analysis. 3. Surface chemistry. 4. Surfaces (Technology)--Analysis. I. Rivière, J. C. II. Myhra, S. (Sverre), 1943- III. Title.

QC173.4.S94H35 2009
620'.44--dc22
                                          2009005505

Visit the Taylor & Francis Web site at
http://www.taylorandfrancis.com

and the CRC Press Web site at
http://www.crcpress.com

Contents

Preface

It is with some trepidation that one approaches the task of producing the second edition of a reference monograph. Many questions come to mind. Did the first edition find a sufficiently large audience and did it meet its objective? Can we improve on the first edition, and has the field moved along to the extent that a second edition, 10 years later, can offer a new and fresh perspective on the field, and new and better insights into the state of the art? Last but not least, can we and our coauthors once more gird our loins and find the time and energy in increasingly busy lives to produce effective and readable accounts of our lives' work? As is often the case, one can answer such questions fully only in hindsight, but we believe nevertheless that there are enough positive indicators to make the project worthwhile.

The first edition was based on our belief that the characterization and analysis of surfaces and interfaces should be done in the context of problem solving rather than being based on the capabilities of one or more individual techniques. If anything, that belief is now held by us even more strongly, and, as it has turned out, recent trends in science and technology appear to have vindicated it. Major instrumental assets are now generally funded and maintained as central facilities that can be accessed as and when potential users make informed decisions that one or more facilities can offer tools appropriate to their analytical problem(s). Industry is increasingly making the assessment that it is more cost effective to contract out analytical services than to maintain in-house facilities of sufficient breadth and expertise. Those are the trends that have motivated this book, and it therefore focuses principally on development of the strategic thinking that should be undertaken by those who decide which facilities to access, and where to subcontract analytical work. While good strategy can win a war, actual battles are fought tactically. Thus this book also attempts to cover most of the major tactical issues that are relevant at the location where the data are being produced.

In this second edition, we attempt to broaden the thinking about the techniques and methods that can be brought to bear on surfaces and interfaces. Thus, there are new chapters that deal with electron-optical imaging techniques and associated analytical methods. Likewise, there is also now an introductory chapter on techniques based on synchrotron sources. Ten years (since the first edition) is quite a long time in science and technology. Even in the traditional surface analytical techniques, there have been significant developments. For instance, most new XPS instruments now have an imaging capability as a matter of course, and parallel multichannel counting is standard. Among the more recent arrivals to the family of techniques the scanning probe group has carved out an even greater role by virtue of convenience and versatility; indeed one of the suppliers has, as of mid-2006, just delivered its 7000th instrument. At the forefront of electron microscopy, high resolution analytical instruments are widely available, while aberration correction is now emerging as the next major advance.

Arguably, the most significant trend in materials science during the last few years has been the upsurge of nanotechnology and nanoscience. In spite of the early hype, and the re-branding of some good, and some not-so-good, "traditional" science and technology, there is now real progress and substance on the meso- and nanoscales (we use "meso" to describe the gray and ill-defined range between "micro" and "nano"). The task of interrogating systems on those scales is presenting considerable challenges to the practitioners of characterization and analysis. Traditional techniques and methods are being pressed into service on a size scale for which they were not intended, and new techniques are being invented, while the practitioners are yet again having to climb new learning curves. The second edition attempts to provide new, relevant material in response to that trend.

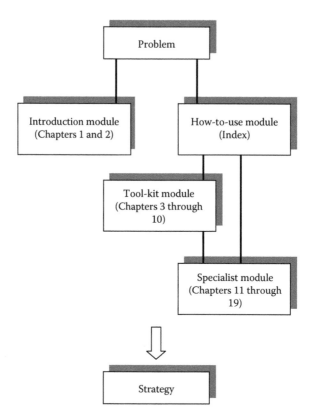

FIGURE 1   Schematic overview of the structure of the book.

The structure of the second edition remains the same as for the first (see Figure 1 for a schematic overview). There are two short introductory chapters dealing with problem solving as a strategic issue, and with guidance on how to locate detailed information on techniques, methods, and materials. Then there are a number of chapters providing the essential physical basis and common modes of operation for groups of techniques. The last half of the volume is concerned principally with exploring the tactical issues for surface and interface characterization and analysis of particular types of materials, or for particular applications of materials.

It goes without saying that this book could not have been written without the time and effort that so many contributors have had to find out of their already overcommitted professional lives. Each chapter represents a distillation of the expertise and experience gained by the contributors as a result of their devotion to one or more aspects of surface science and technology. We would like to set on record our appreciation of that time and effort, and to acknowledge the good-humored acceptance of our (sometimes substantial) recasting and rearranging of contributors' material.

Although success and the accumulation of expertise in science can be a purely individual achievement, for most of us it is largely a cooperative social phenomenon. For that reason we are indebted to all those coworkers, too numerous to mention, who have passed through our research groups over the years and participated in the two-way information transfer process that has added inestimably to our own expertise. We hope that they have found it equally rewarding.

We have both had long-standing connections with what used to be called AERE Harwell, but is now known as the Harwell Laboratory, AEA Technology plc, and we wish to acknowledge that connection. Our special thanks, also, to those individuals at the Harwell Laboratory and elsewhere who have been of particular assistance to us over many years. More recently we have benefited from attachments to the Department of Materials at the University of Oxford, where organizational and collegial support has made it possible to produce the second edition.

Lastly, we would like to express our gratitude to our immediate families. Engagement in a scholarly endeavor may be a source of satisfaction to those directly involved, but tends instead to be an ordeal to those having to live alongside the resultant upheaval.

John C. Rivière
Sverre Myhra
Oxford

Editors

John C. Rivière worked for AEA Technology, Harwell, England, until his retirement. He is the author or coauthor of numerous scholarly papers and chapters that reflect his research interests in both basic and applied surface science and technology. A pioneer in the early development of electron spectroscopies, he was the recipient of the UK Vacuum Council Medal (1989). He received his MSc (1950) from the University of Western Australia, and his PhD (1955) and DSc (1995) from the University of Bristol.

Sverre Myhra is currently a visiting scientist in the Department of Materials at the University of Oxford. Previously, he founded and headed an early scanning probe microscopy group in Australia. He is the author of numerous scholarly papers and chapters in the area of materials science and technology focusing on surface and interface analysis, with an emphasis on recent applications of scanning probe microscopy. He received his PhD (1968) in physics from the University of Utah, Salt Lake City.

Contributors

James E. Castle
School of Engineering
Faculty of Engineering and Physical Science
University of Surrey
Guildford, United Kingdom

David J. Cookson
Science Operations
Australian Synchrotron
Clayton, Victoria, Australia

Alison Crossley
BegbrokeNano
Department of Materials
University of Oxford
Oxford, United Kingdom

Christophe Donnet
University Institute of France
Paris, France

and

Laboratoire Hubert Curien
University Jean Monnet
Saint-Etienne, France

Lara J. Gamble
NESAC-BIO
Departments of Bioengineering
  and Chemical Engineering
University of Washington
Seattle, Washington

Andrea R. Gerson
Applied Centre for Structural
  and Synchrotron Studies
University of South Australia
Mawson Lakes, South Australia, Australia

Daniel Graham
Asemblon, Inc.
Redmond, Washington

Hans J. Griesser
Ian Wark Research Institute
University of South Australia
Mawson Lakes, South Australia, Australia

Thomas Gross
Bundesanstalt für Materialforschung
  und -prüfung
Surface and Thin Film Analysis Working
  Group
Berlin, Germany

Birgit Hagenhoff
Materials Characterization GmbH
Münster, Germany

Gar B. Hoflund
Department of Chemical Engineering
University of Florida
Gainesville, Florida

Marek Jasieniak
Ian Wark Research Institute
University of South Australia
Mawson Lakes, South Australia, Australia

Colin Johnston
BegbrokeNano
Department of Materials
University of Oxford
Oxford, United Kingdom

Reinhard Kersting
Materials Characterization GmbH
Münster, Germany

Peter Kingshott
iNANO
Faculty of Science
University of Aarhus
Aarhus, Denmark

Jean-Michel Martin
University Institute of France
Paris, France

and

Laboratoire de Tribologie et Dynamique des
 Systèmes
École Centrale de Lyon
Écully, France

Sverre Myhra
Department of Materials
University of Oxford
Oxford, United Kingdom

Kevin C. Prince
Sincrotrone Trieste
Trieste, Italy

Derk Rading
Mass Spectrometers GmbH
Münster, Germany

François Reniers
Analytical and Interfacial Chemistry
 Department
Université Libre de Bruxelles
Brussels, Belgium

John C. Rivière
Department of Materials
University of Oxford
Oxford, United Kingdom

Peter M. A. Sherwood
Department of Physics
Oklahoma State University
Stillwater, Oklahoma

Roger St. C. Smart
Applied Centre for Structural
 and Synchrotron Studies
University of South Australia
Mawson Lakes, South Australia,
 Australia

Craig R. Tewell
Sandia National Laboratories
Livermore, California

John M. Titchmarsh
Department of Materials
University of Oxford
Oxford, United Kingdom

Sven Tougaard
Department of Physics and Chemistry
University of Southern Denmark
Odense, Denmark

Wolfgang E. S. Unger
Bundesanstalt für Materialforschung
 und Prüfung
Surface and Thin Film Analysis
 Working Group
Berlin, Germany

John F. Watts
Surrey Materials Institute and Faculty
 of Engineering and Physical Sciences
University of Surrey
Guildford, United Kingdom

Matthias Werner
Nano & Micro Technology Consulting
Berlin, Germany

R. K. Wild
University of Bristol
Bristol, United Kingdom

Vaneica Y. Young
Department of Chemistry
University of Florida
Gainesville, Florida

Zhaoming Zhang
Institute of Materials Engineering
Australian Nuclear Science
 and Technology Organisation
Lucas Heights, New South Wales, Australia

Authors

James E. Castle received his PhD in 1961 for work on the oxidation resistance of the Magnox alloys used in the nuclear power industry. Subsequently, he worked on problems of metal oxidation and corrosion in the power industry before joining the University of Surrey in 1970 where he founded one of the United Kingdom's first surface analysis laboratories, dedicated to understanding the role of surface reactivity in the protection of metals against corrosion, particularly in the industrial context. He is currently the emeritus professor of materials science. He has published more than 200 papers, many dealing with aspects of electron spectroscopy applied to corrosion science and engineering.

David J. Cookson received his PhD in physics in 1989 from Monash University (Australia). After three years in industry, he joined the Australian Synchrotron Research Program in 1992 and has worked as an expatriate scientist/manager in Japan and the United States. He is now the science and beamline development manager at the Australian Synchrotron in Melbourne. His major interests are nanoscale characterization of condensed matter with elastic x-ray scattering, and the investigation of entropy and enthalpy in liquid/solid systems.

Alison Crossley received her PhD in surface science from the University of Liverpool. Since then she has managed the Materials Characterisation Service for AEA Technology plc. In 2002, she was appointed senior research fellow in the Department of Materials, University of Oxford, where she is managing BegbrokeNano—Oxford Materials Characterisation Services—as a commercial interface with industry for other groups within the Department of Materials. She has published more than 55 papers and is a chartered chemist, a member of the Royal Society of Chemistry (RSC), committee member for the RSC Solid Surfaces Group, a fellow of the Royal Microscopical Society and honorary chair of the EM group, and a member of the Institute of Physics Materials and Characterisation Group Committee.

Christophe Donnet received his PhD in analytical chemistry from the University of Lyon, France, in 1990. He has been an associate professor in the French engineering school École Centrale de Lyon up to 2000 and is now a full professor of material science and chemistry at University Jean Monnet, Saint-Etienne, France. He is also member of the University Institute of France. His principal research interests include thin-film deposition (pulsed laser deposition) and characterization, including tribological properties. Recent research has focused on superlow friction of MoS_2 films and diamond-like carbon (DLC), in collaboration with groups at the IBM Research Division (Yorktown Heights, United States) and Argonne National Laboratories (United States). He has recently edited a book dedicated to the tribology of DLC films (Springer, 2008).

Lara J. Gamble is the assistant director of the National ESCA and Surface Analysis Center for Biomedical Problems (NESAC/BIO) at the University of Washington (an NIH-NIBIB-funded biomedical technology resource center). She is also a member of the faculty in the Department of Bioengineering at the University of Washington. Her major interests are in the XPS and ToF-SIMS of biological materials. She received her PhD in physical chemistry from the University of Washington in 1996.

Andrea R. Gerson is the director of the Applied Centre for Structural and Synchrotron Studies (ACeSSS) at the University of South Australia. Her first degrees were from the University of Canterbury (New Zealand) followed by a PhD (1992) from the University of Strathclyde (Scotland).

Andrea completed a postdoctoral fellowship at Kings' College, London University. ACeSSS specializes in the development of advanced techniques for applied applications. Andrea's areas of interest focus on the role of surface and bulk structure in determining reactivity and physical properties.

Daniel Graham received his BS in chemical engineering from Brigham Young University and his PhD in bioengineering from the University of Washington. His PhD thesis work focused on the interpretation of TOF-SIMS spectral fragmentation patterns using multivariate analysis techniques. He has over 10 years of experience in surface modification and characterization. He has worked with a wide range of materials using various analysis techniques including XPS, ToF-SIMS, FTIR, and contact angle. He is currently a founder of, and principal scientist in charge of research and development at, Asemblon Inc in Redmond, Washington.

Hans J. Griesser is professor of surface science at the University of South Australia and deputy director of its Ian Wark Research Institute. He received his PhD in physical chemistry from the Swiss Federal Institute of Technology (ETH) Zürich, Switzerland in 1979. He has been involved in the fundamentals and applications of surface science while employed in industry, in a government laboratory, and in academia. His current research interests are in the areas of surface modification, low-temperature plasma methods, and surface analysis, particularly for bio-interfaces such as protein-resistant and antibacterial coatings.

Thomas Gross received a diploma in 1974 and a Dr. rer. nat. in 1979, both in chemistry, from the Humboldt University, Berlin, Germany. He was a research associate at the Institute of Physical Chemistry of the Academy of Sciences in Berlin, from 1979 to 1991. He then joined the Federal Institute for Materials Research and Testing (BAM), Berlin, in 1992, where he is a scientist in the Surface and Thin Film Analysis Working Group. He is domain expert of the DACH for XPS and the DAP for XPS and ToF-SIMS and was the vice chairman of ECASIA '03. His major interests are electron spectroscopy, secondary ion mass spectrometry, adsorption science applied to catalysis by zeolites, as well as being committed to advances in standardization of surface chemical analysis techniques.

Birgit Hagenhoff received her PhD in 1993 from the University of Münster in Germany. Since then she has worked for the Centre for Manufacturing Technology at Philips in the Netherlands before returning to Münster where she is now the president and CEO of Tascon Gmbh. Her principal research interests are in the area of secondary ion mass spectrometry (SIMS), where she has worked closely with the pioneering group of Benninghoven. She is now fully engaged with applications of SIMS, and with problem solving for industry.

Gar B. Hoflund received his PhD in chemical engineering from the University of California at Berkeley in 1978. He has been a professor of chemical engineering at the University of Florida in Gainesville since 1977. His major interests involve chemical reactions at solid surfaces with applications relating to heterogeneous catalysis and nanotechnology, semiconductor materials and processing, electrochemistry, polymer erosion by oxygen atoms, and UV radiation and tribology. He has contributed to the development of instrumentation, such as atom sources, and methodologies for surface and interface analysis.

Marek Jasieniak has an MS in chemistry and chemical technology from Cracow University of Technology, Poland. Over a number of years at tertiary institutions, he has used extensively a number of instrumental techniques with a major focus on XPS and ToF-SIMS. He is now a staff member of the Ian Wark Research Institute at the University of South Australia. His current interest is in the application of surface analysis to the characterization, development, and performance evaluation of biomedical devices.

Colin Johnston received his PhD in surface science and catalysis in 1987 from the University of Dundee. Since then he has worked at AEA Technology plc in various areas including materials characterization and development of electronic materials, and within the corporate structure. Currently, he is a senior research fellow in the Department of Materials at Oxford University where he is the director of Faraday Advance—the transport node of the materials knowledge transfer network—and coordinator of the Institute of Industrial Materials and Manufacturing, part of the Department of Materials. He is director of HITEN—the network for high-temperature electronics—where he has established a pan-European strategy for high-temperature electronics, and the co-chair of the US High Temperature Electronics biennial conference series. He has published over 80 scientific papers and has edited several books on high-temperature electronics.

Reinhard Kersting received his PhD in 2003 from the University of Münster in Germany. His principal research interests are in applications of Tof-SIMS, with particular focus on organic materials. Since 1997 he has worked with Tascon GmbH, where he does analytical consulting and has responsibilities for customer support and business development.

Peter Kingshott is an associate professor at the Interdisciplinary Nanoscience Centre at the University of Aarhus in Denmark. He has a BSc with honors in chemistry from Murdoch University in Perth, Australia, and a PhD from the University of New South Wales, Sydney. His PhD focused on developing new biomaterial surfaces and on understanding and developing new methods for investigating protein adsorption phenomena. After completing postdoctoral appointments at CSIRO in Melbourne, NESAC/BIO at the University of Washington, and RWTH Aachen, Germany, he became a senior scientist at the Danish Polymer Centre, Risø National Laboratory, where his work has focused on surface functionalization and advanced surface analysis of materials, including medical materials, composites, and materials used in food processing.

Jean-Michel Martin is a graduate in chemistry (1972) and obtained his PhD from the University of Lyon in material science and engineering in 1978. He is presently a full professor at École Centrale de Lyon, France, and a member of the University Institute of France. He has 30 years of extensive experience in fundamental and applied research in tribology of thin films, diamond-like coatings, boundary lubrication, antiwear and extreme-pressure additives, and surface analysis. Along with his colleagues, Christophe Donnet and Thierrry Le Mogne, he discovered experimentally the superlubricity of pure MoS_2 in ultrahigh vacuum. He is currently implementing new additive formulations for metal forming and engine applications, and developing new analytical techniques for tribofilm analyses (XPS/AES/ToF-SIMS, XANES, etc). He has authored or coauthored 130 articles in peer-reviewed journals, 8 patents, 4 book chapters, 195 contributions in conference proceedings, and about 100 oral presentations, including 30 invited talks.

Kevin C. Prince is a research group leader at the Italian National Synchrotron Elettra in Trieste. He received his first degrees from Melbourne University and his DPhil from the Department of Metallurgy and Materials Science (now Department of Materials), University of Oxford, in 1978. He has held posts at Flinders University, the Fritz Haber Institute (Berlin), and the KFA Jülich in Germany before moving to Italy. His interests are in the application of synchrotron work to the understanding of the electronic structure of surfaces, molecular adsorbates, and free molecules.

Derk Rading received his PhD in 1997 from the University of Münster in Germany. His principal research interests are with interactions of energetic particles with surfaces. Since 1998 he has worked with ION-TOF GmbH, where he has responsibilities for analytical customer support, training, and instrument demonstration.

François Reniers received his PhD in chemistry from the Université Libre de Bruxelles (ULB), Belgium, in 1991. He has been professor at ULB since 1999. His main research interests are the areas of modification of surfaces using plasma techniques, and the application of surface analysis techniques (AES and XPS) for the characterization of surfaces, coatings, and buried interfaces. He gave advanced courses on surface characterization at the Chinese University in Hong Kong in 2000 (Croucher Foundation), and was the chairman of the European Conference on the Applications of Surface and Interface Analysis (ECASIA'07), held in Brussels in September 2007. He is currently officer of the IUVSTA, and vice-rector for research at the ULB.

Peter M. A. Sherwood received his PhD in chemistry from Cambridge University in the United Kingdom in 1970. He subsequently worked at Cambridge and at the University of Newcastle-upon-Tyne in the United Kingdom and Kansas State University in the United States before joining the faculty at Oklahoma State University in 2004 where he is currently dean of the College of Arts and Sciences and a Regents professor of physics. He has headed an active and productive research program concerned with inorganic solids and surfaces with particular emphasis on electrode surfaces, corrosion, and carbon fiber surfaces. Many of the projects have been undertaken in response to problems of direct relevance to industry and emerging technologies. He has served with the U.S. National Science Foundation and held several visiting appointments. His contributions were acknowledged by the award of a Regents professorship at Oklahoma State University in 2007, and a university distinguished professorship at Kansas State University in 1997 (where he currently has emeritus status), and the conferring of ScD by the University of Cambridge in the United Kingdom in 1995.

Roger St. C. Smart received his PhD in surface chemistry from the University of East Anglia (United Kingdom) in 1967 and has since worked at Flinders University (Adelaide), University of Papua New Guinea, and Griffith University (Brisbane). At the University of South Australia, he founded the SA Surface Technology Centre in 1987 and was deputy director of the Ian Wark Research Institute from its inception to 2002. Since August 2004, he has been deputy director of a new Applied Centre for Structural and Synchrotron Studies (ACeSSS) at University of South Australia (UniSA). His research interests over 35 years have encompassed surface chemistry and reactions of oxides, sulfides, minerals, ceramics, glasses, polymers, salts, and metals; surface spectroscopy; materials and biomaterials science; leaching and dissolution; nuclear waste disposal; solid-state reactivity; and adsorption in soil mineral systems. In recent years, he has been closely involved with industry in the areas of surface chemistry and modification of mineral and material surfaces in processing.

Craig R. Tewell obtained a PhD in physical chemistry from the University of California, Berkeley in 2002. He is a staff member at Sandia National Laboratories (SNL) in Livermore, California. Prior to joining SNL in 2004, he was a staff member at Los Alamos National Laboratories in New Mexico for two years. His main research interest is understanding helium bubble nucleation and growth in metal tritides.

John Titchmarsh received his PhD from the Department of Materials at Oxford University, where he subsequently carried out postdoctoral work on the development of field-emission scanning transmission electron microscopy and the characterization of defects in III–V semiconductor devices. In 1975, he joined the Fracture Studies Group at the UKAEA Harwell Laboratory where he worked on the development and application of imaging and analytical electron microscopy to studies of nuclear reactor materials, including temper embrittlement, void swelling, irradiation-induced transformation, and corrosion. He returned to academia in 1994 as the Philips professor of analytical techniques at Sheffield Hallam University until 1998, when he returned to the materials department at Oxford University as a Royal Academy of Engineering professor where he continued work on materials problems in the nuclear power field.

Sven Tougaard received his PhD (1979) and Dr. Scient (1987) in theoretical and experimental surface physics at the University of Southern Denmark (SDU), and did his postdoctoral work in the United States and Germany (1978–1984). He has been professor at SDU since 1984, where he founded QUASES-Tougaard Inc. (1994), which develops and sells software for chemical analysis of surface nanostructures by electron spectroscopy. His current research is in experimental and theoretical studies of phenomena induced by the electron–solid interaction at surfaces and in nanostructures, and in the development of practical experimental methods to determine the chemical composition and electronic properties of nanostructures by electron spectroscopy. He has conducted several EU-supported international projects on these subjects, published more than 160 scientific papers, and presented more than 50 invited talks at international conferences and workshops. He serves on the editorial board for *Journal of Electron Spectroscopy, Surface and Interface Analysis, Journal of Surface Analysis*, and *Surface Science Spectra*.

Wolfgang E. S. Unger is the head of the Surface and Thin Film Analysis Working Group at the Federal Institute for Materials Research and Testing (BAM), Berlin, Germany. He is the chairman of the Surface Analysis Working Group at CCQM/BIPM, the German National Representative of the VAMAS Technical Working Area on Surface Chemical Analysis, a convenor of ISO Technical Committee 201 on Surface Chemical Analysis, and was the chairman of ECASIA '03. After studying physics at the Humboldt University in Berlin, he received a PhD in physical chemistry from the Academy of Sciences in Berlin in 1986. His major interests are nanoscale characterization of surfaces by x-ray photoelectron and absorption spectroscopy and by secondary ion mass spectrometry.

John F. Watts is a professor of materials science and director of the Surrey Materials Institute within the Faculty of Engineering and Physical Sciences at the University of Surrey. Applications of XPS, AES, and ToF-SIMS to investigations in materials science have been his principal research interests for more than 30 years. He received his PhD (1981) and DSc (1997) from the University of Surrey for his contributions to fundamental studies of adhesion phenomena using surface analysis methods. He is editor-in-chief of *Surface and Interface Analysis* and is a member of the organizing committees of the biennial ECASIA and SIMS conferences. He has published more than 300 papers in applied surface analysis and is a frequent lecturer in Europe, the Americas, and Asia.

Matthias Werner received his PhD in electrical engineering from the Technical University of Berlin, Germany in 1994. Since then he has acted as a technical consultant for the German engineering foundation VDI/VDE, in the general area of microsystem technologies, and has been the director of the Microtechnology Innovation Team of Deutsche Bank AG. In 2004, he became the managing director of NMTC (a consulting company on nano- and microtechnologies) and has advised many of the leading German players on the exploitation of nanomaterial-related products. He is the author and coauthor of approximately 100 scientific papers, book chapters, and books and has given numerous invited talks. He is coeditor of *mstnews*, the international microsystems magazine, and is a fellow of the Institute of Nanotechnology (IoN), United Kingdom, as well as a member of the National Strategy Group for the UK Micro/Nano Manufacturing Initiative.

Robert K. Wild received a PhD in physics from Reading University in England in 1966. After a brief period at the University of Virginia, Charlottesville, he returned in 1968 to the Research Laboratories of the Central Electricity Generating Board at Berkeley in England. Many of the activities were subsequently transferred in 1992 to the Interface Analysis Centre at the University of Bristol. His principal professional interests are in applying AES, XPS, and SIMS to the study of corrosion of steels, and effects of grain boundary segregation on iron- and nickel-based alloys for which he was awarded a DSc in 1990. He has published texts on the microstructural characterization of materials and the properties of grain boundaries. He was the chairman of the UK Surface Analysis Forum for 10 years from 1985, and in 1993 became secretary of the ISO Technical Committee 201 on surface chemical analysis SC7 XPS.

Vaneica Y. Young received her PhD in analytical chemistry in 1976 from the University of Missouri-Kansas City. She is now a member of the faculty in the Department of Chemistry at the University of Florida. Her major interests are the use of XPS to characterize ion-selective electrode corrosion, the structure of ion-beam-modified surfaces, and the structure of supported rigid rod polymer films.

Zhaoming Zhang received a PhD in applied physics from Yale University in 1993, after which she joined the research staff at the Australian Nuclear Science and Technology Organisation (ANSTO) in Sydney. Since then she has conducted a wide range of research activities, ranging from fundamental studies in surface science and bulk crystallography to more applied topics such as the development of nuclear waste form materials. She is now a principal research scientist in the Institute of Materials Engineering at ANSTO.

1 Introduction

John C. Rivière and Sverre Myhra

CONTENTS

In order to be able to describe the objectives of this book and the audience at which it is aimed, it is first necessary to consider the current situation in surface and interface science and technology.

1.1 SPECTRUM OF PRACTITIONERS AND ACTIVITIES

As in other mainly experimental fields, so in surface and interface science, there is a broad spectrum of approaches and activities. At one end of this spectrum are those whose sole purpose is either the development of new instruments and methods or the improvement of existing ones. Next to them, one finds another group of practitioners who have become expert in the operation and theory of a particular technique, or group of closely related techniques, and their expertise includes application of the techniques to almost any material problem. Moving to the middle, there is a wide and heavily populated band including all those who probably regard themselves more as material scientists than as surface scientists. They would be concerned either with the physical and chemical properties of a class of materials (e.g., polymers, high-tensile steels, etc.) or with a particular class of surface reaction (e.g., catalysis, corrosion, etc.), or with a particular type of application (e.g., nuclear fuel cladding, semiconductor interfaces, etc.). Those techniques would be chosen, which were known or believed to provide the required information and they would not necessarily be restricted to surface-specific techniques. Still further along the spectrum is another sizeable group investigating all types of reactions at interfaces between combinations of vacuum, gas, solids, and fluids, using model systems (e.g., single crystals or other well-characterized surfaces) and carefully controlled experimental variables. An increasingly important subset of that group is concerned with biomaterials and bio-interfaces. Within this group could be included those that are concerned with the development and testing of devices and products. Finally, at the other extreme, are those rare individuals who have managed to achieve a position in which they can devote their entire scientific career to the study of the properties of a single material or group of materials, at the most fundamental levels. They will need to deploy, as appropriate, all the available techniques, and even to invent new techniques to obtain specific information not otherwise forthcoming. The emergent group that falls under the broad heading of nanotechnology will need both to invent new techniques and methods, and to control experimental variables carefully.

As in all continuous spectra, there are no sharp boundaries between the categories listed above, and their activities merge into one another imperceptibly. Such categorization does not carry with it any value judgment. Those working in any one type of activity have much to learn from those in others, and all are necessary for the progress of surface science.

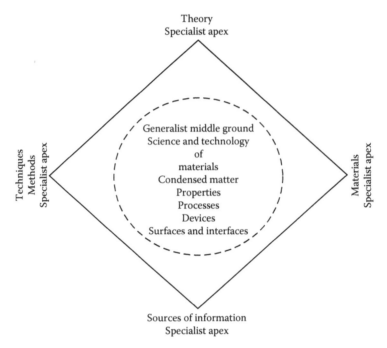

FIGURE 1.1 Schematic compass-point representation of the generalist versus specialist description of the groups of practitioners that are concerned with, and contribute to the knowledge base of surfaces and interfaces. The broken line indicates a region of permeability for two-way flow of information and dialogue. Specialization in analysis and methods, materials, theory, and information reside at the four extreme points of the compass.

For many years now, the majority of those using techniques and methodologies for investigation of surfaces and interfaces have occupied the middle ground of the above spectrum. In Figure 1.1 is shown a compass-point schematic of the potential interactions between the specialists at the four cardinal points, that is, those working at or near the extremes of the above-described spectrum, and the generalists whose expertise, particularly on materials, is much broader than that of the specialists, but which is not as deep on many individual aspects. Thus, on the horizontal axis, information flows into the middle ground from the specialists concerned with techniques and materials, while on the vertical axis, the flow contains information arising from theory and gleaned from sources. The position of the purpose of this volume is therefore on the horizontal $E–W$ axis, to the west, or left of the vertical $N–S$ axis: that is to say, at the point where those specializing in techniques for the characterization of surfaces and interfaces are communicating with those generalists in the broad area of the science and technology of materials. While the principal aim of this volume is to inform the group of practitioners that occupy the middle ground, it is believed that it will also be of considerable value to audiences that reside to the north and the east of the middle ground. Unfortunately, it is often the case that the theoreticians produce outcomes to calculations that cannot be tested experimentally, while the experimentalists tend to produce data for systems that cannot readily be modeled with any degree of confidence. It is also the case that many "specialist" experimentalists prefer to work on model systems under ideal conditions (e.g., ultrapure single crystal Cu at liquid He temperature), while the "generalist" material scientist/technologist has to contend with less ideal and more complex systems under real conditions. The point of these observations is that a secondary objective of this volume is to bridge the gap between the specialist and the generalist. In this sense, the specialist might find the volume useful.

1.2 TRENDS IN SURFACE AND INTERFACE SCIENCE

There are other trends in surface science that need to be considered and accommodated. They can be described loosely as instrumentation, commercialization, and rationalization, and they are all highly interdependent.

Virtually all practitioners in surface and interface science now rely on the instrumentation packages supplied by the manufacturers, although over a period of time, minor in-house modifications and additions will put an individual stamp on any one instrument. This is certainly the situation for the mature surface-specific techniques (e.g., x-ray photoelectron spectroscopy [XPS], Auger electron spectroscopy, and secondary ion mass spectrometry) and the well-established electron-optical microscopies (transmission electron microscopy [TEM], scanning electron microscope [SEM], and the ancillary analytical attachments). Even the more recent scanning probe microscope (SPM) techniques are now based on instrumentation from commercial sources. An important additional trend is that many techniques are located elsewhere at large international facilities (e.g., synchrotron radiation and neutron diffraction sources). Such facilities are becoming increasingly important due to improved access for outside users, their high productivity, and their routine availability of techniques that cannot be supported by local facilities (e.g., extended x-ray absorption fine structure [EXAFS]).

With this achievement of maturity, the technical simplicity has been lost, but there has been a gain in safety, ruggedness, reliability, and automation. For instance, the latest generation of dedicated XPS instruments requires few other experimental skills than being able to load the specimen and use the software. Indeed, some instruments (e.g., SEM and atomic force microscope) can now be made accessible to all-comers, within reason, rather than being under the control of an operator. An important consequence of this is that day-to-day operation of a major instrumental facility may require only the occasional presence of a senior and experienced specialist staff member. Most of the activities can be either under the direction of an operator with background and expertise closer to the technical level, or in the hands of a generalist user of techniques. Intervention by the specialist is then more with matters of interpretation and of the generation of ideas, whenever appropriate. However, the user, who may know very well how to run the instrument, cannot also be expected to know how to deal with every single type of sample that is presented for analysis, in terms of initial preparation and handling, precautions to be taken, range of techniques to be used, and operational parameters for data acquisition. The specialist may not always be available, or may in some laboratories be nonexistent, and in any case, his/her experience is not likely to be encyclopaedic. The user needs a guide that will explain the procedures to be followed when confronted with an unfamiliar material or requirement, based on the experience of others in dealing with such materials, rather than a textbook on techniques *per se*.

The trend toward the deployment of fewer specialists at the operational and instrumental fronts has been given even greater impetus by commercial pressures. With the increasing maturity and sophistication of the techniques, in particular the now-taken-for-granted power of software packages for instrument control, data acquisition, and analysis, has come the possibility of using them in rapid turn-around modes, for troubleshooting and even for online quality control. To be effective in such modes of operation, which become basically part of a production line, the cost per analysis must be minimized. This implies not only high reliability, since both the capital outlay and the cost of maintenance are great, but also low operational costs; thus, the salary and attached overheads of the operator, which form a significant fraction of such costs, need to be kept down. Hence, the increasing reliance on multiskilled generalist staff in the R&D environment. The requirement for budgetary flexibility is another factor. It is becoming increasingly rare that research budgets can underwrite the cost of maintaining permanent staff. It is rather more common that the bulk of the staffing is funded by soft money. Consequently, there will be high turnover of personnel, and less capacity to build up and retain expertise in depth and breadth.

Yet another factor in the cost-and-time-effective equation comes from the need for the rational deployment of techniques and personnel. The growing and desirable tendency to pursue a holistic approach in general material analysis means that analytical activity of sufficient significance, be it a technological research program or an online facility, requires access to a range of techniques in order to obtain complementary information. Decisions then have to be made about which techniques should be used, and whether any or all of them should be available in-house. The delicate balance to be struck is between purchase and operating costs on the one hand, and the cost of subcontracting analysis to one or other of the various fee-for-service analytical centers on the other. Such rational decision-making can be carried out only by those with a basic understanding of a broad range of techniques if an effective choice is to be made. However, also woven into the matrix of this choice is the matter of human resources. The desired minimization of operating costs, as already mentioned, means that the earlier luxury of having an expert attached to each technique or instrument can no longer be afforded. As a result, the available personnel must become multiskilled, that is, able to move from one technique or instrument to another as the analysis requires. There is a price to pay for multiskilling, in that the operator is unlikely to have the deep expertise of the dedicated expert in any one technique, but against that can be set much greater flexibility in deployment over a range of disciplines. Further benefits are that the likelihood of making a correct decision over choice of technique is much enhanced, and that the crossing of disciplinary boundaries will result in improved cross-fertilization of ideas.

Finally, the demographic situation within the broad field needs to be considered. Most of those scientists who grew up with the development of surface analysis and SEM/TEM in the 1960s have either retired or have moved into managerial/executive positions. Hence, there is a generational transition in which a new cadre is entering an already mature field, but is subject to entirely different imperatives. In addition, a new generation has embraced SPM and synchrotron techniques. Part of the rationale for this book is in the area of information transfer from one generation to another. The intention is to speak to members of an audience at the early stages of their careers, rather than to the founding fathers and mothers.

1.3   INTENDED AUDIENCE

The upshot of the foregoing is that there is a large group within the surface science community consisting of those working mostly in industrial laboratories or in analytical service centers, whose job it is to provide reliable and relevant answers to problems involving surfaces and interfaces with rapid turn-around and at relatively low cost. They will know enough about the nature of the information obtainable from some techniques, surface-specific and otherwise, to be able to decide which ones to use or not to use, and will know how to deploy them to best effect. They will not in general have sufficient in-depth background, and certainly not the time, to be able to engage in long-term basic research. Although many might well have expertise in one or more areas of materials science, the nature of their work is such that they are liable to meet at any time a material or a problem from a totally unfamiliar area. It is the members of this group at whom this book is aimed, in an effort to provide them with a guide to the strategies to be adopted when they do come up against something unfamiliar.

Tertiary education institutions are no longer immune to the dicta of efficiency, productivity, and flexibility. Not so long ago, it could be assumed that senior academic staff would devote most of their time and energy to research and the two-way teaching and learning process. As a consequence, research students and postdoctoral researchers would gain insight through lengthy one-to-one interactions in the laboratory with an academic supervisor. Regrettably, most academic supervisors are now increasingly preoccupied with fund-raising, administrative duties, and responding to corporate imperatives. This book is thus intended also as a complement and, in some cases, an alternative to traditional methods of learning for researchers at the formative stages of their careers.

The general thrust of this volume will be a description of the technical, methodological, and phenomenological aspects of surface and interface analysis, in the course of which an attempt will be made to chart the most efficient paths from an initial question to a credible answer, for a series of generic problem areas. There will be less concern with the many techniques as ends in themselves and more as a means to an end. Since this approach is an inversion of the more usual one of first describing a technique and then giving examples of its application, it can be thought of as a top-down approach, which has been shown to be effective in many other areas. In view of the ways in which an increasingly large number of workers in the field operate, the structural inversion seems logical.

From the above, it may be deduced that the structure of the volume is reminiscent of that of an expert system. Starting with some of the most general materials science questions that could arise in surface/interface science and technology, the user is guided to ever more detailed and specific levels of questions, the choice of path between them being based on the experience of experts. There are descriptions of the principal techniques, but not in nearly such great detail as may be found elsewhere; the intention is simply to help the user toward a preliminary choice of techniques. The reader is directed to whichever chapters are appropriate to the particular problem, and find there much essential information, a recommendation for the most productive methodologies, and an indication of the likely answer, in phenomenological terms. Throughout there is an emphasis on the multiskilling approach, by the demonstration in each chapter of how the information provided by one or more of the traditional surface-specific techniques is complemented and reinforced by that either from some of the less common techniques or from the so-called bulk techniques.

The chapters fall neatly into a set of three modules. This and the next chapter are introductory, one to the aims and objectives of the book and the other to the principles of analytical problem-solving. Chapters 3 through 9 describe the tools of the trade, how to use them, when and when not to use them, what information they provide, how to quantify that information, etc. Finally, Chapters 10 through 19 represent an authoritative distillation of the expertise and experience of authors, each of whom is a specialist in the application of surface analytical and complementary techniques to a particular subject area.

Some readers may wish to consult Chapter 2, in order to inform themselves about the strategy and tactics of problem solving. Some may wish to skip the introduction module and go straight to Chapters 3 through 9, others, perhaps of less experience, will benefit from some introduction. Those who already have significant instrumental (i.e., tool-kit) experience will no doubt bypass the tool-kit module and proceed directly to the specialist module. Several variations are therefore possible. Whatever the level of experience, however, and whatever the route chosen, the objective is to provide a practical guide to methods of the solving of problems of surfaces and interfaces. The authors hope that this book will fill the need that they have perceived for such a guide.

2 Problem Solving: Strategy, Tactics, and Resources

Sverre Myhra and John C. Rivière

CONTENTS

2.1 INTRODUCTION

The approach taken in this book assumes that while some readers will already be expert in many aspects of problem solving, others will either be novices or at an early stage on the learning curve. For the latter readership, this chapter sets out the strategic and tactical issues of problem solving, with the focus mainly on traditional surface and interface analysis, but with additional emphasis on the emerging and important field concerned with science and technology on the nanoscale. The successive stages that go to make up the problem-solving route are described together with the associated thought processes, many of which are intuitive for the more experienced practitioners.

2.2 MATERIALS, SIZE SCALES, AND DIMENSIONS

In the interrogation of any system in any scientific discipline to obtain information about that system, at least three components are necessary: a stimulus, an interaction volume in the system that responds to the stimulus, and a means of gathering the information generated within that volume as a result of the stimulus. Since not all the information will necessarily be able to be gathered, there is also an information volume, by definition equal to or smaller than the interaction volume. The shape and size scale of the information volume define the lateral and the depth spatial resolutions, over which there will be some averaging of information. For instance, imaging x-ray photoelectron spectroscopy collects information with microscale lateral resolution, but with a depth resolution of only a few interatomic distances, while scanning transmission microscopy has a lateral resolution in the nanometer range and a depth resolution limited by the thickness of the specimen foil. At one extreme is scanning tunneling microscopy performed in ultrahigh vacuum (UHV) where the information volume approaches that of a single atom in the surface.

In practice, all condensed matter objects, irrespective of the size scale, are bounded by a surface. This surface acts also as an interface due to something else being contiguous. (A number of permutations are possible and relevant, such as solid–solid, solid–liquid, solid–gas, liquid–gas, etc.) Any interactions at the boundary are usually defined by the physical and chemical conditions at the interface. As the size reduces progressively toward the nanoscale region, the interactions are dominated increasingly by surface and interface effects and in the limit converge to those characteristics purely of a surface. With these thoughts in mind, it is instructive to consider the dimensionality of objects in the context of the size scale (Table 2.1).

An important additional category is that of two-dimensional (2D) structures whose overall sizes are on the micro- or mesoscale, but whose surfaces exhibit lateral order on the nanoscale. Examples include arrays of quantum wells, crystalline polymers, zeolites, Langmuir–Blodgett films, etc. A subset includes bio-membranes, which are sometimes referred to as being para-ordered (due to the contiguity of biopolymer chains). The analysis problem is then that of needing to describe lateral differentiation, and structure, on the nanoscale, even though the surface/interface is meso/micro in lateral extent.

In some situations, mostly in basic research, it is the information generated only within the quasi-2D volume itself, for example, a surface layer or an embedded nanoparticle, that is of interest; that is, the analysis is concerned with conditions in that volume in isolation, as it were. When dealing with solids, however, it is not often that the properties of a surface layer, or an embedded

---

TABLE 2.1
Dimensionality and Size Scale

Dimensionality	Size Scales[a]	Analytical Implications and Context
3D	All micro/macro	Bulk analysis of a solid
2D	One nano, two meso/micro/macro	Traditional methods of thin film, and surface and interface analysis/characterization
1D	Two nano, one meso/micro/macro	Traditional techniques require extension and adaption Examples include nano-wires, nanotubes, biomolecular strands, etc.
0D	All nano/(meso?)	Novel techniques are required. Examples include nanoparticles and single quantum wells

[a] The boundaries between regimes of size scales are ill-defined; that is, there is no general rule as to when one departs the micro-regime and enters the nano-regime. Thus, it is useful to use the term "meso" to describe the gray area between nano and micro.

particle, can be considered without reference to the same properties of regions further away, such as a substrate or a surrounding matrix, that is, beyond the shortest (atomic scale) dimension of the information volume. This is so because some properties of the surface depend on what is happening, or has already happened, in those remoter regions, generally termed the "bulk" of the solid. Even in homogeneous solids, there may be internal interfaces of many types such as phase boundaries, grain boundaries, artificially created compositional and structural interfaces, etc. In principle, it is clear that the presence of a surface must imply the existence of an interface, even if the latter is only with the UHV or other environment. The attributes of interfaces are of crucial importance to the behavior of the solid from both surface and bulk aspects, and need to be accessible for analysis. Since the interfaces are themselves effectively quasi-2D volumes, analysis of their attributes should ideally be undertaken by surface-specific methods, or by cross-section electron microscopy. However, they are of course internal and therefore, in current jargon, buried; how to gain access to them for analysis without at the same time causing perturbation leading to ambiguity is still a central question in surface and interface analysis problem-solving. (For purposes of the present discussion, the term "attributes" is being used to describe such things as topography, composition, structure, crystal-chemistry, etc, as opposed to the term "properties", which is reserved for more intrinsic characteristics, such as density, resistivity, etc. There is a gray area between the two terms.)

It should be clear from the above that the traditional surface and interface analysis for 2D objects can be considered to be a mature field, as evidenced by a rich literature and by reasonably well-established codes of best practice and quality assurance. Likewise, the state of the art for 3D objects on the macro- and microscales is well established and proven. The situation is much less satisfactory for lower dimensional objects on the sub-microscales, because in order to attempt to study them, traditional techniques are being pushed beyond their agreed limits of utility. In the limit, on the meso- and nanoscales, new techniques are being explored and practitioners are still feeling their way along the learning curves.

2.3 STRATEGY AND TACTICS OF PROBLEM SOLVING

The logical sequence described below, and summarized in Figure 2.1, is not specific to problem solving by any particular technique or for any particular material, but could be applied to problem

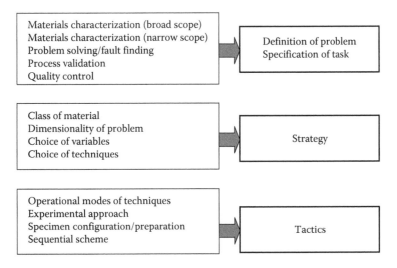

FIGURE 2.1   Schematics of sequence for problem solving.

solving in any branch of material science. In general, it is assumed that material characterization is concerned principally with attributes, as defined above, which may then provide the basis for determination of property.

The block diagram describes the general problem-solving process in a sequential manner. The schematics suggest a linear progression. In most cases, however, it is important to engage in quality control of the process itself, by asking appropriate questions. It might be necessary to backtrack through one or more iterations until one is satisfied that the correct sequence is being followed. The details of the process may differ if the focus is on problem solving as an activity for which a narrowly defined outcome is desired, as opposed to a broader program of investigation, dealing holistically with a particular material, device, or process. There are obviously many common elements, and also some significant differences. Some of the most important issues, alluded to in the block diagram in Figure 2.1, are discussed below. A hypothetical example, assuming a traditional surface analytical approach, has been used for illustration.

2.3.1 IDENTIFICATION OF THE PROBLEM AND FORMATION OF AN INITIAL HYPOTHESIS

A typical example of a problem might be the peeling of Teflon from a pan, for which the initial hypothesis might be failure at the interface. When a technological problem occurs, it will normally be up to the person or the organization suffering the consequences of that problem to identify its nature and to initiate a process of remedy. To do that they, as customers, need to commission one or more analysts to provide both relevant and reliable analytical data to enable them to reach a resolution of the problem via the building up of an overall picture. Preliminary consideration of the nature of the problem may have suggested a hypothesis as to the reasons for its occurrence, and the types of analysis initially chosen are then be based on that hypothesis. On the other hand, there may have been no clues arising from the way that the problem first occurred, and a working hypothesis in that case has to wait for the results from analysis used in either shot-in-the-dark or intuitive manners. With increasing experience, intuition can become surprisingly accurate. Experts in problem solving develop the ability to propound successful hypotheses based either on intuition or on very scant information.

Often the analysts are merely providers of data and it is the customer who works toward the solution. More interesting for the analyst, and usually more rewarding for the customer, are those instances in which the analyst is allowed to play an interactive role in the entire problem-solving sequence. Such interaction can often lead to short cuts, and therefore savings in cost, in the sequence.

2.3.2 IDENTIFICATION OF THE ESSENTIAL VARIABLES

An example of this is the composition at the Teflon/pan interface. Once the problem has been identified and a preliminary hypothesis set up, the nature of the information required, based on that hypothesis, can also be identified. From that, it is a short step to deciding what can actually be measured experimentally to provide that information, that is, the essential variables. Practicalities enter here, and indeed at many other points in the sequence, in that the customer will naturally be seeking the maximum amount of information at minimum cost.

2.3.3 REDUCTION OF THE PROBLEM AS FAR AS POSSIBLE WITHOUT LOSING ESSENTIAL INFORMATION

Examples of this are: disregard bulk composition of substrate, and concentrate on interface composition. Although in basic research on model systems there is usually complete freedom of experimental design, in the real world the problems are too complex to be able to solve in the same complete sense. Life is too short and available finance is strictly limited. In any case, the materials associated with problems, and often the problems themselves, may be ill-defined, with possible hidden variables. It is vital at this stage in the sequence to avoid the temptation to start analysis at once, since the result may simply be to end up with a mass of irrelevant or even misleading data. The most important step in the art of problem solving is pinning down what is the crucial experimental variable that must be measured, and to that end all relevant background information on the system giving rise to the problem should be used. Such information might include the bulk compositions of constituents, any chemical or physical pretreatments, dopant concentrations, ambient atmosphere, etc. It is necessary for the customer and the analyst to have full discussions so that no pieces of the information jigsaw are overlooked.

2.3.4 Selection of the Techniques Likely to Provide the Crucial Information by the Most Reliable and Economic Route

For example, x-ray photoelectron spectroscopy (XPS) might be a good choice, but Auger electron spectroscopy (AES) and static secondary ion mass spectrometry (SSIMS) would be poor choices. Not many institutes or organizations can offer the full range of analytical instrumentation and expertise, and most only a subset of that range. Where there are only a few techniques available in-house, there is an understandable tendency not only to try to apply them regardless of suitability or otherwise, but also to push individual techniques beyond their limits of reliable and efficient function. That tendency must be resisted. In the long run, uncritical application of the same few techniques to the attempted solving of every analytical problem that comes along is counterproductive, in that for many problems the information and data produced are quite inappropriate to the solution. The techniques must be matched to the crucial information, or the whole process becomes nonsensical; if the necessary technique is not available in-house, then it must be sought elsewhere.

2.3.5 Choice of Methodologies Consistent with the Selection of Techniques

Examples of this are: use large sample area for XPS, avoid obscuring contamination, and comparing data from failed components with those from control specimens. Once the choice of necessary techniques has been made, then details such as the way in which the techniques must be used to produce the crucial information, and any precautions that must be taken to avoid ambiguity in interpretation, have to be decided. In many problems, vital clues can be extracted from good vs. bad or before vs. after analytical comparisons, which require control specimens. There is also the ever-recurring and vexed question, referred to earlier, of how to access buried information. Here practicalities enter again. The customer may decide that solution of the problem can be achieved only by such access, but he/she must then be made aware not only that extra time, and therefore money, are involved but also that the mechanics of obtaining access, typically by ion bombardment to removal material, may change the nature of the information irretrievably. There is no firm and final answer to that situation; every sample and problem has to be considered individually, and the pros and cons weighed up.

2.3.6 Acquisition and Processing of Data of Adequate Quantity and Quality

Some examples of this are: energy resolution, signal-to-noise ratio, scan widths required for background subtraction, etc. At this point, the analysis can actually be started. Unless the data are of a minimum quality necessary for correct interpretation, they will not help toward solution of the problem. By this is meant that they must be good enough to disprove a preliminary hypothesis if necessary, and lead to alternative hypotheses, rather than being of just sufficient quality to support the preconceived hypothesis. The level of quality required is a function of the nature of the crucial information sought, and will vary from problem to problem.

2.3.7   INTERPRETATION OF THE DATA

In the interpretation of data typical questions to be asked are: are the results consistent with hypotheses and with other experimental evidence, are they credible, are they reproducible, and if so, at what confidence level? In an ideal situation, there would be abundant data of high quality obtained from several complementary techniques, requiring fairly minimal and straightforward interpretation. In practice, the situation may be constrained to be far from ideal and interpretation must be undertaken carefully and realistically. The more direct and transparent the interpretation, and the greater the extent to which it is compatible with prior information about the problem, the more likely it is that the conclusions from it will be of use to the customer. The main hazards at this stage are that the interpretation is wrong, or irrelevant to the problem.

2.3.8   REVIEW AND EVALUATION, AND ITERATION IF NECESSARY

Examples of these are that the data have been misinterpreted, or in hindsight emphasis should have been put on other variables, or the customer asked the wrong questions in the first place. This is one of the most important stages in the whole sequence, and should be taken seriously. Both individual and institutional credibility are at stake, as is perhaps the long-term viability of the institute or organization in the problem-solving field. A lifetime of carefully nurtured reputation and customer relationships can be undone so easily by one or two shoddy jobs. It is far better, if in doubt, to perform additional measurements and analyses before a report is submitted to the customer.

2.3.9   PRESENTATION

To-the-point reports, false color maps, succinct and adequately descriptive captions, and easy-to-follow conclusions are more useful to the customer than lengthy explanations, particularly if the essential conclusions have to be presented to management. Of course a report must be accurate, factual, relevant, and complete, but its essential messages must be transparent to the relatively nonexpert reader without, for instance, descending into the excessive use of jargon. Nowadays, with so many excellent computer packages available, there is no excuse for submitting a report without substantial visual impact. In addition, it is perfectly possible to assemble data from several different techniques for incorporation into an overall report. With the trend toward full digitization of output, including even high-resolution images, full flexibility of composition, layout, and mode of presentation can be achieved.

The sequence described above suggests that the scope of problem solving is limited to the support of R&D in an industrial setting. However, the ambit of this book is wider than that. For instance, the emphasis in strategic research may be either on a broad characterization of a particular class of material or on the investigation of a generic variety of surface/interface reactions. In both of these, the need is for completeness, which can be achieved only by adopting an integrated approach exploiting complementary techniques and methodologies. Another area in which surface/interface analysis has much to offer is that of quality control. There the emphasis is on demonstrating that the relevant variables are indeed being controlled, and that a process technology remains constant over time. Also, it must be remembered that a significant fraction of surface/interface analysis is carried out within the tertiary education sector by central facilities providing a variety of services for academic customers. The sequence thus includes a component of technology transfer, in the sense that both providers and customers are engaged in a teaching and learning endeavor, as well as a problem-solving process.

2.4   TACTICAL ISSUES IN PROBLEM SOLVING FOR SURFACES AND INTERFACES

Up to now, the discussion and description of the logical sequence of problem solving has emphasized strategic issues. What follows below are some general observations on some of the tactical practicalities that have to be taken into account. Most of the points are described, often in greater depth and breadth, in subsequent chapters, but it is useful at this stage to provide a general overview. The examples given are illustrative, rather than exhaustive, and are framed in the context of surface analysis.

2.4.1 Specimen Handling, Preparation, and Configuration

Whether surface characterization is being used in the context of basic science, in which both the origin and history of the specimen are known, or whether it is applied in problem solving, where the specimen is likely to arrive in a much less well-defined state, the first essential requirement in dealing with the specimen is cleanliness. This should be obvious from the extreme surface specificity of most surface analytical techniques, but must be emphasized nevertheless. Even though specimens exposed to the ambient atmosphere or subjected to various pretreatments carry surface contamination, it is vital not to add to or alter that contamination by manual contact or by any other contact that might cause transfer of material onto the surface. For the same reason, the environment in which the analysis is performed should be such that it cannot contribute to the contamination, that is, the ambient should be oil-free and in the UHV region of pressure.

One of the reasons for avoidance of any alteration to the existing contamination layer is that there will be instances where some or all of the information needed for solving a problem may actually reside in the nature of the contamination itself. These instances might arise, for example, in tribology, adhesion, or corrosion. It is in any case essential to carry out an analysis of the specimen in the as-received condition, if for no other reason than to establish the extent of the contamination and therefore to be able to decide on subsequent procedures.

Specimen preparation and treatment can take place either outside the vacuum envelope of the analysis system, that is, *ex situ* or, within the system, *in situ*. *Ex-situ* preparation could be either in the uncontrolled laboratory ambient or in a controlled atmosphere, while *in-situ* preparation and treatment could be either in the analysis position itself or in an associated chamber.

2.4.1.1 *Ex-Situ* Preparation

In surface and interface problem solving, most specimens will have originated from an external source, that is, the customer, and therefore by definition they will have had some *ex-situ* preparation or treatment, if only as a result of transfer from their original environment to that of analysis. However, that is not what is really meant by *ex-situ* preparation. Many analytical techniques in materials science, those that can be classified as providing bulk analysis, require preparation of the specimen by methods such as polishing, abrasion, and sectioning. These can also be used in the preparation of specimens for certain types of surface analysis, but of course the cleanliness requirements are much stricter than in preparation for bulk analysis. For example, lubricating or cutting liquids cannot be used, because they will leave residues on the surface, while exposure to polar fluids may degrade the surface. In addition, any debris left on the surface after such *ex-situ* methods have been used should be removable by ultrasonic cleaning in a bath of high-volatility solvent. The usual reason for employing such aggressive methods is to expose a buried interface located at a depth too great to be reached by *in-situ* methods.

2.4.1.2 *In-Situ* Preparation

Once inside the analytical system there are many types of preparation or treatment of the specimen that might need to be carried out in the course of any one problem-solving procedure. They fall into the approximate categories of cleaning, depth profiling, interface exposure, and surface treatment.

The most common method of removing surface contamination is by ion beam erosion, normally with energetic Ar^+ bombardment. By adjusting the ion energy and by using a low ion dose (roughly equivalent to about one incident ion per surface site), it is possible to remove just the first one or two atomic layers, which usually consist of contamination. The cycles of alternate erosion and heating to achieve perfect cleanliness of a specimen, used in basic research, are clearly inappropriate to specimens involved in problem solving since the information content would be lost immediately. However, cleaning by gentle ion beam erosion (without heating) is convenient and reliable, and is the most widely used method largely because for the vast majority of specimens there is no alternative. One or more ion guns are therefore mandatory accessories on most surface analytical installations. The principal disadvantages of a technique in which relatively heavy charged particles such as Ar^+ are accelerated to impact on a surface are structural and chemical damage, mentioned in more detail below (see additional discussion in Chapter 10).

In a few special cases, *in-situ* cleaning can be performed by mechanical means. It is possible to fix special tools to the end of an auxiliary manipulator arm so that the operations of abrasion (with a diamond file) or scraping (with scalpel or razor blades), or even cleavage (with razor blade or chisel) can be carried out. When abrading or scraping, care must be taken that material from one surface is not transferred to the next.

As well as for cleaning, as mentioned above, ion beam erosion is used very extensively for the (relatively) controlled removal of material to allow analytical access to regions of the specimen beyond the surface information volume. Erosion and analysis can be either simultaneous (e.g., when using AES) or sequential (e.g., when using XPS); in both cases, the plot of elemental concentration as a function of amount removed is called a depth profile. Examples of depth profiles abound in the literature and there are several in this book. Also in this book in various places (i.e., Chapters 3, 10, 16, and 18), the deleterious effects to be expected from ion bombardment of a surface are described. The latter arise from unavoidable knock-on structural and chemical damage, which disrupts the surface crystal structure and alters the electronic structure. In addition, for a multiatomic specimen, as most are, the removal of surface species by the ion beam is selective, so that some species are enriched and others depleted as the eroded face progresses into the specimen. As a result, there may be selective desorption, interlayer mixing, and the appearance of sub-valent species and non-stoichiometry. For more details, see the above-mentioned chapters. Unfortunately, there is as yet no realistic alternative to ion beam depth profiling, particularly in the realm of problem solving, and the best that can be done is to be aware of the effects and to attempt to account for them in a semiquantitative way. In the case of transmission electron microscopy, focussed ion beam milling is now becoming a widely used tool for preparing specimens (see Chapters 5 and 9).

If internal interfaces are also regions of structural weakness, they can often be exposed for analysis by mechanical methods. The classical example is that of grain boundaries in polycrystalline metals that have been weakened as a result of impurity segregation to them following heat treatment. The boundaries can be exposed by fracture of the metal, usually at low temperatures, but any attempt to measure the nature or level of the impurity by fracture *ex situ* will always be unsuccessful because of the reaction of the impurity with ambient air and of the accumulation of contamination. Thus, there are many available designs of fracture stage for use *in situ*, mostly for impact fracture but some for tensile fracture. The stages can also be used for ceramic fracture, since there are instances in which ceramic materials also lose cohesion through the presence of grain boundary impurities.

In special cases, for example, in adhesion and in thin oxide film studies, it may be possible to expose the interface of interest by peeling techniques (see Chapters 18 and 19). Devices designed to do that are usually constructed individually and are not normally available commercially, unlike fracture stages.

Finally, *in-situ* preparation includes also all those treatments used from time to time to try and create on a specimen the same surface condition as might be found after some technological treatment or other. Since it is normally undesirable to carry out such treatments in the analysis chamber itself, many instruments have an additional chamber, communicating with but isolatable from the analysis chamber, in which the treatments can take place. Indeed, it is increasingly common to incorporate analytical techniques into processing lines, such as those used for chemical vapour deposition (CVD)

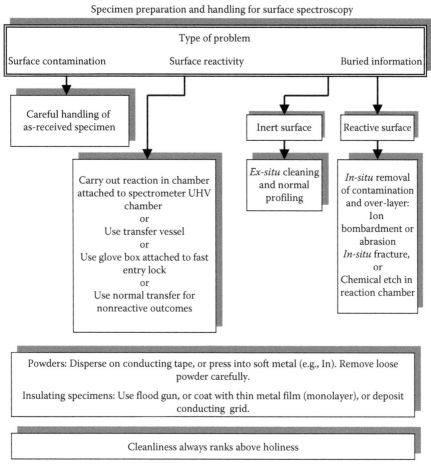

Specimen preparation and handling for surface spectroscopy

FIGURE 2.2 Overview of specimen handling and preparation for surface specific analysis.

and multilayer epitaxial deposition. According to what is required, the additional, or reaction, chamber might therefore be equipped with the means of heating or cooling the specimen, or of exposing it to a gas or mixture of gases via a metered gas handling system, or of depositing thin films of one or more materials on its surface, or of causing alterations in its surface by ion implantation or plasma discharge. Many of these treatments are described in various places in this book.

The main points of specimen handling and preparation for surface-specific analytical techniques are summarized in Figure 2.2. Similar flowcharts of greater or lesser complexity can be constructed for other techniques and types of applications.

2.4.1.3 Specimen Configuration

Just because in some techniques (e.g., scanning tunneling microscope) the quasi-2D interaction volume is exceedingly small does not mean that the specimen configuration itself has to be small. Since specimens have to be positioned and oriented with the naked eye, and handled in so doing with, for example, clean high-quality tweezers, it follows that they are all macroscopic. In addition, many of the surface analytical techniques, when used in routine fashion, are macroscopic in two of the three dimensions of the interaction volume; SSIMS and XPS in their nonimaging forms are good examples.

Despite the above statement, the analyst engaged in problem solving is not infrequently presented with nonstandard specimen configurations, which require ingenuity and inventiveness if a representative analysis is to be achieved. Examples of such difficult specimens are powders, carbon and polymer fibers, biomaterials, and microelectronic circuits; again, these are mentioned in this book in the appropriate chapters, where ways of presenting them for successful analysis are described.

2.4.2 Technique Destructiveness

According to basic physics, any measurement process is irreversible; in that very fundamental sense, then, no measurement technique can claim to be nondestructive. In the world of practical problem-solving, however, such a limitation can be set aside as being too rigid. Given that all the surface analytical techniques can be destructive to greater or lesser extents, a pragmatic approach must be adopted. Depending on the constraints applicable to any particular problem or to a particular type of specimen, a set of functional criteria can be set up to classify techniques and procedures in terms of their potential destructiveness as follows:

2.4.2.1 Functionality Criterion

If the intended functions of a specimen or material are unaffected by the problem-solving process, then the techniques and methodologies required to solve the problem can be said to be nondestructive.

2.4.2.2 Market Value Criterion

Using "market" here in its broadest sense, if the market value of an object is not reduced by the problem-solving process, then the process can be said to be nondestructive. There might even be a gain in value as a result of the analytical procedure, which would be offset against any loss arising from possible destructiveness. Further, if there is a set of nominally identical items, it is possible that the loss in value of those that are analyzed and perhaps irretrievably damaged would be more than made up by the gain in value of the remainder as a result of the information acquired (e.g., due to demonstrable quality control).

2.4.2.3 Sequential Analysis Criterion

When using a multi-technique approach to problem solving, care must be taken that the application of one technique to a specimen does not jeopardize the validity of subsequent analyses by other techniques. Under this criterion, if that validity is unaffected, then the technique first applied can be said to be nondestructive. The criterion puts obvious constraints on the sequence in which techniques should be used, beginning with the least destructive and ending with the most. A corollary of this criterion is that in cases where there are continuing *ex-situ* or *in-situ* specimen treatments requiring periodic analysis then the analytical procedure should not itself affect the course of the treatments.

2.4.2.4 Information Volume Criterion

There is a direct relationship between the quality and quantity of information retrievable from the information volume and the fluence of the stimulus being used. The greater the fluence, the larger will be the number of interactions per unit time, and therefore the better will be the signal-to-background quality of the information. On the other hand, it is the direct, and sometimes indirect, effects of the stimulus that cause destructiveness. Among these effects, for example, might be the rate of energy deposition, or the rate of momentum transfer, or the type of interaction mechanism, or the duration of the analytical procedure. Thus, a balance has to be struck between the quality of information and the level of damage. In practical terms, this may mean having to take longer over an analysis than normal, in order to maintain the rate of energy deposition below a certain critical level, or using an energy of the stimulating probe different from normal, to avoid or minimize a particular interaction mechanism. Where damage is unavoidable, the analysis may reduce simply to having to accept whatever information can be collected before the specimen surface is altered to a predetermined extent.

There are instances in which concerns over destructiveness override all others, as where an object is either irreplaceable or of great value, for example, archaeological artifacts, national treasures, objets d'art, etc. At the other extreme, there are objects that can be replaced at essentially zero cost, compared with the cost of the analysis, for example, mass produced consumables, waste products, common naturally occurring products, etc. Even before analysis starts, it might be the specimen preparation itself that has to be considered because of its destructiveness, for example, if a large item has to be sectioned because it cannot be accommodated in the analysis position.

It is clearly difficult from the above discussion to give fixed criteria for taking destructiveness into account. There is generally a degree of uncertainty with respect to a particular specimen or problem. The most important rule is always to tell the customer beforehand what the potential effects on the specimen might be and to have complete agreement on the procedure to be adopted.

2.4.3 QUALITY ASSURANCE, BEST PRACTICE, AND GOOD HOUSEKEEPING

In an ideal world, the quality of an analytical process, and its outcome, would be guaranteed automatically by scrupulous adherence to the scientific method, independent of the intention of the analysis. That is, it should not matter whether the end user of the information is a research scientist or a customer, nor whether the results are to be used to generate new basic knowledge, improved techniques, better products, or financial gain. In the ideal situation, the quality would indeed be assured, without the need for further formality.

When it comes to the type of contractual relationship that is now commonplace between the supplier of the analytical service and the customer, a more formal guarantee basis is needed. The customer will not necessarily be familiar with the in-principle merits of the scientific method, but will understand a published set of guidelines to which the supplier must adhere. Such a set will include both rules for best practice in all aspects of handling, preparing, and treating specimens, and instructions as to how an analysis must be carried out so that its results have been quantified in an approved way and are traceable back to an agreed international standard. The latter proviso is essential in order that all results from any laboratory using the same technique are intercomparable. Only by general adherence to the same set of guidelines and instructions can quality assurance in an analysis be guaranteed. In many countries, there are now standard organizations that assess analytical supplier laboratories for accreditation in quality assurance, and the number so accredited is growing steadily.

No analytical service can operate efficiently without the internal procedures that can be lumped together under the heading of "good housekeeping". Such procedures are basically commonsense routines that ensure that both personnel and instruments function smoothly and reliably, that proper records are kept and maintained up to date, that specimens are not mislaid, mishandled, or cross-contaminated, that reports are compiled and sent on time, that customers are always fully informed, that safety standards are being met, etc. This book has much to say about the esoteric aspects of surface and interface analysis. It must always be borne in mind, however, that attention to detail is at least as important to the final outcome

2.5   NOTES ON ACRONYMS AND JARGON

Communications within any field of human endeavor, whether it be skateboarding or brain surgery, will sooner or later be conducted with specialized terminology, which to the uninitiated is either just so much alphabet soup or incomprehensible and confusing jargon. The field of surface and interface analysis and characterization is no different. Even to an insider the acronyms and abbreviations may sometimes present barriers to understanding rather than be a means of efficient communication. Regrettably, they have to be used, if for no other reason than for brevity. Other complications arise from redundant terminology where several acronyms are in common use, but refer to essentially the same technique or methodology (e.g., EDS = EDX = EDAX = EDXS). During the editorial process, all reasonable efforts have been made to ensure that acronyms are defined in each chapter when they first appear. Likewise, as far as possible, the use of jargon has been avoided, whenever it is unlikely to be comprehensible to a general scientific audience.

3 Photoelectron Spectroscopy (XPS and UPS), Auger Electron Spectroscopy (AES), and Ion Scattering Spectroscopy (ISS)

Vaneica Y. Young and Gar B. Hoflund

CONTENTS

3.1   INTRODUCTION

The most commonly used surface spectroscopic techniques are x-ray photoelectron spectroscopy (XPS), ultraviolet photoemission spectroscopy (UPS), Auger electron spectroscopy (AES), and ion scattering spectroscopy (ISS). XPS and UPS are similar techniques and can be grouped under photoemission spectroscopy (PES). All four techniques are used widely for the study of solid surfaces both in fundamental scientific studies and in applied studies of polymers, ceramics, heterogeneous catalysts, metals and alloys, semiconductors, nanoparticles, biomaterials, etc. They can provide information about composition, chemical state, electronic structure, and geometrical structure. Detailed reviews have been presented previously [1,2].

3.2   PHYSICAL PROCESSES

In this section, the underlying physical processes are outlined and the nature of the information provided by the resultant spectra is discussed.

3.2.1   PHOTOEMISSION

In PES, either x-rays (XPS) or UV photons (UPS) strike the surface of a sample in an ultrahigh vacuum (UHV) environment. Electrons are emitted as shown schematically in Figure 3.1A. The emission process has been described by Berglund and Spicer [3] in a three-step model, in which the first step involves absorption of an x-ray or UV photon and promotion of an electron from its ground state to the final state above the fermi level. The final state lies within the potential field of the solid

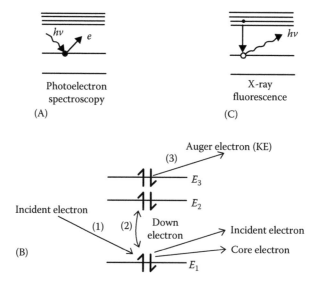

FIGURE 3.1   Schematic diagrams of the (A) photoemission, (B) Auger, and (C) fluorescence processes.

and satisfies the Schrödinger equation for that field. The second step is transport of the electron to the surface, and the third step is escape of the electron into the vacuum. Since the electron is generated within the potential of the solid, its wave function contains contributions from the solid even after it has escaped into the vacuum. In PES, the kinetic energies (KEs) of the emitted electrons are measured using an electrostatic charged-particle energy analyzer, from which their electron binding energies (BEs) can be calculated from the following equation:

$$E_b = h\upsilon - E_k + \Delta\phi \tag{3.1}$$

where

E_b is the electron BE in the solid
$h\upsilon$ is the energy of the incident photon
E_k is the electron KE
$\Delta\phi$ is the difference in work function between the sample and the detector material assuming that there is no charge at the sample surface

Typical XPS and UPS spectra are shown in Figures 3.2 and 3.3, respectively. The characteristics of these spectra are discussed below, as are the photon sources and energy analyzers used to perform PES.

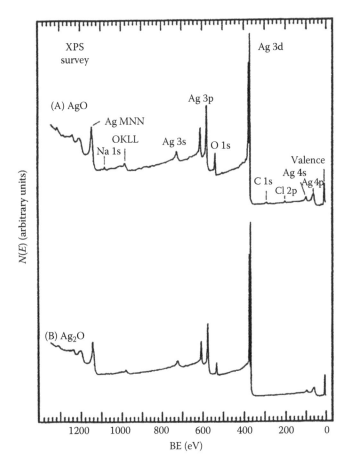

FIGURE 3.2 XPS survey spectra obtained from (A) an AgO sample after a 130°C anneal for 30 min and (B) an Ag₂O sample after a 300°C anneal for 30 min. (From Hoflund, G.B., Hazos, Z.F., and Salaita, G.N., *Phys. Rev. B*, 62, 11126, 2000.)

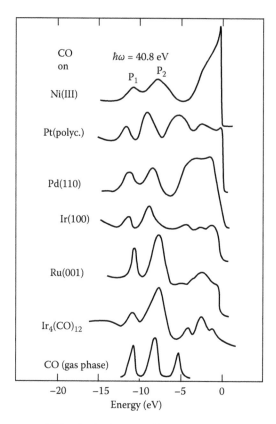

FIGURE 3.3 HeII UPS spectra of CO adsorbed on transition metals. The two peaks below 8 eV, in the group labeled P_2, are induced by CO adsorption. The structure above 8 eV, P_1, which varies from metal to metal, is due to emission mainly from the metal d-orbitals. For comparison, the photoelectron spectra of $Ir_4(CO)_{12}$ and gas-phase CO are shown. (From Gustafsson, T. and Plummer, E.W., *Photoemission and the Electronic Properties of Surfaces*, John Wiley & Sons Ltd., Chichester, U.K., 1978.)

3.2.2 AUGER EMISSION

The process of Auger emission is multistep, as shown in Figure 3.1B. The first step is the production of a core hole by ejection of a core electron as a result of interaction with either incident electrons or photons. The second step involves an electron in a shallower energy level undergoing a transition to fill the core hole. The energy difference is then available to a third electron, which is ejected as the Auger electron. Thus, the AES process involves three different electrons in two or three different energy levels, and the KE of the Auger electron produced from, for example, K, L_1, and L_2 electrons is given by

$$E_{AE} = E_K - E_{L_1} - E_{L_2} - \Delta$$

(3.2)

where E_K, E_{L_1}, and E_{L_2} are the BEs of electrons in the K, L_1, and L_2 energy levels, respectively, and Δ is a complicated term containing both the sample and spectrometer work functions, as well as many-body corrections that account for energy shifts during the Auger process and other electronic effects. The Δ term is usually small (<10 eV) and varies with chemical state. The initial core hole can also decay by x-ray fluorescence, in which an electron in a shallower level drops into the core hole with emission of an x-ray photon as shown in Figure 3.1C. The probability of decay via an Auger

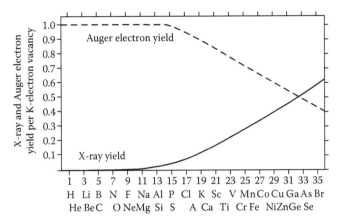

FIGURE 3.4 Auger electron emission and x-ray fluorescence yields for K-shell electron vacancies as functions of atomic number. (From Somorjai, G.A., *Principles of Surface Chemistry*, Prentice Hall, Englewood Cliffs, NJ, 1972.)

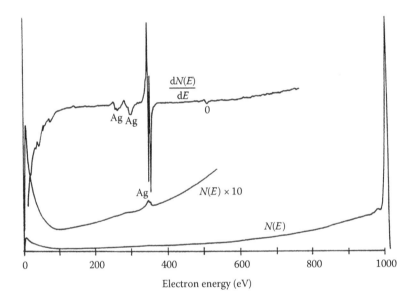

FIGURE 3.5 $N(E)$ and $dN(E)/dE$ electron energy distributions recorded from Ag with a 1 keV primary electron beam. (From Joshi, A., Davis, L.E., and Palmberg, P.W., in Czanderna, A.W. (Ed.), *Methods of Surface Analysis*, Elsevier, Amsterdam, 1975.)

process is greater for light elements than that via fluorescence, as shown in Figure 3.4 for K-shell electrons [6]. Auger processes occur for all elements except hydrogen and helium, which have no or insufficient outer electrons. As can be seen from Equation 3.2, the Auger KE does not depend on the primary beam energy, and the threshold energy for the transition is that required to produce the core-level hole. A typical Auger spectrum in both the $N(E)$ and $dN(E)/dE$ modes, recorded from a poly-crystalline Ag surface, using 1 keV primary electrons, is shown in Figure 3.5 [7].

3.2.3 ION SCATTERING

In ISS, an incident flux of monochromatic inert-gas ions (typically of energies 500–2000 eV) impinges on a solid surface, followed by energy analysis of the ions scattered from the surface at

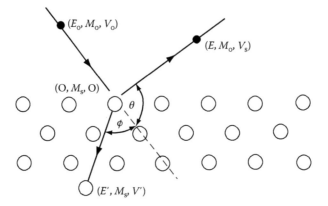

FIGURE 3.6 Schematic representation of the ISS process. The primary ion has energy E_0, mass M_0, and velocity V_0 before scattering, and energy E, mass M_0, and velocity V_s, after scattering. The scattering angle is θ.

some preselected angle, as shown schematically in Figure 3.6. The ion–solid interactions can be approximated as elastic binary collisions between the ions and the individual atoms in the solid. This approximation works quite well because the collision times are short (10^{-15} to 10^{-16} s) compared to the time constant of a characteristic lattice vibration (10^{-13} s). The ion therefore strikes a surface atom and leaves the surface region before the recoiling atom has time to interact with the solid. The conservation of energy and momentum in the binary scattering process can be written as

$$E_0 = E + E' \tag{3.3}$$

$$M_0 V_0 = M_0 V_s \cos\theta + M_s V' \cos\phi \tag{3.4}$$

$$0 = M_0 V_s \sin\theta - M_s V' \sin\phi \tag{3.5}$$

and combined to yield

$$\frac{E_s}{E_0} = \left(\frac{M_0}{M_0 + M_s}\right)^2 \left\{\cos\theta \pm \left[\left(\frac{M_s}{M_0}\right)^2 - \sin^2\theta\right]^{1/2}\right\}^2 \tag{3.6}$$

The symbols in these equations are specified in Figure 3.6. In Equations 3.3 through 3.6, E_0 is the KE of the incident inert-gas ion (set by the ion source), E is the KE of the scattered ion measured with an electrostatic energy analyzer, M_0 is the mass of the primary ion (selected by the choice of inert gas), and θ is the scattering angle determined by the experimental geometry. The variables are all known so that M_s, the masses of the surface atoms, can be determined from the positions of the peaks in the ISS spectrum. If $M_s/M_0 > 1$, then only the plus sign in Equation 3.6 applies and each target mass gives rise to a single peak in the spectrum of scattered intensity as a function of E/E_0. If $M_s/M_0 < 1$, then both signs apply, subject to the constraint

$$\frac{M_s}{M_0} \geq \sin\theta \tag{3.7}$$

and each target mass gives peaks at two energies in the above spectrum. ISS spectra obtained from a higher alcohol synthesis catalyst are shown in Figure 3.7 [8].

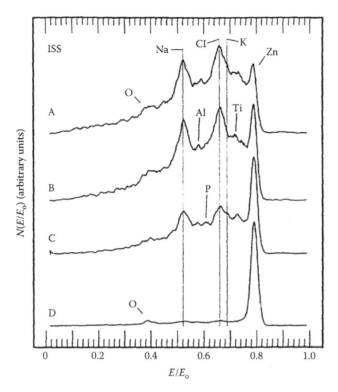

FIGURE 3.7   ISS spectra obtained from an aged catalyst after (A) insertion into the analysis chamber and after sputtering with 1 keV He+ for (B) 5 min, (C) 10 min, and (D) after a further 15 min using a 1:1 He+ and Ar+ gas mixture. (From Hoflund, G.B., Epling, W.S., and Minahan, D.M., *J. Elect. Spec. Rel. Phenom.*, 95, 289, 1998.)

3.3   SPECTRAL FEATURES

3.3.1   PES/XPS

XPS survey (also called wide scan) spectra obtained from Ag_2CO_3 before (a) and then after annealing at (b) 350°C, and (c) 500°C, each for 10 min in vacuum, are shown in Figure 3.8 [9]. In (A), many peaks are present, including the 3s, 3p, 3d, 4s, and 4p Ag peaks, as well as those of O 1s and C 1s. All these peaks arise from direct photoemission processes from core levels as shown in Figure 3.1A. A very small Na 1s peak can also be seen due to a low level of Na contamination. Tables of the BEs of core-level electrons are given in several references [10–12], so if an unknown peak is apparent at a particular BE, it can be identified. Ag and O Auger peaks are also present in Figures 3.2 and 3.8, produced by the process shown in Figure 3.1B, because core holes created by x-rays can also decay by an Auger process. Since the principal electronic shells usually contain electrons of different energies due to multiplet splitting, an Auger feature arising from the ionization of a particular core level consists of multiple peaks. Another feature in Figures 3.2 and 3.8 can be seen at and just above a BE of 0 eV. This is the valence band (VB) photoemission spectrum, due to the valence electrons of each element present. Its position and shape result from the chemical interactions (bonding) of the various elements.

Unmonochromatized x-ray sources contain several x-ray lines, each of which can cause core-hole ionization, giving rise to satellite peaks in an XPS spectrum. These satellite peaks can be removed using a monochromator, but this reduces the x-ray flux and, hence, the photoemission

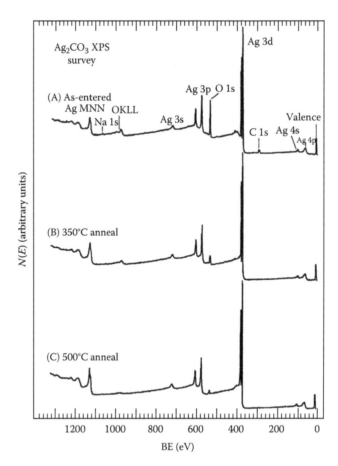

FIGURE 3.8   XPS survey spectra recorded from an Ag_2CO_3 sample (A) as-entered, and after annealing for 10 min at (B) 350°C and (C) 500°C. (From Salaita, G.N., Hazos, Z.F., and Hoflund, G.B., *J. Elect. Spec. Rel. Phenom.*, 107, 73, 2000.)

signal strength. The satellite features appear at BEs lower than those of the primary peaks and are of considerably lower intensity (<10%) [10]. Structure is also apparent at BEs slightly higher (0–50 eV) than those of primary features, due to two types of electron energy loss processes: (1) excitation of plasmons and (2) production of inter- and intra-band transitions.

According to Equation 3.1 for a conductor, a raw spectrum should be shifted in energy by $\Delta\phi$ in order to be able to plot $N(E)$ versus BE. If $\Delta\phi$ were to be known from the work functions of the sample and detector material, then the process would be simple. Where $\Delta\phi$ is not known, which is invariably the case, the calibration shift can be made using a peak that corresponds to a given (known) chemical state. Most samples exposed to air accumulate hydrocarbon contamination resulting in a significant C 1s peak. The magnitude of the shift required is then often determined by assigning the BE of this C 1s peak to a value of 285.0 eV, but it is essential that this assignment be checked against the BEs of other peaks in the spectrum, to ensure they make sense chemically. If they do not, then differential charging is probably occurring, and the data must be retaken using charge compensation techniques. For insulating samples, the surface may charge to some arbitrary potential, which cannot be correlated with any physical or chemical variables.

Damage phenomena must always be considered when using surface techniques. XPS produces the least damage because x-rays interact weakly with a solid, but in conventional XPS, the close proximity of the sample to the source itself can cause thermal damage. Damage can occur by several processes including bond breakage, and emission of particles by photon-stimulated desorption. The possibility of damage can be established by recording sequential spectra as a function of irradiation time. If the spectra change during the sequence, then damage has occurred, and the data collection parameters need to be changed to minimize it.

3.3.2 PES/UPS

In UPS, UV photons, rather than x-rays, are used as the excitation source. In many laboratories, UV sources at fixed energies, for example, HeI and HeII at 21.21 eV and 40.82 eV, respectively, are still used, but more information can be obtained with the help of the variable photon energy provided by a synchrotron (see also Chapter 7). As in XPS, the KEs and signal strengths of the photoelectrons are measured. Similar features are observed using either XPS or UPS to examine the VB, but their relative intensities differ due to variations in the ionization cross-section with photon energy. Another important difference is that the overall intensity of the VB photoemission signal is much greater in UPS than in XPS, because in the former the photon energy is of similar magnitude to that of the energy required for the excitation process (i.e., the Einstein Golden Rule). Since photoelectrons are initially elevated from a filled to an unfilled level, the structure of the density of states (DOS) influences the UPS signal; thus, although UPS occurs by the process shown in Figure 3.1A, and obeys Equation 3.1, the photoemission signal is modulated by the DOS of the unfilled levels just above the fermi level. The UPS VB spectrum therefore reflects a joint DOS. Since the unfilled DOS is much extended at energies far above the fermi level, and is essentially continuous there, such modulation is not a consideration in XPS.

In a synchrotron, the photon energy can be scanned using a monochromator. The initial (i.e., filled) DOS can be recorded by choosing a particular unfilled level (i.e., at a fixed KE) and by scanning the photon energy. The final (i.e., unfilled) DOS, on the other hand, can be obtained by scanning both photon and KE. Figure 3.9 shows the filled and unfilled DOS from the surface of α-quartz [13], recorded in this way. If 21 eV photons had been used, the feature at −10 eV would have been significantly reduced due to the low population of the unfilled DOS at 11 eV. The nature of the orbitals and their related selection rules also influence the intensities in the joint DOS.

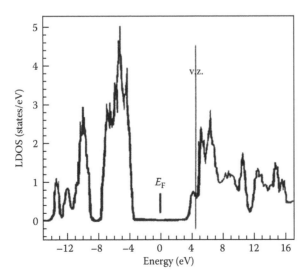

FIGURE 3.9 Filled (below E_F) and unfilled (above E_F) DOS, recorded from an α-quartz surface. (From Garvie, L.A.J., Rez, P., Alvarez, J.R., Busick, P.R., Craven, A.J., and Brydson, R., *Am. Mineral.*, 85, 732, 2000.)

UPS can be used to obtain various types of information, the most obvious being VB electronic structure, as described above. Valence electrons are responsible for the chemical bonding that holds atoms together to form a solid; various quantum mechanical methods can be used to calculate the bonding structure, in particular the VB DOS, thus providing a direct means of comparison of calculated results with experimental data. Both crystalline and amorphous solids can be studied. Using angle-resolved UPS and synchrotron radiation, the complete band structure of single-crystal surfaces can be mapped along different crystal directions. An illustration of this approach is given in Figure 3.10, in which the band structure of a Be(0001) surface has been mapped out in the Δ direction [14]. Comparison is shown with the results of a calculation using an *ab initio* self-consistent pseudo-potential method in the local-density formalism. The agreement between the calculated and experimental results is quite good except near the plasmon threshold. Another important application

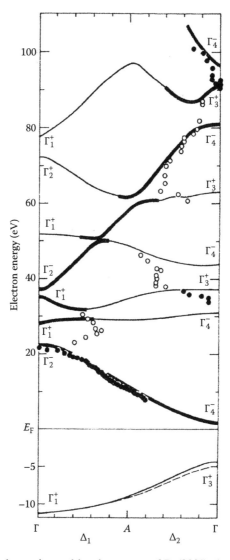

FIGURE 3.10   Calculated and experimental band structure of Be(0001) along Δ. The thicker lines denote final states, which have large plane-wave components along Δ. The dashed line is the corrected initial state used to determine the final state. Note that the vertical scale changes at E_F. (From Jenson, E., Bartynski, R.A., Gustafsson, T., Plummer, E.W., Chou, M.Y., Cohen, M.L., and Hoflund, G.B., *Phys. Rev. B*, 30, 5500, 1984.)

of angle-resolved UPS, using synchrotron radiation, is the determination of bonding orientation of adsorbates on single-crystal surfaces [5,15].

3.3.3 AES

An Auger spectrum, in the derivative, $dN(E)/dE$, mode, recorded from a sputter-cleaned Ni/Cr alloy foil, is shown in Figure 3.11 [16]. In the integral mode, Auger features appear as small peaks riding on a large background of inelastically scattered electrons, as seen in the lower two spectra in Figure 3.5. Differentiating the $N(E)$ spectrum either electronically or numerically accentuates the Auger features relative to the background, which, as can be seen in Figure 3.5, is of much greater intensity. Ni and Cr features are present in Figure 3.11 at both low (<200 eV) and high KEs (>400 eV). The low KE features involve one or two valence levels in the Auger process, while those at high KE arise from three core levels. As with XPS, AES peaks can be identified from their KEs, using reference manuals [12,17–19]. When assigning a particular peak to an element, it is vital to check that all the other Auger peaks belonging to that element are present in the spectrum and with the correct relative intensities.

Damage is a very important consideration in AES since many types of sample are susceptible to electron-beam damage by a variety of processes. Such damage can be minimized by using a low

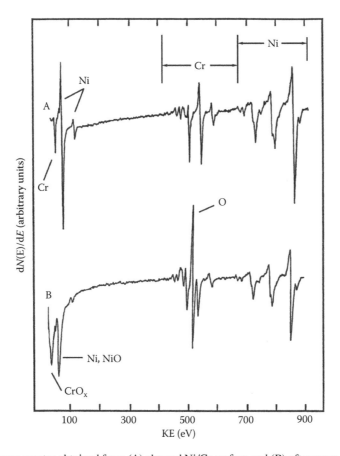

FIGURE 3.11   Auger spectra obtained from (A) cleaned Ni/Cr surface and (B) after room temperature exposure to 100 L of oxygen. (The Langmuir [L] is not an SI unit, but remains in common usage. 1 L is defined as a dose corresponding to one monolayer coverage for a sticking coefficient of unity.) (From Hoflund, G.B. and Epling, W.S., *Chem. Mater.*, 10, 50, 1998.)

electron-beam flux coupled with pulse counting. For a homogeneous sample, the electron beam may be rastered across the surface to minimize beam exposure. As with XPS, damage can be detected in AES by recording sequential spectra and looking for differences, as a function of electron-beam exposure.

3.3.4   ISS

ISS spectra recorded from an higher alcohol synthesis (HAS) catalyst are shown in Figure 3.7. They exhibit a variety of peaks arising from different elements including O, Na, Cl, Zn, Al, P, K, and Ti. Each peak lies at an E/E_o value close to that predicted by Equation 3.6.

Under the usual operating conditions, that is, light incident ion (He$^+$), relatively low incident ion energy (1–2 keV), and low incident ion current (10–50 nA), damage in ISS is negligible. However, when heavier ions are used, for example, Ar$^+$, often at higher energies, then there is a real danger of surface damage, which should be checked in a manner similar to that suggested for XPS and AES.

3.4   DEPTH SPECIFICITY

3.4.1   PES/XPS/UPS

Both XPS and UPS are surface-specific techniques in that the information obtained originates within the outermost 6 nm. In both techniques, the incident photons penetrate deeply, and do not govern the depth specificity. The deciding factor is the attenuation length (λ) of the photoelectrons. The attenuation length is shown in Figure 3.12 as a function of photoelectron KE for elements, inorganic compounds, and organic compounds [20]. The probability of an electron traveling a given distance x in a solid without scattering inelastically is given by

$$P(x) = ke^{-x/\lambda} \tag{3.8}$$

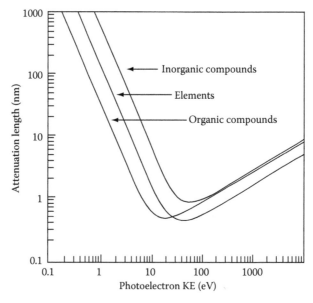

FIGURE 3.12 Dependence of the attenuation length on the photoelectron KE for elements, inorganic compounds, and organic compounds. (From Seah, M.P. and Dench, W.A., *Surf. Interface Anal.*, 1, 2, 1979.)

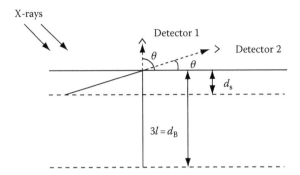

FIGURE 3.13 Schematic diagram illustrating how the collection angle affects XPS depth specificity. If λ of a photo-emitted electron is l, then 90% of the photoelectrons detected by detector 1 are emitted within a distance of $3l(d_B)$ beneath the surface. However, moving the detector to position 2 results in an equivalent detection of electrons emitted within a depth d_s, that is, reduction of the information depth by a factor $\sin \theta$. (From Hoflund, G.B., in Rivière, J.C. and Myhra, S. Eds.), *Handbook of Surface and Interface Analysis*, Marcel Dekker, Inc., New York, 1998.)

where k is a normalization constant. This equation and Figure 3.12 can be used to calculate the contributions to a given elemental peak for varying depths of the element beneath the surface. For any given element, the associated XPS peaks have different KEs and consequently different information depths. Over 90% of photo-emitted electrons originate from a depth less than 3λ, which defines roughly the depth specificity for a particular peak. The spectral features that have the greatest surface specificity are those with KEs near the minimum λ in Figure 3.12. For organic compounds such as polymers, this would be at about 18 eV, where λ is about 0.4 nm and the information depth about 1.1 nm, that is, between 5 and 8 atomic layers for most materials. For photoelectrons at higher KEs, around 1000 eV, from elements or inorganic compounds, λ is about 2.5 nm so that the information depth would be about 7.0 nm. Note that photoelectrons at very low KEs have low depth specificity. Photoelectrons emitted from the outermost atomic layers have the lowest probability of scattering inelastically so that they make the largest contribution to a peak. This contribution decreases exponentially with depth according to Equation 3.8.

XPS and UPS can be performed in angle-resolved modes in order to vary the information depth. For these modes, the only parameter that can be varied is the collection angle, as shown in Figure 3.13. If this collection angle is changed from normal to grazing by rotating the specimen (typically up to 80° off normal), a compositional profile can be obtained, with the help of various numerical procedures [21–23]. In UPS, variation of the collection angle can reveal differences in electronic structure and composition between the surface and the bulk. These differences include band bending and accentuation of some surface electronic states not allowed in the bulk electronic structure.

3.4.2 AES

The Auger electrons generated in AES obey the same scattering rules as do photoelectrons in XPS and UPS, and the Auger electrons in XPS. The difference in the technique is that both the KE and angle of incidence of the primary beam can influence the surface specificity. As the KE of the primary beam is decreased, the beam penetrates less deeply into the surface, thereby increasing the surface specificity. Also, if the incident angle of the primary beam is decreased from normal to grazing, the surface specificity is again increased, because although a grazing beam travels the same distance through a solid as would a normal beam, Auger electrons would then be originating nearer the surface, and the information depth decreases.

As in XPS and UPS, AES can also be used in the angularly resolved mode. There is more flexibility available in AES, because both the incidence and collection angles as well as the primary

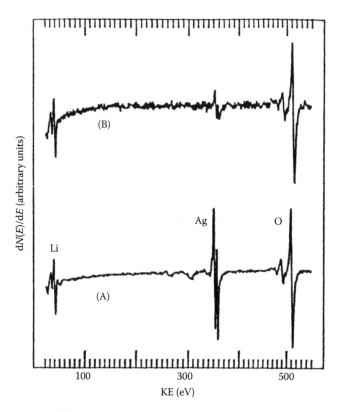

FIGURE 3.14 Angle-resolved AES (ARAES) spectra from a Li-contaminated, polycrystalline, Ag sample after annealing in O_2. Spectrum (A) was recorded using the conventional AES geometry (normal incidence and 42° off-normal collection angle), and spectrum (B) with an angle of incidence of 45° and a collection angle of 7°. (From Davidson, M.R., Hoflund, G.B., and Outlaw, R.A., *J. Vac. Sci. Technol. A*, 9, 1344, 1991.)

beam energy can be altered, and all these influence the surface specificity [24]. An example is shown in Figure 3.14 [25]. The spectra shown there were collected, using a cylindrical mirror analyzer (CMA), from a Li-contaminated Ag sample after annealing in O_2. Spectrum (A) contains three prominent features due to Li, Ag, and O, and was collected in the conventional manner using a primary electron beam at normal incidence and a collection angle of 42° off normal. Spectrum (B) was recorded in a surface-specific mode using angles of incidence and collection of 45° and 7°, respectively, with respect to the surface plane. The Li and O peaks were unchanged but the Ag peak size decreased greatly, indicating that the Li_2O was present as a film covering the Ag.

3.4.3　ISS

Atomic scattering cross sections are in fact larger than indicated schematically in Figure 3.6, which means that a primary ion scattered from an atom in the second atomic layer cannot easily escape without being scattered again by a surface atom. This and the fact that any ion penetrating beneath the surface has a very high probability of being neutralized imply that ISS is almost entirely specific to the outermost atomic layer. However, the situation is not always clear-cut. Figure 3.15 shows two possible atomic configurations of the outermost layer. In Figure 3.15A, ISS spectral features will be present for both elements A and B, regardless of incidence angle or collection angle. On the other hand, in Figure 3.15B features will be present for both A and B only if near-normal incidence and collection angles are used. For any other angles, there will appear a peak only for element A.

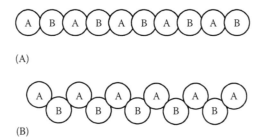

(A)

(B)

FIGURE 3.15   Two schematic models of the outermost atomic layer of a solid.

3.5   COMPOSITIONAL INFORMATION

Since quantification is the subject of Chapter 8, it will not be dealt with here in any detail. In any case, UPS does not provide direct compositional information even though the shapes of features in a UPS spectrum do depend upon composition. XPS and AES can provide good compositional information, as pointed out in Chapter 8, but that usually involves making the assumption that the sample is homogeneous. Except for pure metals, relatively few samples have homogeneous near-surface regions, and the matrix effect (i.e., the spatial distribution of elements in the near-surface region of a homogeneous sample) can have a very large effect on peak intensities [26]. Therefore, any quantification based on the assumption of homogeneity must be viewed with great caution. XPS and AES equipment manufacturers make the homogeneous assumption when programming their software, without providing detailed information, which means that the compositions calculated using the software may be quite far from reality. However, as also described in Chapter 8, study of the shape of the inelastic background under spectra can provide useful information about elemental distributions in subsurface layers in inhomogeneous samples, so the situation is not as difficult as it might seem.

Deriving compositional information from ISS spectra is also challenging for two reasons. The first is that accurate cross-sections are not available for elements as functions of the type of scattering ion and its energy, and the cross sections increase significantly as elemental mass increases [27], so that heavy elements give rise to larger relative peaks than do light elements. The second difficulty involves the neutralization probability, that is, the percentage of ions colliding with an element, that is neutralized, which is very high, ~99%. Thus, small variations from one element to another in this probability can have large effects on quantification. Studies of neutralization using a combination of ISS and the measurement of the scattering of neutrals would be useful, but have not been performed, so the usual assumption is that the neutralization probability is the same for all elements.

Since absolute quantification of XPS, AES, and ISS is not easy, another approach is to use relative quantification, that is, to compare spectra from related surfaces. This could take the form of preparing a number of samples in which just one parameter was varied systematically, and then collecting and comparing spectra. Even small differences in the spectra would then be significant and interpretable. A similar approach would be to subject a sample to various treatments in UHV, and again collect and compare spectra. The treatments could be oxidation, reduction, ion sputtering, annealing, etc., and the sample might oscillate between two distinct states as a result of these treatments. In a study of $TiO_2(001)$ [28], sequential reduction, oxidation, and sputtering was applied, with analysis after each treatment by XPS, AES, and ISS, and it was found that the chemical state at the surface changed systematically and reproducibly with the various treatments, in a cyclical manner. The study demonstrated that very small changes in features or lineshapes in spectra can be significant.

A useful way of reporting comparative XPS, AES, and ISS data is by way of peak-area ratios (in the case of AES, in the undifferentiated $N(E)$ vs. E spectra). Such ratios contain both compositional and matrix effects, of course, but the use of ratios tends to minimize those effects, and the resultant relative compositions are more accurate than those based on the assumption of homogeneity.

3.6　ELEMENTAL SENSITIVITY

The question often arises as to what are the elemental sensitivities in XPS, AES, and ISS. The answer is complex because the elemental sensitivity depends upon the technique, the element, the nature of the sample, including compositional and matrix effects, the instrument used, and the parameters used for data collection. Each of these is discussed below.

The three techniques XPS, AES, and ISS have different information depths, dependent in the first two on experimental conditions. ISS collects information from the outermost surface layer only, while in AES the information depth is governed by the incident and collection angles, and the primary beam energy; in XPS, it is determined by the incident photon energy and the collection angle. Thus, the observed sensitivity of a technique to any one element will depend on the spatial distribution of that element. If an element is not located in the outermost couple of layers, but is distributed in the immediate subsurface layers, then it will not appear in ISS spectra, but may yield large XPS and AES features. Again, an element spread uniformly over a surface will give rise to a larger signal for all three techniques compared to the same amount of that element present as clusters [26]. In a multielement sample, there may be overlapping peaks, which would affect sensitivity, since less prominent, but well-separated, peaks might have to be used.

The relationship between elemental sensitivity and the nature of the physical process is also different in each of the techniques. It is simplest for ISS, in that the scattering cross section increases monotonically with mass. In AES, the Auger electron yield per K-electron vacancy is higher than the x-ray yield for lighter elements, but becomes lower for heavier elements [6]. Relative sensitivity factors are provided by equipment suppliers for XPS and AES since the factors will depend on the analyzers used, and can vary by factors of 30–100. Thus, some elements are easier to detect than others. Using both AES and XPS can be advantageous because a given element may have a large sensitivity factor in one technique and a small one in the other. Carbon is an example, having a relatively high sensitivity factor in AES but a low one in XPS. Thus, a C-contaminated surface may appear to be relatively clean when using XPS but not when using AES.

Hydrogen is an element that is virtually impossible to detect with these methods. This is unfortunate because surface hydrogen is often present and can determine the chemical behavior of a surface. Surface hydrogen has been observed directly using ISS [29,30] but only at extremely small scattering angles that are not accessible with most ISS systems. Hydrogen has no core-level electrons so it cannot be observed directly with either XPS or AES, but it can sometimes be observed indirectly by XPS because of its presence as part of a surface group. For example, hydroxyl groups on oxide surfaces yield O 1s peaks at higher BEs than the oxide O 1s peaks. If it is necessary to establish that hydrogen, or an H-containing molecule, is definitely present, then either one of the SIMS family of techniques, or electron stimulated desorption (ESD), must be used.

All the various instruments available for XPS and AES produce different signal intensities for a given element for reasons such as differences in sample area analyzed, in electron optics and analyzer transmission function, and in type of electron detection. These topics are discussed in detail later in this chapter. The data collection parameters used can also have a great influence on elemental detectability. Elemental detectability can always be improved via the data collection parameters; for example, a small peak can be amplified by making multiple (N) scans over the energy range around the peak for a long time. The number of counts increases with N, but the background noise increases with $\sqrt{N}$. Using a lower energy resolution setting also increases the signal strength, but at the cost of poorer spectral resolution.

3.7　CHEMICAL-STATE INFORMATION

Of these techniques, XPS offers the best ability to provide chemical-state information. In this context, chemical state is often taken to mean that of an element in one or other of its known various valence states. What XPS actually measures is the influence on the BE of a given orbital of the electronic structure surrounding an element, whether the element is in a well-defined valence state or not. This is a more general definition of chemical state that is appropriate to the interpretation of XPS data. For example, on a completely oxidized surface, the metallic constituent may indeed be in one well-defined valence state. However, bombardment with energetic inert-gas ions will remove selectively oxygen atoms and ions, leaving an oxygen-depleted surface (i.e., a sub-valent oxide, see Chapter 10). The metallic element may then be reduced to a range of intermediate valence states, which is usually reflected in the XPS spectra [31].

Following their discovery in 1954 [32] that photoemission spectra could be used for elemental identification, Siegbahn et al. went on to demonstrate [33] that spectral shifts could be used to provide chemical-state information. Their classical example [34] is that of ethyl trifluoroacetate, whose C 1s spectrum is shown in Figure 3.16. Each of the four carbon atoms in this molecule has a different chemical environment, resulting in four different C 1s BEs, as seen in Figure 3.16. In most cases, however, BE shifts are not as obvious as in Figure 3.16 and may be difficult to detect.

This point is illustrated by the high-resolution spectra recorded from Kapton (a polyimide polymer used for coating spacecraft in orbit at altitudes of 200–700 km) [35]. The structure of Kapton is shown in Figure 3.17; it can be seen that the C atoms are in six different chemical positions and the O atoms in two. Survey spectra taken before and after exposure to atomic oxygen (AO) and to air give the compositions (calculated on the basis of a homogeneous distribution) shown in Table 3.1.

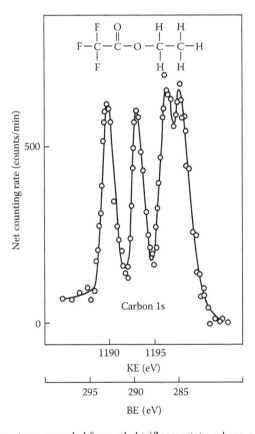

FIGURE 3.16 C 1s XPS spectrum recorded from ethyl trifluoracetate, whose molecular structure is shown at the top. The four C atoms are each in a different chemical environment, giving rise to peaks at four different BEs. (From Siegbahn, K., Nordling, C., Fahlman, A., Nordberg, R., Hamrin, K., Hedman, J., Johansson, G., Bergman, T., Karlson, S., Lindgren, I., and Lindberg, B., *Electron Spectroscopy for Chemical Analysis—Atomic, Molecular and Solid-State Structure Studies by Means of Electron Spectroscopy*, Almquist & Wiksells, Stockholm, 1967.)

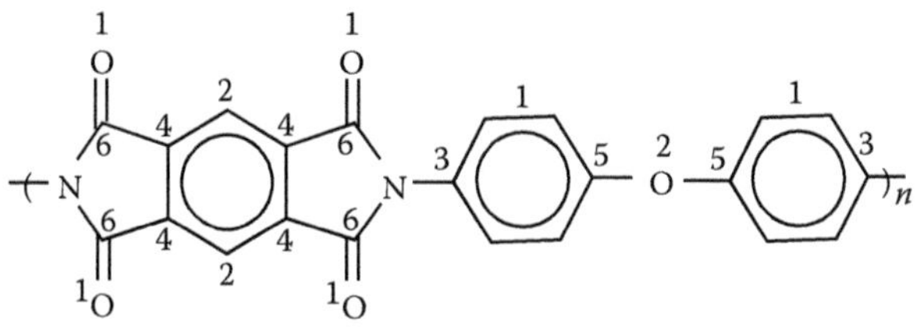

FIGURE 3.17 Molecular structure of poly(ether imide) (Kapton HN) with labeled atomic sites. The C atoms are in six different chemical environments, and the O atoms in two. (From Grossman, E., Lifshitz, Y., Wolan, J.T., Mount, C.K., and Hoflund, G.B., *J. Spacecraft Rockets*, 35, 75, 1999.)

TABLE 3.1
Near-Surface Composition Determined from XPS Spectra from Solvent-Cleaned, O-Atom, and Air-Exposed Kapton

Surface	Composition (%)		
Sample treatment	O	C	N
Theoretical	17.2	75.9	6.9
Solvent cleaned	18.1	77.7	4.2
20 min O-atom exposure	14.4	78.4	7.2
24 h O-atom exposure	9.2	83.0	7.8
3 h air exposure following 24 h O-atom exposure	17.9	78.2	3.9
RF plasma	28.4	64.9	6.7
Low-earth orbit environment	22.2	70.8	7.0

Source: Grossman, E., Lifshitz, Y., Wolan, J.T., Mount, C.K., and Hoflund, G.B., *J. Spacecraft Rockets*, 35, 75, 1999.

The as-received Kapton surface had a composition close to the theoretical, while on AO treatment, the O 1s peak intensity decreased significantly compared to that of the C 1s. However, the high-resolution C 1s, O 1s, and N 1s spectra in Figure 3.18 provide much more information. In the as-received state (a), the C 1s spectrum contains one large and rather broad peak, corresponding to unresolved peaks due to the C atoms labeled 1–5 in Figure 3.17, and a smaller peak at higher BE that can be attributed to the C atom in position 6, bonded to carbonyl. With increasing AO exposure, the smaller peak diminished, thought to be a result of surface reaction to form CO_2, which desorbed. At the same time, the peak in the O 1s spectrum due to carbonyl also decreased, but to a greater extent than the C(6) peak, so that the O/C compositional ratio also decreased. If this interpretation is correct, then the N 1s spectrum should also have altered, since all the N atoms in Kapton are bonded to carbonyl groups, and indeed that is what was found, as can be seen from the N 1s spectrum. With AO exposure, the predominant N 1s peak diminished as a new one formed at a lower BE and increased in intensity.

In some previous studies on Kapton [36,37], it was found that the surface oxygen content increased with AO exposure, contrary to the above observations. In those studies, the samples were exposed to air after the AO exposure and before taking XPS data. After the 24 h AO exposure, the sample described above was exposed to air for 3 h and then analyzed again, whereupon the O content was found to have nearly doubled. This demonstrates that it is important to carry out such experiments *in situ* in order to understand the processes occurring during AO exposure, since air exposure alters the results.

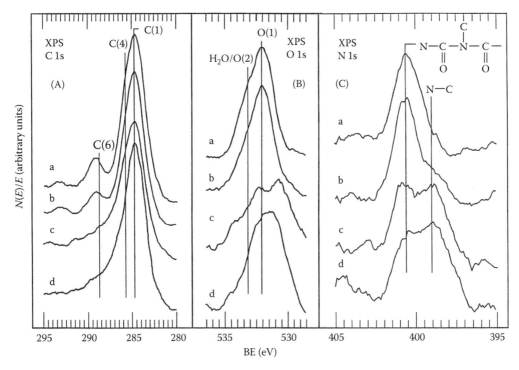

FIGURE 3.18 High-resolution C 1s (A), O 1s (B), and N 1s (C) XPS spectra from a solvent-cleaned Kapton film after (a) insertion into the vacuum system, (b) 20 min, (c) 24 h exposure to the hyperthermal O-atom flux, and (d) 3 h air exposure following the 24 h O-atom exposure.

In some cases, air exposure between treatment and XPS analysis does not affect the results, but this should always be tested and never assumed. An example is shown in Figure 3.19 [38]. In this study, Pd metal, anhydrous PdO, and hydrous PdO powders were given various treatments in a catalytic reactor operating at 1 atm, exposed to air, and then analyzed. The spectrum shown in (a) was taken from Pd metal powder that had been reduced in hydrogen in the reactor, and consisted of only the metallic Pd 3d features, indicating that the reduction was complete, and that air exposure at room temperature did not result in oxidation of Pd. Spectrum (b) was recorded after exposing the reduced Pd metal powder to methane oxidation conditions in the reactor, transferring the sample in air to the XPS system, and then collecting XPS data. Features due to both Pd metal and PdO were clearly present, which is interesting because the reaction mixture consisted of methane, which is a reducing species, and oxygen, an oxidizing species. This spectrum thus indicates that a layer of PdO had formed over the Pd metal, showing that PdO was the catalytically active species under the conditions.

The spectra shown in (c) and (d) were from fresh anhydrous PdO before and after reduction, respectively, in hydrogen in the reactor. The treatment converted the near-surface PdO back to metallic Pd, but there was peak broadening on the high-BE side due to the presence of subsurface PdO. A similar treatment on hydrous PdO gave similar results, as shown in (e) and (f). A mild sputter treatment with Ar ions also reduced hydrous PdO to Pd metal as shown in (g).

In order to understand the behavior of a complicated system, the different elemental spectra from all the elements involved should be compared, and consistency attained. Analysis of the spectrum from just one element may give a partial explanation, but a more complete explanation can be found only by analysis of all of the relevant peaks. This also provides a check on the consistency of the analysis, since if it is found that changes in all the spectra are not self-consistent, then the interpretation must be rethought. For example, consider the thermal decomposition of silver carbonate (Ag_2CO_3) as studied by XPS [9]. A fresh silver carbonate powder sample was analyzed before and

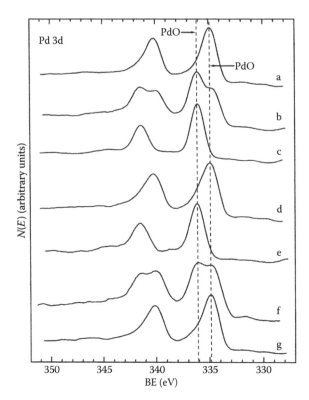

FIGURE 3.19 High-resolution Pd 3d XPS peaks from (a) and (b) Pd powder after hydrogen and methane/oxygen oxidation, respectively, (c) and (d) anhydrous PdO before and after reduction, respectively, (e) and (f) hydrous PdO before and after reduction, respectively, and (g) PdO after ion sputtering. The specimens were exposed to air between each treatment. (From Hoflund, G.B. and Hagelin-Weaver, H.A.E., to be published.)

after annealing at 350°C and then at 500°C (survey spectra shown in Figure 3.8). The resultant Ag 3d spectra are shown in Figure 3.20A, in which the peaks recorded from the fresh sample exhibited a single peak at a BE characteristic of either Ag_2CO_3 or AgO. Although the sample was nominally Ag_2CO_3, the presence of AgO could not be eliminated. After annealing at 350°C, another single peak was present with a BE characteristic of Ag_2O, suggesting that Ag_2CO_3 decomposes by losing a CO_2 molecule to form Ag_2O. Annealing at 500°C resulted in another single Ag 3d peak at the BE characteristic of Ag metal, due to desorption of O_2.

Now consider the corresponding O 1s and C 1s spectra shown in Figure 3.20B and C, respectively. Again, the spectra labeled (a) in these figures were obtained from fresh Ag_2CO_3. The predominant O 1s peak was due to either Ag_2CO_3 or $AgHCO_3$. The presence of a significant shoulder due to hydroxyl groups or possibly water suggests the presence of bicarbonate. No O 1s features due to Ag_2O or AgO were present, which indicates that these species were absent and the Ag was present as Ag_2CO_3. The corresponding C 1s spectrum exhibits two well-defined and well-separated peaks due to carbonate/bicarbonate species and adsorbed hydrocarbons, respectively, the latter being present on all air-exposed surfaces. Annealing at 350°C produced the spectra shown in (b). A C 1s peak was not positively identifiable in the survey spectrum, and the O 1s peak was reduced by a factor of about 3 as the Ag_2CO_3 was converted to Ag_2O, consistent with the O 1s spectrum (b) in Figure 3.20B, which exhibits a well-defined Ag_2O feature as well as a feature due either to hydroxyl groups or to adsorbed water. The spectrum in the C 1s region was essentially noise. After annealing at 500°C, the O 1s peak in the survey spectrum decreased greatly, while the

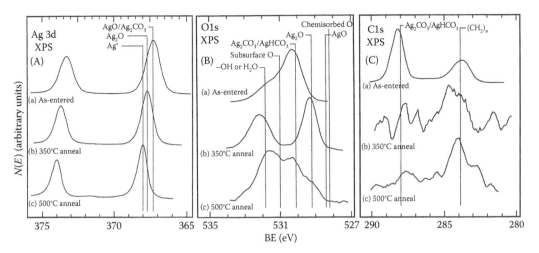

FIGURE 3.20 High-resolution Ag 3d, O 1s, and C 1s XPS spectra from an Ag_2CO_3 sample (a) as-entered, and after annealing at (b) 350°C for 10 min and at (c) 500°C for 10 min. (From Salaita, G.N., Hazos, Z.F., and Hoflund, G.B., *J. Elect. Spec. Rel. Phenom.*, 107, 73, 2000.)

high-resolution O 1s spectrum indicated that the low level of oxygen remaining was present as hydroxyl groups/water, subsurface dissolved oxygen, and $Ag_2CO_3/AgHCO_3$. The C 1s feature was consistent in that a small amount of $Ag_2CO_3/AgHCO_3$ was present as well as some hydrocarbons. This group of spectra demonstrates that the chemical behavior of the system was much more complex than that indicated by the Ag 3d spectra on their own.

In principle, chemical-state information can also be derived using AES, by observing changes in peak shapes and positions, but the complexity of the Auger process means that interpretation is usually too difficult. No chemical-state database has yet been formulated for AES. There are, however, special cases where AES can provide useful chemical-state information.

3.8 SPECTRAL RESOLUTION

It can be seen in the above examples that peaks recorded at high resolution often overlap. Ideally, peaks would be very narrow and well separated. In practice, they are broadened by three factors: (1) the full width at half-maximum (FWHM) of the x-ray line, ΔE_x, (2) the width of the analyzer energy window, ΔE_A, and (3) the natural line width of the orbital in the atom, ΔE_2. The overall energy resolution (ΔE in units of eV) is given by

$$\Delta E = \left(\Delta E_x^2 + \Delta E_A^2 + \Delta E_2^2 \right)^{1/2} \tag{3.9}$$

ΔE_2 cannot be altered, while the analyzer energy window width can be set so that it does not limit resolution. Thus, it is the x-ray line-width that usually limits the overall energy resolution in XPS, which is the critical factor in the unambiguous extraction of chemical-state information. Mg K_α x-rays have a natural line-width of 0.70 eV, while line widths from other possible anode metals can be as broad as several electron volts (Table 3.2). Reduction in x-ray line-width, and hence improvement in resolution, can be achieved by the use of a monochromator. For example, the Mg and Al K_α widths can be reduced to about 0.35 eV. Monochromatization also removes the x-ray satellite lines and most of the Bremsstrahlung background, leading to cleaner spectra. The disadvantage of a monochromator is that the incident x-ray flux is reduced significantly (by as much as a factor of 40), thereby reducing the photoemission signal strength. Improvements in signal detectability, for example, multichannel detection, are beginning to overcome this problem. When a spectrum contains

TABLE 3.2
X-Ray Sources

Source[a]	Energy (eV)	Line Width (eV)
Zr M_ξ	151.4	0.77
Mg K_α	1253.6	0.70
Al K_α	1486.7	0.85
Mono-Al K_α	1486.7	0.26
Mono-Si K_α	1739.9	0.35
Zr L_α	2042.4	1.7
Au M_α	2122.9	
Ag L_α	2984.3	1.3
Mono-Ti K_α	4510.0	2.0
Mono-Cr $K_{\beta1}$	5946.7	2.0
Mono-Cu K_α	8047.8	2.4

Sources: Patthey, F. and Schneider, W.-D., *J. Electron Spectrosc. Relat. Phenom.*, 81, 47, 1996; Beamson, G., Haines, S.R., Moslemzadeh, N., Tsakiropoulos, P., Weightman, P., and Watts, J.F., *Surf. Interface Anal.*, 36, 275, 2004; Moslemzadeh, N., Beamson, G., Haines, S.R., Tsakiropoulos, P., Watts, J.F., and Weightmam, P., *Surf. Interface Anal.*, 38, 703, 2006; Diplas, S., Watts, J.F., Morton, S.A., Beamson, G., Tsakiropoulos, P., Clark, D.T., and Castle, J.E., *J. Electron Spectrosc. Relat. Phenom.*, 113, 153, 2001; Vargo, T.G. and Gardella, J.A., Jr., *J. Vac. Sci. Technol. A*, 7, 1733, 1989; Wagner, C.D., *J. Electron Spectrosc. Relat. Phenom.*, 15, 518, 1978.

[a] "Mono" refers to use of a monochromator.

several closely spaced and overlapping peaks whose separation and identification is essential for the analysis, then it is worth compromising on the loss in signal strength caused by using a monochromator.

According to Equation 3.1, only core levels with BEs less than the energy of the x-ray line can be ionized, which, with the conventional Mg and Al sources, can occasionally be a limitation. Photoelectrons from levels with higher BEs can of course be excited by x-rays of higher energy. One source that has been found useful for that application is Ti, whose K_α line energy is at ~4510 eV. However, the line width is 2.0 eV, so that the energy resolution, and hence the ability to acquire chemical-state information, is much reduced. The situation is not helped by the fact that chemical shifts in core levels at higher BEs are usually small.

The cross-section for photoemission of an electron in a core or molecular level is dependent on the photon energy. An important consequence of this statement relates to the VB electrons that have large cross-sections for photoemission by UV light but very small ones for photoemission by x-rays. A good choice of anode for XPS VB photoemission is therefore Zr, for which the M_ξ line has a photon energy of 151.4 eV and a line width of ~0.77 eV. With Zr M_ξ radiation the VB cross-sections are large, the spectral resolution is not significantly limited by the photon line-width, and electrons in core levels with BEs between 20 and 145 eV can also be excited. The latter are not accessible with most UV sources and are useful because they are quite sensitive to chemical state, even if a compositional analysis cannot be made.

Curve-fitting techniques to separate the contributions of the various species are widely used, and manufacturers usually supply programs for them, but, caution should be exercised when using them.

It is all too easy to allow a program free rein to produce a fit involving many component peaks, some of which may have no chemical or physical meaning. If possible, it is better to have available a set of peak shapes and positions of all the species that may be present, and then to add these together in the appropriate proportions to obtain an envelope that is the best fit to the experimental one. The use of an x-ray monochromator for recording both the experimental envelope and the spectra of the standards yields the best results.

3.9   DEPTH PROFILING

Most samples encountered are layered, either naturally or intentionally, or are otherwise spatially inhomogeneous, over depths greater than the information depths available with these techniques. The layer structure or the inhomogeneity can in principle be uncovered by depth profiling using inert-gas ion sputtering [45]. Surface analysis performed after each sputtering dose then provides some sort of elemental depth profile. However, there are several processes that may distort the compositional profiles so obtained, and in addition there is the problem of ion-beam damage leading to chemical reduction of some species. Nevertheless, under carefully controlled conditions it can prove useful. Analysis of buried information is discussed in Chapter 10.

3.10   MODULAR INSTRUMENTATION

The four techniques described here generally share a similar instrumental platform and have a functional modularity, but of course the actual components used for a module vary with the technique. In this section, upgrades to existing modules, as well as new modules that have been introduced in the last 10 years, are discussed. In the following sections, the modules for each of the techniques are discussed individually, concentrating on the commercially available instruments and their figures of merit, and also on emerging developments in the types of samples analyzed and the conditions under which they can be analyzed.

3.10.1   EXCITATION SOURCES

The primary excitation sources are a flux of vacuum ultraviolet (VUV) light for UPS, of x-rays for XPS, of electrons for AES, and of ions for ISS. Ideally, all the particles making up any one flux are identical, that is, the photon and the electron fluxes would be mono-energetic, and the ion flux would consist of ions of a single mass, energy, and charge.

3.10.1.1   UV Sources

In conventional UPS, UV photons are produced using an inert gas, usually He but occasionally Ne, in a plasma-discharge lamp. If He is being used, then, according to the pressure of He in the discharge capillary, either the HeI line at 21.21 eV or the HeII at 40.82 eV can be selected. Control of the gas pressure in the discharge zone is crucial, particularly when trying to maximize the HeII or NeII line intensity, and it is normal to leave a lamp operating for a long time continuously once it has started, to save time. A system of differential pumping in the source housing ensures that UHV can be maintained in the experimental chamber. The line-widths of the resonant lines are narrow (~20 meV), so that, unlike the x-ray sources for XPS, there is negligible contribution to instrumental broadening from the UV source. The most important application of UPS is in the angle-resolved mode, in which the angular orientation of the specimen, as well as that of the analyzers, can be changed, so that all emission angles are covered.

As discussed in Section 3.3.2, UPS can also be performed using radiation from a synchrotron, and indeed much additional and valuable information can be gathered by so doing. For a full description of the design of, and techniques available with, synchrotrons, see Chapter 7.

3.10.1.2   X-Ray Sources

X-rays are generated by accelerating electrons emitted from a heated filament onto a metal anode with an applied voltage of 10–15 keV. The x-ray spectrum so generated consists of a continuous radiation band (Bremsstrahlung) on which discrete lines characteristic of the anode material are superimposed. For a standard x-ray source using a dual Mg/Al anode source, a thin aluminum foil is interposed between the anode and the sample, to reduce the Bremsstrahlung background, to prevent stray electrons from hitting the sample, and to maintain the UHV in the system. Since the imaginary component of the atom scattering coefficient decreases monotonically with increasing energy both before and after an absorption edge, the thin aluminum foil functions as an imperfect high-energy cutoff filter. Without monochromatization, seven different Mg or Al K-emission lines are produced: the unresolved $K_{\alpha 1,2}$ and the $K_{\alpha 3}$, $K_{\alpha 4}$, $K_{\alpha 5}$, $K_{\alpha 6}$, and K_{β} lines [10]. Of these, the $K_{\alpha 1,2}$ lines are the most intense. Thus, an XPS spectrum will contain peaks excited not only by $K_{\alpha 1,2}$ but by the other five emission lines as well. The peaks excited by the minor emission lines are called satellites. The $K_{\alpha 3}$ and $K_{\alpha 4}$ source satellites are of sufficient intensity (3%–8% of the principal lines) for many vendors to provide a software algorithm that allows them to be stripped from the observed spectra.

For complete removal of satellite lines, and nearly all the Bremsstrahlung radiation, coupled with a valuable improvement in line width, an x-ray monochromator should be used. It so happens that, for Al x-rays, diffraction from the $(10\bar{1}0)$ plane of a quartz crystal at a Bragg angle of 78.5° selects the larger component of the K_{α} doublet, and rejects all other wavelengths. If then the x-ray source anode is placed at a particular point on a focussing sphere (often of diameter 0.5 m), the quartz crystal at another, and the sample at a third, x-rays will be dispersed by the crystal and refocussed on the sample. In this way, the line width of the Al K_{α} radiation is reduced from 0.85 to 0.35 eV. Another advantage of a monochromator is that the hot anode is removed from the near vicinity of the sample, thereby eliminating any possible degradation of the sample by heating. However, a disadvantage is that the x-ray intensity is reduced by a factor of up to 40. In principle, it is also possible to use Ag L_{α}, Ti K_{α}, and $Cr_{\beta 1}$ radiations in monochromators, since their photon line-energies are multiples of Al K_{α}, which means that a quartz crystal could again be used for dispersion. The use of higher energy x-rays for special applications is becoming more common, and many vendors sell sources with anodes other than Mg and Al. Table 3.2 is a tabulation of x-ray sources in current use. They are all commercially available except for monochromated Si K_{α} [39] and Cu K_{α} [40,41]. The former uses a quartz crystal at a Bragg angle of 56.85°, but requires special procedures to prepare the Si anode. The latter uses an LiF(220) crystal.

Several different figures of merit can be used to characterize optical sources, and, in principle, similar ones may be used to compare XPS sources. One such figure of merit is the spectral radiance (L_{λ}), defined as the radiance (power per unit area per steradian) per unit wavelength with SI units of W m^{-2} sr^{-1} m^{-1}, although the units W m^{-2} sr^{-1} nm^{-1} are usually used [46]. Since $h\nu$, and not λ, is the parameter usually used in XPS, the analogous figure of merit might be $L_{h\nu}$, with units of W m^{-2} sr^{-1} eV^{-1}. This would be a very useful parameter for comparing sources with different anode materials operating at the same source power, but, unfortunately, is not provided by suppliers, so that research workers simply have to make qualitative comparisons based on the source power (in W), as determined by the maximum filament current and the maximum voltage that may be applied to the anode. On output this energy appears mainly as thermal energy, which is usually removed by cycled chilled water. Only about 1% appears as spectral energy. Radiance can be calculated from the source power provided that the area of the source, the solid angle of emission, and the efficiency of conversion of supply power to photon power are known. To determine the spectral radiance, the actual spectral distribution is needed. Several factors determine the spectral distribution of the x-ray source. One is the geometry, as shown in Figure 3.21, for the x-ray photon intensity distributions of three different source geometries [47]. A second is the target material of the source for a fixed geometry and electron power, as shown in Figure 3.22 [47]. In Figure 3.23, it can be seen that the intensities of the

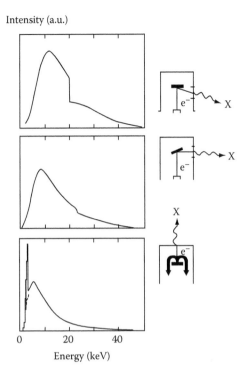

FIGURE 3.21   Measured x-ray source spectral continuum distributions: effect of source geometry. (From Broll, N. and de Chateaubourg, P., *Adv. X Ray Anal.*, 41, 393, 1999. With permission.)

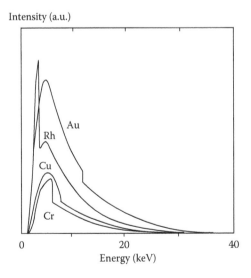

FIGURE 3.22   Measured x-ray source spectral continuum distributions: effect of target material. (From Broll, N. and de Chateaubourg, P., *Adv. X Ray Anal.*, 41, 393, 1999. With permission.)

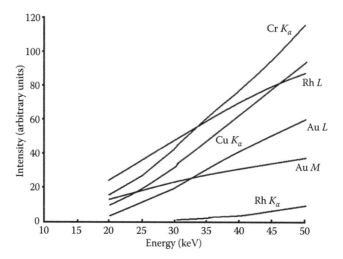

FIGURE 3.23 Measured x-ray source line intensities as a function of electron power. (From Broll, N. and de Chateaubourg, P., *Adv. X Ray Anal.*, 41, 393, 1999. With permission.)

x-ray lines as a function of electron power all increase monotonically, but the slopes depend on the individual material [47]. In order to evaluate experimental differences in photoionization cross-sections for different sources, the measured intensities must be referenced to a standard level recorded using all sources.

One new source that is not yet commercially available, but has great potential, is the table-top synchrotron [48], in which the radius of the electron orbit is a mere 15 cm. Under the operating conditions described in Ref. [48], it has a spectral radiance similar to that of a rotating anode, that is, a factor of about 10 brighter than that of a fixed anode. The spectral radiance can be increased by a judicious choice of target. However, the real potential of a synchrotron lies in its ability to tune the emergent x-rays over an energy range of several keV up to 20 MeV. Furthermore, the spectral distribution function is almost flat over this range, that is, nearly ideal. Many important photoemission studies that cannot be performed with classical x-ray sources can be with synchrotron radiation. The use of radiation from large diameter (tens of meters) synchrotrons also allows chemical imaging at a lateral resolution of around 20 nm using a photoelectron microscope. Unfortunately, there are only about 50 operational synchrotron facilities worldwide, and there is greater demand than time available. Although the table-top synchrotron, at its present level of development, will not achieve 20 nm lateral resolution, it is likely to be reasonably affordable compared to large synchrotrons, thereby allowing greater numbers of research workers to carry out experiments of greater sophistication.

3.10.1.3 Electron Sources

Electron sources for AES and SAM consist of an emitter, either thermionic or field emission, a set of electron lenses for beam focus and transport, and an optional electron energy selector for high-energy resolution. Almost all electron guns provided by suppliers come with a four-pole electrostatic deflector, which allows the beam to be moved in any direction perpendicular to the optical axis; the deflection system for scanning Auger microscopy (SAM) is more complex. The beam stability is determined by the power supply. Figures of merit used for these electron sources are radiance, spot size, and energy spread, and a comparison of them for the various sources is shown in Table 3.3. The suppliers normally make this information available.

Field-emission sources have radiances that are similar to those of synchrotron sources; for example, a synchrotron source operated at 3 GeV and 300 mA stored beam current, an x-ray conversion efficiency of 0.01, a solid angle of $2\pi \times 10^{-3}$ sr, and an area of 0.1 mm^2, has a radiance of

---

TABLE 3.3

Comparison of Electron Sources for AES/SAM

Source	Radiance (W mm^{-2} sr^{-1})	Spot Size	Energy Spread ΔE (eV)
W	2×10^7	30–100 μm	1.0–3.0
Thoriated W[a]	4×10^8		
LaB$_6$	2×10^8	5–50 μm	1.2
Field emission			
Cold	2×10^{10}	<5 nm	0.3
Thermal	2×10^{10}	<5 nm	1.0
Shottky	2×10^{10}	15–30 nm	0.3–1.0

[a] Calculated at 1900 K using the Richardson–Dushman equation and the maximum brightness equation [49].

~2×10^{11} W mm^{-2} sr^{-1} [50], to be compared to those quoted in Table 3.3 for field emission. The energy spread in field-emission sources rivals those of x-ray sources used for XPS. They are used in the more sophisticated SAM instruments, where lateral resolution in the nanometer range is currently the best achieved in dedicated laboratory instruments. The lifetimes of the various types of filament are determined by the material properties and the operating conditions. On the horizon is a new field-emission cathode, based on an array of carbon nanotubes. It has been demonstrated that a single carbon nanotube with a tip diameter of ~20 nm has a brightness that is a factor of 10 greater than that of conventional field emitters [51]. Field-emission carbon-nanotube cathodes have been used to make miniature x-ray sources with brightnesses comparable to classical ones [52]. However, the x-ray line-widths are much too large for photoemission.

3.10.1.4 Ion Sources

Ion sources for ISS comprise a beam of ions, electrostatic lenses for focus and ion transport, an optional mass filter, and an optional energy filter/selector. They are available with or without deflectors, the latter being used mostly for sputter cleaning. The most commonly used ion is He$^+$, but Ne$^+$ and Ar$^+$ have also been employed; the ions are generated from the gases by electron impact ionization. Alkali metal ions, that is, Li$^+$, Na$^+$, and K$^+$, produced by the surface ionization of directly heated alkali aluminosilicate plugs are also used. Figures of merit for ion sources include the ranges of beam energy, beam current, and spot size, and the spread in the beam energy; beam current and spot size are adjustable independently. Typical operating parameters for a noble-gas ion source for ISS are as follows: beam energy 10 eV–5.0 keV, beam current 1–50 nA, spot size 1–20 mm, and energy spread <5 eV at low current. For an alkali ion source, the corresponding figures would be as follows: beam energy 50 eV–5.0 keV, beam current 1–100 nA, spot size 1–10 mm, and energy spread 0.4 eV (calculated for thermal spread).

The typical energy range for primary ions used in ISS is in fact 1–2 keV, which means that the energy spreads for the sources are acceptable, and an energy selector would be an unnecessary expense. Mass filters, which actually filter according to mass/charge ratio, remove multiply charged ions from the transported beam. When using He, the most commonly used gas, multiply charged ions are not a problem, and so a mass filter is not normally required.

A less common type of ion source used in ISS is the duoplasmatron, a schematic of which is shown in Figure 3.24; it is used when negatively charged primary ions are to be scattered from surfaces. A gas discharge is generated between a cathode and an anode, and the plasma so formed

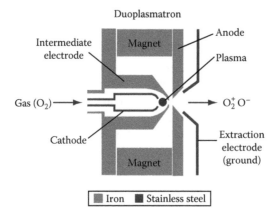

FIGURE 3.24   Schematic diagram of a duoplasmatron ion source. (From Evans Analytical Group, Online SIMS Instrumentation Tutorial. With permission.)

contains positive ions, negative ions, and neutrals. Either positive ions or negative ions can be extracted, but neutrals with velocities parallel to the source axis and moving toward the exit will also enter the next stage. Deflection is therefore used to separate ions from neutrals, followed by a mass filter to select ions with the desired mass/charge ratio. Duoplasmatron sources are bright sources with energy spreads of the order of 15 eV, and spot sizes of 40–300 μm. A duoplasmatron-like ion source that produces a microbeam with a spot size less than 0.1 μm is being developed [54].

3.10.2 ENERGY ANALYZERS

An energy analyzer consists of an energy dispersing element, entrance and exit slits, and (optionally) entrance and exit lenses. Many commercial instruments use an entrance lens, but only a few an exit lens. In XPS, μXPS, iXPS, AES, and SAM, energy dispersion is achieved by deflection in an electrostatic field. The three types of electrostatic energy analyzer commonly in use are the CMA, the double-pass CMA (DPCMA), and the concentric hemispherical analyzer (CHA). An electrostatic energy analyzer is also the commonly used method of energy dispersion in ISS, but time-of-flight (TOF) analyzers are also used, in which the primary ion-beam is pulsed, and the time taken for a scattered ion to reach a detector, at a known distance from the surface, is measured. In the DPCMA, better energy resolution is achieved by placing two CMAs in series. A schematic of a CMA is shown in Figure 3.25. The analyzer consists of a pair of coaxial cylinders with entrance and exit apertures at the front and back ends, respectively, of the inner cylinder. To reach the detector, a particle must travel along a trajectory that takes it from the entrance aperture, through the space between the two cylinders, through the exit aperture, and into the detector. An ideal CMA has a circular entrance aperture of zero width. The position of the entrance slit relative to the sample position, S in Figure 3.25, determines the entrance angle, α. In the figure, α is the angle between the cylinder axis and the central trajectory shown beginning at S. The distance L from S to the focal point on the axis, F in Figure 3.25, depends on α. To discriminate charged particles on the basis of their energies, an electric field is set up between the two cylinders. The inner cylinder is usually grounded, and for the analysis of electrons a negative potential, $-V$, is used to generate an electrical field between the two cylinders. Only those electrons with energy E satisfying the condition

$$E = K_o eV \left[\ln\left(\frac{R_2}{R_1}\right) \right]^{-1} \tag{3.10}$$

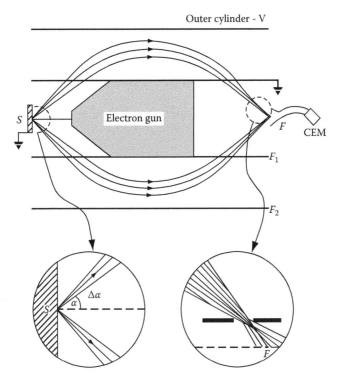

FIGURE 3.25   Schematic of a CMA. (From Seah, M.P., *Methods of Surface Analysis*, Cambridge University Press, Cambridge, U.K., 1989. With permission.)

are focussed at F, where

K_o is a constant
e is the elementary charge
V is the magnitude of the applied potential
R_1 is the radius of the inner cylinder
R_2 is the radius of the outer cylinder [55]

The same equation applies for univalent positive ions (charge = Ze, with $Z = 1$), but a positive potential, $+V$, is applied to the outer cylinder. For $\alpha = 42° \ 18.5'$, the analyzer becomes a second-order focussing device, since the first-order spherical aberrations vanish, and then $K_o = 1.3099$. In practice, the entrance aperture has a finite width, which allows electrons with entrance angles of $\alpha - \Delta\alpha$ to $\alpha + \Delta\alpha$ to enter into the space between the two cylinders. For small values of $\Delta\alpha$ on either side of $\alpha = 42° \ 18.5'$, electrons of energy E will be brought to the same focal point, therefore all CMAs are constructed using this particular acceptance angle. However, an unavoidable consequence of a finite width entrance slit is that electrons with energies slightly smaller and slightly larger than E will also be brought to focus very close to F. This leads to an uncertainty in the value of ΔE for any given applied deflecting voltage. Notice that all the electron trajectories cross in a small region of space within the inner cylinder just before they reach the detector. This crossing region is called the circle of least confusion, and is shown magnified below the analyzer in Figure 3.25.

If there is no ring slit at the circle of least confusion, the relative energy resolution is given by

$$\frac{\Delta E}{E} = \frac{C_S (\Delta\alpha)^3}{D_E + C_C (\Delta\alpha)} \tag{3.11}$$

where

D_E is the specific linear energy dispersion
C_S is the spherical aberration coefficient
C_C is the chromatic aberration coefficient
$\Delta\alpha$ is the in-plane half-angle of acceptance into the analyzer (in radians)

If a ring slit of width W is placed at the circle of least confusion, the relative energy resolution is then given by

$$\frac{\Delta E}{E} = \frac{0.9W}{R_1} + \frac{1}{4}\left[\frac{C_S(\Delta\alpha)^3}{D_E + C_C(\Delta\alpha)}\right] \tag{3.12}$$

where R_1 is the radius of the inner cylinder. An appropriate choice of W for any given set of concentric cylinders will result in improved resolution, but at the expense of a reduction in transmission. The ring slit is not normally adjustable by the user, which represents a consideration that must be addressed at the time of purchase. Although the energy resolution of a CMA is adequate for classical AES and ISS, it is not adequate for XPS or for lineshape analysis in AES. In order to obtain better energy resolution, the DPCMA was developed, and an example is shown in Figure 3.26A. Two hemispherical, retarding grids are placed in front of the sample. The nearer hemispherical grid is at the same potential as the sample, usually ground. When used as an electron energy analyzer, a negative potential, $-V_1$, is applied to the second hemispherical grid in order to decelerate the electrons to a selected pass energy E_p. The same potential is applied to the inner cylinder. A potential $-V_2$, such that V_2 is greater than V_1, is applied to the outer cylinder, the difference $V_2 - V_1$ determining the pass energy. A spectrum is recorded by ramping the potential on the inner cylinder while maintaining the potential difference between the outer and inner cylinders. The analyzer in Figure 3.26A has four individually selectable entrance/exit slits in the inner cylinder, and two apertures that lie in planes perpendicular to the cylinder axis, one at the midpoint and one at the end of the inner cylinder. During the first pass, electrons are focussed into a region that is smaller than the entrance region. This becomes a source for the second pass CMA, which overall has the effect of reducing $\Delta\alpha$. Because the pass energy is fixed, so is ΔE throughout the spectrum. The improvement in energy resolution comes with a loss in analyzer transmission.

A CHA consists of two concentric hemispheres, of radii R_1 for the inner and R_2 for the outer. The trajectory along which a charged particle moves from the entrance to the exit slits of the analyzer lies on a hemispherical surface with radius R_o, between the two hemispheres. R_o is equal to ½$(R_1 + R_2)$. An electron of energy E can move from the entrance slit to the exit slit if the condition

$$E = \left(\frac{R_1 R_2}{R_2^2 - R_1^2}\right)e(V_2 - V_1) \tag{3.13}$$

is satisfied, where $-V_1$ and $-V_2$ are the potentials applied to the inner and outer hemispheres, respectively, with V_2 greater than V_1 (i.e., the inner hemisphere is more positive than the outer hemisphere) [55]. A spectrum is recorded by scanning the applied potential difference across the concentric hemispheres. For a CHA used without a transport lens, the relative energy resolution is given by the recursive relationship

$$\frac{\Delta E}{E} = \frac{W_1 + W_2}{2} + \frac{C_S(\Delta\alpha)^2}{D_E + C_C(\Delta\alpha) + C_E\left(\dfrac{\Delta E}{E}\right)} \tag{3.14}$$

where

W_1 and W_2 are the entrance and exit slit widths, respectively
C_E is the nonlinear energy dispersion coefficient

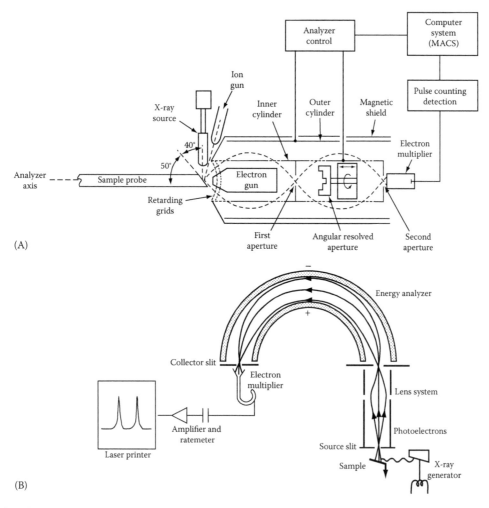

FIGURE 3.26 Schematics of (A) a DPCMA and (B) a CHA. (From Davis, L.E., MacDonald, N.C., Palmberg, P.W., Riach, G.E., and Weber, R.E., *Handbook of Auger Electron Spectroscopy*, Physical Electronics Industries Inc., Eden Prairie, MN, 1976; Hughes, A.E. and Sexton, B.A., *J. Electron Spectrosc. Relat. Phenom.*, 46, 31, 1988.)

The entrance and exit slit widths are adjustable by the user, from outside the instrument, by means of feedthroughs, although with some instruments they are not independently adjustable, in which case $W_1 = W_2$. As with the CMA, there will be a trade-off between transmission and resolution, but at least the user can control this trade-off. For any given set of parameters, the relative energy resolution is constant, so that the absolute energy resolution will increase with increasing E. In order to obtain a constant absolute resolution, a transport lens placed between the sample and the analyzer input slit is used to decelerate or accelerate electrons to a selected pass energy, in the range 5–100 eV. The potential difference across the concentric hemispheres is therefore fixed, and spectra are recorded by scanning the transport lens.

Even though the DPCMA is competitive with the CHA-transport lens system with respect to energy resolution, the latter combination is superior with respect to the versatility of sample positioning, the ability to add additional components near the sample position, and the potential of achieving further improvements by increasing the number of elements in the transport lens. For these reasons, the demise of the DPCMA had been anticipated. However, if the claims made by STAIB Instruments for the so-called super-CMA and double-pass super-CMA, which they introduced in 2004 [57], are true, there may be a revival in the use of the CMA-type analyzer. These

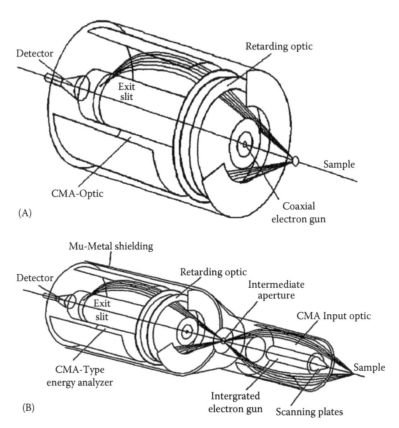

FIGURE 3.27 Schematics of (A) Super-CMA and (B) double-pass super-CMA. (From Staib Instruments manufacturers' literature: www.staibinstruments.com/english/products/desa/)

analyzers are shown in Figure 3.27A and B, respectively. The super-CMA differs from the standard CMA in that it does not have an entrance slit machined in the front end of the inner cylinder, but instead, electrons from the sample are introduced directly into the front space between the two cylinders, after being retarded by high-precision retarding field optics. This method of introducing the electrons into the analyzer allows for a much larger working distance between the sample and the analyzer than in the standard CMA. The retarding field optics serve the same function as the hemispherical grids used in a standard DPCMA. For the double-pass super-CMA, a CMA optical system and an aperture are placed between the sample and a super-CMA. The additional optics consist of a small CMA, which also has no slits machined in the inner cylinder. Instead, electrons enter the front space between the two cylinders directly, and exit from the space between the two cylinders at the far end. Then they enter a separate, cylindrical aperture, from which they enter the second stage super-CMA. It is claimed that these analyzers retain all the advantages of the classical CMA and DPCMA, but overcome the disadvantages of short working distance, need of fine control on sample positioning, and a small acceptance area on the sample.

The super-CMA shares the following features with CHA analyzers [57]:

- Simultaneous analogue and pulse counting modes
- Constant energy resolution or constant resolving power modes
- Electronically controlled energy resolution, from <100 meV up to 6 eV

The development of new types of energy analyzer and the improvement of older ones are still active areas of research, and some of these developments and improvements may appear on the market in the next 10 years. Just as miniaturization has been important in other fields, so is it also important in

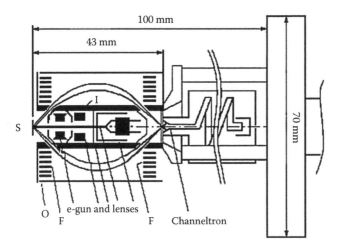

FIGURE 3.28 Miniature CMA with coaxial electron gun and channeltron, mounted on a 70 mm flange: S, sample; O, outer CMA cylinder; I, inner CMA cylinder; and F, fringing field correction rings. (From Grzelakowski, K., Man, K.L., and Altman, M.S., *Rev. Sci. Instrum.*, 72, 3362, 2001. With permission.)

surface analysis. Miniaturization of the CMA and the DPCMA has been achieved by Grzelakowski et al. [58], who have designed a CMA that fits onto a 70 mm diameter flange. It was constructed by mounting a miniature electron gun equipped with a cerium hexaboride filament coaxially inside the analyzer, and placing a channeltron detector at the image position (Figure 3.28). Using the FWHM of the elastic peak in a spectrum recorded from contaminated W(100), the developers calculated a relative energy resolution of 0.9%, indicating that there was no significant loss in resolution on miniaturization. An AES spectrum from contaminated W(100) in the analogue detection mode is shown in Figure 3.29, illustrating that spectra of high quality may be obtained with small devices. A miniature

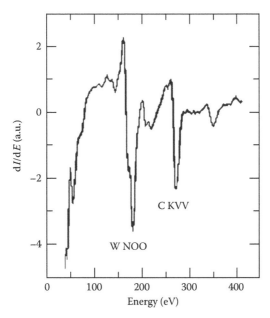

FIGURE 3.29 AES spectrum from contaminated W(100) using a miniature CMA. (From Grzelakowski, K., Man, K.L., and Altman, M.S., *Rev. Sci. Instrum.*, 72, 3362, 2001. With permission.)

DPCMA has also been developed for recording photoemission spectra from gas-phase molecular beams, or from molecular cluster beams under conditions of higher density [59].

Another new type of energy analyzer based on cylindrical electrodes is the TCMA (tri-CMA), which consists of three coaxial cylindrical electrodes, as shown in Figure 3.30 [60]. It has a theoretical relative energy resolution of 0.03% at $\alpha = 33° 55'$ and is being developed for the purpose of AES lineshape analysis, in order to provide chemical-state information. A high-resolution carbon KLL spectrum obtained by it from a carbon-contaminated Si(111) surface is shown in Figure 3.31, exhibiting characteristic satellite lines.

Almost all commercial instruments that employ a CHA as the energy analyzer also use a transport lens between the sample and the analyzer entrance. Such a lens is used not only to focus an area of the sample surface onto the entrance slit of the CHA, but also to retard the electrons or ions so as to control the absolute energy resolution ΔE.

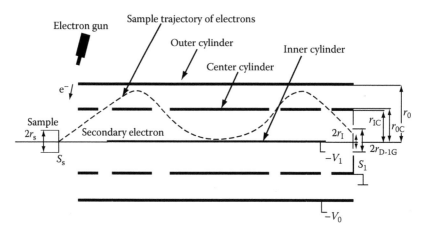

FIGURE 3.30 TCMA. (From Kościelniak, P., Kaszczyszyn, S., and Szuber, J., *Vacuum*, 63, 361, 2001. With permission.)

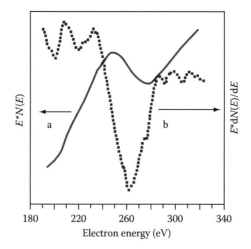

FIGURE 3.31 Carbon KLL Auger spectrum in both $N(E)$ and $dN(E)/dE$ modes from contaminated Si(111) using a TCMA. (From Kościelniak, P., Kaszczyszyn, S., and Szuber, J., *Vacuum*, 63, 361, 2001. With permission.)

The most commonly used transport lenses are multielement and electrostatic, employing four or five elements, although a magnetic immersion lens can also be used (e.g., the axis instrument). With a four-element lens, the magnification of the image, and the energies of the electrons forming it, in the entrance plane of the analyzer, may be kept constant as the energies of the electrons emitted from the sample are increased or decreased. If the electron energies in the sample image in the entrance plane of the analyzer are constant and independent of the applied, scanned, accelerating, or decelerating voltages, then the position of the image in the detector plane will also be fixed. With a five-element lens, the capability of making the size of the output angle at lens element 5 the same size as the input angle at lens element 1 is also available.

When a transport lens is operated in the magnification mode, its magnification M and the slit width into the analyzer determine the acceptance area, which defines the lateral resolution for μXPS. Thus, M is an important figure of merit for the analyzer. The acceptance area is the analyzer entrance slit width divided by M. The smallest available slit width and the largest value for M will therefore determine the theoretical minimum lateral resolution for the instrument. In the magnification mode, the acceptance angle into the analyzer is usually controlled by a further aperture, called the iris, located inside the lens at the first diffraction plane. The function of the iris can probably best be illustrated by some graphics taken from the trade literature. The multi-element lens used in the particular instrument being discussed may be operated in any one of three magnification modes, that is, $M = 10$, 5, or 2. In Figure 3.32, the trajectories are shown for two different sets of 1000 eV electrons being transported through a lens with $M = 10$. In each case, the electrons have been retarded to a final energy of 100 eV. The first set of trajectories start on the lens axis at angles of $0°$, $\pm1°$, and $\pm2°$ relative to the axis, while the second set of trajectives start at the same angles, but 0.5 mm off the lens axis. At the first diffraction plane, the two sets of parallel trajectories focus at different positions. The iris aperture is located at this plane. Using the largest area entrance slit ($7 \times 20\,mm^2$) and $M = 10$, the acceptance area is $0.7 \times 2\,mm^2$. When the intensities of the Ag and Cu XPS peaks from a sample consisting of a silver foil covered with a 10 mm diameter copper aperture were plotted as a function of iris diameter, the result shown in Figure 3.33 was obtained. The iris aperture was varied continuously up to 50 mm (wide open). Both Cu and Ag were observed until the iris aperture had decreased to 25 mm, because the acceptance angle decreased at constant acceptance area. For an aperture of 25 mm

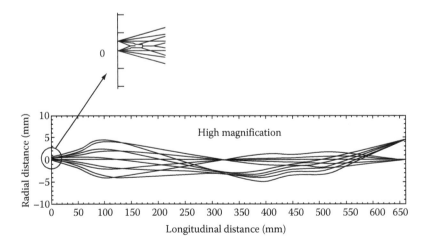

FIGURE 3.32 (See color insert following page 396.)   Trajectories of 1000 eV electrons through a transport lens set at $M = 10$ with retardation of the electrons to a final energy of 100 eV. Black lines indicate trajectories from a point on axis at angles of $0°$, $\pm1°$, and $\pm2°$ to the axis, and blue lines the same from a point 0.5 mm off axis. (From SPECS literature: www.specs.com. With permission.)

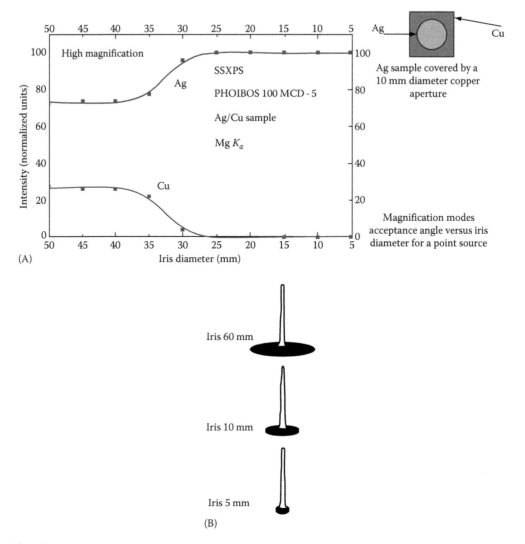

(A)

Iris 60 mm

Iris 10 mm

Iris 5 mm

(B)

FIGURE 3.33 (A) Copper and silver intensities as a function of the diameter of the iris aperture of the electron transfer optics referred to in page 53 for a silver sample covered by a copper aperture. The acceptance area is centered within the silver region. (B) Distribution of intensity as a function of position for three values of the iris aperture diameter. (From SPECS literature: www.specs.com. With permission.)

diameter, the acceptance angle would have been about ±8°. Thus, the lens settings may affect the energy resolution through their effect on $\Delta\alpha$. The explanation for the results illustrated in Figure 3.33A can be established by examining the distribution of intensity as a function of position. The distribution is cylindrically symmetric, and is Gaussian at small distances from the center, but has a very wide low intensity tail. The iris aperture's function is to truncate the tail, as illustrated in Figure 3.33B.

The transmission of an energy analyzer is an important parameter, because the sensitivity of a technique for any given element will depend on it. Dubé and Roy [62] have developed a generalized approach for calculating the transmission function, but the equation derived by them depends critically on the initial and boundary conditions. Procedures for measuring this function have been published in the literature [63], so it may be found empirically for some instruments. A more useful figure of merit is called the étendue, the product of transmission and slit area. The user should measure this function for the conditions appropriate to a specific application, particularly in cases where stringent quantification measurements are to be made.

As has been mentioned at the beginning of this section, a TOF analyzer is sometimes used in ISS instruments. In a TOF analyzer, bunches of charged particles with different energies enter a flight path that has a fixed length to a particle detector. The particles are discriminated on the basis of their velocities, that is, the time taken to traverse the flight path, which depends on their masses and energies. In conventional ISS, all the incident particles have ideally the same mass, charge, and energy. Through collisions with a sample surface, the energy of a particle is reduced, and usually the charge also. Thus, the collision process converts a mono-energetic particle beam to a multiple-energetic beam, and the TOF analyzer measures the energy distribution of the latter. A TOF analyzer can analyze positive, negative, and neutral beams, although in ISS the scattered beam is always positive. It may also provide higher energy resolution (ΔE) than an electrostatic energy analyzer [64]. An excellent review of TOF analyzers, as well as electrostatic analyzers, for ISS, may be found in a recent text by Rabalais [65].

The electrostatic analyzers discussed above may all be used for simultaneous rather than single energy detection since they disperse a band of energies at the image position in the analyzer exit plane. If the exit slit is removed, then the whole energy band is available for detection. However, the range of energies is small. For example, a CHA operated in the fixed analyzer transmission (FAT) mode will disperse a range of energies given by the interval $\pm 0.1 E_{p}$, where E_{p} is the pass energy, imaged in the analyzer detector plane. For $E_{p} = 100\,\text{eV}$, this results in a band of width $20\,\text{eV}$, which is often the width used to record a single spectral peak in XPS (e.g., C 1s, O 1s, etc.) in the high-resolution mode. An entire spectrum may be collected over some appropriate time interval using a position-sensitive detector in the exit plane of the analyzer.

Alternatively, a number (e.g., 3, 5, etc.) of single channeltrons may be positioned near the image point for E_{p}, and the signals from them summed as the analyzer is energy-scanned in the normal fashion. In this case, some energy resolution is sacrificed for a large gain in intensity. The latter allows peaks from elements present at low surface concentrations, peaks with low-ionization cross sections, or both, to be recorded in a reasonable time, particularly for low-intensity x-ray sources such as those that are monochromated. The ability to record an extended range of energies simultaneously also introduces the possibility of carrying out real-time surface analysis.

New analyzers that can be used for simultaneous energy analysis over a large range are under development. One of these is the hyperbolic field analyzer (HFA), shown schematically in Figure 3.34 [66]. With this lensless analyzer, an energy range of $E_{\max}/E_{\min} \approx 34$ (compared to 1.2 for the CHA) may be achieved.

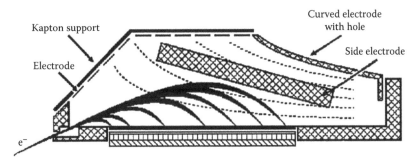

FIGURE 3.34   Schematic diagram of the HFA. (From Kale, A., PhD thesis, The University of York, 2003. With permission.)

3.10.3 Detectors

The workhorse detector for instruments using single-channel detection is still the channeltron. The technology itself is old, but manufacturers continue to develop methods of extending the dynamic range, improving detection efficiency at higher masses, and increasing the lifetime of the device. The three parameters—dynamic range, detection efficiency, and lifetime—along with the transit time, are its figures of merit. The transit time is the elapsed time from the initiation of an event until the termination of the resulting electron avalanche; it determines the temporal resolution of the device, and thus places a limit on the maximum count rate. The characteristic curve of a channeltron consists of a plot of gain versus applied collector voltage, and is supplied by the manufacturer with the device. However, the characteristic applies to incident electrons, and since the detector may also be used to detect ions or even neutrals, it should be determined experimentally on installation of the instrument. The resulting plots then serve as standards for comparison of the instrument performance over time.

Detector efficiency is important because it is a parameter in the mathematical expression for spectral peak intensity. For electrons, it is a nonlinear function of the electron KE, while for ions it depends on both KE and mass. Unfortunately, there is no universal mathematical curve for the detector efficiency as a function of particle energy. Thus, for quantitative analysis, the FAT mode should be used to collect data since the detector efficiency will then be the same for all peak energies. Some suppliers include generic information in their handbooks, which are usually free. For example, Burle's *Channeltron Electron Multiplier Handbook* is an excellent choice [67].

Instruments using multichannel detection employ either multiple channeltron detectors (MCD-n, where n is the number of channeltrons), micro-channel plates (MCP), charge coupled device (CCD) detection systems, or delay line detectors (DLD). In the case of the MCD-n, the channeltrons are spaced at regular intervals across the detector plane, which corresponds to increments in the energy dispersion, with a reference channeltron centrally positioned. The experimental spectrum is then collected by shifting the spectrum of each channeltron to put it in register with that of the reference channeltron, and then adding the intensities for each energy increment. This is a way of exploiting the multiplex advantage without an increase in the time required to record a spectrum. Theoretically, it is possible to do this without degrading resolution, but in practice, there is always some loss in resolution because the shifts to registry are not perfect; however, the loss in absolute energy resolution does not usually exceed 0.1 eV.

An MCP stack may be used to obtain data for particles at a fixed polar angle, but widely varying azimuthal angles. For example, in ISS with a scattering angle of ~180°, the azimuthal angle of the scattered primary ion is in the interval 0°–360°. The MCP stack intercepts the scattered ions and generates a signal, which is sent to an amplifier. MCPs used in this configuration are called detection-quality MCPs. Such detectors are useful when TOF analyzers are used. The performance figures of merit for the detection quality MCP are the same as those for a channeltron.

CCD detection systems are used in imaging instruments, and consist of an MCP stack, a phosphor screen, a demagnifying lens, and a digital camera (Figure 3.35). The phosphor is an electron-to-light transducer, and as such, needs to respond very quickly to changes in the electron current impinging on the screen. The demagnifying lens is used to enlarge the image on the phosphor so that it fills the input aperture of the camera. The performance figures of merit for the detection system depend on the individual figures of merit. The resolution of an MCP stack is inversely proportional to its pitch, which is the center-to-center distance between adjacent pores, and the pore size determines the magnitude of the electron avalanche beam leaving the stack. As the beam leaves the stack it diverges, the degree of divergence depending on the distance it has to travel to hit the plate and on the accelerating voltage. By keeping the distance between the MCP stack and the phosphor plate short, and adequately controlling the accelerating voltage, divergence can be minimized. Each avalanche beam creates a tiny light source whose intensity is determined by the number of electrons in the beam. These multiple light sources are focussed onto the CCD, which has its own resolution determined by

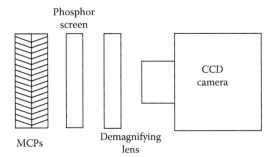

FIGURE 3.35   Schematic of a CCD detector system.

the size of a single pixel in its photodiode array. If the pixel size is smaller than that of the pore size of the MCP stack, then the MCP stack will determine the overall detector spatial resolution, and most CCD detectors are constructed in that way. Ideally, the spatial resolution of the detector should be selected so that it is negligible compared to that of the multichannel dispersion analyzer.

Another figure of merit is the temporal resolution, which is the time between successive images and is the reciprocal of the frame rate, the parameter specified by the vendor. A further figure of merit is the number of camera bits. This refers to the number of bits in the analog-to-digital converter that determines the number of intensity levels in the digitized image. For example, a 12-bit camera would provide $2^{12} = 4096$ intensity levels. Cameras with 12, 14, or 16 bits are available.

The latest innovation in detectors is the DLD. A standard DLD consists of an MCP stack and two mutually perpendicular delay-line windings at the exit from the stack. Its operation can be described with reference to Figure 3.36. The perpendicular line shows a single event being detected by an MCP stack. Below the stack a winding path delay-line for the x-direction is shown. The avalanche electron generated by the event hits the winding and produces an electrical pulse that propagates to the two end stations, from which two signals register. By measuring the difference in signal arrival time between the stations, the event's x position can be determined. Similarly, a perpendicular y-winding delay-line (not shown) allows the y position to be determined. The summed signal gives the amplified pulse count for the event, and hence a parallel image with pulse-counted data is recorded.

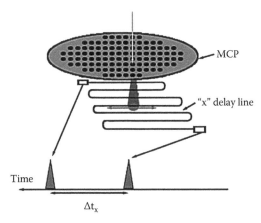

FIGURE 3.36   Simplified diagram of the DLD, designed to show its operation. The vertical line represents an incoming particle. A winding delay line for the x-direction is shown. A similar one would exist for the y-direction. (From Kratos literature: www.kratos.com. With permission.)

3.11   SPECIAL APPLICATIONS OF XPS

3.11.1   High Pressure XPS

XPS has proved to be a powerful tool for, amongst many other applications, the characterization of heterogeneous catalysts and of corrosion processes on materials. However, such studies have been carried out almost entirely on before-and-after specimens, in order that the high-vacuum requirements of the traditional instruments can be maintained. In such cases, it is the surface/vacuum interface that is actually studied. It has always been the goal of research workers to be able to study surface reactions under conditions that more closely represent the native reaction environments. For example, in order to characterize a catalyst–gas interface or a material–liquid interface, instrumentation that allows the sample to be immersed in a gas or a liquid, but does not degrade spectrometer performance, is needed. Now, two relatively high-pressure XPS instruments are commercially available. Both are based on differential pumping; a four-stage arrangement is shown in Figure 3.37 [69]. An alternative high-pressure monochromatic XPS/EELS (electron energy loss spectroscopy) system has three-stage differential pumping, and is equipped with a high-pressure cell [70].

The vacua maintainable in the various differentially pumped stages are determined by the pumping speeds of the pumps used at each stage, and are functions of the nature, and the equilibrium pressure, p_0, of the particular gas used in the high-pressure cell in the sample chamber. The upper limit for p_0 in the cell is 1020 Pa, so to maintain a pressure of no more than about 10^{-5} Pa at the analyzer requires a pressure reduction factor of $\sim 10^8$. The pressure in each stage may be expressed in terms of p_0 by the function $p_i = C_i(p_0) \cdot p_0$, where i indexes the individual stage, and C is the reduction factor. The three-stage instrument uses turbomolecular pumps with the following pumping speeds: $TP_1 = $ 330–680 L/s, $TP_2 = 105$–180 L/s, and $TP_3 = 105$–180 L/s. The results of tests relating the reduction factors to the pressure in the reaction cell are shown in Figure 3.38. For p_0 in the range 100–1020 Pa, the factors $C_i(p_0)$ are constant, with respective values $C_1 = 6.4 \times 10^{-4}$, $C_2 = 3.0 \times 10^{-6}$, and $C_3 = 2.1 \times 10^{-8}$ for the first, second, and third differential pumping stages. Thus, at the highest working pressure in the cell of 1020 Pa, the pressure at the analyzer could be maintained at $\sim 2 \times 10^{-5}$ Pa, adequate for analysis. Since the goal of investigators using high-pressure XPS is to work at the highest sample chamber pressure possible, the reduction factors are important parameters characterizing the system. The lower their values, the higher p_0 can be for a given target value of the analyzer pressure.

In high-pressure XPS, the maximum pressure possible in the sample chamber is determined in part by the mechanical stability of the x-ray source window and the maximum window thickness

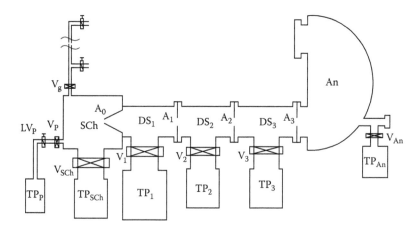

FIGURE 3.37   Schematic of a high-pressure XPS instrument with multistage differential pumping. TP, turbomolecular pump; Sch, sample chamber; DS, differential pressure chamber; An, analyzer; V, valve. (From Kleimenov, E., PhD thesis, Technical University of Berlin, 2005. With permission.)

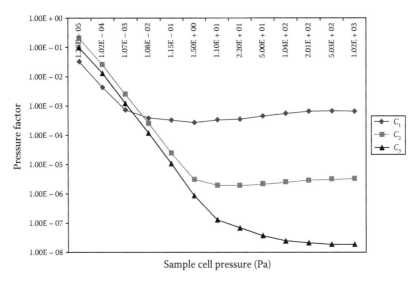

FIGURE 3.38 First, second, and third pressure reduction factors versus pressure in the sample cell for the instrument shown in Figure 3.37. The cell gas was air. Above about 100 Pa, the factors are almost constant. (From Kleimenov, E., PhD thesis, Technical University of Berlin, 2005.)

that can be tolerated without severely degrading the source intensity at the sample. For example, Kleimenov [69] calculated that a 100 nm thick silicon nitride window could withstand a pressure difference up to 10^3 Pa. It can be concluded that there will be some lower limits for the above factors, beyond which there will be no further improvement possible for a given source.

The ability to obtain spectra from samples subjected to *in-situ* high pressures is not without sacrifice. There will be a consequent pressure-dependent loss in source intensity at the sample surface due to the x-ray transmission properties of the introduced gas. In addition, photoelectrons ejected from the sample undergo scattering off the gas molecules so that the reduction in signal depends on both the nature of the gas and its pressure. Since the scattering cross sections of the photoelectrons are energy dependent, the overall transmission function of the spectrometer changes, which affects data analysis. Correction of spectra for charging encountered with semiconductor or insulator samples is more complicated because the surface charging is also influenced by both gas composition and pressure. The highest pressure that has been reported when using a conventional x-ray source is about 100 Pa [70].

3.11.2 Automated XPS

In industry, XPS analysis is often carried out in the context of quality control. For the industrial laboratory, a compact instrument with high sample throughput, low downtime, adaptation for special sample dimensions, and minimum operator intervention is required. For example, XPS is widely used for characterizing the surfaces of silicon wafers, which can have diameters of up to 300 mm.

Most of the instruments aimed at the quality control market are based on the CHA. However, one manufacturer has opted for a Dupont-type band-pass filter analyzer, a schematic of which is shown in Figure 3.39. The grid and absorber electrodes on the left of the diagram function as a low-pass reflective filter. The curved grid electrode is biased at $-V$ volts; photoelectrons with energies lower than eV are reflected by the grid, while those with higher energies are absorbed by the absorber electrode. The reflected electrons travel through a field-free region toward the opposite end of the analyzer, where they encounter a high-pass transmission filter. Thus, electrons with a small energy bandwidth with respect to the nominal electron energy are detected. Typically, the bandwidth (ΔE)

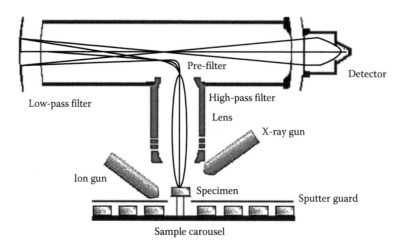

FIGURE 3.39 Schematic of the Dupont-type band-pass filter analyzer. (From Kratos literature: www. kratos.com. With permission.)

is 0.8 eV at 1000 eV. The system has a fully automated multi-sample carousel, and both data acquisition and analysis may be performed entirely under computer control.

Some instruments now allow full remote control and service via the Internet, and the only required manual operation is the loading of the sample holder into the transfer system. Additional information on this class of XPS instrumentation may be found in Refs. [22,32,37].

3.11.3 DEDICATED XPS

Dedicated here means state-of the-art instruments that have been optimized for XPS analysis, but that may have other techniques, such as AES, available on them. They include both small-spot and imaging instruments. Within this category, there are instruments, based on a CHA and a combination of electrostatic and magnetic lenses, that have been specifically designed for small-spot analysis, in which the minimum spot size is 30 μm. Of course, traditional, i.e. so-called large area, XPS can also be performed, as can large-area imaging by means of stage mapping. One manufacturer of imaging instruments has based its design on a 127 mm radius CHA, with either an MCD-3 (3 channeltron array) detector or an MCD-5 detector, and with automated multi-sampling as an option. Other manufacturers offer similar designs, with the main difference between them being the size of the CHA.

The so-called NanoESCA probably leads the field in iXPS at the moment, with a claimed lateral resolution of <200 nm when operated at a synchrotron beamline. A more conventional design uses a monochromatic Al K_α source in which a raster-scanned focussed electron-beam generates micro-beam x-rays with a spot size adjustable from ~9 to 200 μm. Thus, small-spot XPS is available at ≤10 μm, and imaging XPS with a lateral resolution of 9 μm. Another design can perform small-spot and angle-resolved XPS using monochromatic Al K_α radiation and a 110 mm radius CHA, and uses optics that allow angular information and electron energy information to be imaged in parallel on a two-dimensional detector. Ninety-six angular channels may be accessed without having to tilt the sample. It is ideal for the nondestructive depth profiling of very thin films, and a single angular-resolved XPS experiment may be performed very quickly. Yet another instrument, based on a CHA, has a 2-stage, 11-element electrostatic lens, and a detector system consisting of an MCP with a fast fire wire pulse transmission CCD camera. The spatial resolution for imaging XPS is 10 μm, and angle-resolved XPS may also be performed. Additional information on these products may be found in Refs. [68,71–75].

3.11.4 MULTI-TECHNIQUE INSTRUMENTS

These are instruments designed to accommodate several surface analysis techniques in a single-sample analysis chamber and may not be optimized for any of them, although they usually are for XPS. Current-generation models from the major manufacturers represent evolutionary developments along the following lines:

- *Small-spot and imaging XPS.* Depending on the model, available spot size ranges from a few to ~100 μm, while the state-of-the-art in lateral resolution for iXPS is in the low micrometer range.
- *Parallel data acquisition.* Parallel detection is now the rule, rather than the exception, by one or other of the technologies, that is, based on DLD, MCD, or MCP with camera.
- *Monochromatic excitation sources and greater choice of sources.* In addition to the traditional dual-anode Mg K_α/Al K_α and monochromatic Al K_α, single anode Ag L_α, Zr L_α, and Au M_α sources are available.
- *Better integration of multi-technique attachments.* Most instruments can be tailor-made with a great range of optional, additional techniques. These include the following: SAM with a spot size of <100 nm using either Schottky field emission or LaB_6 electron sources; routine SEM with a secondary electron detector; ISS using rasterable, electron-impact-generated noble-gas ion beams; SIMS using a quadrupole mass analyzer; and UPS using a He discharge lamp.

Additional technical information on multi-technique instruments may be found in Refs. [62,69,72–75].

3.12 AES CONFIGURATIONS

3.12.1 DEDICATED INSTRUMENTS

Because lineshape analysis can now be used to obtain chemical speciation data in AES, there are fewer dedicated instruments, and they are all research-oriented scanning instruments with electron-beam spot sizes in the low nanometer range. As a rule, the instruments use Schottky field-emission electron guns, while both CMAs and CHAs are on offer. The claimed lateral resolution ranges from 8 to 12 nm for the top-of-the-range models. Detection is based on either MCD or MCP technology. Although these instruments are designed specifically for AES and SAM, they all allow some modules to be attached for other techniques (e.g., XPS, SIMS, EDS, and ISS). Additional information may be found in Refs. [72,75].

3.12.2 OTHER INSTRUMENTS

Instruments designed primarily for low-energy electron diffraction, EELS, or scanning probe microscopy often offer AES/SAM as an option. The SAM specifications are then comparable to those found for dedicated AES/SAM instruments.

3.13 ISS

3.13.1 DEDICATED INSTRUMENTS

ISS analysis using instruments designed primarily for XPS is limited to the determination of the energy loss spectrum of scattered primary ions. The energy losses are related to the masses of the collision partners, and thus the qualitative and quantitative elemental composition of the top monolayer of a solid can be determined, provided that the ion flux is low enough. A new instrument design for ISS/LEIS was introduced by Brongersma in 1985 [76], allowing both energy-resolved and angle-resolved analysis of scattered primary ions. The design has subsequently become commercially available as an ISS low energy ion scattering (LEIS) instrument [77]. It uses an electron-impact noble-gas ion source with an energy range of 0.5–5 keV and a spot size of ≈0.3 mm. The spatial resolution is 10 μm, the absolute energy resolution ΔE is 10% of the pass energy, and the mass resolution, $M/\Delta M$, is in the range 10–20. The analyzer consists of a four-element electrostatic zoom lens and a double toroidal electrostatic analyzer, with a scattering angle of 145° and an azimuthal angle that ranges from 0° to 320°. A TOF option may be added, allowing TOF scattering and recoiling spectroscopy analysis to be carried out, but this requires that a second ion gun be used. It is also possible to modify the instrument to perform TOF-CAICISS (coaxial impact collision ISS [78]. An instrument optimized for CAICISS is under development [79]. Another ISS variation is mass-resolved ion scattering spectrometry. In this technique, the incident ion beam is mono-energetic, but not mass-selected, and the backscattered ions are mass and energy analyzed [80,81]. Just as the commercialization of XPS and AES instruments has led to their tremendous growth and maturation, so the commercialization of dedicated ISS instruments has positioned ISS to mirror the growth and maturation of its related surface spectroscopic techniques.

REFERENCES

1. G. B. Hoflund, Spectroscopic techniques: X-ray photoelectron spectroscopy (XPS), Auger electron spectroscopy (AES) and ion scattering spectroscopy (ISS), in *Handbook of Surface and Interface Analysis*, J. C. Rivière and S. Myhra (Eds.), Marcel Dekker, Inc., New York, 1998, Chapter 4, pp. 57–158.
2. G. B. Hoflund and J. C. Rivière, Less commonly used techniques for analysis of surfaces and interfaces, in *Handbook of Surface and Interface Analysis*, J. C. Rivière and S. Myhra (Eds.), Marcel Dekker, Inc., New York, 1998, Appendix 3, pp. 885–888.
3. C. N. Berglund and W. E. Spicer, *Phys. Rev. A*, 136 (1964) 1030.
4. G. B. Hoflund, Z. F. Hazos, and G. N. Salaita, *Phys. Rev. B*, 62 (2000) 11126.
5. T. Gustafsson and E. W. Plummer, in *Photoemission and the Electronic Properties of Surfaces*, B. Feuerbacher, B. Fitton, and R. F. Willis (Eds.), John Wiley & Sons, Ltd., Chichester, U.K., 1978, Chapter 12.
6. G. A. Somorjai, *Principles of Surface Chemistry*, Prentice Hall, Englewood Cliffs, NJ, 1972.
7. A. Joshi, L. E. Davis, and P. W. Palmberg, Auger electron spectroscopy, in *Methods of Surface Analysis*, A. W. Czanderna (Ed.), Elsevier, Amsterdam, the Netherlands, 1975, Chapter 5, p. 159.
8. G. B. Hoflund, W. S. Epling, and D. M. Minahan, *J. Elect. Spec. Rel. Phenom.*, 95 (1998) 289.
9. G. N. Salaita, Z. F. Hazos, and G. B. Hoflund, *J. Elect. Spec. Rel. Phenom.*, 107 (2000) 73.
10. C. D. Wagner, W. M. Riggs, L. E. Davis, J. F. Moulder, and G. E. Muilenberg (Eds.), *Handbook of X-Ray Photoelectron Spectroscopy*, Perkin-Elmer Corp., Physical Electronics Division, Eden Prairie, MN, 1979.
11. J. F. Moulder, W. F. Stickle, P. E. Solol, and K. D. Bomben, in *Handbook of X-Ray Photoelectron Spectroscopy*, J. Chastain and R. C. King (Eds.), Physical Electronics, Inc., Eden Prairie, MN.
12. G. Beamson and D. Briggs, *High Resolution XPS of Organic Polymers: The Scienta ESCA 300 Database*, Wiley, New York, 1992.
13. L. A. J. Garvie, P. Rez, J. R. Alvarez, P. R. Busick, A. J. Craven, and R. Brydson, *Am. Mineral.*, 85 (2000) 732.
14. E. Jenson, R. A. Bartynski, T. Gustafsson, E. W. Plummer, M. Y. Chou, M. L. Cohen, and G. B. Hoflund, *Phys. Rev. B*, 30 (1984) 5500.
15. J. W. Davenport, *Phys. Rev. Lett.*, 36 (1976) 945.
16. G. B. Hoflund and W. S. Epling, *Chem. Mater.*, 10 (1998) 50.
17. L. E. Davis, N. C. MacDonald, P. W. Palmberg, G. E. Riach, and R. E. Weber, *Handbook of Auger Electron Spectroscopy*, 2nd edn., Physical Electronics Industries Inc., Eden Prairie, MN, 1976.
18. G. E. McGuire, *Auger Electron Spectroscopy Reference Manual*, Plenum, New York, 1979.
19. American Vacuum Society (Ed.), *Surface Science Spectra*, American Institute of Physics, New York.
20. M. P. Seah and W. A. Dench, *Surf. Interface Anal.*, 1 (1979) 2.
21. P. C. McCaslin and V. Y. Young, *Scan. Electron Microsc.*, 1 (1987) 1545.

22. C. S. Fadley, R. J. Baird, W. Siekhaus, T. Novakov, and S. Bergström, *J. Electron Spectrosc. Relat. Phenom.*, 4 (1974) 93.
23. C. S. Fadley, *Prog. Surf. Sci.*, 16 (1984) 275.
24. G. B. Hoflund, D. A. Asbury, C. F. Corallo, and G. R. Corallo, *J. Vac. Sci. Technol. A*, 6 (1988) 70.
25. M. R. Davidson, G. B. Hoflund, and R. A. Outlaw, *J. Vac. Sci. Technol. A*, 9 (1991) 1344.
26. G. B. Hoflund and D. M. Minahan, *J. Catal.*, 162 (1996) 48.
27. A. C. Miller, in *Treatise on Analytical Chemistry*, Part 1, Vol. 11, J. D. Winefordner (Ed.), Wiley, New York, 1989, p. 253.
28. G. B. Hoflund, H.-L. Yin, A. L. Grogan, Jr., D. A. Asbury, H. Yoneyama, O. Ikeda, and H. Tamura, *Langmuir*, 4 (1988) 346.
29. R. Germans, PhD thesis, Eindhoven University of Technology, 1996.
30. F. Shoji, K. Kashihara, K. Oura, and T. Hanawa, *Surf. Sci.*, 220 (1989) L719.
31. J. L. Sullivan, S. O. Saied, and T. Choudhury, *Vacuum*, 43 (1992) 89.
32. K. Siegbahn, *Alpha, Beta and Gamma-Ray Spectroscopy*, North-Holland, Amsterdam, the Netherlands, 1954.
33. C. Nordling, E. Sokolowski, and K. Siegbahn, *Ark. Fys.*, 13 (1958) 483.
34. K. Siegbahn, C. Nordling, A. Fahlman, R. Nordberg, K. Hamrin, J. Hedman, G. Johansson, T. Bergman, S. Karlson, I. Lindgren, and B. Lindberg, *Electron Spectroscopy for Chemical Analysis-Atomic, Molecular and Solid-State Structure Studies by Means of Electron Spectroscopy*, Almquist & Wiksells, Stockholm, Sweden, 1967.
35. E. Grossman, Y. Lifshitz, J. T. Wolan, C. K. Mount, and G. B. Hoflund, *J. Spacecraft Rockets*, 35 (1999) 75.
36. M. A. Golub, T. Wydeven, and R. D. Cormia, *Polym. Commn.*, 29 (1988) 285.
37. S. L. Koontz, K. Albyn, L. J. Leger, D. E. Hunton, J. B. Cross, and C. J. Hakes, *J. Spacecraft Rockets*, 32 (1995) 483.
38. G. B. Hoflund and H. A. E. Hagelin-Weaver, unpublished data.
39. F. Patthey and W.-D. Schneider, *J. Electron Spectrosc. Relat. Phenom.*, 81 (1996) 47.
40. G. Beamson, S. R. Haines, N. Moslemzadeh, P. Tsakiropoulos, P. Weightman, and J. F. Watts, *Surf. Interface Anal.*, 36 (2004) 275.
41. N. Moslemzadeh, G. Beamson, S. R. Haines, P. Tsakiropoulos, J. F. Watts, and P. Weightmam, *Surf. Interface Anal.*, 38 (2006) 703.
42. S. Diplas, J. F. Watts, S. A. Morton, G. Beamson, P. Tsakiropoulos, D. T. Clark, and J. E. Castle, *J. Electron Spectrosc. Relat. Phenom.*, 113 (2001) 153.
43. T. G. Vargo and J. A. Gardella, Jr., *J. Vac. Sci. Technol. A*, 7 (1989) 1733.
44. C. D. Wagner, *J. Electron Spectrosc. Relat. Phenom.*, 15 (1978) 518.
45. G. B. Hoflund, Depth profiling, in *The Handbook of Surface Imaging and Visualization*, A. T. Hubbard (Ed.), CRC Press, Boca Raton, FL, 1995, Chapter 6, p. 63.
46. A. D. McNaught and A. Wilkinson (Eds.), Compendium of chemical terminology, in *The Gold Book*, 2nd edn., Blackwell Science, 1997.
47. N. Broll and P. de Chateaubourg, *Adv. X Ray Anal.*, 41 (1999) 393.
48. H. Yamada, *Nucl. Instrum. Meth. Phys. Res. B*, 199 (2003) 509.
49. F. Reniers and C. Tewell, *J. Electron Spectrosc. Relat. Phenom.*, 142 (2005) 1.
50. G. P. Williams, *Rep. Prog. Phys.*, 69 (2006) 310.
51. N. de Jonge, Y. Lamy, K. Schoots, and T. Oosterkamp, *Nature*, 420 (2002) 393.
52. G. Z. Zue, Q. Qiu, B. Gao, Y. Cheng, J. Zhang, H. Shimoda, S. Chang, J. P. Lu, and O. Zhou, *Appl. Phys. Lett.*, 81 (2002) 355.
53. Evans Analytical Group SIMS Tutorial. http://www.eaglabs.com/tutorials/sims_theory_tutorial/
54. Y. Ishii, R. Tanaka, and A. Isoya, *Nucl. Instrum. Meth. Phys. Res. B*, 113 (1996) 75.
55. M. P. Seah, in *Methods of Surface Analysis*, J. M. Walls (Ed.), Cambridge University Press, Cambridge, UK, 1989.
56. A. E. Hughes and B. A. Sexton, *J. Electron Spectrosc. Relat. Phenom.*, 46 (1988) 31.
57. Staib Instruments manufacturers' literature: www.staibinstruments.com/english/products/desa/
58. K. Grzelakowski, K. L. Man, and M. S. Altman, *Rev. Sci. Instrum.*, 72 (2001) 3362.
59. C. M. Teodorescu, D. Gravel, E. Rühl, T. J. McAvoy, J. Choi, D. Pugmire, P. Pribil, J. Loos, and P. A. Dowben, *Rev. Sci. Instrum.*, 69 (1998) 3805.
60. P. Kościelniak, S. Kaszczyszyn, and J. Szuber, *Vacuum*, 63 (2001) 361.
61. SPECS literature: www.specs.com
62. D. Dubé and D. Roy, *Nucl. Instrum. Meth.*, 201 (1982) 291.

63. L. T. Weng, G. Vereecke, M. J. Genet, P. Bertrand, and W. E. E. Stone, *Surf. Interface Anal.*, 20 (1993) 179.
64. M. Draxler, S. N. Markin, M. Kolíbal, S. Průša, T. Šikola, and P. Bauer, *Nucl. Instrum. Phys. Res. B*, 230 (2005) 398.
65. J. W. Rabalais, *Principles and Applications of Ion Scattering Spectrometry: Surface Chemical and Structural Analysis*, John Wiley & Sons, Inc., Hoboken, NJ, 2003, Chapter 3.
66. A. Kale, PhD thesis, Quantitative Microanalysis Using the Hyperbolic Field Analyzer, The University of York, 2003.
67. Burle's *Channeltron Electron Multiplier Handbook*: www.burle.com
68. Kratos literature: www.kratos.com
69. E. Kleimenov, PhD thesis, High-pressure x-ray Photoelectron Spectroscopy Applied to Vanadium Phosphoross Oxide Catalysts Under Reaction Conditions, Technical University of Berlin, 2005.
70. J. Pantförder, S. Pöllmann, J. F. Zhu, D. Borgmann, R. Donecke, and H.-P. Steinrück, *Rev. Sci. Instrum.*, 76 (2005) 14102.
71. VG SCIENTA literature: www.vgscienta.com
72. Thermo Electron literature: www.thermo.com
73. JEOL literature: www.jeol.com
74. Omicron literature: www.omicron.com
75. Phi literature: www.phi.com
76. G. J. A. Hellings, H. Ottevanger, S. W. Boelens, C. L. C. M. Knibbeler, and H. H. Brongersma, *Surf. Sci.*, 162 (1985) 913.
77. AnaLEIS literature: www.calipso.nl and private communication
78. B. Moest, PhD thesis, Temperature Dependent Segregation of Stepped Surfaces, Eindhoven Technical University, 2004.
79. R. N. Fujita, S. Hayasi, and S. Makoto, *Shimazu Criticism*, 55 (1999) 259.
80. K. Wittmaack, *Surf. Sci.*, 3459 (1996) 110.
81. K. Franzreb, A. Pratt, S. Splinter, and P. A. W. van der Heide, *Surf. Interface Anal.*, 26 (1998) 597.

4 Ion Beam Techniques: Time-of-Flight Secondary Ion Mass Spectrometry (ToF-SIMS)

Birgit Hagenhoff, Reinhard Kersting, and Derk Rading

CONTENTS

4.1 INTRODUCTION: THE ANALYTICAL QUESTION

A universally applicable surface analysis technique for the characterization of chemical composition should be able to (a) locate points and areas of interest on the surface as well as in deeper layers, (b) identify any elements and molecules found there, and (c) determine their concentrations with an overall sensitivity of at least parts per million (ppm) for surfaces and parts per billion (ppb) for the bulk.

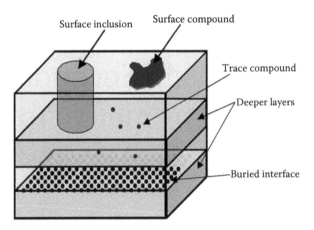

FIGURE 4.1 Definition of surface components.

Figure 4.1 shows schematically the various types of external and internal structures that an analyst might want to characterize. Unfortunately, no known analytical technique can satisfy all these requirements. Nevertheless, the performance of a technique in the three areas of location, identification, and quantification may be used as a measure of its usefulness in routine analysis. Time-of-flight secondary ion mass spectrometry (ToF-SIMS), to which this chapter is dedicated, has confirmed its position as a sought-after tool for surface characterization amongst the better-known analytical techniques. This position has been achieved over the last 20 years because of its supreme capabilities in the areas of location (imaging and depth profiling) and identification (spectrometry). Not only can elements be detected but molecules as well, usually simultaneously. Lateral resolution has improved to below 100 nm, while depths down to several micrometers can be analyzed at the same time as the lateral distributions of species (three-dimensional [3D] microarea analysis) are probed. It is therefore a tool well suited to failure analysis and sample screening. New options, in particular for the analysis of organic materials, have arisen with the advent of cluster-ion sources for the polyatomic bombardment of surfaces. The latter approach has improved the detection limit for organic materials by several orders of magnitude, and the first steps into organic depth profiling have been taken.

Given its strong positions in the areas of location and identification, the inferior performance of ToF-SIMS with respect to quantification is unfortunate but can often be overcome by the use of, for example, internal and external standards. Additionally, the last few years have shown that semiquantitative information can be obtained without much effort for use in failure analysis.

4.2 DEFINITIONS OF IMPORTANT TERMS

In the following paragraphs definitions are given, and explained; these are either used later in the chapter or are commonly used in the SIMS community. It serves as reference and can be skipped at first reading.

Secondary ion yield $Y_i(X_q)$ (general)
$Y_i(X_q)$ = number of detected secondary ions X in charge state q per number of primary ions, where Y_i is dependent on the particular instrumental conditions used (e.g., transmission and detection efficiency). For measurements performed with the same instrument under near-identical operational conditions, the yield determination is the easiest way to compare different spectra quantitatively.

Useful yield $Y_u(X_q)$ (general)
$Y_u(X_q)$ = number of detected secondary ions X in charge state q per number of sputtered species M. Y_u is also dependent on instrumental parameters, as above. For determination of Y_u the total number of sputtered species must be determined by some independent technique.

Energy distribution (general)

As a consequence of ion bombardment, elements and molecules are desorbed with different energy distributions. For elements the maximum in the distribution is at about 5–10 eV, with a slow decrease toward higher energies (proportional to $E/(E + E_B)^3$, where E is the energy of the desorbing species, and E_B the surface-binding energy [1]). The maximum for molecules is at lower energies, ca. 1 eV, and the distribution decreases exponentially with increasing energy (proportional to $\exp(-E/E_m)$, where E_m is the maximum in the distribution [2,3]). This spread in energy must be compensated for in mass analyzers in order to achieve high mass resolution.

Angular distribution (general)

For primary ion bombardment perpendicular to the target surface, the sputtered material is ejected in a cosine distribution [4], $dY/d\Omega \propto \cos v$, where v is the emission angle with respect to the surface normal. For other angles of incidence the cosine distribution shows an anisotropy in the direction of the reflected primary ions [5]. Just as for the energy distribution of the desorbed particles, the angular distribution must also be taken into account in the design of suitable mass analyzers. For example, it influences the analyzer performance with respect to topographically demanding samples (e.g., wires, balls, and edges).

Disappearance yield $Y_D(M)$ (organic materials)

$Y_D(M)$ = number of surface species M having disappeared per number of primary ions. This quantity refers to those surface species that have disappeared from the surface as a result of primary ion bombardment, irrespective of their fate (i.e., due to sputtering, fragmentation, thermal desorption, etc.)

Disappearance cross-section σ (organic materials)

σ is defined as the area from which surface species have disappeared under primary ion bombardment, irrespective of their fate; the magnitude of σ is a function of the particular species M being analyzed (i.e., $\sigma = \sigma(M)$). Typical values are 10^{-13} to 10^{-14} cm^2. It is used mainly in connection with organic monolayers.

Static limit (organic materials)

In general, the probability P of collecting data from an already damaged area must be kept very small, that is,

$$P = \sigma_{tot}/A \ll 1 \tag{4.1}$$

where
   A is the bombarded area
   σ_{tot} is the total area of damage

If P is sufficiently small, then the bombardment processes can be assumed to be independent of each other, i.e.,

$$\sigma_{tot} = \sum \sigma = \text{PID} \times \sigma \tag{4.2}$$

where
   σ is the disappearance cross-section
   PID is the primary ion dose

Then

$$P = \text{PID} \times \sigma/A = \text{PIDD} \times \sigma \ll 1 \tag{4.3}$$

where PIDD (=PID/A) is the primary ion dose density. Thus, for $P = 0.01$ and $\sigma = 10^{-13}$ to 10^{-14} cm^2, the consequence is that $(PIDD)_{max} = 10^{11}$ to 10^{12} ions/cm^2, while for $P = 0.1$, $(PIDD)_{max} = 10^{12}$ to 10^{13} ions/cm^2. It must be emphasized that the static limit $(PIDD)_{max}$ depends on σ, which itself depends on the particular species.

Transformation probability $P(M \rightarrow X_q)$

$P(M \rightarrow X_q)$ = number of emitted secondary ions X_q per number of sputtered surface species $M = Y_i(X_q)/Y_D(M)$. $P(M \rightarrow X_q)$ is the probability of a surface species M being transformed into an ion X in charge state q. The transformation probability can be determined by evaluation of a plot of $N_d(t)$ versus t if $N(M)$ is known. $P(M \rightarrow X_q)$ is used mainly for the description of monolayers. It is assumed that desorbed ions are stable. For metastable ions, measured values of $P(M \rightarrow X_q)$ are too low.

Number of detectable ions N_d (organic materials)

$$N_d(t) = N_d(0)\exp(-(\sigma I/Ae)t) \qquad (4.4)$$

where
  $N_d(0)$ is the number of detectable ions at $t = 0$
  I is the primary ion current
  A is the bombarded area
  σ is the disappearance cross-section

From a plot of $N_d(t)$ versus t, σ can be found. $N_d(0)$ itself is given by

$$N_d(0) = N(M)P(M \rightarrow X_q)TD \qquad (4.5)$$

where
  $N(M)$ is the number of molecules M present at the surface
  T is the transmission of the analyzer
  D is the detection efficiency
  $P(M \rightarrow X_q)$ is the transformation probability (see previous paragraph)

Ion generation efficiency $E(X_q)$ (organic materials)

$$E(X_q) = Y_i(X_q)/\sigma \qquad (4.6)$$

$(=N(M)\, P(M \rightarrow X_q)\, TD$, for monolayers)

where
  $Y_i(X_q)$ is the secondary ion yield
  σ is the disappearance cross-section

$E(X_q)$ describes the maximum number of secondary ions that can be detected from the unit surface area (typically 1 cm^2) if the whole area were to be consumed. It therefore refers to the effectiveness with which ions can be generated. $E(X_q)$ depends on the primary ion species and is also used for the analysis of multilayers.

Useful lateral resolution Δl (organic materials)

Estimation of Δl is based on the area σ that has to be damaged to detect n secondary ions:

$$(\Delta l)^2 = \sigma = n/E(X_q)$$

$$\Delta l = (n / E(X_q))^{1/2} \qquad (4.7)$$

The dependency of Δl on $E(X_q)$ is such that a good lateral resolution requires very efficient secondary ion generation independent of the physical properties of the analytical instrument (e.g., focus of the primary ion gun, transmission of the analyzer).

4.3 PHYSICAL EFFECTS OF ION-INDUCED SPUTTERING

SIMS is based on the initially unexpected observation that the bombardment of a surface with ions of kiloelectronvolt energy leads to the emission (sputtering) of a secondary species characteristic of the surface chemical composition, an effect that was first discovered by Thomson in 1910 [6]. A small fraction of the sputtered particles was found to be charged (secondary ions).

Even now, almost 100 years later, the physical effects that lead to the eventual desorption* of secondary particles are not yet completely understood. Whereas the desorption of secondary neutrals from elemental targets can be described fairly well, an explanation of the emission of charged particles and, in particular, of the emission of molecular species is still being sought. Nevertheless, there is a consensus that sputtering is based primarily on the formation of a collision cascade in the target caused by the impinging primary ion [7,8].

4.3.1 SPUTTERING

When a primary ion hits a solid, it loses its energy and momentum by elastic and inelastic collisions with target atoms, which are then displaced. These moving atoms (i.e., primary recoils) themselves induce the movement of further target atoms (i.e., secondary recoils), so that the paths of the primary ion and of the highly energetic primary recoil atoms are surrounded by a cloud of recoil particles of lower energies (i.e., a collision cascade; see Figure 4.2). Recoil atoms can leave the solid if their

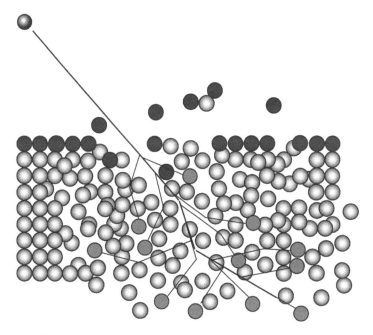

FIGURE 4.2 Schematic of the sputtering process.

---

* Although the term "desorption" is now used mainly to describe the spontaneous disappearance of a species from a surface, in this chapter it will also be used to describe sputter-induced particle emission from surfaces.

momenta are directed at least partially toward the surface, and if their energies are sufficient to over-come the surface binding energy.

The greatest number of recoils is produced near the end of the collision cascade and their aver-age energy is consequently comparatively low. Only recoils originating in the outermost monolayer of the solid can overcome the surface energy barrier and contribute to the sputtered flux. The low-energy of the secondary recoils is sufficient to cause the ejection of not only atoms into the gas phase but also intact molecules (see Section 4.3.3). The total sputtered flux is therefore characteristic of both the elemental and molecular composition of the uppermost monolayer of the bombarded sur-face (see Refs. [2,7–11] for detailed descriptions).

4.3.2 Ionization

By far the greatest proportion of the sputtered flux has zero charge, i.e., consists of neutrals. The fraction desorbed as ions is only 10^{-1} to 10^{-6}. The number of sputtered ions per incident primary ion (i.e., the secondary ion yield) can also vary tremendously with the chemical environment in the near-surface layers, a phenomenon known as the matrix effect. Examples of the secondary ion yield enhancement, which in some cases can reach several orders of magnitude, are described in Refs. [12,13] for positive metal ions in oxidic environments.

With such yield variations direct quantification of surface species based on the number of des-orbed secondary ions (i.e., from the SIMS data) is difficult (see Section 4.5.1). Sputtered neutrals suffer much less from changes in the chemical matrix. However, the analytical technique, secondary neutrals mass spectrometry (SNMS), that makes use of these neutrals, has so far been applied far less often than SIMS for the simple reason that mass analysis can be achieved only if the species is charged. Sputtered neutrals must therefore be ionized prior to mass analysis (i.e., postionization). This process not only complicates the experimental setup but also reduces the number of surface species that can be detected to such an extent that only for specific applications does the SNMS approach the levels of sensitivity (in terms of minimum detectable surface species) of SIMS.

4.3.3 Formation of Molecular Species

Whereas the desorption of atoms and small molecular clusters by ion bombardment can be described in a satisfactory way by the formation of collision cascades, it is surprising that large molecular spe-cies, and even molecular constituents that are thermally unstable and cannot be vaporized, are also emitted from the first monolayer of a surface as intact neutral particles and as positive or negative secondary ions. This phenomenon provides the basis for the mass spectrometric analysis of molecu-lar surfaces first described by Benninghoven [14,15]. An early summary of emission models for the desorption of organic species can be found in Ref. [16].

As mentioned, the entire energy of the impinging primary ion is transferred in the collision cascade. Those areas affected by the cascade therefore undergo dramatic changes in their com-position, including the breaking and reformation of chemical bonds. If information on the molecular structure of a surface is required, collection of data from already damaged areas must be avoided. From a statistical point of view this can be achieved if the PIDD (i.e., the number of primary ions per bombarded area) is reduced to such an extent that a maximum of only 1% of the atoms in the surface suffer primary knock-on collisions. For such an operational mode the term static SIMS (SSIMS) is used in contrast to dynamic SIMS (DSIMS). In the dynamic versions the operational conditions are such that successive atom layers are removed com-pletely and rapidly under the primary ion bombardment, and the sputtered flux is used to obtain a "depth profile" of the elements present [2]. The static limit, i.e., the highest PIDD that can be tolerated without causing significant surface damage, is of the order of 10^{11} to 10^{13} ions/cm^2, depending on the size of the species analyzed (see also Section 4.2).

During the last 10 years it has transpired that the efficiency with which secondary ions of organic materials can be generated (secondary ion formation efficiency E; see Section 4.2) is strongly dependent on the nature of the primary ion. In particular, the use of polyatomic primary ions, or clusters, has lead to dramatic increases in formation efficiency. As the formation efficiency influences not only the ultimate detection limit, but also the achievable lateral resolution (useful lateral resolution Δl, see Section 4.2), the advent of cluster ion sources has allowed the expansion of analytical applications in organic analysis.

4.3.4 Polyatomic Primary Ion Bombardment

Polyatomic bombardment refers to a cluster of atoms jointly reaching the target and thus allowing cooperative phenomena in the substrate. Currently, SF_5 [17,18], Au_x [19], Bi_x [20], as well as C_{60} [21–23], are in use.

Whereas the emission behavior of elemental ions remains almost unchanged when switching from monatomic to polyatomic bombardment, the ion generation efficiency E for organic materials is enhanced by up to three orders of magnitude. In Figure 4.3, a systematic study of the emission behavior of the polymer additive Irganox 1010 (mass 1176 amu) under bombardment with different primary ion species is shown. The additive was prepared as a monolayer on low-density polyethylene (LDPE). As a result of this method of preparation, secondary ions from the additive can originate only in the uppermost monolayer, and therefore disappearance cross-sections (necessary for the determination of E; see Section 4.2) can be measured exactly. Furthermore, the physical effects leading to the emission of secondary ions take place entirely in an organic substrate.

From Figure 4.3a it can be concluded that the formation efficiency increases with the mass of the primary ion for monatomic bombardment. When using Au^+ or Bi^+ instead of Ga^+ the efficiency increases by more than a factor of 10. An even more dramatic increase is observed when switching to polyatomic bombardment. Compared to the efficiency of Ga^+, the efficiencies of Bi_3^+ or C_{60}^+ are higher by more than three orders of magnitude. As the useful lateral resolution has a square root dependency on $1/E$ (see Section 4.2), the enhancement in ion formation efficiency leads to a corresponding improvement in the useful lateral resolution of more than one order of magnitude. Whereas under Ga^+ bombardment an area with a diameter of at least 8 μm must be removed in order to generate four intact molecular ions, a diameter of 200 nm is sufficient under polyatomic bombardment. Thus, for the first time, organic imaging in the submicrometer range becomes possible [24].

Apart from enhancing the ion formation efficiency in the uppermost monolayer, polyatomic bombardment also leads to reduced damage in the deeper layers of an organic substrate. Therefore, not only does the uppermost monolayer contribute to the organic secondary ions signals, but signals from deeper layers can also be used. In some cases it is even possible to obtain stable secondary ion signals of organic species under prolonged primary ion bombardment, allowing the monitoring of the distribution of organic material as a function of sputtering depth (organic depth profiling) [25–27]. Several factors seem to influence the possibility of organic depth profiling, including the target chemistry [28,29] and the surface temperature [30,31].

The underlying physical and chemical effects are currently a topic of intense discussion. Explanations considered include a deposition of collision energy closer to the surface [32] and cooperative phenomena in overlapping collision cascades [33]. The first review on the current status in cluster ion bombardment is given in Ref. [34]. However, the field is still in development and new insights and developments can be expected over the next years.

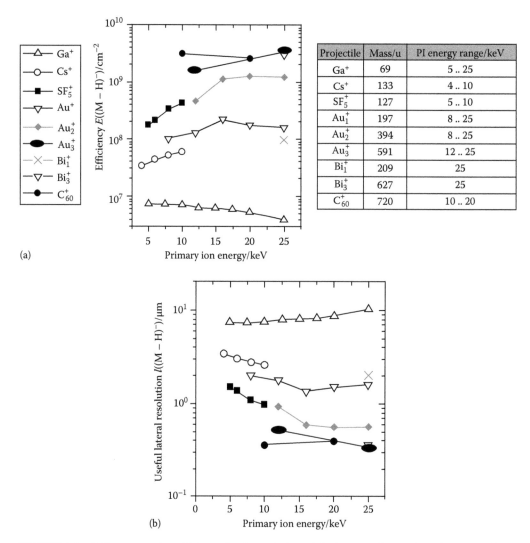

FIGURE 4.3 (a) Ion generation efficiency for the emission of the intact molecular ion of Irganox 1010 prepared as a monolayer on LDPE, as a function of primary ion species and primary ion energy. (b) Useful lateral resolution for detection of four intact molecular ions of Irganox 1010, as a function of primary ion species and primary ion energy.

4.4 INSTRUMENTATION

In general the instrumentation for a SIMS experiment can be divided into two parts, (I) and (II), as follows:

(I) Primary ion generation (primary ion column)
↓
(Secondary ion generation, target)
↓
(II) Mass analysis and detection (secondary ion column).

In the primary ion column the primary ions are generated, focused, and transported toward the target; in the secondary ion column the desorbed species are extracted, sorted by mass, and detected. Mass detection in ToF analyzers requires pulsed operation of the primary ion column. The primary ions in fact reach the target in the form of ion packages with well-defined arrival times, a prerequisite for the measurement of the flight time of the generated secondary ions toward the detector.

4.4.1   PRIMARY ION GENERATION

In general, primary ion guns can be distinguished between those used for the analysis of the uppermost monolayer (analysis guns) and those used for subsequent sputter removal of surface species (sputter guns). The selection of a suitable ion gun for SIMS depends on the type of application intended and the particular operational mode (see Section 4.5).

4.4.1.1   Ion Guns for Surface Analysis

A primary ion gun for surface analysis by ToF-SIMS should provide high lateral resolution, high mass resolution, and fast measurement times. For high mass resolution the pulse length of the primary ion packages must be very small (<1 ns). Good lateral resolution requires good focus ability in the ion column, while the measurement time is influenced by the rate at which data can be acquired, which is dependent principally on the achievable primary ion current. Unfortunately, the fundamental laws of physics decree that in any one ion optical arrangement the three decisive parameters, pulse length, focus diameter, and data rate, cannot all be optimized at the same time. The possible options fall within a parameter triangle as shown in Figure 4.4. For example, having a high mass resolution and a high lateral resolution will be possible only at the cost of data rates, i.e., measurement time, while, on the other hand, fast production of ion images with high lateral resolution requires a compromise with respect to mass resolution (i.e., primary ion pulse length). Depending on the ion gun used, either the whole parameter triangle can be accessed or special fixed points can be predefined.

Today, the two groups of ion guns most used in ToF-SIMS are liquid metal ion guns (LMIG) and gas-phase electron impact (EI) ionization guns. These two types of ion guns will now be discussed briefly.

In an LMIG a small needle is wetted by a liquid metal (e.g., Ga, Au, Bi). A positive voltage of some 10 kV is applied to the needle and a spray process (ion formation by field emission) occurs at the needle tip. This field-induced stream of positively charged ions (or jet) is used as the primary ion beam [35–37]. Because the ion production volume at the tip is very small, and the energy distribution in the emission solid angle is very narrow, LMIG sources have high brightness, and consequently the ion beam can be focused to a small area with high current density. The necessary use of liquid metals means that chemical reactions at the surface cannot be completely excluded, although they have not been reported so far. Generally, LMIGs allow fast switching between monatomic and polyatomic primary ions (e.g., Au_1^+, Au_3^+, Au_5^+; Bi_1^+, Bi_3^+, ..., Bi_7^+) and are therefore well suited to both elemental and molecular surface analyses. They are easy to use and the pulse length can be kept very short. They are currently the "universal workhorse" guns in ToF-SIMS surface analysis.

Before the advent of LMIGs, gas-phase ionization guns were the most used in surface analysis, usually operated with noble gas ions such as Ar^+ or Xe^+, and almost always involving EI. In an EI gun gas atoms are ionized by collisions with electrons of some 10 eV energy (i.e., five to six times the ionization potential) attracted to an anode from a heated filament. The gas ions so generated are

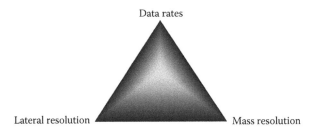

FIGURE 4.4   Parameter triangle for primary ion guns.

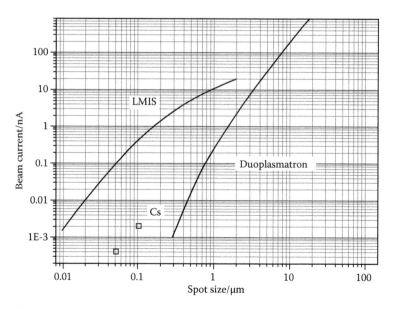

FIGURE 4.5 Relationship between ion gun beam current and spot size for ion guns with LMIS and gas-phase ionization (i.e., EI) ion guns (here: duoplasmatron as a member of this family), respectively. Values for the Cs gun in an Cameca NanoSIMS have been added for comparison. (From Briggs, D.A. and Seah, M.P., *Practical Surface Analysis (Second Edition). Volume 2: Ion and Neutral Spectroscopy*, Chichester, New York, 1992; Product Brochure Cameca NanoSIMS 50, http://www.cameca.fr/doc_en_pdf/ns50_instrumentation_june2007_web.pdf)

accelerated toward an extraction electrode and can then be focused onto the target [2,38,39]. The use of noble gases offers the advantage that there are no chemical reactions between the primary ion and the target species that could affect the correct determination of the surface composition. Although EI guns can deliver very high data rates, the focus quality and the achievable pulse length are limited. As a result, they are now not used as often as LMIGs. Recently, EI guns have found new interest because they allow the production and application of large polyatomic cluster ions such as C_{60}^+ and SF_5^+.

Figure 4.5 summarizes the performance data of primary ion guns used for analysis. It can be seen clearly that at a given beam current the focusing capability of LMIGs (liquid metal ion source [LMIS] in Figure 4.5) is more than one order of magnitude better than that of gas-phase ionization guns (duoplasmatron: the gas-phase ionization gun with the best focus quality [40]). Additionally, data have been added for Cs primary ions produced in a surface ionization gun (for operating principles see Ref. [2]). Please note that the data refer to unpulsed operations of the guns. Pulsing will reduce further the focus quality, making it difficult to reach a submicrometer focus quality with EI guns.

4.4.1.2 Ion Guns for Sputtering

For surface erosion in ToF-SIMS, either EI (O_2^+) or surface ionization ion guns (Cs^+) are used. These ions are used because, on the one hand, O_2^+ stabilizes the chemical environment for the emission of positively charged secondary ions and, on the other, Cs^+ does the same for negatively charged secondary ions (see Chapter 10). Their use thus guarantees high secondary ion yields even from deep layers. Recently, C_{60}^+ and SF_5^+ have also been used (gas-phase ionization) in context with organic depth profiling (see Section 4.5.3).

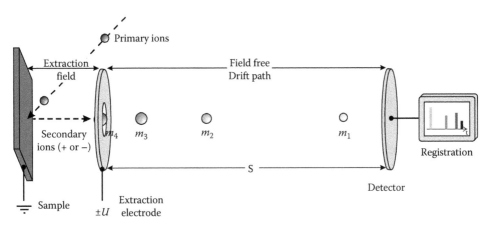

FIGURE 4.6   Schematic of a ToF analyzer. The flight path length is s and the extraction voltage U.

4.4.2   MASS ANALYZER

In a ToF-SIMS experiment, a ToF analyzer is used for mass separation of the desorbed secondary ions. In a ToF analyzer (Figure 4.6), the ions desorbed are accelerated to a common energy qU (where q = ion charge and U = applied voltage), and the time t taken by each type of ion to reach a detector, after traveling along a field-free drift path of given length s, is measured, according to

$$ms^2/2t^2 = qU$$

(4.8)

which can be rearranged to

$$t^2 = ms^2/2qU \propto m/q$$

(4.9)

The m/q ratio can then be calculated from the flight time. Because a very well-defined start time is required for the flight time measurement, the primary ion gun has to be operated in a pulsed mode in order to be able to deliver discrete primary ion packages [41]. Electric fields (using, e.g., ion mirrors [41,42] or electrical sectors [43,44]) can be introduced into the drift path in order to compensate for different incident energies and angular distributions of the ions (see also Section 4.2).

The main features of a ToF analyzer are

- Quasi, simultaneous detection of all masses of one polarity
- High geometrical transmission
- High mass resolution

ToF analyzers therefore offer excellent sensitivity and have become the analyzer of choice for organic analysis and for screening purposes.

4.4.3   CHARGE COMPENSATION

The primary ion bombardment leads not only to the desorption of sputtered elements and molecules but also to the emission of further secondary species, among the electrons. The number of emitted electrons per primary ion (the so-called ion-induced secondary-electron coefficient) can be as large as 10 depending on the particular material. Ion bombardment of insulating surfaces therefore leads to the build-up of positive charge at the surface that can severely disturb the secondary ion extraction. The charge can be compensated by auxiliary electron bombardment. Low-energy electrons are generally used (<20 eV) because, firstly, at those energies the electron-induced secondary-electron coefficient is <1 (if the electron-induced secondary-electron coefficient is less than unity, then every incoming electron leads to less than one outgoing electron and an existing positive charge can be compensated) and, secondly, because sensitive organic molecules are hardly affected by the additional bombardment. For the latter reason high-energy electrons (e.g., >3 keV, for which the electron-induced secondary-electron yield coefficient would also be below unity) are less suitable for charge compensation purposes.

When ToF analyzers are used for mass determination, low-energy electrons can easily be injected into the target zone in the period between secondary ion extraction and the next primary ion pulse. During that period all high voltages near the target zone, which might deflect low-energy electrons, can be switched off. With this arrangement, stable spectra of both polarities can be obtained even for extremely insulating materials, including powders (for details see Ref. [45]).

4.5 OPERATIONAL MODES

Depending on whether or not the primary ion beam is focused and rastered across the surface, and on whether or not the surface is continuously eroded during the SIMS experiment, four operational modes can be distinguished. These modes are pictured in Figure 4.7, and described below.

In surface spectrometry (see Section 4.5.1), an area of interest is bombarded by the primary ion beam and the mass spectra are recorded for all secondary ions originating from the entire bombarded area. The mass spectrum so obtained then contains information on the chemical composition in the area of interest. Such an experiment is normally performed under static or near-static conditions, i.e., the information originates from the uppermost monolayers of the sample.

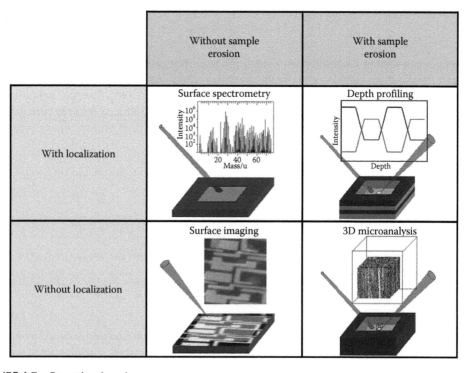

FIGURE 4.7 Operational modes.

In surface imaging (Section 4.5.2) the primary ion beam is focused to a very small diameter (typically <1 µm) and is rastered across the surface point-by-point. The whole bombarded area is divided into so-called pixels that are interrogated step-by-step by the primary ion beam (ion probe mode of imaging). As a result, a mass spectrum is obtained from every pixel addressed, and therefore chemical maps can be acquired for any secondary ion of interest [46]. Alternatively, when operating a ToF analyzer with imaging capabilities, an unfocused primary ion beam is directed onto the sample surface, and a position-sensitive detector is used to deduce the point of origin of the secondary ions (ion microscope mode of imaging [43]). In general, the microprobe mode is used more often due to the better lateral resolution and higher data rates that can be obtained.

In depth profiling (Section 4.5.3) the PID density is increased to the point at which continuous sample erosion occurs during the SIMS experiment; the masses of the desorbed secondary ions are then analyzed as a function of sputter time. Since the sputter time is correlated with the eroded depth, a mass spectrum as a function of depth is recorded.

In 3D microarea analysis (Section 4.5.4) the combination of surface imaging with depth profiling allows any position within a 3D volume (voxel) to be addressed, thus providing access to the chemical compositions in the bombarded and eroded volumes.

4.5.1 SURFACE SPECTROMETRY

Surface spectrometry is one of the oldest operational modes. Since all other modes require the detection of a spectrum in any case, this mode is described in detail below.

4.5.1.1 Typical Characteristics of SIMS Spectra

Figure 4.8 shows the positive secondary ion mass spectrum ("positive" refers to the detection of positively charged secondary ions) from an isolated oligomer of polydimethylsiloxane (PDMS) prepared as a monolayer on an etched silver substrate (solution: 0.1 mg/mL in toluene, 1 µL spread over 80 mm^2). This particular material is used here as an example since it allows most of the typical spectral features of SIMS spectra to be described. In contrast to polymer preparations containing more than one oligomer, spectra of isolated oligomers, though technically irrelevant, do not suffer from peak interferences with other oligomers, and the fragments can arise from only one type of molecule instead of from the whole molecular weight distribution. The chosen molecule was kindly provided by Wacker Chemie AG (München, Germany). PDMS molecules, known also as silicone oils or rubbers, are among the most troublesome contaminants in adhesion problems (see Chapter 18).

The spectrum can be divided into three general regions: elements and fragments (1–500 amu), quasimolecular ions (1150–1350 amu), and an intermediate region, in which almost no secondary ions can be detected.

The term "quasimolecular ions" refers to those ions that are formed either by the attachment of low-molecular-weight cations and anions (e.g., salt and metal ions) to the parent mass, or by the loss of small fragments (mostly functional groups such as CH_3, OH, etc.). The most intense group of peaks among the quasimolecular ions can be attributed to the attachment of silver to the intact molecule (i.e., $(M + Ag)^+$) confirmed by a comparison of the calculated and measured isotopic patterns. Almost the same holds for the peak at mass 701.1 amu. Here, cationization has been achieved by the attachment of two Ag ions. Consequently, the ion is doubly charged and appears at half the mass of the parent molecule (1402.2 amu) (cf. detection of m/q). (The low intensity of this peak shows that formation of a doubly charged species is not a dominant process in SIMS.) Further, quasimolecular ions can be attributed to $(M–CH_3)^+$ M^+, and $(M + H)^+$ (the isotopic patterns of M^+ and $(M + H)^+$ overlap), as well as $(M + Na)^+$ and $(M + K)^+$. The differences in the intensities of these ions, compared to that of the $(M + Ag)^+$ peak, demonstrate the high formation probability of the cations. If a complete oligomer weight distribution were to be analyzed, the form of the distribution could be derived directly from the distributions of the respective quasimolecular ions.

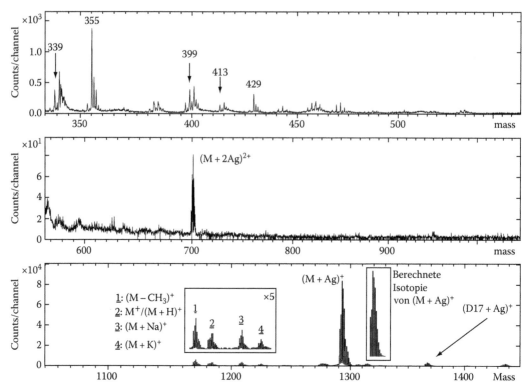

FIGURE 4.8 Positive secondary ion mass spectrum of D16 (PDMS oligomer consisting of 16 repeat units), prepared as a monolayer on silver.

The mass range up to 500 amu is dominated by element and fragment ions. Na, Si, and Ag are detected as elements. Na is due to a salt contamination, Si originates from the oligomer backbone, and Ag is a substrate ion. Four organic fragment series can be identified starting at masses 73, 103, 117, and 133 amu, respectively. All further fragments can be explained by the addition of the PDMS repeat unit (74 amu). The ion structures are given in Figure 4.9. With the exception of the primary ionization process, fragments in SIMS are nearly all formed according to the conventional rules of EI mass spectrometry (i.e., α and β cleavages, rearrangement processes [47]). As other polymers give rise to other fragment patterns, the type of polymer can be determined by evaluation of the low-mass fragments alone.

In the intermediate mass range, where the PDMS does not give rise to any characteristic secondary ions, large fragments can be observed originating from single or multiple main-chain scissions. In many cases these fragments are neutral and can therefore be observed only as $(F + Ag)^+$ (where F is an appropriate fragment) from monolayer preparations on noble metals. From the mass separation between two fragment peaks the mass of the polymer repeat unit can be derived.

Of course, negatively charged secondary ions can also be formed, as quasimolecular ions (e.g., $(M - H)^-$) as well as fragments. Small, negatively charged fragments are often more characteristic of a molecule than those positively charged, because charge stabilization is usually related to oxygen, a heteroatom having a profound influence on chemical behavior. From PDMS, for example, no negative quasimolecular ions are formed, but a strong fragment series starting at mass 75 amu (CH_3SiO_2) is observed.

In summary, SIMS spectra provide not only evidence of all the elements present but also detailed insight into the molecular composition. Quasimolecular ions can be desorbed intact up to 3,500 amu

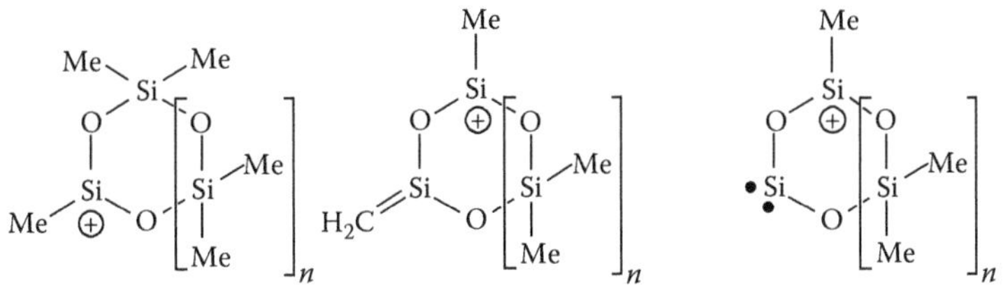

$n = 0$	133	117	103
$n = 1$	207	191	177
$n = 2$	281	265	251
$n = 3$	355	339	325
$n = 4$	429	413	399
$n = 5$	—	—	473

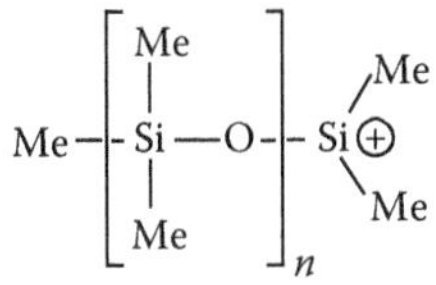

$n = 0$	73
$n = 1$	147
$n = 2$	221

FIGURE 4.9   Fragment structures of PDMS produced as a result of ion bombardment.

and sometimes even up to 15,000 amu, depending on the particular molecule [48] and on whether or not an effective ionization mechanism is present. Larger molecules tend to fragment under the SIMS excitation conditions due to stronger intra- and intermolecular bonds.

If no quasimolecular ions are formed, due to steric hindering during desorption (large molecules), or hindered ionization (e.g., thick molecular overlayers without contact with cations), small-fragment ions still appear, because they can be formed even if strong intra- and intermolecular bonds exist, and because they carry an intrinsic charge. The observed fragment peak patterns are characteristic of the particular molecules and of the mass range up to 500 amu; the term "fingerprint region" is thus used. Whereas quasimolecular ions are comparatively easy to identify, the interpretation of the fingerprint region requires much experience and should best be left to the expert. Meanwhile, some spectral libraries have been published [49]. However, the number of substances represented in these libraries is still comparatively small. To date no material has been reported from which SIMS spectra could not be obtained, provided that it is stable in a vacuum environment, making SIMS a universal analysis technique.

4.5.1.2 Quantification of SIMS Spectra

Generally, the matrix effect, i.e., the dependence of the secondary ion yields, transformation probability, and ion formation efficiency, on the nature of the chemical environment complicates quantification of SIMS data or makes it even impossible. Nevertheless, many cases exist where at least semiquantitative data can be obtained. The three most common ones will be discussed in the following because it is essential for the user of the technique to be aware of the options and limitations in this respect.

4.5.1.2.1 Use of Internal Standards

Internal standards are elements or molecules added to the analyte in known concentrations. If chemically identical (i.e., isotopically labeled) or chemically similar (i.e., same substance class) materials are used, the analyte and standard react in the same way to changes in the chemical environment, which means that the respective transformation probabilities change in the same ratios. The evaluation of I(analyte)/I(standard), with I being the peak areas of the corresponding characteristic peaks, therefore allows quantification of the amount of analyte material if a calibration curve had previously been constructed from which the (fixed) ratios of the transformation probabilities could be derived [50]. Although the use of internal standards can be a very powerful approach for absolute quantification, there are several limitations to it in SIMS:

- Choice of a suitable standard: The standard (S) must mix homogeneously with the analyte (A). Both S and A must show the same behavior with respect to segregation or crystallization. If a sample has to be prepared specially prior to analysis, both materials must show the same behavior in all preparatory steps. All these requirements can be fulfilled for isotopically labeled materials, which are of course chemically identical, but problems can occur when using chemically similar materials. On the other hand, the latter are generally more readily available, more stable, and cheaper, compared to isotopically labeled materials.
- As A and S must be mixed prior to the measurement, the use of internal standards will usually be limited to solutions and subsequent droplet depositions onto substrates.
- Best results can be achieved when S and A are present in approximately equal amounts, i.e., prior knowledge is required about the expected amount of A.

4.5.1.2.2   (Sub)monolayer Coverages

If the analyte is present on a substrate at a submonolayer or single-monolayer coverage, quantification can be achieved by normalizing an analyte peak area to a substrate peak area. In this approach linear relationships can be found between the surface coverage and the analyte signal normalized in this way. Absolute quantification is possible if at least one coverage value can be determined by independent quantitative techniques.

As an example, Figure 4.10 compares normalized analyte signals with known coverages of sputter-deposited standards [51]. The quantification refers to Ni on a Si wafer. Ni can cause severe problems in semiconductor production and must be quantified even for extreme submonolayer coverages. The advantage of ToF-SIMS lies in the possibility of analyzing small areas with good detection limits.

For submonolayer coverages, quantification is possible because in such dilute systems the matrix of an analyte species is always that of the neighboring substrate, and therefore $P(M \rightarrow X_q)$ is stable. Nevertheless, it would be expected that deviations from linearity would occur when approaching monolayer coverage. The finding that the linear behavior extends even to monolayer coverage can

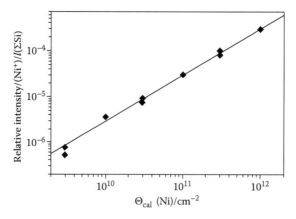

FIGURE 4.10   Plot of Ni peak areas normalized to the Si substrate peak areas, as a function of the Ni surface coverage. (From Schnieders, A., Möllers, R., Terhorst, M., Cramer, H.-G., Niehuis, E., and Benninghoven, A., *J. Vac. Sci. Technol. B*, 14, 2712, 1996. With permission.)

be explained in a simple way by the fact that desorption in SIMS is caused by the collision cascade, which in this case takes place entirely in the substrate and "pushes" the species from below the surface. Interactions in the vertical direction (between the layers) are thus much more important for desorption and ionization than interactions in the horizontal direction (within a layer). As long as the maximum coverage does not exceed one monolayer, a stable ion formation process (i.e., stable $P(M \rightarrow X_q)$) can be expected. This argument implies that severe problems could occur in quantification during the transition from monolayer to multilayer. In that case the collision cascade could take place partially in the substrate and partially in the analyte itself, which would affect principally the sputter yield. In addition, the ion formation process could also change during the transition. Coverage effects on the secondary ion yield can be observed even for those cases where a second monolayer has just started to be formed (i.e., submonolayer coverage of the second monolayer on a complete first monolayer); the effect on $(M + H)^+$ formation is an example. For the same reason quantification is also unreliable for isolated multilayer clusters ("island" nucleation) on an otherwise uncovered substrate.

4.5.1.2.3   Organic Multilayers

Quantification of organic multilayer systems is surprisingly successful in view of the enormous chemical varieties of organic materials. For quantification purposes a typical analyte molecule A (mostly a fragment from the "fingerprint" area) is first chosen, and its intensity is then normalized to either

- That of a fragment of a second compound or that of the sum of both compounds (in binary mixtures), or
- The total spectral intensity, or
- A sum over the intensities of peaks (including all relevant compounds), or
- That of an uncharacteristic hydrocarbon fragment, $C_xH_y^{\pm}$

An example is given in Figure 4.11, which shows the quantification of polybutadiene (PBD) in a binary system of PBD and polystyrene (PS). Both polymers are unsaturated hydrocarbons and quantification by other techniques is not easy. A peak typical of PBD is the $C_4H_6^+$ fragment at mass 54 amu. A linear relationship can be obtained if the peak area of this fragment, normalized to the total spectral intensity, is plotted against the PBD concentration in the particular sample. The quantification is achieved directly at the surface, allowing the monitoring of surface phenomena such as segregation, preferential adsorption, and diffusion.

Quantification in such a system is possible mainly because the chemical environment consists only of hydrocarbons, which from a physical (in contrast to a chemical) point of view do not vary much (only C, H, and few heteroatoms are present). If fragment ions are used, then the matrix is formed partially by the (larger) molecule itself, which helps to stabilize $P(M \rightarrow X_q)$ further.

However, this quantification approach requires considerable experience with SIMS. First of all, the choice of a fragment is not always straightforward and not all characteristic peaks are suitable. In particular, the most intense peaks are not necessarily the most successful candidates for quantification purposes. Second, a suitable normalization procedure must be chosen. In reality it might prove necessary to try several combinations of fragments and normalization procedures while at the same time using any additional foreknowledge of the sample. The selection should then, in principle, be tested on a model system of known composition in order to establish adequate reliability. Fortunately, statistical procedures based on multivariate statistical modeling can help to select suitable normalization candidates. In particular, the use of partial least squares fitting has shown promising results [52–54]. Furthermore, reference spectra can be consulted in order to identify characteristic fingerprint peaks. Absolute quantification in multilayer systems is only rarely achieved, but semiquantitative information can be obtained in many cases.

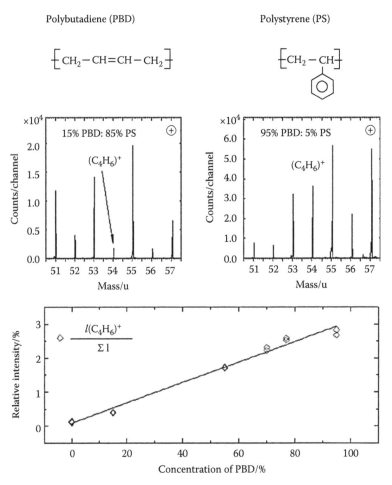

FIGURE 4.11 Positive ToF-SIMS spectra of PBD and PS. The lower part shows the calibration curve derived from the results.

In summary, it can be concluded that quantification of surface coverages is possible much more often than might be expected given the matrix effect in SIMS. Internal standards should be used whenever possible, since they give the most reliable results. When working with (sub)monolayer coverages, quantification by normalization to a substrate-related peak is always worth the effort. If the matrix is genuinely stable, then information on the surface coverage can be derived. If the matrix is varying (e.g., due to oxygen treatment in a plasma or from UV/ozone exposure), at least some new insight into the matrix effect itself might be gained. In this approach even a change in oxidation state can be monitored. For organic multilayers it is well worthwhile to attempt quantification. If there are no initial clues as to which peaks and normalization procedures might be suitable, then a start could be made by normalizing identified fragment ion intensities to those of uncharacteristic hydrocarbon fragments (e.g., $C_3H_5^+$, $C_3H_2^-$), which are always present on samples that have been in contact with ambient air. This is surprisingly successful in many cases and easy to apply. Further progress could then be made from spectral libraries and eventually multivariate statistical modeling. Quantification remains difficult for systems where there are only a few monolayers (one to five) cover the substrate.

It should be emphasized that the determination of surface concentrations and coverages is only one aspect of quantification. SIMS can also be used successfully for the determination of molecular weight distributions and average molecular weights in polymers, for the determination of diffusion constants (vertical and horizontal diffusion), and for the determination of surface coverages via the evaluation of species portions from ion images. Some examples will be shown in Section 4.7.2.

4.5.2 SURFACE IMAGING

In general, three factors are essential for the successful location of surface species, and they therefore govern the overall performance in imaging (i.e., the level of spatial resolution). They are the

1. Physics of the analysis process
2. Lateral resolution provided by the instrumentation
3. Achievable sensitivity of the particular surface analytical technique

In SIMS, secondary ions are formed in a collision cascade following the impact of the primary ion. The diameter of this collision cascade therefore determines the ultimate limit of the accuracy with which the original position of a surface species can be located. Depending on the type of primary ion, on the bombardment energy, and on the particular surface species, the collision cascade diameter is about 2–5 nm [55,56].

The lateral resolution offered by the instrumentation is determined mainly by the spot diameter of the primary ion beam. When LMIGs are operated in a static mode, i.e., not pulsed, a focus diameter of 10 nm can be achieved [57]. However, when modern, sensitive ToF analyzers are used for mass determination and ion detection, the static operational mode of the LMIG only allows the acquisition either of total secondary ion images or of total ion-induced secondary-electron images. Mass information cannot be obtained. The total secondary ion and secondary-electron images nevertheless do give valuable topographical information. The image quality nowadays is close to that of dedicated SEM instruments. If mass information is required, then the LMIG must be pulsed, resulting in an inferior focus diameter. With only a purely nominal mass resolution, however, 50–80 nm can still be achieved. A focus diameter of 200 nm allows the acquisition of spectra with a reasonable mass resolution of $m/\Delta m = 5000$, which is sufficient to differentiate elements from organic species reliably. For maximum mass resolution, an LMIG with a focused diameter of ca. 1 μm has to be used.

A small focus diameter, however, is not of much help if not even one secondary ion can be detected from the bombarded area due to insufficient sensitivity. This is a problem mainly in the analysis of organic materials. However, the use of polyatomic primary ions has improved the situation considerably (see Section 4.3.4). Figure 4.3b shows that intact organic species can indeed be detected with submicrometer pixel sizes.

As an example of the lateral resolution obtainable with a ToF-SIMS instrument, Figure 4.12 shows negative secondary ion images of a Ba halide crystal coated with Sr. The image resolution is better than 100 nm (with Bi_3^{2+} as the primary ion).

However, it is not only the small fields of views that are of interest with respect to problem solving. ToF-SIMS instruments can also allow imaging of large fields of view, up to almost $10 \times 10 \, cm^2$. In this case, either the stage is moved continuously under the primary ion beam and the resulting data stream is reduced to pixels (stage scanning), or several adjacent images of maximum field of view of the particular ion gun are assembled or "stitched" together. Important applications include screening for unknown contaminants on glasses, polymers, and wafers.

In summary, it can be stated that imaging performance has improved dramatically over the last decade. Fields of view ranging from $10 \times 10 \, \mu m^2$ to $10 \times 10 \, mm^2$ can be addressed routinely. The LMIG has developed into the ion gun of choice for ToF-SIMS imaging because it shows excellent focus quality, high ion currents, and can be pulsed in very short intervals, while at the same time being able to produce polyatomic primary ions. Nevertheless, the learning curve for this operational mode is still steep and more progress can be expected in the future.

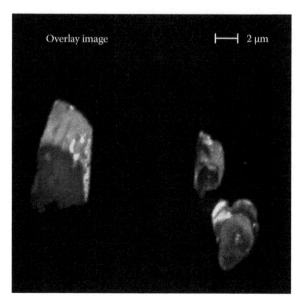

FIGURE 4.12 (See color insert following page 396.) Positive ion mass resolved ToF-SIMS images of Ba halide crystals covered by Sr halide. The Ba signal is in red, and the Sr signal in green.

4.5.3 DEPTH PROFILING

Depth profiling, i.e., the sputter erosion of a surface with the purpose of analyzing the sample composition as a function of depth (DSIMS), is not the main subject of this chapter. A more detailed description can be found in Chapter 10. Nevertheless, in the last decade the number of depth profiling applications with ToF-SIMS instruments has increased continuously for three reasons:

- The possibility of detecting all masses simultaneously allows the screening of samples without the necessity of having prior knowledge of the sample composition.
- Depth profiles can be acquired easily from insulating materials.
- The sputtering ion energy can be lowered to <200 eV, improving the achievable depth resolution toward atomic dimensions.

For ToF-SIMS depth profiling a dual beam mode is used, i.e., two ion guns, one for sputter erosion and one for analysis of the crater center. An electronic timing scheme switches between the two ion guns automatically. This approach allows the optimization of sputter and analysis conditions independently from each other. The sputter gun can be operated at ultra-low energies for good depth resolution, while the primary ion species can be selected for optimum ion yields (e.g., O_2^+ for enhancement of electropositive elements, Cs^+ for enhancement of electronegative elements, and cluster ions (C_{60}^+, SF_5^+) for organic materials). For the analysis gun an LMIG is the most frequently used because of its excellent performance characteristics. The sputter rate of the analysis beam should be at the most 0.5% of that of the sputter beam, in order to avoid degrading the depth resolution. Normally, the field of view of the analysis gun is chosen to be smaller than that of the sputter gun, because such an arrangement automatically ensures that crater edge effects are minimized. For a detailed description of ToF-SIMS depth profiling see Ref. [58] and references therein.

Applications are wide ranging and include semiconductor materials, ceramics, and glass. As mentioned in Section 4.3.4, even depth profiles of organic materials are possible if polyatomic primary ions are used. An example is presented in Figure 4.13, which shows a ToF-SIMS depth profile through a layer system consisting of polyvinylpyrolidone (PVP)/Irganox 1010 (polymer additive)/polymethylmethacrylate (PMMA). As can be seen there, for the polymers the principal constituents

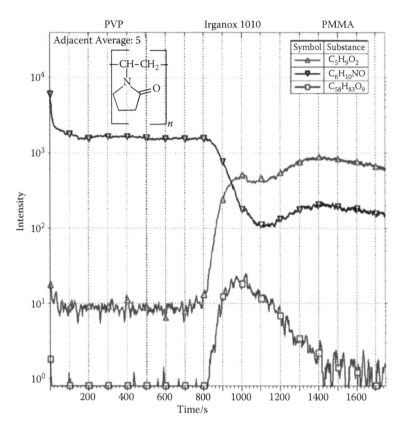

FIGURE 4.13   ToF-SIMS depth profile of the layer system PVP/Irganox 1010/PMMA. Experimental conditions: sputtering; C_{60}^{++} (20 keV) on $300 \times 300 \ \mu m^2$; analysis: Bi_3^{++} (50 keV) on $100 \times 100 \ \mu m^2$. Marker ion for PVP: $C_6H_{10}NO^+$, for PMMA: $C_5H_9O_2^+$ and for Irganox 1010 the large fragment ion $C_{56}H_{83}O_9^+$.

of the repeat units could be detected with significant signal intensities. For the Irganox 1010, the largest positively charged fragment could be used. Such results are promising for the future. But as mentioned in Section 4.3.4, organic depth profiling is not always successful and the underlying physical and chemical effects are still under discussion. Recently, organic depth profiling was successfully reported even without the use of polyatomic primary ions but with low-energy Cs bombardment [59]. The field is still emerging and much progress can be expected in the near future.

4.5.4   3D MICROAREA ANALYSIS

Three dimensional ToF-SIMS analysis is the simultaneous combination of imaging and depth profiling, in which it is possible to analyze a volume of $300 \times 300 \times 3 \ \mu m^3$ (with the smallest dimension in the z-direction). For any voxel within this volume a complete ToF-SIMS spectrum is written to the computer disk and, via retrospective analysis of the raw data, the following types of information are available:

- A spectrum for any lateral position and depth
- A depth profile for any lateral position and mass
- An image for any depth and mass

It is therefore well suited to screening applications. An example is given in Figure 4.14, which shows a diffusion study in a metal oxide with a Cr + Fe + Y overlayer. The 3D distribution of the substrate compound Mg is shown on the left and the distribution of Y from the overlayer is shown on the right.

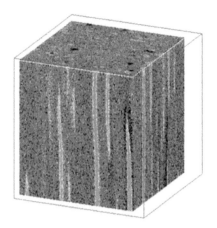

FIGURE 4.14 (See color insert following page 396.) Three-dimensional ToF-SIMS study of diffusion. The sample consisted of a polycrystalline metal oxide $O_{2.8}$ $Mg_{0.2}$ $Ga_{0.8}$ $Sr_{0.2}$ $La_{0.8}$ with a $Cr+Fe+Y$ overlayer. Left, the distribution of Mg^+, and, right, that of Y^+; a color coded map is used to represent ion intensities with blue representing high intensity and red representing low intensity, respectively. Experimental conditions: sputtering; O_2^+, 1 keV on 150×150 μm^2; analysis; Ga^+, 25 keV on 60×60 μm^2. The sample was kindly provided by Professor Martin, RWTH Aachen.

From the data it can be concluded that diffusion of the overlayer metals into the substrate occurred along the grain boundaries, and not through the bulk. In addition, organic materials have been analyzed. For 3D imaging of single biological cells, see Ref. [60].

4.6 PERFORMANCE SUMMARY

Table below summarizes the performance data at the time of writing. Instrumental performance (e.g., mass resolution, lateral resolution, etc.) is always likely to improve progressively with time.

Principle of measurement	Excitation of the surface with high-energy ions; determination of the mass of desorbed secondary species (atoms, molecules)
Obtained information	Chemical (elemental/molecular) composition for the surface; detection of all elements and stable isotopes; detection of molecules (intact for $m < 10,000$ amu)
Excitation depth	Some 10 nm
Information depth	Outermost monolayer
Operational modes	Spectrometry, imaging, depth profiling, 3D microarea analysis
Maximum field of view	ca. 10×10 cm^2
Lateral resolution	<100 nm (slightly higher for molecules)
Maximum depth	10 μm (depth profiling); 500 μm (imaging of cross-sections)
Depth resolution	<1 nm (depth profiling); 100 nm (imaging of cross-sections)
Mass resolution	$>10^4$
Quantification	Only using standard reference samples; semiquantitative information can be obtained without standards
Detection limits	A few 10 ppb (elements), fmol (molecules)
Identification of compounds	Yes (including screening for unknown compounds)
Sample requirements	Compatible with vacuum conditions, including insulators and powders

4.7 PROBLEM SOLVING

It will be shown here, taking four selected examples, how ToF-SIMS can be used to solve surface-related problems. All examples have either originated directly from a production site or are related closely to development and production problems.

4.7.1 DEFECTS IN CAR PAINT

As one of the last steps in car production, visual checks are performed with respect to the quality of the car paint layers. If defects are observed, then that car cannot be processed further and paint repair is necessary. For quality and cost reasons it is therefore essential to find the cause of paint defects occurring at various stages of the production process, even if that means cutting samples from the (rather expensive) cars themselves.

One example of such a defect is described in Figure 4.15. On the left-hand side a sum image of all paint-related secondary ions is shown. Image intensities have been translated into a gray-scale color and height code. As can be seen, in the crater center hardly any paint-related ions were detected. Bearing in mind the information depth of ToF-SIMS of one monolayer, this means that at the crater position the uppermost monolayer must have consisted of something other than paint. On the right-hand side a sum image of the ions characteristic of a perfluorinated polyether is shown, indicating the presence of this material in the uppermost monolayer at the crater center. The identification was possible because perfluorinated polyethers have a characteristic ToF-SIMS spectrum consisting of $C_xF_yO_z$ fragment ions. Such polyethers are used, for example, as high-performance temperature-stable lubricants. Detailed discussions with the line engineers responsible for the painting process revealed that the lubricant originated from the transport belts used in the production. Lubricant droplets can fall into the paint bath and prevent adhesion of the paint to the metal. Tests in which the metal was deliberately contaminated by the lubricant proved that typical craters were indeed produced.

ToF-SIMS was decisive in solving the crater problem because it gave organic information about the various species present. It was thus possible to exclude polysiloxane as the cause of the defect although it often gives similar problems. Furthermore, the sensitivity was sufficient to detect sub-monolayer coverages of the lubricant at the crater bottom. The lubricant concentration was so low that neither EPMA nor SEM, imaging techniques normally applied to these kinds of problems, was able to detect the fluorine in the crater.

4.7.2 CHLORINE DIFFUSION IN POLYMER MATERIALS

Nowadays, products ranging from car bumpers to CD players to shavers consist of plastics covered by (colored) organic lacquers. One of the plastics generally used is polypropylene (PP). Unfortunately, not every lacquer adheres well to PP-based substrates. In order to overcome the problem either the PP surface can be modified (e.g., by corona or plasma treatment) or adhesion

FIGURE 4.15 ToF-SIMS images of a paint crater. Left: The sum of paint-related ions; right: the sum of characteristic ions of a perfluorinated polyether; field of view: $500 \times 500 \, \mu m^2$. Image intensities have been converted into gray-scale color coding (high intensities: bright areas) and height (high intensities: large height).

promoters can be used. For PP, chlorinated polyolefines (CPO) have proved to be very useful for adhesion improvement. However, not much is known about the underlying adhesion mechanism. Furthermore, experience shows that the improvement is also dependent on the nature of the PP substrate.

One mechanism proposed is the diffusion of CPO into the PP substrate [61]. Differences in adhesion behavior based on this mechanism should then arise from different diffusion constants for CPO in the various PP substrates. In order to test this hypothesis, a 50 μm CPO layer was deposited onto two different substrates (the ethylene-propylene-diene terpolymers Keltan and Hifax). From the samples prepared in that way cryosections were cut, and the PP–CPO interfaces (as revealed by the surfaces of the sections) were analyzed with imaging ToF-SIMS. The chlorine distribution (originating from the CPO molecules) in the interface was monitored as a function of time in order to observe possible diffusion processes. Figure 4.16 shows the Cl⁻ image of the CPO-treated Hifax interface after 57 days. Broadening of the interface can be seen clearly, a first hint of the occurrence of diffusion. Linescans across this image, and across another taken after 8 days, are shown in Figure 4.17, demonstrating the broadening of the interface with storage time. A mathematical evaluation of the interface broadening based on diffusion theory (i.e., determination of the distance at which the original Cl concentration in CPO had decreased to one-tenth of its value as a function of the storage time, according to Fick's second law) for the two substrates (Figure 4.18) showed not only that diffusion of CPO into the PP had indeed occurred but also that the diffusion constant in Keltan was significantly greater than in Hifax. This was in very good agreement with the empirical finding that lacquer adhesion was better on Keltan than on Hifax [61].

Imaging ToF-SIMS was used to investigate the adhesion mechanism because the values of the diffusion constants were not known in advance, and it was feared that collisional mixing, as occurs unavoidably in sputter depth profiling, could have destroyed the effect to be measured. Sectioning of samples instead of sputter depth profiling is a very useful methodology for the determination of depth distributions in those cases where the collision cascade would destroy sensitive material (e.g., organic layer systems).

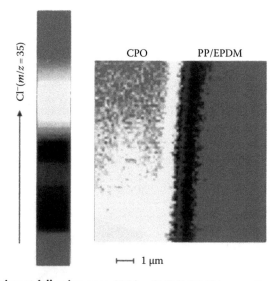

FIGURE 4.16 (See color insert following page 396.)   Cl-ToF-SIMS image of a cryosection of a CPO layer deposited on a PP substrate (diffusion time: 57 days). A color code is used to represent image intensities (see left color bar; pink and red correspond to low and high intensities, respectively). (From Rulle, H., PhD thesis, Münster, Germany, 1996.)

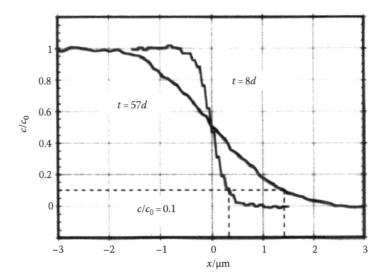

FIGURE 4.17 Cl^- line images across the image from Figure 4.16 and also across a similar image after 8 days of CPO diffusion time: The data were recorded by adding up all ion intensities over 10 lines in the center of the image in Figure 4.16. (From Rulle, H., PhD thesis, Münster, Germany, 1996.)

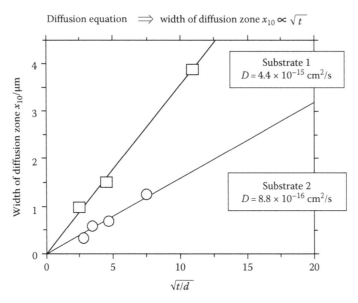

FIGURE 4.18 Cl diffusion profile (10% definition) as a function of sample storage time (room temperature) for the PP substrates Keltan (substrate 1) and Hifax (substrate 2). From the slopes of the curves the diffusion constants could be determined. (From Rulle, H., PhD thesis, Münster, Germany, 1996.)

4.7.3 Residues on Glass

This example is taken directly from a production site for consumer electronics. The product in question consists mainly of plastic materials and glass held together by various glues. The product also contains moving and rotating metal parts that need to be lubricated. Unfortunately, the process technology of the product is confidential and therefore cannot be shown, but the example demonstrates how ToF-SIMS can be used to solve real production problems.

By optical inspection of the glass parts of products that had failed in a lifetime test, a contamination layer was detected, which in some areas was as thick as a few micrometers. In order to characterize the contaminant, infrared (IR) and x-ray photoelectron spectroscopy (XPS) measurements were performed. ToF-SIMS was not chosen in the first place because the problem seemed neither to require very sensitive analysis nor to be related in any way to the outer monolayer of the glass. The analyses showed that the material must be organic in origin, most probably based on aliphatic hydrocarbons (single Cls peak in the XPS spectrum, CH vibrations in the IR spectrum), but no further information could be obtained (no loss features or chemical shifts in the XPS spectra, no typical bonds in the IR spectra). The diagnosis "aliphatic hydrocarbon" included several glues and fats as possible contamination sources.

More information on the organic structure was therefore required. With the prior information from XPS and IR spectroscopies, it was clear that to record a ToF-SIMS spectrum directly from the material on the underlying glass would not be advisable. First of all the large amount of contamination present could pollute the vacuum system of the ToF-SIMS instrument (SIMS is sensitive to readsorbed submonolayer coverages and can therefore suffer severely from memory effects), and, second, aliphatic hydrocarbons need an effective ionization mechanism in order to be visible in a SIMS spectrum. Preparation was therefore carried out simply by rubbing a freshly etched silver substrate over the contamination, outside the vacuum system, thus transferring about a monolayer of the material to the silver. Noble metal substrates are well suited to this type of analysis because organic materials adsorbed on them can be ionized intact in the form of $(M + NMe)^+$, with M being the molecule of interest and NMe the noble metal. This ionization mechanism works almost independently of the molecule chemistry [62]. For reference purposes the procedure was repeated for a component that had not failed.

The respective spectra are compared in Figures 4.19 and 4.20. Whereas the spectrum from the contaminated product showed a distinct peak pattern above 500 amu, this pattern was completely absent from the spectrum of the properly working product (note the enlarged section in the latter spectrum). The peaks in the spectrum of the contamination have a mass separation of 140 amu,

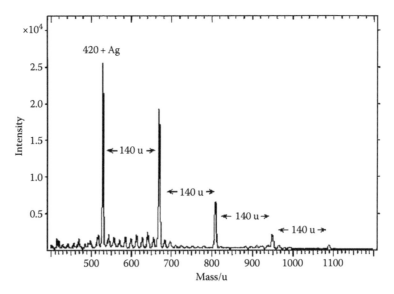

FIGURE 4.19 Positive ToF-SIMS spectrum from a reference glass product with visible residues; the sample was rubbed onto a freshly etched silver substrate.

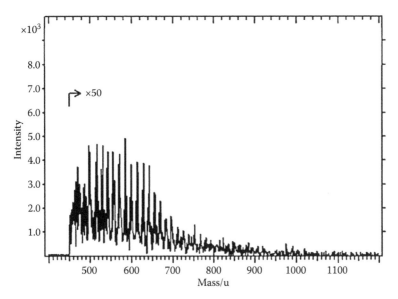

FIGURE 4.20 Positive ToF-SIMS spectrum from a reference glass product without visible residues.

indicating some kind of polymer-like material. The isotopic pattern of the peaks is mainly that of silver, indicating that the spectra originated from neutral molecules cationized by silver from the substrate (i.e., $(M + Ag)^+$ formation). From evaluations of the absolute masses it could be concluded that the repeat unit of the material must have the base composition $C_{10}H_{20}$. The first peak of the distribution would then correspond to a trimer of the repeat unit with hydrogen end groups, so that the contamination must have been a purely aliphatic hydrocarbon.

After further analysis of six glues and three fats used in the production (all dissolved in suitable solvents and deposited as monolayers on silver substrates), one particular fat was found to have the same characteristic peak pattern as the contamination. The fat formula is based on polyalphaolefines (Figure 4.21 shows the spectrum of the extracted base oil). Intact molecules cationized by silver produced the characteristic peak pattern in the spectrum. By comparing Figures 4.19 and 4.21 it can be concluded that the contamination must have occurred via the gas phase, because the contamination and the fatty base oil have significantly different molecular weight distributions. Low-mass molecules ($n = 3, 4$) are more pronounced in the contaminant compared to the base oil. This difference is caused by the fact that low-molecular-weight molecules can desorb more easily into the gas phase than those of higher molecular weights, and the shifted distribution is found after readsorption of the desorbed molecules on the glass. The mechanism was confirmed by storing some fat material together with a clean silver substrate in a closed environment for some hours. The base oil indeed desorbed into the gas phase and readsorbed on the silver, where it could be detected by ToF-SIMS, showing the shifted molecular weight distribution. After the fat was replaced, the number of products failing the lifetime test was significantly reduced, and no further optically visible contamination of the glass surfaces was observed.

4.7.4 Monitoring of Cleaning Efficiencies

Molecular information from the outermost monolayer of a solid is decisive in the monitoring of cleaning efficiencies because most cleaning agents and many contaminants are organic in origin. The following two examples show how ToF-SIMS can provide useful information in this context.

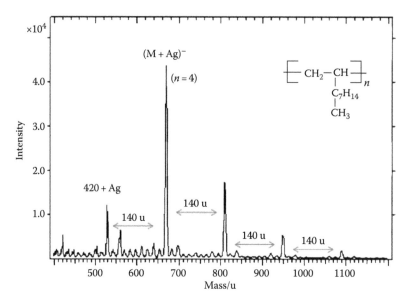

FIGURE 4.21 Positive ToF-SIMS spectrum from the base oil extracted from one of the fats employed. The oil was prepared from a 1 mg/mL solution in hexane (about 3 μL deposited onto 80 mm² of an etched silver substrate).

Figure 4.22 shows ToF-SIMS data from a metal surface. The upper row shows the spectra before cleaning. On the left-hand side, positively charged secondary ions are shown and on the right-hand side negatively charged secondary ions. Clearly, the surface contamination by mineral oil and fatty acid residues can be seen, originating from the metal forming processes. Using ToF-SIMS it was possible not only to determine the average molecular weight of the mineral oil, which is related to its viscosity, but the various fatty acid residues could be distinguished as well. In the negative spectra, the most prominent ion, at mass 255 amu, is that originating from a palmitate residue. After cleaning, the peaks from the mineral oil and the fatty acid residues are diminished significantly, proving the success of the cleaning process. The peak around 485 amu belongs to the used surfactant (the active ingredient in the surfactant cleaning agent).

The use of surface analytical techniques such as ToF-SIMS allows not only the monitoring of the removal of surface contaminants, but also the visualization of the presence of cleaning residues. As an example, Figure 4.23 shows ToF-SIMS images of a Ni surface that had been treated using a water-based multibath cleaning step. All cleaned surfaces showed bad adhesion behavior in the next step in the manufacture, which was soldering. Even the additional use of a plasma cleaning step did not improve the adhesion strength.

The lateral distribution of some characteristic substrate ions is presented on the left-hand side of Figure 4.23. The right-hand side shows the lateral distribution of the cleaning agent dodecylbenzene-sulphonic (DDBS) acid. Taking into account that the information in ToF-SIMS is coming exclusively from the outermost one to three monolayers of the surfaces under investigation, it can be concluded that the Ni surface was contaminated by spot-like remnants of the cleaning agent, with a local coverage in the multilayer range. It is not surprising, therefore, that the additional plasma treatment was unsuccessful because it could not remove the locally thick cleaning remnants. The problem was solved, however, by optimization of the rinsing steps in the water-based cleaning cascades.

4.8   SUMMARY AND OUTLOOK

ToF-SIMS is a very powerful tool for the chemical characterization of surfaces. It is able to identify the presence of elements and molecules, it is sensitive down to the ppb and fmol levels, it is very

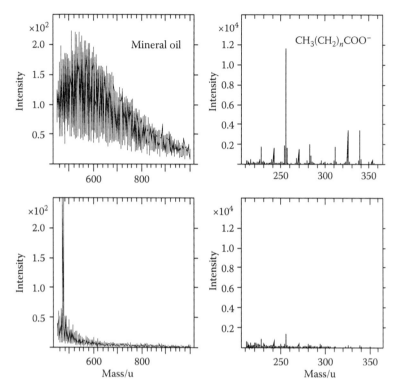

FIGURE 4.22 Monitoring of cleaning efficiency by ToF-SIMS. Upper row: Data before cleaning. Lower row: Data after cleaning and rinsing. The amounts of mineral oil (left, positive polarity) and fatty acid residues (right, negative polarity) are clearly diminished after cleaning.

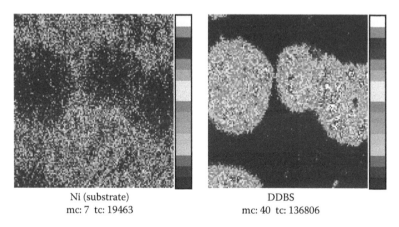

Ni (substrate)
mc: 7 tc: 19463

DDBS
mc: 40 tc: 136806

FIGURE 4.23 Mass resolved ToF-SIMS images of a Ni surface after cleaning. Left: Substrate (Ni). Right: Signals characteristic of DDBS acid. mc: number of counts in the brightest pixel; tc: total number of counts; the field of view was $500 \times 500 \ \mu m^2$.

surface specific (information depth = the outermost monolayer), and it allows the determination of the lateral distributions of elements and molecules with submicrometer resolution. Furthermore it can be applied to almost any material in general without any special preparation (i.e., as-received). It should therefore be the method of choice in those cases where molecular information is requested or where the material available for analysis is very limited. With ToF-SIMS, screening of unknown samples is possible; i.e., information on all elements and molecules present on the surface can be obtained simultaneously without any need for prior sample knowledge. The in-depth distribution of elements, and increasingly often molecules, can be probed with a depth resolution in the nanometer range. Combining imaging and depth profiling yields information on the 3D distribution of the species of interest. Here again, the screening capabilities of ToF-SIMS make it a sought-after tool, for example, production control and failure analysis.

On the negative side it must be said that SIMS is inherently a non-quantitative technique, although in many analytically important cases at least semiquantitative information can be obtained. The high sensitivity of surface mass spectrometry can sometimes be troublesome when dealing with production problems. Submonolayer coverages of possible contaminants are present virtually on all surfaces originating from production sites, and careful thought must always be given to whether or not a molecule that has been detected is really the one causing the problem. Working with surface mass spectrometry still requires considerable experience, although the use of the technique continues to expand rapidly and is becoming more user-friendly. Currently, more than 300 commercial instruments are installed worldwide, 60% of which are used in industrial laboratories. A German idiom says: "If the only tool you have is a hammer, you tend to think of every problem as a nail." In this sense ToF-SIMS is not always the most suitable analytical tool for solving surface-related problems. Sometimes cheaper techniques (e.g., measurement of contact angle) can give information faster and more easily. Nevertheless, ToF-SIMS is currently the only technique available for the unambiguous identification of those organic molecules present at surfaces that so often cause problems in adhesion, wetting, corrosion, etc. It must therefore be regarded as a powerful tool that contributes several large and important pieces to the general analytical jigsaw puzzle that has to be solved by a combination of several techniques when dealing with complex surface-related problems.

ToF-SIMS is still developing very quickly. Whereas the last decade saw the development of dedicated depth profiling options, and the improvement of the performance of ion guns for imaging purposes, currently the use of polyatomic primary ions for applications in organic depth profiling and biological analysis is the main point of focus. Additionally, the level of automation for measurement and data evaluation is constantly increasing, making the technique available for high-throughput and routine applications. With the improvement in ion formation efficiency, for the first time organic imaging with submicrometer resolution has become possible, thus making ToF-SIMS a real nano-analytical tool.

REFERENCES

1. M. W. Thompson, *Phil. Mag. 18*, 377 (1968).
2. A. Benninghoven, F. G. Rüdenauer, and H. W. Werner, *Secondary Ion Mass Spectrometry*, Wiley, New York, 1987.
3. R. D. Macfarlane, *Nucl. Instrum. Method. 198*, 75 (1982).
4. E. Dennins and R. J. MacDonald, *Radiat. Eff. 13*, 243 (1972).
5. F. J. Hennequin, *J. Phys. 29*, 957 (1968).
6. J. J. Thomson, *Phil. Mag. 20*, 752 (1910).
7. P. Sigmund, *Phys. Rev. 184*, 383 (1969) and *187*, 768 (1969).
8. P. Sigmund, in: *Sputtering by Particle Bombardment I* (R. Behrisch, ed.), Springer, Berlin, 1981, p. 9.
9. B. Garrison and N. Winograd, *Science 216*, 805 (1982).
10. P. Williams, *Surf. Sci. 90*, 588 (1979).
11. H. M. Urbassek, Status of cascade theory, in: *ToF-SIMS: Surface Analysis by Mass Spectrometry* (J. Vickerman, ed.), IM Publications, Huddersfield, 2001, p. 139.
12. A. Benninghoven, E. Löbach, and N. Treitz, *J. Vac. Sci. Technol. 9*, 600 (1972).
13. K. H. Müller, P. Beckmann, M. Schemmer, and A. Benninghoven, *Surf. Sci. 80*, 325 (1979).
14. A. Benninghoven, D. Jaspers, and W. Sichtermann, *Appl. Phys. 11*, 35 (1976).
15. A. Benninghoven and W. Sichtermann, *Anal. Chem. 50*, 1180 (1978).
16. S. J. Pachuta and R. G. Cooks, *Chem. Rev. 87*, 647 (1987).
17. G. Gillen and S. V. Roberson, *Rapid Commun. Mass Spectrom. 12*, 1303 (1998).

18. F. Kötter and A. Benninghoven, *Appl. Surf. Sci. 133*, 47 (1998).
19. N. Davies, D. E. Weibel, P. Blenkinsopp, R. Hill, J. C. Vickerman, and N. P. Lockyer, *Appl. Surf. Sci. 203/204*, 223 (2003).
20. F. Kollmer, *Appl. Surf. Sci. 231–232*, 153 (2004).
21. D. Weibel, S. Wong, N. Lockyer, P. Blenkingsopp, R. Hill, and J. C. Vickerman, *Anal. Chem. 75*(7), 1754 (2003).
22. S. C. C. Wong, R. Hill, P. Blenkinsopp, N. P. Lockyer, D. Weibel, and J. C. Vickerman, *Appl. Surf. Sci. 203–204*, 219 (2003).
23. D. E. Weibel, N. Lockyer, and J. C. Vickerman, *Appl. Surf. Sci. 231–232*, 146 (2004).
24. R. Kersting, B. Hagenhoff, A. P. Pijpers, and R. Verlaek, *Appl. Surf. Sci. 203–204*, 561 (2003).
25. C. M. Mahoney, S. V. Roberson, and G. Gillen, *Anal. Chem. 76*, 3199 (2004).
26. C. Szakal, S. Sun, A. Wucher, and N. Winograd, *Appl. Surf. Sci. 183*, 231 (2004).
27. J. Cheng, A. Wucher, and N. Winograd, *J. Phys. Chem. B 110*, 8329 (2006).
28. M. S. Wagner, *Surf. Interface Anal. 36*, 42 (2004).
29. M. S. Wagner, *Anal. Chem. 77*, 911 (2005).
30. R. Möllers, N. Tuccitto, V. Torrisi, E. Niehuis, and A. Licciardello, *Appl. Surf. Sci. 252*, 6509 (2006).
31. C. M. Mahoney, A. Fahey, G. Gillen, C. Xu, and J. D. Batteas, *Anal. Chem. 79*, 837 (2007).
32. N. Winograd, *Anal. Chem. 143A* (2005).
33. B. J. Garrison, *Appl. Surf. Sci. 252*, 228 (2006).
34. A. Wucher, *Appl. Surf. Sci. 252*, 6482 (2006).
35. R. Levi-Setti, G. Crow, and Y. L. Wang, in: *Secondary Ion Mass Spectrometry V* (A. Benninghoven, R. J. Colton, D. S. Simons, and H. W. Werner, eds.), Springer, Berlin, 1986.
36. J. Orloff, *Rev. Sci. Instrum. 64*, 1105 (1993).
37. L. W. Swanson, *Nucl. Instrum. Method 218*, 347 (1983).
38. F. A. White and G. M. Wood, *Mass Spectrometry-Applications in Science and Engineering*, Wiley, New York, 1986.
39. H. E. Duckworth, R. C. Barber, and V. S. Venkatasubramanian, *Mass Spectroscopy*, 2nd ed., Cambridge University Press, Cambridge, 1986.
40. C. D. Coath and C. V. P. Long, *Rev. Sci. Instrum. 66*, 1018 (1995).
41. E. Niehuis, T. Heller, H. Feld, and A. Benninghoven, *J. Vac. Sci. Technol. A5*, 1243 (1987).
42. V. I. Karataev, B. A. Mamyrin, and D. V. Shmikk, *Sov. Phys. Techn. Phys. 16*, 1177 (1972).
43. B. W. Schueler, *Microsc. Microanal. Microstruct. 3*, 119 (1992).
44. T. Sakurai, T. Matsuo, and H. Matsuda, *Int. J. Mass. Spectrom. Ion Phys. 63*, 273 (1985).
45. B. Hagenhoff, D. van Leyen, E. Niehuis, and A. Benninghoven, *J. Vac. Sci. Technol. A7*, 3056 (1989).
46. J. Zehnpfenning, H. G. Cramer, T. Heller, U. Jürgens, E. Niehuis, J. Schwieters, and A. Benninghoven, in: *Secondary Ion Mass Spectrometry VIII* (A. Benninghoven. K. T. F. Janssen, J. Tümpner, and H. W. Werner, eds.), Chichester, New York, 1992, p. 501.
47. F. W. McLafferty and F. Turecek, *Interpretation of Mass Spectra*, 4th ed., University Science Books, Mill Valley, 1993.
48. D. van Leyen, B. Hagenhoff, E. Niehuis, A. Benninghoven, I. V. Bletsos, and D. M. Hercules, *J. Vac. Sci. Technol. A7*, 1790 (1989).
49. At the moment, there is only one commercial spectra library on the market. Information can be found at http://www.surfacespectra.com/simslibrary/index.html. Additionally, the ToF-SIMS instrument manufacturers and some service laboratories offer libraries with their systems or services.
50. B. Hagenhoff, Quantification in molecular SIMS in: *Secondary Ion Mass Spectrometry X* (A. Benninghoven, B. Hagenhoff, and H. W. Werner, eds.), Wiley, New York, 1997, p. 81.
51. A. Schnieders, R. Möllers, M. Terhorst, H.-G. Cramer, E. Niehuis, and A. Benninghoven, *J. Vac. Sci. Technol. B14*, 2712 (1996).
52. K. R. Beebe and B. R. Kowalski, *Anal. Chem. 59*, 1007A (1987).
53. A. Chilkoti, B. D. Ratner, and D. Briggs, *Anal. Chem. 65*, 1736 (1993).
54. B. Tyler, G. Rayal, and D. G. Castner, *Biomaterials 28*, 2412 (2007).
55. H. J. Whitlow, M. Hautala, and B. U. R. Sundqvist, *Int. J. Mass Spectrom. Ion Proc. 78*, 329 (1987).
56. G. Betz and F. Rüdenauer, *Appl. Surf. Sci. 51*, 103 (1991).
57. R. L. Kubena, J. W. Ward, F. P. Statton, R. J. Joyce, and G. M. Atkinson, *J. Vac. Sci. Technol. B9*, 3079 (1991).
58. E. Niehuis and T. Grehl, Ultra-shallow depth profiling, in: *ToF-SIMS: Surface Analysis by Mass Spectrometry* (J. Vickerman, ed.), IM Publications, Huddersfield, 2001.
59. N. Mine, B. Douhard, J. Brison, and L. Houssiau, *Rapid Commun. Mass Spectrom. 21*, 2680 (2007).

60. D. Breitenstein, C. E. Rommel, R. Möllers, J. Wegener, and B. Hagenhoff, *Angew. Chem. Int. Ed. Engl* *46*, 5251 (2007).

61. H. Rulle, PhD thesis, Münster, Germany, 1996.

62. B. Hagenhoff, Optimisation methods: Noble metal cationisation, in: *ToF-SIMS: Surface Analysis by Mass Spectrometry* (J. Vickerman, ed.), IM Publications, Huddersfield, 2001.

63. D. A. Briggs and M. P. Seah, *Practical Surface Analysis (Second Edition). Volume 2: Ion and Neutral Spectroscopy*, Chichester, New York, 1992.

64. Product Brochure Cameca NanoSIMS 50, http://www.cameca.fr/doc_en_pdf/ns50_instrumentation_june2007_web.pdf

5 Surface and Interface Analysis by Scanning Probe Microscopy

Sverre Myhra

CONTENTS

5.1 INTRODUCTION

The first member of the scanning probe microscope (SPM) family, the scanning tunneling microscope (STM), saw the light of day in 1982 [1], and earned the two protagonists, G. Binnig and H. Rohrer, a Nobel Prize. Shortly thereafter, the atomic force microscope (AFM) was demonstrated [2]. Since then the technologies and ideas that underpinned the initial developments have provided the inspiration for numerous additional related techniques, each capable of functioning in several different operational modes.

The emergence of SPM as a versatile and powerful group of techniques for visualizing and manipulating surfaces and interfaces has been given additional impetus by two coincident trends in science and technology. Numerically intensive computation (NIC), beginning with the advent of digital computers, has provided increasingly powerful tools for modeling and simulation of complex systems and processes. More recently nanotechnology, indeed anything with the prefix "nano," has become the most topical, and most rapidly expanding, area of scientific and technical endeavor. (There is a great deal of confusion about the designation and use of size scales; e.g., when does one leave the micro-regime and enter the nano-world? Throughout the chapter hard and fast definitions will be avoided, but "meso" will be used to refer to the gray area between micro and nano.) SPM has proved to be an invaluable and timely tool of the trade as support, and driver, of the two trends. The developments are illustrated in schematic form in Figure 5.1. The relative maturity and increasing topicality of SPM is reflected in the appearance of dedicated monographs [3–6].

5.1.1 Essential Elements of SPM

An SPM system, shown schematically in Figure 5.2, consists of a surface having structural and physicochemical attributes, a tip also with structural and physicochemical attributes, that can be located and controlled in the spatial and temporal domains, and interactions with characteristic strengths and ranges that depend on the respective attributes of surface and tip. The latter allows a two-way transfer of "information" between tip and surface. On the basis of this description one can make the observation that, given sufficient knowledge about any two of the three elements, one can in principle gain knowledge about the third element. Therein is the richness of information inherent in SPM techniques.

An abbreviated tree of the SPM family is shown in Figure 5.3. More will be said later in this chapter. The salient characteristics of some members of the SPM family are sketched in Figure 5.4. In the case of scanning thermal microscopy (SThM) the tip is essentially a local thermal radiation sensor (i.e., a thermocouple or a thermistor). The underlying mechanism of scanning tunneling microscopy

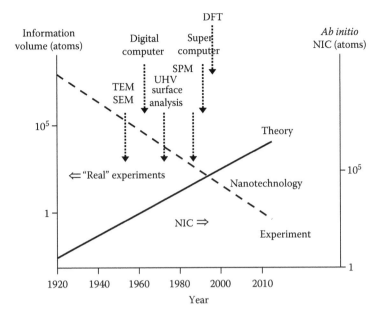

FIGURE 5.1 Schematic illustration of trends in NIC (as defined by the number of interacting atoms that can be subjected to *ab initio* numerical calculation), and of the ability to do "real" experiments on increasingly small information volumes (where the limit for "real" experiments arises from the spatial resolution of the probe). The two trends are correlated with the emergence of theoretical and experimental tools (DFT = density functional theory). The cross over of the two trends has defined the arrival of nanotechnology. (However one chooses to define nanotechnology, it emerged as a potentially practical proposition during the 1990s.)

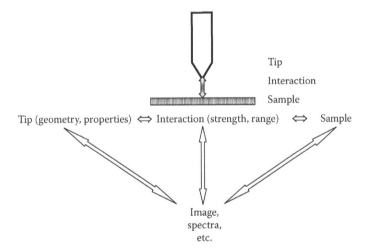

FIGURE 5.2 Schematic representation of the essential elements of an SPM system.

inverse photoemission (STMIP) is that of the emission of low energy photons in response to tunnel electrons being injected into unoccupied surface states.

5.1.2 WHY SPM?

Nearly all traditional techniques for the characterization of surfaces and interfaces have emerged from the technical exploitation of significant discoveries, e.g., x-ray diffraction from the discovery of x-rays,

SPM family tree (abbreviated)

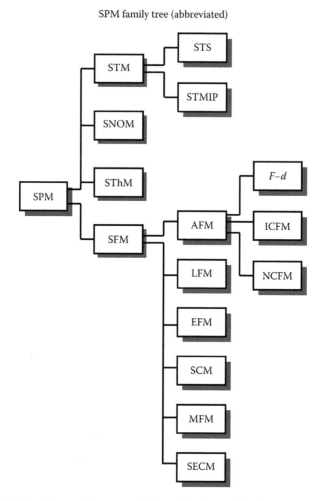

FIGURE 5.3 The SPM family tree. The most widely used and user-friendly techniques for characterization of surfaces and interfaces are shown in italics. Those techniques that are the most important research tools, in terms of usage, are shown underlined. The other techniques are less widely used, and then usually by seasoned practitioners. (STS, scanning tunneling spectroscopy; STMIP, scanning tunneling microscopy inverse photo-emission; SNOM (NSOM), scanning near(-field) optical microscopy; SThM, scanning thermal microscopy; SFM, scanning force microscopy; *F–d*, force versus distance (analysis); ICFM, intermittent contact (= tapping) force microscopy; NCFM, noncontact force microscopy; LFM, lateral force microscopy; EFM, electrostatic force microscopy; SCM, scanning capacitance microscopy; MFM, magnetic force microscopy; SECM, spreading electrical current microscopy. *Note:* other writers may use different acronyms.)

surface spectroscopies from the then mature field of atomic spectroscopy, Raman spectroscopy from the discovery of the Raman effect, etc. SPM, on the other hand, arrived unannounced by any significant precursor discovery. Nevertheless, the family of local probe techniques instantly caught both popular and professional imagination, and became mainstream in less than a decade. Some of the reasons are described in the following subsections.

5.1.2.1 Cost Effectiveness

The capital cost of a multitechnique multimode SPM instrument is a small fraction of that of a UHV-based surface technique or a state-of-the-art analytical SEM or TEM. The exception is a UHV-STM facility. Likewise, the day-to-day cost of operation of SPM techniques compares favorably, with the principal cost being that of the probes.

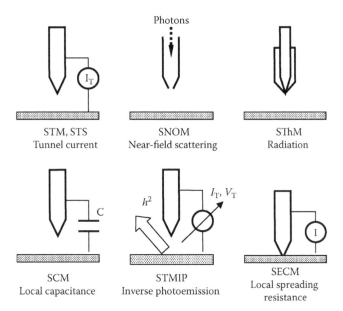

FIGURE 5.4 Schematic description of some SPM techniques.

5.1.2.2 Platform Flexibility

An SPM instrument is essentially a tool-kit that can be configured readily to support the great majority of local probe techniques and their respective operational modes [3]. Other techniques can be added to an existing platform with modest capital outlays.

5.1.2.3 Ambient Tolerance

The great strength of SPM techniques resides in the relative insensitivity of performance to almost any ambient environment ranging from biocompatible fluids, via either laboratory ambient or a controlled gaseous environment, through to high and ultra-high vacuum (UHV).

5.1.2.4 User-Friendliness

Analysis in ambient or controlled atmospheres is consistent with minimal specimen preparation and rapid turnaround, comparable to SEM. Obtaining a usable AFM image in the contact mode requires only marginally greater training than that for a standard SEM; mounting and aligning the probe is probably the most taxing task.

5.1.2.5 Ease of Interpretation

SPM images have the great advantage of providing information in "real" space, as opposed to techniques that are based on diffraction and/or phase contrast. In the case of $F\text{–}d$ (force versus distance) analysis, the nature of the information is readily apparent, although its relationship to mechanical or electronic properties may require considerable theoretical insight.

5.1.2.6 Unique Capabilities

The SPM family has had a great impact on recent progress in surface and interface science and technology due to its broad range of capabilities, some of which are unique. True "real" space single atom and single molecule resolution for STM and scanning force microscopy (SFM), respectively, was the initial focus of attention. Subsequently, SPM has provided unique opportunities for studies in nano-tribology, nano-biology, and nano-mechanics, and has provided the means for simultaneous manipulation and visualization of systems on the meso- and nanoscale.

5.2 TECHNICAL IMPLEMENTATION

There are several essential elements that are common to all instrumental SPM platforms, and which merit brief descriptions.

5.2.1 SPATIAL POSITIONING AND CONTROL

Spatial positioning and control of the probe with respect to a location on the surface, in combination with stability and temporal response, is a key factor in the implementation of SPM techniques. Three spatial regimes spanning micro- to nanoscale positioning need to be considered, viz.

- Coarse lateral movement in order to locate promising regions of the surface within the dynamic range of the available field of view. The present generation of instruments achieves that objective either with stepper-motors driving differential screw arrangements, or with an "inch-worm" arrangement (an inch-worm drive is a piezoelectric device consisting of an expandable body and two clamps).
- Coarse and fine movements in the z-direction of the tip with respect to the surface in order to bring the probe into the dynamic range of the gap control loop (the tip approach sequence). A stepper-motor method is the most widely used. All commercial instruments have placed the final approach under software control.
- Lateral rastering of the tip relative to the surface either by displacement of the sample with respect to a stationary tip (usually the case for SFM instruments), or vice versa (usually the case for UHV-STM instruments).

Piezoelectric materials have the property that a dimensional change is proportional to the local **E**-field. In the case of a beam (or a tube), the change in length is given by

$$\Delta x = \frac{d_{31} V \ell}{h} \tag{5.1}$$

where
     d_{31} is the piezoelectric coefficient
     V is the applied potential
     ℓ is the length in the direction of expansion
     h is the spacing between the two electrodes (usually the smallest dimension of a beam, or the wall thickness of a tube)

Other figures of merit relate to mechanical eigenmodes of the devices, coefficient of thermal expansion, creep, hysteresis, linearity, orthogonality, and aging. The two devices most widely used for positional control, the tube scanner and the tripod, are shown in Figure 5.5. The tube scanner has four external segmented electrodes and a common electrode inside. The tube can undergo either lateral bending or extension/contraction in the z-direction. Nonideal characteristics can mostly be overcome by clever engineering, by software correction routines, and by periodic calibration. Routine calibration is usually carried out with semiconductor grids of known dimensions on the μm scale in the x–y plane and along the z-direction. Suitable standards can now be obtained from most of the manufacturers of SPM instrumentation. On the nano-scale, a freshly cleaved face of highly oriented pyrolytic graphite (HOPG) is a widely used and convenient standard.

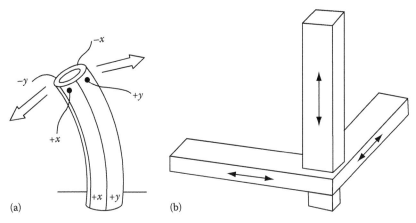

FIGURE 5.5 Schematic illustration of the two most widely used positional devices: (a) the tube scanner and (b) the tripod scanner.

Some instruments are based on single scanners with fields of view in excess of $100 \times 100 \, \mu m^2$, which can be subdivided by D/A converters and low-voltage modes to achieve bit-size resolutions of 0.01 and 0.1 nm, in the z- and x/y-directions, respectively. Other instruments have exchangeable stages that cover the high and low spatial resolution ranges.

5.2.2 FEEDBACK CONTROL LOOP AND FUNCTIONAL BLOCK DIAGRAMS

The electronic control unit (ECU) closes the loop consisting of surface, interaction, and tip. The essential objective is to ensure the most accurate tracking by the tip of a contour (e.g., of constant tunnel current, attractive force, etc.) in space, where the contour is defined by a set point. In order to appreciate fully the power of the technique one should be mindful of the phase shifts and gains in the mechanical components in the loop and of the functional dependence of the interaction on gap dimension. The ECU must thus have facilities not only for filtering, proportional gain, integral and differential time constants, and phase shifting, but also for generating and mixing in other signals for spectroscopy, F–d analysis, tapping mode imaging, etc. These parameters and functions are under software control, but are accessible to the experimenter so that optimum conditions can be set up. Current generation instruments tend to use digital techniques supported by software and A/D and D/A interfacing for instrument control and data acquisition.

Block diagrams in Figures 5.6 and 5.7 show typical STM and SFM configurations and in Figure 5.8 the arrangement whereby the PSPD (position-sensitive photodiode detector) can sense the response of a force-sensing lever to out-of-plane and in-plane force components.

5.2.3 THE PROBE

All the science accessible to SPM analysis takes place at the interface between the probe tip and the surface. Accordingly the choice of probe and its mechanical and physicochemical properties are crucially important. Specific requirements will be described in subsequent sections, but it is useful to summarize the main requirements.

5.2.3.1 STM

The probe, with a length of a few millimeters, is prepared from a Pt/Ir or W wire, typically of 0.3 mm diameter. A sharp apex is normally produced by electrochemical etching (an early description is

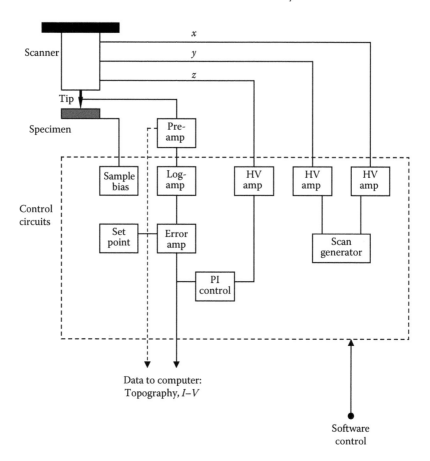

FIGURE 5.6 Block diagram layout for a typical STM configuration. The purpose of the log-amp is to linearize the tip current, which has an exponential dependence on the tunnel gap distance. The error amplifier senses deviation from the set point, and then provides the input for the PI controller (proportional and integral treatment of the error signal). A scan generator drives the raster via high voltage amplifiers. The tunnel gap is maintained by the high voltage input to the z-stage scanner in response to the output from the PI controller.

given in Ref. [7], and similar recipes can be found on the Internet). If the surface to be analyzed is atomically flat, then the aspect ratio is not relevant, and one can rely on the fact that not only will the tunneling probability be significant only between the closest pair of atoms, but also there will be one atom at the apex that will protrude beyond all other atoms. That atom will *de facto* be the apex. If the surface is corrugated with steep slopes, then the overall aspect ratio will be important, in order to maximize access by the tip to all parts of the surface. Contamination and oxide, in the case of a W tip in UHV, must be removed, usually by field emission, in order to eliminate any surface barrier, and to ensure stability of the apex.

5.2.3.2 SFM

Several factors, additional to those that are relevant for STM probes, affect the choice and performance of SFM probes and are discussed below. An early attempt to specify "best practice" for determination of parameters for SFM probes can be found in the literature [8].

- The normal force constant, k_N, refers to deflection of the lever due to force components normal to the surface. It must be matched to the operational mode and to the effective force constant

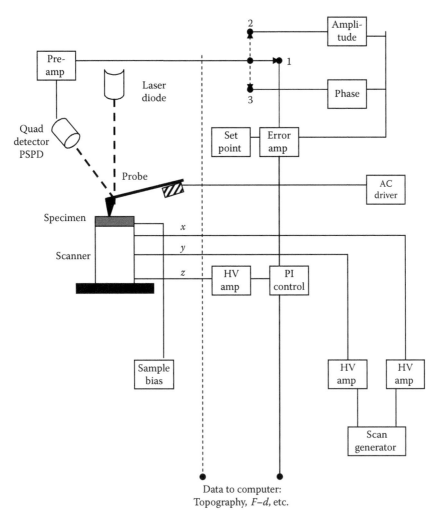

FIGURE 5.7 Block diagram layout for a typical SFM configuration. Output from a laser diode is incident on the lever at the location of the tip (tip plus lever constitutes the probe); the reflected light is sensed by a position-sensitive photodiode (PSPD). The resultant optical lever arrangement has an angular deflection sensitivity in the range 10^{-6} to 10^{-7} rad. The lever can be stimulated to oscillate at, or near, its fundamental eigenmode by an AC signal driving a piezoelectric actuator at the point of attachment of the lever, thus allowing operation in the noncontact or intermittent contact modes.

of the surface being probed. "Soft" surfaces (e.g., organics, polymers, biomembranes, etc.) will be deformed and/or damaged unless a "soft" lever is being used ($k_N < 0.1\,\text{N/m}$). On the other hand, contact mode imaging of "hard" surfaces tends to produce better outcomes with a "stiff" lever (>1 N/m). AC mode imaging requires a much stiffer lever, typically with $k_N = 3$–$10\,\text{N/m}$, in order to have a principal resonance frequency above 100 kHz. The lateral torsional force constant, k_T, is relevant for lateral force imaging and for nano-tribological analysis. Beam-shaped probes have simpler bending modes, and are therefore preferred. Sensitivity to in-plane force components is enhanced by high ratios of tip length to length of lever, and by low normal force constant (see below). On the other hand, a V-shaped probe is preferable if it is desirable to suppress torsional response.

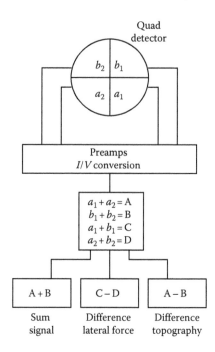

FIGURE 5.8 Signals from the quadrants of the PSPD are manipulated by an arithmetic circuit. The sum signal is used to optimize the intensity of the reflected light from the lever during the alignment procedure. In the topographic imaging mode the A − B signal is compared to a set-point offset, thus generating an error signal; the latter constitutes the input to the control loop. In the F–d mode the A − B signal is proportional to the normal deflection of the lever, and is thus a measure of the strength of the out-of-plane force component acting on the tip. The C − D signal represents a measure of the torsional distortion of the lever in response to in-plane force components perpendicular to the long axis of the lever acting on the tip.

- The resonance frequency needs to be no lower than that of the instrumental eigenmode in order to ensure decoupling from mechanical noise (preferably above 10 kHz), and in order to allow high scan rates. For AC mode operation a high resonance frequency ensures a better Q-value for the resonance envelope and thus better response of the control loop. (The envelope is essentially a plot of vibrational amplitude of the lever, at constant excitation power, as a function of excitation frequency.) The envelope will have a maximum at the eigenfrequency of the lever; its width depends on dissipation and is measured by the Q-value, where Q = quality).
- The length of the lever affects the sensitivity of the optical beam method of detection, being proportional to $\Delta z/L$, where Δz is the deflection of the lever and L its length. A lever length of 100 μm will result in z-resolution of better than 0.1 nm for an optical path length of 1 cm. Lengths of levers range from 100 to 300 μm, with the other dimensions being ca. 30 μm width and ca. 1 μm thickness.
- The optical reflectivity of the top face of the lever is important in order to improve the signal-to-noise ratio of reflected light intensity and to ensure that the thermal stability of the assembly is not affected by absorbed energy. Thus levers are often coated with a thin layer of Al or Au.
- The aspect ratio of the tip (A_r, height to half-width at base) determines the extent to which high aspect ratio features on a surface can be accessed by the tip. Also, a high aspect ratio will minimize the effects of tip-to-surface "convolution" in the image. Routine probes may have tips of pyramidal shape with A_r of order unity. Sharpened tips may have A_r in region of 10, while special purpose tips with attachment of carbon whiskers, or carbon nanotubes, will have even greater A_r values (although they are rather more expensive).

- The radius of curvature of the tip, R_T, at the apex is an important parameter that will affect the image quality. Any feature in a surface with a radius of curvature less than that of the tip will produce an image of the tip (known as reverse imaging [9]). Nonstandard tip shapes are also likely to produce image artifacts that can be misleading or lead to wrong conclusions. Routine tips may have an R_T of ca. 50 nm, while tips for higher resolution can have R_T values smaller by a factor of 10.
- The surface chemistry of the tip is of critical importance for F–d analysis (e.g., measurement of adhesion, investigation of biomolecular bonding, protein unfolding, analysis of double layer interactions, etc.). As well, the surface chemistry will affect friction measurements in the lateral force microscopy (LFM) mode. An as-received clean tip, microfabricated from doped Si or nominally stoichiometric Si_3N_4, will be covered by a native Si oxide, and will thus present a hydrophilic surface. However, atmospheric exposure will render the tip hydrophobic due to adsorbed hydrocarbon contamination. There is increasing interest in studies requiring specific and known tip functionalities. These can be engineered through silane coupling to the oxide layer [10], or through thiol coupling to an Au coating [11]. Several commercial suppliers offer AFM probes with tailored surface chemistries.
- Special purpose probes are required for some applications and particular operational modes. For instance, an magnetic force microscopy (MFM) probe is commonly produced by magnetizing a thin coating of Co on a standard tip. Scanning capacitance microscopy (SCM) analysis requires that the tip be a "metallic" conductor. Spreading electric current microscopy (SECM) needs to make an ohmic contact with the surface being probed. SThM scanning is carried out with a tip that is essentially a thermometer (either a thermocouple junction or a thermistor device).

5.2.3.3 SFM Probe Calibration

SPM in general and SFM in particular are relatively immature by the standards of the more "traditional" techniques for surface and interface analysis. Thus, even though generically derived assumptions and information about the characteristics and properties of probes, such as those provided by a supplier, have sufficed for many past and present investigations, it is becoming apparent that each probe may need to be calibrated before, after, and even during an experiment in order to demonstrate "quality assurance." Failure to do so, and ignorance of the limitations inherent in the image formation process(es), could invalidate what might otherwise be a carefully designed and executed study. The topic of probe calibration is extensive, and cannot be covered in detail in a broad description of SPM techniques and applications. Nevertheless, significant issues are dealt with below, with an emphasis on SFM probes, and mainly by references to the literature.

5.2.3.3.1 Normal Spring Constant

In many cases the adoption of the manufacturer's nominal values for k_N will suffice, with due recognition of uncertainties. However, the actual value for a particular probe may be needed. The range of available procedures is summarized in Table 5.1.

5.2.3.3.2 Lateral Spring Constant

The lever responds to in-plane force components, leading to torsional or longitudinal bending, as well as to out-of-plane components. In the case of a single-beam diving-board configuration, and if the long axis of the lever is perpendicular to the fast scan direction, then only the torsional mode will be stimulated, to a good first-order approximation. If the torsional spring constant, and the relationship between the torsional angle at the tip and the PSPD response are known, then lateral forces can be quantified. Therein is the principle of LFM (also known as chemical force microscopy). In the so-called multi-asperity regime, where friction is independent of contact area, the lateral force component, F_L, due to surface chemistry, can be written as

TABLE 5.1

Methods for Determination of k_N

Methods	Refs.	Accuracy (%)	Comments/Merits/Demerits
Dynamic response methods			
Resonance frequency with added mass	[12]	≈ 10	Positioning and calibration of load difficult; potentially destructive
Thermal fluctuations	[13]	10–20	Temperature control essential; only suitable for soft levers; requires analysis of resonance curve
Simple scaling from resonance frequency	[14]	5–10	Depends on dimensional accuracy and determination of effective mass. Convenient, often implemented in instrument software
Theoretical methods			
Finite difference method	[15]	>10	Depends on dimensional accuracy and Young's modulus
Parallel beam approximation	[16,17]	>10	Depends on dimensional accuracy and Young's modulus
Static response methods			
Static deflection with added mass	[16]	15	Positioning and calibration of load difficult; potentially destructive
Response to pendulum force	[16]	30–40	Complex and time-consuming procedure
Static deflection with external standard	[18]	15–40	Requires accurate external standard

$$F_L = \mu * (F_N + SI_A)$$ (5.2)

where

   $\mu*$ is an equivalent coefficient of friction
   F_N is the normal force
   S is the area of contact
   I_A is the force of adhesion per unit area

The normal force is $F_N = k_N \Delta z$, with k_N in the range 10^{-2} to 10^2 N/m. Since the z-resolution of a typical SFM is better than 0.1 nm, the minimum detectable normal force is of order 10^{-12} N. Assuming a maximum compliance of the lever of 1 µm, the maximum normal force imposed by the lever will be 10^{-4} N. The corresponding approximate angles of deflection, for a lever of length 100 µm, will be 10^{-6} rad (resolution limit) and 10^{-2} rad (maximum compliance). The lateral force is also given by $F_L = k_L \Delta x$ (k_L is the lateral spring constant related to the torsional spring constant and the length of the tip), and Δx is the displacement of the apex of the tip perpendicular to the long axis of the lever. The angle of torsional rotation will be $\Delta x/h$ where h is the length of the tip. Given a limit on resolution of 10^{-6} rad, and a tip length of 5 µm, the minimum detectable Δx will be ca. 5 pm. Disregarding the effects of adhesion, the effective coefficient of friction is then

$$\mu* = \frac{F_L}{F_N} = \frac{k_L}{k_N} x \frac{\Delta x}{\Delta z}$$ (5.3)

Values of $\mu*$ down to 10^{-3} can be measured with standard probes. Some manufacturers will provide an estimate of the lateral spring constant. Analytical expressions have been derived for k_L for V-shaped levers [15]. However, finite element analyses [19] suggest that the bending modes of such levers can be more complex than those that can be modeled analytically. Beam-shaped levers in the long thin beam approximation have much simpler modes, and can be modeled with reasonable accuracy by the simple expression

$$k_L = \frac{Et^3 w}{6L(1+v)h^2}$$ (5.4)

where

t, w, and L are the thickness, width, and length of the lever, respectively
E is Young's modulus
v is Poisson's ratio
h is the length of the tip

The so-called wedge calibration provides a direct experimental method for determining k_L [20,21]. A lateral electrical nano-balance for determination of lateral force constants has been described recently [22].

5.2.3.3.3 Tip Parameters

Aspect ratio, radius of curvature at the apex, and tip length are the three most critical geometrical parameters of the probe tip. The parameters affect the quality of images, particularly in the contact mode, and are the principal source of artifacts ([23–25] and references therein). All three can be determined most conveniently by a method known as reverse imaging. The principle is simple. If an actual tip is used to image a feature that is known to approximate to a delta-function spike, then the image of the spike will be a true representation of the tip shape [26]. An example of the application of reverse imaging is shown in Figure 5.9.

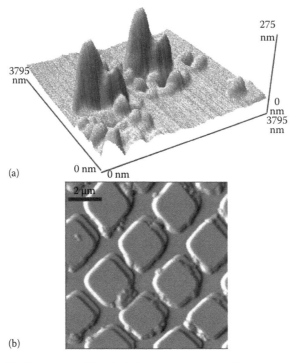

FIGURE 5.9 (a) A double tip is revealed by obtaining reverse images of two carbon whiskers and (b) shows the resultant spatially correlated artifact, an apparent ledge, for a semiconductor grid.

5.3 STM/STS

5.3.1 Physical Principles: Brief Theory

Electron tunneling is the elementary process that accounts for the operation of STM/STS. The approach adopted by Tersoff and Hamann [27] remains an intuitive and useful tool for interpretation of data. Additional material can be found in the review literature [28,29]. The energy level scheme of a tunnel barrier is shown in Figure 5.10. The respective wavefunctions outside and inside the barrier are sinusoidal and exponential; the latter is particularly relevant to the problem at hand. The tunnel current is given by a summation over elastic tunneling channels (hence the presence of $\delta(E2 - E1)$ in the expression for the tunnel current, where E refers to energy as a variable, and the subscripts are defined below) linking occupied states (as described by the Fermi–Dirac function f) on one side of the barrier, with unoccupied states $(1 - f)$ on the other side, as shown in Equation 5.5.

$$I = \frac{2\pi e}{h} \sum_{1,2} f\left(E_2\right)\left[1 - f\left(E_1 + eV_\mathrm{T}\right)\right]\left|M_{1,2}^2\right|\delta\left(E_2 - E_1\right) \tag{5.5}$$

The distributions of states in the tip and sample are displaced by eV_T, where V_T is the tunnel voltage. The diagram illustrates the case of tunneling between two metal surfaces in close proximity, from occupied states below E_F (in metal 1) to unoccupied states above E_F (in metal 2) (where E_F is the Fermi energy). The matrix element $M_{1,2}$ is the transfer-Hamiltonian of Bardeen [30] and is a weakly varying function.

The simple one-dimensional case of two identical free-electron metals with identical work functions, Φ, provides useful insight. Applying boundary conditions, the wavefunctions in the gap may be written as

$$\begin{aligned}
\psi_1 &= \psi_1^0 e^{-kz} \\
\psi_2 &= \psi_2^0 e^{-k(s-z)}
\end{aligned} \tag{5.6}$$

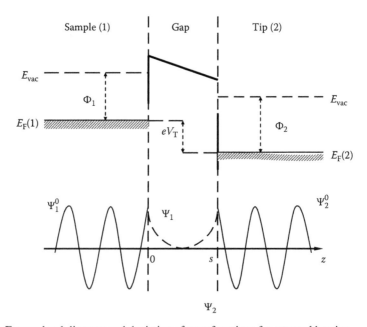

FIGURE 5.10 Energy level diagram and depiction of wavefunctions for a tunnel barrier.

where

s is the width of the tunnel barrier

k is a real number (typically 10 nm^{-1}), and is given by $k = (2m\Phi/h^2)^{1/2}$

It is straightforward to show, since M is a weakly varying function and if $eV_T < \Phi$ and $eV_T < E_F$, that

$$I \propto \sum_{1,2} \left|\psi_1^0\right|^2 \left|\psi_2^0\right|^2 e^{-2ks} \tag{5.7}$$

The expression gives qualitative insight into the tunneling process, irrespective of tip shape, if the wavefunction anchored in the tip can be approximated by an s-wave. Furthermore, if the overlaps of the respective wavefunctions within the barrier are small, and if the tails of the two functions are of similar shape, then

$$I \propto \sum_{i} \left|\psi_i\right|^2 \delta\left(E_1 - E_F\right) \tag{5.8}$$

It can now be recalled from solid state physics that the local density of states in the sample surface at the position of the tip can be written as

$$\rho(E_F, r) = \sum_{1} \left|\psi_1(r)\right|^2 \delta(E_1 - E_F) \tag{5.9}$$

with r being a spatial coordinate in direct space. This description allows a number of qualitative statements about the STM/STS system to be made, as set out below. More sophisticated approaches to the interpretation of STM images have been reviewed in the literature [29].

- The exponential dependence of the tunnel current I on the width of the tunnel gap accounts for the extreme spatial resolution in the z-direction; a change in s of 0.1 nm will change I by a factor of e^2.
- The model shows that an STM image, formed by mapping constant tunnel current, actually represents a map of constant density of states. Thus, unless there is a one-to-one spatial correlation between the positions of atoms, i.e., nuclei, and maxima/minima in the density of states map, then it is the electronic structure that is being revealed. Conversely, the symmetries in the image must reflect the surface structural symmetries, if there is a one-to-one correspondence.
- The lateral resolution in the x–y plane is due, in part, to the exponential dependence of I on the barrier width. As well, the lateral decay of the s-state located in the apex atom of the tip will confine the interaction laterally to a radius of ca. 0.1 nm. Thus there will be single atom resolution due to the lateral spatial decay of the density of states function.
- Of equal significance is the fact that the elastic tunnel process picks out particular states in the sample and tip, as a consequence of energy conservation. Thus the tunnel voltage, V_T, sets an energy window, with a width of a few k_BT. Consequently, a tunnel voltage can be chosen for maximum resolution, i.e., an energy at which the corresponding electron states have the greatest spatial variation. These deductions account for the relative ease with which "atomic" resolution can be obtained for covalent materials (e.g., Si); this is a consequence of the strong spatial dependence of the density of states across the unit cell at the extrema of bands. Conversely, similar resolution is much more difficult to obtain for metals where the spatial variations are more gentle.
- The ability to "tune" the energy window of the tunneling process by changing eV_T is the basis for STS. Successive maps of the surface at different values of V_T will reveal the real-space locations of the corresponding equi-energy contours of the density of states.

Local spectroscopy can be carried out by fixing the lateral and vertical position of the apex of the tip with respect to the surface and recording an I_T versus V_T curve. The I_T–V_T data will not be interpretable as a simple plot of the density of states function; however, since there will be an exponential dependence of the tunnel probability with V_T from the matrix element M. This and other effects can be eliminated, in part, by measuring $(dI/dV)/(I/V)$ which is a useful dimensionless quantity related to the local band structure. As well as being local, STS has the additional merits of extreme surface specificity—literally the first monolayer—and of being able to probe both valence and conduction band states by the simple expediency of reversing the polarity of V_T (and thus reversing the tunnel process). The pioneering work by Wolkow and Avouris [31] is an excellent example.

- Investigations of localized surface states are particularly useful and rewarding with STS. In the context of the "traditional" surface spectroscopies, one should note that STS is complementary to UPS/IPES. The additional merits and convenience of STS vis-à-vis UPS/IPES make this an attractive proposition.

- From the description of the tunnel process it can be seen that the tunnel current is exponentially dependent on k, as well as on barrier width. The decay constant, k, is a function of the effective barrier height, and is therefore related to the work function of the surface being probed. Thus it is possible to extract the relative spatial variation of the work function from the STM map.

- It will also be apparent that STM/STS can be applied only to materials that are tolerably good conductors. The criterion for "goodness" is that the effective sample resistance, R_S, must be such that $I_T R_S \ll V_T$. Hence the technique will work well for clean surfaces of metals, alloys, semimetals, and doped semiconductors. The presence of thin (<1 nm) insulating surface barriers (e.g., oxide layers) can be accommodated by allowing them to act as tunnel barriers. Because of the extreme surface specificity of STM/STS, a UHV environment is necessary when the surface is reactive. One non-UHV area of application in which the STM has much to offer is that of a fluid environment. In particular, electrochemical STM, where the tip can be a local electrode as well as a local probe, has revealed a great deal of detailed information [32].

5.3.2 OPERATIONAL MODES: STM

It is useful to distinguish between techniques that are members of the SPM family (e.g., STM, AFM, etc.), and the operational modes of particular techniques (e.g., constant current versus constant height imaging in the case of STM; constant force versus constant height imaging in the case of AFM).

5.3.2.1 Imaging at Constant Tunnel Current

The feedback loop senses the difference signal between the tunnel current and a set point; the loop then adjusts the z-height of the scanner in order to minimize the difference signal. The topographical map is derived from the relative z-stage excursion at each pixel location in the x–y field of view. The dynamic range (the depth of focus) is set by the limits of extension/contraction of the z-stage of the scanner. The mode is appropriate for relatively rough surfaces, and for large fields of view.

5.3.2.2 Imaging at Constant Height

The z-stage is essentially disabled by increasing the time constant of the feedback loop, and the z-height information is derived from the change in tunnel current at each pixel location. The mode is preferred for flat surfaces (cleaved crystal faces, epitaxial films, etc.), and allows rapid scanning over small fields, thus minimizing the effects of thermal drift.

5.3.2.3 Error Signal Mapping

The error signal mode is a variation on the constant height mode. In this case the difference signal arising from the inability of the feedback loop to respond to sudden changes in slope, with the loop enabled, is sensed and plotted. Changes in the slope of the topography, e.g., edges, will appear more prominently in the image, while more gentle changes in topography will be correspondingly washed out.

5.3.2.4 *I–V* Spectroscopy

The tip is positioned at a particular location in the x–y plane. The feedback loop is momentarily disabled while V_T is swept over the range of interest. The effect is to shift the Fermi edge of the tip with respect to energy states in the surface at the location of the tip. While the tunnel current is exponentially dependent on the voltage, additional contributions to the current will occur when the Fermi edge coincides with either the energy of a surface state, or a band edge, in the electronic structure in the surface. The result is shown schematically in Figure 5.11. The effect can be enhanced by plotting dI/dV_T, or $d(\ln I)/dV_T$) versus tunnel voltage.

5.4 SFM

A description of an SFM system cannot be carried out with the same degree of generality, neatness, and unity as for STM/STS. Neither can the image formation process be understood at the same level of detail and physical insight. Continuum theories will provide useful guidance, but phenomenologically only a rough description of the system can be provided, which may, however, be adequate for most purposes.

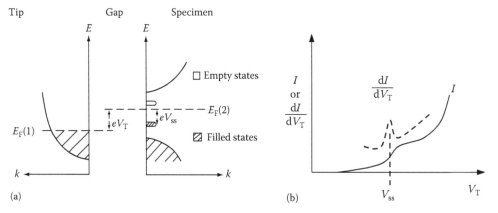

FIGURE 5.11   Schematic illustration of STM *I–V* spectroscopy (eV_{ss} is the energy of a surface state with respect to the Fermi energy). The schematics shows (a) densities of occupied and unoccupied states as functions of k-vector, and (b) the STS response in terms of tunnel current, I, or its derivative, dI/dV_T, as a function of tunnel bias, V_T.

5.4.1 Physical Principles

It is illustrative to consider some of the general characteristics of interatomic interactions, as shown schematically in Figure 5.12. The short-range repulsive interaction, due to nonclassical exchange forces, combines with the longer-range attractive ones, due to dispersion forces of the van der Waals type, to produce a potential well. There will be a location of lowest energy, the binding energy E_b, where the attractive and repulsive force components are balanced. Due to the shape of the potential well, the force constants in the repulsive part of the well are much greater than the force constants in the attractive part. The latter observation is highly relevant to the design and characteristics of SPM instrumentation. In the repulsive contact mode, where the force constant is high, of order 100 N/m, the demands on spatial control in the z-direction are relatively modest, and good performance can be obtained in the DC control mode. In the attractive noncontact mode, on the other hand, the force constant is low, of order 10 N/m. In practice, it becomes necessary to control the z-direction spatial position in the AC mode.

A Lennard-Jones type of potential function is usually adopted as a workable approximation for the contact mode. However, a full description would require extensive modeling (such as a full simulation by molecular dynamics [33]). In the noncontact mode, the van der Waals interaction plays a major role, but other types of interaction may need to be considered. Relevant geometries and useful expressions are summarized in Table 5.2. Attempts to reconcile theory and experiment are commonly based on the work by Israelachvili [33].

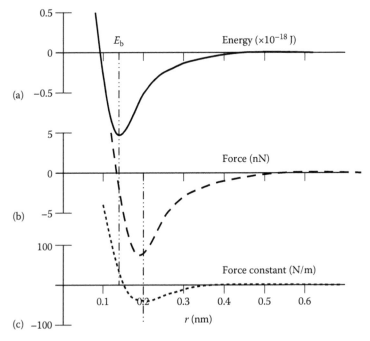

FIGURE 5.12 (a) The attractive and repulsive forces acting on the tip combine to produce a potential well; (b) resultant net force as a function of tip-to-surface distance; and (c) plot of force constants reveals the high constants in the repulsive regime ($E < E_b$), and the lower constants in the attractive regime ($E > E_b$). The force arising from the deflection of the lever maintains a quasi-static equilibrium.

TABLE 5.2
Tip-to-Surface Interactions *(D ≪ R)*

Interaction	Geometry	Expression for Force
Capacitance	Sphere-to-flat	$\dfrac{\varepsilon\varepsilon_0\pi R V^2}{D}$
	Cone-to-flat	$\varepsilon\varepsilon_0\pi V^2\tan^2\theta\ln\left(\dfrac{1}{D}\right)$
Charge versus fixed dipole	Charge-to-flat	$\pm\left(\dfrac{\pi\rho\mu}{2\pi\varepsilon\varepsilon_0}\right)\ln\left(\dfrac{D+t}{D}\right)$
Charge versus "free" dipole	Charge-to-flat	$-\dfrac{\pi\rho q^2\mu^2}{3(\pi\varepsilon\varepsilon_0)^2 kT}\left(\dfrac{1}{2D^2}-\dfrac{1}{(D+t)^2}\right)$
Charge-induced dipole	Charge-to-flat	$-\dfrac{\pi\rho q^2 a}{3(\pi\varepsilon\varepsilon_0)^2}\left(\dfrac{1}{2D^2}-\dfrac{1}{(D+t)^2}\right)$
Capillary	Sphere-to-flat	$-4\pi R\gamma_{LV}\cos\theta + 4\pi R\gamma_{SL}$
van der Waals	Sphere-to-flat	$-\dfrac{HR}{6D^2}$
	Cone-to-flat	$-\dfrac{H\tan^2\theta}{6D}$
Fixed dipoles	Sphere-to-flat	$\pm\beta[(D+2t)\ln(D+2t)+D\ln D-2(D+t)\ln(D+t)]$
	Cone-to-flat	$\pm\eta[(D+2t)^2\ln(D+2t)+D^2\ln D-2(D+t)^2\ln(D+t)]$
Patch charges	Sphere-to-flat	$-\dfrac{\delta}{(D+A)^2}+\dfrac{\xi}{(2D+A+B)^2}$

Source: Adapted from Burnham, N.A., Colton, R.J., and Pollock, H.M., *Nanotechnology*, 4, 64, 1993. With permission.
Note: R, radius of curvature of tip; V, potential; D, tip-to-surface separation; a, length of dipole; q, point charge; θ, half-angle of cone, or contact angle of meniscus; $\varepsilon_0\varepsilon$, permittivity of free space and relative permittivity, respectively; ρ, charge density; μ, dipole moment; t, layer thickness; β, polarizability; $\gamma_{LV,SL}$, surface energy (tension); LV and SL = liquid-to-vapor and solid-to-vapor, respectively; H, Hamaker's constant; ξ,β,η,δ = constants (see Ref. [34]); A, location of charge on tip; B, radius of curvature of tip (for patch charge model).

5.4.2   SFM Operational Modes

5.4.2.1   AC Modes: Noncontact and Intermittent Contact

The probe is stimulated to oscillate at resonance as a free-running oscillator (well away from the surface). The shape of the resonance peak of the oscillating probe is shown in Figure 5.13. When the tip enters the force field of the surface, the system becomes similar to that of a classical damped and driven anharmonic oscillator, described by the following equation.

$$\frac{d^2 z}{dt^2}+\frac{\gamma\,dz}{dt}+\omega_0^2 z=\frac{A_0}{m*}\exp(i\omega t) \tag{5.10}$$

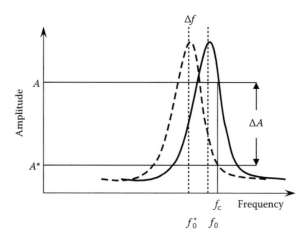

FIGURE 5.13 The principle of AC control for noncontact and intermittent contact imaging. A is the set-point amplitude at which the system is maintained, while ΔA is the (exaggerated) error signal.

where

> $\gamma = b/m^*$, with m^* being the effective mass and b the drag coefficient
> $\omega_0 = (k/m^*)^{1/2}$ is the free-running resonance frequency with k being the spring constant
> A_0 is the amplitude of the driving force
> ω is the driving frequency

The system can then be described completely for various conditions [35]. However, it is sufficient for a qualitative description to note that the resonance frequency of a free oscillator is

$$f_0 = \frac{1}{2\pi}\left(\frac{k_N}{m^*}\right)^{1/2} \tag{5.11}$$

where k_N is the "normal" spring constant of the lever. During the closest approach to the surface, the tip will sense an attractive force with an effective force constant k^*. The shift in the characteristic frequency of the lever is then given by

$$f_0^* = \frac{1}{2\pi}\left(\frac{k_N - k^*}{m^*}\right)^{1/2} \tag{5.12}$$

The situation is shown schematically in Figure 5.13. The dependence of the characteristic frequency on the strength of tip-to-surface interaction can be exploited in mapping the topography of the surface. There are three methods of control and detection.

5.4.2.1.1   Slope Detection (Also Known as Amplitude Detection)
The probe is driven at a frequency f_c resulting in an amplitude A, below that of the free-running resonance amplitude. The conditions constitute the set point. An increase, or decrease, in strength of tip-to-surface interaction will give rise to a frequency shift, and thus a further decrement, or increment, in amplitude to A^*. The decrement/increment is then fed back to the control-loop, which generates an error signal. The error signal adjusts the z-height of the scanner, thus restoring the set-point conditions.

5.4.2.1.2   Phase Detection: Topographic Imaging
There is rapid change in the phase of the oscillating lever, with respect to the phase of the driving signal, at the peak of the resonance envelope, in response to topographic excursions in the z-direction. Thus the system can be phase locked and controlled at a fixed phase change; the fixed phase will then

constitute the set point for the feedback loop, while the phase increment/decrement constitutes the control signal being fed back to the loop. Most current instruments have adopted amplitude (slope) detection as the preferred method for topographical imaging in the intermittent contact mode.

5.4.2.1.3 Phase Detection: Mapping of Surface Stiffness

In this mode, the feedback loop is controlled by the amplitude/slope detection method. The damping term determines the phase increment (or decrement) of the oscillating lever with respect to the driving signal, and the damping is a function of the stiffness of the surface (i.e., more or less energy is being transferred from the oscillator to the surface). Thus lateral variations in surface stiffness can be sensed and mapped with the aid of a phase-locked loop. This way of mapping is particularly useful for a two-phase surface, where one phase has a lower elastic modulus than the other phase (e.g., a two-phase polymer, or a composite consisting of ceramic particles embedded in a polymer matrix). It must be borne in mind that the topographic signal and the phase signal are not completely decoupled. Thus the method is most reliable for phase mapping, when the surface is relatively flat.

5.4.2.1.4 FM Detection

The tip-to-surface interaction sensed during the closest approach by the tip in the AC mode gives rise to a frequency shift, Δf, of the oscillating probe with respect to the fixed frequency of excitation. The shift increases with increasing strength of interaction. Thus the feedback loop can be controlled by a signal proportional to the deviation from a set point that is determined by a given frequency shift while maintaining constant amplitude of oscillation. In this manner a surface map, based on constant strength of interaction, can be generated. The FM AC mode may be preferable to slope detection for high Q resonance modes, i.e., in vacuum, and for relatively flat surfaces.

5.4.2.2 LFM and Friction Loop Analysis

In-plane as well as out-of-plane force components will act on the tip at the point of contact with the surface. The former will exert a friction-like force on the tip in the direction of travel. If the fast-scan raster direction is perpendicular to the long axis of the lever, then the lateral friction force will cause a torsional deformation of the lever that can be sensed by the signal on the left–right PSPD segments. Detection of lateral forces can be used to generate an image based on "friction" contrast. Quantitative investigations of local friction require that the friction loop be analyzed [33]. The latter is illustrated in Figure 5.14. The torsional force constant of a lever, in the long and thin beam approximation, can be written as

$$k_{T} = k_{N} \left[\frac{2L^{2}}{3(1+\mu)h^{2}} \right] \tag{5.13}$$

where the form of the expression shows that k_{T}, and thus the sensitivity to in-plane forces, depends on the normal force constant, the length of the lever, and the length of the tip. μ is Poisson's ratio.

To a first approximation, the lateral force effect is similar to that of macroscopic friction between two objects in sliding contact (the lateral force is independent of relative speed, and independent of the "contact" area). The relationship between lateral force, F_{F}, being sensed, and normal force, F_{N}, being imposed by the lever, is given by

$$F_{F} = \mu * (F_{N} + F_{A}) \tag{5.14}$$

where
  $\mu*$ is the coefficient of friction
  F_{A} is the force of adhesion between tip and surface

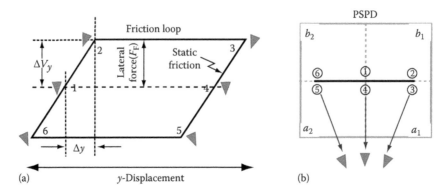

(a)   y-Displacement   (b)

FIGURE 5.14 Schematic representation (a) of the friction loop. The response of the lever to lateral forces acting on the tip is shown (in exaggerated form). The lateral "friction" force F_F can be quantified from the signal ΔV_y, while the displacement Δy is due to static "friction." The corresponding positions of reflected light incident on the PSPD are illustrated in (b). The top–bottom differential signal, sensed by the control loop, is the output $(b_1 + b_2) - (a_1 + a_2)$ from the arithmetic unit, and the left–right differential lateral signal is the output $(b_2 + a_2) - (b_1 + a_1)$.

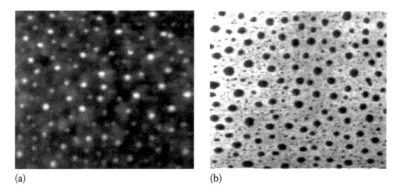

(a)   (b)

FIGURE 5.15 Topographic (a) and lateral force (b) images of a phase-separated polymer blend film (PMA/PMMA at 95:5 nominal weight fraction). The images show that the phase structure is more clearly delineated in the LFM than in the topographic imaging mode. (Adapted from Li, W.-K., *Polym. Test.*, 23, 101, 2004. With permission.)

AFM in the lateral force mode is now a routine tool for investigations of nano-tribology and nano-mechanics (see review literature, e.g., Ref. [37]). In the single asperity regime, when a sharp tip is sliding across a hard surface, the relationship is more complex [38]. A surface may be laterally differentiated by virtue of variation in surface chemistry (the differentiation may not manifest itself in the topographical contrast). Chemical contrast will result from chemical differentiation that manifests itself as a change in adhesive force and/or a change in in-plane force components. Hence the LFM mode is often called chemical force microscopy (or friction force microscopy [FFM]) [39]. The LFM imaging mode is particularly useful for delineating phase-separated polymer surfaces, as shown in Figure 5.15 [40], and for investigating the dependence of adhesion and friction on interface modification [41].

5.4.2.3 *F–d* Analysis

Force versus distance analysis is rapidly gaining in importance and popularity in comparison with imaging modes. *F–d* curves can provide information on local materials properties, such as elasticity, hardness, Hamaker constant, adhesion, and surface charge. *F–d* analysis has had a major impact on

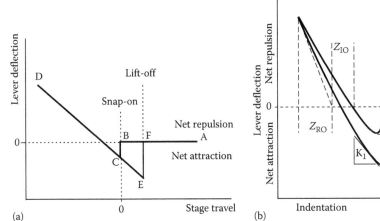

FIGURE 5.16 Outcomes of F–d analysis. (a) The system consists of an incompressible tip and a surface, with the surface being covered by an adsorbed aqueous film. (b) The surface is compliant so that the tip indents the surface. The surface undergoes deformation in response to the force being applied by the lever, in combination with partial elastic recovery.

the understanding of fundamental interactions in colloid science, and has shed new light on nanoscale friction and lubrication. A recent review covers F–d exhaustively and provides a near-encyclopaedic list of references [42].

Quasi-static F–d analysis can be undertaken by holding the tip at a particular x–y location far away from the surface. The sample is then driven toward the tip at a rate that is slow in comparison with the mechanical response of the system. The net force is sensed during the approach, contact, and retraction parts of the cycle. Two idealized response curves are shown in Figure 5.16, with stage travel and lever deflection plotted on the horizontal and vertical axes, respectively. The vertical scale can be converted into force sensed/applied by the lever by the simple expediency of multiplying the deflection by k_N. The curve in Figure 5.16a represents the case when both tip and surface are incompressible, and when the surface is covered with a thin adsorbed aqueous film. The various segments represent

1. Approach half-cycle
   a. AB—tip and surface are well-separated, no interaction.
   b. BC—the tip senses the attractive interaction from the meniscus layer, the force constant of interaction exceeds k_N, and the tip snaps into contact with the "hard" surface. There is a regime of instability due to the force constant of interaction being greater than that of the lever.
   c. CD—tip and surface are incompressible, and the stage travel distance must therefore be equal to the lever deflection.
2. Retract half-cycle
   a. DE—the system retraces itself since all deformations/deflections are elastic.
   b. EF—the meniscus interaction has increased due to capillary action, and there will be greater vertical spacing at the lift-off instability/discontinuity (than for the snap-on).
   c. FA—return to large separation and no interaction.

This kind of curve is representative of events for an air-ambient instrument due to adsorbed moisture. The meniscus interaction is generally a nuisance feature in that it will mask other surface mechanical effects. The meniscus can be eliminated by carrying out F–d analysis under water (or some other fluid ambient). Alternatively, in the case of an instrument operated within a vacuum envelope the aqueous phase will be pumped away. The "hard" surface F–d curve is used to calibrate the detection system so that a measurable detector response can be related accurately to the lever deflection. Applications of F–d analysis in most areas of science and technology are now ubiquitous. The impact on the study of the nano-mechanical properties of soft materials (e.g., organics, polymers, and biomaterials) has been particularly significant.

The schematic curve in Figure 5.16b shows lever deflection as a function of tip indentation of the specimen surface, or separation from the surface (i.e., the difference between stage travel and lever deflection), and thus illustrates other surface mechanical aspects of the system which are accessible to F–d analysis. The shape of the curve is based on the assumption, not ever strictly realized in practice, that the force constant of attraction always is less than k_N (hence instabilities at snap-on and lift-off are suppressed). The information content of the F–d curve in Figure 5.16b may be summarized as follows:

- The force constant of interaction, k_i, is simply the slope of the curve, $k_N z_L / |z_d - z_L|$, where z_d and z_L refer to stage travel and lever deflection, respectively.
- The forces at "contact" on approach and retract, F_A and F_R, may be defined at the inflection points where the force constants of interaction are the greatest.
- The snap-on and lift-off forces, F_{SO} and F_{LO}, are measured at the points of greatest net attraction. The latter is generally taken to be the force of adhesion.
- The distances z_{RO} and z_{IO} are measures of the elastic recovery and the plastic indentation, both at zero lever loading.
- The extent of hysteresis in the system is given by the area enclosed by the approach and retract curves.

The parameters defined above cannot readily be related to the familiar macroscopic definitions of mechanical properties, e.g., hardness, adhesion, Young's modulus, tensile strength, flexural strength, etc., unless the system can be specified further (e.g., tip shape, contact area, surface free energy of tip, etc.) and unless additional assumptions are made (homogeneity, isotropy, surface topography, etc.). Details of the procedure can be found in the literature [43].

A consequence of the discussion above is the need to match the force constant of the lever to that of the interaction being investigated in order to extract maximum information. If there is mismatch, then either the lever will be the only compliant element, and no information is obtained about the surface, or the surface will be the only compliant element, and the deflection of the lever is not measurable.

5.4.2.3.1 Colloidal Probe Analysis

This method of carrying out F–d analysis was initially adopted for measurement of colloidal forces in an aqueous fluid arising from surface charges, e.g., in Ref. [44]. A growing number of AFM-based studies of biomolecular interactions have been carried out by sensing forces in the sub-nanonewton range versus distance (i.e., standard F–d analysis), between a Si or Si_3N_4 tip prepared with a particular (bio)chemical functionality and a functionalized surface (e.g., in order to investigate the interaction between antibody and antigen molecular species, where the species have been attached to the tip and surface, respectively). More often, however, a bead (of diameter from less than 1 to several μm) is attached at the location of the tip, so as to produce a "colloidal" probe; an arrangement that has advantages for investigations of intermediate- and short-range interactions [45]. The geometry of the probe bead is then known with greater relative certainty than in the case of the tip. Likewise, the surface chemistry of a bead can be prepared with greater flexibility and reliability than in the case of a tip; a wide range of well-characterized microspheres/beads with known surface (bio)chemistries can be obtained from several suppliers. Also, the much greater surface area will effectively amplify the strength of interaction and thus improve the signal-to-noise ratio. On the other hand, there will be greater uncertainty as to the number of interacting species, and lateral spatial resolution is substantially degraded, in comparison with a sharp probe tip. In the case of a functionalized tip, with a radius of curvature less than 50 nm, the number of interacting species may be less than 10, thus allowing single event binding forces to be estimated from a histogram of the data.

Typical F–d curves for different systems, and associated expressions describing the dependence of force on probe-to-sample separation are shown in Figure 5.17 [46].

The F–d operational mode is now used for nano-mechanical analysis throughout science and technology. An early description [47] provides a good introduction to the basic ideas. The merits of

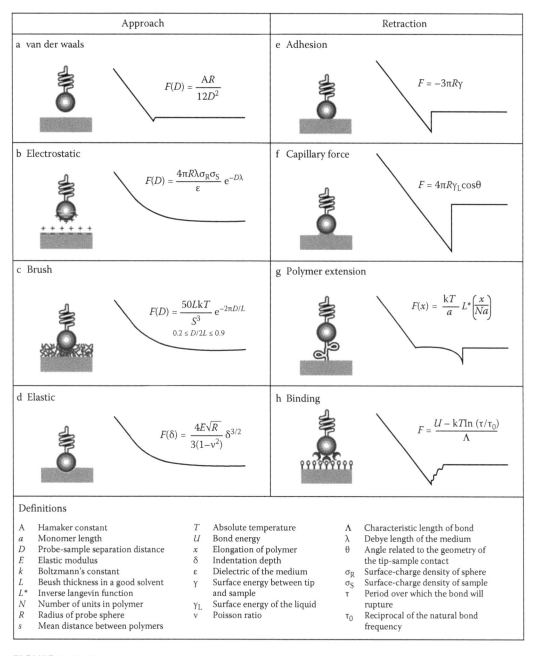

FIGURE 5.17 Some examples of F–d curves and the expressions that describe the dependence of force on tip-to-surface distance. (a–d) represent the approach cycle, and (e–h) represent the retract cycle. (From Heinz, W.F. and Hoh, J.H., *Nanotechnology*, 17, 143, 1999. With permission.)

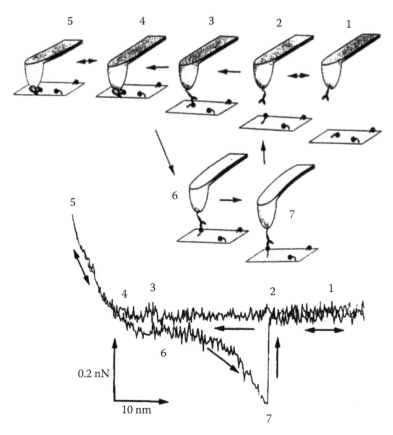

FIGURE 5.18 The schematic sequence shows (1–5) approach of the antibody and attachment to the antigen, followed by (6) retraction of the probe and stretching of the tether and bond. Finally, the antibody–antigen bond is ruptured at (7) due to the lever-imposed force of ca. 200 pN. (From Hinterdorfer, P., Baumgartner, W., Gruber, H.J., Schilcher, K., and Schindler, H., *Proc. Natl. Acad. Sci. USA*, 93, 3477, 1996. With permission.)

F–d are particularly apparent in studies of biomolecular binding [48,49] and protein (un)folding. An example of the former is shown in Figure 5.18, where both the location and the strength of a single antibody–antigen recognition event have been demonstrated [50]. Actual data in the form of an F–d curve are shown below.

Every long-chain molecule has its own unique folded native structure, which will depend on the environment and on the stage of the biomolecule in its functional life-cycle. Due to the enormous number of degrees of freedom, it is a daunting task to calculate its minimum energy configuration. Nevertheless, a polypeptide chain will find its folded global minimum energy configuration in a remarkably short time (Levinthal's paradox [51]). Explanations currently favored revolve around the existence of an identifiable directed pathway through the multidimensional potential landscape. One possible approach to the gaining of insight into the problem is to reverse-engineer the folding process, namely to induce unfolding.

It has been shown that an AFM operated in the F–d mode can shed considerable light on the problem [52]. A folded protein is attached to a substrate without being denatured (i.e., without undergoing change from its "natural" condition). The probe tip, often prepared with a particular chemical surface functionality, is first attached to a reactive site, through trial and error, and then withdrawn slowly, while a force is applied, causing extension initially, and then subsequently leading to a sequence of unfolding stages. A typical outcome is shown in Figure 5.19.

Studies of dynamics are gradually beginning to take advantage of the F–d operational mode. A nice example is that of an investigation of spontaneously beating cardiomyocytes (i.e., heart cells) [53],

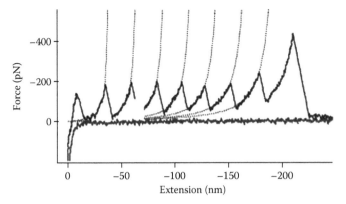

FIGURE 5.19 Force versus extension graph of an octameric TI 127 poly-protein being pulled by an AFM probe. The upper curve represents a sequence of continuous extension curves (rising portions of the trace) and sudden unfolding events. The lower trace represents the initial approach. (From Best, R.B. and Clarke, J., *Chem. Commun.*, 183, 2002. With permission.)

where the force sensing lever simply responds to expansions and contractions at various locations of a single cell. Sequences of mechanical pulses are shown correlated with the location in Figure 5.20. Other reports, e.g., Refs. [54,55], describe time-resolved drug-induced cytoskeletal changes of single cells.

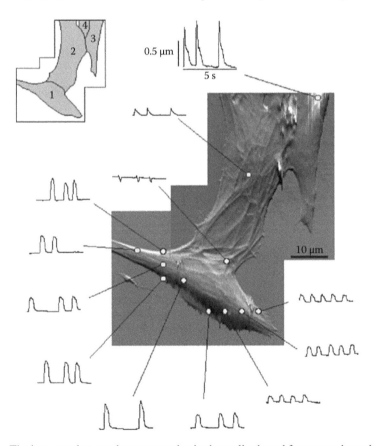

FIGURE 5.20 The image and traces demonstrate that both amplitude and frequency depend on the location within a single cardiomyocyte cell. The sketch (upper left) shows the boundaries of single cells. The numbers refer to adjacent cells. (Adapted from Domke, J., Parak, W.J., George, M., Gaub, H.E., and Radmacher, M., *Eur. Biophys. J.*, 28, 179, 1999. With permission.)

5.5   SCM

5.5.1   Principles and Implementation

SCM is a relatively recent and specialized technique, but is gaining in importance and popularity for semiconductor device characterization in response to the shrinking dimensions of such devices. Under favorable conditions, surface conductivity can be mapped with a lateral resolution of 5 nm [56]. Its principal merits are those of being nondestructive and of having lateral spatial resolution in the 5–10 nm range. As shown in Figure 5.21, the configuration of tip and specimen becomes that of a metal oxide semiconductor (MOS) capacitor. The equivalent circuit has a resonance frequency, ca. 1 GHz, determined by the impedance of the transmission line (arising from a capacitance of a few picofarads and an inductance of a few nanohenries). If a frequency shift of 1 kHz can be detected, then in principle the sensitivity is of the order of 10^{-18} F.

5.5.2   Capacitance Mapping

The method for mapping lateral variations in capacitance is illustrated in Figure 5.22. In essence, a control point is established at a frequency corresponding to the steepest part of the resonance envelope. As the local capacitance changes the resonance envelope will shift, and thus gives rise to a difference in output voltage for a fixed set-point frequency. Examples are shown in Figures 5.23 and 5.24 [57,58].

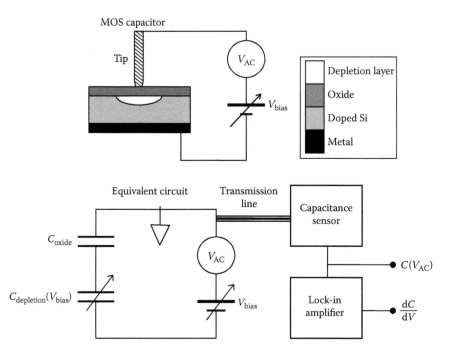

FIGURE 5.21   In SCM the tip/specimen system can be modeled as a MOS device where the metallic tip constitutes the gate electrode. Extent of depletion below the gate dielectric layer is controlled by V_{bias}. The equivalent circuit consists of a grounded tip in series with two capacitors also in series. In combination with the transmission line the system will have a resonant frequency of ca. 1 GHz.

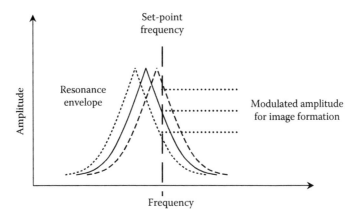

FIGURE 5.22 The resonance envelope shifts in response to change in local capacitance, giving rise to a corresponding change in amplitude at the set-point frequency.

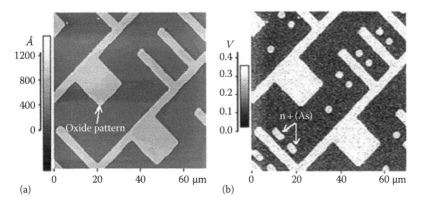

FIGURE 5.23 (a) Contact mode topographic and (b) SCM images of a patterned semiconductor surface. Regions with bright contrast in the topographic image represent a thermally grown SiO_2 pattern of height 70 nm. The additional circular and rounded rectangular features evident in the SCM image correspond to regions heavily doped by ion implantation (50 keV As$^+$ ions and 10^{14} ions/cm^2 dose density). (From veeco.com/library/nanotheater_detail.php?type = application&id = 277&app_id = 10)

5.5.3 MAPPING DIFFERENTIAL CAPACITANCE

The applied AC signal, in combination with the DC bias, causes a time-dependent variation in the extent of depletion, and thus a change in local capacitance. If the corresponding change in amplitude, or phase, is detected by a lock-in amplifier, at the AC set-point frequency, then the differential capacitance, dC/dV, can be mapped. The effect of bias voltage on capacitance and differential capacitance is illustrated in Figure 5.25.

5.6 SNOM

It is well known that the spatial resolution of optical imaging in the far-field region is limited by diffraction to slightly less than half the wavelength of the incident light. The possibility of imaging in the near-field regime was recognized more than 70 years ago [59], but was given serious consideration as a practical proposition only in the early 1980s with the advent of SPM technologies [60–63].

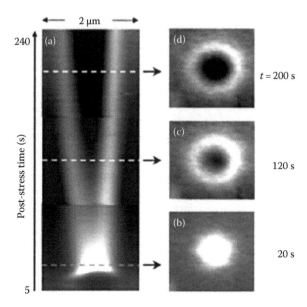

FIGURE 5.24 Charge redistribution mapped as a function of time. The charge was trapped initially on the surface of a 6 nm thick silicon oxide on p-doped Si by a voltage stress (−8 V applied by the tip). The SCM images were obtained with the instrument controlled in the contact topographic mode, while an AC modulation signal of 50 mV at 50 kHz was applied to the tip. The fields of view in (b–d) are $2 \times 2 \mu m^2$, and the bright/dark contrast refers to net positive/negative charge. (From Mang, K.M., Kuk, Y., Kwon, J., Kim, Y.S., Jeon, D., and Kang, C.J., *Europhys. Lett.*, 67, 261, 2004. With permission.)

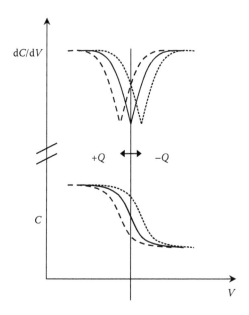

FIGURE 5.25 Effect on capacitance and differential capacitance of local accumulation or depletion of charge as a function of applied voltage. Addition or subtraction of charge causes the characteristic curves to shift left or right, respectively.

5.6.1 Physical Principles

While SFM techniques have many unique attributes, they cannot provide analytical or vibrational spectroscopic information. As well, the current generation of SFM instrumentation has relatively poor temporal resolution. SNOM extends the spatial resolution of classical optical microscopy by a factor of 10 or more, in combination with modest spectroscopic capability and excellent temporal resolution. SNOM exploits the properties of evanescent fields, which can be confined on structures much smaller than the wavelength of the incident light. Evanescent fields do not propagate into the far-field region; thus local probe techniques can be adapted either to illumination of the sample or to the detection of the evanescent fields emitted by the sample. Commercial instruments are based on the former principle, whereby the field is emitted from, and defined by, an aperture of dimensions of ca. $\lambda/20$.

An object illuminated by an external field will become polarized, and will reemit some of the absorbed radiation. The scattered radiation fields have evanescent components that are bound to the scatterer, and non-evanescent components that can propagate into the far field. Conversely, polarization of matter will also occur when an external field interacts with an extremely small volume of matter. In the latter case, the evanescent field in the vicinity of the scatterer will contain information about the geometry of the scatterer.

The currently favored arrangement is that of the generation of an optical near-field at a small aperture at the apex of a tapered fiber. If a sample is located within a few nanometers of the aperture, then the sample will be submerged in, and be coupled to, near-field radiation. The sample will act as a scatterer of near-field radiation, resulting in information being available in the far-field, in transmission or reflection modes. It is important to recognize that the light injected into the fiber will be subject to interactions with the whole system, i.e., during transmission along the fiber, by the aperture, and from coupling to the sample. Thus detailed modeling must take account of the whole system. Much work has been done on theoretical investigations of the transmission coefficient of the fiber. It turns out that the main limitations are the size of the aperture and the angle of the tapered cone. Coefficients of transmission in the range 10^{-3} to 10^{-6} are required for practical exploitation. Calculation of the actual field distribution at the aperture, when a sample is present, is nontrivial, even though it can be carried out within the framework of classical electrodynamics. The state of the art has been described in several excellent reviews, e.g., Refs. [64,65].

5.6.2 Technical Details

A schematic of a SNOM instrument is shown in Figure 5.26. It is usually based on an inverted microscope platform where the existing conventional optics are adapted to collect transmitted and reflected intensity in the far-field region. Likewise, conventional optical components can be introduced into the optical paths in order to filter, disperse, or polarize scattered light. Other components that are specific to the SNOM technique, and which affect its performance, need further description.

5.6.2.1 Shear-Force Detection

Unlike the more widely used SPM techniques, the "strength" of interaction, e.g., tunnel current in the case of STM and interatomic forces in the case of AFM, cannot easily be used to control the z-position of the SNOM probe. Instead shear-force detection has been adopted as the most widely used method. The fiber tip is stimulated to oscillate laterally with an amplitude of 1–5 nm at its fundamental resonance by a tuning fork arrangement driven by an exciting signal at the requisite frequency. In the vicinity of the surface, shear forces will cause damping, resulting in an amplitude decrement and a change in phase. Several mechanisms can be the source of damping, e.g., interaction with the adsorbed moisture film, intermittent contact with the surface, and/or

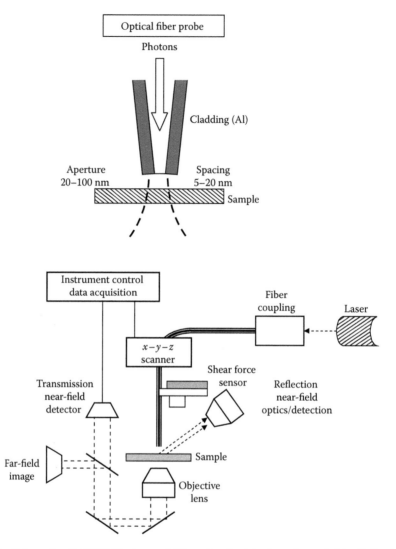

FIGURE 5.26   Schematic of a SNOM instrument. The optics for detection of near-field radiation in transmission or reflection are based on a standard inverted microscope platform. Polarization contrast can be implemented by insertion of polarizing elements in the optical path before and after scattering by the specimen. Spectroscopic information in transmission can be obtained by inserting a dispersive component or a notch filter into the far-field optical path.

electrostatic image forces. Normally the phase change is monitored and constitutes the control signal that allows the feedback loop to maintain the probe at a constant distance from the surface. Thus a SNOM can function in an AFM imaging mode, albeit at a somewhat lower z-resolution.

5.6.2.2   Optical Fiber Probe

The final probe is the outcome of a two-step processing of a standard optical fiber. The first step consists of preparing a transparent tapered conical section with a sharp apex. Two methods are used, either local heating and pulling [66], or chemical etching in HF (the so-called Turner's technique) [67,68]. The former is relatively labor intensive and lacks reproducibility, but leaves a smooth surface finish. The latter is more convenient, lending itself to higher throughput parallel fabrication, but may not result in an optimum surface finish. The second step consists of deposition of a metal coating and

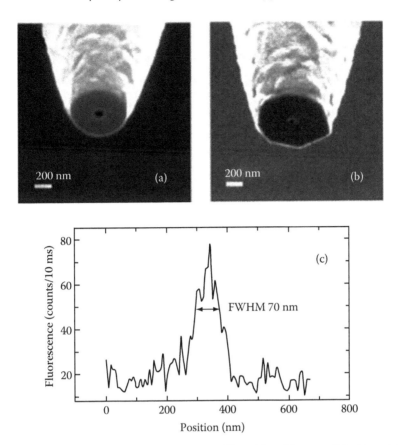

FIGURE 5.27 Aperture obtained by focused ion-beam milling. (From Veerman, J.A., Otter, A.M., Kuipers, L., and van Hulst, N.F., *Appl. Phys. Lett.*, 72, 3115, 1998. With permission.)

creation of an aperture at the apex. Angle-resolved vapor deposition can be used to form a self-aligned aperture at the apex. More recently, focused ion-beam micro-machining has been used to produce reproducible apertures of high quality [69,70]. A typical outcome is shown in Figure 5.27.

5.6.3 OPERATIONAL MODES

The most widely used modes include transmission imaging and fluorescence detection. Imaging in the reflection mode can also be carried out, although at lower signal-to-noise ratio due to the smaller angle of detection and the need to work with an objective lens at lower numerical aperture. In either case, features in the optical image can be correlated with those in a shear-force topographical image during parallel imaging, see Figure 5.28 [71].

The most recent addition is that of near-field Raman spectroscopy, where in practice SERS enhancement is required in order to compensate for the low incident power (typically less than 100 nW).

5.6.3.1 Transmission Imaging

The operational mode is illustrated by the images in Figure 5.28.

5.6.3.2 Fluorescence

Fluorescence analysis is particularly useful for localizing labeled compounds within larger bio/organic-structures. An example is shown in Figure 5.29. The technique is gaining in popularity.

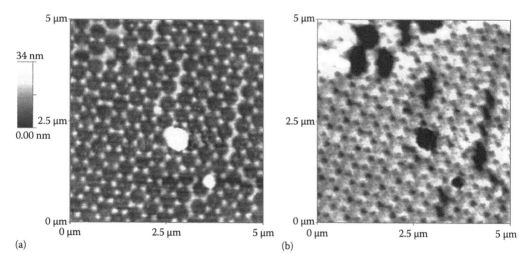

(a) (b)

FIGURE 5.28 (a) Shear force and (b) optical transmission images of a standard SNOM specimen. The specimen was produced by metal evaporation onto a flat substrate with close-packed polystyrene spheres as a shadowmask (the spheres were subsequently removed). (Courtesy of Dr. D. Higgins, Kansas State University.)

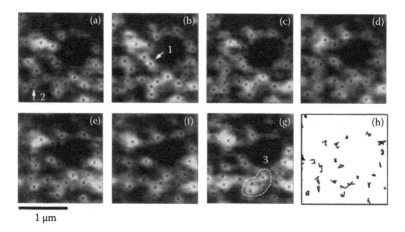

FIGURE 5.29 Consecutive SNOM fluorescence images, obtained at 13 min intervals, of single molecules (Rhodamine-6G) embedded in polyvinylbutyral. The sequence (a–g) illustrates lateral mobility of individual molecules at ambient conditions. The resultant trajectories are shown in (h). (From Bopp, M.A., Meixner, A.J., Tarrach, G., Zschokke-Gränacher, I., and Novotny, L., *Chem. Phys. Lett.*, 236, 721, 1996. With permission.)

5.6.3.3 Near-Field Raman Spectroscopy and Mapping

Raman spectroscopy is of great value as a tool for structural fingerprinting. In the microprobe version, conventional far-field Raman mapping can resolve and identify structures with spatial resolution of ca. 1 μm. However, the "normal" Raman process is relatively inefficient; only 1 in 10^7 incident photons will undergo the characteristic scattering. SERS enhancement can improve the yield by several orders of magnitude, thus making Raman a practical proposition as a SNOM operational mode. The technique is currently being developed, and can be implemented with existing instrumentation. Several examples have been described in the literature, e.g., Ref. [65]. As well, the dependence on polarization can be investigated in a fluid environment [72].

A recent study based on aperture-less near-field SNOM suggests that SERS enhancement can be induced by the near-field itself [73]. In the aperture-less mode a metal probe is illuminated, thus stimulating near-field emission at its apex. High spatial resolution of SERS-enhanced Raman analysis is illustrated in Figure 5.30. The technique appears to have potential for further development and wider utility.

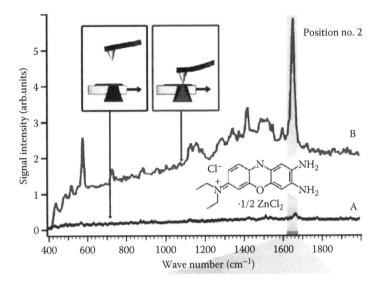

FIGURE 5.30 Raman analysis by aperture-less SNOM in the transmission mode, illustrating the merit of surface enhancement (SERS). The AFM probe is silver-coated and illuminated by 488 nm radiation at 5 mW incident power. The sample consisted of a thin layer of BSB (brilliant cresyl blue with formula repeat unit shown above) on a transparent substrate. The lower and upper traces represent data collection with the probe lifted off, and in contact with the surface, respectively. The SERS enhancement, when in contact, was ca. 2000, at an estimated lateral spatial resolution comparable to the size of the probe, ca. 50 nm. (From Stöckle, R.M., Suh, Y.D., Deckert, V., and Zenobi, R., *Chem. Phys. Lett.*, 318, 131, 2000. With permission.)

5.7 SECM

SECM is an extension of conventional spreading current analysis with the additional advantages of greater lateral spatial resolution, <5 nm in favorable cases, and the ability to correlate I–V measurements with features in a high-resolution topographic map.

5.7.1 Physical Principles

The technique is based on contact mode AFM, with a sharp conducting tip making a point contact with a surface at a particular location in the x–y plane. The lateral extent of the point contact will depend on the shape of the tip at the apex, the normal force loading, and the stiffness of the surface. The technique can function as a straight spreading resistance probe, if the circuit is ohmic, or as a tunneling probe, if there are one or more tunnel junctions in the circuit. Likewise, local defect states can be imaged as functions of applied bias. (The method will then essentially be similar to STS analysis, but with the difference that the states are buried. Thus the quality of the analysis will not depend on an UHV environment.) High sensitivity current sensing is then required, typically in the fA range, in order to obtain tunneling I–V data for thin dielectric films. The current, due to either direct or Fowler–Nordheim tunneling, will depend on the thickness and dielectric strength of the film, leakage paths, and charge traps [74].

5.7.2 Technical Details and Applications

In the case of an ohmic circuit and with the tip in contact with a semi-infinite homogeneous solid, the resistance of the solid can be inferred from the simple expression $R = \rho/2d$, where R is the measured resistance (from V/I), ρ is the resistivity, and d is the diameter of the (circular) contact junction. The underlying assumption is that the probe-to-surface contact resistance is negligible.

A variety of materials have been used in order to ensure ohmic contact, durability of the probe, and the prevention of the formation of rectifying and/or insulating interface layers; they include highly doped Si, diamond, Pt/Ir, Co/Ir, Au, etc. Also, it is important that the measurements be undertaken on a flat, <0.5 nm RMS roughness, surface, in order to ensure that features in a current map are not obscured by variations in contact area.

The most widespread application of the technique is in the area of characterization of electronic devices and materials; for example, a recent study of an silicon on insulator (SOI) device [75]. A number of case studies have been discussed by De Wolf et al. [74]. A sequence of current images on a SiO_2 film, Figure 5.31, demonstrates the voltage-dependent leakage due to embedded defects.

The effect of tip conditions, in the context of an investigation of direct and Fowler–Nordheim tunneling through insulating surface barriers on oxidized Si and a diamond-like carbon (DLC) film, has been demonstrated [76]. The results are shown in Figure 5.32.

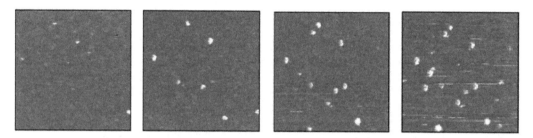

FIGURE 5.31 Tunneling current images, revealing embedded defects in an SiO_2 film at increasing bias of 1, 2, 3, and 4 V, from left to right. The field of view was $1 \times 1\ \mu m^2$. (From De Wolf, P., Brazel, E., and Erickson, A., *Mater. Sci. Semiconductor Process.*, 4, 71, 2001. With permission.)

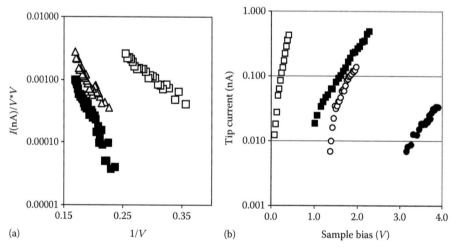

FIGURE 5.32 (a) *I–V* data plotted to emphasize a fit with a Fowler–Nordheim tunneling model. The logarithmic vertical axis refers to tip current in nA divided by the square of the sample bias. From left to right the sets of data refer to: as-received n-type tip with native oxide against n-type Si with 2.4 nm oxide; etched tip against Si with 2.4 nm oxide; and as-received tip against DLC. (b) *I–V* characteristics for the following combinations (left to right): Au-coated tip against DLC; etched tip against DLC; oxidized tip against Au; oxidized tip against DLC. The semilog plot emphasizes the exponential dependence due to direct tunneling. (From Myhra, S. and Watson, G.S., *Appl. Phys. A*, 81, 487, 2005. With permission.)

5.8 FUTURE PROSPECTS

5.8.1 INCREASED SPATIAL RESOLUTION FOR SFM

It has long been received wisdom that only UHV-STM can deliver true single atom resolution. There are a few examples of genuine single atom resolution for contact mode AFM, but only for chemically inert surfaces under exacting conditions [77], while similar resolution was reported for the imaging of more reactive surfaces, i.e., Si(111) (7 × 7), in the small-amplitude noncontact mode [78]. Recently, great progress in UHV-SFM has been reported [79,80]. Indeed subatomic resolution has been reported through the adoption of very stiff levers ($k_N > 10^3$ N/m) which allows noncontact operation at amplitudes of ca. 0.1 nm. The information content can be increased further by detection of higher harmonics. Comparative STM and AFM (higher-harmonic detection) images are shown in Figure 5.33. While these techniques cannot be described as routine, they do offer directions for future improvements in instrumentation.

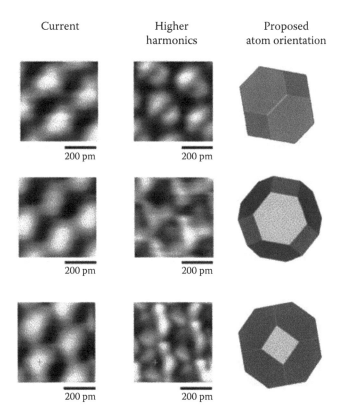

Current   Higher harmonics   Proposed atom orientation

200 pm   200 pm

200 pm   200 pm

200 pm   200 pm

FIGURE 5.33 Constant-height UHV-STM (left column) and higher-harmonic UHV-AFM images (central column) of graphite, obtained with a W tip. The right-hand column shows the proposed orientation of the W tip atom. The W atom is represented by its Wigner–Seitz unit cell, which reflects the full symmetry of the bulk. The bonding symmetry of the adatom is assumed to be similar to the bonding symmetry of the bulk. Top row: the higher harmonics show a twofold symmetry, resulting from a [110] orientation of the front atom. Second row: the higher harmonics show threefold symmetry, as expected for a [111] orientation. Third row: the symmetry of the higher-harmonic signal is approximately fourfold, as expected for a tip in a [001] orientation. The imaging was carried out at 4.9 K with k_N = 1800 N/m, amplitude = 0.3 nm, resonance frequency = 18.0765 kHz, and with Q = 20,000. (From Giessibl, F.J., *Mater. Today*, May, 32, 2005. With permission.)

5.8.2 SINGLE ATOM CHEMICAL IDENTIFICATION BY AFM

Although the SPM system has turned out to be a rich source of information, it has been unable to provide direct compositional information. However, advances in small-amplitude intermittent contact imaging in which the combination of long-range van der Waals interaction and the short range exchange force is being sensed have allowed genuine single atom spatial resolution imaging by SFM techniques (see Section 5.8.1). Those techniques have been extended to acquisition of $F–d$ data over a range of less than 1 nm, and at a force resolution in the low piconewtons range, in the regime where the atom at the apex of the tip is making an intermittent "bond" with the closest atom in the surface [80]. The technique is based on the shape of the $F–d$ curve within 0.2 nm of the greatest force of attraction unique to a particular atom species. The map in Figure 5.34 illustrates the specificity of identification of atomic species in a field of view where simple imaging cannot discriminate any differences [81]. The data were obtained in an UHV instrument at room temperature in the frequency modulation (FM) (see Section 5.4.2.1.4) control mode.

The technique places considerable demands on spatial control and stability. In principle a moderate vacuum should be sufficient to achieve the required conditions (i.e., a high Q resonance mode and low thermal drift). However, in contrast to the ultra-high resolution mode described in Section 5.8.1, the data can be obtained at room temperature. Thus there is considerable scope for implementing SFM as a true analytical technique.

5.8.3 FASTER SCAN RATES

The principal reason for increasing scan rates is to track surface and interface dynamics in real time and thus increase the knowledge base for the thermodynamics and kinetics of processes. TV scan rates are likely to emerge in next-generation instruments when smaller probes with higher resonance frequencies become available. In addition, the mechanical stiffness of the instrumental closed-loop must increase correspondingly, in unison with faster electronics for control and data acquisition.

(a)

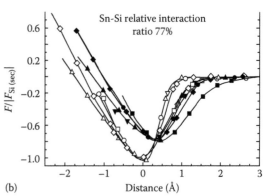

(b)

FIGURE 5.34 (See color insert following page 396.)   (a) False color image of surface alloy composed of Si (red), Pb (green,) and Sn (blue) atoms deposited on a Si(111) surface. The field of view is 4.3 × 4.3 nm². (b) Normalized $F–d$ curves show the distinct difference in interaction between the tip and Si and Sn surface atoms. Curves of different color refer to data obtained with different probes. The curves demonstrate that the assignments are independent of the tip material. The $F–d$ data have been normalized to those obtained for the Si species ($F/F_{Si\text{-set}} = 1$ at distance = 0). (From Sugimoto, Y., Pou, P., Abe, M., Jelinek, P., Pérez, R., Morita, S., and Custance, Ó., *Nature*, 446, 64, on-line05530, 2007. With permission.)

Increased scan rates will have the added benefits of ensuring greater convenience in the laboratory and greater throughput in the factory.

5.8.4 GREATER INTEGRATION AND SPECIALIZATION

There is likely to be a trend toward integrating SPM with complementary techniques, e.g., LV-SEM (LV = low voltage), HRTEM (HR = high resolution), and surface analytical techniques. Conversely, while the multimode multi-technique modularity of SPM instrumentation will continue to lead the market, there will be increasing emphasis on producing SPM instrumentation that has been optimized for niche markets, e.g., the electronic device industry and the life sciences.

5.8.5 APERTURE-LESS SNOM

Greater use of aperture-less illumination and enhanced spectroscopic capabilities is likely to emerge in next-generation instruments. The development will lead to a better integration of SNOM with other SPM techniques.

ACKNOWLEDGMENTS

Much of the insight and knowledge I have gained, and on which this chapter is based, has come from work in several laboratories, and from interactions with many industrious and able colleagues. I owe a great deal to former coworkers in the SPM Laboratory at Griffith University, namely Drs. G. S. and J. A. Watson, S. A. Holt, and C. T. Gibson. Many of the activities undertaken in that laboratory have benefited from funding from many sources: the Australian Research Council, CSIRO, US DOD DARPA, and Griffith University. Many projects benefited from collaboration with research groups at AEA Technology (Harwell) and the University of Oxford, where Drs. J. A. A. Crossley and C. Johnston were important contributors to success. A great deal is owed to Dr. Helen Bergen who over many years has with grace and forgiveness let me spend long hours in the laboratory or hunched over a keyboard.

REFERENCES

1. G. Binnig and H. Rohrer, *Helv. Phys. Acta*, **55**, 726 (1982).
2. G. Binnig, C. F. Quate, and C. Gerber, *Phys. Rev. Lett.*, **56**, 930 (1986).
3. E. Meyer, H. J. Hug, and R. Bennewitz, *Scanning Probe Microscopy—the Lab on a Tip*, Springer, Berlin, 2004.
4. R. Wiesendanger, E. Meyer, and S Morita, *Noncontact Atomic Force Microscopy*, Springer, Berlin, 2002.
5. R. Wiesendanger (ed.), *Scanning Probe Microscopy,* Springer, Berlin, 1998.
6. K. S. Birdi, *Scanning Probe Microscopes: Applications in Science and Technology*, CRC Press, Boca Raton, FL, 2003.
7. P. K. Hansma and J. Tersoff, *J. Appl. Phys.,* **61**, R1 (1987).
8. C. T. Gibson, G. S. Watson, and S. Myhra, *Scanning*, **19**, 564 (1997).
9. L. Hellemans, K. Waeyaert, and F. Hennau, *J. Vac. Sci. Technol. B*, **9**, 1309 (1991).
10. E. P. Plueddemann, *Silane Coupling Agents*, 2nd edn., Plenum Press, New York, 1991.
11. A. Ulman (ed.), *Self-Assembled Monolayers of Thiol*, Academic Press, San Diego, CA, 1998.
12. J. P. Cleveland, S. Manne, D. Bocek, and P. K. Hansma, *Rev. Sci. Instrum.*, **64**, 403 (1993).
13. J. L. Hutter and Bechhoefer, *Rev. Sci. Instrum.*, **64**, 1864 (1993).
14. J. E. Sader, I. Larson, P. Mulvaney, and L. R. White, *Rev. Sci. Instrum.*, **66**, 3789 (1995).
15. J. M. Neumeister and W. A. Ducker, *Rev. Sci. Instrum.*, **65**, 2527 (1994).
16. T. J. Senden and W. A. Ducker, *Langmuir*, **10**, 1003 (1994); H. J. Butt, P. Siedle, K. Seifert, K. Fendler, T. Seeger, E. Bamberg, A. L. Weisenhorn, K. Goldie, and A. Engel, *J. Microsc.*, **169**, 75 (1992).
17. J. E. Sader, *Rev. Sci. Instrum.*, **66**, 4583 (1995).
18. C. T. Gibson, G. S. Watson, and S. Myhra, *Nanotechnology*, **7**, 259 (1996).
19. M. Labardi, M. Allegrini, M. Solerna, C. Frediani, and C. Ascoli, *Appl. Phys. A*, **59**, 3 (1994).

20. D. F. Ogletree, R. W. Carpick, and M. Salmeron, *Rev. Sci. Instrum.*, **67**, 3298 (1996).
21. R. W. Carpick, D. F. Ogletree, and M. Salmeron, *Appl. Phys. Letts.*, **70**, 1548 (1997).
22. P. J. Cumpson, J. Hedley, and C. A. Clifford, *J. Vac. Sci. Technol. B*, **23**, 1992 (2005).
23. C. Odin, J. P. Aimé, Z. El Kaakour, and T. Bouhacina, *Surf. Sci.*, **317**, 321 (1994).
24. T. O. Glasbey, G. N. Batts, M. C. Davies, D. E. Jackson, C. V. Nicholas, M. D. Purbrick, C. J. Roberts, S. J. B. Tendler, and P. M. Williams, *Surf. Sci.*, **318**, L1219 (1994).
25. F. Atamny and A. Baiker, *Surf. Sci.*, **323**, L314 (1995).
26. L. Montelius, J. O. Tegenfeldt, and P. van Heeren, *J. Vac. Sci. Technol. B*, **12**, 2222 (1994).
27. J. Tersoff and D. R. Hamann, *Phys. Rev. B*, **31**, 805 (1985).
28. L. E. C. van Leemput and H. van Kempen, *Rep. Prog. Phys.*, **55**, 1165 (1992).
29. W. A. Hofer, *Prog. Surf. Sci.*, **71**, 147 (2003).
30. J. Bardeen, *Phys. Rev. Lett.*, **6**, 57 (1961).
31. R. Wolkow and Ph. Avouris, *Phys. Rev. Letts.*, **60**, 1049 (1988).
32. A. A. Gewirth and B. K. Niece, *Chem. Rev.*, **97**, 1129 (1997).
33. J. N. Israelachvili, *Intermolecular and Surface Forces*, 2nd edn., Academic Press, San Diego, CA, 1992.
34. N. A. Burnham, R. J. Colton, and H. M. Pollock, *Nanotechnology*, **4**, 64 (1993).
35. D. Sarid, *Scanning Force Microscopy with Applications to Electric, Magnetic and Atomic Forces*, Oxford University Press, New York, 1991.
36. C. T. Gibson, G. S. Watson, and S. Myhra, *Wear*, **213**, 72 (1997).
37. B. Bhushan, *Wear*, **259**, 1507 (2005).
38. R. W. Carpick and M. Salmeron, *Chem. Rev.*, **97**, 1163 (1997).
39. A. Noy, C. D. Frisbie, L. F. Rozsnyai, M. S. Wrighton, and C. M. Lieber, *J. Am. Chem. Soc.*, **117**, 7943 (1995).
40. W.-K. Li, *Polym. Test.*, **23**, 101 (2004).
41. V. V. Tsukruk and V. N. Blizniuk, *Langmuir*, **14**, 446 (1998).
42. H.-J. Butt, B. Capella, and M. Kappi, *Surf. Sci. Rep.*, **59**, 1 (2005).
43. J. A. Blach, G. S. Watson, W. K. Busfield, and S. Myhra, *Polym. Intl.*, **51**, 12 (2001).
44. W. A. Ducker, T. J. Senden, and R. M. Pashley, *Nature*, **353**, 239 (1991).
45. G. S. Watson, J. A. Blach, C. Cahill, S. Myhra, D. V. Nicolau, D. K. Pham, and J. Wright, *Coll. Polym. Sci.*, **282**, 56 (2003).
46. W. F. Heinz and J. H. Hoh, *Nanotechnology*, **17**, 143 (1999).
47. N. A. Burnham and R. J. Colton, *J. Vac. Sci. Technol. B*, **7**, 2906 (1989).
48. J. Zlatanova, S. M. Lindsay, and S. H. Leuba, *Prog. Biophys. Mol. Biol.*, **74**, 37 (2000).
49. O. H. Willemsen, M. M. E. Snel, A. Cambi, B. G. de Groot, and C. G. Figdor, *Biophys. J.*, **79**, 3267 (2000).
50. P. Hinterdorfer, W. Baumgartner, H. J. Gruber, K. Schilcher, and H. Schindler, *Proc. Natl. Acad. Sci. USA*, **93**, 3477 (1996).
51. C. Levinthal, *J. Chim. Phys.*, **65**, 44 (1968).
52. R. B. Best and J. Clarke, *Chem. Commun.*, 183 (2002).
53. J. Domke, W. J. Parak, M. George, H. E. Gaub, and M. Radmacher, *Eur. Biophys. J.*, **28**, 179 (1999).
54. G. R. Bushell, C. Cahill, F. M. Clarke, C. T. Gibson, S. Myhra, and G. S. Watson, *Cytometry*, **36**, 254 (1999).
55. C. Rotsch and M. Radmacher, *Biophys. J.*, **78**, 520 (2000).
56. D. Álvarez, J. Hartwich, M. Fouchier, P. Eyben, and W. Vandervorst, *Appl. Phys. Lett.*, **82**, 1724 (2003).
57. veeco.com/library/nanotheater_detail.php?type=application&id=277&app_id=10
58. K. M. Mang, Y. Kuk, J. Kwon, Y. S. Kim, D. Jeon, and C. J. Kang, *Europhys. Lett.*, **67**, 261 (2004).
59. E. H. Synge, *Phil. Mag.*, **6**, 356 (1928).
60. D. W. Pohl, W. Denk, and M. Lanz, *Appl. Phys. Lett.*, **44**, 651 (1984).
61. D. W. Pohl and L. Nowotny, *J. Vac. Sci. Technol. B*, **12**, 1441 (1994).
62. S. Kirstein, *Curr Opin. Coll. Interface Sci.*, **4**, 256 (1999).
63. B. Hecht, B. Sick, U. P. Wild, V. Deckert, R. Zenobi, O. J. F. Martin, and D. W. Pohl, *J. Chem. Phys.*, **112**, 7761 (2000).
64. C. Girard, C. Joachim, and S. Gauthier, *Rep. Prog. Phys.*, **63**, 893 (2000).
65. C. Girard, *Rep. Prog. Phys.*, **68**, 1883 (2005).
66. E. Betzig, J. K. Trautmann, T. D. Harris, J. S. Weiner, and R. L. Kostalek, *Science*, **251**, 1468 (1991).
67. D. Turner, US Patent 4,469,554 (1984).
68. P. Hoffmann, B. Dutoit, and R.-P. Salathé, *Ultramicroscopy*, **61**, 165 (1995).
69. S. Pilevar, K. Edinger, W. Atia, J. Smolianinov, and C. Davis, *Appl. Phys. Lett.*, **72**, 3133 (1998).

70. J. A. Veerman, A. M. Otter, L. Kuipers, and N. F. van Hulst, *Appl. Phys. Lett.*, **72**, 3115 (1998).
71. M. A. Bopp, A. J. Meixner, G. Tarrach, I. Zschokke-Gränacher, and L. Novotny, *Chem. Phys. Lett.*, **236**, 721 (1996).
72. J. Grausem, B. Humbert, and A. Burneau, *Appl. Phys. Lett.*, **70**, 1671 (1997).
73. R. M. Stöckle, Y. D. Suh, V. Deckert, and R. Zenobi, *Chem. Phys. Lett.*, **318**, 131 (2000).
74. P. De Wolf, E. Brazel, and A. Erickson, *Mater. Sci. Semiconductor Process.*, **4**, 71 (2001).
75. D. Álvarez, J. Hartwich, M. Fouchier, P. Eyben, and W. Vandervorst, *Appl. Phys. Lett.*, **82**, 1724 (2003).
76. S. Myhra and G. S. Watson, *Appl. Phys. A*, **81**, 487 (2005).
77. F. Ohnesorge and G. Binnig, *Science*, **260**, 1451 (1993).
78. F. J. Giessibl, *Science*, **267**, 68 (1995).
79. F. J. Giessibl, *Mater. Today*, May, 32 (2005).
80. S. Hembacher, F. J. Giessibl, and J. Mannhart, *Science*, **305**, 380 (2004).
81. Y. Sugimoto, P. Pou, M. Abe, P. Jelinek, R. Pérez, S. Morita, and Ó. Custance, *Nature*, **446**, 64 (2007).

6 Transmission Electron Microscopy: Instrumentation, Imaging Modes, and Analytical Attachments

John M. Titchmarsh

CONTENTS

6.1 INTRODUCTION

Transmission electron microscopy (TEM) and the related scanning transmission electron microscopy (STEM) have developed over several decades into a huge field with wide ranging areas of application in both the physical and life sciences. Although many of the basic instrumental requirements are common to both areas of application, there are important differences in sample preparation methods, the sensitivity to irradiation by high-energy electrons, the contrast mechanisms, and the dimensions of features of interest. The contents of this chapter and of Chapter 10 describe TEM and STEM methods for the characterization of boundaries and interfaces only in inorganic materials. Even so, the topic is still so vast that, in the space available, only a general overview is possible. Several comprehensive texts [1–7] covering EM are listed in the references, should the reader wish to delve more deeply into the subject. The topic has been divided into two chapters. This chapter describes instrumentation, basic theory, and techniques. Such aspects underlie all types of analysis and their description is essential before considering specific interfacial application, examples of which are covered in Chapter 9.

Although this chapter includes current developments and recent results, it also covers the more established characterization methods found on both new and relatively old instruments, because the latter are still widely used in materials science. This is because many areas of materials science still rely on simple diffraction contrast (DC) imaging, selected area diffraction (SAD), and energy dispersive x-ray analysis (EDX) rather than more demanding methods such as high resolution electron microscopy (HREM) imaging, electron energy-loss spectroscopy (EELS), and high-angle annular dark-field (HA-ADF) imaging. The most modern instruments now incorporate aberration correction, monochromators, holography, energy-filtered imaging, and HA-ADF, but these advances are not yet widely available to many materials scientists. However, these methods will become increasingly available in the coming years, and so are included in this chapter in anticipation that the content will remain or become more useful to materials scientists in the future.

An internal interface in a sample can be investigated by TEM provided that an electron beam can be transmitted through the sample and collected with sufficient intensity and spatial resolution by an imaging system. Because they are charged particles, electrons interact strongly with matter compared to other particle beams such as x-rays or neutrons that are also employed for internal structural studies. The disadvantage of this is that the strong interaction causes significant scattering out of the incident electron-beam direction in even the thinnest samples and the scattering increases with atomic number. Penetration by an electron beam is, therefore, much smaller than for an energetic x-ray or neutron beam. TEM will generally be useful when the sample thickness $t < {\sim}150$ nm. For the highest spatial resolution, much thinner samples are necessary. As the electron energy is increased, penetration improves but the atoms within the sample are then increasingly likely to be displaced from their positions and even ejected from the sample. Such processes interfere with the integrity of both imaging and analytical measurements. Hence, although the accelerating voltage, E_0, of TEMs was increased from <100 keV to 1 MeV, or more, in the 1960s, the vast majority of instruments are now designed to operate with E_o = 200–300 keV, because this range is considered to be the optimum compromise between TEM size and the cost of increasing E_o, useful penetration, spatial resolution, sample damage, and analytical sensitivity.

Abbreviations and acronyms are used extensively throughout the chapter in the interest of economy of space. Definitions are summarized below as an aid to the reader.

6.1.1 List of Symbols (Units), Acronyms, and Abbreviations

a_i^a	partition function for ith shell of element a
A^a	atomic weight of element a
$A(\alpha)$	focused probe aperture function
$\boldsymbol{b}$	dislocation Burger's vector (m)
B	electron source brightness (current emitted per unit area per steradian)
$c_{a,b,\ldots}$	fractional weight concentrations of elements a, b, ...
C_c	chromatic aberration coefficient (m)
C_s	spherical aberration coefficient (m)
d	crossover diameter (m)
d_{hkl}	interplanar spacing (m)
e	electronic charge (C)
E_o	accelerating voltage/energy of the electron beam (eV)
E_t	temporal coherence envelope function
E_B	continuum x-ray energy (eV)
E_x	characteristic x-ray energy (eV)
E_i^a	energy of ith shell x-ray from element a (eV)
$F(\mathbf{K})$	Fourier transform of crystal potential
F_g	structure factor for diffraction vector $\boldsymbol{g}$ (m)
$\boldsymbol{g}_{hkl}$	reciprocal lattice diffraction vector (m^{-1})
h	Planck's constant
i, I_o	probe current (A)
I^{BF}, I^{DF}	bright field, dark field image intensity
I_i^a	number of x-rays from ith shell of element a
$I_p(\boldsymbol{r})$	probe intensity at position $\boldsymbol{r}$
k	Boltzmann's constant (J/K)
k_{ab}	factor scaling ratio of x-ray counts to weight ratio of elements a and b
$\boldsymbol{k}$	wave vector (m^{-1})
$\mathbf{K}$	wave vector (m^{-1})
$\boldsymbol{k}_o$	incident beam wave vector (m^{-1})
$\boldsymbol{k}'$	diffracted beam wave vector (m^{-1})
L	effective camera length of diffraction pattern
m	electron mass (g)
n	integer index
N_A	Avagadro's number
P_o	probability of electron suffering no inelastic scattering
q_o	limit of illumination angle at specimen (rad)
$\boldsymbol{r}$	real space position vector (m)
r_c	radius of coherent illumination at specimen (m)
s	deviation vector (m^{-1})
$S_i^A(W, \beta)$	energy loss signal intensity integrated over window width, W, and collection angle, β, above the ith ionization edge of element A
t	sample thickness (m)
T	temperature (K)

V_o	beam accelerating voltage (V)
V_c	volume of crystal unit cell (m^3)
z	coordinate of the optic axis (m)
Z	atomic number
α	semi-angle of probe convergence (rad)
β	scattering angle collected by aperture or detector (rad)
χ	path difference (m): $2\pi\chi\lambda$ = phase angle (rad)
$\chi(\alpha)$	aberration function
Δf	probe defocus distance (m)
ΔE	energy range (window) in EELS spectrum (eV)
$\varepsilon(E)$	x-ray detector efficiency for x-ray energy E
ϕ	azimuthal angle (rad)
ϕ_o	mean inner potential inside sample (eV)
$\phi(r)$	crystal potential distribution (eV)
$\phi_P(r)$	projected crystal potential distribution normal to beam (eV)
λ	electron wavelength (m)
λ_{inel}	total inelastic scattering mean free path (m)
$(\mu/\rho)_i^j$	mass absorption coefficient of x-ray from element i by atomic mixture j (kg/m^2)
$\Psi(r)$	electron wave function at a point, r, in the $x–y$ plane
ρ^a	density of sample of element a (kg/m^3)
σ_K	K-shell ionisation cross-section (m^2)
σ_i^a	ionization cross-section for ith shell of element a (m^2)
τ	analysis time (s)
θ	diffraction angle (rad)
θ_B, θ_{hkl}	Bragg diffraction angle for (hkl) planes (rad)
θ_i, θ_o	inner and outer angular limits of HA-ADF detector (rad)
ω	angular frequency (rad)
ω_i^a, ω_K	fluorescent yields for ith, Kth shells
ξ_g	extinction distance of diffraction vector $\mathbf{g}$ (m)
ξ_g^{eff}	effective extinction distance of diffraction vector $\mathbf{g}$ (m)
ψ	eletron wave function
$\psi_o(r), \psi_e(r)$	incident, exit surface wave functions at position r
$\psi_p(r)$	probe wave functions at position r
ψ_i, ψ_i^*	image wave functions and its complex conjugate
ADF	annular dark field
BF	bright field (image)
BFP	back focal plane of the objective lens
CA	contrast aperture
CC	camera constant
CCD	charge-coupled device
CL	condenser lens
CTEM	convergent beam TEM
CTF	contrast transfer function
DC	diffraction contrast (image)
DF	dark field (image)
DOS	density of states
DP	electron diffraction pattern
EDX	energy dispersive x-ray analysis
EELS	electron energy-loss spectrometer
EFTEM	energy filtered TEM
EHT	electrostatic high tension
ELNES	energy-loss near-edge structure

EPMA	electron probe microanalyser
ESI	electron spectrum image
FEG	field-emission gun
FT	Fourier transform
FWHM	full-width half-maximum
HA-ADF	high-angle annular dark-field detector
HREM	high-resolution electron microscopy
IL	intermediate lens
IP	image plane
MAC	mass absorption coefficient
OA	objective aperture
OL	objective lens
PL	projector lens
SAD	selected area diffraction
SADA	selected area diffraction aperture
SEM	scanning electron microscope
SNR	signal-to-noise ratio
SSD	single scattering distribution
STEM	scanning transmission electron microscopy
TDS	thermal diffuse scattering
TEM	transmission electron microscopy
WB	weak beam
WPOA	weak-phase object approximation
ZLP	zero loss peak

6.2   ELECTRON MICROSCOPE INSTRUMENTATION

6.2.1   CONVENTIONAL TEM

The TEM system is pumped to a low pressure (the 10^{-5} Pa range or lower in the case of FEG [field-emission gun]) to ensure stable electron emission from the source, to minimize scattering of the electron beam between source and detectors, to prevent ion damage in the sample and source, and to minimize contamination deposits on the sample. It follows that great care must always be undertaken during maintenance or repair to use clean and dry gas whenever the system is vented to atmospheric pressure, and any components introduced into the evacuated system should always be degreased, dried, and never handled without gloves. Figure 6.1 shows, schematically, the essential electron-optical components of a TEM instrument used for DC and HREM imaging, which can be found in many materials science laboratories. Different sections of the column are often differentially pumped because the electron gun and sample stage sections demand lower pressure than other regions. The maximum tolerable gun pressure depends on the type of electron emitter (described in the following section). All the lenses are electromagnetic, except for those of the electron gun, which are electrostatic, and have axial symmetry. Deflection coils, necessary for accurate alignment of the electron optical column, and astigmatism correctors are excluded from Figure 6.1.

The electron gun provides a source of electrons at the point S1 (Figure 6.1), which is focused onto the specimen by two or more condenser lenses (CLs) plus an objective mini-lens. The sample is positioned on the optical axis (OA) in the sample stage inside the strong magnetic field of the objective lens (OL). The sample holder, of which there are various designs for different types of experiment, is inserted into the stage through an airlock. Samples can, therefore, be exchanged without venting the column to atmospheric pressure. The OL produces a magnified image of part of the sample and a series of intermediate lens (IL) and projector lens (PL) further magnify the image and transfer it to viewing chamber for assessment and recording.

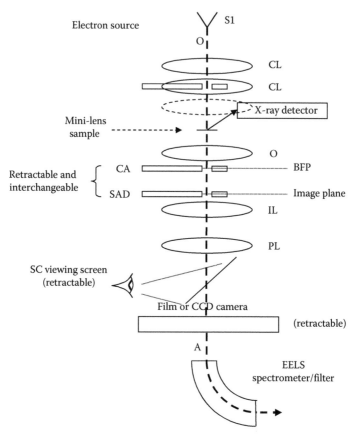

FIGURE 6.1 Major components of a TEM. Explanations of the acronyms can be found in Section 6.1.1.

6.2.2 ELECTRON SOURCES

Various types of electron source are used and all employ electrostatic extraction and acceleration. The source is held at a high negative-potential, E_o, with respect to the grounded column and sample. An important characteristic of the source is its brightness, B. The thermionic tungsten hairpin filament is still used in older TEMs, but more recent instruments use a heated, pointed, single crystal of LaB_6 [8] as the electron emitter because it is brighter and has a much longer lifetime than the W hairpin. Increasingly, the Schottky FEG [9,10] is found in the most modern high-performance instruments. The FEG depends on electron tunneling from a single crystal of oriented W wire, etched to a fine point (~50 nm radius at the apex) and coated with an appropriate material to lower the potential barrier to electron emission. The Schottky source is brighter than the simple thermal sources but it still needs to be heated to keep the tip clean and to give thermal assistance to the tunneling process. The brightest source is the cold FEG that operates at ambient temperature [11]. However, FEGs require a much lower operational pressure (of order 10^{-9} Pa) than simple thermal guns (10^{-5} Pa) and, consequently, FEG-TEMs are considerably more expensive. The electron gun, whatever the type, generates an effective source focused at a crossover, S1, on the optical axis, which forms the object for the condensing lens system.

The brightness, B, is defined as the current emitted per unit area per steradian:

TABLE 6.1

Comparison of Electron Source Parameters

Type	Operating Temperature (K)	Operating Pressure (Pa)	Maximum Brightness (A/mm²/sr)	Life (h)	Energy Spread (FWHM eV)
Thermal W	2500–2800	10^{-3}	10^3	25	1–2
Thermal LaB$_6$	1800	10^{-5}	10^4	500	1–2
Schottky	1000–1500	10^{-8}	10^6	10^4	0.8–1.0
Cold FEG	300	10^{-9}	10^7	10^4	0.3–0.5

$$B = \frac{4i}{\pi^2 d^2 \alpha^2} \tag{6.1}$$

where

i is the probe current
d is the effective crossover diameter
α is the semi-angle of probe convergence

Provided aberrations are negligible, B is constant when measured at any focused beam crossover in the TEM column. Table 6.1 lists typical values of operating parameters for the sources discussed above. The FEG is a factor of about 10^3 brighter than a thermal source. Compared to a thermal source, the spatial coherence of the FEG beam is much greater at the specimen for a given current density and area of illumination. Coherence is important for HREM imaging and is discussed in Section 3.4. High B is equally important for analytical work because the current in a very small focused probe is much greater for an FEG. The brightness of any source tends to fall gradually during its operational lifetime.

6.2.3 ILLUMINATION SYSTEM

The effective source position (S1 in Figure 6.1) and the sample height are fixed in the column, so the CLs can transfer either a defocused image of S1 (i.e., an approximately parallel beam) to the sample plane, for TEM imaging and diffraction, or a focused probe at the sample, for convergent beam diffraction and spot analysis using EDX or EELS. The three CLs allow, in principle, independent control of the three beam parameters, i, d, and α. However, the condenser mini-lens is often coupled with CL2, rather than operated in a completely independent manner for TEM imaging, and so independent control of all three parameters is not always readily available to the user. The beam size at the sample is controlled by changing the strength of CL1. However, this requires corresponding adjustment to CL2 to maintain focus at the sample and, from Equation 6.1, α will increase. In practice, an interchangeable aperture in CL2 is used to adjust α, so that changing CL1 with CL2 results in d and i changing while α remains constant. Instrument manufacturers usually provide a number of fixed settings for CL1. However, when a free lens control facility is available, it is possible to adjust CL settings to allow continuous adjustment of α and d, which may have advantages for focused beam diffraction experiments.

The CL2 aperture can be regarded as the effective source of electrons and such a source can be considered to be completely incoherent in the sense that individual electrons are seen by the sample as originating at distinct points within the CL2 aperture at different times.

6.2.4 SAMPLE STAGE

In older TEMs, the sample holders were of the top-entry type, whereby the sample was inserted into the column through an airlock above the OL and then lowered through the OL upper pole piece. This gave improved positional stability for HR imaging compared with the side-entry stages found on lower resolution, analytical TEMs. However, in modern systems, improved stage design for side-entry holders now provides equivalent performance without the mechanical complexity of top-entry. The sample is inserted horizontally, directly between the pole pieces of the OL. Sample tilting is possible about two axes that are approximately orthogonal over a typical range up to ±45° from the horizontal position, which allows considerable flexibility for examining crystal features under a wide range of orientations and diffraction conditions. TEM sample dimensions must usually be no more than 3 mm in diameter and ~0.2 mm thick. It is important to remember that, in any TEM image, the structural features appear as a 2D (two-dimensional) projection of the real 3D structure and that greater insight can be gained by tilting to several orientations. There is currently much interest in the development of tomography by combining series of images acquired with tilt increments [12,13], leading to the development of stages that permit tilting up to ±80°. An HREM will probably have a reduced tilting range, typically ±25°, because the pole-piece gap is lower than for a general purpose TEM, in order to reduce resolution-limiting aberrations of the OL. Special holders can be purchased for *in-situ* studies of heating, cooling, straining, magnetization, chemical reactions, etc. [6], and these often require extensive changes to the stage design.

6.2.5 OBJECTIVE LENS

The OL is the most important imaging lens in the TEM and is designed to minimize the spherical and chromatic aberration coefficients that limit the spatial resolution. Aberrations are discussed in Section 3.4.3.

Image and diffraction pattern formation in the OL are shown, schematically, in Figure 6.2 using the concept of ray optics. Although the sample is, in practice, immersed in the field of the OL, it is still sufficiently accurate for many purposes to regard the OL as thin, with the object (sample) outside the lens and close to the principal focus of the lens. It is assumed in Figure 6.2 that all rays incident on the sample are parallel to OA, i.e., there is an incident plane wave, and those which are

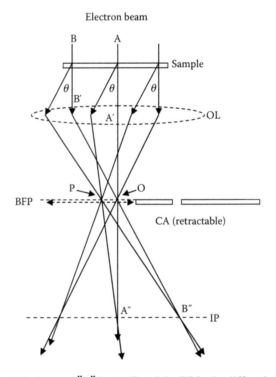

FIGURE 6.2   Formation of the image A″B″ in the IP and the DP in the diffraction plane (BFP) by the OL.

transmitted through the sample but not deflected, such as AA′ and BB′, are brought to a focus, O, in the back focal plane (BFP) of the OL on OA. All rays that are diffracted through a particular angle, θ, are brought to a common, off-axis point, P, in the BFP. The intensity distribution in the BFP is the diffraction pattern (DP). Continuation of the pairs of rays A′O and A′P and B′O and B′P result in intersections at points A″ and B″, respectively, in the image plane (IP) of the OL. Points A and B in the sample are imaged at A″ and B″, respectively, in IP. Every point in the sample contributes intensity to every part of the DP and every point in the DP subsequently contributes to the image intensity in the IP. When the sample is very thin and crystalline, the DP intensity is concentrated into an array of spots, with relatively low intensity between the spots. What cannot readily be shown in Figure 6.2 is that there is a rotation about OA between corresponding directions in the image and diffraction planes.

Close to the BFP, the operator can insert and position an objective aperture (contrast aperture [CA]), which is essential for performing DC imaging. Without the insertion of a CA, the final image is a superposition of all the diffracted images from each point in the BFP. When the CA is inserted and positioned on OA to allow only the directly transmitted spot to pass through, the resulting image is called the bright-field (BF) image. (The term "undiffracted" beam is sometimes used as a convenience to describe the directly transmitted beam. However, diffraction also occurs in the directly transmitted beam direction along OA and so it is more accurate to call the undif-fracted spot the forward-diffracted spot.) A dark-field (DF) image is formed by positioning the CA around a single off-axis region of the DP. For a crystalline sample, the aperture is usually centered round a single diffraction spot. A tilted DF image is obtained by tilting the incident beam through an angle such that the selected DF spot is moved onto the OA to reduce the blurring caused by OL aberrations. If the diameter of the CA is large relative to the separation of diffraction spots in the BFP, or no CA is inserted, then many DF images, together with the BF image, are superposed in the image plane. An HREM image is formed in this manner. HREM image formation is discussed in Section 3.4.

A retractable SAD aperture (SADA) can be inserted in the IP (see Figure 6.1). By definition, the IP is optically conjugate with the sample plane so that the SADA appears sharply focused in the image. The SADA can be accurately positioned, usually in the center of the image, and any sample region of interest can be moved until it appears within the SADA. The insertion of the SADA, therefore, is equivalent to placing an area-selecting aperture on the sample itself. Figure 6.2 shows how the area of sample, diameter AB, contributing to the DP, is restricted to the area of sample apparently defined by the SADA, diameter A″B″. Hence, the SADA can be used to collect diffraction information from, for example, a relatively small particle contained in a matrix, or to determine the local orientation in a bent crystal.

6.2.6 PROJECTION AND RECORDING

The excitation of lenses IL and PL (Figure 6.1), which follow the OL, can be changed so that a magnified image of either the BFP or the IP of the OL can be transferred to the fluorescent viewing screen (SC), which is viewed through a glass window. The operator can toggle between the two modes using a single button and the magnification of either image or DP can be separately changed to suit the experiment.

Recording of images was, until recently, performed using photographic emulsion on cellulose plates in a camera located below the retractable SC (Figure 6.1). However, it is now commonplace to use a solid-state charge-coupled device (CCD) camera, the output from which can be read directly into a computer memory, from which real-time viewing and processing of images and DPs can be performed to assess resolution and aberrations [14]. The only limit on the number of recorded images is the capacity of the computer hard drive. This is a great advantage for HREM studies, where large numbers of exposures are often required for through-focal series. CCD cameras with up to 4k × 4k pixel arrays are now becoming affordable, providing a pixel density close to that of photographic emulsion.

Although the CCD camera is better than emulsion, it still lacks the sufficient dynamic range necessary for some measurements. Image plates provide a superior recording medium in this respect [15,16]. The image plate is exposed in the same manner as photographic emulsion such that the electron intensity is stored as a charge distribution across the plate. The plate must be removed from the microscope before the intensity distribution can be read and stored in a computer. Hence, image plates suffer from the same slow pre-pumping and exchange of plate as conventional photographic emulsion. The dynamic range of the image plate, however, is several orders of magnitude greater than that of a CCD camera and this makes image plates suitable for certain specialist applications, for example, for measuring quantitative data from DPs where the intensity at the center of bright spots is often many orders of magnitude greater than the background.

6.3 ELECTRON–SAMPLE INTERACTION

A high-energy electron travels as an isolated wave packet with a wavelength, λ, much smaller than the typical interatomic distance and is determined by the electron energy:

$$\lambda = \frac{1.226}{E_0^{1/2}(1+0.97845\times10^{-6}E_0)^{1/2}} \tag{6.2}$$

where
  λ is in nanometers
  E_0 in volts

For example, $\lambda = 2.51\,pm$ for an electron accelerated through 200 keV. A plane wave is represented as

$$\psi = \exp\left[i(\omega t - 2\pi k \cdot r)\right] \tag{6.3}$$

traveling in the direction of k, where r is the (vector) distance from a chosen origin and $|k| = 1/\lambda$, although the time varying factor, ωt, where ω is the angular frequency, is usually omitted for convenience. (It is worth noting that some texts use the definition of $|k| = 2\pi/\lambda$, so Equation 6.2 must then be modified appropriately.) The physical extent, typically tens of nanometers, of the wave packet in the direction, z, parallel to OA, is usually much smaller than the average separation of electrons, typically tens of micrometers, in beams used for both HREM and analysis. The manner in which the electron beam interacts with the sample can be described either as the propagation of a continuous plane or as a spherically converging monochromatic wave, even though the electron wave-packets are discrete, or as a stream of energetic particles (e.g., Ref. [17]). The former is more convenient when discussing image and DP formation while the latter is often more useful when considering analytical modes such as EDX and EELS.

6.3.1 DIFFRACTION PATTERN

When a sample is placed in the electron beam, the fast electron wave-field interacts with the electric field of the atomic nuclei and their surrounding electrons. The wave function ψ_e, of the fast electron that emerges at the exit surface of the sample, is no longer planar (or spherically converging, depending on the incident illumination condition), and this causes a variation of amplitude with direction of propagation. The angular distribution of the wave function, viewed in a plane at a large distance, L, the camera length, from the sample, is called the Fraunhofer electron DP. In the TEM, the DP is observed by magnifying the electron distribution in the BFP of the OL at the detector. In Figure 6.2, all rays emerging from the sample at a common angle to OA are focused by the OL to a single point in the BFP. For an amorphous sample, there will be a continuous distribution of intensity in the BFP. However, the DP is

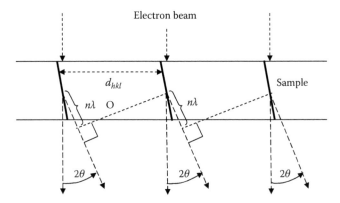

FIGURE 6.3 Diffraction from a set of planes.

most conveniently described by considering a section through a simple crystalline sample with just one set of diffracting planes of spacing d_{hkl}, tilted at the Bragg angle, θ_{hkl}, to the incident beam (Figure 6.3). The phase of ψ_e then varies periodically in the x–y plane, and constructive interference occurs in certain directions, $2\theta_{hkl}$ from the incident direction, that are derived using the well-known Bragg equation:

$$\lambda = 2d_{hkl}\sin\theta_{hkl} \tag{6.4}$$

A bright spot is observed in the BFP for every set of lattice planes with a normal lying close to the x–y plane (i.e., normal to OA) at a location where the path difference between diffracted intensity from adjacent planes is a whole number of wavelengths (shown as $n\lambda$ in Figure 6.3). The symmetry of the spots in the DP will reflect the symmetry of the projected crystal lattice in the plane perpendicular to OA. An extra constraint on the formation of any diffraction spot is that the structure factor [1] of the sample unit cell must be nonzero, otherwise systematic absences occur. For certain crystals, however, double diffraction by two different sets of planes can lead to additional spots that are kinematically forbidden.

The Ewald sphere construction (Figure 6.4) provides a convenient, geometric method of illustrating, in reciprocal space, the relationship between the incident beam direction, with a wave vector k_o, where $|k_o| = 1/\lambda$, and a diffracted beam with wave vector k', also with $|k'| = 1/\lambda$. The vector k_o is

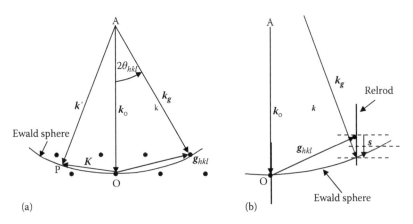

(a) (b)

FIGURE 6.4 Ewald sphere construction with (a) the Bragg condition satisfied exactly and (b) with the crystal tilted positive from the Bragg condition.

drawn from a point, A, in the direction AO, parallel to the incident electron wave, to the point O. A sphere is drawn with radius $AO = 1/\lambda$, centered on A, part of which is shown as a circular arc in the 2D section of reciprocal space in Figure 6.4a. The diffraction vector $OP = K = k' - k_o$ has a magnitude $|K| = 2|k_o| \sin \theta$.

The locations of possible Bragg reflections are represented by points, located at the end of vectors, g_{hkl}, drawn from O and that form a 3D lattice in reciprocal space. As the sample is tilted, the array of spots rotates with respect to the direction of k_o so that the surface of the sphere sweeps through the array of reciprocal lattice points. At specific tilt angles of the beam, the sphere will pass exactly through a reciprocal lattice point when $k' = k_g$, as shown in Figure 6.4a for the point g_{hkl}, where $|g_{hkl}| = 1/d_{hkl}$, signifying that the (hkl) planes satisfy the condition for Bragg diffraction, Equation 6.3, such that $K = g_{hkl}$ and the angle between k' and k_o is $2\theta_{hkl}$. Other reciprocal lattice points that do not lie on the sphere are not seen in the DP.

A more accurate construction of the reciprocal lattice requires that the points are elongated into rods, called relrods, in a direction normal to the top surface of the sample (i.e., the entry surface of the electron beam), and by a length that is inversely proportional to the thickness of the sample. Hence, a very thin sample has very long relrods. The implication of this is that when the Ewald sphere intersects a relrod, a spot is generated in the DP, even though there is a deviation from the precise Bragg orientation by the deviation vector, s, parallel to k_o, as shown in Figure 6.4b, so that $k_g = k_o + g_{hkl} + s$. In Figure 6.4b, g_{hkl} lies inside the Ewald sphere and $s > 0$. For the values of λ used in TEM, the Ewald sphere is almost planar and intersects many relrods, so many spots are then visible in a DP from a thin EM sample.

A DP from a thin crystal of Inconel 600, a face-centered cubic metal alloy, oriented with a <110> zone axis almost exactly parallel to the electron beam, is shown in Figure 6.5. Several dozen spots are seen due to the extension of the relrods and the curvature of the Ewald sphere. The spots in the pattern have been indexed using a right-handed axis notation and assuming that the outward direction from the page is [109], i.e., anti-parallel to the beam.

More rigorous treatment of electron diffraction in standard texts [18] shows that the DP is given by the Fourier transform (FT) of the 3D distribution of the sample potential. Because the potential generally consists of a distribution of spatial-frequency components, $\phi(r)$, related to the distribution of nuclear and electronic charges in the sample, the DP intensity $F(K)$ at some point, K, in the DP is given by

$$F(K) = C\int\left[\phi(r)\exp(-2\pi iK \cdot r)\right]dr \tag{6.5}$$

where C is a constant.

The DP can be used to determine d_{hkl} if L for the TEM is known. The magnification of the DP and, hence, L can be changed by altering the setting of the projector lenses, in a manner analogous to changing the image magnification. Typical values of L are in the range 50–1000 mm. Equations

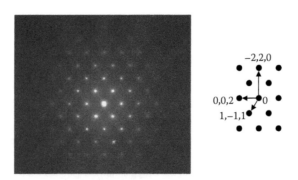

FIGURE 6.5 DP from a thin crystal of Inconel 600, a face-centered cubic metal alloy, oriented with the [109] zone axis almost exactly parallel to the electron beam.

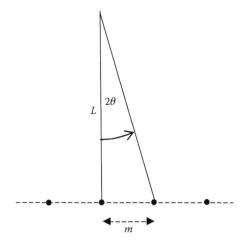

FIGURE 6.6 Geometry related to the calculation of d_{hkl} from m and L.

6.6 and 6.7 and Figure 6.6 show how the modulus, m, of $\mathbf{g}$, measured from the recorded DP, is related to L and to its Bragg angle (in the small angle approximation):

$$m \propto 2\theta L \tag{6.6}$$

Combining Equations 6.3 and 6.5 gives

$$md_{hkl} \propto \lambda L \tag{6.7}$$

The product λL is called the camera constant (CC) and, for quantitative measurements, must first be accurately calibrated at each magnification of the DP for the appropriate E_0 (i.e., λ) using a sample of known crystal structure. The d_{hkl} spacings of an unknown structure are then easily calculated from measurements of m and Equation 6.7. In using Equation 6.7, it is important to ensure that the dimensions are correct: if L and m are measured in millimeters and λ in nanometers, then the units of d_{hkl} will be in nanometers.

Unless the sample is very thin, the DP also contains intensity between the spots arising from inelastic phonon scattering (thermal diffuse scattering, TDS) whereby electrons are scattered over all angles with very small energy losses, proportional to kT (~0.025 eV at ambient temperature). Low intensity is observable in Figure 6.5 between the spots. In thicker samples, when electrons undergo both phonon and diffraction scattering, a pattern of lines, known as the Kikuchi pattern [19], appears in the DP, an example of which is shown in Figure 6.7 from a thicker region of the same sample of Inconel as in Figure 6.5, but at a random incident-beam orientation away from a zone axis. The Kikuchi lines appear in pairs, one brighter (excess, E), and the other darker (deficient, D), than the background intensity, with the angular separation $2\theta_{hkl}$ of the particular set of (hkl) planes giving rise to the contrast and a line direction perpendicular to $\mathbf{g}_{hkl}$. The excess line of any pair is always the further from the central BF spot in the DP. When the incident beam is exactly parallel to a set of planes, then the two lines have equal intensity and form a Kikuchi band with the central spot lying between and equidistant from both lines. As with diffraction spots, additional higher order Kikuchi lines are often seen lying parallel to, and at an angular distance of θ_{hkl}, outside, the principal lines.

The Kikuchi pattern is extremely useful because, like the DP spots, it displays the same symmetry as the sample. Major zone axes are defined by the intersection of two or more bands. As the sample is tilted, the DP spots remain fixed in position but change in intensity. When large changes in tilt are made, some spots will disappear while others appear at new positions when a new zone axis, with new $\mathbf{g}_{hkl}$ vectors, comes close to OA. However, while the DP is changing, the Kikuchi pattern

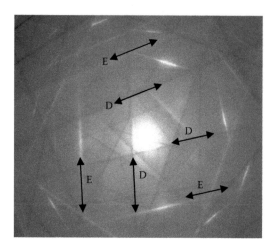

FIGURE 6.7   Kikuchi pattern from a sample of Inconel 600, showing lines with excess (E) and deficient (D) intensities.

intensity stays roughly constant and the lines move across the DP, as the sample is tilted, as though fixed to the sample. The Kikuchi pattern, therefore, can be regarded as a road atlas of the crystal orientation and is used to tilt the sample to specific zone-axis orientations or to select specific diffraction conditions for defect analysis, with an accuracy as high as 0.05 mrad.

The clarity of the Kikuchi pattern is degraded when the sample is bent or contains crystal defects. When such degradation occurs when using a SADA, it is better to remove the SADA and change the illumination to a convergent probe. The area of sample generating the DP is then greatly reduced, as is the range of bending, and it is then often possible to select a defect-free region.

6.3.2   DIFFRACTION CONTRAST IMAGING

DC imaging is considered to be a low-resolution imaging mode because the contrast arises from changes on a scale significantly larger than typical interatomic distances. The name HREM imaging is used to denote the imaging of the atomic structure in a crystalline sample and is considered separately in Section 3.4. While DC imaging can be used in thin samples under the so-called kinematic conditions where the undiffracted spot is much brighter than any diffracted beam, the lack of a Kikuchi pattern impedes accurate sample tilting. DC is more often applied in thicker samples than those used for HREM. Dynamic scattering occurs in thicker crystals such that the intensity in the diffracted beams is as strong, or stronger, than in the undiffracted beam, and where the Kikuchi pattern is visible. Scattering then occurs between the various diffracted beams because each becomes an effective illumination direction within the sample.

The most common type of DC image is the two-beam BF image, briefly mentioned in Section 2.5. Figure 6.3 shows how diffraction occurs from a single set of planes oriented exactly at the Bragg condition ($s = 0$, Figure 6.4a), i.e., when the diffracting planes are inclined at exactly the Bragg angle to the incident beam. At such an orientation and when no other set of planes lies close to its Bragg diffracting orientation, then the DP contains only two strong spots: the undiffracted spot and the spot from the diffracting planes. Figure 6.8a shows such a DP, where only one strong spot, g_{200}, is excited. The sample was too thin in this area to generate visible Kikuchi lines. The positions of the much weaker spots, present because of their extended relrods, still show the symmetry of a <110> zone axis, tilted a few degrees away from the incident beam direction around the <100> direction parallel to g_{200}. The BF and DF images in Figure 6.8b and c are formed by placing the CA around the lower

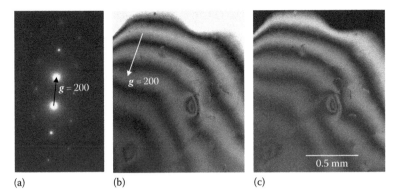

(a)   (b)   (c)

FIGURE 6.8   Thickness fringes in (a) BF, and (b) DF, images of Inconel 600 close to a two beam, $g = 200$ Bragg orientations. (c) Corresponding two-beam DP.

and upper spots, respectively, in Figure 6.8a. Note that the difference in the direction of g_{200} arrowed in Figure 6.8a and b is due to the rotation between image and DP mentioned in Section 2.5.

Under two-beam conditions, a perfect lattice of uniform sample thickness, t, will generate a BF or DF image of uniform intensity. If t varies, however, the intensities in the BF or DF images will increase or decrease in a complementary manner as a function of t. A wedge sample, therefore, will show intensity contours in both images [20,21] as in Figure 6.8. When the crystal is bent, however, the symmetric two-beam condition shown in Figure 6.3 occurs only along a line in the image and the intensity will then vary in an oscillatory manner on each side of this line to generate a bend contour. In practice, the two-beam BF image is often recorded with $s > 0$. This reduces the contrast from thickness and bending variations, thereby improving the visibility of microstructural features.

When a lattice defect such as a dislocation is present in the crystal, then some of the lattice planes in a small region around the core are distorted. Under two-beam conditions, a local change of intensity will be seen along the line of the dislocation core, projected onto the image plane, whenever a distorted set of lattice planes is diffracting. The magnitude of the defect contrast can be calculated using numerical calculations, the details of which can be found in standard texts [1,4]. The analysis of the dislocation Burgers vector, b, is performed by systematically recording a series of BF images, using a range of different g-vectors, to find those in which the dislocation is rendered invisible, i.e., to identify lattice planes that are not distorted by the defect. Invisibility requires that the vector product $g \cdot b = 0$ so, by finding three such noncoplanar g-vectors, a unique set of indices can be derived for b. In practice, it is found that a residual image can still be present when $g \cdot b = 0$ and then it is necessary to compare experimental images with calculated images [21]. This is especially so when analyzing the dislocations present in a grain boundary or interface [22] and specific examples of this are described in Chapter 9. An example of dislocation invisibility is shown in Figure 6.9 for a face-centered cubic nickel alloy. In Figure 6.9a, the dislocation line images labeled AA' have strong contrast when $g = 220$, whereas only residual contrast remains in Figure 6.9b for which $g = 113$, when $b = a/2(1,-1,0)$ and $g \cdot b = 0$.

6.3.3   WEAK-BEAM IMAGING

The width of a dislocation image in a two-beam BF or DF image is approximately $\xi_g/3$, where the extinction distance ξ_g, Equation 6.8, is a characteristic thickness difference between two adjacent intensity thickness contours in the BF or DF image at the exact Bragg orientation:

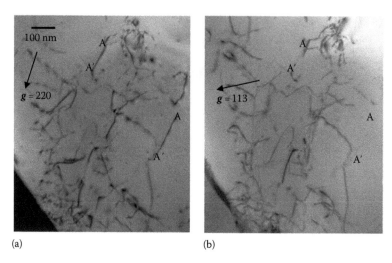

(a)   (b)

FIGURE 6.9 Dislocations in Inconel 600: (a) dislocations such as AA′ are clearly visible when g = 220, whereas (b) they show only very faint residual contrast in g = 113 because $g \cdot b$ = 0.

$$\xi_g = \pi V_c \cos \theta_B /(\lambda F_g) \qquad (6.8)$$

where

V_c is the volume of the crystal unit cell
θ_B is the Bragg angle
F_g is the structure factor for g

Calculation of ξ_g using atomic scattering amplitudes [23] at E_o = 100 keV yields values in the range 15–200 nm, increasing as $|g|$ increases and atomic number, Z, decreases, so that a dislocation will have an image line width of approximately 5–50 nm, severely limiting individual defect discrimination in many materials. However, ξ_g decreases with increasing $|s|$, such that the effective value, ξ_g^{eff}, is

$$\xi_g^{\text{eff}} = \xi_g /(1+|s|^2 \xi_g^2)^{1/2} \qquad (6.9)$$

This is exploited when weak-beam (WB) DF imaging is employed [24,25], the diffraction conditions for which are shown in the Ewald sphere construction in Figure 6.10.

For WB imaging, the sample is tilted about an axis normal to the selected g_{hkl} so that the Ewald sphere intersects close to a higher order spot, ng_{hkl}, along the systematic row in which g lies. In Figure 6.10, the $5g$ spot is shown intersected by the Ewald sphere, although n need not be integer, and the WB DF image is then recorded by placing the CA around a spot such as $+g$ (as shown in Figure 6.10) or $-g$, for which s is large ($>\sim 0.2$ nm^{-1}). Prior calculation is necessary to determine the value of n for a given λ and g suitable for generating a WB image. However, because the image intensity falls inversely with $|s|^2$, it becomes increasingly difficult to focus on the image. A dislocation image width, approximately $\xi_g^{\text{eff}}/3$, is then as small as ~ 1 nm and closely spaced dislocations are resolved much more clearly, as in Figure 6.11 [26]. Because the WB image is always a DF image, the dislocation cores are imaged as thin bright lines on a dark background. It is also important to use tilted illumination so that the DF spot used for imaging lies on OA and aberrations are minimized. This is indicated in Figure 6.10 by the incident electron wave vector, k_o, inclined to the optical axis AO. The WB image resolution is such that in many materials it is then possible to image the partial dislocations formed when perfect dislocations dissociate, to image the stacking-fault ribbon (labeled SF in Figure 6.11) between the partials, and even to estimate limits for the stacking-fault energy [27].

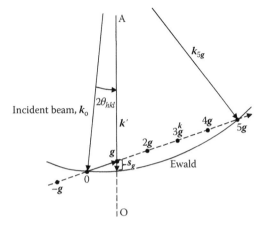

FIGURE 6.10 Ewald sphere construction of diffraction conditions used for WB DF imaging. For the conditions shown, the incident illumination has been tilted such that the $+g$ diffracted beam lies on the optical axis and the B image is formed by placing the CA around the $+g$ spot in the DP.

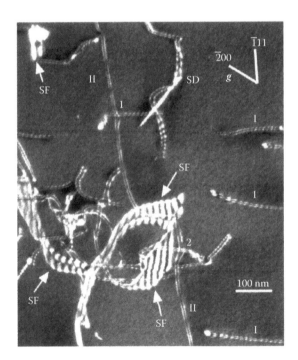

FIGURE 6.11 WB DF image of a $(NiFe)_3Ge$ alloy showing the dissociation into closely spaced partial dislocations (I, II) and stacking-fault ribbons (SF). (From Dirras, G.F. and Douin, J., *Phil. Mag. A*, 81, 467, 2001. With permission.)

6.3.4 HREM IMAGE FORMATION

HREM imaging provides a means for exploring crystal structures and imperfections on the atomic scale when the sample is tilted such that a low-order zone axis is closely parallel to the incident beam, and columns of atoms are viewed end-on by the beam. However, a direct correspondence between atom columns and image features does not generally occur. In particular, the contrast will vary with OL excitation (C_1), t, and the convergence angle, α, of the illumination. In order to interpret HREM image contrast, a comparison is usually necessary between a series of experimental images, recorded as C_1 is systematically changed by a known increment, with images generated by computer simulation for which the parameters listed above are also systematically varied. Even when an accurate match is obtained, it is commonly observed that the experimental values of image contrast are significantly lower (typically by a factor of roughly 3) than computed values [28].

In this section, the issues relevant to HREM imaging, both experimental and computational, are described. These include sample illumination, the propagation through the sample, the distortion introduced by OL aberrations and defocus, C_1, and the stability of the microscope power supplies.

6.3.4.1 Spatial Coherence

First, consider the nature of the illumination at the sample. It was noted earlier that the sample sees electrons as emanating sequentially in time from different points within the CL2 aperture, of angular radius q_0, and that the electron wave packets are spaced far apart in the z direction parallel to OA; thus, the probability of electron overlap in the sample is very small when using the nearly parallel illumination required for HREM. Any image contrast, therefore, must arise from self-interference within individual electron wave packets and, across the area of interest, the wave front must therefore be spatially coherent [29].

On the wave front, ψ_0, of any incident wave packet, a small path difference arises between two points in the sample, separation r, which increases with $|r|$. The incident wave is spatially coherent over $|r|$ provided that this path difference is negligible compared with λ. Clearly, electrons emitted at the periphery of the CA generate the greatest path difference across the sample. Hence, as q_0 increases, the fraction of electrons in the beam that coherently illuminates any small area of the specimen falls. The spatial coherence length, r_c, is a measure of the lateral distance over which the phase of the incident wave packet is coherent at the sample. For a wave packet with incident illumination angle, α, at the sample,

$$r_c = \lambda/2\pi\alpha$$

(6.10)

Even at very high magnification and with $q_0 \sim 0.5$ mrad, the field of view is usually larger than r_c because $q_0 > \lambda/2\pi r_c$, and the illumination is then partially coherent because any area of the sample will be illuminated by a mixture of both coherent and incoherent illuminations [30,31].

For atomic imaging by HREM through the interference of diffracted beams, it follows that r_c needs to be larger than the interatomic spacing for a significant fraction of the incident wave packets, so that the phase of the atomic scattering accurately contributes to the wave function at the exit surface, ψ_e. Atomic resolution is usually required over a field of view typically tens of nanometers in diameter. When the illumination is spread over the sample to become more parallel, α tends to zero and r_c increases. However, the intensity of illumination also falls as q_0 and α are reduced. The need for both spatial coherence and sufficient intensity of illumination for HREM imaging is readily achieved with an FEG source and even with LaB_6. (Although partial coherence is possible with a simple heated W source, the intensity of illumination is weak, so HREM imaging becomes more difficult.) Even though q_0 is small (~ 0.5 mrad or less), computer image simulation still requires that the CL2 illumination aperture area is divided into many small elements, with each element being treated as a parallel beam at a specific incident angle, α, to OA. A separate image calculation is then made for each element, and the contributions from all the elements added incoherently, weighted according to the intensity distribution across the CL2 aperture, to give a single image.

6.3.4.2 Electron Beam Propagation through the Sample

Although HREM is primarily concerned with imaging atomic structure, first consider propagation through a homogeneous continuum sample of thickness t, inside which the potential of the fast electron is uniformly raised by a few volts, ϕ_0, the mean inner potential, relative to vacuum. Within the sample, therefore, the electrons travel faster than in vacuum and the refractive index is increased. A plane wave incident on the sample will emerge with a path difference, χ, relative to the same wave traveling through vacuum. The phase difference, $2\pi\chi/\lambda$, can be derived from the dependence of λ on E_o, Equation 6.2, as

$$2\pi\chi/\lambda = \pi\phi_o t/\lambda E_o = \sigma\phi_o t \tag{6.11}$$

where $\sigma = 2\pi m e \lambda/h^2$ [1]. Hence, an incident plane wave of unit amplitude, given by Equation 6.3, exits with a wave function, ψ_e, of the form

$$\psi_e = \exp(-2\pi i\, \boldsymbol{k}\cdot\boldsymbol{r}) \cdot \exp(-i\sigma\phi_o t) \tag{6.12}$$

i.e., the phase of the wave is modified by the term $\sigma\phi_o t$. As only the variations of phase introduced by the sample are of interest, Equation 6.12 is usually abbreviated to

$$\psi_e = \exp(-i\sigma\phi_o t) \approx 1 - i\sigma\phi_o t \tag{6.13}$$

provided $|\sigma\phi_o t| \ll 1$. This is known as the weak-phase object approximation (WPOA) and assumes that the scattered intensity is much smaller than the incident beam intensity.

When a crystalline sample is viewed down a zone axis, then the phase change, $\phi_P(\boldsymbol{r})$, at a point, $\boldsymbol{r}$, in the exit surface of the sample will be determined by the integrated path difference, along the z direction in the sample. Thus, along the path AB in Figure 6.12, which passes close to an atomic nucleus represented by the black dots, $\phi_P(\boldsymbol{r})$ will increase with depth faster than $\phi_P(\boldsymbol{r})$ along the path CD, which passes between nuclei, as the path difference increases. The exit wave function, ψ_e, therefore, in the WPOA will vary with position $\boldsymbol{r}$ and Equation 6.12 can be rewritten as

$$\psi_e(\boldsymbol{r}) = \exp\left[-i\sigma\phi_P(\boldsymbol{r})\right] \approx 1 - i\sigma\phi_P(\boldsymbol{r}) \tag{6.14}$$

Equation 6.14 shows that $\psi_e(\boldsymbol{r})$ has two terms, the first term, i.e., "1," is the undiffracted beam amplitude, while the scattering term $\sigma\phi_P(\boldsymbol{r})$, which is complex and has both an amplitude and phase variation across the sample, contains all the information about the crystal structure. The DP is given by the FT of $\psi_e(\boldsymbol{r})$.

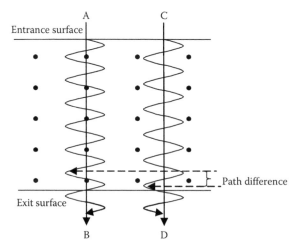

FIGURE 6.12 Wavelength variation and path difference as a function of position at a crystal zone-axis orientation. AB and CD denote electron paths.

As t increases, the WPOA breaks down because the scattered electron intensity in the diffracted beams is no longer small compared with the incident wave intensity and multiple elastic scattering occurs between the diffracted beams. In practice, the WPOA breaks down even when the sample has a low Z and is only a few nanometers thick. Computer simulation must then be used to model the wave propagation through a known crystal structure to calculate $\psi_e(r)$ [32–36]. Such calculations must be performed at discrete thicknesses within a likely range of t when comparing with experimental images.

6.3.4.3 Objective Lens Aberrations

It has been known for many years that round electromagnetic lenses suffer from aberrations that cannot be eliminated. When a plane wave is transmitted through an ideal lens, it is transformed into a spherical wave, represented by the solid curves in Figure 6.13 that converge to an exact focus, P, on OA. Hence, an ideal OL changes the phase along each ray path in an incident plane wave such that all the rays have identical phase at the point of interference, P. In a real lens, the converging wave front is no longer spherical but becomes increasingly distorted by various lens aberrations as α increases, indicated by the dashed curves in Figure 6.13. Different parts of the wave front converge to different points along OA. Additional, azimuthal, distortion arises through astigmatism, when the focusing field departs from the ideal cylindrical shape, and through coma, when the incident illumination is inclined to OA. The effect of the various aberrations (the wave aberration function or path difference), $\chi(\alpha)$, can be represented as a power series in α and the azimuthal angle, ϕ [37], i.e.,

$$\chi(\alpha) = A_0\alpha \cos(\phi - \phi_{11})$$
$$+ 1/2A_1\alpha^2\cos 2(\phi - \phi_{22}) + 1/2C_1\alpha^2$$
$$+ 1/3A_2\alpha^3\cos 3(\phi - \phi_{33}) + 1/3B_2\alpha^3\cos(\phi - \phi_{31})$$
$$+ 1/4A_3\alpha^4\cos 4(\phi - \phi_{44}) + 1/4S_3\alpha^4\cos 2(\phi - \phi_{42}) + 1/4C_3\alpha^4 + \cdots \qquad (6.15)$$

The coefficients corresponding to the different aberrations are defined as follows: A_1, first-order (twofold) astigmatism; C_1, defocus; A_2, threefold astigmatism; B_2, coma; A_3, fourfold astigmatism; S_3, star distortion; C_3, threefold spherical aberration, etc. ϕ is the azimuthal angle and the various angles ϕ_{ij} refer to the azimuthal values of reference planes.

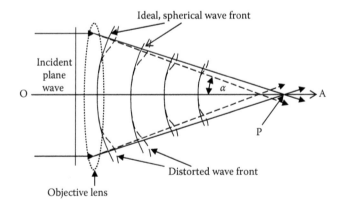

FIGURE 6.13 Distortion of a converging spherical wave front by objective lens aberrations. P is the focal point for undistorted spherical waves. Waves distorted by aberration come to a focus at other points on OA.

All microscopes are fitted with correctors to minimize first-order astigmatism, A_1, while coma, B_2, can be effectively eliminated by careful alignment of the illumination in a modern HREM. However, the defocus setting, C_1, of the OL is a user-controlled parameter that is optimally non-zero (see below) and, hence, is an important aberration. The other major aberration that affects χ is spherical aberration, C_3. Note that, although the IL and PL further magnify the image by several orders of magnitude, the aberrations in these lenses are not significant because electrons travel through the OL at much greater angles to OA than in the IL and PL. Consequently, the pole-piece bore and separation in the IL and PL can be much larger than in the OL and their aberrations still be insignificant.

Each spatial frequency of the crystal potential, $\phi_p(r)$, generates intensity in the DP at the corresponding reciprocal lattice position, K (Figure 6.4), which is transferred through the OL at its unique angle, α, to OA. Hence, the aberrations of the OL change the phase of each diffracted beam by an amount related to α and to the azimuthal angle of propagation, ϕ, before the beams interfere (Figure 6.13), thereby distorting the structural information in the HREM image. Because the phase varies with C_1, the image intensity at any point varies with OL focus. When computing images, it is necessary to know the value of C_3 and to vary systematically the value of C_1 for every value of α and t.

The phase distortion, $2\pi\chi/\lambda$, of a diffracted beam from the ideal spherical wave, described in terms of its scattering vector K, is given by [5]

$$2\pi\chi(K)/\lambda = \pi C_1 \lambda K^2 + \pi C_3 \lambda^3 K^4/2 \tag{6.16}$$

Constructive interference between the central, undiffracted beam and a scattering direction, K, at some value of C_1 will enhance the image contrast of that spatial frequency. Other spatial frequencies, however, may be destroyed by destructive interference in that image. Under the WPOA and the linear imaging approximation [38], whereby scattering between diffracted beams is assumed to be negligible, the contrast transfer function (CTF) is related to the gradient of the phase function given in Equation 6.16 by

$$CTF = \sin\left[(2\pi/\lambda)\cdot d\chi(K)/dK\right] \tag{6.17}$$

The effect of partial spatial coherence (Section 3.4.1) resulting from focused illumination with maximum semi-angle q_0 requires multiplication of the CTF in Equation 6.17 by an exponential function that attenuates the CTF at high K. The form of the CTF is then given by

$$CTF = \exp\left[-\pi^2 q_0^2 (C_1 \lambda K + C_3 \lambda^3 K^3)^2\right] \sin\left[(2\pi/\lambda)\cdot d\chi(K)/dK\right] \tag{6.18}$$

The CTF describes how the amplitude of the transmitted spatial frequency, K, depends on the defocus C_1, λ, and C_3. CTFs are shown in Figure 6.14 for a 200 keV TEM with C_3 of 0.5 mm, $q_0 = 1$ mrad, and at two defocus values, $C_1 = -35$ and -48 nm.

Astigmatism is assumed to have been corrected and all other higher order aberrations ignored. At very small K, CTF ≈ 0 but it then falls to -1 over a range, ΔK, before oscillating between $+1$ and -1. This first wide band, ΔK, in the curves in Figure 6.14, where the CTF ≈ -1, implies that ψ_e is faithfully transferred by the OL over ΔK, and that direct interpretation of the image intensity is possible for this range of K up to the point where the CTF first crosses the K-axis yields. This intersection point of K defines the point resolution, d ($d = 1/K$), given by [39]

$$d = 0.707\,(C_3 \lambda^3)^{1/4} \tag{6.19}$$

The corresponding value of C_1 is called the Schertzer defocus value and is calculated as $C_1 = -(C_3\lambda)^{1/2}$ [39].

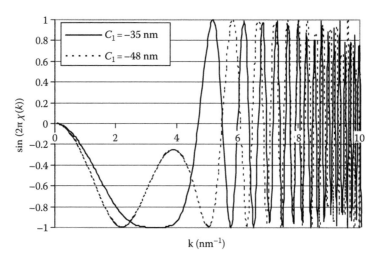

FIGURE 6.14 CTFs for two values of defocus, −35 and −48 nm, for $q_o = 1$ mrad, $C_3 = 0.5$ mm, and $E_o = 200$ keV.

In Figure 6.14, the curve for $C_1 = -35$ nm is close to Schertzer defocus and gives a point resolution of ~0.47 nm^{-1} (or ~0.21 nm). At higher K, however, contrast reversal occurs such that alternate passbands of K are transferred with reversed phase. Comparison of the curves in Figure 6.14 shows that the positions of these passbands have a sensitive dependency on C_1, implying that contrast reversal of image features are induced by small changes in C_1, preventing direct image interpretation at higher K. Equation 6.19 suggests that the resolution can be improved by reducing λ (i.e., by increasing E_o) or C_3 through OL design or C_3 aberration correction. A detailed discussion of the passbands can be found in Ref. [5].

6.3.4.4 Temporal Coherence

In addition to the inherent aberrations included in Equation 6.15, the image is further degraded by chromatic aberration and limited temporal coherence. Temporal coherence is determined by the mechanical and electrical instabilities in the microscope system that cause changes in C_1 during the image recording time. Such instabilities are always present and manufacturers are constantly trying to minimize fluctuations in power supplies and vibrational resonances. (Protection from stray external acoustic and electromagnetic sources is of prime concern in the selection of a site for a TEM.) Assuming that these instabilities occur independently, their effect in blurring the image is combined using the central limit theorem to give a defocus distribution $F(C_1)$ [40]:

$$F(C_1) = (1/\pi^{1/2}\Delta) \exp(-C_1^2/\Delta^2) \tag{6.20}$$

where,

$$\Delta = 2^{1/2} C_c \left[\frac{\sigma^2(V_o)}{V_o^2} + 4\frac{\sigma^2(I_o)}{I_o^2} + \frac{\sigma^2(E_o)}{E_o^2} \right]^{1/2} \tag{6.21}$$

There are two voltage-dependent terms in Equation 6.21: $\sigma^2(V_o)$ is the variance of the electronic fluctuations of the EHT power supply and $\sigma^2(E_o)$ is the variance of the natural spread of energy of the electrons from the source (given in Table 6.1) that are always present even if the EHT were perfectly stable; $\sigma^2(I_o)$ is the variance of the OL current; and C_c is the chromatic aberration coefficient of the OL. The factor 4 occurs because the OL focal length is inversely proportional to the square of the OL current. The effect of temporal incoherence is to attenuate the transfer of higher spatial frequencies, described mathematically by another exponential envelope function, E_t, given by [41,42]

$$E_t = \exp\left[-\pi^2 \Delta^2 (0.5\lambda K^2)^2\right] \tag{6.22}$$

The effect on the CTF curves of Figure 6.14, following multiplication by E_t, is shown in Figure 6.15. The temporal incoherence leads to much greater attenuation of the CTF at high K compared with spatial incoherence under normal HREM imaging conditions. The value of K at which the envelope function falls from unity to $1/e$ defines the information limit, an important parameter in HREM [5]. Even if the point resolution, d, Equation 6.19, could be made very small, the information limit will usually determine the minimum K that can be usefully recorded by HREM. In Figure 6.15, a value of $\Delta = 5$ eV has been assumed, consistent with $C_c = 1$ mm, and instabilities in power supplies of 1 part in 10^6. Great efforts are being made continually to reduce instabilities further (e.g., Ref. [43]).

Following careful alignment of the TEM column, correction of astigmatism, and orientation of the sample to a selected zone axis, the HREM experimental procedure is to record a series (10–20) of images while systematically varying C_1 over a range of values around Schertzer defocus. The incremental changes in C_1 are, typically, a few nanometers. The experimental images are then compared in detail with an appropriate series of computed images to look for correspondence. Figure 6.16 illustrates the need for careful C_1 adjustment. The four images of the same sample area are from a more extensive through-focal series from the complex oxide Nb16W18O94 projected along the [010] direction, recorded using a 300 keV FEG TEM (JEOL 3000F) [44]. Fine detail is observed to vary rapidly for focus changes of only a few nanometers; focus values are marked in the top left of each image (negative implies underfocus). Because there is clear, periodic structure in all images, every image contains structural information and it is impossible to state that any particular image is the one which is correctly focused. (Note that the sample is covered by a very thin amorphous film of contamination, which can be seen in each image at the edge of the sample.)

The expressions for the envelope functions of the CTF imposed by spatial and temporal coherence limits are applicable only in the WPOA and the linear imaging approximation. No scattering between one diffracted beam and another is permitted in this approximation. As t increases, scattering between diffracted beams increases and nonlinear imaging becomes increasingly important. The

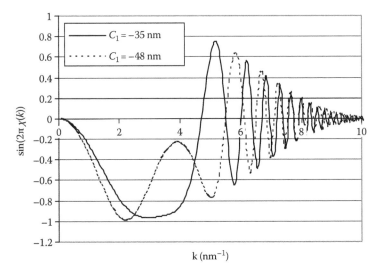

FIGURE 6.15 CTF curves in Figure 6.14 after multiplication by the temporal coherence term in Equation 6.22, using $\Delta = 5$ eV.

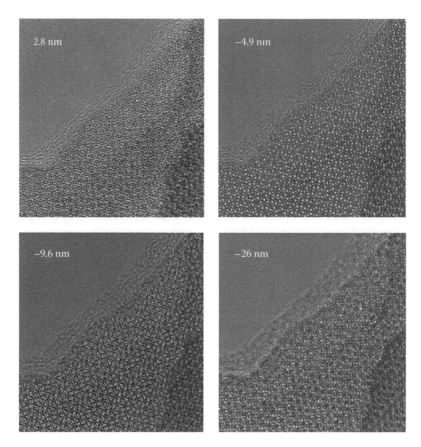

FIGURE 6.16 HREM through-focal image series of the complex oxide Nb16W18O94 projected along the [010] direction, recorded using a 300 keV FEG TEM (JEOL 3000F). Focus values are marked (negative underfocus) in the top left of each image. The fine detail changes rapidly for focus changes of only a few nanometers. (From Kirkland, A.I., unpublished data, 2007.)

envelope functions, Equations 6.18 and 6.22, are modified by nonlinear imaging such that the point resolution and information limit change [45].

6.3.4.5 Phase Problem

ψ_e is a complex function, Equation 6.14, and the phases of the K components, containing the crystal structural information, are distorted by the OL before the image, ψ_i, is formed. Moreover, the detector records only the wave intensity, $\psi_i\psi_i^*$, in the image plane, in which case all the phase information is lost. This is the phase problem of HREM.

It can be seen from Figure 6.15 that the point resolution, which defines the limit of the zero-order passband, ΔK, within which all spatial frequencies are transferred to the image with roughly the same phase and efficiency, varies with C_1, as do the higher passbands of K. One possibility for imaging a known crystal structure is to adjust C_1 so that all the spatial frequencies present at a selected zone axis lie in a single passband above the point resolution, or lie in different passbands that all have the same polarity in the CTF. Then, all K at that zone axis are correctly combined into the image, even though the point resolution is exceeded. The problem is that small focus fluctuations can easily reverse the phase of some of the higher frequency passbands because the passbands are relatively narrow. The method is, therefore, unsuitable for unknown structures and the information limit still imposes a fundamental limit that cannot be overcome by this approach.

Schiske [46] has suggested that it should be possible to recover both the phase and amplitude of ψ_e by recording and subsequently analyzing a series of HREM images at appropriate C_1 values, such that the whole range of K up to the information limit would be transferred strongly in at least one of the images. Other groups have developed algorithms to achieve this [47–49], and it is now routinely performed in a number of laboratories. Further advantages, using series of images for which both C_1 and α [50] are systematically changed, have also been demonstrated [51].

Equation 6.19 indicates that increasing E_0 and reducing C_3 increases the point resolution. However, there is a limit to the smallest C_3 that can be produced in an OL. An alternative approach has been to develop correctors, based on non-round lenses, to correct and eliminate aberrations up to third order, including C_3 [52–57]. The through-focal series reconstruction methods can still be used. The point resolution can be moved much closer to the information limit in a C_3-corrected TEM, as illustrated in Figure 6.17, in which the CTFs for C_3 values of 0.5 mm ($C_1 = -35$ nm) for a conventional OL, and 5 µm ($C_1 = -3.5$ nm) for a C_3-corrected OL, are compared for $\Delta = 5$ eV, $q_0 = 1$ mrad, and 200 keV. For the latter, the spatial coherence has been improved to the extent that the CTF oscillations have been completely suppressed, and the structure can be interpreted directly, out to the information limit $K \approx 7.5$ nm^{-1}.

The development of monochromation [58–62] for FEG guns provides a means of pushing the information limit to higher spatial frequencies by reducing $\sigma^2(E_0)$ in Equation 6.21. However, the development of chromatic aberration correctors [63,64] offers the most dramatic means of improving the information limit by reducing C_c, again in Equation 6.21. In combination with C_3 correctors, a point and information limit close to 0.05 nm should ultimately be obtainable, with acceptable image intensity, when C_1 can be corrected. Figure 6.17 also shows a CTF for $\Delta = 1$ eV to simulate such a microscope. At present, C_c correctors are still under development.

It is desirable to use a TEM that has both point resolution and information limits that allow direct interpretation of an HREM image, and that is particularly true when characterizing an interface on the atomic scale. The information limit of a TEM can be measured using the Young's fringe method [65] with an amorphous sample of metal, typically Ge or Au, evaporated on to a holey C support film. Two images of the same area are sequentially recorded with a small lateral displacement. The images are then added and the power spectrum is formed by computing the FT (readily performed online when a CCD detector and computer image storage are available). An example is

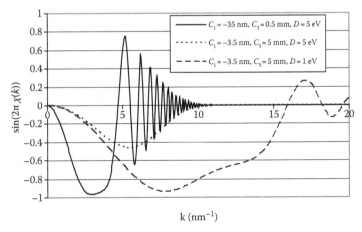

FIGURE 6.17   Comparison of CTFs. Full line: $C_1 = -35$ nm, $C_3 = 0.5$ mm, $\Delta = 5$ eV; dotted line: $C_1 = 3.5$ nm, $C_3 = 0.5$ mm, $\Delta = 5$ eV; and dashed line: $C_1 = 3.5$ nm; $C_3 = 0.5$ mm, $\Delta = 1$ eV. $E_0 = 200$ keV and $q_0 = 1$ mrad.

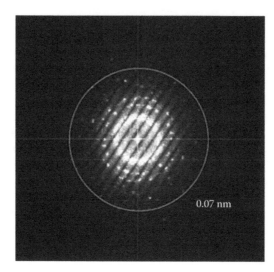

FIGURE 6.18 Young's fringe pattern from an amorphous carbon film, recorded using an aberration-corrected 200 keV TEM (JEOL 2200MCO) showing information transfer to 0.07 nm. (From Kirkland, A.I., unpublished data, 2007.)

shown in Figure 6.18 for an amorphous C sample using an aberration-corrected TEM (JEOL 2200MCO) at 200 keV with very stable EHT and lens power supplies. The Young's fringes have uniform intensity close to the center (the OA) but with gradually fading intensity as the distance from the OA increases. The superimposed circle marks the position of the 0.07 nm information limit, to which the fringes extend. The shape of the temporal coherence envelope is determined from the intensity attenuation with radius (proportional to K) in this power spectrum.

6.3.4.6 Implications for HREM Imaging of Interfaces

In order to apply HREM imaging to the study of the atomic displacements at a grain boundary, it is necessary (1) to orient the sample such that the axis about which one grain is rotated into the orientation of the other is parallel to OA and (2) that both grains are viewed along a low-index zone axis parallel to the rotation axis. For many general grain boundaries and interphase interfaces in metals, it is unlikely that the common rotation axis will coincide with a low-index direction. HREM imaging is limited to the study of simple tilt boundaries, a few special boundaries, and the interfaces between coherent and semi-coherent epitaxial layers. Fortunately, there are many important, nonmetallic, structures that contain such interfaces, including semiconductor devices, giant magnetoresistant structures, ceramic alloys, metal alloys, incommensurate minerals, etc.

6.4 SCANNING TRANSMISSION ELECTRON MICROSCOPY

In both DC and HREM conventional imaging modes described in Section 6.3, all parts of the sample are simultaneously illuminated by the electron beam, and the intensity in the image is integrated in parallel by the detector at each point (or pixel). STEM provides an alternative imaging method, which has both advantages and disadvantages compared with convergent beam TEM (CTEM), and many TEMs now offer a choice of both CTEM and STEM modes. There are also dedicated STEM instruments that allow imaging only in the scanning mode. The so-called HA-ADF STEM mode is an incoherent high-resolution image mode that is particularly powerful for imaging atomic structures as an alternative to conventional HREM. The same requirements and restrictions on zone-axis orientation and range of interfaces apply to both imaging modes.

6.4.1 STEM IMAGING

An STEM image is formed in a manner analogous to that from a solid sample in a scanning electron microscope (SEM). The incident illumination has the form of a cone of semi-angle of convergence q_o, defined by the diameter of the aperture of the probe-forming OL. The probe is focused to a fine crossover at the entrance surface of the thin foil sample, and then scanned as a raster over a selected area of the sample. In a modern STEM, the scanning is performed digitally such that the probe is held for a selected dwell time sequentially at each pixel. A detector on the exit side of the sample intercepts part of the transmitted intensity (cf. SEM) over the selected angular range determined by the detector geometry, and the electron current is integrated and stored at each pixel, sequentially in time, as the probe is scanned over the image area. Ideally, STEM requires an FEG to ensure sufficient intensity in a very small probe, Equation 6.1, for atomic resolution. When STEM facilities are interfaced with CTEMs with LaB_6 sources, resolution is limited to a few nanometers.

6.4.2 STEM INSTRUMENTATION

A typical imaging configuration for a dedicated STEM is shown in Figure 6.19. The electron probe is focused on, and transmitted through, the thin sample. A retractable fluorescent screen is inserted to view the DP and a retractable annular detector is positioned symmetrically around the OA to record the annular dark-field (ADF) image: the ADF detector collects electron intensity scattered over a range defined by its inner (θ_i) and outer (θ_o) cutoff semi-angles. Usually, there are no lenses between the sample and the ADF detector in a dedicated STEM, so the physical size and ADF detector position determine θ_i and θ_o. (When an STEM facility is incorporated into a CTEM, the ADF detector is located within the intermediate/projector lens system such that lenses between the sample and the detector can be used to change the effective values of θ_i and θ_o at a detector of fixed physical dimensions.) A separate, on-axis, disk detector is used to collect the BF and low-order diffracted beams up to a maximum semi-angle of $\beta (\leq \theta_i)$, determined by the collector aperture. When an EELS spectrometer is installed, the undiffracted beam passes through the spectrometer (described later in Section 5.2) before reaching the BF detector, as shown in Figure 6.19. The difference between ADF and HA-ADF imaging lies simply with the range of θ_i and outer θ_o, as discussed in more detail in the following section.

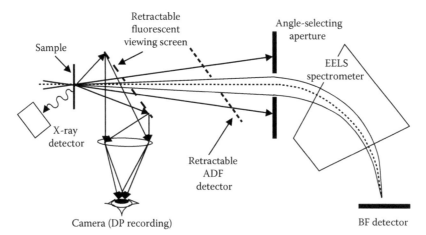

FIGURE 6.19 Dedicated STEM: the configuration of the probe, sample, detectors, EDX, and EELS.

STEM offers significant, and increasingly exploited, advantages over CTEM. These include the following:

- Image of higher resolution can be formed using incoherent HA-ADF than can be formed using coherent HREM under equivalent electron-optical conditions (E_o, OL aberrations, optimum focus, etc.) [66–70].
- More than one detector can be used to record simultaneous multiple images. Thus, BF and ADF electron images can be acquired together with chemical compositional maps, for example.
- Information using EELS, EDX, and diffraction can be acquired with a spatial resolution similar to the probe diameter using a stationary probe.
- No changes in lens excitation are required when changing from a scanning (imaging) mode to an analytic (stationary probe) mode. Although, in principle, in a CTEM, it should be possible to switch from an imaging mode to a static-focused probe at a selected point, in practice, it is always necessary to make minor adjustments to stigmators, C_1, and probe position due to hysteresis in the power supplies. Such adjustments can be time consuming and increase damage and contamination.

Potential disadvantages of STEM compared with HREM include the following:

- Very slow image recording times because pixels are recorded sequentially.
- Image distortions and drift over extended acquisition times limit the accuracy of atomic column positions, for example, at interfaces and around crystal defects.
- Greater difficulty in performing DC imaging and working with DPs in STEM.

6.4.3 STEM IMAGING MODES

The formation of the CTEM image and DP was shown in Figure 6.2 using simple ray optics. In the STEM, the incident illumination is a cone (Figure 6.20) and every diffracted beam emerges as a cone. The axes of the cones lie at $2\theta_B$ with respect to the axis, OA, of the forward scattered beam. Figure 6.20a shows just one diffracted beam such that $q_o < \theta_B$ and the diffraction disks do not overlap. However, when $q_o > \theta_B$ (Figure 6.20b), the disks overlap. A ray from point E in the incident cone can reach point Q in the disk overlap region, without diffraction, passing through the point P in the

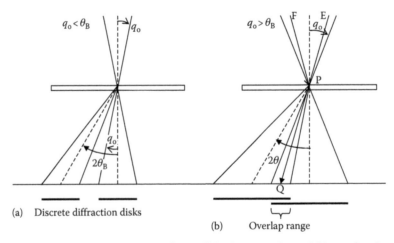

(a) Discrete diffraction disks

(b) Overlap range

FIGURE 6.20 DP disk geometry when (a) $q_o < \theta_B$ and disks do not overlap and (b) $q_o > \theta_B$, where disk overlap and coherent STEM BF imaging is possible.

sample, while a ray from point F in the incident cone can be diffracted through $2\theta_B$ at P and then travel to Q. If the source disk is coherent, then the two rays traveling along PQ will interfere. As the probe position, P, is translated across columns of atoms in a thin sample, to generate a scanned image, the phase of the diffracted ray at Q will change, depending on the probe position within the unit cell of the crystal. However, this would not be observed if β were made very large (say $\pi/2$) such that the BF detector always collected all the transmitted intensity.

In practice, the DP from point P in a thin sample will contain many disks, with overlap between adjacent disks. Also, β is 10–20 mrad and q_0 is ~10 mrad, so the BF detector integrates the intensity over relatively few regions of disk overlap, together with the areas within disks where there is no overlap. As the probe is scanned, the intensity in the coherent overlap regions integrated by the BF detector changes with the periodicity of the projected crystal potential, and an HR image is generated, i.e., the fraction of signal lost to the BF detector varies as the probe is scanned. The principle of reciprocity [71,72] shows that a CTEM HR image will be identical to the corresponding STEM BF image provided (1) the beam convergence angle in CTEM is equal to the STEM BF detector angle, β, and (2) the STEM probe convergence angle, q_0, is equal to the angular diameter of the OL CA used to form the CTEM BF image. This is true both for elastic scattering and for inelastic scattering with small energy losses, such as TDS. In practice, HREM imaging in CTEM uses an illumination angle q_0 of ~0.5 mrad and a CA that allows many diffracted beams to contribute to the image. When an STEM BF β of only 0.5 mrad is used, the image is extremely noisy and, consequently, STEM BF is rarely used for HREM. It is important to stress that, for application of the principle of reciprocity, the STEM BF HREM image must have the same spatial and temporal coherence as the corresponding HTEM image and, therefore, depends on defocus, aberrations, and power supply instabilities as described in Section 3.4.4. Contrast reversals in the image will still occur as the C_1 changes and the limitations of the WPOA will apply (Section 3.4.2).

It follows that a DF image, $I^{DF}(r)$, formed by collecting the signal lost to the BF detector with an ADF detector (Figure 6.19) that extends from β to $\pi/2$ will be complementary to the BF image, $I^{BF}(r)$, i.e.,

$$I^{DF}(r) = 1 - I^{BF}(r) \tag{6.23}$$

Contrast will, therefore, still arise by coherent interference between adjacent diffraction cones. Coherent STEM ADF lattice fringes were first demonstrated by Cowley [73]. However, the intensities of diffracted beams fall with increasing angle and, above a few tens of milliradians, the scattered signal arises largely through quasi-elastic TDS, manifest in the Kikuchi pattern in the CTEM DP (Figure 6.6). Such scattering is believed to be largely incoherent. The principle of reciprocity requires that a large STEM collection angle corresponds to a large illumination angle in CTEM. In Section 3.4.1, it was noted that the coherence of the CTEM illumination falls as the CL2 aperture angle, q_0, increases and, for very large angles, the spatial coherence distance, Equation 6.10, will eventually become too small for coherent HREM. Hence, for an ADF detector with collection limits of θ_i and θ_0, the DF image will become increasingly incoherent as θ_i increases. Typically, when $\theta_i > $ ~40 mrad, corresponding to the TDS vibrational amplitude [74], it is possible to form an ADF image that is essentially incoherent and this is known as an HA-ADF image. Such images have low intensity, even when θ_i is very large (~200 mrad).

When a cold FEG is used, an OL aperture of semi-angle 10 mrad can be filled with coherent illumination, and the probe intensity distribution at the sample can simultaneously be made very small by appropriate demagnification of the source. The probe wave function, $\psi_p(r)$, and intensity, $I_p(r) = \psi_p(r)^* \cdot \psi_p(r)$, are mathematically defined by the FT of the aperture disk function, $A(\alpha)$, convoluted by the OL phase function, $\chi(\alpha)$, i.e.,

$$\psi_p(r) = \text{FT}\left[A(\alpha) * \chi(\alpha)\right] \quad \text{and} \quad I_p(r) = |\psi_p(r)|^2 \tag{6.24}$$

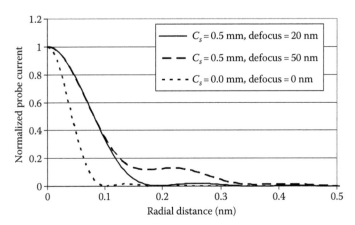

FIGURE 6.21   Theoretical STEM probe current distributions for $E_o = 200$ keV, showing the effects of Δf and C_s correction.

$A(\alpha) = 1$ if $\alpha < q_o$, and $A(\alpha) = 0$ if $\alpha > q_o$. Three theoretical probe current distributions are shown in Figure 6.21 for 200 keV electrons, and these take the approximate shape of a Bessel function, with a large central peak and weaker oscillations as r increases. The two curves for which $C_s = 0.5$ mm and $q_o = 8$ mrad demonstrate the way in which the tail in the distribution changes as the defocus changes from 20 to 50 nm, while the third curve is for an aberration-corrected STEM where C_s and Δf are zero and $q_o = 15$ mrad. The benefits of C_s-correction are obvious; even though q_o is almost doubled, giving an increase of almost four times the current in the probe, the FWHM (full-width half-maximum) of the central peak is almost halved. The optimal probe shape approaches a Gaussian distribution for $C_1 = -0.75(C_3\lambda)^{1/2}$ and $\alpha = 1.27(\lambda/C_3)^{1/4}$ [75]. It must be stressed, however, that such probe current distributions are modified immediately when the probe enters the sample, leading to channeling of intensity along atom columns at zone-axis orientations. Although the HA-ADF signal intensity is often described as having a power law dependence on atomic number, Z, of the form Z^n, where $1.7 < n < 2$ [67], in thicker alloy samples the simple relationship between intensity and Z no longer applies for atomic resolution. Channeling and scattering between the atom columns occur, leading to more complex dependence on both location and crystal thickness [76,77]. Currently, there is much interest in atomic imaging by combining STEM with EELS to use chemically specific signals generated by inner shell ionization. Such experiments will be complicated because recent calculations show that it is possible, when the probe is positioned on a specific atom column, to cause ionization in adjacent atom columns that might contain completely different atoms [78–81].

The incoherent image is defined by the convolution of $I_p(r)$ with the object function, which can be represented as an array of scattering objects at the positions of the atom columns (when the sample is oriented with a zone axis parallel to OA). A detailed analysis [5,82] suggests that the form of the object function depends on θ_i. The image is then an array of bright spots close to the positions of the atomic columns, provided the probe-forming aperture is filled with coherent illumination. However, unless the image is completely incoherent (i.e., $\theta_i > \sim 3q_o$), the apparent separation of adjacent atom columns can be inaccurate. Kirkland [35] has written a computer code to calculate the STEM ADF image using an alternative, multi-slice approach to the Bloch wave analysis of Ref. [82].

The first atomic images in STEM were reported by Pennycook and Boatner [83]. An example of an HA-ADF image is shown in Figure 6.22 for a GaAs sample at a <110> zone axis [84]. The great advantage of the incoherent HA-ADF image is that there are no contrast reversals with focus; the image can be interpreted directly with confidence, unlike the CTEM HR image, i.e., the optical transfer function does not have the reversals shown in Figures 6.15 and 6.17 but tends to decay out to the information limit, and the point resolution is extended to higher K compared to the HREM image obtained using the same E_o, C_3, and C_c.

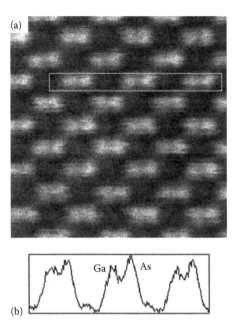

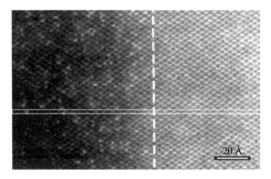

FIGURE 6.22   HA-ADF image from a GaAs sample at the <110> zone axis. (From Nellist, P.D. and Pennycook, S.J., *Ultramicroscopy*, 78, 111, 1999. With permission.)

The first images of single atoms in the electron microscope were obtained using a STEM with $E_o = 30$ keV by displaying the image as the ratio of the inelastic to elastic signals [85,86]. The atoms were those of U, i.e., with high Z, on a low-Z supporting film. Such work stimulated much of the subsequent development of STEM and FEG. The ratio method of Crewe is not now commonly used because it is relatively straightforward to generate HA-ADF images of atom columns in many crystal structures. However, the imaging of low-Z atoms remains a challenge, as is the imaging of a single substitutional atom in a lattice of another element. The imaging of Sb atom complexes and even single Sb atoms in Si, Figure 6.23 [87], has demonstrated the power and potential of HA-ADF imaging. The application of HA-ADF imaging to interfaces is described in Chapter 9.

FIGURE 6.23   HA-ADF STEM image showing the locations of atomic columns in a <110> cross section through an undoped Si substrate, to the right of the dashed line, and a Sb-doped epilayer. The sample thickness increases from left to right. The atom column intensity is almost constant in the substrate but, in the doped layer, some columns appear much brighter than neighboring columns, consistent with the presence of one or more heavier Sb-dopant atoms located in such columns. (From Voyles, P.M., Grazul, J.L., and Muller, D.A., *Ultramicroscopy*, 96, 251, 2003. With permission.)

6.5 CHEMICAL ANALYSIS METHODS

6.5.1 COMPOSITIONAL ANALYSIS BY EDX

EDX [88,89] has been used to determine chemical composition in the TEM and STEM for at least three decades. Although EELS [3] is a more sensitive method and can give more detailed information at higher spatial resolution than EDX, the relative simplicity of EDX still ensures that it is the method of choice for simple compositional measurement for many materials scientists using TEM. In this section, the principles of the EDX method are outlined, together with its advantages, limitations, and potential pitfalls. Examples of its detailed application to the measurement of compositional changes at interfaces and grain boundaries are described in Chapter 9.

6.5.1.1 Principles of EDX

When a fast electron passes through the cloud of bound electrons around an atom in a TEM sample, it can lose energy and generate two kinds of x-ray. If the atom is ionized, then it can return to an equilibrium state by emitting either a characteristic x-ray or an Auger electron. (The analysis of surface composition using Auger electrons is described in Sections 3.2.2 and 3.3.3.) The probability of X-ray emission is called the fluorescent yield, ω ($0 < \omega < 1$), the value of which varies with Z and ionized shell. For example, ω is greater for L-shells than for K-shells and increases with Z in a nonlinear manner. These x-rays have precise energies, E_X, that are characteristic of the atom within which they were generated and are emitted isotropically from the sample. Alternatively, the fast electron can transfer energy to the atom without causing ionization, yet the energy is still emitted in the form of an x-ray. Such a continuum, or Bremsstrahlung, x-ray has energy, E_B, between 0 and E_o. Continuum x-ray emission is anisotropic, being strongly peaked in the forward direction.

In electron probe microanalysis (EPMA) [89,90] of a bulk sample, E_X is measured using x-ray diffraction by a large single crystal. Spectral accumulation over a range of E_X is sequential in time, although it provides greater sensitivity and much better energy resolution than EDX. The EDX method is based on the detection and energy measurement of individual x-rays, whatever the order of detection, and was developed from the same technology used by the nuclear industry to measure gamma radiation. A full description of the EDX method can be found in reference texts [88–90] so only the broadest outline is given here. Currently, the only type of x-ray detector interfaced to (S) TEM instruments is the Li-drifted Si type; a few Ge crystal detectors have been supplied and the same operational principles and analytical procedures apply. The detector is cooled by liquid nitrogen to reduce artifacts from electronic noise.

The detected x-ray undergoes complete absorption in a Si single-crystal detector by the photoelectron process. The emitted photoelectron has a limited range in the Si and soon loses energy through further ionization events involving successively less tightly bound electrons, finally resulting in the generation of electron–hole (e–h) pairs. On average, one e–h pair is produced for approximately every 3.8 eV of E_X or E_B. An electric field of about 200 V/mm is applied across the Si crystal to separate the e–h pairs, and the resulting charge pulse is amplified, shaped, and measured by sophisticated electronics. This whole process must be concluded in time to be repeated when the next x-ray arrives and hence the time constant of the amplifier determines the maximum measurement rate. If a second x-ray is absorbed too soon, then the system will detect it and prevent it from interfering with the measurement of the first. However, the second x-ray will not be measured, and its absorption delays the return of the amplifier to its quiescent state in readiness for the measurement of the x-ray after that. The greater the frequency of overlap then the larger the dead time the system spends in a state of unreadiness. The microscopist can vary the x-ray generation rate by changing the electron probe current or, possibly, the sample thickness. A dead time <25% is recommended. The total time for analysis, τ, is selected to allow accumulation of enough counts in the

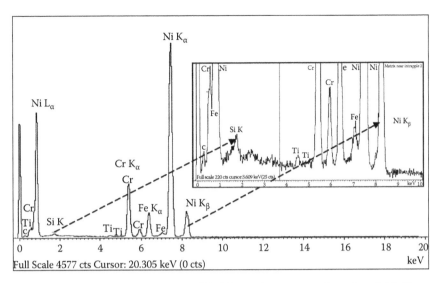

FIGURE 6.24 EDX spectrum from Inconel 600. The main spectrum from 0 to 20 keV contains the characteristic peaks of the main constituent elements, Ni, Fe, and Cr. The inset shows an enlarged section of the range 0–10 keV, enhancing the visibility of the Si peak, and showing the typical shape and noise of the continuum.

spectrum for analysis with the required level of precision. The user must base this on previous experience and appropriate calculation. The accuracy of the analysis will depend on the care taken with the calibration of the system and measurement of a number of parameters that are described below.

A typical spectrum is shown in Figure 6.24 from a sample of Inconel 600 containing Fe, Cr, and Ni as the major alloying elements, with smaller concentrations of Si and Mo. The range of the energy scale is from 0 to ~20 keV, divided into 1024 channels of 20 eV/channel. Characteristic peaks from the major elements have a peak height much larger than the continuum intensity and are readily visible. The peak from Si is much smaller and noisier, as seen in the enlarged inset section between 0 and 10 keV. The continuum, also noisy, has a maximum intensity typically between 1 and 2 keV and then falls gradually as the energy increases. The width of the characteristic peaks is a measure of the detector resolution. Although a specific characteristic x-ray has a natural linewidth less than 1 eV, the measured spectrum peak-width is much larger, reflecting the stochastic processes that cause different numbers of e–h pairs to be generated by x-rays of the same energy. Hence the peak width increases with x-ray energy because the variance of the numbers of e–h pairs increases with E_X or E_B. The capacitance and electronic noise of the detector and amplifiers also contribute to the peak width but, as this cannot be known before fabrication, the resolution and price of similar detectors can vary considerably. A good-quality detector will provide an FWHM resolution of ~60 eV at the C K peak at ~0.28 keV and an FWHM value about 130 eV at the Mn K_α peak at about 5.9 keV.

The shape of the continuum is the first feature that the microscopist should always examine when performing EDX. The actual generation of the continuum x-ray intensity is predicted to rise steeply as the x-ray energy falls [91,92] and is roughly proportional to Z^2. However, the detector has an energy-dependent detection efficiency factor, $\varepsilon(E_X)$. Low energy x-rays are always strongly absorbed within the sample, in the protective window in front of the detector, and in the Au contact on the front surface of the detector, so the intensity always falls to zero at $E_X = 0$ eV. Within the energy range 3–10 keV, almost all x-rays are absorbed within the Si crystal, so $\varepsilon(E_X) \approx 1.0$. As the energy increases further, there is an increasing probability that the x-ray will pass right through

TABLE 6.2

Example of Measured Detector Efficiency Factors at Several Peak Energies for a Windowless Detector

L-line	Energy (keV)	Efficiency, ε	K-Line	Energy (keV)	Efficiency, ε
Fe	0.70	0.53	Fe	6.40	1.00
Ni	0.85	0.66	Ni	7.48	1.00
Ge	1.19	0.82	Ge	9.88	1.00
Mo	2.29	0.97	Mo	17.48	0.81
Sn	3.44	1.00	Sn	25.25	0.45

Source: Paterson, J.H., PhD thesis, University of Glasgow, 1988.

the detector crystal and be lost to the analysis, so $\varepsilon(E_X)$ begins to fall. Table 6.2 lists the values of $\varepsilon(E_X)$ measured for a windowless detector interfaced to an FEG-STEM at the K_α and L_α x-ray energies of several elements, to illustrate the magnitude of the variation [94].

There should, therefore, be a maximum in the measured continuum intensity, the position of which depends on (1) the type of detector window and (2) the extent of self-absorption of x-rays within the sample itself. The peak is usually between 1 and 2 keV and moves to higher energies for Be windows compared to ultrathin polymeric windows or windowless detectors, as the self-absorption (i.e., sample thickness) increases. However, the peak will also move to higher energies when there is a strong contribution from stray x-rays generated by thick material in the vicinity of the sample, or when there is not a complete line of sight between the electron probe position and the whole of the detector window.

6.5.1.2 Protocol for EDX Spectrum Acquisition

Although it is apparently very straightforward to acquire an EDX spectrum, high accuracy in quantitative analysis is difficult to achieve. The following protocol should alert the potential analyst to some of the many potential pitfalls:

- Detector should be reconditioned within the recommended period prior to use, otherwise preferential absorption of low energy x-rays can occur and k-factors used for quantification can be erroneous.
- Liquid nitrogen dewar should contain sufficient coolant to last longer than the duration of the experiment.
- Detector and amplifier should be under power for several hours before use in order to ensure stability of the pulse-processing electronics, amplifiers, and the energy scale in the spectrum.
- Low-background holder and sample-retaining screw made from Be should be used to minimize spectral contributions from x-rays generated outside the sample by stray scattered electrons.
- Ensure that the sample and holder are both clean, ideally by plasma cleaning, before insertion into the TEM. Wherever possible, ensure that the thin regions of the sample to be analyzed are close to the center of the holder and remote from the supporting cup or grid bars, which can generate x-rays by stray scattered electrons.
- Ensure that the probe size is small enough to provide the desired spatial resolution. (In a TEM, it is possible to image the probe directly. This is more difficult to ensure in a STEM.)

- Select the analysis position and first acquire a spectrum from a nearby location.
- Check that the dead time is ≤25% and, if too high, reduce the beam current or the probe size.
- Confirm that the sample is not significantly damaged or contaminated during the trial analysis. If it is, then try using a smaller beam current or spread the beam over a larger area (if the desired spatial resolution permits this). If contamination persists, then it might be reduced by flooding the analysis area with a high intensity of electrons for several minutes. To do this, retract the detector from the TEM column, select the largest probe size and largest CA and defocus the illumination to spread the beam over a diameter of 10–20 μm. Illuminate for ~5 min before returning the TEM to the normal analysis conditions.
- Examine the shape of the continuum to ensure that no gross absorption is present. If it is, then tilt the sample holder further toward the detector and repeat until it seems that a clear line of sight is present between the point of analysis and the detector. If the sample is supported on a grid, then ensure that the analysis region is not shadowed by a grid bar. When analyzing an interface, this might be difficult because an additional geometrical constraint must be also met, i.e., to keep the interfacial plane parallel to the optic axis.
- Commercial analysis packages assume that the sample is homogeneous and amorphous. Hence, if the sample is crystalline at the point of analysis, it is important to ensure that the OA is not parallel to a low-index zone axis and is not strongly exciting any diffracted beams, i.e., that a kinematic diffraction condition operates, rather than dynamic scattering. Strong electron channeling operates when dynamic scattering is present and this can result in large variations in acquired counts [95,96]. In addition, false, coherent, Bremsstrahlung peaks can arise in the spectrum, which cause inaccuracies in the analysis of a number of elements, if present in small concentrations [97,98].
- Ensure that the characteristic peaks are accurately located at the correct positions on the energy scale. (Some systems require the collection of additional spectra using a range of counting rates to calibrate or check this.)
- If possible, collect a spectrum with the probe positioned in the center of a large hole near the analysis point using the same probe configuration to be used for the actual analysis. If the hole-count spectrum contains appreciable counts, then check that the CA and SADA are fully retracted and the hard x-ray aperture, if fitted on the TEM, is inserted and accurately centered. A badly positioned hard x-ray aperture can generate large stray x-ray signals instead of removing them.
- Collect spectra as the probe is moved from a hole and into the sample, noting the distance over which the counts increase from zero to a maximum. When this distance is much larger than the probe diameter, then there is significant intensity in an extended tail of the probe that will affect accuracy when high spatial resolution is required. A possible solution is to reduce the probe convergence angle by changing the CA diameter.
- Collect and store the required spectra.
- Perform experiments to measure the sample thickness and tilt with respect to the detector for use in correction for self-absorption of x-rays in the sample.
- Collect spectra from other samples to be used as standards for quantification.

6.5.1.3 EDX Spectrum Processing

The calculation of elemental concentrations from an EDX spectrum has been described in several references [99]. The number, $N_i^a(\text{ion})$, of ionization events in shell i, in a sample composed of one pure element, a, can be calculated from

$$N_i^a(\text{ion}) = (i/e)\,\tau\rho^a N_A t\sigma_i^a / A^a \tag{6.25}$$

where

    i is the probe current

    e is the electron charge

    τ is the analysis time

    t is the sample thickness

    ρ^a is the sample density

    N_A is the Avogadro's number

    σ_i^a is the total ionization cross section of shell i in element a at the given electron accelerating voltage

    A^a is the atomic weight

Only a fraction of these events, $\omega_i^a \, a_i^a$, generate characteristic x-rays, where ω_i^a is the fluorescent yield and a_i^a is the partition function [100–103], i.e., the fraction of the x-rays that arise in the specific sub-shell, α, β, etc., chosen for analysis. If all the lines in a sub-shell are combined for the analysis, then $a_i^a = 1$. Only those x-rays emitted into the solid angle subtended by the detector, Ω, have a chance of measurement, so the fraction of such x-rays is $\Omega/4\pi$. Finally, the number actually measured is further reduced by the relevant detector efficiency factor, $\varepsilon(E_i^a)$, for the specific x-ray energy. Hence, the expression for the number of counts actually detected, I_i^a, becomes

$$I_i^a = (i\,/\,e)\tau\rho^a N_A t\sigma_i^a \, \omega_i^a \, a_i^a \, \Omega\varepsilon(E_i^a)/(A^a 4\pi) \tag{6.26}$$

In a binary alloy containing elements a and b, with weight fractions c_a and c_b, the number of x-rays from element a is given by the expression in Equation 6.26 multiplied by c_a. An equivalent expression gives the number of x-rays collected from shell j of element b:

$$I_j^b = c_b(i\,/\,e)\tau\rho^b N_A t\sigma_j^b \, \omega_j^b \, a_j^b \, \Omega\varepsilon(E_j^b)/(A^b 4\pi) \tag{6.27}$$

The two expressions finally yield the relationship between the weight fractions and the collected counts:

$$c_a/c_b = (k_{ab}I_i^a)/I_j^b, \quad \text{where } k_{ab} = \left[\sigma_j^b \, \omega_j^b \, a_j^b \, \varepsilon(E_j^b)/A^b\right] \Big/ \left[\sigma_i^a \, \omega_i^a \, a_i^a \, \varepsilon(E_i^a)/A^a\right] \tag{6.28}$$

Knowledge of the k-factors, k_{ab}, also known as the Cliff–Lorimer factors [104,105] permits quantification of spectra.

Spectra can be processed to yield a chemical composition using either a standardless procedure or using standard samples to determine the k-factors, k_{ab}. For both methods, it is necessary (1) to identify the characteristic peaks present and (2) to remove the underlying continuum to extract the number of counts, I_i^a, I_j^b, etc., in the peaks. Automated routines are provided by system manufacturers to remove the background, which can be problematic when peak overlap occurs. Digital spectrum filtering or fitting of theoretical or experimentally recorded standard peaks to the spectrum are usually employed.

The use of k-factors measured from standard samples of known composition provides the most accurate results provided suitable standards are available. Standards must be homogeneous on the nanoscale, be stable, and not damaged or oxidized. Moreover, appropriate correction may be required for self-absorption (see below) that can be tedious to compute. Unlike EPMA, where many bulk standards can be stored inside the microscope column and readily accessed and analyzed at precisely known geometry, sample exchange in the (S)TEM is slow and analysis is tedious. Hence, standard spectra acquired on the same or a similar (S)TEM at an earlier time are sometimes used. Standardless analysis requires calculation of the k-factors from first principles using values for σ, ω, a, ε, and A, the last of which presents no problem. Values for ω and a are available in the literature

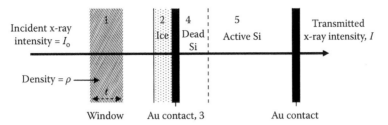

FIGURE 6.25 Schematic section through a detector showing the absorbing layers that have to be penetrated by an x-ray.

[100,101], and either parametric expressions (National Bureau of Standards [NBS] special technical publication)] or numerically computed values [106,107] are used for σ. The detector efficiency, ε, is perhaps the most difficult parameter to calculate, especially for low x-ray energies where it varies quickly and is sensitive to the precise thickness of detector components. The absorbing layers in the detector that influence the efficiency are shown in Figure 6.25 and include (1) the window, (2) possible ice and contamination, (3) the Au contact in the front of the detector, (4) an inactive (dead) layer of Si just below the Au contact, and (5) the active Si crystal.

The thickness, density, and mass absorption coefficient (MAC) (a function of absorbing material Z and x-ray energy) must be known for all five layers of Figure 6.25. The detector efficiency is then computed as

$$\varepsilon(E_X) = I/I_0 = \left\{1 - \exp[-(\mu/\rho)_5 \rho_5 t_5]\right\} \cdot \Pi \exp[-(\mu/\rho)_i \rho_i t_i] \tag{6.29}$$

Here, the multiplication operation, denoted by Π, is for the factors of layers 1–4, and the parameter $(\mu/\rho)_i$ is the MACs for the x-ray energy in the absorbing material in layer i. Parameterized expressions for MACs for the whole periodic table can be found in the literature [108]. Detector manufacturers can provide data on construction materials and thicknesses of the layers (apart from layer 2), which, hopefully, can be eliminated by periodic conditioning of the detector. In practice, the system software should include appropriate expressions for the detector efficiency for use with standardless analysis.

Whether measured or standardless k-factors are used, the EDX system software derives the composition of the sample at the probe position, along with the standard error of each element. If the sample is thick or the angle between the plane of the sample and the direction to the center of the detector crystal (the takeoff angle) is small, then a correction may be necessary for the absorption of x-rays within the sample. This is especially important when one element in the sample, present in a significant concentration, strongly absorbs x-rays of another element. The expression in Equation 6.26 for the detected x-ray intensity, I_i^a, must be scaled by the factor

$$\left[1 - \exp(-x_i)\right]/x_i \tag{6.30}$$

where $x_i = (\mu/\rho)_i \rho_i y$ and $y = t/(\sin\theta + \cos\theta\cos\phi\tan\beta)$ [109]. The term $(\mu/\rho)_i$ is the effective MAC of the x-ray by the sample itself. The geometric parameters θ, ϕ, β, and the distance y are shown in Figure 6.26. The upper surface of the wedge sample is inclined at an angle β from the horizontal plane down its slope of steepest descent. The probe travels along TB, the local sample thickness, t, within the sample, generating x-rays uniformly along the trajectory. Depending on the generation depth, the x-rays have to travel through different distances before emerging from the sample. Figure 6.26 shows the distance, y (=BE), to be used in Equation 6.30; it equals the distance traveled by x-rays generated at the point B. The direction of the x-ray detector is at an angle θ above the horizontal and at an azimuthal angle, ϕ, to the plane containing TB and the normal to the plane of steepest descent.

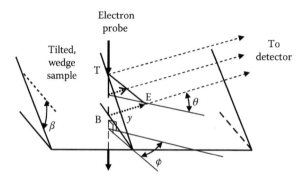

FIGURE 6.26 Geometric parameters θ, ϕ, β, and y required for the self-absorption correction.

As the mass absorption $(\mu/\rho)_i$ depends on the sample composition, whose measurement is required, an initial calculation of composition is performed without correction for self-absorption. This first estimate is used to generate MACs for the relevant x-rays in the specimen, for example, $(\mu/\rho)_a^{\text{spec}}$ for element a, defined in Equation 6.31, allowing a first-absorption correction to be made:

$$(\mu/\rho)_a^{\text{spec}} = c_a(\mu/\rho)_a^a + c_b(\mu/\rho)_a^b + c_c(\mu/\rho)_a^c + \cdots \tag{6.31}$$

where
    c_i is the estimate of the concentration of a
    $(\mu/\rho)_i^j$ is the MAC of the x-ray from element i by element j

In practice, the analyst is required to measure values for t, θ, ϕ, and β at each analysis point, and to store details of the detector structure and geometry in the computer, which also stores values for MACs in look-up tables.

6.5.2 ELECTRON ENERGY-LOSS SPECTROSCOPY

EELS is a very powerful and versatile technique when coupled to either a TEM or STEM [3]. The following kinds of information can be found using EELS:

- Chemical composition: point, line profile, or map
- Valence and electronic bonding
- Local foil thickness
- Dielectric function
- Mapping of oxidation states and variations in bonding
- Atomic site location of impurities in ordered crystals (ALCHEMI) [110]
- Measurement of the bandgap of semiconductor materials [111]

In this section, the basic instrumentation and implementation of EELS techniques are presented. Detailed examples of applications to the characterization of interfaces appear in Chapter 9. The spatial and energy resolution limits in EELS have recently been reassessed by Egerton [112].

6.5.2.1 EELS: Principles and Instrumentation

EELS provides information from the distribution of energy loss, ΔE, by the electrons in the beam during interaction with the atoms in the thin foil sample. There are two types of spectrometer: the

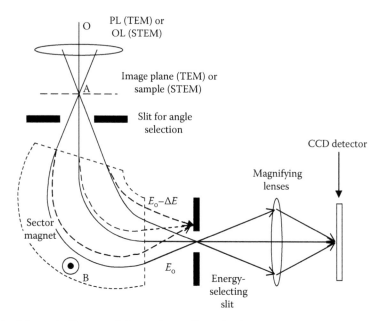

FIGURE 6.27 Schematic diagram of electron beam dispersion in a sector magnet spectrometer. B shows the direction of the magnetic field.

in-column Ω-filter and the post-column sector-magnet filter. Both types of spectrometer generate EELS spectra for ΔE up to ~3 keV, above which the intensity falls to very low levels over the angular range over which the signal can be collected. At the present time, the post-column sector-magnet systems made by Gatan Inc. are more common [113,114]. The sector-magnet system can be fitted to both the TEM and the dedicated FEG-STEM, and the basic design is illustrated schematically in Figure 6.27.

The incident electrons are scattered and undergo energy losses, ΔE, in passing through the sample. An aperture limits the angular range of the scattered intensity that passes into the dispersing magnetic field with a direction normal to the plane of this page. All electrons are deflected along a curved trajectory within a drift tube located inside the sector magnet. Figure 6.27 shows how the electrons with energy E_o (solid lines), and electrons with energy-loss ΔE (dashed lines), are focused by the sector magnet. An energy-selecting slit is inserted to limit the transmitted range of energy loss when forming an energy-filtered TEM image (EFTEM). The slit is removed and the magnifying lenses adjusted when the spectrum is viewed. The CCD array detector, of size 1 K $\times$ 1 K pixels or larger, is located inside the vacuum and is bombarded directly by the high-energy electron beam. The integrated charge at the CCD is periodically read into a computer memory for display as a spectrum or image, and successive readouts are added to achieve the required statistical accuracy, typically for times of up to tens of seconds. A spectrum can be shifted along the direction of dispersion by application of a voltage to the drift tube, in order to choose a range of the spectrum for investigation. Electromagnetic lenses control energy dispersion at the detector. Second-order aberrations in the magnetic field are reduced by shaping the entrance and exit surfaces of the sector magnet and further alignment and astigmatism correction is achieved by incorporating additional compensating coils that are not shown in Figure 6.27.

The simple spectrometer in Figure 6.27 is positioned below the standard viewing chamber in a TEM, so that the viewing screen can be used to select the area for analysis and then retracted to allow passage of the beam into the spectrometer. In a dedicated FEG-STEM, where there are no imaging lenses, the spectrometer is placed after the ADF detector and post-sample deflection coils (Figure 6.19). BF imaging is achieved using all, or part of, the forward-diffracted intensity that has passed through the spectrometer. The spectrum is acquired with the beam focused on a selected point in the STEM (or TEM operated in a STEM mode) or with a more parallel beam in a TEM.

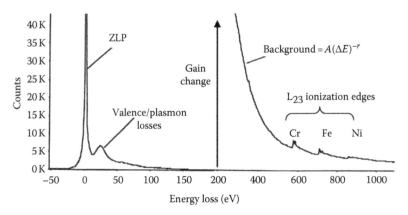

FIGURE 6.28 EELS spectrum showing the ZLP, plasmon loss, gain change, and ionization edges containing fine structure (ELNES).

A typical EELS spectrum is shown in Figure 6.28. A large zero-loss peak (ZLP) marks the origin of the energy-loss scale, and the width of the ZLP is a measure of the energy resolution of the combined spectrometer/microscope system. The highest probability of any electron losing energy is by phonon excitation, which, at ambient temperature, requires ~0.025 eV energy transfer to the lattice and this is too small, currently, to be detected by spectrometers interfaced to EMs. Hence, electrons contributing to the ZLP include electrons that have suffered both zero loss and phonon losses. Apart from the ZLP, most of the spectral intensity is concentrated in the first few tens of electron volts, often peaking in the range 15–30 eV. This is the plasmon-loss region where the energy losses arise through cooperative excitations of loosely bound electrons, and ionization of valence electrons. As the sample thickness increases, the ZLP intensity falls, while the chance of multiple plasmon losses increases, leading to overlapping plasmon-loss distributions that decay into a fairly smooth intensity distribution as ΔE increases further. In a few materials such as Al alloys, the plasmon peaks are much narrower (in the range 1–2 eV FWHM).

Superimposed on the decaying background are ionization edges at values of ΔE corresponding to characteristic electron ionization energies of the elements that compose the sample. At high-energy resolution, many edges are observed to contain fine structure, and the precise value of ΔE at the onset of the edge can shift over a few electron volts, depending on the valence state of the excited atom, and the nature and configuration of the surrounding atoms.

Many studies have been reported in which experimental fine structure has been compared with modeling predictions in attempts to understand the local atomic environment and interatomic bonding, and this is becoming increasingly important for studying atomic arrangements at interfaces. For a fast electron to contribute to an ionization edge, it must impart energy to a bound electron in the sample, such that the bound electron is transferred to an empty state above the Fermi level. Hence, the fine structure at an edge is proportional to the product of the distribution of the density of vacant states and the probability of an electron transition from the occupied inner-shells to empty states, initiated by the fast electron. This product must then be convoluted by the spectrometer/detector response function and the energy spread of the incident electron beam to generate the theoretical spectral details.

In theory, the energy resolution in the spectrum is determined by a number of factors. Clearly, one limit is imposed by the physical separation of the diodes in the array detector and the dispersion. However, the dispersion can be made as large as necessary to overcome this constraint. The low-order aberrations in the spectrometer have now been largely overcome by improvements in design so that, for a small angular range of electrons close to the OA, it is the energy spread in the incident electron beam (see Table 6.1) that then limits energy resolution in many instruments. The advent of monochromators [58–60] overcomes even this constraint, so that the fundamental limit imposed by the natural line width of the electron transitions during ionization is now achievable with a few sophisticated EMs. Currently, a ZLP resolution of ~0.15 eV is possible with adequate intensity for spectroscopy.

In practice, however, many EELS experiments are still performed using an LaB_6 or FEG source without monochromation, and in spectrometers in which higher order aberrations are still present. Also, it is not always possible to eliminate the degrading effects of stray fields and vibrations, which can change on a daily basis due to factors beyond the control of the analyst. An energy resolution around 1 eV is typically achievable and acceptable for some of the purposes listed at the beginning of this section. In addition, there is usually a compromise to be made between energy resolution and SNR (signal-to-noise ratio). Because the electrons undergo momentum change during energy loss, they are scattered over a range of angles from the OA. As the collection angle at the entrance to the spectrometer increases, the signal intensity will increase. However, the resolution will degrade due to distortion by higher order aberrations, which increase with angle.

6.5.2.2 Protocol for Collection of EELS Spectrum

It is assumed that the microscope site is adequately protected from stray fields, mechanical vibration, and earth loops. During spectrum acquisition, no movement of metallic objects, such as chairs or noise generation, e.g., talking, should occur.

- Ensure the sample is clean because EELS spectra are very sensitive to C contamination. Select a region of the sample for use during alignment close to the region to be analyzed.
- Decide which coupling mode is to be used: the image mode requires a DP on the viewing screen and the diffraction mode an image on the screen [115]. The image mode is more accurate for compositional analysis. The diffraction mode might be more convenient for analyzing a very small particle using a convergent beam. The projector lens crossover located at the exit aperture of the final PL is close to the object point for the sector magnet (Figure 6.27). In image mode, there is a very low magnification image close to the crossover and in diffraction mode, there is a very low magnification DP close to the crossover (hence the names of the operating modes).
- Angular scattering range at the sample, β, collected by the spectrometer is selected using one of the interchangeable spectrometer entrance apertures located below the final viewing screen, together with adjustment of the DP camera length. An alternative method is to use the CA in the OL to limit β.
- β should be measured for later calculation of partial differential scattering cross sections. When the OL CA is used, β can be measured directly from a DP. When the spectrometer aperture defines β, then an estimate is obtained by dividing the aperture radius by the distance from the aperture to the crossover point in the final PL.
- Chosen spectrometer entrance aperture must provide the desired energy resolution in the spectrum. A sector magnet is equivalent to a lens. A larger aperture implies a larger angular range of electrons passing through the sector magnet and a greater influence of aberrations on the spectrum resolution. The best energy resolution, therefore, is achieved only when using the smallest available spectrometer entrance aperture (unless the image of the OL CA diameter is smaller at the spectrometer entrance aperture), but the signal intensity is reduced.
- Sample orientation should ideally be set to avoid strongly diffracting conditions when measuring composition, otherwise ordered alloys can generate orientation-dependent compositional data. When strongly diffracting conditions are unavoidable or chosen deliberately, as for a zone-axis orientation, then the angle-defining aperture should be symmetrically located with respect to the spots in the DP [116].

- Incident illumination convergence angle, α, at the sample should be recorded. The theory underlying the conversion of an ionization edge intensity counts ratio into an atomic concentration ratio assumes that α is zero (i.e., parallel illumination). When $\alpha \approx \beta$, then a significant correction factor might be necessary to ensure compositional accuracy [3].
- Align the spectrometer. Nearly all sector-magnet systems have been provided by Gatan Inc. and aligned during initial installation to ensure that the OA of the microscope closely matches that of the spectrometer. However, further small adjustments are essential at the beginning of each session, and then should be checked periodically throughout an experiment. The procedure should be described in the user manual.
- Fine alignment is initially performed using the ZLP formed through a hole in the sample. Stigmators and compensators are adjusted to optimize the sharpness of the ZLP such that the ZLP does not move along the energy scale as the focus control is changed over a wide range about the optimum setting. The ZLP position on the CCD array is selected by deflection coils and major offsets are made by adjustment of the voltage on the drift tube. The sample is then moved under the beam.
- Ensure that the ZLP does not saturate the CCD during the data acquisition period by reducing the acquisition period, or the beam current, or both.
- Trial spectrum should be examined to ensure that there are no unexplained artifacts present, such as regions where the slowly changing background intensity rises with increasing ΔE, rather than falling. The cause of any such artifact should be found, which might be difficult and even demand cessation of any further analysis until it can be rectified.
- Energy scale must be accurately calibrated. The energy scale is determined by the dispersion of the sector magnet and subsequent magnification (in $\mu m/eV$), selected by the user, before projection on the CCD array. The channel energy width in the spectrum is determined by dividing the physical spacing of the pixels by the dispersion. Calibration is performed automatically by the system software following the identification of two known edges, or the ZLP and one known edge, in a spectrum. (Note: Calibration can be performed post-acquisition on stored spectra.)
- Choice must be made about which ΔE is to be focused optimally. Although the initial alignment checks are most easily performed on the ZLP because it is so intense, ionization edges will then be progressively more blurred as ΔE increases. When energy-loss near-edge structure (ELNES) information of a particular edge is important, then the spectrometer should be focused on the selected edge rather than on the ZLP. The ZLP will then be blurred.
- Because the intensity of most ionization edges is so much lower than that of the ZLP, it is common practice to offset the energy scale of the spectrum so that the ZLP is not recorded. Long acquisition times are then feasible for ionization edges without saturation of any spectral channels. The analyst must decide on the extent of the offset voltage applied to the drift tube. Some further experimentation with acquisition times might then be necessary in order to optimize data recording at the edges of interest.
- Sample is then moved to enable recording of spectra from the required areas. Sufficient time must be allowed for any sample drift to subside and, if the CCD has recently been exposed to a ZLP, for any afterglow in the CCD to decay.
- With an older spectrometer, it might be necessary to record a dark current spectrum, with the beam deflected away from the spectrometer, to measure the background noise in the spectrum. Modern systems should automatically record such data with every spectrum acquisition.
- Spectrum containing the unsaturated ZLP and the adjacent low-loss region of the spectrum should be separately recorded. This spectrum can later be used for deconvolution of the multiple scattering contribution in the spectra containing the ionization edges, using the Fourier-ratio method [117,118].

- Each individual pixel of the CCD array has an individual gain efficiency. Ideally, a calibration of the CCD pixel response should be made for correction of spectra. If such a calibration is not performed automatically, then the user manual should provide information on how to measure and apply this scaling correction.

6.5.2.3 Sample Thickness Measurement Using EELS

The sample thickness can be determined from the low-loss region of the EELS spectrum. The single scattering distribution (SSD) is the energy-loss distribution arising from an infinitely thin sample such that no electron undergoes more than one inelastic event. Almost all the SSD intensity occurs within ~100 eV of the ZLP because plasmon, single electron excitations, and ionization of weakly bound core electrons with energy losses of only a few tens of electron volts have much larger cross-section (i.e., probabilities) than for ionization of the more tightly bound core electrons. The total inelastic scattering mean free path, λ_{inel}, is the average distance traveled by a fast electron in the sample between any two inelastic scattering events. The low-loss region of the EELS spectrum from a very thin sample displays the SSD. As the sample thickness, t, gradually increases from zero, intensity in the ZLP is gradually transferred to the SSD. However, the probability of multiple scattering increases with t, and the low-loss region then includes contributions from multiple convolutions of the SSD. The probability of a fast electron undergoing multiple scattering is given by the Poisson distribution, from which it can be shown that the probability of an electron suffering no inelastic scattering, P_o, is given by

$$P_o = I_o/I_t = \exp\left(-t/\lambda_{inel}\right) \tag{6.32}$$

where I_o and I_t are, respectively, the intensity in the ZLP and in the total EELS spectrum, including the ZLP. Even when multiple scattering is present, almost all the total spectrum intensity lies within the first 100 eV, so that integration of counts in the ZLP and up to 100 eV provides experimental values of I_o and I_t. t/λ_{inel} can be therefore be calculated rapidly. Figure 6.29 illustrates, schematically, the integration ranges for the ZLP and the low-loss region. Provided λ_{inel} is known, t can be derived.

The simple theory assumes that all the emergent electron intensity is collected in the spectrum. In practice, collection is limited over an angular range, β, as already described earlier, so that a high proportion can be lost by high-angle elastic scattering (diffraction). It is reasonable to assume that the lost intensity is divided equally between the ZLP and the inelastic intensity, so that Equation 6.32 remains valid. However, λ_{inel} then becomes a function of β, as well as E_o and atomic number Z. An empirical formula for λ_{inel} as a function of β, E_o, and Z [119] has been shown to generate values of λ_{inel} with an accuracy of ~10% over a wide range of the three parameters, and can be used with Equation 6.32 to yield values of t that are sufficiently accurate for many purposes.

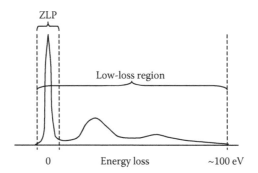

FIGURE 6.29   Schematic illustration of the energy ranges of I_o (ZLP) and I_t (low-loss region) for calculation of t.

It is recommended (cf. EELS protocol) that measurements of t/λ_{inel} be made whenever EELS experiments are performed. When $t/\lambda_{\text{inel}} \geq 0.5$, the results will contain multiple scattering contributions that can distort ELNES peaks and lower the accuracy of compositional measurements unless they are removed by deconvolution [117,118].

6.5.2.4   Measurement of Chemical Composition by EELS

EELS provides an alternative to EDX for compositional measurement. The combinations of elements present in the sample often determine which method is preferable. EDX is generally suitable for quantitative analysis of all elements with $Z \geq 10$ (x-ray energy ≥ 1 keV) even when peaks overlap; most of the k-factors lie within a factor of 3, and thicker samples are more readily analyzed. EELS, however, has a huge advantage for low-Z analysis. Although elements such as B, C, N, and O have large σ_K, their very low x-ray fluorescent yields, ω_K, high absorption in both sample and detector window, and the small solid angle of collection lead to very low EDX detection rates. EELS is not affected by ω_K or absorption in the sample, and a high fraction of events is detected, so that these low-Z elements, and even He, Li, and Be are very easily detected and quantified. Count rates for many edges are orders of magnitude larger than for peaks in EDX so that precision and detection limits are generally much better in EELS [3].

Ideally, a single spectrum should be acquired containing all peaks of interest. However, when the ionization edges are too widely separated, it might be necessary to acquire consecutive spectra over different energy ranges, with consequently increased chance of changes due to drift, damage, and contamination, or to changes in beam current and electron-optical parameters. Ideally, any multiple scattering should be deconvoluted to yield the SSD; however, for $t < \lambda_{\text{inel}}$, failure to deconvolute is unlikely to degrade accuracy by more than a few percent.

The background intensity, B, must first be determined and subtracted from under each ionization edge. The simplest method of background removal is to fit a power law of the form

$$B = A(\Delta E)^{-r} \tag{6.33}$$

where A and r are constants. This is applicable only over regions where B is varying smoothly, so that the low-energy region of the spectrum ($\Delta E \leq 100$ eV) and regions on the tails of preceding ionization edges must be avoided. Even then, in practice, r increases as ΔE increases, so that the fitting function is only approximate. Figure 6.30 shows a section of an EELS spectrum from a stainless steel sample displaying the L_{23} edges of Cr, Fe, and Ni. The variation of intensity in the energy window A has been used to extrapolate a background under the Fe and Ni edges, which has then been subtracted to generate the lower spectrum. The intensity in a window such as B is then used to quantify the Fe content of the sample.

More complex methods of background fitting to improve accuracy have been assessed in Ref. [120]. Subtraction of B reveals the ionization edges that extend over the range $I_i^A < \Delta E < E_0$, where I_i^A is the ith shell ionization energy for element A. For quantification, the edge intensity, $S_i^A(W, \beta)$, is found by integration over a selected window of width, W. The same procedure is repeated for all edges of interest. The edge intensity, measured in this way, is proportional to the probability that the fast electron is scattered through an angle no larger than β while losing energy in the range I_i^A to $I_i^A + W$, namely the partial differential cross section, $\sigma_i^A(\beta, W)$. As with EDX, it is usual to measure ratios of elemental concentration, so that the atomic ratio c^A/c^B of two elements, A and B, is given by

$$c^A/c^B = S^A \sigma_j^B (\beta, W) \big/ \left[S^B \sigma_i^A (\beta, W) \right] \tag{6.34}$$

The validity of this formula has been discussed in Ref. [3] and some of the assumptions relating to crystalline sample and probe convergence were covered in Section 6.5.2.2.

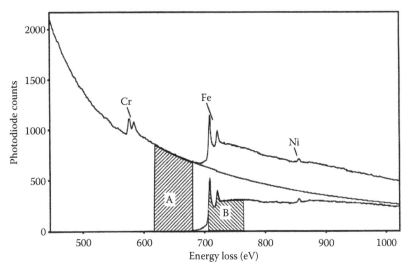

FIGURE 6.30 Background extrapolation and subtraction under the Fe-L$_{23}$ edge in a spectrum from a stainless steel sample.

6.5.2.5   Ionization Edge Shape and ELNES

Spectrometers typically transfer electrons that have undergone losses of no more than 3 keV. Within the energy range 0–3 keV, most elements have ionization edges that can be detected by EELS [121], although the intensities fall rapidly as ΔE increases. Edge shapes can be grouped broadly into different types:

- Type 1. Hydrogenic-like, with a large step at the onset, followed by a gradual decay (K-edges for all elements)
- Type 2. A sharp peak, or peaks, at the onset, superposed on a hydrogenic-like edge (L$_{2,3}$-edges for K–Cu and Rb+)
- Type 3. A delayed edge showing a gradual rise from the onset and then a gradual decay (L$_{2,3}$-edges for P–Ar, Zn–Br, and most M- and N-edges)

An energy resolution of ~1 eV in EELS spectra is often sufficient to reveal small shifts in the onset energy, and changes in the fine structure related to differences in the unoccupied density of states (DOS) with coordination and bonding. Good experimental agreement has been obtained between XANES, using synchrotron irradiation, and EELS edges, for crystals with known structure. Figure 6.31 illustrates how the shape and fine structure of the O–K edge varies within iso-structural transition metal chromites [122]. Changes in the ratio of the L$_2$/L$_3$ peak intensities are sensitive to the valence state of 3d-transition elements [123], and the ELNES in many materials varies with orientation of the beam with the crystal sample [124].

Theoretical ELNES structure can be predicted for periodic crystal structures by self-consistent calculations using density functional methods [125] to derive the DOS, both occupied and unoccupied. The Wien2K code [126] is suitable for large crystal structures as it is based on a band-structure approach. However, this requires super-cells to be constructed in the computer from large numbers of atoms, which is computationally demanding. An alternative approach, also using self-consistent density functional theory, but requiring much less computing power, is to use a multiple scattering code such as FEFF8 [127,128] to derive the equilibrium configuration of an atomic cluster and its associated DOS. As the cluster size is increased, good agreement with experiment can be achieved [129]. Once

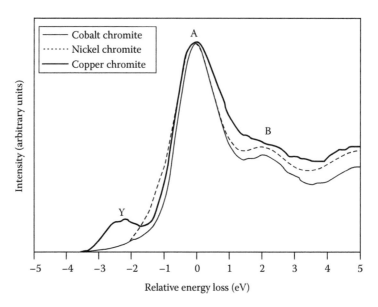

FIGURE 6.31 Changes in the peak shape and ELNES fine structure at the O-K ionization edge in different iso-structural transition metal chromites. (From Docherty, F.T., Craven, A.J., McComb, D.W., and Skakle, J., *Ultramicroscopy*, 86, 273, 2001. With permission.)

the DOS has been calculated, the theoretical EELS spectrum is generated from the product of the probability of the excitation of occupied states below the fermi level to a specific unoccupied state and the square of the appropriate transfer matrix elements. Convolution is then required with an appropriate function to account for the spectrometer resolution and the spread of energy in the electron beam.

Figure 6.32 illustrates the experimental ELNES Mg-L_{23} ionization edge in MgO with corresponding calculated fine structure. There is excellent qualitative agreement between the two with respect to the presence and positions of all the major spectral peaks [130], in which all the observed fluctuations in the ELNES are reproduced with high accuracy by the calculations. Self-consistent approaches using density functional theory are essential to predict accurately the ELNES close to the edge onset. A number of less rigorous calculations based on non-self-consistent calculations have been reported over a number of years, but the level of agreement with experiment has generally been poorer.

As computing power increases, it is becoming feasible to model the ELNES measured from the disordered region of an interface. In addition to generating the DOS, such calculations will also yield the most energetically favorable positions of the atoms. For special interfaces, such as tilt boundaries and coherent interphase interfaces, the atomic structural information derived from HREM and HA-ADF images can be directly used as input into the calculations.

6.5.2.6 Energy Filtered Imaging in the TEM

So far, the description of EELS has concentrated on spectroscopy using the sector-magnet spectrometer. The analyzer system has been developed further by Gatan into the Gatan imaging filter [113] by the inclusion of additional lenses between the spectrum plane and the detector, as shown schematically in Figure 6.33. The object plane of the spectrometer is shown as a DP plane, where the spots contain electrons with a range of energy losses determined by the sample thickness and composition. The direction of the field in Figure 6.33 is out of the plane of the page. Inside the sector magnet the electrons are dispersed, the deflection increasing with the energy loss. The width and position of the energy-selecting slit determines the range of energies used to form the final image.

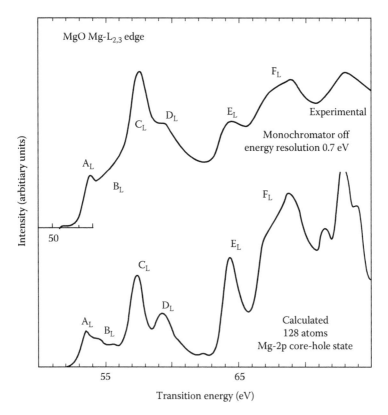

FIGURE 6.32 Comparison of ELNES experiment and theory for the Mg-L$_{23}$ edge in MgO. (From Mizoguchi, T., Tatsumi, K., and Tanaka, I., *Ultramicroscopy*, 106, 1120, 2006. With permission.)

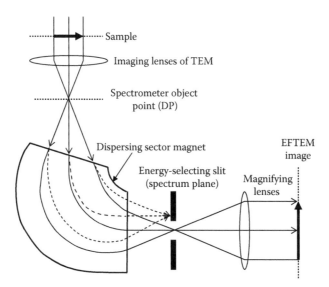

FIGURE 6.33 Schematic of an imaging filter illustrating how the paths of electrons with two specific energies differ in the magnetic field (the direction of which is out of the plane of the page) and how the slit controls the energy range of an EFTEM image.

The dashed lines represent the paths of electrons with energies lower than those of electrons with ray paths represented by the solid lines.

In practice, the slit position is fixed and E_o is changed to form images with selected energy loss. The EELS computer can be programmed to record an image for each selected incremental value of E_o, thereby generating a series of energy-loss images. For a specific image from a selected ionization edge, three energy-loss images are usually recorded, two with energies below the identified ionization edge and one with energy just above the edge. For each corresponding pixel in the images, the intensities in the two images below the edge are used to calculate and subtract the background, B, from the pixel intensity in the image above the edge [3]. When the sample thickness is constant, the intensity at that pixel is then proportional to the atomic concentration at that pixel, so that an elemental map is generated. A thickness map can also be made from two further energy-loss images recorded from (1) the ZLP and (2) the low-loss region, including the ZLP, from which t/λ_{inel} is calculated at each pixel using Equation 6.32. The elemental maps can then be normalized to remove the effects of any t variation.

Figure 6.34 shows an EFTEM example from an aged Fe–Cu alloy in which small spherical Cu-rich precipitates are formed in the Fe matrix. Several Cu precipitates with diameters of only a few nanometers are imaged in both the Cu-L_{23} and Fe-L_{23} maps [131]. Precipitates appear as bright objects in Figure 6.34a because they are Cu-rich, and as dark objects in Figure 6.34b because they are Fe-deficient. Intensity profiles along the diameter of an object were shown to correspond to a sphere, and the diameter of the sphere, calculated from the profile, was equal within experimental error to the diameter measured independently, from the total Fe-deficient signal of the whole object for radii >~3 nm. Below this size signal broadening prevented accurate measurement from line profiles. However, blurred images of particles with radii as small as 1 nm, measured from the total Fe-depletion, were still detected [131].

The energy slit width, W, and image acquisition time, τ, must be adjusted to collect enough counts to generate images with suitable SNR. Typically, τ is of order 10 s, even when the illumination at the sample is focused to a relatively small area. An LaB$_6$ source generates images with higher SNR than an FEG except at the very highest magnifications. Although it is possible to generate EFTEM images with W about 1 eV in order to select specific features in the ELNES [132–134], the SNR is very low and the smaller energy spread in the FEG beam cannot readily be exploited. When a series of many images is acquired, a 3D EFTEM data-cube can be recorded (x, y, ΔE). An alternative way of generating such data is to use an FEG-STEM with a conventional sector magnet so that a full spectrum is recorded at every pixel with both the highest possible spatial and energy resolutions

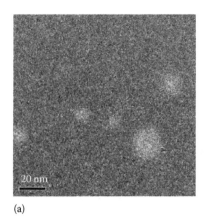 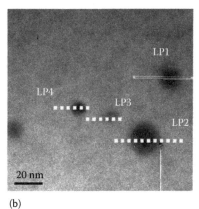

(a)   (b)

FIGURE 6.34 EFTEM images of spherical Cu-rich precipitates in Fe. (a) Cu-L_{23} map and (b) Fe-L_{23} map, in which Cu particles show reduced intensity. (From Lozano-Perez, S., Titchmarsh, J.M., and Jenkins, M.L., *Ultramicroscopy*, 106, 75, 2006. With permission.)

available. This approach is called electron spectroscopic imaging (ESI) [135]. Systematic back-ground subtraction is then performed for all peaks at every pixel in order to generate energy-filtered images, as for EFTEM. The potential for imaging with specific ELNES features using ESI is much greater than with EFTEM, although it can take tens of minutes to record full data sets and significant image distortion due to drift is likely.

6.5.2.7 In-Column Filters

An alternative to the sector magnet for EELS and EFTEM is the in-column filter, first introduced by Zeiss into a TEM column [136,137]. The in-column filter, located between the intermediate and the projector lenses, as shown in Figure 6.35, is often known as the omega filter because of its shape. Four sector magnets, M1–M4, are located symmetrically about the mirror plane AB. Figure 6.35a shows the ray paths in the $x–z$ plane for two electrons entering along the same trajectory but with an energy difference of ΔE. The electrons emerge from the spectrometer at different angles, i.e., a spectrum is generated. A slit located below the filter allows a small energy range of the spectrum to be used for filtering the image or for DP. Figure 6.35b shows ray paths, also in the $x–z$ plane, of three electrons with the same energy diverging from a point in the plane C1. The C1 plane corresponds to the image plane of the IL, so that either an image or a DP can be selected for filtering. In practice, a circular aperture is inserted between C1 and M1 in order to limit the angular range of electrons entering the filter. The rays in Figure 6.35b are parallel as they cross the plane AB but are subsequently focused to a point in the plane C2, located symmetrically with respect to C1.

In Figure 6.35c, three rays in the $y–z$ plane through the filter are shown diverging from the same point in C1 as depicted in Figure 6.35b. (Note that the sector magnets are still shown in this plane as they would appear in the $x–z$ plane to allow positional correspondence to be made.) These rays pass through a focus in the plane AB and are again focused at the same point in C2 as in Figure 6.35b. When the emerging rays in Figure 6.35a are linearly projected back into the filter, they intersect at a point where the height defines the achromatic image plane of the filter in M4, labeled AP in Figure 6.35a. Hence, in the plane C2, a spectrum is formed, apparently emanating from a point in AP. This

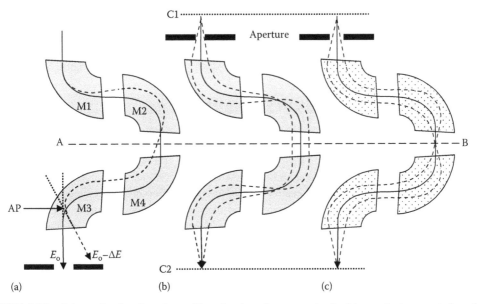

FIGURE 6.35 Schematic of an in-column filter showing electron paths for (a) two electrons entering along the same trajectory but with an energy difference, ΔE, (b) three electrons of equal energy but different angles of entry in the $x–z$ plane, and (c) three electrons of equal energy but different angles of entry in the $y–z$ plane.

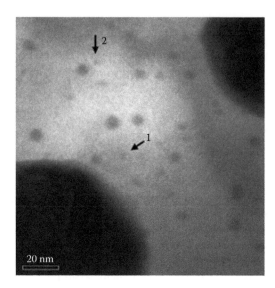

FIGURE 6.36 EFTEM image formed using an in-column EELS filter and the Fe-L$_{23}$ edge to reveal very small spherical Cu-rich precipitates in Fe (diameter <5 nm). The sample is from the same material as shown in Figure 6.34. (From Lozano-Perez, S., Titchmarsh, J.M., and Jenkins, M.L., *J. Mater. Sci.*, 41, 4394, 2006. With permission.)

spectrum is in the form of a thin line, very bright at one end where the high-intensity ZLP is located and then smoothly fading to almost zero intensity at the other end when the range of ΔE is large. The whole spectrum is imaged by the following PL and recorded using a CCD detector. Alternatively, an energy-selecting slit can be inserted across the spectrum line in the plane of C2, as indicated in Figure 6.35a, to transmit an EFTEM image. In practice, as with the post-column filter, the slit is permanently positioned on the optic axis and E_o adjusted to select the desired ΔE.

The symmetric construction of the filter ensures that low-order aberrations generated in the first half are canceled exactly in the second half (provided the mechanical construction and alignment is perfect). Because the image formed by the IL in the plane C1 is transferred to plane C2 with unit magnification, the filter is essentially invisible to the TEM operator in standard imaging and diffraction modes. An example of an EFTEM image obtained with an in-column filter of the same sample material shown in Figure 6.34, but recorded with a post-column spectrometer, is shown in Figure 6.36 [138].

6.6 CURRENT INSTRUMENTAL DEVELOPMENTS

Currently, there are several major developments in TEM instrumentation in progress, the most notable concerning aberration correction and monochromation.

The effect of aberrations in HREM imaging has been described in Section 3.4.3. Aberrations in the OL adversely affect the direct interpretation of HREM images in CTEM (Equation 6.15) and enlarge and distort the focused STEM probe (Equation 6.24). Chromatic effects also limit resolution (Equation 6.21). Early in the history of EM, it was recognized that the aberrations of round electromagnetic lenses could be corrected by combinations of multipole lenses [39]. Developments in manufacturing precision, computer control of multiple power supplies, and in real-time analysis of aberrations have recently allowed correctors to be fitted by all major EM instrument manufacturers. The principle of aberration correction is straightforward but difficult to realize. A multipole lens, such as a hexapole, introduces aberrations into an image, but the relative components of the aberration function, Equation 6.15, will be different from those of a round lens. When two identical multipole lenses are combined symmetrically with a round OL and excited in a complementary manner, then many of the aberrations introduced by the multipoles cancel exactly while others remain and are doubled. The magnitude of these remaining aberrations can be adjusted selectively and the polarity controlled to reduce, eliminate, or even reverse the polarity of an aberration such as C_3 in a round OL. Other remaining aberrations introduced by the multipoles must themselves be canceled by the introduction of further multipoles. Clearly, as the lower order coefficients in Equation 6.15 are successively minimized or eliminated, then the relative importance of the higher order coefficients becomes more important and will determine the point resolution and Schertzer defocus value. Correctors of increasing complexity will be necessary to reduce simultaneously higher and higher orders of aberration.

Two types of aberration correctors have been developed for control of C_3. One of these [51–54] uses a combination of hexapoles and dipoles and is now increasingly incorporated into the OL of TEMs. A similar design of corrector has also been incorporated into the probe-forming optics of a CTEM, in addition to the OL, to generate probes in the range <0.1 nm [139]. A second type of corrector, using quadrupoles and octupoles, was designed for retrofitting to the dedicated HB501 STEM sold by Vacuum Generators for several decades [55,56], also to generate sub-0.1 nm probes. A point resolution below 0.1 nm is now achievable in both CTEM and STEM instruments. The possibility of using core-loss signals to form energy-filtered atomic images [78] is currently being attempted in several laboratories.

Unfortunately, the introduction of C_s correction increases the chromatic aberration, C_c, significantly, so that the information limit is made worse than in the corresponding non-corrected instrument. Equations 6.20 and 6.21 show that the information limit can be extended by (1) reducing the energy spread in the incident beam, (2) improving the stability of power supplies, and (3) correcting C_c. The first of these can be achieved by including a monochromator in the electron gun and this has recently been realized [59,60]. A monochromator disperses the electron beam before it is accelerated to high energy, in such a way that an aperture can be used to permit only part of the energy distribution to proceed down the column. The design and operating principles of a monochromator are similar to those of an EELS system. The disadvantages of monochromation include a significant reduction in beam current and an increase in the minimum focused probe diameter at the sample. However, a huge advantage is that the achievable resolution in an EELS spectrum can be reduced to ~0.15 eV, leading to significant advantages for ELNES applications. The second in the above list is a never-ending aim of the instrument manufacturers, and instabilities are now typically at a level of less than one part in 2×10^6. The third is currently one of the aims of a major instrument design program based in the United States [140] and should be realized very soon.

6.7 SUMMARY

In this chapter, a description has been presented of many of the features of EM as currently applied to a wide range of materials science issues. The spectroscopic methods for compositional and bonding characterization on the nanoscale are becoming increasingly important, and these are driving instrumental development just as hard as the desire for improved image resolution. Space limitation has prevented any description of some specialized but important areas for interfacial characterization, and the reader is referred to the specialist literature for such information. In particular, electron holography is developing as an alternative means to HREM image series deconvolution for the recovery of the phase of the exit-surface wave [141,142]. The study of magnetic materials and domain boundaries [143] requires specific methods and instrument modification, and so has not been directly included. Examples of the application of the methods described in this chapter to the study of grain boundaries and interfaces are described in Chapter 9.

REFERENCES

1. P. B. Hirsch, A. Howie, R. B. Nicholson, M. J. Whelan, and D. W. Pashley, *Electron Microscopy of Thin Crystals*, 2nd edition, 1977, Krieger, Huntingdon, New York.
2. L. Reimer. *Transmission Electron Microscopy: Physics of Image Formation and Microanalysis.* Springer Series in Optical Sciences, Vol. 36, 4th edition, 1997, Springer, Berlin/New York.
3. R. F. Egerton, *Electron Energy-Loss Spectroscopy in the Electron Microscope*, 2nd edition, 1996, Plenum Press, New York.

4. D. B. Williams and C. B. Carter, *Transmission Electron Microscopy*, 1996, Plenum Press, New York/London.
5. J. C. H. Spence, *High-Resolution Electron Microscopy*, 3rd edition, 2003, Oxford University Press, Oxford, United Kingdom.
6. P. L. Gai, *In-Situ Microscopy in Materials Science*, 1997, Kluwer Academic Publishers, Boston/Dordrecht/London.
7. P. W. Hawkes and J. H. C. Spence, *Science of Microscopy*, 2007, Springer-Verlag GmbH & Co, New York.
8. A. N. Broers, *J. Appl. Phys.* 38 (1967) 1991.
9. M. T. Otten and W. M. J. Coene, *Ultramicroscopy* 48 (1993) 77.
10. T. Honda, T. Tomita, T. Kaneyama, and Y. Ishida, *Ultramicroscopy* 54 (1994) 132.
11. A. V. Crewe, D. N. Eggenberger, J. Wall, and L. M. Welter, *Rev. Sci. Instrum.* 39 (1968) 576.
12. P. A. Midgley, M. Weyland, J. M. Thomas, and F. G. Johnson, *Chem. Commn.* 10 (2001) 907.
13. I. Arsan, J. R. Tong, and P. A. Midgley, *Ultramicroscopy* 106 (2006) 994.
14. O. L. Krivanek, P. E. Mooney, G. Y. Fan, M. L. Leber, and Y. Sugimoto, *J. Electron Microsc.* 40 (1991) 290.
15. N. Mori, T. Oikawa, Y. Harada, and J. Miyahara, *J. Electron Microsc.* 39 (1990) 433.
16. J. M. Zuo, M. R. McCartney, and J. C. H. Spence, *Ultramicroscopy* 66 (1996) 35.
17. R. H. Dicke and J. P. Wittke, *Introduction to Quantum Mechanics*, 1963, Addison-Wesley, Reading, MA.
18. J. M. Cowley, *Diffraction Physics*, 1975, North-Holland, Amsterdam/Oxford.
19. S. Kikuchi, *Jpn. J. Phys.* 5 (1928) 83.
20. R. D. Heidenreich, *J. Appl. Phys.* 20 (1949) 993.
21. P. B. Hirsch, A. Howie, and M. J. Whelan, *Phil. Trans. Roy. Soc.* A252 (1960) 499.
22. C. T. Forwood and L. M. Clarebrough, *Electron Microscopy of Interfaces in Metals and Alloys*, 1991, Adam Hilger, Bristol/Philadelphia/New York.
23. G. H. Smith and R. E. Burge, *Acta Cryst.* 15 (1962) 182.
24. D. J. H. Cockayne, I. L. F. Ray, and M. J. Whelan, *Phil. Mag.* 20 (1969) 1265.
25. I. L. F. Ray and D. J. H. Cockayne, *Proc. Roy. Soc. (London)* A325 (1971) 543.
26. G. F. Dirras and J. Douin, *Phil. Mag. A* 81 (2001) 467.
27. D. J. H. Cockayne, *Ann. Rev. Mat. Sci.* 11 (1981) 75.
28. C. B. Boothroyd, *J. Microsc.* 190 (1998) 99.
29. W. O. Saxton, *Optik* 49 (1977) 51.
30. P. W. Hawkes, *Advances in Optical and Electron Microscopy*, Vol. 7, Eds. V. E. Cosslett and R. Barer, 1978, Academic Press, London, United Kingdom, p. 101.
31. C. J. Humphreys and J. C. H. Spence, *Optik* 58 (1981) 125.
32. J. M. Cowley and A. F. Moodie, *Acta Cryst.* 10 (1957) 609.
33. J. M. Cowley and A. F. Moodie, *Acta Cryst.* 12 (1959) 353.
34. J. M. Cowley and A. F. Moodie, *Acta Cryst.* 12 (1959) 360.
35. E. J. Kirkland, *Advanced Computing in Electron Microscopy*, 1998, Plenum Press, New York.
36. P. A. Stadelmann, *Ultramicroscopy* 21 (1987) 131.
37. K. Ishizuka, *Ultramicroscopy* 55 (1994) 407.
38. E. J. Kirkland and B. M. Siegel, *Ultramicoscopy* 6 (1981) 169.
39. O. Scherzer, *J. Appl. Phys.* 20 (1949) 20.
40. P. Fejes, *Acta Cryst.* A33 (1977) 109.
41. J. Frank, *Optik* 38 (1973) 519.
42. R. H. Wade and J. Frank, *Optik* 49 (1977) 81.
43. M. A. O'Keefe, C. J. D. Hetherington, Y. C. Wang, E. C. Nelson, J. H. Turner, C. Kisielowski, J. O. Malm, R. Mueller, J. Ringnalda, M. Pan, and A. Thust, *Ultramicroscopy* 89 (2001) 215.
44. A. I. Kirkland, unpublished data, 2007.
45. K. Ishizuka, *Ultramicroscopy* 5 (1980) 55.
46. P. Schiske, Zur Frage der Bilderkonstruktion durch Fokusreihen, *4th European Conference on Electron Microscopy*, 1969, Rome, Italy, p. 145.
47. E. J. Kirkland, *Ultramicroscopy* 17 (1984) 151.
48. W. O. Saxton, *Scanning Microsc.* 2 (1988) 213.
49. W. Coene, G. Janssen, M. Op de Beeck, and D. van Dyck, *Phys. Rev. Lett.* 69 (1992) 3743.
50. W. Dowell, *Optik* 20 (1963) 535.
51. A. I. Kirkland, W. O. Saxton, K. L. Chau, K. Tsuno, and M. Kawasaki, *Ultramicroscopy* 57 (1995) 355.
52. M. Haider, G. Braunshausen, and E. Schwan, *Optik* 99 (1995) 167.
53. M. Haider, S. Uhlemann, E. Schwan, H. Rose, B. Kabius, and K. Urban, *Nature* 392 (1998) 768.

54. M. Haider, H. Rose, S. Uhlemann, E. Schwan, B. Kabius, and K. Urban, *Ultramicroscopy* 75 (1998) 53.
55. O. L. Krivanek, N. Dellby, and A. R. Lupini, *Ultramicroscopy* 78 (1999) 1.
56. P. E. Batson, N. Dellby, and O. L. Krivanek, *Nature* 418 (2002) 617.
57. M. Lentzen, B. Jahnen, C. Jia, A. Thust, K. Tillmann, and K. Urban, *Ultramicroscopy* 92 (2002) 233.
58. P. C. Tiemeijer, J. H. A. van Lin, and A. F. de Jong, *Microsc. Microanal.* 7(2) (2001) 1130.
59. P. E. Batson, *Ultramicroscopy* 78 (1999) 33.
60. B. Freitag, S. Kujawa, P. M. Mul, J. Ringnalda, and P. C. Tiemeijer, *Ultramicroscopy* 102 (2005) 209.
61. M. Mukai, T. Kaneyama, T. Tomita, K. Tsuno, M. Terauchi, K. Tsuda, M. Naruse, T. Honda, and M. Tanaka, *Microsc. Microanal.* 11(Suppl. 2) (2005) 2134.
62. P.C. Tiemeijer, *Ultramicroscopy* 78 (1999) 53.
63. P. W. Hawkes, *Nucl. Instrum. Method A* 519 (2004) 1.
64. H. Rose, *Nucl. Instrum. Method A* 519 (2004) 12.
65. J. Frank, *Optik* 44 (1976) 379.
66. S. J. Pennycook and D. E. Jesson, *Phys. Rev. Lett.* 64 (1990) 938.
67. J. Silcox, P. Xu, and R. F. Loane, *Ultramicroscopy* 47 (1992) 173.
68. P. D. Nellist and S. J. Pennycook, *Adv. Imaging Electron Phys.* 113 (2000) 148.
69. P. D. Nellist and S. J. Pennycook, *Ultramicroscopy* 78 (1999) 111.
70. R. F. Loane, P. Xu, and J. Silcox, *Ultramicroscopy* 47 (1992) 121.
71. A. P. Pogany and P. S. Turner, *Acta Cryst.* A24 (1968) 103.
72. J. M. Cowley, *Appl. Phys. Lett.* 15 (1969) 58.
73. J. M. Cowley, *Bull. Mater. Sci.* 6 (1984) 477.
74. A. Howie, *J. Microsc.* 117 (1979) 11.
75. C. Mory, C. Colliex, and J. M. Cowley, *Ultramicroscopy* 21 (1987) 171.
76. L. J. Allen, S. D. Findlay, M. P. Oxley, and C. J Rossouw, *Ultramicroscopy* 96 (2003) 47.
77. C. Dwyer and J. Etheridge, *Ultramicroscopy* 96 (2003) 343.
78. E. C. Cosgriff, M. P. Oxley, L. J. Allen, and S. J. Pennycook, *Ultramicroscopy* 102 (2005) 317.
79. M. P. Oxley, E. C. Cosgriff, and L. J. Allen, *Phys. Rev. Lett.* 94 (2005) 203906.
80. L. J. Allen, S. D. Findlay, M. P. Oxley, C. Witte, and N. J. Zaluzec, *Ultramicroscopy* 106 (2006) 1001.
81. J. C. H. Spence, *Rep. Prog. Phys.* 69 (2006) 725.
82. P. Nellist and S. J. Pennycook, *Adv. Imaging Electron Phys.* 113 (2000) 148.
83. S. J. Pennycook and L. A. Boatner, *Nature* 336 (1988) 565.
84. P. D. Nellist and S. J. Pennycook, *Ultramicroscopy* 78 (1999) 111.
85. A. Crewe, J. Wall, and J. Langmore, *Science* 168 (1970) 1338.
86. M. Ohtsuki, *Ultramicroscopy* 5 (1980) 325.
87. P. M. Voyles, J. L. Grazul, and D. A. Muller, *Ultramicroscopy* 96 (2003) 251.
88. J. C. Russ, *Fundamentals of Energy Dispersive X-Ray Analysis*, 1984, Butterworths, London, United Kingdom.
89. D. B. Williams, J. I. Goldstein, and D. E. Newbury (Eds.), *X-Ray Spectrometry in Electron Beam Instruments*, 1995, Plenum Press, London/New York.
90. J. I. Goldstein, D. E. Newbury, P. Echlin, D. C. Joy, A. D. Romig, C. E. Lyman, C. Fiori, and E. Lifshin, *SEM and X-Ray Microanalysis*, 1992, Plenum Press, London/New York.
91. J. N. Chapman, C. C. Gray, B. W. Robertson, and W. A. P. Nicholson, *X-ray Spectrom.* 12 (1983) 153.
92. C. C. Gray, J. N. Chapman, W. A. P. Nicholson, B. W. Robertson, and R. P. Ferrier, *X-ray Spectrom.* 12 (1983) 163.
93. J. H. Paterson, PhD thesis, 1988, University of Glasgow, UK.
94. J. H. Paterson, J. N. Chapman, W. A. P. Nicholson, and J. M. Titchmarsh, *J. Microsc.* 154(Pt. 1) (1989) 1.
95. C. J. Rossouw, P. S. Turner, T. J. White, and A. J. O'Connor, *Phil. Mag. Lett.* 60 (1989) 225.
96. C. J. Rossouw, C. T. Forwood, M. A. Gibson, and P. R. Miller, *Micron* 28 (1997) 125.
97. J. H. C. Spence, J. M. Titchmarsh, and N. Long, *Proceedings of the 20th Annual Conference MAS*, Ed. J. T. Armstrong, 1985, San Francisco Press Inc., San Francisco, CA, p. 349.
98. K. S. Vecchio, and D. B. Williams, *J. Microsc.* 147 (1987) 15.
99. J. I. Goldstein, *Introduction to Analytical Electron Microscopy*, Eds. J. J. Hren, J. I. Goldstein, and D. C. Joy, 1979, Plenum Press, New York, Chapter 3.
100. M. O. Krause, *J. Phys. Chem. Ref. Data* 8 (1979) 307.
101. J. H. Schofield, *At. Data Nucl. Data Tables* 14 (1974) 122.
102. W. Bambenyk, B. Crasemann, R. W. Fink, H-U. Freund, H. Mark, C. D. Swift, R. E. Price, and P. V. Rao, *Rev. Mod. Phys.* 44 (1972) 716.
103. A. Lagneberg and J. Van Eck, *J. Phys. B* 12 (1979) 1331.

104. G. Cliff and G. W. Lorimer, *J. Microsc.* 103 (1975) 203.
105. T. P. Schreiber and A. M. Wims, *Ultramicroscopy* 6 (1981) 323.
106. C. J. Powell, *Rev. Mod. Phys. A* 48 (1976) 33.
107. M. Inokuti, *Rev. Mod. Phys.* 43 (1971) 297.
108. K. F. J. Heinrich, *Proceedings of the 11th ICXOM*, Eds. J. D. Brown and R. H. Packwood, 1987, University of Western Ontario, London, Canada, p. 67.
109. N. J. Zaluzec, *Introduction to Analytical Electron Microscopy*, Eds. J. J. Hren, I. Goldstein, and D. C. Joy, 1979, Plenum Press, New York, Chapter 4.
110. J. C. H. Spence and J. Taftø, *J. Microsc.* 130 (1983) 147.
111. S. Lazar, G. A. Botton, M.-Y. Wu, F. D. Tichelaar, and H. W. Zandbergen, *Ultramicroscopy* 96 (2003) 535.
112. R. F. Egerton, *Ultramicroscopy* 107 (2007) 575.
113. O. L. Krivanek, A. J Gubbens, N. Delby, and C. E. Meyer, *Microsc. Microan. Microstr.* 3 (1992) 187.
114. A. J Gubbens, H. A. Brink, M. K. Kundman, S. L. Friedman, and O. L. Krivanek, *Micron* 29 (1998) 81.
115. P. Kruit and H. Shuman, *Ultramicroscopy* 17 (1985) 263.
116. N. J. Zaluzec, J. Hren, and R. W. Carpenter, *38th Annual Proceedings of the EMSA*, Ed. G. W. Bailey, 1980, Claitor's Publishing, Baton Rouge, LA, p. 114.
117. D. W. Johnson and J. H. C. Spence, *J. Phys. D (Appl. Phys.)* 7 (1974) 771.
118. R. F. Egerton, B. G. Williams, and T. G. Sparrow, *Proc. R. Soc. (London)*, A398 (1985) 395.
119. T. A. Malis, S. C. Cheng, and R. F. Egerton, *J. Electron Microsc. Tech.* 8 (1988) 193.
120. R. F. Egerton and M. Malac, *Ultramicroscopy* 92 (2002) 47.
121. C. C. Ahn and O. L. Krivanek, *EELS Atlas*, 1983, Gatan Inc. and ASU HREM Facility, Warrendale, USA.
122. F. T. Docherty, A. J. Craven, D. W. McComb, and J. Skakle, *Ultramicroscopy* 86 (2001) 273.
123. D. H. Pearson, C. C. Ahn, and B. Fultz, *Phys. Rev. B* 47 (1993) 8471.
124. R. D. Leapman, P. L. Fejes, and J. Silcox, *Phys. Rev. B* 28 (1983) 2361.
125. P. Rez, J. R. Alvarez, and C. Pickard, *Ultramicroscopy* 78 (1999) 175.
126. P. Blaha, K. Schwarz, G. K. H. Madsen, D. Kvasnicka, and J. Luitz, *WIEN2k, an Augmented Plane Wave + Local Orbitals Program for Calculating Crystal Properties*, 2001, Karlheinz Schwarz Technical Universität, Wien, Austria.
127. A. L. Ankudinov, B. Ravel, J. J. Rehr, and S. D. Conradson, *Phys. Rev. B* 58 (1998) 7565.
128. M. S. Moreno, K. Jorissen, and J. J. Rehr, *Micron* 38 (2007) 1.
129. S. Lazar, C. Hébert, and H. W. Zandbergen, *Ultramicroscopy* 98 (2004) 249.
130. T. Mizoguchi, K. Tatsumi, and I. Tanaka, *Ultramicroscopy* 106 (2006) 1120.
131. S. Lozano-Perez, J. M. Titchmarsh, and M. L. Jenkins, *Ultramicroscopy* 106 (2006) 75.
132. D. A. Muller, Y. Tzou, R. Raj, and J. Silcox, *Nature* 366 (1993) 725.
133. W. Grogger, P. Warbichler, F. Hofer, T. Lang, and W. Schintlmeister, *Proceedings of the ICEM14*, Eds. H. A. Calderon Benavides and M. J. Yacaman, 1998, Institute of Physics, Bristol, p. 215.
134. G. A. Botton and M. W. Phaneuf, *Micron* 30 (1999) 109.
135. C. Jeanguillaume, P. Trebbia, and C. Colliex, *Ultramicroscopy* 3 (1978) 237.
136. S. Lanio, H. Rose, and D. Krahl, *Optik* 73 (1986) 56.
137. A. Berger, J. Mayer, and H. Kohl, *Ultramicroscopy* 55 (1994) 101.
138. S. Lozano-Perez, J. M. Titchmarsh, and M. L. Jenkins, *J. Mater. Sci.* 41 (2006) 4394.
139. J. L. Hutchison, J. M. Titchmarsh, D. J. H. Cockayne, R. C. Doole, C. J. D. Hetherington, A. I. Kirkland, and H. Sawada, *Ultramicroscopy* 103 (2005) 7.
140. U. Dahmen, *Microsc. Microanal.* 13 (2007) 1150.
141. D. Gabor, *Nature* 161 (1948) 777.
142. M. Lehmann and H. Lichte, *Microsc. Microanal.* 8 (2002) 447.
143. J. N. Chapman, *J. Phys. D (Appl. Phys.)* 17 (1984) 623.

7 Synchrotron-Based Techniques

Andrea R. Gerson, David J. Cookson,
and Kevin C. Prince

CONTENTS

7.1 INTRODUCTION

Synchrotron radiation was first generated at the General Electric Research Laboratory (Schenectady, New York) in 1945 [1] using a 70 MeV electron synchrotron. Subsequent experiments by the General Electric team and by scientists using the Cornell synchrotron [2] revealed the unique combination of spectral and polarization properties of the radiation. However, it was not until 1961 that sufficient interest was generated in the scientific community for an experimental program to be set up at the (then) National Bureau of Standards. In the early 1960s, synchrotron radiation research blossomed with the establishment of many first-generation parasitic synchrotron radiation facilities built onto accelerators whose primary purpose was high energy or nuclear physics. In 1968, the first synchrotron spectrum was generated from an electron storage ring [3]. However it was not until 1981 that the first purpose-built, so-called second-generation, synchrotron radiation facility was commissioned at the Daresbury Synchrotron Radiation Source (UK).

In the late 1970s, with the advent of more sophisticated synchrotron-based experimentation, research workers called for greater x-ray radiation brightness rather than just increased flux. This is a requirement for high spatial and spectral resolution, and is a function of the size and divergence of the electron beam in the storage ring. Third-generation synchrotron radiation facilities are those designed to maximize brightness through the application of undulator and wiggler multi-magnet insertion devices. Bend magnets, as used within the second-generation sources, result in lower intensity and degree of polarization, and are based on a single magnet. Undulators provide much more intense radiation and generally have high degrees of polarization, which may be linear, circular, or elliptical, depending on the type of device. The first of these third-generation sources were the European Synchrotron Radiation Facility (ESRF, 6 GeV, Grenoble), the Advanced Light Source in Berkeley, California, and Elettra, Trieste, Italy, all of which provided their first user access in 1994. Since then many more have followed, such as the Canadian Light Source and the Australian Synchrotron.

Why use light from a synchrotron rather than from a laboratory source? The reason is that synchrotron light has many advantages which make it much more powerful for some applications than laboratory-based sources. These advantages are high brightness (for third-generation sources greater than 10^{18} photons/s/mm^2/mrad^2/0.1% bandwidth, Figure 7.1), low emittance, high collimation, wide range of selectable energies, and polarization. There are also some disadvantages. Synchrotrons are very large, nationally owned, devices and there are only a limited number (~50) around the world. An experimenter wishing to use one might have to travel a considerable distance. In addition, synchrotron light sources typically run 24 h a day for many days at a time, with a few intervals of unavailability, and beam line times have to be scheduled on a rota basis. Access is by application and the experimenter has to have everything ready to be able to perform the measurements within the allocated, and limited, period of time. Consequently the timescales for planning and preparing experiments are rather long.

William Henry Bragg, and his son William Lawrence, are credited with being the first to observe elastic scattering of x-rays from matter, in 1913 [4]. They found that single crystals deflected beams of x-rays at very specific angles, which depended on the wavelength of the incident radiation and the atomic lattice spacings in the crystal. This is now justifiably known as Bragg diffraction and is an important special case of the more general phenomenon of elastic x-ray scattering. For elastic x-ray scattering the interaction between photon and electron (or atom) is a single-step process where photon energy is conserved [5]. There is a well-defined phase relationship between the incident and scattered photon which makes it a coherent process. A process where the phase relationship is lost is called incoherent and often (but not always) implies a change in photon energy—resulting in inelastic scattering.

Over the last 20 years the range of applications of synchrotron x-ray techniques to the study of surfaces has grown enormously. In this brief review an attempt will be made to provide an overview of the capabilities and applications of some synchrotron radiation techniques without recourse to a high level of technical detail. Electromagnetic radiation (light) generated with a wavelength of around 0.1 nm is generally called "hard" x-ray radiation. Such x-rays are photons with an energy greater

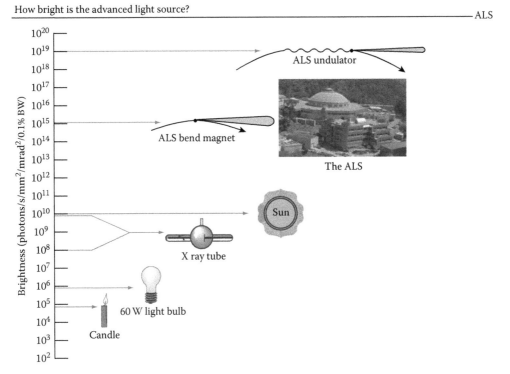

FIGURE 7.1 Comparison of the brightness of bend magnets and undulators at the Advanced Light Source at the E. O. Lawrence Berkeley National Laboratory, with other types of light source. Brightness (sometimes also called brilliance) is measured by the flux of light emitted over a narrow range of wavelengths from a small source area and within a narrow cone of light. Note that the sun emits a huge amount of light but over a range of wavelengths and in all directions, so it is not as bright as a synchrotron light source. (Courtesy of Advanced Light Source, Berkeley, CA.)

than about 5 keV, and are distinguished from "soft" x-rays by the fact that they can penetrate a meter of atmosphere at sea-level pressure. A brief introduction will be presented to the soft x-ray technique, and variations thereof, of photoemission spectroscopy, x-ray absorption spectroscopy, which may be carried out using either hard or soft x-rays, and small angle x-ray scattering and reflectometry, typically carried out with hard x-rays. In particular, the analytical possibilities arising from the application of synchrotron x-radiation, as opposed to conventional cathode-tube based x-ray sources, are emphasized.

7.2 SYNCHROTRON PHOTOEMISSION

7.2.1 INTRODUCTION

As described in Chapter 3, in photoemission spectroscopy, an electron is ejected from a solid after the absorption of an incident photon of well-defined energy. The energy of the ejected electron is measured and is characteristic of the element and energy level from which it came. Electrons in solids are described as either strongly bound (core) or loosely bound (valence), the latter being responsible for the chemical bonding of solids. It has also been shown in Chapter 3 how core level spectral intensities can be used for quantitative elemental analysis, and for the characterization of the chemical state of elements in a sample via the core level shift. Valence band (VB) spectroscopy can also provide valuable information about the nature of the chemical bonding.

The apparatus used for photoemission spectroscopy at a synchrotron is very similar to that used for laboratory x-ray photoelectron spectroscopy (XPS), namely, an ultra-high vacuum (UHV) chamber, a sample manipulator, an electron energy analyzer, and various preparation facilities (for sputtering, heating, cleaving, etc.).

7.2.2 CORE LEVEL SPECTROSCOPY

7.2.2.1 Varying Surface Specificity by Varying Photon Energy

The attenuation length (AL) of electrons in most materials is shortest at about 20–50 eV kinetic energy (KE) (see Figure 3.12 in Chapter 3), and has a value of a few tenths of a nanometer. In conventional XPS, the KE of an outgoing electron from a particular core level of a particular element is fixed because the x-ray line energy of the source is fixed at one or other of very few values. Thus the surface specificity cannot be varied easily. However, with synchrotron light, it is possible to choose the photon energy so that the KE of the photoelectron lies in the range of maximum surface specificity. It is also sometimes useful to be able to reduce the surface specificity for a particular peak, by choosing the photon energy so that it is near, or well above, threshold, thus verifying the assignment of that peak to either a surface or a bulk species. This method is particularly useful for observing surface core level shifts, by enhancing the intensity of surface peaks.

Figure 7.2 shows an example of a spectrum of iron sulfide (pyrite) at two different photon energies. In the lower curve, at 449 eV x-ray energy, the kinetic energy is high and the spectrum is dominated by two peaks due to the spin-orbit split $2p_{1/2}$ and $2p_{3/2}$ sulfur core levels of the bulk material. Only weak extra shoulders are visible. At 206 eV x-ray energy (upper curve), the photon energy has been adjusted in order to maximize surface specificity. The bulk peak is still the strongest, but now two peaks due to surface sulfur atoms are visible, as well as two weak (spin-orbit split) peaks due to oxidized sulfur. Similar spectra have been published and discussed in Refs. [6,7].

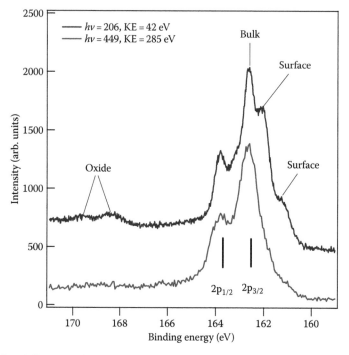

FIGURE 7.2 S 2p XPS spectra from pyrite (FeS$_2$) at two photon energies. (Reprinted from Plekan, O., Feyer, V., Šutara, F., Skála, T., Švec, M., Cháb, V., Matolín, V., and Prince, K.C., *Surf. Sci.*, 601, 1973, 2007. With permission.)

7.2.2.2   Variation of Cross-Section

Another application uses variation of cross-section to separate neighboring or overlapping peaks, and to identify them. In the simplest case, spectra are recorded as a function of x-ray energy, the intensities are compared, and the assignment of the peaks is obtained by reference to published cross-sections [8].

An important application occurs where an element being analyzed has a Cooper minimum in the cross-section. Such minima occur for core levels where the angular momentum quantum number l is less than the principal quantum number n, that is, for core levels labeled (n, l), where $l < n$. Examples are 2s, 3s, 3p, 4d, 5f, but not 1s, 2p, 3d, etc. These minima can be very deep, with reductions of cross-section of as much as two orders of magnitude. This very large change makes identification of the core level easy, and may reveal peaks due to dilute species masked by emission from the matrix for energies away from the Cooper minimum.

In Figure 7.3 the example of a dilute alloy of Al in Cu is shown. At 106 eV x-ray energy, near the Cooper minimum of Cu 3p, the spin-orbit split Al 2p peaks are clearly visible, while the Cu 3p peaks are almost invisible. At higher x-ray energy, 170 eV, the cross-section of Cu 3p has increased while that of Al 2p has decreased, so that the 3p states become visible along with the 2p peaks. At still higher x-ray energy, 430 eV, the Cu 3p and Al 2p peaks have about equal peak heights, while at the energy of Al K_α, i.e., that used in traditional XPS, the Al 2p peaks have become almost invisible in the low binding energy tail of the Cu 3p states.

When varying the cross-section by means of the photon energy, it must be remembered that the KE of the electrons, and therefore the surface specificity, also changes. Thus, in the example given, the information depth does not change significantly between 106 and 170 eV photon energies, and the Cu and Al core levels can be distinguished and assigned. However, at higher energy, the signal is dominated by bulk photoemission.

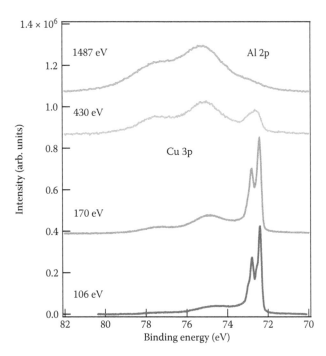

FIGURE 7.3   Cu 3p and Al 2p XPS spectra from a Cu-9% Al alloy at the photon energies shown. The crystal has been treated to segregate Al to the surface as an oxide. (Courtesy of S. Nemšak.)

The Cooper minimum technique can also be applied to VB emission. For the VB, atomic orbitals may be hybridized with other states, and the calculated energies of Cooper minima may be more or less shifted from the values calculated for pure atomic orbitals. Generally, for higher orbital angular momentum quantum numbers, more atomic character is retained. Thus 4d and 4f bands are rather atomic like and calculations of the Cooper minimum energy for atomic orbitals are reliable, while ns and np orbitals are more distorted by bond formation.

7.2.2.3 High Resolution

X-ray energy resolution can be specified as the bandwidth of the incident photons, $\Delta h\nu$, usually expressed in meV or eV, or the resolving power, $h\nu/\Delta h\nu$, at a particular x-ray energy. Typically, laboratory-based XPS achieves a resolution of about 0.85 eV using non-monochromatized Mg K_α radiation, or about 0.3–0.4 eV for monochromatic Al K_α radiation, corresponding to resolving powers of about 1500 and 5000, respectively. Undulator beamlines at synchrotron radiation sources can routinely achieve a resolving power of 10,000 at lower energies, so that the absolute energy resolution is much better. This high resolution is not always necessary, because for many applications it is sufficient to have resolution that is better than the natural line-width of the core level being examined. In this case resolution can be traded for flux, by opening the slits of the beamline monochromator.

Figure 7.4 shows an example of high resolution photoemission from Pb on Rh(100) [9]. The Pb 5d core levels give rise to sharp peaks, and undergo small shifts depending on the ordered structures they form on the surface. Using high resolution, it is possible to detect quite small shifts in energy, in this case 160 meV, and changes in line shape.

In the example shown in Figure 7.4, after adsorption of Pb at room temperature, a structure labeled α-c(2 × 2) was formed, in which Pb atoms were adsorbed above the substrate surface. On heating to 470 K the peaks began to shift and after heating to 520 K, the binding energy increased by 160 meV. The diffraction pattern remained the same, c(2×2), but the surface had different chemical properties, so the structure is labeled β. The Pb atoms are believed to be embedded in the surface in a substitutional alloy.

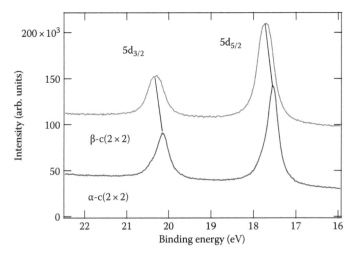

FIGURE 7.4 Pb 5d core level XPS spectra from 0.5 monolayers of Pb adsorbed on Rh(100), in two different ordered structures. The shift in the 5d peaks is 160 meV. (Reprinted from Švec, M., Dudr, V., Šutara, F., Tsud, N., Skála, T., Cháb, V., Matolín, V., and Prince, K.C., *Surf. Sci.*, 601, 5673, 2007. With permission.)

7.2.3 Valence Band Spectroscopy: Band Structure

As noted in Section 7.2.1, the VB electrons are responsible for the bonding of molecules and solids. The electronic states in the VB can be probed directly by photoemission; for example, in Figure 7.5, the VBs of single crystals of (110) oriented Ag, Cu, and Pd are shown. These metals have the atomic configurations $4d^{10}5s$, $3d^{10}4s$, and $4d^{9.75}s^{0.3}$, respectively. The VB spectra show strong peaks due to the high cross-section d electrons, marked by arrows. For Ag and Cu, there is a weaker band of sp electrons between the d band and the Fermi level (in the solid state the s electrons hybridize to become sp like), while in Pd the sp band is below the d band. It is immediately obvious that the d bands of Ag and Cu are full (no density of d states at the Fermi level), while that of Pd is not full (high density of d states at the Fermi level). This in turn accounts for the higher catalytic activity of Pd, because the d electrons can more easily participate in chemical bonds than those of Ag or Cu.

The simple interpretation of the VB spectrum in terms of a direct projection of the density of states (DOS) is a reasonable approximation for polycrystalline solids, and for spectra integrated over a large angle of emission. However for single crystal samples, the situation is more complicated. The electrons in solids exist in bands, and generally have well-defined momenta. A simple picture of the photoemission process is the three-step model [10]: in the first step, photoexcitation, the momentum of the electron is conserved (to within a reciprocal lattice vector) and it is excited to a higher energy state. In the second step, the electron travels to the surface without change of energy or momentum. In the third step, it is emitted from the surface, and at this stage the momentum can change. Momentum is only conserved in a solid because it is periodic in three dimensions; the surface is periodic in the two dimensions parallel to the surface, but not perpendicular to it, so an electron can lose or gain momentum in that direction.

In an experiment, the energy of an emitted electron is measured, which provides its total momentum through the following relation:

$$p = m_e v = \sqrt{2 m_e E_k} \tag{7.1}$$

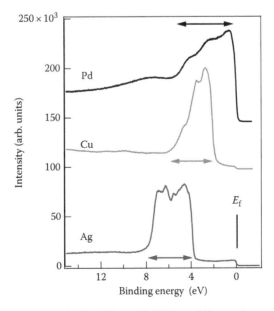

FIGURE 7.5 VB spectra from Ag(110), Cu(110), and Pd(110) at different photon energies. The Fermi level, E_f, is indicated. The horizontal arrows indicate the ranges of the d bands. (From Prince, K.C., unpublished data.)

Equation 7.1 becomes, after converting momentum p to wave number $k(k = p/\hbar)$, where $\hbar$ is Planck's constant, inserting the mass of the electron m_e, and writing the energy E in electron volts,

$$k = \hbar^{-1}\sqrt{2m_e E_k} = 5.1\sqrt{E_k} \qquad (7.2)$$

where
  the units of k are inverse nm (nm^{-1})
  E_k is the KE in eV

In electron band theory, the units are usually inverse Å, but nm^{-1} are used here to be consistent with Sections 7.3 and 7.4. Note that the constant is different from that in Equation 7.6 because here eV are used as the unit of energy, whereas in Equation 7.6 keV are used.

In this chapter, the wave number k and the wave vector $\boldsymbol{k}$ of both electrons and photons will be used frequently. As the names imply, these two quantities emphasize the wave nature of the particles under discussion, but unfortunately confusion can arise about their respective significance. The dimensions of k are inverse length, since $k = 2\pi/\lambda$, where λ is the wavelength of the photon, or the de Broglie wavelength of a particle with mass. In optical spectroscopy, wave number expressed in cm^{-1} is used as a unit of energy; indeed, cm^{-1} are sometimes referred to as wave numbers. Strictly speaking, the energy of a photon is $\hbar kc$, where c is the velocity of light. In the theory of scattering of x-ray photons, neutrons, and electrons, k is sometimes referred to as momentum, although again strictly speaking, momentum is actually $\hbar k$. It is more correct to refer to "the momentum associated with the wave number k" or "the vector momentum associated with the wave vector $\boldsymbol{k}$." However, for brevity, common practice will be followed, and the reader should remember that it is necessary to multiply by $\hbar$ to convert k to momentum, or $\hbar c$ to convert photon k to energy.

The vector momentum $\boldsymbol{k} = (k_x, k_y, k_z)$, where the z-axis is along the surface normal, is calculated from k and the angle of emission. The components of the momentum parallel to the surface, (k_x, k_y), are the values in the crystal. However the normal component, $k_z = k \cdot \cos \theta$, where θ is the angle from the normal, is the momentum in vacuum and its value in the crystal is not known. There are various methods of determining the value of this momentum. The simplest approximation is to assume a free electron final state, in which the relation between the momentum and the energy is parabolic [11]. A better approximation is to calculate the band structure completely for the occupied and unoccupied states and compare it with measurement [12]. Indeed, it is possible to calculate the spectra exactly using the single-step model, in which the entire process of photoexcitation and emission is included, and the two steps are not artificially separated. There are also precise experimental methods, such as triangulation [13], which do not require calculation. All of these depend on the tunability of the light, and are therefore almost exclusively the domain of synchrotron radiation.

The power of mapping the band structure of solids is one of the most successful applications of synchrotron radiation. Other methods, such as de Haas-van Alphen measurements, can map the electronic structure at the Fermi level, but not at higher binding energy. Optical measurements can provide partial information on band structure. Synchrotron radiation band mapping, however, provides a broad picture of the whole VB in energy and momentum space.

The data in Figure 7.6 [14] were recorded in order to determine the band structure of PbS. The spectra were taken in the normal direction, and as the photon energy changed, the structure of the VB also changed dramatically. From these data, the binding energies of peaks, and the momenta of the electrons, were determined, and from them the band structure.

Figure 7.7 shows an example of surface band mapping, in this case of oxygen adsorbed on a Rh(100) single crystal [15]. This is a purely two-dimensional system, so only the momentum parallel

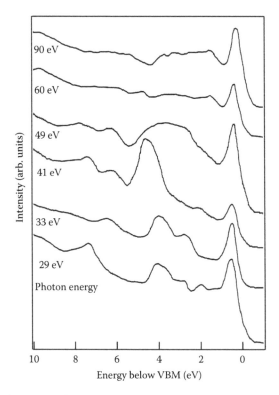

FIGURE 7.6 Photoemission spectra from a lead sulfide (PbS) single crystal along the normal to the surface. As the photon energy changes, the spectrum changes markedly. (From Santoni, A., Paolucci, G., Prince, K.C., and Christensen, N., *J. Phys.: Condens. Matter*, 4, 6759, 1992. With permission.)

to the surface is relevant. A fixed photon energy and fixed total momentum were used, and the parallel momentum was varied by measuring the spectrum at different angles of emission. As the angle varied, the binding energies of the peaks varied (Figure 7.7a). The binding energy of the oxygen-induced peak is plotted in Figure 7.7b against the parallel momentum, $k \cdot \sin \theta$. The band dispersion is observed, and repeats in higher Brillouin zones.

7.2.4 Determination of Symmetry

Synchrotron light is naturally polarized; the polarization may be linear, circular, or elliptical, and the degree of polarization may be as high as 100%. If a substrate has a well-defined and sufficiently high symmetry (usually a single crystal), linearly polarized light can be used to determine the symmetry of the valence orbitals observed in photoemission. This is performed by applying the powerful techniques of group theory. If a system has an axis of rotation or a plane of reflection symmetry, then the molecular orbitals or VBs must be either symmetric or antisymmetric under this rotation or reflection. Similarly, the experimental setup can be so arranged that the light is symmetric or antisymmetric under rotation or reflection. For instance, if the electric vector of the light lies in a reflection plane, it is purely symmetric. If it is perpendicular to a reflection plane, it is antisymmetric.

To determine the nature of an orbital or band, the sample, linearly polarized light, and the electron energy analyzer are set up in various geometries. If a peak is observed, it is "allowed," but if it disappears, it is "forbidden." The above symmetry considerations and the Fermi Golden Rule are applied to determine in which of these geometries the peak should be allowed or forbidden. The details of how this is carried out in practice and many examples have been discussed by various authors [11,16–18].

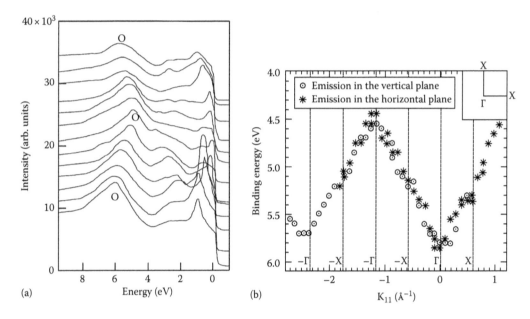

FIGURE 7.7 Surface band mapping: oxygen adsorbed on Rh(100) in the p4g structure. (a) VB photoemission spectra taken at different angles of emission referred to the surface normal. The peaks marked "O" are induced by the adsorption of oxygen. Other peaks are Rh 4d states. (b) A dispersion plot: the binding energies of the peaks are plotted against momentum (using the formula above). G and X represent high symmetry points in the Brillouin zone. (Reprinted from Zacchigna, M., Astaldi, C., Prince, K.C., Sastry, M., Comicioli, C., Rosei, R., Quaresima, C., Ottaviani, C., Crotti, C., Antonini, A., Matteucci, M., and Perfetti, P., *Surf. Sci.*, 347, 53, 1996. With permission.)

7.2.5 RESONANT SPECTROSCOPY

Resonant spectroscopy is a special case of the use of cross-section to obtain chemical and physical information. At resonance, the photoabsorption cross-section is enhanced over a narrow energy interval with respect to the usual calculated cross-section, because transitions occur to an unoccupied orbital. To find and identify resonances, tunable radiation is necessary, and the technique of x-ray absorption is used (Section 7.3). The photon energy is tuned to the resonance energy and the photoemission spectrum is measured. As an example, resonant photoemission from titanium dioxide [19], Figure 7.8, will be considered.

In this experiment, the photon energy is tuned through the threshold for Ti 2p photoemission (see Ref. [18] for the near edge x-ray absorption fine structure [NEXAFS] spectrum). Peaks in the absorption cross-section occur when the 2p core electron is promoted to an empty 3d state, and, taking account only of the orbitals directly involved, the transition can be written as

$$2p^6 3d^n \rightarrow 2p^5 3d^{n+1} \tag{7.3}$$

This excited state can decay in various ways, including

$$2p^5 3d^{n+1} \rightarrow 2p^6 3d^{n-1} + e \tag{7.4}$$

where e is a photoelectron. This configuration, $2p^6 3d^{n-1}$, is also the final state reached in direct photoemission:

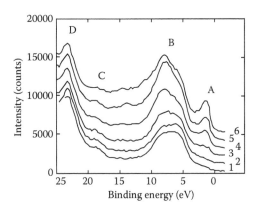

FIGURE 7.8 VB photoemission spectra of TiO$_2$ at photon energies of (1) 454, (2) 455, (3) 456, (4) 458.6, (5) 459.2, and (6) 460.2 eV. A: defect states. B: valence band. C: final states with two ligand holes, and one 3d electron. D: oxygen 2s emission. (Reprinted from Prince, K.C., Dhanak, V.R., Finetti, P., Walsh, J.F., Davis, R., Muryn, C.A., Dhariwal, H.S., Thornton, G., and van der Laan, G., *Phys. Rev. B*, 55, 9520, 1997. With permission.)

$$2p^6 3d^n \rightarrow 2p^6 3d^{n-1} + e \qquad (7.5)$$

If two final states can be reached by different paths, they interfere and the intensity changes—it may increase or decrease, and in this case it increases substantially. The resonance is known as a Fano resonance, and the process as resonant photoemission (other names are resonant Auger spectroscopy and de-excitation spectroscopy). Various types of information can be extracted from the experimental spectrum in Figure 7.8. Firstly, the nominal configuration of Ti in this oxide is 3d^0, and the VB is nominally composed of oxygen 2p derived states. However if there were no d character in the valence orbitals ($n = 0$), then they would not resonate, because the configuration 2p^{6}3d^{n-1} could not be reached. However the peak B does resonate, so it can be concluded that Ti 3d character is mixed into the VB. This admixture covers the whole VB and has a maximum at around 8 eV. Secondly the peak A is assigned to defect states, nominally 2p^{6}3d^1, which are present at low concentration. They are almost invisible at low energies, below the resonance, but become quite strong on resonance. This is another useful property of resonant photoemission; since the increase in intensity can be as much as one or two orders of magnitude, it is possible to observe otherwise undetectable states.

7.3 X-RAY ABSORPTION SPECTROSCOPY

7.3.1 INTRODUCTION

There have been a considerable number of useful fundamental texts and reviews of the theory and application of x-ray absorption spectroscopy (XAS). The topic is very large so here only a brief, nontechnical, introduction with pointers to suitable further reading will be attempted. The most comprehensive text to date remains that by Koningsberger and Prins [20]. The aim here is to provide the reader with a basic conceptual framework by which to approach surface-sensitive XAS measurements. There exist many modes by which XAS can be measured, generally with accompanying acronyms. In the following sections, the basic interpretation of both x-ray absorption near edge structure (XANES) and extended x-ray absorption fine structure (EXAFS) and the most common methods by which surface XAS can be measured will be discussed.

7.3.2 REGIONS OF THE XAS SPECTRUM

XAS (alternatively called XAFS, x-ray absorption fine structure) enables the local structure around a specific element to be determined regardless of the nature of the long range structure of the sample. This means that structural information can be derived from samples unsuitable for crystallographic analysis, for instance, liquids, glasses, amorphous metals, multicomponent systems, surfaces, trace impurities, etc.

XAS measurements generally require the energy of the incident x-ray beam to be scanned, by means of a monochromator, across the energy required to excite an electron from a specific orbital of a specific element. In addition, the intensity of synchrotron radiation allows processes to be followed in real time *in situ* [21]. For rapid *in situ* measurements, an energy dispersive setup may be used so that the broadband-transmitted radiation is dispersed as a function of energy, and then measured using an area detector. This type of measurement, while enabling extremely fast data collection, is not generally applicable to surface-specific measurements. Both types of XAS measurement require an x-ray source with a broad and intense spectral range, which in practical terms, only a synchrotron can supply.

The excitation edge, which manifests itself as a sharp transition in the XAS spectra (Figure 7.9) is defined as the energy required to excite an electron from a core or valence orbital (or band for solid materials) to a continuum (i.e., delocalized or unbound) state. The exact edge energy (E_0) is

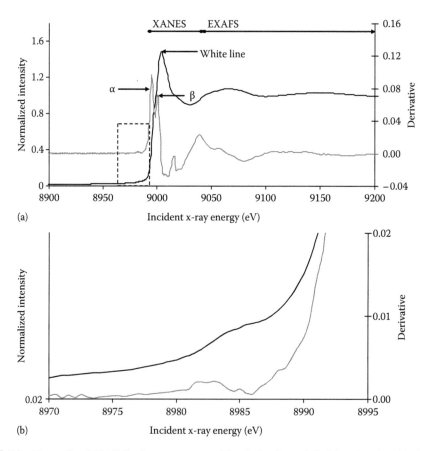

FIGURE 7.9 Normalized XAS Cu 1s spectrum, and its derivative, of $CuSO_4$, showing (a) the XANES and EXAFS regions. The portion of the spectra indicated in (a) by a broken-line is enlarged in (b).

defined as the first maximum in the first derivative of the measured absorption with respect to incident x-ray energy. Note that E_0 defined in this manner does not correspond to the Fermi energy. Pre-edge features, at energies less than E_0 but greater than the Fermi energy, may occur on electron excitation to empty bound states. The XAS spectrum incorporating the pre-edge features and the region immediately above the absorption edge is termed the XANES region. Alternatively, for lower x-ray energy applications, the terminology NEXAFS is commonly used.

Above the XANES region lies the EXAFS region. There is no exact demarcation between the XANES and EXAFS regions. However, the transition between them is generally taken to be where the electron wavelength is approximately equal to the distance to the nearest neighboring atom, at about 40–50 eV above the absorption edge (Chapter 2 of Ref. [20]).

7.3.3 EXAFS OR XANES?

There are several advantages of XANES analysis as compared to EXAFS [22]:

- The spectra are generally more intense and it is therefore possible to analyze lower concentrations of the element of interest.
- The spectra appear over a smaller incident x-ray energy range and it is therefore possible to carry out either a greater number of measurements or spend longer on any given measurement.
- The spectra are sensitive to chemical/electronic structure, particularly conduction band or unoccupied states.
- The spectra are more amenable to empirical analysis where more than one chemical state is present.

However, XANES data have been, and to date, remain less amenable to fundamental structural analysis than those of EXAFS. In addition, if x-ray absorption edges from the material of interest fall within the EXAFS range of the edge being studied, the measurement may be limited to only the XANES region.

7.3.4 INTRODUCTION TO THE INTERPRETATION OF XANES

Historically, XANES has been interpreted as providing a straightforward methodology for probing unoccupied electronic states, both localized (bound) and delocalized (unbound). The excitation from the core or VBs to the conduction bands is required to obey the dipole selection rules as defined by the Pauli exclusion principle, i.e., $\Delta l = \pm 1$, $\Delta s = 0$ so that $\Delta j = \pm 1$ where l is the angular momentum, s is the electron spin angular momentum, and j is the total angular momentum quantum number, containing both l and s. Typical transitions are therefore from s to p states or p to d states.

The original analytical method for the identification of XANES features, and it remains a useful practice, was the comparison of measured spectra with those of standards. The best introductory account, with emphasis on empirical analysis, is that by Bare [22]. It can be summarized as follows:

- The position of the absorption edge is sensitive to local environment and to a first approximation to elemental oxidation state. As the oxidation state increases, the incident x-ray energy required to cause electron excitation also increases. For instance, George and Gorbaty [23] demonstrated an overall edge shift for sulfur of 11 eV within an extensive series of sulfur-containing cyclic aromatic compounds.
- Examination of pre-edge features can be used as a means of identifying the local coordination. An example of this is provided by Lytle et al. [24], who identified the significant pre-edge feature in spectra of the tetrahedrally coordinated transition element oxides of Ti^{4+}, V^{5+}, Cr^{6+}, Mn^{7+}, and Fe^{3+} as being due to a 1s to 3d transition. The decrease in intensity

observed across this series is due to the progressive filling of the 3d band. In contrast, the pre-edge feature is virtually absent in octahedral oxide coordination.

- Empirical analysis of the "white line" height and peak shape is also useful. The "white line" is defined as the peak immediately above the x-ray absorption edge (Figure 7.9). The term derives from the now historical use of x-ray film where strong x-ray absorption appeared as a white line. The peak height, relative to the pre-edge and post-edge regions, can be indicative of the relative occupancy (as for pre-edge features) of the orbital/band to which electron excitation is occurring. The "white line" of the group VIII metals increases in intensity from Au to Re ($2p_{3/2}$ XANES) as the occupancy of the 5d orbital decreases [25].

- Quantitative speciation analysis of the XANES region by the measurement of standard spectra is frequently carried out. The spectra of the "unknown" and the standards are normalized so that the step height between the pre-edge and post-edge regions is one and the standards are then fitted as a linear combination to the unknown.

An example of some of the XANES features discussed is provided in Figure 7.9. A pre-edge feature for $CuSO_4$ is shown in Figure 7.9a and enlarged in 9b. It has been attributed to a number of possible phenomena [26], but is present only for Cu^{2+} containing species. The two major peaks in the derivative of the XANES portion of the spectra, referred to as α and β respectively, result from the distortion of Cu coordination octahedra on elongation of the two axial bonds. The greater the energy separation of these two components, the greater is the degree of distortion.

Strictly speaking, XANES spectra cannot be regarded simply as a ground-state projected DOS, although this has been used as a basis for interpretation through the application of quantum chemical codes such as CASTEP [27], and X_α [28]. Rather the spectra are more indicative of a transition state, i.e., the projected DOS of the excited state. This more sophisticated approach is available through STOBE [29], VASP [30], and Gaussian 03 (using time-dependent Hartree–Fock [HF] and density functional theory [DFT] [31]) which enable core-hole formation and/or excited state configuration to be taken into account explicitly. However, these approaches do not account for energy-dependent self-energy effects, final state effects, or multiple scattering, and thus data interpretation remains limited.

In recent years, significant advances have been made in the theoretical interpretation of XANES measurements with a growing awareness of the role that multiple scattering plays in determining the XANES profile. For both EXAFS and XANES, the photoelectron can be backscattered from the surrounding atoms to interact either constructively or destructively with both the outgoing waves and other backscattered photoelectron waves. Whether the interference is constructive or destructive depends on the energy of the photoelectron and the local atomic geometry. Hence, by changing the energy of the photoelectron by varying the energy of the incident x-ray radiation, an interference pattern can be obtained. For both XANES and EXAFS, the photoelectron wave undergoes a phase shift on backscattering that must be taken into account during analysis.

Multiple scattering of the excited photoelectron wave in the XANES region is now recognized as being an even more significant process than in the EXAFS region. The low kinetic energy of the outgoing photoelectron wave in the XANES region results in strong backscattering from the neighboring atoms, and this scattering tends to be from valence and/or conduction bands due to their large photoelectron cross sections. As the KE of the photoelectron increases, that is, as the incident x-ray energy increases into the EXAFS region, the scattering becomes weaker and is less sensitive to the surrounding electronic structure, since the scattering tends to be more from the core levels of the neighboring atoms. For this reason, EXAFS data are used for geometrical interpretation of the local structure whereas XANES data contain more electronic state information and are particularly sensitive to the excitation and relaxation processes inherent in photoelectron generation. Therefore calculation of scattering potentials and their impact on the x-ray wave phase shift in the XANES region is complicated and not effectively modeled by spherical potentials whereas in the EXAFS region spherical scattering potentials are sufficient.

Sophisticated analytical methodologies now exist, and are being constantly developed, in which the wave functions used in ground state quantum chemical applications are replaced by a one-particle Green's function to enable a "quasi-particle" approach. This enables final state effects encompassing both inelastic losses resulting in final state broadening, and peak shift effects, to be determined. In addition, energy-dependent self-interaction effects prevalent in the XANES region and core-hole potentials can be accounted for. This approach has been implemented in a number of packages, the most recent of which is the FEFF8 code [32] available from http://feff.phys.washington.edu/feff/. It is also implemented in the ARTEMIS module [33] of the IFEFFIT freeware suite [34] available at http://cars9.uchicago.edu/~ravel/software/. Useful reviews of theoretical approaches to XANES (and EXAFS) analysis are those by Rehr [35] and Rehr and Albers [36].

7.3.5 Introduction to the Interpretation of EXAFS

Fundamental understanding of the physics of the EXAFS process is more complete than for XANES, due primarily to the lower sensitivity of the EXAFS spectra to the electronic environment. However, unlike XANES, where considerable information may be derived from empirical or comparative studies, even basic EXAFS analysis requires mathematical computations. Analysis of the EXAFS interference fringe allows determination of the structure surrounding the atom, from which the electron has been excited, to within, ideally, ±0.002 nm. Where the data are of sufficient quality the secondary interactions, i.e., those between atoms from which the photoelectron wave is backscattered, can be taken into account during the analyses and the surrounding structure can be determined in three dimensions (shown schematically in Figure 7.10). When either the quality of the data or the nature of the material being examined preclude this, then analysis will yield only distances between the element from which electron excitation has taken place, and its neighbors.

Prior to modeling the EXAFS data, background subtraction must be undertaken, typically involving two stages, the first of which is the removal of the pre-edge background from the entire spectrum. This is performed by fitting the pre-edge region to an appropriate function, and then removing the extrapolated contribution of this function across the entire energy range of the data. After that the post-edge background, which in theory involves the removal of the contribution from the smooth monotonically decreasing edge cross section, can be removed. It is not, however, directly measurable, and the background can be affected by the nature of the pre-edge subtraction and other system variables. This has traditionally been undertaken by fitting a spline to this region (as in Figure 7.10). However, there is no certainty that the background fitted is correct and the use of a spline consisting of too many individual polynomials will result in the EXAFS oscillations being subtracted as well.

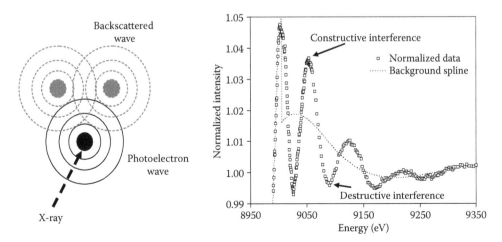

FIGURE 7.10 Schematic of the EXAFS process and the resulting constructive/destructive interference.

EXAFS data are presented typically as a function of k (nm^{-1}), the modulus of the photoelectron wave vector:

$$k = 10 \times \hbar^{-1} \sqrt{15.2 m_e (E - E_0)} = 162 \sqrt{E - E_0} \quad (7.6)$$

where

m_e is the rest mass of an electron

E (keV) is the energy of the incident radiation

E_0 is the absorption edge (this variable was also described in Section 7.2.3)

The first-order interference fringe, i.e., ignoring secondary interactions between backscattering atoms, may by constructed by considering the sum of sinusoidal contributions, described by Ref. [37]:

$$\chi(k) = \sum_i \frac{N_i A_i(k)}{k R_i^2} e^{\frac{-2R_i}{\lambda}} e^{-2\sigma_i^2 k^2} \sin\left[2kR_i + \phi(k)\right] \quad (7.7)$$

where there are i shells, A is the scattering amplitude function, λ is the magnitude of the photoelectron attenuation length, ϕ is the phase function, and σ is the Debye–Waller factor, which accounts for atomic disorder. It is more common than not to use algorithms that take into account multiple scattering and these have been implemented widely. R_i is the distance between the photoelectron emitter and the backscattering atom. The oscillation intensity, $\chi(k)$, is frequently weighted by k^3 in order to increase the amplitude of the oscillations at high k.

One way of assessing whether a background subtraction is likely to be realistic is to examine the Fourier transform of the background-subtracted data, which provides a pseudo-radial distribution function ($R+\Delta$). A high magnitude of the Fourier transform at low $R+\Delta$ is physically unrealistic, and the flat region at low $R+\Delta$ should be maximized by manipulation of the background spline. An alternative, and much easier approach from the experimenter's point of view, is implemented in the analysis routines ARTEMIS and ATHENA [33]. The background in this case is refined from the data by analyzing and removing the low frequency components of the Fourier transform [38], and is usually carried out automatically during the data processing.

The resulting background-subtracted interference pattern is then analyzed by constructing a theoretical atomic model and refining the model to give a best fit to the data measured. The specific neighboring atomic sites taken into account within the model are described as "shells". Where a three-dimensional analysis is not being undertaken, each shell consists of a distinct interatomic distance between the atom from which the photoelectron has been excited and a particular element from which it has been backscattered. Where a three-dimensional analysis is being carried out, each shell will consist of a specific neighboring element that has a similar geometrical relationship to the emitter atom. In either case, the occupancy (N) of any one shell may be more than unity and is dependent on coordination geometry.

In practice, to fit a particular set of data, σ, N, and various coordinates defining spatial relationships are refined for each shell. In addition, E_0 and an overall amplitude factor must be refined. The closer together the emitter and the backscattering atoms are, the higher is the frequency of scattering oscillations in k-space. Also, larger N gives greater oscillation amplitude, while large σ leads to more rapid damping of the oscillation as k increases. If on refinement, σ becomes negative for one or more shells then the probability is that the model is not reliable. A useful introduction to EXAFS analysis is that by Garrett and Foran [39].

In Figure 7.11a, there is an example of Cu 1s EXAFS resulting from the presence of Cu(I) on the surface of sphalerite (ZnS), deposited under the same conditions as detailed in Ref. [40]. The sphalerite grains were of size less than 20 μm, and the Cu was adsorbed at a coverage of less than a monolayer from 10^{-5} M $Cu(NO_3)_2$ solution at pH 5.5; the conditions precluded Cu hydroxide or oxide precipitation. The measurement was carried out in the fluorescence mode (Section 7.3.6).

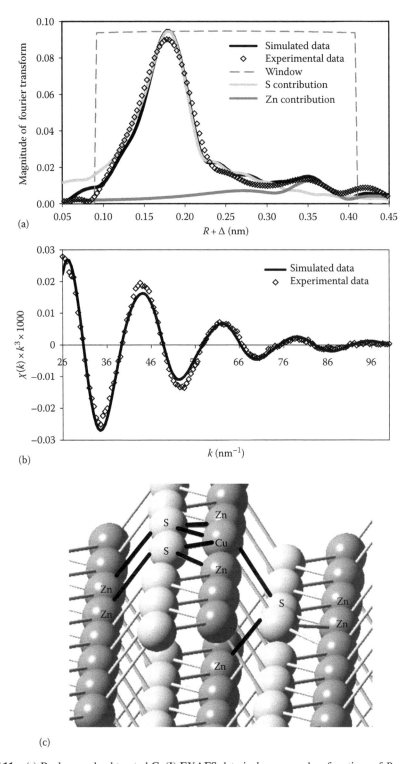

FIGURE 7.11 (a) Background-subtracted Cu(I) EXAFS data in k-space and as functions of R, subsequent to Fourier transformation. The simulated data resulting from refinement of the model in $(R+\Delta)$-space are also shown in both (a) and (b). In (b) the contributions from the S and Zn containing shells are given. (c) A small section of the sphalerite (110) surface. The nearest shell containing S and the second shell containing Zn are indicated by labels. The minimal distortion required within the surface to accommodate the Cu atom is apparent.

The EXAFS data in k-space appear smoothly sinusoidal indicating that they arose predominantly from a single shell. On simulation of the data in $(R+\Delta)$-space, this was indeed found to be the case, with the majority of the profile in Figure 7.11b being contributed by an S shell. The Cu–S bond length was refined to be 0.009 ± 0.001 nm shorter than the Zn–S bond length in sphalerite, with a Cu to Zn distance 0.013 ± 0.006 nm longer than the Zn to Zn distance in sphalerite. The occupancies of the S and Zn shells were fixed at 3 and 7, respectively, and not refined. A possible structure, commensurate with these bond lengths, is given in Figure 7.11c. The three Cu–S bond lengths are slightly shortened while the S to Zn lengths are elongated by pushing the Cu–S$_3$ structure slightly up and out of the surface.

7.3.6 Surface-Specific Modes of Measurement

The method of obtaining XAS data determines the analytical information depth, which can vary from that of the bulk to that of surface-specific analysis, probing only the top few nm.

One of the most common modes is the measurement of the intensity I of the x-rays transmitted through a sample, which is then compared to the incident intensity, I_0, to give a plot of absorbance versus incident photon energy. This is effectively a bulk measurement with the depth of measurement being limited only by the depth of penetration of the incident x-rays into the sample. Where the element of interest is present only on the surface of the sample, measurements may be carried out in transmission mode but sensitivity is likely to be lost due to x-ray absorption in the underlying bulk, thus reducing the signal-to-noise ratio. In the case where a core electron is excited in a relatively heavy element, and the underlying structure is light, this mode of measurement may indeed be viable. Depending on the energy of the incident x-rays, such measurements may be carried out in air or an environmental cell. Transmission XAS is particularly sensitive to the homogeneity of a sample and in particular to pin holes in the sample, which can cause "leakage" of the incident beam through the sample, thus corrupting the transmitted XAS intensities.

XAS spectra may be measured via other processes which also reflect the x-ray absorption characteristics. These include direct measurement of the intensities of the emitted electrons, or of processes related to the subsequent relaxation events. On excitation of an electron from a core orbital, both x-ray fluorescence and Auger electron emission may occur (see also Chapter 3). Fluorescence results in the emission of an x-ray as an electron from an orbital with a higher (or lower) angular momentum ($\Delta j = \pm 1$) electron relaxes into the core level hole created. For fluorescence XAS the information depth is governed by the distance that the emitted x-rays can travel through the sample, and is thus determined by their energy and by sample absorption, both of which can be calculated readily. As for transmission measurements, fluorescence XAS measurements may be carried out in air, or an environmental cell may be used if the energy of the emission is sufficiently high.

Surface EXAFS measurements are generically referred to as SEXAFS and involve the measurement of electron emission. In the surface spectroscopies Auger electron spectroscopy (AES), x-ray photoelectron spectroscopy (XPS), ultraviolet photoelectron spectroscopy (UPS) etc., the required information about the nature and number of surface and subsurface species is obtained from those features in a spectrum that arise from elastically scattered electrons produced in Auger and photoemission events. According to the number of different surface species, such a spectrum may contain many features, or peaks, each characteristic of the element in which the primary event occurred. Apart from the use of the shape and magnitude of the underlying inelastic background to determine the depth distribution of the various species, as described by Tougaard in Chapter 8, little attention is paid in those techniques to the inelastically scattered electrons.

For SEXAFS, on the other hand, attention is focused on one particular Auger excitation event, and on *both* the elastic and inelastic events that follow the excitation. Thus, in SEXAFS, measurement takes the form of integration of the electron yield, over a chosen range of KEs, as a function of the variable incident x-ray photon energy. It is this integrated yield that displays oscillatory XAS behavior.

The range of KE chosen does not have to include the discrete features of interest in AES and XPS, but often does. This type of measurement can be contrasted with that in XPS and x-ray excited Auger electron spectroscopy (XAES), where the incident x-ray energy is kept fixed.

Elastically scattered Auger electrons, i.e., those of interest in AES and XAES, have fixed KEs, whereas the KEs of the electrons resulting from photoemission vary as a function of the incident x-ray energy, and will at some stage overlap those of the elastically scattered Auger electrons as the x-ray energy is varied. This results in intensity modulations that are not related to XAS processes and limit the incident x-ray energy range available. The measurement of elastic Auger yield (EAY) results in good signal-to-noise ratios but small signals, and must be carried out in a photoemission analyzer with the energy range centered on the KE of the Auger peak. If, for the material being analyzed, the EXAFS measurement is made problematical by an x-ray absorption energy falling just below the edge of interest (<200 eV), EAY measurement may still be made by choosing an Auger emission at a higher KE than that anticipated for Auger emission resulting from the lower energy absorption edge.

Integration of other regions of the spectrum leads to other yield modes. For example, partial Auger yield (PAY) measures the integrated electron intensities of both the elastically scattered Auger electrons and the higher energy region of the inelastically scattered Auger electrons. The cutoff energy (usually 5–10 eV below the Auger energy) above which PAY electron intensities are measured can be chosen to avoid interference by direct photoemission. Secondary yield (SY) measurements are made by integrating the low energy region of the inelastic electron emission spectrum (i.e., below the 5–10 eV cutoff). Finally, in the total yield (TY) mode the intensities of all electrons emitted, consisting of Auger and photoelectrons and those inelastically scattered subsequent to the excitation event, are measured. The TY is usually dominated by the intensities of inelastically scattered Auger electrons, so that the elastically scattered photoemission electrons, which do not necessarily exhibit EXAFS modulations, do not interfere significantly.

At the moment of excitation of an Auger event, all electrons, regardless of depth within the sample, will start out with the same KE. But almost at once inelastic scattering begins as the Auger electrons are scattered from the neighboring atomic electron clouds, resulting in a loss of KE. Thus the deeper within a solid the original Auger event has occurred, the greater the probability that the Auger electron will suffer energy loss through successive scattering events, before it emerges from the surface. It follows that, for any one Auger event, the elastically scattered Auger electrons must originate closer to the surface than those inelastically scattered, in order to be observable at the characteristic Auger energy. Since in SEXAFS spectral structure both elastic and inelastic scattering can be recorded, information as a function of depth can be obtained. Thus the SY and TY modes have the lowest, and the PAY and EAY the highest, surface specificity.

The choice between the modes of measurement (including fluorescence) must be made on the basis of both total signal-to-noise and signal-to-background ratios, which are best determined experimentally. EAY, PAY, and SY may be measured using an electron energy analyzer or a biased detector, in order to select the desired electron energy range. TY may also be measured by an electron detector (with no bias), or by drain current, i.e., the current required to cancel the current of electrons emitted from the surface. Measurements from light elements, regardless of the nature of the data collection, must also be carried out under vacuum due to attenuation in air of the incident x-ray beam. Information depths may vary from <5 nm to 50–200 nm for low to higher incident x-ray energies, respectively (e.g., for K-shell excitation from Cu at 8980 eV) [20].

A further method for carrying out surface-specific measurements is that of total reflection EXAFS (ReflEXAFS). For this mode, the incident x-ray reflection at angles less than the critical angle for total reflection is measured, resulting in both high reflectivity (>50%) and low penetration depth (<3 nm) [20]. However, for these measurements to be effective, the sample must be extremely flat and smooth, and a high degree of control of sample orientation is required. In addition, the footprint of the incident x-ray beam on the sample is elongated due to the low angle of incidence (typically <0.5°), so that the samples need to be several tens of millimeters long and homogeneous over the entire length [41]. ReflEXAFS may be measured in electron yield, fluorescence yield, or x-ray absorption modes.

7.4 SMALL ANGLE X-RAY SCATTERING AND REFLECTOMETRY

7.4.1 INTRODUCTION

7.4.1.1 X-Ray Diffraction, Reflection, and Refraction

The term "diffraction" refers to the interaction between a propagating wave (e.g., of photons or neutrons) and a structure with features of length scales similar to the wavelength. By contrast, reflection and refraction (Figure 7.12) are phenomena observed at the interface of two different media—with no requirement that either the media or interface possess any "fine" structure.

In practice, there is little point in using x-ray reflectivity unless one is interested in some kind of fine structure with length scales between 0.1 and 100 nm. Deviations from "ideal" specular reflection are often caused by the same elastic scattering mechanisms that give rise to diffraction.

7.4.1.2 Reflectivity, SAXS, GISAXS, and Synchrotron Radiation

Figure 7.13 shows typical experimental arrangements for x-ray reflectivity, small angle x-ray scattering (SAXS), and glancing incidence small angle x-ray scattering (GISAXS) [42]. In all three types of measurement, the incident beam is usually monochromatic and collimated with a series of mirrors and slits.

These three techniques have a number of things in common:

1. They are all manifestations of elastic scattering, in that the incident and scattered x-ray energies (and hence wavelength) are the same.
2. They all involve relatively small deflections (generally less than about 10°) of all or part of an incident x-ray beam.
3. The higher the measurable angle of deflection or scatter, the smaller the real-space length scales that can be examined. Measurement is usually limited by the low intensity of scattering signal that becomes even lower as angle increases.

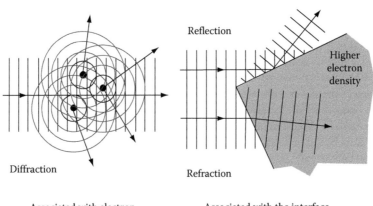

Diffraction	Reflection · Refraction
Associated with electron density fluctuations on 0.1–1000 nanometer scale	Associated with the interface between two media with different electron densities

FIGURE 7.12 Diffraction, reflection, and refraction shown schematically as resulting from differences in electron density. In the case of reflection and refraction, at least one well-defined interface must be present.

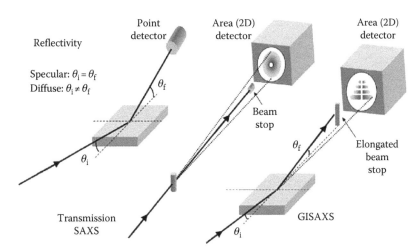

FIGURE 7.13 Reflectivity, SAXS, and GISAXS shown in their simplest configurations. The incident x-ray beam is assumed here to be monochromatic and well collimated. Reflectivity here is shown from the surface of a specimen—it could also be from the interface of two media such as immiscible liquids. (From Luo, G., Malkova, S., Yoon, J., Schultz, D.G., Lin, B., Menon, M., Benjamin, I., Vanysek, P., and Schlossman, M.L., *Science*, 311, 216, 2006.)

4. Unlike techniques such as AFM, SEM, and the fluorescence microprobe, the region of interest is not inherently limited by the smallest measurable length scales. This allows bulk averaging of nm-scale data over macroscopic volumes and areas.

The high flux of a synchrotron beam is important for all these techniques, as small angle scattering and reflectivity intensities tend to decrease as the fourth power of the scattering angle [43]. As a result, intensities must often be measured over many orders of magnitude, placing high demands on the linearity and dynamic range of detector systems. A single-channel, or point detector, in conjunction with attenuator foils, provides the greatest dynamic range possible, and is often the detector system of choice for reflectivity. SAXS and GISAXS require area detectors, which usually have a restricted dynamic range—necessitating the use of a beam stop to prevent the detector from being damaged by the main transmitted or specularly reflected x-ray beam.

7.4.1.3 What Sort of Interfaces Do Reflectivity and SAXS See?

X-ray reflectivity is built around the idea of refraction and reflection at a single well-defined interface between materials with different average electron densities. The local electron density at this interface may have a rich substructure that extends some micrometers away from the plane of reflection, but all real-space features can be described in terms of their proximity and orientation to the interface (Figure 7.14).

Transmission SAXS (as opposed to GISAXS) samples a large number of interfaces with a range of orientations to the incident beam. Often these interfaces are actually the boundaries of particles suspended in some liquid or embedded in a solid matrix. In other cases, they simply delineate material phases of different density, as in the case of block copolymers or polymer blends. SAXS samples a large population of such interfaces, thus making it a bulk-averaging technique.

In SAXS, every sub-volume (or particle) presents the incident beam with a range of interface orientations. In short, all elastic x-ray scatters come from the interfaces between media of different electron density. In a material with no electron density variation, no interfaces exist, and no diffraction or reflection is observable. Perfect single crystals show no small angle scatter, but at higher

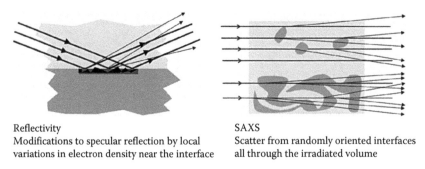

Reflectivity
Modifications to specular reflection by local
variations in electron density near the interface

SAXS
Scatter from randomly oriented interfaces
all through the irradiated volume

FIGURE 7.14   Types of region probed by reflectivity and SAXS.

scattering angles, where smaller length scales are probed, very strong scatter can be seen. This is because on a sufficiently small length scale, all matter starts to look very "grainy" with electrons clustered around atomic nuclei.

7.4.1.4 Momentum Transfer Vector q

In matter, an incident electromagnetic wave will cause free (or nearly free) electrons to vibrate, due to the oscillating electric field of the incident wave [44]. This produces secondary or scattered waves which interact with each other to cause diffraction and reflection. Refraction is caused by the interaction of the secondary waves with the primary beam [5].

A key concept in explaining these phenomena is that two waves with the same wavelength propagating in the same direction will interfere, either constructively or destructively, depending on their relative phase. This can be calculated easily if the wavelength and the difference in path length that each wave front travels are known.

Figure 7.15 shows how a wave front that moves in the direction $\hat{k}_0$ and generates a secondary wave from the electron at P, needs to travel a further distance $|AP'|$ in order to generate another

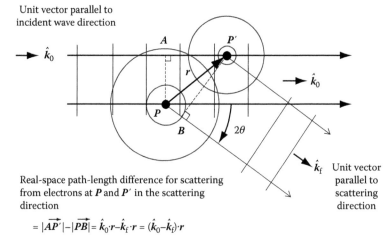

Unit vector parallel to
incident wave direction

Real-space path-length difference for scattering
from electrons at P and P' in the scattering
direction

$$= |\overrightarrow{AP'}| - |\overrightarrow{PB}| = \hat{k}_0 \cdot r - \hat{k}_f \cdot r = (\hat{k}_0 - \hat{k}_f) \cdot r$$

FIGURE 7.15   Vector representation of path-length difference for a wave scattered from electrons at P and P'.

secondary wave at P'. After the secondary waves are generated, the wave from P in the direction $\hat{k}_f$ will have to travel further than that from P' by an amount $|PB|$. This path-length difference is given by $(\hat{k}_f - \hat{k}_0) \cdot r$. The overall phase delay between the secondary waves scattered from P and P' is therefore given by:

$$\text{Phase difference} = \frac{2\pi}{\lambda} \cdot \text{path length difference} = q \cdot r \tag{7.8}$$

where

$$q = \frac{2\pi}{\lambda}(\hat{k}_0 - \hat{k}_f) \tag{7.9}$$

The dot product $q \cdot r$ is seen in much of the theory used to describe elastic scatter and reflectivity. The vector q is known as the momentum transfer vector because photons have inherent momenta* which must be conserved in an elastic interaction. The magnitude of q is given by the important relation

$$|q| = q = \frac{4\pi \sin \theta}{\lambda} \tag{7.10}$$

Equation 7.10 is a near-universal expression since q is a parameter widely used in x-ray, neutron, and electron scattering. It has units of inverse distance, most commonly nm^{-1}. The advantage of this more "universal" quantity is that it allows easier comparison of scattering data taken at different wavelengths (energies) and scattering angles.

7.4.1.5 Measurable q and Real-Space Length Scales

The average phase difference $<q \cdot r>$ between scattered waves from all the possible combinations of charge/volume elements separated by r in a sample must become larger if the average magnitude of either q or r also increases. The larger the average phase difference, the more destructive the interference in the secondary scatter, and the weaker the scatter at a momentum transfer of q.

The implication of this is profound. At a constant wavelength, scatter or reflectivity at higher angles contains information relating to smaller features. At the highest scattering or reflection angles, it is possible to obtain information relating to the smallest atomic length scales.

7.4.2 REFLECTIVITY

7.4.2.1 Data Collection: Specular Reflectivity and Diffuse Scattering

Figure 7.16 shows the basic geometry used for x-ray reflectometry. When the surface to be studied is a liquid (i.e., subject to the leveling effect of gravity), some way must be found to steer the x-ray beam vertically downward. This is a nontrivial technical challenge when using a synchrotron, and usually involves the use of a separate beam-steering crystal [45].

X-rays encountering an interface between the medium in which they are propagating and another medium of higher electron density will reflect and refract to a degree dependent on their wavelength

---

* This is a finding that comes from special relativity, which says that although photons have no rest mass, they do in fact have a momentum given by h/λ (cf. Section 7.2.3) where h is Planck's constant and λ the photon wavelength. "Real" photon momentum was experimentally verified by the observation of radiation "pressure" of intense light beams on reflective surfaces (cf. *The Astrophysical Journal*, **17**, 315 (1903)).

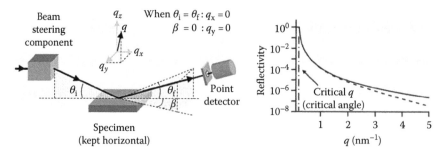

FIGURE 7.16 X-ray reflectometry experiment. Specular reflectivity is measured when $\theta_i = \theta_f$ and $\beta = 0$. When neither of these conditions is met (i.e., $\theta_i \neq \theta_f$ or $\beta \neq 0$) the intensity measured is due to diffuse scattering.

and angle of incidence on the interface. If the angle of incidence is less than the critical angle (i.e., their q is less than the critical q) such x-rays will undergo total external reflection.* Above this angle, the amount of reflected intensity generally drops off rapidly while an increasing fraction is transmitted through the interface. For specular reflectivity measurements, the in-plane incident and reflected angles (θ_i and θ_f) are equal and the out-of-plane scattering angle (β) is zero. X-ray photons measured under these conditions have suffered only a change of momentum normal to the surface (in the q_z direction).

A perfectly smooth flat surface will exhibit Fresnel reflectivity, giving a profile shown by the solid line in the plot to the right of Figure 7.16. This response curve is flat with a maximum possible reflectivity of 1.0 below the critical angle (and critical q) while at higher values of q it soon starts to decay as q^{-4}. Surfaces of materials such as silica (glass) that can be fabricated with an atomically smooth surface routinely demonstrate Fresnel reflectivity. Coating these surfaces with thin layers of dense metal (such as vapor-deposited platinum) greatly increases their critical angle. Such surfaces form the basis of one of the most important optical components used in synchrotrons—the x-ray mirror.

In the plot to the right in Figure 7.16 the Fresnel reflectivity of a perfectly flat and smooth surface (solid line) is compared to the reflectivity from pure water at room temperature (dashed line) [46,47]. The roughness of the liquid water surface is typically defined as a Gaussian height distribution with $\sigma = 0.27$ nm at room temperature.

By contrast, a reflectivity measurement of the surface of pure water will not give a perfect Fresnel profile. This is due to capillary waves, which are microscopic ripples generated thermally by Brownian motion on the surface of the liquid [47]. Such roughness reduces the effective reflectivity which becomes more noticeable (Figure 7.16) at higher q values. The intensity lost to the specularly reflected beam appears as diffuse scatter which can be measured by allowing the conditions $\theta_i \neq \theta_f$ and $\beta \neq 0$, which probes inhomogeneities in electron density parallel to the surface or interface.

7.4.2.2 Interfacial Structure and the Reflectivity Profile

While roughness and nonuniformity in electron density parallel to the interface can reduce the Fresnel reflectivity, it is also possible to increase and modulate the observed reflectivity with uniform electron density variation normal to the surface. Figure 7.17 shows the measured reflectivity from gold nanoparticles (6 nm diameter) floating on the surface of water [48]. The thin and uniform layer of highly

---

* Total external reflection for x-rays is equivalent to total internal reflection for visible light. In both cases the photons are traveling in a medium with a higher refractive index and striking at a shallow angle an interface to a medium with lower refractive index. The key difference is the fact that higher electron density for visible light means higher refractive index, whereas for x-rays, the reverse is true.

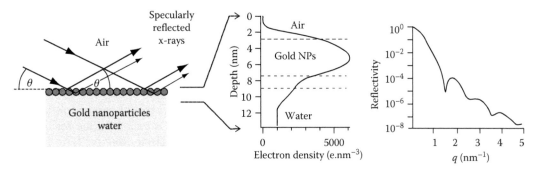

FIGURE 7.17 X-ray reflectometry measurement of a gold-nanoparticle monolayer on the surface of water. Layer thickness is about 6 nm as seen in the reconstructed electron density curve fitted to the experimental reflectivity profile (far right). Electron density is given relative to that of pure water at room temperature.

electron-dense gold atoms gives rise to pronounced oscillations in the reflectivity curve,* which can be used to reconstruct a profile, with fractional-nm resolution, of electron density normal to the water/gold and gold/air interfaces.

This technique can be used in a more generalized way to study highly aligned molecule monolayers on the surface of some liquid, usually (but not necessarily) water. A liquid handling device known as a Langmuir trough [49] can be used to corral and compress thin layers of molecules floating on such a surface.

7.4.3 SMALL ANGLE X-RAY SCATTERING

7.4.3.1 Data Collection

Figure 7.18 shows the simplest SAXS geometry (sometimes known as pinhole geometry) used to obtain a two-dimensional SAXS pattern. The flight path between the sample and the detector must be evacuated, as scattering from air molecules would overwhelm the scattered intensity from the

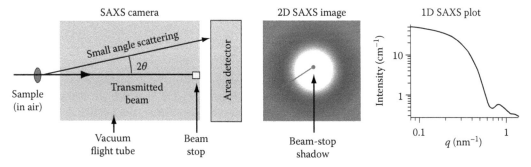

FIGURE 7.18 SAXS instrumentation. The 2D SAXS image of highly monodispersed iron nanoparticles suspended in toluene, with an average diameter of 14 nm, was azimuthally integrated (averaged over 360°) and then plotted against q. Note that q is derived from the 2θ scattering angle, which is in turn calculated from the radial distance from the center of the scattering pattern and the sample-to-detector distance.

---

* A smooth surface covered by a thin, even layer of material with different electron density will produce evenly spaced reflectivity oscillations which decay smoothly with increasing q. These are called Kiessig fringes and arise from interference between reflected waves from the top and bottom interfaces.

sample. The specimen in this case was a suspension of iron nanoparticles, which scatter isotropically (i.e., without preferred orientation). Data from such measurements are usually presented as log I versus log q plots.

7.4.3.2 Form Factor

In Section 7.4.1.4 it was seen that a single free electron will elastically reemit a certain fraction of the amplitude of an incident electromagnetic wave. If this fraction is called A_e, the scattering power of any particle of matter can be related to that of a single electron using the form factor $f(q)$. The intensity of scatter from any particle is equal to the square of the amplitude, which means that

$$\text{Scattered intensity} = I(q) = \left| f(q) \cdot A_e \right|^2 \tag{7.11}$$

In the case of single atoms, the form factor is called the atomic form factor and has units of "electrons per atom." This particular form factor is often written as

$$f_a(q, E) = f_a^0(q) + f_a'(E) + i f_a''(E) \tag{7.12}$$

and is dominated by the $f_a^0(q)$ term, which at $q = 0$ is exactly equal to Z, the number of electrons surrounding the atomic nucleus. The second two terms, $f_a'(E)$ and $i f_a''(E)$, are energy-dependent terms known as dispersion corrections. They arise because the electrons around an atom are not truly free, and they are generally difficult to calculate. Fortunately, as long as x-ray absorption edges are avoided the second two terms can usually be ignored. In addition, for the low q-values (low angles) routinely probed in SAXS the $f_a^0(q)$ term differs little from Z. As a result, when calculating the form factor for a particle containing many atoms, it is possible to dispense with individual atomic form factors, and instead talk about electron density.

In short, the form factor describes fully the x-ray scattering of a given particle. In theory it can be calculated exactly if a complete inventory of all the atoms and their positions within a particle is available. This is usually difficult to know, except in the case of biological macromolecules where tens of thousands of atomic coordinates can be documented. For the most part, it is more common to assume that the particles can be approximated by simple geometric solids such as spheroids, rods, and disks [50,51]. The $f(q)$ for these shapes are well known, and the collective SAXS profile for N such particles in an irradiated volume can be predicted to be

$$I(q) = N \left| f(q) \cdot A_e \right|^2 \tag{7.13}$$

Figure 7.19 shows the theoretical scattering curves calculated for several size distributions of spherical particles [52]. The intensities are given as reciprocal centimeters (cm^{-1}), which are also known as absolute intensities because they relate only to the inherent sample material, and not to the thickness of the specimen. It can also be seen that the $I(q)$ versus q curve for the monodispersed (identically sized) spheres has strong oscillations that become "washed out" for the polydispersed samples where the spheres have a range of diameters.

This highlights an important general feature of SAXS scattering profiles. A collection of particles with well-defined shapes and a narrow size distribution will often give rise to a SAXS profile with clear oscillations.

7.4.3.3 General Features of a $I(q)$ versus q SAXS Profile

In most materials, inherent variations in particle size and shape remove almost completely all the form-factor oscillations, giving profiles that look more like that shown in Figure 7.20.

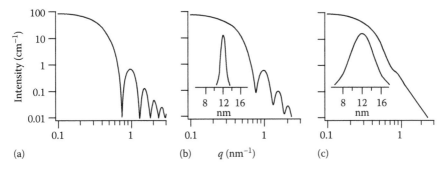

FIGURE 7.19 Solid iron spheres (diameter ~12 nm) suspended in toluene with (a) zero polydispersity, (b) ~10% polydispersity, and (c) ~20% polydispersity. The size distributions of spheres used to calculate (b) and (c) are shown as inset histograms. (Available from www.xor.aps.anl.gov)

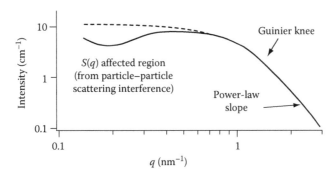

FIGURE 7.20 SAXS profile of 5 nm diameter Cu nanoparticles implanted in silica. The dashed line shows how the profile would look without particle–particle scattering interference. The estimated nearest neighbor distance is about 8 nm. The power law slope gradient is −4.0 on the log–log curve, making it a Porod slope (see text) and indicating that the particle–matrix interfaces are smooth.

In such profiles, a number of other important features are usually apparent—hopefully within the scattering q-range available to the experiment. These include [50] the power law slope at high q, the Guinier knee at lower q, and sometimes a correlation peak. These features are indicated in Figure 7.20 and can be summarized as follows:

1. The power law slope: the high-q part of the SAXS profile where intensity decreases as q^{-p} and p can take values from 1 to 4. This high-q behavior is indicative of what the interfaces within the sample look like on the smallest length scales. If these interfaces are smooth and well defined, then $p = 4$, and the slope is called a Porod slope after Gunther Porod, the first person to show that smooth interfaces in a solid give rise to a q^{-4} decay in scattered intensity [53]. Smaller values of p can indicate rough interfaces, or a more complex particle geometry.
2. The Guinier knee: a change in gradient described by the relationship: $I(q) \propto \exp\left(\dfrac{-R_g^2 q^2}{3}\right)$

   where R_g is the radius of gyration and can be related to some limiting length scales, such as the radius of a spherical particle.
3. Particle–particle scattering interference: arises when the concentration of similar-sized particles in a solution is sufficiently high for the scattering from one particle to interfere with the scatter from its nearest neighbors. This gives rise to a structure factor that modifies Equation 7.13 to

$$\text{Scattered intensity} = I(q) = \left| f(q) \cdot A_e \right|^2 \cdot S(q) \qquad (7.14)$$

A significant structure factor can make analysis of solution SAXS difficult, which is why it is generally best to measure SAXS from the most "dilute" samples possible. It should be noted, however, that powder and single crystal diffraction are possible only because the structure factor $S(q)$ becomes highly significant when assembling millions of atoms or molecules on a periodic lattice.

7.4.3.4   Particle versus Interface Picture of Condensed Matter

In many solids that show strong SAXS there are no discernible particles. As mentioned in Section 7.1.4, it is the interfaces between regions of different electron density that are important to SAXS. Voids or low-density inclusions in a high-density matrix can be just as effective at scattering x-rays as high-density inclusions or particles in a low-density matrix. This is known as Babinet's principle [43].

7.4.3.5   Some Comments about GISAXS

GISAXS is a powerful technique [54] that has come into its own since synchrotron radiation became widely available (Figure 7.13). It incorporates all three elastic scattering phenomena including diffraction, reflection, and refraction. The advantage of this technique is that it allows the measurement of diffuse scatter (as for reflectivity with $\theta_i \neq \theta_f$ and $\beta \neq 0$) over a large range of q_x, q_y, and q_z simultaneously by using a 2D area detector. This allows rapid measurement of surfaces that may be changing with time, or that have a high degree of texture and richness of structure.

The main disadvantage of GISAXS lies in the fact that scattering at different angles from different parts of the specimen can be difficult to de-convolute from the final image. In addition to this, any data analysis [55] must take into account refraction effects from scatter originating below the sample surface.

REFERENCES

1. H. C. Pollock, *Am. J. Phys.*, *51*, 278 (1973).
2. P. L. Hartman, *Synch. Rad. News*, *1*, 4, 28 (1988).
3. K. Codling, *J. Synch. Rad.*, *4*, 316 (1997).
4. W. L. Bragg, *Proc. Camb. Philos. Soc.*, *17*, 43 (1914).
5. J. Als-Nielsen and D. McMorrow, *Elements of Modern X-Ray Physics*, John Wiley & Sons, New York (2001).
6. A. G. Schaufuß, H. W. Nesbitt, I. Kartio, K. Laajalehto, G. M. Bancroft, and R. Szargan, *J. Electron Spectrosc. Relat. Phenomena, 96*, 69 (1998).
7. O. Plekan, V. Feyer, F. Šutara, T. Skála, M. Švec, V. Cháb, V. Matolín, and K. C. Prince, *Surf. Sci., 601*, 1973 (2007).
8. J. J. Yeh and I. Lindau, *Atomic Data Nucl. Data Tables, 32*, 1 (1985).
9. M. Švec, V. Dudr, F. Šutara, N. Tsud, T. Skála, V. Cháb, V. Matolín, and K. C. Prince, *Surf. Sci., 601*, 5673 (2007).
10. C. N. Berglund and W. E. Spicer, *Phys. Rev. A, 136*, 1030, 1044 (1964).
11. E. W. Plummer and W. Eberhardt, *Adv. Chem. Phys., 49*, 533 (1982).
12. T. Strasser, C. Solterbeck, W. Schattke, I. Bartoš, M. Cukr, P. Jiříček, C. S. Fadley, and M. A. Van Hove, *Phys. Rev., B, 63*, 195321 (2001).
13. R. Courths, B. Cord, H. Wern, and S. Hüfner, *Physica Scripta, T4*, 144 (1983).
14. A. Santoni, G. Paolucci, K. C. Prince, and N. Christensen, *J. Phys.: Condens. Matter, 4*, 6759 (1992).
15. M. Zacchigna, C. Astaldi, K. C. Prince, M. Sastry, C. Comicioli, R. Rosei, C. Quaresima, C. Ottaviani, C. Crotti, A. Antonini, M. Matteucci, and P. Perfetti, *Surf. Sci., 347*, 53 (1996).
16. K. C. Prince, G. Paolucci, and A. M. Bradshaw, *Surf. Sci. 175*, 101 (1986).

17. K. C. Prince, M. Surman, Th. Lindner, and A. M. Bradshaw, *Solid State Comm., 59*, 71 (1986).
18. V. Pérez-Dieste, J. F. Sánchez-Royo, J. Avila, M. Izquierdo, L. Roca, A. Tejeda, and M. C. Ascenzio, *Surf. Sci., 601*, 742 (2007).
19. K. C. Prince, V. R. Dhanak, P. Finetti, J. F. Walsh, R. Davis, C. A. Muryn, H. S. Dhariwal, G. Thornton, and G. van der Laan, *Phys. Rev. B, 55*, 9520 (1997).
20. D. C. Koningsberger and R. Prins (ed.), *X-Ray Absorption, Principles, Applications, Techniques of EXAFS, SEXAFS and XANES*, Volume 92, Chemical Analysis Monographs, John Wiley & Sons, New York (1988).
21. M. M. Hoffman, J. G. Darab, S. M. Heald, C. R. Yonker, and J. L. Fulton, *Chem. Geol., 167*, 89 (2000).
22. S. R. Bare, XANES measurements and interpretation, EXAFS data collection and analysis course, APS, July 26–29, http://xafs.org/Workshops/APS2005 (2005).
23. G. N. George and M. L. Gorbaty, *J. Am. Chem. Soc., 111*, 3182 (1989).
24. F. W. Lytle, R. B. Greegor, and A. J Panson, *Phys. Rev. B, 37*, 1550 (1988).
25. G. Meitzner, G. H. Via, F. W. Lytle, and J. H. Sinfelt. *J. Phys. Chem., 96*, 4960 (1992).
26. A. J. Berry, A. C. Hack, J. A. Mavrogenes, M. Newville, and S. R. Sutton, *Am. Mineralogist, 91*, 1773 (2006).
27. G. U. von Oertzen, R. T. Jones, and A. R. Gerson, *Phys. Chem. Min., 32*, 255 (2005).
28. K. Tanaka, J. Kawai, and H. Adachi, *Phys. Rev. B, 52*, 733 (1995).
29. R. G. Wilks, J. B. MacNaughton, H.-B. Kraatz, T. Regier, and A. Mowes, *J. Phys. Chem. B, 110*, 5955 (2006).
30. Y.-A. Jeon, Y. S. Kim, S. K. Kim, and K. S. No, *Solid State Ionics, 177*, 2661 (2006).
31. R. K. Pandey and S. Mukamel, *J. Chem. Phys., 124*, 094106 (2006).
32. A. L. Ankudinov, B. Ravel, J. J. Rehr, and S. D. Conradson, *Phys. Rev. B, 58*, 7565 (1998).
33. B. Ravel and M. Newville, *J. Synch. Rad., 12*, 537 (2005).
34. M. Newville, *J. Synch. Rad., 8*, 322 (2001).
35. J. J. Rehr, *Rad. Phys. Chem., 75,* 1547 (2006).
36. J. J. Rehr and R. C. Albers, *Rev. Mod. Phys., 72*, 621 (2000).
37. E. A. Stern, *Phys. Rev. B, 10*, 3027 (1974).
38. M. Newville, P. Līviņš, Y. Yacoby, J. J. Rehr, and E. A. Stern, *Phys. Rev. B, 47*, 14126 (1993).
39. R. F. Garrett and G. J. Foran, EXAFS, in *Surface Analysis Methods in Materials Science*, O'Connor, D. J., Sexton, B. A., and Smart R. St. C. (Eds.) Springer-Verlag, pp. 347–373, Berlin, Heidelberg, New York (2003).
40. A. R. Gerson, A. G. Lange, K. E. Prince, and R. S. C. Smart, *Appl. Surf. Sci., 137*, 207 (1999).
41. V. López-Flores, S. Ansell, D. T. Bowron, and S. Díaz-Moreno, *Rev. Sci. Instrum., 78*, 013109 (2007).
42. G. Luo, S. Malkova, J. Yoon, D. G. Schultz, B. Lin, M. Meron, I. Benjamin, P. Vanysek, and M. L. Schlossman, *Science, 311*, 216 (2006).
43. O. Glatter and O. Kratky, *Small Angle X-Ray Scattering*, Academic Press Inc., London (1982).
44. E. Hecht and A. Zajac, *Optics*, Addison-Wesley, London (1974).
45. M. L. Schlossman, D. Synal, Y. M. Guan, M. Meron, G. Shea-McCarthy, Z. Q. Huang, A. Acero, S. M. Williams, S. A. Rice, and P. J. Viccaro, *Rev. Sci. Instrum., 68*, 4372 (1997).
46. S. K. Sinha, E. B. Sirota, S. Garoff, and H. B. Stanley, *Phys. Rev. B, 38,* 2297 (1988).
47. A. Braslau, P. S. Pershan, G. Swislow, B. M. Ocko, and J. Als-Nielsen, *Phys. Rev. A, 38*, 2457 (1988).
48. D. G. Schultz, X.-M. Lin, D. Li, J. Gebhardt, M. Meron, P. J. Viccaro, and B. Lin, *J. Phys. Chem. B, 110*, 24522 (2006).
49. V. M. Kaganer, H. Mohwald, and P. Dutta, *Rev. Mod. Phys., 71*, 779 (1999).
50. A. Guinier and G. Fournet, *Small-Angle Scattering of X-Rays*, John Wiley & Sons, New York (1955).
51. L. A. Feigin and D. I. Svergun, *Structure Analysis by Small-Angle X-Ray and Neutron Scattering*, Plenum Press, New York (1987).
52. I vs q plots calculated using Gaussian distribution of spherical form factors as implemented in the IRENA-2 software, available from www.xor.aps.anl.gov
53. G. Porod, *Kolloid Zeit., 124*, 83 (1951)
54. G. Renaud, R. Lazzari, C. Revenant, A. Barbier, M. Noblet, O. Ulrich, F. Leroy, J. Jupille, Y. Borensztein, C. R. Henry, J.-P. Deville, F. Scheurer, J. Mane-Mane, and O. Fruchart, *Science, 300*, 1416 (2003).
55. R. Lazzari, *J. Appl. Cryst., 35*, 406 (2002).

8 Quantification of Surface and Near-Surface Composition by AES and XPS

Sven Tougaard

CONTENTS

8.1 INTRODUCTION

Auger electron spectroscopy (AES) and x-ray photoelectron spectroscopy (XPS) are applied widely to problem solving in the analysis of surfaces and interfaces. The underlying principles, the physical mechanisms, and the technical basics of the instrumentation are described in Chapter 3. Here the focus will be on the practical application of AES and XPS to the quantitative characterization of surfaces on the nanometer depth scale and of near-surface elemental distributions. The reason that AES and XPS are so powerful for surface analysis on the nanometer scale is that the typical distance λ between inelastic scattering events for electrons of energies between 100 and 2000 eV (i.e., typical for Auger and photoelectrons in AES and XPS, respectively) is of the order of 1 nm. Measured emitted electron intensities $I(z)$ are then attenuated exponentially as a function of distance $z/(\lambda \cos \theta)$ to the surface according to

$$I(z) = I_0 e^{\frac{-z}{\lambda \cos\theta}} \tag{8.1}$$

where θ is the angle of emission with respect to the surface normal. With $\lambda \sim 0.5$–2 nm, AES and XPS are ideal techniques for the study of the outermost few nanometer.

When the practical working analyst sets out to solve a particular problem, he/she should first consider what information about the surface composition is needed in order to solve the problem, because the spectra that should be acquired, as well as the subsequent nature of the analysis, depend on the specific aim of the analysis. Is mere identification of the presence or not of a given element in the near-surface region (ca. 5 nm) sufficient? Is it necessary to know the precise number of atoms in this region? Is it important to know the chemical state of the atoms? Is the depth distribution over the outermost 5–10 nm important? Perhaps it is of crucial importance to know whether atoms that were deposited intentionally on the surface have stayed there or have moved inside over distances of a few atomic layers, whether they have agglomerated and formed islands or spheres on the surface, or if they have covered the surface completely as, e.g., a protective layer. It is amazing that XPS (and to a lesser extent AES) can provide answers to these vastly varied questions [1].

The basic principle of XPS is explained in Chapter 3. In summary, the sample surface is irradiated by monoenergetic photons, produced usually by electron bombardment of an Al or Mg anode, which causes the emission of soft x-rays with principal energies of 1486.7 eV for Al K_α, and of 1253.6 eV for Mg K_α. The photon energy is absorbed by atom core-electrons (with binding energy E_{bind}) in the surface region, and a photoelectron is emitted with kinetic energy $E_{kin} = h\nu - E_{bind} - \Delta$ where Δ accounts for the work function of the spectrometer and for corrections due to the nature of the coupling to the surrounding electrons. The latter depends on the chemical bond of the atom and E_{kin} therefore contains valuable information on the chemical state (see Chapters 3, 14, and 16 through 19).

The basic principle of AES is also explained in Chapter 3. Again in summary, the sample surface is irradiated by electrons of 2–30 keV energy which excite a core electron of binding energy E_1. In the Auger process, the core hole so produced is filled by an electron with binding energy E_2 and the released energy is then given up to a third electron with binding energy E_3 which is the Auger electron. Its kinetic energy is $E_{123} = E_1 - E_2 - E_3 - \Delta$. The core hole needed to trigger the Auger process may also be created by photon excitation and therefore an XPS spectrum also exhibits Auger peaks. Since electrons E_2 and E_3 are often from the valence band, which has a width of several electron volts, Auger peaks are inherently much wider than XPS peaks. Another difference is that the backscattered primary and secondary electrons create a huge background. These effects usually make the analysis of electron-stimulated AES more complex as compared to XPS. Most early AES spectrometers provided the spectrum in a differentiated form, i.e., instead of giving the number $N(E)$ of electrons detected at energy E, EdN/dE was recorded. Even today, although modern spectrometers record $N(E)$, many analysts numerically differentiate the spectrum because Auger peaks are small and on a strongly sloping background of inelastically scattered electrons, and the latter can be removed to a large extent by differentiation.

In this chapter, the discussion will be mostly about XPS and photon-excited AES. However, the principles of the analysis apply also to electron-stimulated AES. Although Auger electrons can be excited with x-rays, the term AES is normally taken to refer to Auger spectra excited by an electron beam. The main advantage of AES is that the electron beam can be focused with high precision and high flux, thus providing the ability to form images with lateral resolution in the nanometer range. During the past decade XPS imaging has become more common, largely due to the advances in the electron optical and multichannel detection capabilities of analyzers, and spatial resolution of ~5–10 μm is now provided by many commercial instruments [2,3] (see also Chapter 10). In addition, the new generation of synchrotrons has extremely high brightness, allowing x-ray spot sizes of <100 nm diameter to be obtained routinely [4]. However, synchrotrons are still not widely used because of their size, cost, and limited availability.

8.1.1 XPS SPECTRUM

Figure 8.1a shows an Al K_α-excited XPS spectrum of a PET (polyethylene terephthalate) sample. This polymer contains only two elements, carbon and oxygen, and it can be seen that distinct

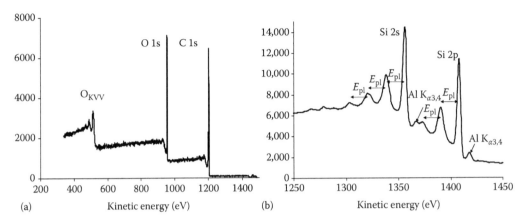

FIGURE 8.1   (a) Monochromatized Al K_α-excited XPS spectrum of PET and (b) Al K_α-excited XPS spectrum of Si.

photoelectron peaks at energies corresponding to the binding energies for C 1s (285 eV) and O 1s (531 eV) dominate the spectrum. The peak structure corresponding to the O_{KVV} Auger transition at ~510 eV can also be seen, and there would be a similar C_{KVV} Auger peak structure at ~275 eV. These transitions are initiated by the O 1s and C 1s core holes, followed by the involvement of two weakly bound valence electrons; the Auger peaks therefore appear at energies 10–20 eV lower than the binding energies of the corresponding 1s levels. Figure 8.1b shows an Al K_α-excited XPS spectrum from pure silicon in the Si 2s and Si 2p energy range. The peaks with a separation of $E_{pl} \approx 17.5$ eV are due to multiple plasmon excitations, while the $\alpha_{3,4}$ satellite can also be seen (see Figure 8.1b).

Figure 8.2 shows an Al K_α-excited XPS spectrum of an iron sample that had been exposed to a corrosive maritime environment. Peaks corresponding to Fe, O, C, Mg, and Ca atoms can be seen, and from this it can be concluded that these atoms are present somewhere within the outermost ~5 nm.

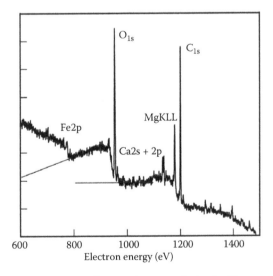

FIGURE 8.2   Al K_α-excited XPS spectrum from an iron sample that had been exposed to a corrosive maritime environment. (From Tougaard, S., *QUASES: Software Package for Quantitative XPS/AES of Surface Nanostructures by Peak Shape Analysis,* Ver. 5.1 (1994–2005). See www.quases.com. With permission.)

The signal-to-noise ratios in Figures 8.1a and 8.2 are typical for fast survey spectra recorded with data acquisition times of less than a couple of minutes. If minor details in the spectrum need to be analyzed the signal-to-noise can be improved by using longer acquisition times for selected narrow energy ranges. On the low kinetic energy side of each peak, there is an increase in the background intensity which originates from electrons, excited at greater depths in the solid, that have undergone one or more inelastic collisions before they reach the surface. The spectrum in Figure 8.1a was recorded with monochromatized Al K_α radiation, while unmonochromatized x-rays were used in Figures 8.1b and 8.2. Unmonochromatized Al K_α or Mg K_α x-ray sources have a number of weaker satellite x-rays associated with them and each of these will produce photoelectron peaks shifted by characteristic energies relative to the dominant $\alpha_{1,2}$ lines. The prominent $\alpha_{3,4}$ photolines with intensities ~5%–10% of the main $K\alpha_{1,2}$ line, and located ~10 eV to higher kinetic energy (see Table 8.1), can be identified easily (see Figure 8.1b). The other photolines further away from the main line can cause difficulties in interpretation. Thus, although their intensity is low, when these satellite x-ray lines excite atoms at high concentration, the resultant photoelectron intensity may be comparable to that from elements at low concentrations excited by the main $K\alpha_{1,2}$ line. For example, the Mg K_β satellite of the C 1s peak is at almost the same energy as the Mg K_α-excited Mo 3d peak. In this case, one should look for the existence of the corresponding Mo 3p peak. If the latter also occurs in the spectrum, then the former is indeed the Mo 3d, but if it does not, then the former probably arises from Mg K_α-excited C 1s electrons; confirmation would come from the presence of a very strong Mg K_α-excited C 1s peak. Table 8.1 gives a list of satellite energies and their relative intensities, which is useful in the interpretation of minor peaks originating from excitation by unmonochromatized x-rays.

Often a twin anode capable of producing either Al K_α (at 1486.7 eV with FWHM ≈ 0.85 eV) or Mg K_α (at 1253.6 eV with FWHM ≈ 0.70 eV) radiation is used. Since the kinetic energies of Auger peaks are independent of the excitation, such an anode can be used to identify Auger peaks and also to separate the positions of XPS and Auger peaks if they are overlapping. A twin anode is formed by coating the two opposing sides of a V-shaped water-cooled Cu block with Mg and Al, respectively. Each of the two sides has its own W filament for electron bombardment (to excite the x-rays), so arranged that neither can bombard the opposite side, thus avoiding mutual excitation. However, the arrangement is never perfect and there is usually a very small amount of crossover of intensity. This can lead to the appearance of small peaks which are always shifted in energy by 233 eV (i.e., the difference between the main Mg and Al K_α photon energies).

Other unwanted radiation may also occur from the anode. For example, after extended use, particularly in poor vacuum conditions, the anode material may become oxidized, resulting in weak oxygen x-rays. In extreme cases, e.g., if the instrument has been operated under very poor vacuum conditions for a long time, the anode material may be sputtered away and the underlying Cu base material exposed, thus producing Cu x-rays. The result is small peaks all displaced by a characteristic energy. Such "ghost" peaks may also arise if the anode has been operated at too high

TABLE 8.1

X-Ray Satellites and Their Relative Intensities

	$\alpha_{1,2}$	α_3	α_4	α_5	α_6	β	O K_α	Cu L_α
Al displacement (eV)	0	9.8	11.8	20.1	23.4	69.7	961.7	556.9
Relative height	100	6.4	3.2	0.4	0.3	0.6	[a]	[a]
Mg displacement (eV)	0	8.4	10.2	17.5	20.6	48.7	728.7	323.9
Relative height	100	8.0	4.1	0.55	0.5	0.5	[a]	[a]

[a] The O and Cu x-ray intensities depend on the condition of the anode. For a "good" anode the intensities should be zero.

a temperature for too long, i.e., when the water cooling has been inadequate. Such over-heating can cause migration of Mg onto the Al side, and vice versa, with the result that the spectrum then contains peaks due to excitation by both Mg and Al. To check one's instrument for the existence and actual intensities (which will be instrument dependent and likely to change with time) of these ghost peaks, it is good laboratory practice to record a spectrum from a material possessing a simple XPS spectrum (i.e., with a few sharp peaks) such as PET or Si, and look for the existence and intensities of any unexpected peaks in the spectrum.

Most instruments manufactured today are equipped with an Al x-ray source combined with a focusing quartz crystal monochromator. The monochromator removes the satellite peaks and the Bremsstrahlung background, and by separating the main K_α doublet at 1486.7 eV, provides a much improved line width, i.e., 0.26 eV. The focusing effect also enables XPS to be performed with a resolution of 5–10 μm.

8.1.2 XPS Peak Shapes

The observed shape of a photoelectron peak in an XPS spectrum is determined by a convolution of four factors: the true shape of the excited peak obtainable with an ideal monochromatic light source, the energy distribution of the light from the x-ray source, changes caused by the inelastic scattering events photoelectrons undergo before they escape from the surface and enter the electron energy analyzer, and the energy resolution of the analyzer. Since an understanding of the photoelectron peak shape is crucial for quantification, these factors are now discussed briefly.

The peak shape of a photon-excited core level is inherently Lorentzian, with a width which is the inverse of the core-hole lifetime, but the photoelectron peak is also greatly influenced by excitations of electrons in the local environment of the atom that are triggered by the sudden appearance of the potential from the core hole, leading to electron-hole pair and plasmon excitations of the valence sea of electrons [5–8]. Since they are an intrinsic part of the photoexcitation transition, they are usually called intrinsic excitations or shake-up excitations, to emphasize the nature of the resulting final state in which electrons other than the photoelectron are excited into higher energy states. These excitations extend ca. 20–40 eV to the low-energy side of the main peak [7,8], usually with a broad distribution, but in some cases distinct peaks can be observed, which may be either intrinsic plasmon excitations, or the result of shake-up from occupied valence states to unoccupied states, in which case they contain valuable chemical bond information because the involved energy levels depend on the chemical state (see Chapters 3 and 7).

For energies of less than a few eV, where the valence band density of states may be considered roughly constant, the combined effect of these electron-hole pair excitations and the lifetime broadening can be described by the Doniach and Sunjic formula [5]. Although this formula is usually valid only for the first couple of eV from a peak position, it is frequently applied to a much larger energy range. Although such application is incorrect, it does provide a consistent (but not accurate) way to deal with certain aspects of the peak shape, and is a useful model for the identification of differences between the electronic structures of different samples. Other functions can also be applied in practice, such as a mixed Gaussian–Lorentzian function with an empirical exponentially decaying intensity added to represent the peak asymmetry [9,10]. These procedures are not supported by physical models, but as long as they are used in a consistent way when comparing different samples, the observed differences between the samples give valuable information on the differences in the chemical states of the atoms. For insulators it is not possible to excite electron-hole pairs with an energy smaller than the band gap energy. The effect is therefore absent in the near-peak region and the peaks are symmetric. The C 1s peak at a binding energy of ~285 eV will nearly always be present since even nominally carbon-free samples will exhibit the C 1s peak due to exposure to atmospheric contaminants, and after prolonged exposure to any residual hydrocarbons in the ultra high vacuum (UHV) atmosphere of the spectrometer. For saturated hydrocarbons, the peak is highly symmetric because no low-energy shake-up electronic excitations are possible.

In graphitic carbon, where small energy electron-hole excitations are possible, the C 1s peak is clearly asymmetric.

On their way out of the solid, some electrons undergo inelastic scattering processes, and the typical energy loss in a single-scattering event is between 10 and 30 eV [11,12]. This results in a fairly broad background with weak features corresponding to plasmon and interband transitions, but from a few solids (e.g., Al, Si, SiO_2) clear plasmon peaks can be observed [12] (see Figures 8.1b and 8.5). The inelastic processes lead to a distortion of the energy distribution as compared to the original distribution at the point of excitation in the solid [13–15]. The distance between inelastic scattering events is only ca. 1 nm [16,17] and the resultant shape of the background in a wide energy range, 50–100 eV below the characteristic peaks, therefore depends critically on the atom depth distribution on the nanometer scale. This phenomenon can be used to enhance the accuracy of XPS quantification and to provide information on the depth distribution [15,18–20]. To interpret XPS spectra, it is important to have a good feeling for the importance of this effect and to be able intuitively to identify information on the depth distribution from simple visual inspection of the shape of the background associated with the characteristic peaks. This topic will be discussed again later, as well as the quantitative analysis of the effect.

The most popular type of spectrometer is the hemispherical analyzer operated in the fixed pass energy mode, which is the energy to which the electron is decelerated before it passes through the analyzer energy filter. The analyzer energy resolution is then essentially constant (often ~1% of the pass energy), and is independent of the original photoelectron energy. It is important for the practical working analyst always to keep in mind that there is a trade-off between analyzer resolution and signal-to-noise in the acquired spectra. Thus, in order for the analyst to save machine time, it is important not to acquire spectra with an analyzer resolution better than that needed for the particular analysis. For example, for a standard determination of the composition, an energy resolution of 2 eV is often adequate, which suggests using an analyzer pass energy of 200 eV. To be able to distinguish the variations in binding energy corresponding to different chemical states, a better energy resolution is required. The optimum analyzer resolution (pass energy) depends on the x-ray source used. For an unmonochromatized source, where the widths of the principal lines in the photon spectrum are ca. 0.7–0.8 eV, there is no advantage in using an analyzer resolution better than ca. 0.5 eV (corresponding to a pass energy of ca. 50 eV), while for the monochromatized Al K_α x-ray source (with line width of ca. 0.26 eV) it makes sense to use a pass energy of 15–20 eV. Using lower pass energy values merely reduces the signal-to-noise ratio without introducing any noticeable improvement in the energy resolution. To optimize the performance and efficiency of the use of a particular instrument, it is good laboratory practice to record a few spectra of a sharp peak (e.g., Si 2p or Au 4f) with different pass energies, and to determine the minimum pass energy below which a further reduction in pass energy does not reduce the width of the measured peak. To save data acquisition time, the energy step should also be adjusted to the maximum possible without losing information, e.g., 3–5 energy points over the width of the instrument resolution.

8.1.3 BACKGROUND CORRECTION

Quantification of sample composition usually relies on the measurement of peak area. The most common methods for correcting measured spectra for the inelastically scattered electron background, and thus determine the peak area are the linear, Shirley, and Tougaard background subtraction. They are available in the software that comes with most commercial instruments today, and will be discussed briefly here. Tougaard and Jansson have compared their accuracies [21]. For a linear background, a straight line is drawn from a point close to the low-energy side of a peak, E_{min}, to a point on its high-energy side, E_{max}, and subtracted from the peak. One of the problems with this method is that it is not well defined since the choice of low- and high-energy points is entirely subjective. The Shirley background [22] subtraction assumes that the background intensity at any energy point is proportional to the total spectral area to higher energies, or, in other words, that for each photoelectron at a particular energy there is a constant background to lower energies. If $J(E)$ is the observed intensity at E (which includes the background), and if the operator-chosen upper and lower energy limits are E_{max} and E_{min}, respectively, then the background corrected intensity in this range is given by

$$F^n(E) = J(E) - k_n \int_E^{E_{max}} F^{n-1}(E') \cdot dE' \qquad (8.2)$$

where $F^0(E) = J(E)$ and k_n is found from the requirement that $F^n(E_{min}) = 0$. Since the "true" intensity appears also in the integral the method is iterative, but the series converges rapidly after 3–4 iterations $F^n(E) \sim F^{n-1}(E)$. The method is fairly independent of E_{max}, but does depend on E_{min}, so the Shirley background is thus also not well-defined. In the left-hand panel of Figure 8.3, the Shirley method is used to remove the background from the same spectrum but with different E_{min} values. When it is applied over a small energy range (Figure 8.3a), the peak areas derived are quite similar to those found using the straight line method. For wider energy ranges, it can give strongly misleading results and the resulting peak areas may even be negative as in Figure 8.3c. To minimize the uncertainty, it is important to be consistent in picking E_{min} and E_{max} by using the same criteria (although also entirely subjective and based on visual observation) for all peaks [21]. One should be very careful to avoid situations in Figure 8.3b and c. This is a particular problem when automatic data-handling is used.

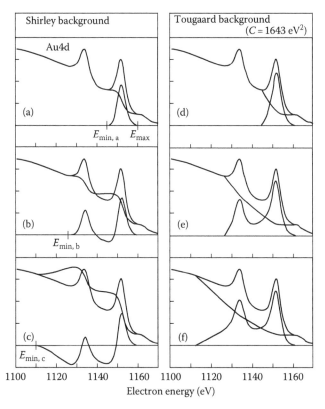

FIGURE 8.3 Illustration of the result obtained for Au 4d spectra using the Shirley and Tougaard backgrounds (Equations 8.2 and 8.3, respectively) with different values for E_{min}. (From Tougaard, S., *Surface Analysis by Auger and X-Ray Photoelectron Spectroscopy*, IM Publications, Chichester, West Sussex, U.K., 2003, 243–295.)

Another background correction method was suggested originally in Refs. [15,23], and has subsequently become known as the Tougaard background. It relies on a quantitative description of the physical processes that lead to the background. The algorithm is

$$F(E) = J(E) - B_1 \int_E^{E_{max}} J(E') \frac{E' - E}{\left(C + (E' - E)^2\right)^2} \cdot dE' \tag{8.3}$$

where $C = 1643\,eV^2$. The factor B_1 is adjusted to give zero intensity in a region between 30 and 50 eV below the characteristic peak structure. For polymers, and other materials (such as Si and Al) with sharp plasmon structure, the three-parameter Tougaard-background algorithm is more accurate, viz

$$F(E) = J(E) - B_1 \int_E^{E_{max}} J(E') \frac{E' - E}{\left(C - (E' - E)^2\right)^2 + D(E' - E)^2} \cdot dE' \tag{8.4}$$

where C and D depend on the material [12,20]. Figure 8.4 shows the full Al K_α-excited spectrum from a Cu foil analyzed by Equation 8.3. As can be seen, the inelastically scattered electron background is described with good accuracy over the full 1000 eV energy range. Note also that peaks extend to ca. 30 eV on the low-energy sides of the characteristic peaks. These arise from the intrinsic (or shake-up) electrons discussed above.

When this method is applied to a single peak, a straight line is first fitted to the spectrum, starting on the high-energy side of the peak, and then subtracted from a region of the spectrum so that the peak, originating from a given core level, can be isolated. Such straight line backgrounds are shown in Figure 8.2 for the O 1s and Fe 2p peaks. Figure 8.5 shows the Si 2p and 2s spectra from SiO_2 analyzed by Equation 8.3 (with $C = 1643\,eV^2$) (upper), and by Equation 8.4 (using $C = 542\,eV^2$ and $D = 275\,eV^2$, valid for SiO_2 [12]) (lower). SiO_2 has a narrow plasmon at ca. 23 eV, and while

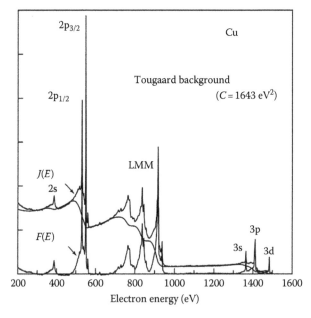

FIGURE 8.4 Experimental Al K_α-excited spectrum $J(E)$ of pure copper corrected for the analyzer transmission function, and the primary excitation spectrum $F(E)$ determined from Equation 8.3 with $C = 1643\,eV^2$ and $B_1 = 3010\,eV^2$. (From Tougaard, S., *Surf. Sci.*, 216, 343, 1989.)

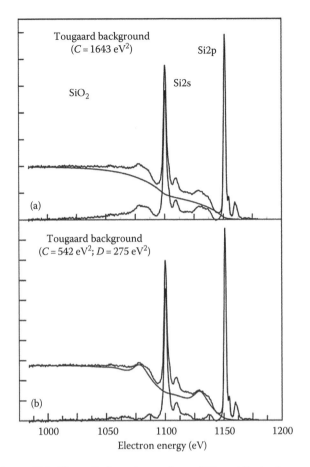

FIGURE 8.5 Experimental Mg K_α-excited spectrum of pure SiO_2 and the background-corrected spectrum after using Equation 8.3 with $C = 1643\,eV^2$ and $B_1 = 3510\,eV^2$ and Equation 8.4 with $C = 542\,eV^2$, $D = 275\,eV^2$, and $B_1 = 306\,eV^2$, respectively.

both Equations 8.3 and 8.4 give consistent and good correction beyond some 30 eV from the Si 2s peak, Equation 8.4 accounts better for the intensity in the plasmon energy loss structure. Strictly speaking, Equations 8.3 and 8.4 are valid only for homogeneously and exponentially distributed depth profiles. When the profiles differ considerably from these, more involved algorithms give a better description of the background. However, even in this case, Equation 8.3 gives quite a robust peak area determination (see Figure 8.3d through f where it has been applied to the same energy range as for the Shirley method in Figure 8.3a through c.). Seah [24] has suggested using smaller values of C in Equation 8.3. Although part of the intrinsic shake-up intensity is then removed from the peak area, Seah found that the ratios of the peak intensities so determined were reasonably independent of the value of C. This has the advantage that the width of the peak region appears to be smaller, as seen in Figure 8.6, where different model spectra (corresponding to different depth distributions of Au atoms in a solid) are analyzed by Equation 8.3 with $C = 200\,eV^2$. The resulting peak areas are similar to those derived by the Shirley method for very narrow peaks, although the method is more robust than that of Shirley and does not give the sort of erroneous results seen in Figure 8.3b and c. Provided one is careful to avoid such situations, the Shirley and straight line methods, although they are not based on physical models, still give a reasonably accurate relative measure for the peak area and may be applied when the peak areas are normalized against the peak area recorded from a reference sample. The two methods have the advantage that only the main

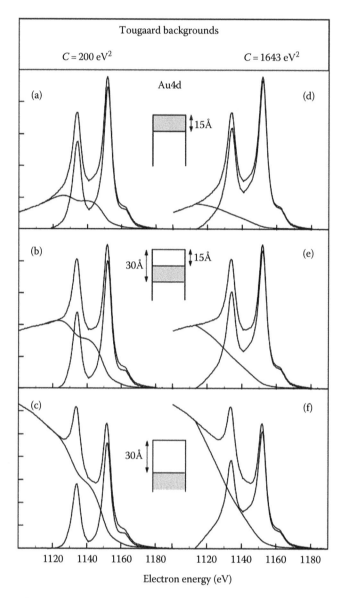

FIGURE 8.6 Comparison of the application of the Tougaard background subtraction method to model spectra from three different depth distributions of Au atoms, using Equation 8.3 with, on the left, $C = 200\,eV^2$, and on the right $C = 1643\,eV^2$. The three model spectra were calculated using the QUASES software [25] for different in-depth distributions of Au atoms, with a value of $\lambda = 1.5\,nm$. The spectra have been normalized to the same maximum intensity. (From Tougaard, S. *Surface Analysis by Auger and X-Ray Photoelectron Spectroscopy*, IM Publications, Chichester, West Sussex, U.K., 2003, 243–295.)

peak region (typically of width <5 eV) needs to be analyzed, while proper use of the Tougaard background requires an energy range of 30–50 eV. In practice, the Shirley and straight line backgrounds are still widely used for relative peak area determination. Since the Tougaard background with small C gives similar peak areas but is more robust than the Shirley background it may be a better choice, especially for automatic data analysis where situations like Figure 8.3b and c may occur unnoticed.

8.2   QUANTIFICATION

AES and XPS are such powerful tools for the analysis of surface composition because the mean free path between inelastic scattering events in a solid is ca. 0.5–2 nm for 100–2000 eV electrons [16,17]. If the distribution of atoms is uniform within a depth of about 5 nm, then the measured peak area is to a good approximation proportional to the atomic concentration. The factor of proportionality depends on the surface composition which is taken into account by a matrix factor. With these assumptions, a simple algorithm can be derived, to be discussed below. The model, which relies on measured peak areas, is usually the preferred starting point for quantification and is available with all commercial instruments. However, it is important to note that it assumes a homogeneous depth distribution, whereas solids that are of interest for surface analysis are hardly ever homogeneous to depths of several nanometers. It is precisely because samples are inhomogeneous on the nanometer depth scale that analysis is performed with XPS or AES, rather than with other well-established but less surface-specific techniques.

Figure 8.7 shows the dependence on depth distribution of model spectra of Cu 2p peaks, cor-responding to four different distributions of Cu atoms on and in a gold matrix [19]. The XPS peak intensity from all four solids is identical even though the concentration in the immediate surface varies between 0% and 100%, while the actual amount of Cu within the surface region can be any-where between the equivalent of 0.11 nm (as in type (a)), or 1.0 nm (as in type (c)), or even greater (as in type (d)). From this it is clear that peak areas can give only qualitative compositional infor-mation unless the atom depth distribution is known, in which case it is straightforward to correct the measured intensities (see below) by application of Equation 8.1. For accurate quantification,

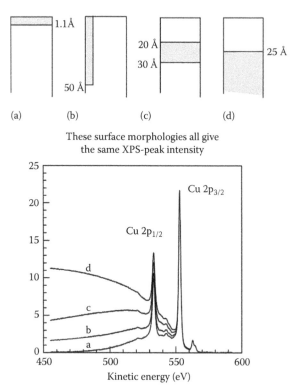

FIGURE 8.7   Four widely different surface and near-surface distributions of Cu atoms in and on Au, that give identical Cu XPS peak intensities, but quite different inelastic backgrounds. (From Tougaard, S., *J. Vac. Sci. Technol. A*, 14, 1415, 1996.)

"the task of determining the atomic concentration cannot therefore be decoupled from the task of determining the depth distribution."

Figures 8.6 and 8.8 show other examples in which the shape of the inelastic background associated with the Au 4d peaks can be seen to vary with the depth distribution of the Au atoms.

It is clear that the background contributions to the spectra over a wide energy range below the Cu peaks depend strongly on the depth distribution of the element. It is thus very easy experimentally to distinguish between the depth distributions contributing to the backgrounds of the four spectra in Figure 8.7, over a 50 eV energy region. Much more accurate quantification can thus be achieved when the dependence of background on surface morphology is taken into account in the analysis. This is the idea behind the formalism, developed by Tougaard et al. [13–15,18–20,25], which provides quantitative information on the surface and near-surface elemental distribution by analysis of XPS or AES backgrounds.

Because the effect is so strong, even quite simple and approximate models enhance the accuracy of quantification significantly and models with different levels of sophistication have been developed.

Just by simple visual inspection of the Fe 2p spectrum in Figure 8.2, it is evident that the depth distribution is of type (d) in Figure 8.7 rather than of any of the other types. It can thus be concluded that Fe was primarily in the bulk, while the O 1s background is of type (b) or (c), indicating that the oxygen atoms were confined mostly to the outermost atomic layers. It can be concluded immediately that qualitatively the sample consisted of an iron substrate covered with a fairly thick oxide layer.

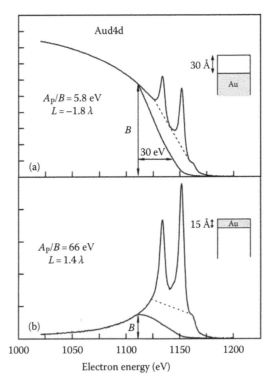

FIGURE 8.8 Two examples of the application of the A_p/B and decay length (L) methods, respectively. The two model spectra were calculated for (a) a gold substrate covered with a 3.0 nm overlayer and (b) a 1.5 nm gold film on top of a substrate. $\lambda = 1.5$ nm in both cases. (From Tougaard, S., *J. Vac. Sci. Technol. A.*, 14, 1415, 1996. With permission.)

Thus, even a quick look at an XPS survey spectrum can give a rough picture of the depth distributions of the various elements. However, a more detailed numerical analysis is required to determine the quantitative extent of these distributions (see below).

8.2.1 MEASURED INTENSITY

The instrumental efficiency, D_{instr}, depends on the characteristics of the x-ray source and the electron energy analyzer, which can be separated into two factors, $D_{instr} = DT(E)$, where $T(E)$ is the energy-dependent part (often called the analyzer transmission function), and D is the energy independent part. $T(E)$ is a smooth function of E and is often approximated by E^{-m} where $0.5 < m < 1$, and m depends on E and the pass energy. Procedures to determine $T(E)$ have been developed by Seah [26]. Consider now photoelectrons excited from atoms of type A resulting in a peak centered at energy E_A with peak intensity (area) $I_A^{measured}$ determined by one of the methods discussed above. Correction of the measured intensities by the energy-dependent part of the detector efficiency results in the expression $I_A = I_A^{measured} /T(E_A)$.

Take the situation where atoms A are in a thin layer of thickness dz at a depth z in a solid. Assume that the x-ray irradiated surface area is larger than the area A_0 from which photoelectrons are accepted into the analyzer. The intensity from the layer is then

$$dI_A(z) = \left[D \frac{A_0}{\cos \theta} \right] \cdot \left[\frac{d\sigma_A(h\nu)}{d\Omega} W_A(\gamma, \beta_A) \right] \cdot \left[N_A CF_A(z, \theta, E_A) \right] \cdot \exp\left(-z/(\lambda \cos \theta)\right) dz \Delta\Omega \quad (8.5)$$

where

$\Delta\Omega$ is the solid acceptance angle of the analyzer

$\sigma_A(h\nu)$ is the photoionization cross section for the particular core level in atom A for photons of energy $h\nu$

N_A the atomic density of A atoms

γ is the angle between the incident x-ray beam and the electron detector

$W_A(\gamma, \beta_A) = 1 + \frac{1}{2}\beta_A\left(\frac{3}{2}\sin^2(\gamma) - 1\right)$ is the anisotropy of the photoemission

σ_A and the asymmetry parameter β_A for the core level A can be taken from theoretical or experimental tables [27–30]. For the magic angle $\gamma = 54.7°$, W_A is independent of β_A. Many spectrometers are designed to operate close to this geometry.

The factor $CF_A(z, \theta, E_A)$ accounts for the effects of elastic electron scattering [31,32]. By definition, $CF = 1$ when elastic scattering is negligible, i.e., when the electron moves along a straight line from the point of excitation to the detector entrance. This factor was introduced by Jablonski and Tougaard as a practical way of accounting for elastic electron scattering effects. They made extensive Monte-Carlo simulations, and found that it was possible to express CF in a simple parameterization, which becomes particularly simple for the experimental geometry where $\theta < 30°$ and $45° < \gamma < 65°$, which is typical for most instruments. For this geometry, they derived the expression

$$CF(z, \theta) \cong \exp(-0.157764\tau - 1.25132) + \exp(-0.0562417\tau^2 + 0.00698849\tau - 0.201962) \quad (8.6)$$

where $\tau = z(\lambda + \lambda_{tr} /\lambda\lambda_{tr})$, λ_{tr} being the transport mean free path [33,34]. Thus CF is independent of θ for $\theta < 30°$. For larger values of θ, they found a more general parameterized expression [32]. This provides a simple method for correcting angle-resolved XPS (ARXPS) measurements for large emission angles. For those cases where the majority of the photoelectrons come from atoms at depths <1–2λ, CF is generally a small correction (~ 0.9–1.1). But in situations where all the photoelectrons to be used for analysis originate from depths $>2\lambda$ (e.g., a substrate with a thick overlayer), CF can be substantial [31,32]. For homogeneous materials, $CF \approx 1$ because electrons from the shallower layers dominate the measured peak intensity.

8.2.2   Quantification of Homogeneous Samples

Although, as discussed above, samples analyzed by AES and XPS are not normally homogeneous in the outermost few nanometers, the assumption of homogeneity is still the usual starting point in the analysis programs provided by most commercial instruments. This procedure will therefore be discussed here. Assume that the sample composition is homogeneous to depths of $\sim 3\lambda$, and that the concentration of element A is being sought. First calculate the intensity of A from pure A, since this is often used as a reference. From Equation 8.5, by integration over all depths, the intensity of the photoelectron current from A in the pure material is given by

$$I_A^p = \int dI_A^p(z) = \left[D_R A_0\right] \cdot \left[\frac{d\sigma_A(h\nu)}{d\Omega} W_A(\gamma)\right] \cdot \left[N_A CF_{\text{eff}}^A \lambda_A(E_A)\right] \cdot \Delta\Omega \tag{8.7}$$

where D_R is the efficiency of the spectrometer used to measure the reference spectra, and where $CF_{\text{eff}}^A \approx 1$ because pure A is homogeneous. In Equation 8.7, I_A^p is written as a product of three factors where the first is characteristic of the instrument, the second of the atom and the last of the material. Now, the intensity when A atoms are distributed homogeneously in a matrix m with atomic fraction X_A is

$$I_A^m = DA_0 \cdot \left[\frac{d\sigma_A(h\nu)}{d\Omega} W_A(\gamma)\right] \cdot \left[X_A N_m CF_{\text{eff}}^m \lambda_m(E_A)\right] \cdot \Delta\Omega \tag{8.8}$$

where CF_{eff}^m is the weighted average of CF. In fact, this is the same as the factor Q which was introduced by Jablonski [35]. Jablonski also showed that elastic scattering can be approximately accounted for in W_A by using a β_{eff} instead of β. Tabulations of these two factors are available [34]. Equation 8.8 can be rewritten as

$$X_A = \frac{I_A^m}{DA_0} \cdot \frac{1}{\dfrac{d\sigma_A(h\nu)}{d\Omega} W_A(\gamma)} \cdot \frac{1}{N_m CF_{\text{eff}}^m(E_A)\lambda_m(E_A)} \cdot \frac{1}{\Delta\Omega} \tag{8.9}$$

Using Equation 8.7 to eliminate the photoionization cross section by multiplication and rearrangement, and neglecting a small possible effect of dependence on γ in W_A if the two spectrometers have different geometries, Equation 8.9 becomes

$$X_A = \frac{I_A^m}{I_A^p} \cdot \frac{D_R}{D} \cdot \frac{N_A CF_{\text{eff}}^A(E_A)\lambda_A(E_A)}{N_m CF_{\text{eff}}^m(E_A)\lambda_m(E_A)}$$

or

$$X_A = \frac{I_A^m}{I_A^p} \cdot \frac{D_R}{D} F_{Am}$$

where the matrix factor is

$$F_{Am} = \frac{N_A CF_{\text{eff}}^A(E_A)\lambda_A(E_A)}{N_m CF_{\text{eff}}^m(E_A)\lambda_m(E_A)} \tag{8.10}$$

The dependency on D and D_R can be eliminated by normalizing, using $\sum_i X_i = 1$ where i runs over all elements in the matrix m, from which can be obtained

$$X_A = \frac{F_{Am}\left(I_A^m / I_A^p\right)}{\sum_i F_{im}\left(I_i^m / I_i^p\right)} \tag{8.11}$$

The matrix factor F_{Am} should not be neglected since it may vary over the range ~ 0.3–3.

For homogeneous materials, $CF_{eff} \approx 1$ and is often neglected. The inelastic mean free path can be calculated from the practical TPP-2M formula [16,17] (freeware for this can be downloaded at www.quases.com) or can be taken from an extensive database [33].

If the depth distribution is known, it is straightforward to apply Equation 8.5 and correct for the attenuation effect of the overlayer, thus arriving at an accurate concentration. One should be aware of the need to include the CF_{eff} correction if the majority of the detected electrons in a peak originate from depths $>2\lambda$, or if the angle of detection is $>60°$ from the surface normal [32].

As stated above, this formalism assumes a homogeneous depth composition of the sample, which is rarely the case in practical analysis. The uncertainty is therefore large. There are situations in which it is known that the detected atoms are likely to be on the surface in a thin layer. These include carbonaceous contamination and thin layers of an oxide or a polymer and they will now be discussed.

8.2.3 QUANTIFICATION OF UNIFORM THIN FILMS

Assume that the surface is completely covered with a film of A atoms of uniform thickness d on a substrate of B atoms. From Equation 8.5 the peak intensities are then

$$I_A^{film}(d) = \int_0^d dI_A(z) = I_A^p \cdot \left(1 - e^{\frac{-d}{\lambda_A(E_A) \cdot \cos\theta}} \right) \tag{8.12}$$

and

$$I_B^{substr}(d) = \int_d^\infty dI_B(z) = I_B^p \cdot e^{\frac{-d}{\lambda_B(E_B) \cdot \cos\theta}} \tag{8.13}$$

where the correction CF_{eff} has been ignored. Similar equations can be derived easily for other depth distributions. The film thickness d can then be determined easily from the measured intensities. The major limitation to practical application of this method is that the depth distribution must be known in advance, which is not usually the case. The above analysis is not valid if, for example, the film has formed islands or if atoms from the film have diffused into the substrate.

8.2.4 METHODS FOR ESTIMATING THE ATOM DEPTH DISTRIBUTION

Several methods are in use for the determination of the depth distribution of atoms. For depths >5–$10\,nm$, sputter depth profiling, where atoms are removed by bombardment with energetic inert gas ions (usually Ar^+), is a widespread and effective technique. This is described in detail in Chapter 10. It is a destructive technique and effects such as the preferential sputtering of one type of atom compared to another, intermixing, and radiation enhanced diffusion, combine to limit the resultant depth resolution. In addition, there is the problem of the reduction of some species to lower oxidation states. Other techniques are therefore used to achieve accurate analysis of the surface composition of the outermost few nanometers. ARXPS has been used for more than 30 years as a technique for nondestructive analysis of surface structures [36], and a facility for this analysis is available in most software. It relies on the angular dependence of the peak intensity; in other words, the greater angle of take-off into the analyzer with respect to the surface normal, the shallower the depth z from which photoelectrons are accepted (see Equation 8.1). The acronyms ARXPS and ARAES (angle-resolved AES) have been coined for these procedures. In an excellent paper by Cumpson [37], the limitations, the problems, and the accuracies that can be achieved with this method have been investigated systematically from a theoretical point of view. Parallel experimental investigations to determine the limitations of ARXPS have unfortunately not been performed except for specific systems such as SiO_2 on Si [38]. Another technique, developed by Tougaard et al. [13–20], relies on the fact (see Figures 8.6

through 8.8) that the inelastic background in the energy distribution of emitted electrons depends strongly on the depth concentration profile. The range of validity of this method has been studied extensively both theoretically and experimentally [20].

8.2.5 ANGLE-RESOLVED XPS

The ARXPS formalism is founded on a simple expression (derived from Equation 8.5) that relates the measured photoelectron intensity to the concentration profile, $c(z)$, viz

$$I_A(\theta) = I_0 \int_0^\infty CF(z,\theta)c(z)\exp\left(-\frac{z}{\lambda\cos\theta}\right)dz$$

$$\cong I_0 s \int_0^\infty c(z) \exp\left(-\frac{z}{\lambda\cos\theta}\right)dz \tag{8.14}$$

where $I_0 \lambda \cos\theta$ is the intensity recorded from a reference sample with $c(z) \equiv 1$. In the last expression on the right, elastic collisions have been neglected so that $CF = 1$. Elastic scattering may be important for $\theta > 60°$ or if the majority of detected electrons originate from layers $z > 2\lambda$. A correction for this is implemented in some software.

By measuring the intensity for different values of θ, the depth profile $c(z)$ can in principle be determined by comparison with Equation 8.14. Several numerical procedures have been suggested [36] for the inversion of Equation 8.14 and for the direct determination of $c(z)$, but they tend to be unstable and in practice a simple trial and error procedure is usually preferred, where $c(z)$ is changed until a good match of Equation 8.14 to experiments recorded for a few values of θ is obtained.

The most serious limitation of this method is that it works only for very flat surfaces, because if the surface is not flat, there is a shadowing effect for large θ. In general, the use of large values of θ cannot be avoided because $\cos\theta$ varies only slightly with θ for small θ. It is therefore necessary to include measurements at θ larger than about 50° to obtain good depth information. So for rough surfaces the interpretation of ARXPS is complicated because the angular variation of the XPS peak intensity will depend on the surface roughness. Even for ideally flat substrates, ARXPS analysis of laterally inhomogeneous surface structures grown on the surface is quite complex. The reason is that at large θ neighboring nanoclusters can cast shadows, which affect the observed XPS peak intensity. The effect depends on both the shape and distribution of clusters on the surface. ARXPS analysis may then become quite unreliable. The problem was addressed recently in detail [39,40]. For crystalline solids, measurements in high symmetry directions should be avoided because forward photoelectron focusing effects can lead to 20%–30% variations in peak intensity which (if they are included in the analysis) lead to large errors.

Quantification of elemental depth distribution by ARXPS is however straightforward and quite accurate for perfectly flat surfaces of amorphous solids. It is applied widely to the study of thin SiO_2 layers on Si, and has been shown to work very well for this particular system [38] because these surfaces can be made extremely flat. Furthermore the light elements Si and O are weak elastic scatterers so that accurate correction for elastic electron scattering is not so important. ARXPS is also useful in quantification of thin polymer films and carbonaceous contamination layers.

8.2.6 QUANTIFICATION BASED ON COMBINED ANALYSIS OF PEAK-INTENSITY AND -SHAPE

Since 1983, methods of varying degrees of complexity for the extraction of quantitative information from the large variation of the inelastic background with atom depth distribution, have been suggested, and will be discussed briefly here. All information is derived from the analysis of a single spectrum and is therefore also valid for rough surfaces when the take-off angle is close to the surface normal.

8.2.7 PEAK AREA-TO-BACKGROUND RATIO A_p/B

The simplest quantitative description of the variation in peak shape and background with depth is to take the ratio of the peak area A_p to the increase in background height B at a chosen energy below the peak energy. This ratio is very sensitive to the in-depth distribution because A_p and B vary in opposite directions as a function of the depths of the atoms in a solid. For a homogeneous distribution of atoms, it has been shown that this ratio, D_0, is almost constant (ca. 23 eV), independently of material and peak energy. Deviations from this value can then be used to estimate the depth distribution of atoms [13,41].

The algorithm can be defined from Figure 8.8. A_p is the peak area (of the doublet in this case) determined after a linear background has been subtracted (dashed line) from the measured spectrum. The upper energy point to be chosen for the straight line background is taken to be at that energy at which the spectral intensity is 10% of that at the peak energy, while the low-energy point at the other end of the straight line is defined as being at the same distance below the peak energy as the high energy point is above it [13]. B is the increase in intensity measured 30 eV below the peak energy. (In the case, as here, of a doublet peak, the geometrical weighted centroid of the peak structure is used as reference energy). A quick estimate of the in-depth distribution of atoms can then be found from the rules in Table 8.2. For a given system, the method may be fine-tuned by calibrating D_0 against A_p/B determined from the analysis of a sample known to have a homogeneous atom distribution. An example of its application is also shown in Figure 8.8, where the values A_p/B are seen to be consistent with the rules in Table 8.2. See also the example in Table 8.4. Other examples of its practical application may be found in Refs. [42,43].

8.2.8 DECAY LENGTH L AND AMOUNT OF SUBSTANCE $(AOS)_{3\lambda}$

Another simple algorithm was first suggested in 1990 [44], and later improved [45], and its validity has been tested [46]. It provides quantitative information on the atom depth distribution as well as on the amount of substance $(AOS)_{3\lambda}$, i.e., the number of atoms per unit surface area at depths z between 0 and 3λ. In this method, all depth distributions are approximated as exponential, i.e., $\exp(-z/L)$ where L is a characteristic decay length for the profile. The theoretical basis to the method is that the simple Tougaard background is an exact solution to the analysis of the background for all depth profiles of exponential form, and application of this to a general (not necessarily exponential) profile therefore determines the "best" exponential profile fit to the actual depth profile. If most atoms are at shallow depths, L will turn out to be small and positive, while if most atoms are at large depths, L will be small and negative. To be specific, the method is as follows: the measured spectrum, $J(E)$, where E is the electron energy in eV, is first corrected by a standard Tougaard background, i.e.,

---

TABLE 8.2
Rules for Estimating the Depth Profile from A_p/B

A_p/B	Depth Distribution
≈23 eV	Uniform
>30 eV	Surface localized
<20 eV	Subsurface localized

If the same peak from two samples has the values $D_1 = (A_p/B)_1$ and $D_2 = (A_p/B)_2$ then

if $30 eV < D_1 < D_2$	Atoms are surface localized in both samples and are at shallower depths in sample 2 than in sample 1
if $D_1 < D_2 < 20\ eV$	Atoms are primarily in the bulk of both samples and at deeper depths in sample 1 than in sample 2

$$F(E) = J(E) - B_1 \cdot \int_E^{E\,\mathrm{max}} J(E') \frac{E' - E}{\left(1643 + (E' - E)^2\right)^2} \,\mathrm{d}E' \tag{8.15}$$

B_1 is adjusted to give zero background at a point 30 eV below the peak centroid (see the solid background lines in Figure 8.8) and from this, the decay length

$$L = \frac{B_1}{B_0 - B_1} \cdot \lambda \cos\theta \tag{8.16}$$

is determined. In Equation 8.16, B_0 is the value of B_1 determined from analysis, by Equation 8.15, of the spectrum from a homogeneous sample. In practice, $B_0 \approx 3000$ eV^2 for most materials [20]. The depth distribution is then estimated from the rules in Table 8.3 [45,46]. A negative value of L corresponds to a depth distribution that increases with depth within the analyzed depths (i.e., for z up to ca. 5λ). For a given system, the value B_0 may be fine-tuned by replacing B_0 by B_1^H where B_1^H is determined by analysis, from Equation 8.15, of a spectrum from a sample that is known to have a homogeneous atom distribution.

The decay length method has been applied to the spectra in Figure 8.8, and from Table 8.3 and the values of L determined to be $L = -1.8\lambda$ and $+1.4\lambda$, it was concluded that the atoms in (a) are strongly bulk localized and in (b) strongly surface localized, in good agreement with the actual distributions.

The absolute AOS within the outermost layer of thickness ca. 3λ is [45]

$$(\mathrm{AOS})_{3\lambda} = \frac{L + \lambda\cos\theta}{1 - e^{-3\frac{\lambda\cos\theta + L}{L\cos\theta}}} \cdot A_p \frac{c_H}{A_p^H} \cdot \left(1 - e^{-3\lambda/L}\right) \tag{8.17}$$

where the peak area is

$$A_p = \int_{E_{p-30\,\mathrm{eV}}}^{E_{\mathrm{max}}} F(E)\,\mathrm{d}E \tag{8.18}$$

and A_p^H is the peak area from a solid with a homogeneous atomic distribution of density c_H. If the objective is to find the relative $(\mathrm{AOS})_{3\lambda}$ in a set of samples, it is not necessary use a reference sample and determine c_H/A_p^H. It is convenient to define an equivalent film thickness

$$d = \frac{\mathrm{AOS}_{3\lambda}}{c_H} \tag{8.19}$$

TABLE 8.3

Rules for Estimating the Depth Profile from L

L	Depth Distribution
$6\lambda < |L|$	Almost uniform
$-3\lambda < L < 0$	Most atoms are at depths $>1\lambda$
$0 < L < 3\lambda$	Most atoms are at depths $<1\lambda$
If the same peak from two samples has values L_1 and L_2, then	
if $0 < L_1 < L_2 < 3\lambda$	Atoms are surface localized in both samples and are at shallower depths in sample 1 than in sample 2
if $-3\lambda < L_1 < L_2 < 0$	atoms are primarily in the bulk of both samples and at deeper depths in sample 2 than in sample 1

which is the thickness of the material if it were distributed as a uniform film with the same atom density as in the reference. If c_H is in atoms/nm^3 and λ is in nanometers, then AOS is in atoms/nm^2 and d is in nanometers.

The validity of the algorithm in Equation 8.17 has been tested and found to give a good measure of the AOS, as well as a reliable depth sectioning based on both model and experimental spectra [44–46]. The accuracy of $(AOS)_{3\lambda}$ is typically better than ~15% when this method is applied to a wide range of depth distributions. The algorithm is suitable for automation and can also therefore be applied to mapping where several thousand spectra at a time must be analyzed. The practical applicability of this algorithm for three-dimensional surface mapping has recently been demonstrated successfully [47]. It gives an image with much more quantitative information compared to the simple assumption that the concentration is proportional to the peak area (which is the standard method), because in the latter case the peak areas, and hence the resulting maps, will not necessarily reflect either the actual surface concentration or the AOS in the outermost atomic layers.

8.3 QUANTIFICATION BY DETAILED ANALYSIS OF THE PEAK SHAPE

The above quantification procedures are easy to apply in practice. However, with greater effort and more elaborate algorithms it is possible to obtain an even more accurate and detailed analysis of the near-surface elemental distribution. The underlying algorithms for this were published more than a decade ago and are summarized in Ref. [19]. The validity of the technique has been established through a series of systematic experiments, some of which have been reviewed in Ref. [20]. These studies showed that quite detailed information on the in-depth atom distribution on the nanometer depth scale can be extracted. The algorithms are fairly complex and a user-friendly software package, which provides tools to do the full analysis from raw spectra to the resulting distribution, was developed to make this type of analysis available for nonspecialists [25]. The method is now widely used and has been applied to the study of a wide range of systems and physical phenomena, including thin-film growth mechanisms and subsurface elemental distributions of films, nucleation, island formation, diffusion, etching, etc. [48–57]. Since the method is nondestructive, it also allows study of the changes in the surface morphology of a given surface atomic structure during surface treatment, as in chemical reaction or gradual annealing. Recently, other software for similar peak shape analysis has been described [58].

As an example of the practical application of the method the analysis of the XPS spectrum of the corroded iron sample in Figure 8.2 will be described. Using the QUASES software [25], the Fe 2p and O 1s energy regions are first isolated by subtracting straight lines fitted to the intensity on the high-energy sides of the peaks as shown in Figure 8.2. The peaks isolated in this way are shown in Figure 8.9 after a 5-point Savitzky–Golay smooth. The spectra are then analyzed with the QUASES software by varying the assumed depth profile, and the corresponding background-subtracted spectra are calculated and plotted. In the upper panel of Figure 8.9 Fe is assumed to be distributed homogeneously and it is clear that such a depth distribution does not account for the strong increase in background intensity. On the other hand, in the middle panel the assumption is that the Fe is distributed from a depth of about 3.5 nm to infinite depths, and that is clearly a much better model for the peak shape and background over a wide energy region. The lower panel shows a similar analysis of the O 1s peak shape. The peak shape and background can be described well by assuming that the O atoms are distributed with constant concentration between depths of 2.5 and 8.0 nm. The C 1s, Mg KLL and Ca 2s and 2p peaks overlap somewhat in energy and an analysis of them would require a spectrum with a better signal-to-noise ratio. The conclusion from the analysis is then that the sample consists of an Fe substrate covered first with a layer of oxide about 5.5 nm thick, on top of which is a mainly carbonaceous layer of thickness about 2.5 nm. All information was extracted from the rather noisy survey spectrum in Figure 8.2.

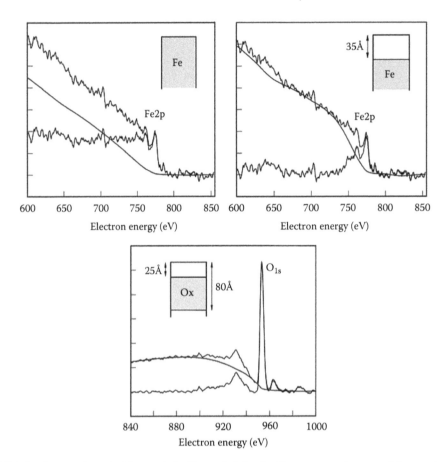

FIGURE 8.9   Spectra from the Fe sample in Figure 8.2 that had been exposed to a maritime environment. The Fe 2p and O 1s peaks have been analyzed with the QUASES-Tougaard software [25] to determine the depth distribution of the O and Fe atoms. (From Tougaard, S., *QUASES: Software Package for Quantitative XPS/AES of Surface Nanostructures by Peak Shape Analysis,* Ver. 5.1 (1994–2005). See www.quases.com. With permission.)

The spectrum in Figure 8.2 was also analyzed by the simple A_p/B and decay length methods, and the results are shown in Table 8.4. Comparing the A_p/B values with the rules in Table 8.2 shows that Fe and O are present in the subsurface region and that the Fe atoms are at depths deeper than those of the O. Comparing the decay lengths L with the rules in Table 8.3, shows that most Fe and O atoms are at depths $>1\lambda$ and that Fe atoms are at greater depths than O. These results are in full agreement with the more detailed analysis in Figure 8.9.

TABLE 8.4
Results from Analysis of the O 1s and Fe 2p Spectra in Figure 8.9

Method of Analysis	Fe	O
Detailed peak shape	3.5–100 nm	2.5–8.0 nm
A_p/B	3.9 eV	14.8 eV
Decay length L	-1.54λ	-1.95λ

8.4 SUMMARY

XPS and AES are powerful techniques for the study of the composition of surfaces on the nanometer depth scale. Analysis of spectra is often carried out on the assumption that the surface composition is homogeneous within the outermost few nanometers, in which case the composition would be proportional to the intensity of the corresponding peaks. Samples are however rarely homogeneous within a few nanometer, and this limits the accuracy of the analysis. Improvements to the analysis can be made easily if the depth composition of the sample is known prior to analysis. But this is seldom the case in practice and the determined compositions are therefore rather uncertain. The techniques can be made more quantitative if both the peak intensity and the shape of the background of the associated inelastically scattered electrons are taken into account, because these quantities depend strongly on the depth distributions of the atoms emitting the XPS or AES electrons. From a simple visual inspection of the peak shapes over a wide (ca. 50 eV) energy region one can obtain immediate information on whether the atoms are located primarily in the near-surface region or deeper in the bulk. More quantitative information on the composition as well as on the in-depth distributions of atoms can be provided by algorithms that have been developed for this purpose. With the most elaborate algorithms, XPS can determine the amount of material within the outermost 10 nm of the sample to an accuracy of better than about 10 at.%, and it can also determine the distributions of these atoms with nanometer depth resolution. XPS can thus be applied to the study of a wide range of systems and physical phenomena, including thin-film growth mechanisms and subsurface elemental distributions of films, nucleation, island formation, diffusion, etching, etc. Since XPS is nondestructive, it also allows study of the changes in the surface morphology of a given surface atomic structure during surface treatment, as in chemical reaction or gradual annealing.

REFERENCES

1. S. Tougaard, in: *Surface Analysis by Auger and X-Ray Photoelectron Spectroscopy*, Eds. D. Briggs and J. T. Grant (IM Publications, Chichester, West Sussex, UK, 2003), pp. 243–295.
2. Y. J. Chen, E. T. Kang, K. G. Neoh, and K. L. Tan, *Macromolecules* 34, 3133 (2001).
3. F. Reniers and C. Tewell, *J. Electr. Spectrosc.* 142, 1 (2005).
4. S. Günther, B. Kaulich, L. Gregoratti, and M. Kiskinova, *Progr. Surf. Sci.* 70, 187 (2002).
5. S. Doniach and M. Sunjic, *J. Phys. C Solid State Phys.* 3, 285 (1970).
6. P. M. van Attekum and G. K. Wertheim, *Phys. Rev. Lett.* 43, 1896 (1979).
7. D. R. Penn, *Phys. Rev. Lett.* 40, 568 (1978).
8. S. Tougaard, *Phys. Rev.* B34, 6779 (1986).
9. R. O. Ansel, T. Dickingson, A. F. Povey, and P. M. A. Sherwood, *J. Electroanal. Chem.* 98, 79 (1979).
10. N. Fairley, in: *Surface Analysis by Auger and X-Ray Photoelectron Spectroscopy*, Eds. D. Briggs and J. T. Grant (IM Publications, Chichester, West Sussex, UK, 2003), pp. 397–420.
11. S. Tougaard, *Solid State Comm.* 61, 547 (1987).
12. S. Tougaard, *Surf. Interface Anal.* 25, 137 (1997).
13. S. Tougaard, *Surf. Sci.* 162, 875 (1985).
14. S. Tougaard, *J. Vac. Sci. Technol.* A5, 1230 (1987).
15. S. Tougaard, *Surf. Interface Anal.* 11, 453 (1988).
16. S. Tanuma, C. J. Powell, and D. R. Penn, *Surf. Interface Anal.* 21, 165 (1993).
17. S. Tanuma, C. J. Powell, and D. R. Penn, *Surf. Interface Anal.* 35, 268 (2003).
18. S. Tougaard and H. S. Hansen, *Surf. Interface Anal.* 14, 730 (1989).
19. S. Tougaard, *J. Vac. Sci. Technol.* A 14, 1415 (1996).
20. S. Tougaard, *Surf. Interface Anal.* 26, 249 (1998).
21. S. Tougaard and C. Jansson, *Surf. Interface Anal.* 20, 1013 (1993).
22. D. A. Shirley, *Phys. Rev.* B 5, 4709 (1972).
23. S. Tougaard, *Surf. Sci.* 216, 343 (1989).
24. M. P. Seah, *Surf. Sci.* 420, 285 (1999).
25. S. Tougaard, *QUASES: Software Package for Quantitative XPS/AES of Surface Nanostructures by Peak Shape Analysis,* Ver. 5.1 (1994–2005). See www.quases.com

26. M. P. Seah, *Surf. Interface Anal.* 20, 243 (1993).
27. J. H. Scofield, *J. Electr. Spectrosc.* 8, 129 (1976).
28. R. F. Reilman, A. Msezane, and S. T. Manson, *J. Electr. Spectrosc.* 8, 389 (1976).
29. M. P. Seah, I. S. Gilmore, and S. J. Spencer, *Surf. Interface Anal.* 31, 778 (2001).
30. I. M. Band, Y. I. Kharitonov, and M. B. Trzhaskovskaya, *At. Data Nucl. Data Tables* 23, 443 (1979).
31. A. Jablonski and S. Tougaard, *Surf. Interface Anal.* 26, 17 (1998).
32. A. Jablonski and S. Tougaard, *Surf. Interface Anal.* 26, 374 (1998).
33. NIST SRD 71 Electron Inelastic Mean Free Path Database, Ver 1.1. National Institute of Standards and Technology (NIST), Gaithersburg, MD 20899 (2001).
34. NIST SRD 65 Electron Elastic Scattering Cross Section Database, Ver 2.0. National Institute of Standards and Technology (NIST), Gaithersburg, MD 20899 (2002).
35. A. Jablonski, *Surf. Interface Anal.* 23, 29 (1995).
36. P. Cumpson, in: *Surface Analysis by Auger and X-Ray Photoelectron Spectroscopy*, Eds. D. Briggs and J. T. Grant (IM Publications, Chichester, West Sussex, UK, 2003), pp. 651–675.
37. P. J. Cumpson, *J. Electr. Spectrosc.* 74, 25 (1995).
38. M. P. Seah, *J. Vac. Sci. Technol. A* 22, 1564 (2004).
39. P. Kappen, K. Reihs, C. Seidel, M. Voertz, and H. Fuchs, *Surf. Sci.* 465, 40 (2000).
40. A. I. Martín-Concepción, F. Yubero, J. P. Espinós, and S. Tougaard, *Surf. Interface Anal.* 36, 788 (2004).
41. S. Tougaard, *J. Vac. Sci. Technol. A* 5, 1275 (1987).
42. L.-S. Johansson, J. M. Campbell, K. Koljonen, M. Kleen, and J. Buchert, *Surf. Interface Anal.* 36, 706 (2004).
43. K. Idla, L.-S. Johansson, J. M. Campbell, and O. Inganäs, *Surf. Interface Anal.* 30, 557 (2000).
44. S. Tougaard, *J. Electr. Spectrosc. A* 8, 2197 (1990).
45. S. Tougaard, *J. Vac. Sci. Technol. A* 21, 1081 (2003).
46. S. Tougaard, *J. Vac. Sci. Technol. A* 23, 741 (2005).
47. S. Hajati, S. Coultas, C. Blomfield, and S. Tougaard, *Surf. Sci.* 600, 3015 (2006).
48. H. F. Winters, D. B. Graves, D. Humbird, and S. Tougaard, *J. Vac. Sci. Technol. A* 25, 96 (2007). In press.
49. J. P. Espinós, A. I. Martín-Concepción, C. Mansilla, F. Yubero, and A. R. González-Elipe, *J. Vac. Sci. Technol. A* 24, 919 (2006).
50. M. Zhou, C. Wu, P. D. Edirisinghe, J. L. Drummond, and L. Hanley, *J. Biomed. Mater. Res.* 77A, 1 (2006).
51. D. Eon, G. Cartry, V. Fernandez, C. Cardinaud, E. Tegou, V. Bellas, P. Argitis, and E. Gogolides, *J. Vac. Sci. Technol. B* 22, 2526 (2004).
52. F. Gracia, F. Yubero, J. P. Espinós, and A. R. González-Elipe, *Appl. Surf. Sci.* 252, 189 (2005).
53. E. Johansson and L. Nyborg, *Surf. Interface Anal.* 35, 375 (2003).
54. J. Zemek, P. Jiricek, A. Jablonski, and B. Lesiak, *Appl. Surf. Sci.* 199, 138 (2002).
55. E. R. Fuoco and L. Hanley, *J. Appl. Phys.* 92, 37 (2002).
56. M. D. Re, R. Gouttebaron, J. P. Dauchot, P. Leclère, R. Lazzaroni, M. Wautelet, and M. Hecq, *Surf. Coat. Technol.* 151–152, 86 (2002).
57. C. Mansilla, F. Gracia, A. I. Martín-Concepción, J. P. Espinós, J. P. Holgado, F. Yubero, and A. R. González-Elipe, *Surf. Interface Anal.* 39, 331 (2007).
58. W. S. M. Werner, W. Smekal, and C. J. Powell, NIST Database for the Simulation of Electron Spectra for Surface Analysis, SRD 100, Version1.0, NIST, Gaithersburg, 2005.

9 Structural and Analytical Methods for Surfaces and Interfaces: Transmission Electron Microscopy

John M. Titchmarsh

CONTENTS

9.1 INTRODUCTION

Chapter 6 outlined electron microscope instrumentation and methods, giving some of the theoretical background to imaging by diffraction contrast (DC), high-resolution electron microscopy (HREM), and the high-angle annular dark field (HAADF) method, in the scanning transmission electron microscopy (STEM). Chemical characterization by energy dispersive x-ray (EDX) analysis and electron energy-loss spectroscopy (EELS) was also discussed. Such information is necessary in order to appreciate the specific application of transmission electron microscopy (TEM) to the characterization of a grain boundary (GB) or an interface (IF), which is the particular subject of this chapter. After a brief summary of the methods available to generate electron-transparent foils from bulk samples and the TEM parameters that require calibration, procedures are described for the determination of misorientation across a GB/IF using electron diffraction. The imaging and analysis of interfacial dislocations are discussed. For GBs with suitable orientation, HREM and HAADF now enable the direct imaging of atomic structure, interfacial dislocations, and the structural units from which the GB is composed. In the space available it is not possible to acknowledge the many excellent contributions to this subject reported in the literature. However, specific examples are described in order to illustrate the current state of characterization at the leading edge of this field. The established area of characterization of interfacial segregation by EDX is then described before considering the rapidly expanding area of EELS characterization by spectroscopy, energy-filtered mapping (energy-filtered TEM [EFTEM]), spectroscopic imaging in STEM (electron energy-loss spectroscopy and imaging [EELSI]), and absorption edge fine structure analysis (energy-loss near-edge structure [ELNES]). Space limitation prevents specific description of more specialized areas of TEM IF analysis, such as magnetic domain imaging, in situ experiments, and electron holography.

9.2 SAMPLE PREPARATION METHODS

The preparation of the sample for TEM characterization is important and sometimes difficult. Most methods have been in use for many years and details of these are available in several texts [1–4]. In recent years, new methods based on mechanical polishing and ion-beam erosion have become more common and, in particular, the use of focused ion-beam (FIB) methods for site-specific foil preparation has had a huge impact on the sectioning of semiconductor devices. The most common methods, together with the categories of material for which the methods are suitable, are outlined below. Details of equipment, procedures, consumables, and conditions for specific materials, can be found in many papers in the literature. The microscopist might well have to spend considerable effort in developing variations of these basic procedures before finding an appropriate way of generating electron-transparent areas.

9.2.1 CRUSHING AND DISPERSION IN SOLVENT

This method is used for brittle, nonconducting materials such as minerals, ceramics, and glasses. The sample is ground to a fine powder and dispersed in a suitable solvent such as ethanol. Small fragments of the sample can then be collected on a holey carbon film mounted on a standard 3 mm diameter metal (usually Cu) grid. Some of the fragments may be electron transparent, particularly around the edges or when the material fractures into parallel-sided plates. The most suitable particles are those suspended over the edges of the holes in the carbon.

9.2.2 EVAPORATION ONTO HOLEY CARBON FILMS

Direct evaporation of metals onto a holey carbon film is suitable for making standard samples used for calibration of microscopes or for testing performance. The evaporated material may have an amorphous structure or be nanocrystalline. Heavy metals such as Au are commonly used for measuring the information limit in the TEM (see Section 10.3 of Chapter 10). Evaporated Al in a nanocrystalline form has often been used to generate "amorphous" rings to calibrate diffraction pattern (DP) camera length and circularity. Evaporated films can sometimes be grown on a crystalline substrate such as rock salt, which is then dissolved in a solvent to leave a free-standing film, or one which can be supported directly on a metal grid without an additional carbon film support. Grain boundaries are produced in the evaporated film by controlled heating.

9.2.3 CHEMICAL POLISHING

Chemical polishing is the slow, controlled, dissolution of a material by a suitable chemical. An example of the use of this method is the thinning of silicon by simple immersion in dilute hydrofluoric acid (HF). Prior mechanical dimpling ensures that, when the acid finally perforates the sample, the area along the edge of the hole is electron transparent. A variation of this method is to direct a fine jet of acid from both sides at the center of a disc of material that has been cut or drilled to a diameter of 3 mm and mechanically ground to a thickness of ca. 0.1 mm. Such a method was once commonly used to prepare Si, Ge, and many III–V semiconductor materials, some ionic crystals (e.g., NaCl using jets of water), and metals.

9.2.4 ELECTROPOLISHING

Electropolishing procedures are used to prepare metal and alloy samples. An electrochemical cell is constructed with the sample as the anode, a Pt cathode and a suitable acid electrolyte. Passing direct current through the cell causes removal of the anode until perforations appear. As with simple chemical polishing, the method has been refined by using jets of electrolyte with precut, thin (ca 0.1 mm) sample discs. Commercial apparatus is readily available with sensors to detect the moment of perforation, at which the current is stopped. The microscopist is required to experiment with the type of electrolyte, the jet speed, cell temperature and DC voltage in order to optimize the polishing.

9.2.5 ULTRAMICROTOMY

Ultramicrotomy is a preparation method most commonly used for soft materials of interest to the life science community. However, a number of attempts have been made to adapt the method for harder materials, including semiconductors [5], minerals [6], and metals [7]. A sharp glass or diamond knife is used to cut many slices of uniform thickness off a large piece of sample material. The area of the slice is larger for softer materials. Hard materials, e.g., many metals, damage the knife blade too quickly. Trials with soft metal alloys of Al reveal that the high-speed plastic deformation that occurs during cutting induces a large defect density and microstructural modifications. Other methods outlined in this section are usually much more appropriate for hard materials.

9.2.6 ION-BEAM MILLING

Ion-beam milling, or thinning (IBT), was developed in the 1970s as an alternative means of thinning semiconductor samples, particularly for the generation of cross-sections through epitaxial layers and devices, so that the near-surface regions of the cross-section could be made electron transparent. It is used extensively today, not only for semiconducting and magnetic devices but also for preparing TEM samples of nonconducting or partially nonconducting materials such as ceramic hard coatings on metals or bulk ceramics, for which the methods previously described are not viable. All samples require initial mechanical cutting and polishing. For a bulk ceramic sample this might involve slicing (thickness of ca. 1 mm) with a diamond-impregnated cutting wheel, grinding to ca. 0.2 mm thickness, ultrasonically cutting 3 mm diameter discs from the ground slice, and then using a dimpling apparatus to thin selectively the center of the disc to 10–20 μm. Only then is the sample inserted into the ion-milling apparatus for controlled removal of material and perforation through the center of the disc. Depending on the ceramic and the type of TEM characterization required, the ion-thinned sample might then require the deposition of an evaporated carbon film to prevent charging during TEM examination.

The preparation of cross-sectional samples from brittle semiconductors demands extremely careful handling. Usually, the sample containing the device is stuck, face-to-face, with a similar sample using strong adhesive. Pieces are then sliced perpendicularly to the interface from the resulting "sandwich." Mechanical grinding in stages using successively finer grades of polishing compounds (SiC, diamond, and alumina) is applied until the sandwich is about 10–20 μm thick. Only then is IBT used in the final stage.

9.2.7 MECHANICAL POLISHING

The idea of mechanically grinding a sample until part of it became electron transparent, without inducing extensive damage, was considered completely impractical in the early years of TEM. However, it is now routinely used for the preparation of cross-sections through semiconductor layered structures. Samples are prepared as above for IBT. Particular care is required to polish both sides of the sandwich so that no mechanical damage is introduced and propagated through the successive mechanical polishing stages. The final stages of polishing on both sides of the sample require the use of polymeric paper impregnated with submicrometer particles of diamond. For the final stage on the second side, the sample is polished with a slight taper angle (about 1°) using a calibrated tripod. For the most demanding characterization an 18-stage preparation protocol has been developed so that single-embedded dopant atoms can be detected [8].

9.2.8 FOCUSED ION-BEAM MILLING

The FIB mill, combining ion-machining and imaging, is particularly useful for preparing electron-transparent samples from specific locations in a flat sample [9–11]. Older FIBs are "single beam," while the newest ones are "dual beam." Single-beam instruments generate an image by collecting the secondary electrons emitted during the scanning of a beam of Ga ions, accelerated through some 30 keV and focused to a diameter of 5–10 nm, over the sample surface, analogous to scanning the electron beam in an SEM. Atoms are simultaneously sputtered from the sample surface. Any microstructural feature of interest exposed on the flat surface, such as an individual component of an integrated circuit or the trace of a GB, can be selected. The ion beam is then moved under computer control to machine away material, sectioning through the feature, to generate a thin membrane, typically 10 mm × 10 mm × 50 nm in size. Various machining procedures have been reported. For some sample geometries membrane(s) are left attached to the original bulk sample, which then provides the support necessary for handling and mounting a membrane. For others, the membrane is cut away from the bulk sample and micromanipulation is performed to pick up the membrane and place it on a standard holey carbon film on a grid, or on some other appropriate means of support. Figure 9.1 shows, schematically, the stages in the fabrication of such a TEM sample membrane containing a boundary between two grains (1 and 2). The boundary area is selected from the image of the polished surface (Figure 9.1a) and reference identification markers are ion-etched into the adjacent area. A strip of heavy metal, usually Pt, is then selectively deposited over a length of ca. 15 mm to protect the surface from ion erosion (Figure 9.1b) during the subsequent ion-machining (Figure 9.1c). Two trenches with triangular cross-sections are machined in such a way that a thin membrane is left between them. Final thinning of the membrane is performed using a less-energetic beam to minimize microstructural damage and Ga ion implantation. A spatial accuracy of 5–10 nm is possible and a membrane thickness of ca. 50–100 nm is readily achievable. The membrane is cut away using the ion beam and carefully collected for mounting (Figure 9.1d). Figure 9.2 [12] shows an image of a membrane before removal from the original sample and inspection in the TEM. It is also possible to deposit films of materials such as Pt over small sample regions to protect the surface and edges of the specific sites of interest.

A major problem with the single-beam FIB is that there is continual erosion of the scanned area during imaging. This causes unwanted damage while deciding the geometry of machining and during checks to monitor machining progress. A dual beam instrument has a separate electron beam for imaging, in addition to the ion beam. It is now also possible to add an EDX system for chemical analysis. However, the cost of a dual beam FIB is similar to that of an FEG-TEM and it can take many hours to machine even a very small sample, thus specimen preparation using FIB is very

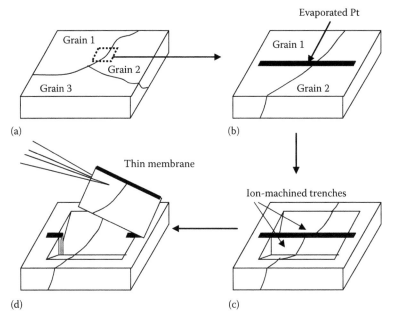

FIGURE 9.1 Schematic of FIB sample generation. (a) Selection of feature; (b) deposition of protective Pt strip; (c) machining of trenches with membrane between; (d) cutting and removal of membrane.

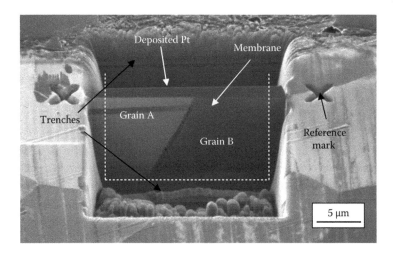

FIGURE 9.2 Secondary electron image of a thin membrane TEM sample containing a GB, fabricated in an FIB and viewed from an oblique angle before cutting free and removing on to a support grid. The dashed lines indicate the approximate lines along which cutting is yet to be performed. (From Lozano-Perez, S., unpublished work, 2003.)

expensive compared to other methods. FIB is increasingly the only means whereby TEM can be used to investigate structures relevant to nanotechnology.

9.3 CALIBRATION OF TEM

Alignment and calibration of the TEM is essential for reproducible measurements and for HR (S) TEM. Mechanical alignment of the gun and lenses in a modern microscope is performed by the manufacturer in the factory and further checked after site installation. The manufacturer's manual will provide full details of the additional alignment procedures that the user needs to apply and check on a regular basis to optimize performance. An EELS spectrometer, where fitted, is an extension of the column containing several additional lenses and apertures, which also require careful interfacing to the microscope and regular alignment checks, again in accordance with the manufacturer's guidelines. In this section it will be assumed that the microscope alignment has been performed and attention will be concentrated on the calibrations required for accurate measurements from images and DPs relating to the characterization of interfaces. Although the magnification (M) and camera length (L) are usually continuously displayed on the EM console during use and then recorded with the data, the displayed values are only nominal. Standard samples can either be purchased or made specifically for periodic checks of important parameters or remeasurement, after the column has been reassembled following repair or maintenance. Any sample should always be positioned at a reference height within the magnetic field of the objective lens (OL), usually the eucentric height of a side-entry tilt holder. The OL current is recorded for accurate focusing at that reference height so that the sample height can be physically readjusted to refocus, if necessary, following sample tilting around the noneucentric axis. The main calibrations are

- Low magnification: A replica from a mechanically scribed optical diffraction grating has traditionally been used for $M < \sim 20$ K. The grating is coated with evaporated metal that provides fine structure in addition to an array of lines with a spacing of ca. 0.5 μm.
- High magnification: The HREM or HAADF atomic image of a known crystal structure at a known zone axis allows direct measurement for $M >$ ca. 300 K.
- Intermediate magnification (20–300 K): Extrapolation of the low M and high M calibrations is made using fine and coarse features, respectively, in the two samples above. An alternative is a commercially available sample containing latex spheres of uniform diameter, typically a few tens of nanometers, supported on a holey carbon film.
- Camera constant (CC): The diffraction ring patterns from evaporated amorphous metals, e.g., Al, allow the radius in reciprocal space to be calibrated and checked for radial uniformity. Astigmatism in the IL (intermediate lens) is detected and corrected. Calibration is performed over the whole range of CC.
- Camera length (L): The nominal E_o and, hence, wavelength, λ, of a TEM can have considerable error. The CC = λL (L is the camera length) and L, rather than CC, is required for accurate measurement of angles in a Kikuchi pattern. The easiest way to calibrate E_o (or λ) is to exploit the fact that, in a Kikuchi pattern, the angular relationships between lines depend on the specific crystal structure and, therefore, are independent of E_o and L, while the separations of lines in the DP depend on E_o and L [13]. An orientation is found such that two zone axes are present from a DP of a known crystal structure at the CL (condensing lens) to be calibrated. The angle, θ, between the two zones is calculated from the known crystal structure and related to the distance, m, between the two zones measured in the DP. Because $L = m/\theta$, then λ (or E_o) = (CCθ)/m. A specific example is described in Section 9.4.2.
- Rotation angle, ϕ, between a direction in the DP and the same direction in a TEM image. The formation of the DP and the TEM image are shown in Figure 6.2 in Chapter 6. In practice there is a rotation, ϕ, of the image around the optic axis with respect to the DP which varies with M. When calculating the crystallographic directions of image features in order to characterize the misorientation across a GB, described in the following section, it is essential to correct for this rotation by measuring ϕ for every combination of CL and M. The standard procedure is to use a sample containing MoO_3 crystals on a holey carbon support. The orthorhombic crystals ($a = 0.3962$ nm; $b = 1.385$ nm; $c = 0.3697$ nm) grow as platelets with the (100) direction parallel to the longer side of crystals when viewed down the [010] zone axis, which is the orientation in which they tend to lie on the holey carbon film. A rectangular array of spots is, therefore, usually present in the DP with the [100] direction corresponding to the larger spacing. A double exposure of the DP and the image is recorded for each required M and CC, from which ϕ is measured. The 180° lens inversion

ambiguity can be checked by observing features in the shadow image in the DP spots as the OL is changed through focus [14].

- Illumination convergence angle, α: A CBDP (convergent beam DP) is recorded from a known crystal structure at a known zone axis, for the required CL2 aperture size and E_o. The separation of the indexed discs in the CBDP yields the value of L, and $2\alpha L$ equals the disc diameter, that is also measured from the CBDP, allowing calculation of α.

9.4 STRUCTURAL CHARACTERIZATION OF GRAIN BOUNDARIES AND INTERFACES

9.4.1 LATTICE MISORIENTATION

An interface defines the plane where two separate assemblies of atoms meet. In the most general case, each separate assembly can be either crystalline or amorphous, and the two assemblies have completely different compositions and atomic coordination. An interface is a region where the energy of the structure is raised by the increased atomic disorder relative to that inside the grains away from the interface. The atoms at the interface will try to adopt configurations that minimize the total interfacial energy. In materials with metallic bonding, energy minimization is achieved by maximizing the atomic coordination and, in alloys, this could involve the preferential segregation of specific atom species to the interface. In ionic and covalent materials, charge neutrality and bond hybridization are the principal factors that determine local atomic structure. The text by Sutton and Balluffi [15] is recommended for the reader requiring a comprehensive description of GBs and IFs. There is currently much interest in characterizing both the arrangements and the chemical nature of atoms at interfaces using TEM to enhance understanding and exploitation of mechanical and physical properties of materials. In this section we shall outline procedures for the structural characterization of an interface.

The simplest type of interface is a GB between two crystal grains, A and B, of a single-phase material when one grain is rotated with respect to the other. For such a GB it is always possible to find an axis, p, with a common crystallographic direction, $[H, K, L]$, in both grains about which a rotation by $\theta°$ will transform the orientation of A into that of B. It is conventional to define a low-angle GB as one for which $\theta < 15°$ and a high-angle GB one for which $\theta > 15°$. The plane of the GB interface is specified by the three crystal indices that define the direction normal to the GB at any point in the GB; the interface plane, therefore, can be described by two alternative sets of indices, one for each grain. Clearly, for a given misorientation, (p, θ), a continuous variation of interface plane orientation is possible and a total of five parameters (degrees of freedom) are required to describe a general GB. The degrees of freedom can be found by TEM.

A useful concept when describing a GB is the idea that the lattices of the two grains are allowed to extend and interpenetrate on both sides of the interface. This is the basis of the coincident site lattice (CSL) formulation. For high-angle GBs, certain specific rotational transformations (p, θ) may result in some fraction $(1/\Sigma)$ of lattice points in A being exactly coincident with lattice points in B, where Σ is given by [16,17]:

$$\Sigma = n^2 + m^2 (H^2 + K^2 + L^2) \tag{9.1}$$

In Equation 9.1, n and m are positive integers. The value of Σ is usually used to denote these special orientations. In a cubic system Σ always has an odd value of 3 or greater. The CSL points, however, are only a subset of a more general three-dimensional (3D) set of lattice points called the "O-lattice" [18], defined in the following manner. For a specific CSL orientation, only the CSL points in A and B are coincident while many more O-lattice points may not be coincident. However, a rigid translation by an O-lattice vector of A relative to B, without further rotation, causes a new set of O-lattice points in A to become coincident with a different set of O-lattice points in B and, hence, these form the new CSL sites. Both configurations have the same Σ value and are alternative descriptions for the same rotation (p,θ). The translation is not required to lie in the GB plane. The set of three smallest nonplanar vectors that define the O-lattice are called DSC (displacements which are symmetry-conserving) vectors [19]. DSC vectors can be identical to lattice vectors of A and B but, generally, they are smaller in magnitude. The volume of the O-lattice cell and, hence, the magnitude of the DSC vectors, have an inverse relationship with Σ [20]. The Burgers vectors of interfacial dislocations are described in terms of DSC vectors.

9.4.2 DETERMINATION OF GRAIN BOUNDARY MISORIENTATION

The TEM experimental procedure for the determination of GB misorientation has been detailed in Refs. [21,22]; it has been most widely applied to materials with cubic symmetry but the methods can be extended to other crystal systems and to phase interfaces between two different crystal systems. Careful calibration of the microscope (Section 9.2) is assumed, especially the image magnification, the angular rotation calibration between DP and image, the camera length and electron wavelength that determine the angular magnification of the DP.

High accuracy can be achieved only when the sample is not bent. This requirement often prevents the characterization of GBs near the thin edge of a sample in many metal alloys. Bending is detected and measured within each grain by observing lateral shifts in the Kikuchi pattern as either grain is moved under the beam or, in severe cases, by the presence of bend contours in DC images. Once a GB has been selected in a flat area, the sample is tilted to a two-beam diffraction condition in Grain A. The DP from Grain B at this sample tilt is most likely to be kinematic, i.e., no strongly diffracted beams are present. However, by alternately viewing the DP from the two grains as the sample is slowly tilted along the Kikuchi band seen in the DP from Grain A, a Kikuchi band will eventually be found in Grain B and two-beam conditions established simultaneously in both grains. The DPs from both grains are then recorded, together with an image of the GB. The sample is then tilted along the Kikuchi band in Grain B, which tilts Grain A away from its initial two-beam orientation, until a new Kikuchi band appears for Grain A. Fine tilt adjustment is made until both grains are at two-beam orientations, before recording DPs and GB image again. In principle, (p,θ) can be determined from just these two pairs of Kikuchi patterns. However, several pairs of data permit multiple derivations and averaging of (p,θ) values so, for high accuracy, the process described above is repeated several times over as large a range of sample tilt as is feasible.

The CC used for the above experiment should be sufficiently small to contain pairs of Kikuchi lines from several planes in order to allow determination of the beam direction indices at each sample tilt for both grains. An example of the procedure to determine the beam direction, O, from a Kikuchi pattern is shown in Figure 9.3 in which corresponding pairs of Kikuchi lines have been indexed, and zone axes calculated, at the intersections of pairs of prominent bands. The [1,–2,0], [4,–7,–1], [5,–8,–1], and [3,–5,0] zone axes are labeled for a right-handed set of axes and with the directions upward with respect to the plane of the paper. With knowledge of the CC, the precise angular distances of the incident beam along and perpendicular to one of the zone axes are measured and the beam direction is marked on a stereographic projection. In Figure 9.3 the beam direction is found by rotating the crystal 3.17° (clockwise) from [1,–2,0] about the [0,0,1] direction toward [1,–1,0], and then further clockwise rotation by 3.88° from that point about the [4,2,0] axis. It has indices close to [7,–12,–1].

For each orientation of the sample, 1, 2, 3, ..., pairs of directions, a_i and b_i, of the incident beam, indexed with respect to Grains A and B, can be determined, as just described. (Here, a_i and b_i each represent the indices of unit vectors in the crystal directions parallel to the beam at sample orientation i.) From an experimental set of data pairs: $a_1, b_1; a_2, b_2; a_3, b_3;$ ..., it is required to determine unique values of p and θ which relate the orientations of Grains A and B. The derivation can be made by stereographic projection or using vector algebra.

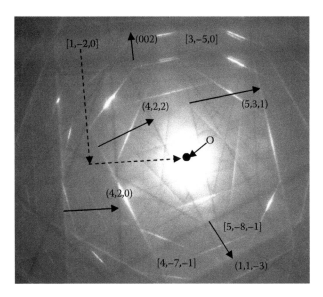

FIGURE 9.3 A Kikuchi pattern from a grain of Inconel 600. The precise orientation of the grain with respect to the incident beam direction, O, is measured from the direction and magnitude of the displacement of the zero-order beam from the indexed directions of the Kikuchi bands.

The stereographic method is shown in Figure 9.4, in which the points a_1 and b_1 indicate the directions of the incident beam at sample tilt position 1, referenced to the crystallographic indices of Grains A and B, respectively. A great circle (zone), EE′, is constructed through a_1 and b_1 and the rotation angle, θ, between the two points is measured directly along the zone. The rotation axis, p, must lie somewhere on the zone, FF′, which intersects EE′ at the midpoint between a_1 and b_1. A similar procedure is followed for the second pair of beam directions, a_2 and b_2, at sample tilt position 2. The great circle, GG′, is constructed through a_2 and b_2 and the angular separation again provides a value for θ which should be the same as measured previously on EE′, within experimental accuracy. A zone,

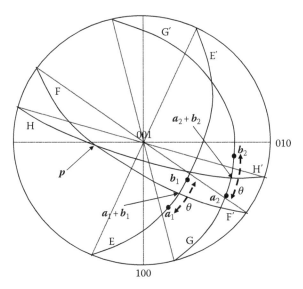

FIGURE 9.4 Stereographic projection illustrating how the rotation axis, p, and rotation angle, θ, between Grains A and B are found from two-beam directions, 1 and 2.

HH′, is constructed perpendicular to GG′, passing through the midpoint between a_2 and b_2. The rotation axis, p, is required to lie on HH′ and, therefore, is found from the intersection of zones FF′ and HH′.

The accuracy of p and θ depends on the accuracy of the stereographic construction and the errors in the data pairs. When zones FF′ and HH′ intersect at a shallow angle the error in p can be large, which is why it is recommended to reduce errors by repeating the construction for several pairs of data to find average values.

The indexing of the first pattern in any grain can be chosen arbitrarily but, thereafter, the indexing must be self-consistent with the initial choice when indexing all other beam directions recorded for that grain. Depending on the crystal symmetry, a number of equivalent sets of indices could be used for any grain. For example, for a simple cubic lattice, two Kikuchi bands in the first pattern of grain A, indexed as (h_1, k_1, l_1) and (h_2, k_2, l_2), might just as easily have been indexed as $(-h_1, k_1, l_1)$ and (h_2, k_2, l_2), or as any of the other combinations of equivalent crystallographic indices. If the initial reference set of axes used to define the directions in Grain A is changed to a symmetry-related alternative, then that does not change the actual physical misorientation between A and B. However, the points in Figure 9.4 will be relocated and therefore completely different values will be determined for p and θ, which will be related to the first set of values through the crystal symmetry. Thus, a specific ΣCSL GB in the cubic system might be described by up to 24 different combinations of p and θ [23]. Detailed, worked examples showing the derivation of (p, θ) values are described in Ref. [21].

The algebraic derivation of p and θ is described using the same beam direction vector pairs shown in Figure 9.4. The axis of the zone, EE′, containing a_1, b_1 (and their midpoint, $a_1 + b_1$) is defined by $a_1 \times b_1$. The zone FF′, therefore, has a zone axis direction given by $[a_1 + b_1] \times [a_1 \times b_1]$. In a corresponding manner, the zone, HH′, is derived from: $[a_2 + b_2] \times [a_2 \times b_2]$. Hence, the direction, p, is given by

$$p = \{[a_1 + b_1] \times [a_1 \times b_1]\} \times \{[a_2 + b_2] \times [a_2 \times b_2]\} \tag{9.2}$$

The rotation angle, θ, is calculated from

$$\cos \theta = [(p \times a_1) \cdot (p \times b_1)]/[|p \times a_1| \cdot |p \times b_1|] \tag{9.3}$$

As with the stereographic method, accuracy is improved by averaging values derived from combinations of all the possible pairs of beam directions $(a_i, b_i; a_j, b_j)$. A matrix method for estimating the most likely values of p and θ from multiple measurements has been described by Mackenzie [24].

The values of p and θ for GBs in cubic crystals are relatively easy to calculate, as described above, because the lattice constants are equal and the axes are orthogonal. The derivation for a general interface between two crystals of different systems is more complicated. However, even for the triclinic system with the lowest symmetry of all, it is possible to transform any crystal direction, a_1, to a direction, a_1', in a chosen, reference lattice with orthogonal axes, by multiplication with an appropriate matrix, T. Thus, $a_1' = T \cdot a_1$. A second orthogonal reference lattice is chosen for the second phase, so that a direction, b_1, is transformed into $b_1' = Sb_1$. If the two reference lattices are then scaled to have identical lattice parameters, which can be all equal (i.e., cubic) for convenience, the misorientation across an interface can be described by p and θ in a manner equivalent to that for a GB in the cubic system described above. As with cubic crystals, the symmetry of two phases will probably permit various equivalent indices to be allocated to the Kikuchi lines in a DP, leading to different possible indices for a beam direction and alternative, but corresponding, values of p and θ.

9.4.3 DIFFRACTION EFFECTS

The DP acquired with the selected area diffraction aperture (SADA) placed over a GB/IP (grain boundary and phase boundary) contains the spots and Kikuchi lines from both grains but it might also show extra spots which are not present in either of the DPs recorded separately from the

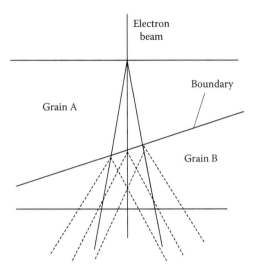

FIGURE 9.5   Sketch of inclined interface geometry and double diffraction beams.

grains. When one grain overlaps the other (Figure 9.5), each diffracted beam (solid line) generated in the upper grain acts as an incident beam when it enters the lower grain. Each beam from Grain A generates a new DP in Grain B, indicated by the dashed lines in Figure 9.5. This double diffraction causes the DP from the lower grain to be a superposition of similar DPs, each centered on every spot from the upper grain. (A similar situation occurs when a DP is formed from a matrix containing a second-phase particle.) It is possible to identify double diffraction spots by comparing DPs recorded separately from the two grains. When double diffraction spots lie close to the "undiffracted" beam and are included within the CA (contrast aperture) used to form a BF (back focal) image, Moiré fringes are seen in the GB/IP running along the length of the GB/IP, parallel to the intersections of the GB/IP and the sample surfaces. Such fringes can be falsely identified as interfacial dislocation images.

In Section 6.3.1 of Chapter 6 it was noted that diffraction spots are, in reality, "relrods" extended in reciprocal space in a direction perpendicular to the entrance surface of the sample. In the case of the DP from an inclined GB (Figure 9.6), when the electron beam passes through Grain A and enters Grain B through the GB, a relrod of a DP spot from Grain B is generated in a direction perpendicular to the GB plane. However, a relrod is also generated from the nonoverlapped region of Grain B perpendicular to the entrance surface. Hence, two relrods are generated by Grain B. The intersection of the Ewald sphere with the relrods generates two slightly separated DP spots from Grain B (Figure 9.6b).

In the following section, the nature of GB dislocations will be described. Often, such dislocations are present as parallel arrays with a constant spacing. The strain fields of such defects generate a periodic modulation of the crystal potential seen by the electron beam and, hence, act as a 1D lattice with a planar spacing equal to the defect separation. Because the dislocations are located in the inclined GB, a weak line of additional relrods and, hence, diffraction spots is generated in a direction perpendicular to the projected dislocation direction and passing through the extra, displaced spot from Grain B associated with the relrod of Grain B from the inclined GB (Figure 9.6c). Note that, in Figure 9.6b and c, the relrods are coplanar with g. In practice, the relrods from the inclined GB and dislocations are likely to be inclined to the plane containing g and the relrod associated with the entrance surface. Any such extra DP spots from a GB dislocation array are usually very faint and can be masked by fogging of the recording medium from the much stronger spot from Grain B. If visible, the separation of the spots provides a measurement of the average dislocation spacing in the array.

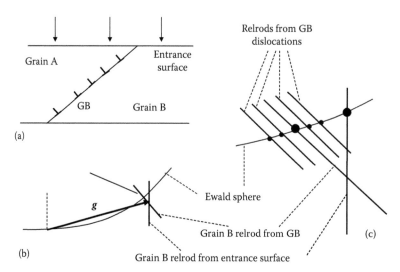

FIGURE 9.6 Diffraction effects from an inclined GB: (a) schematic showing inclined a GB plane containing dislocations; (b) the intersection by the Ewald sphere of the relrods for a diffraction vector, g, in Grain B arising from the entrance surface of Grain B and the inclined GB plane; (c) an enlarged view of the intersection of the Ewald sphere and the relrods. Additional relrods, generated by the array of parallel GB dislocations, are shown lying parallel to the relrod of g from Grain B associated with the inclined GB.

9.4.4 Grain Boundary Dislocations

A GB/IF plane is often observed to contain dislocations. The Burgers vector of a perfect lattice dislocation, b, is a lattice vector of the parent crystal structure. A low-angle GB can be described by a set of perfect dislocations aligned in a plane. An ideal low-angle tilt boundary contains an array of edge dislocations with roughly equal spacing and a common b, while an ideal low-angle twist boundary contains only arrays of screw dislocations each with a common b. Combinations of more than one b form more complex low-angle GBs that are mixed (i.e., both tilt and twist components) in nature. A flat section of a GB (i.e., a facet) has uniform spacing of all the dislocation sets while curvature of a boundary is accommodated by local changes in the spacing of one or more of the dislocation sets.

In a manner analogous to the description of a low-angle GB in terms of lattice dislocations and Burgers vectors, it has been established, at least for the case of cubic crystals, that a planar section of a CSL high-angle GB can be described in terms of up to three arrays of dislocations each with a different, noncoplanar Burgers vectors [21]. However, the three Burgers vectors in the case of a CSL are either base vectors of the DSC lattice, or linear combinations of the three base DSC lattice vectors. The line direction and spacing of the three sets of dislocations are constant in any facet of an interface and, analogous to lattice Burgers vectors in a low-angle GB defining the total rotation between the two grains, an array of dislocations with the same DSC lattice Burgers vector and a fixed spacing describe a specific rotation, θ', "from a precise CSL interface which has (p,θ) close to that of the actual GB." Curvature of such a CSL GB is again accommodated by variation of the spacing of the sets of GB dislocations.

The total Burgers vector, B, of all the dislocations crossing a distance, x, in the GB interface plane is [25]:

$$B = 2 \, (\sin (\theta/2) \, (x \times p) \tag{9.4}$$

For a low-angle GB ($\theta < 15°$) B is an aggregate of lattice Burgers vectors, b, while for a high-angle GB, B is the aggregate of DSC vectors and θ is replaced by θ'.

The general validity of Equation 9.4 has been demonstrated (Figure 9.7) both for low-and high-angle GBs in f.c.c. metals using specially fabricated bicrystals of Ag and Au, in which twist boundaries

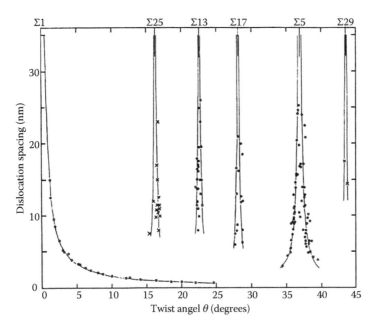

FIGURE 9.7   Comparison of dislocation spacing calculated using Equation 9.4 and measured by TEM as a function of rotation angle in twist boundaries in Au. (Reprinted from Babcock, S.E. and Balluffi, R.W., *Phil. Mag. A*, 55, 643, 1987. With permission.)

of predetermined rotations about a chosen crystal direction were made, and which generated arrays of screw dislocations [26–28]. Corresponding confirmation for tilt GBs in Au has also been reported [29]. Figure 9.8 shows an example of interfacial DSC screw dislocations with $b = a/10<310>$ in $a <100>$ twist boundary close to S5 (36.9° rotation) in Au [27].

In a TEM DC image, two adjacent grains will generally show different transmitted intensities unless the diffraction conditions in both grains are equivalent. The projection of the GB is a band of varying width, contrast, and direction, defined by the intersection of the GB with the upper and lower sample surfaces. A GB in Inconel 600, an f.c.c. nickel-based alloy, is shown in Figure 9.9. The misorientation was determined using the methods described above to be within 2° of $\Sigma35a$ for which, nominally, $p = <510>$ and $\theta = 119°$ [23]. Fringes are observed running along the band of the projected GB image, and two sets of closely spaced parallel dislocation lines are present. The GB plane is curved and the dislocation spacing changes with the inclination.

The analysis of interfacial dislocations at a low-angle GB with matrix Burgers vectors can be performed using the method described in Chapter 6, which requires seeking diffraction vectors, g, which render the dislocations "invisible" when $g \cdot b = 0$. However, as θ increases, the spacing of the dislocations decreases, their strain fields begin to overlap, and it becomes more difficult to resolve the individual defects and to determine b, especially when there are two or three sets of intersecting dislocations. Analysis then requires the computation of dislocation images imaged under a range of different diffraction conditions, using different values of b equal to possible DSC vectors, which are then compared with corresponding experimental images to seek a match. For large Σ, when the DSC Burgers vectors are small, and the spacing is too small to permit the resolution of individual dislocation images, experimental dislocation analysis is impossible using DC methods.

On the atomic scale, it has been proposed [30,31] that GBs are composed of a limited number of atomic polyhedra, equivalent to the Bernal polyhedra proposed for models of liquid structure [32]. An earlier proposal was that basic atomic structural units could be used to construct a GB, and that the frequency of these units within the GB plane could be changed to vary angular offsets from a given GB that, itself, was constructed from different structural units with a much larger spacing

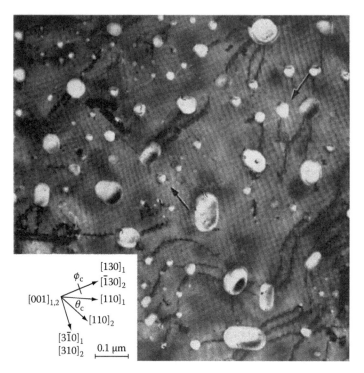

FIGURE 9.8 Interfacial DSC screw dislocations with $b = a/10<310>$ in a $<100>$ twist boundary close to S5 (36.9° rotation) in Au. (Reprinted from Schober, T. and Balluffi, R.W., *Phil. Mag.*, 21, 109, 1970. With permission.)

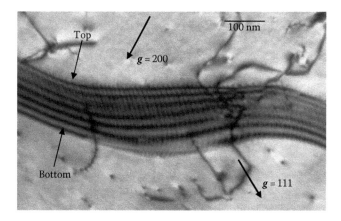

FIGURE 9.9 A GB in Inconel 600 imaged under kinematic diffraction conditions. The GB is close to Σ35. The intersections of the GB plane with the top and bottom sample surfaces are labeled. Dislocations are observed lying in the boundary plane.

[33–37]. The precise atomic arrangement at the GB could be computed using density functional theory to seek the minimum energy configuration of assemblies of atoms [15]. Such calculations have demonstrated that GB interfaces are, indeed, composed of repeated arrangements which are constructed from combinations of a limited number of polyhedral atomic arrangements. A repeat unit might consist of one, two, or more of the different polyhedra and a particular structural polyhedron can occur several times within a repeat unit. Changes in θ or in the boundary orientation (H, K, L) are accommodated by changes in the combinations of structural units within the repeat units.

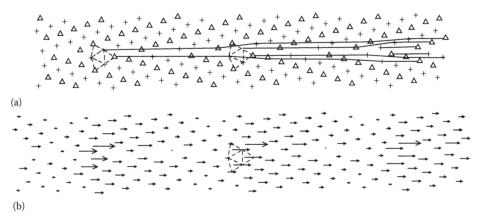

FIGURE 9.10 Theoretical structure of a relaxed (229) symmetric tilt boundary in Al, showing correspondence of structural units (a) and terminating {115} planes of atoms (b). (Reproduced from Sutton, A.P. and Vitek, V., *Phil. Trans. R. Soc. Lond. A*, 309, 1, 1983. With permission.)

It was stated earlier that such changes in boundary geometry could be accommodated by varying the spacing and densities of three sets of GB dislocations which had DSC Burgers vectors. Direct correspondence between the computed repeat units and the positions of the cores of DSC GB dislocations was established by Sutton and Vitek [38–40] in computations of the GB structures for a series of symmetric $p = <110>$ tilt GBs in Al, as θ was systematically changed between 31.59° (corresponding to a {115} GB plane) and 50.48° (a {113} plane). Figure 9.10 shows one of these <110> GBs, lying on a {229} plane ($\Sigma = 49$, $\theta = 34.89°$) [38]. The repeat units contain capped trigonal prisms, indicated by the dotted lines, and three repeated combinations of an irregular pentagon and a tetrahedron that are not outlined for the sake of clarity. An extra {115} half-plane of atoms in both grains terminates at each trigonal prism, so that $b = (2a/27)$ [115] for each DSC dislocation. As θ varied, the structure of the repeat units changed, as did the indices of the extra half-planes terminating (and, hence, b for the DSC dislocations) at the new structural repeat unit. Hence, there is a direct equivalence between the description of a GB in terms of structural polyhedra and by dislocations with DSC Burgers vectors.

9.4.5 DETERMINATION OF DISLOCATION LINE DIRECTIONS AND INTERFACIAL PLANE INDICES

The interfacial plane is independent of (p, θ) and can, in principle, have any orientation. The GB planar indices are defined with respect to either grain. The experimental determination of the indices requires the imaging of two, nonparallel, linear features (i.e., the trace of the GB with the sample surface, an interfacial dislocation, or a GB triple point intersection) at two (or more) widely separated orientations. In practice, the GB must be flat over the analysis region. The TEM images contain only projected images of the linear features. The true, 3D directions, t_1 and t_2, of the two features are initially unknown and are to be determined. The indices of the beam direction, B_i, and the projected direction of one linear feature, pt_1, at orientation i, are determined by stereographic construction from the Kikuchi lines in the DP recorded for each image. The plane, $P_i = B_i \times pt_1$, which contains both the beam direction and the projected image line, also contains the real direction of the linear feature, t_1. A second plane, $P_j = B_j \times pt_1$, at orientation j, are then found, which also contains t_1, which is then found from the vector product: $t_1 = P_i \times P_j$. The procedure is repeated for the second linear feature with direction t_2. The GB/IP is the plane that contains both t_1 and t_2, and so the GB plane normal is determined from their vector product, $t_1 \times t_2$.

Interfacial dislocations can never be accurately analyzed in a high-angle GB simply using the $g \cdot b = 0$ invisibility criteria that are employed for standard lattice dislocations in a low-angle GB or in a single crystal. Any rigid body displacement, R, between the two grains complicates the dislocation analysis. R is detected from the presence of fringes in the GB image acquired using a diffraction vector, g, common to both grains, i.e., by selecting planes that have the same orientation in both grains and are, therefore, continuous across the GB, apart from the displacement R. This is directly equivalent to the imaging and analysis of stacking faults [41,42]. However, determination of the magnitude and direction of R for a general GB is not always feasible. Forwood and Clarebrough [43] have shown that $\Sigma 3\{112\}$ and $\Sigma 3/\{213\}$ GBs in b.c.c. Fe containing C and N are symmetric and only have a component of R normal to the boundary plane. However, the experimentally determined R was significantly greater than that predicted from a hard sphere atomic model, and it was concluded that segregation of C would explain the additional contribution to R. The only safe way to confirm Burgers vectors and R is to match dislocation images recorded in a minimum of three noncoplanar diffraction vector pairs, with corresponding computer-generated images. It is beyond the scope of this chapter to describe the full procedure for such computations but these have been described in detail in Ref. [15].

9.5 ATOMIC IMAGING OF INTERFACES

It is a time-consuming activity to analyze completely the orientation and dislocation structure of a random GB, and it becomes an impossible task using DC imaging when the dislocation spacing becomes too small to be resolved. The whole concept of GB representation by dislocation arrays is then questionable. Even when a complete description of the GB is obtained, the information becomes useful only when it can be related to other information such as the chemical environment of the GB, the correlation with precipitate nucleation, and the emission or blocking of dislocations, in explaining plastic deformation behavior, etc. The whole GB analysis must then be repeated for many individual boundaries in order to gather statistically meaningful data. GB characterization is much more useful when the sample contains a very limited range of interfaces, because the physical or mechanical properties of the sample depend on fewer variables. For such interfaces with special geometry, direct imaging of atomic structure parallel to the interfaces plane using HREM and STEM-HAADF has become important both in research and in the quality assurance assessment of devices.

Many semiconducting and magnetic devices are fabricated by growing homo- and hetero-epitaxial layers on large single-crystal planar substrates. Examples include Si/SiGe layered structures, high-density IC chips, and giant magnetostriction devices for read–write heads on computer disc drives. The physical abruptness, the chemical diffusion of the component chemical species, and the changes in bonding at the interface, are important parameters that determine device performance. Great care in sample preparation is crucial for successful characterization of these structures. Samples are prepared in the cross-section orientation using ion-beam methods for thinning. As the lateral dimensions of components become smaller, FIB thinning, rather than a broad ion beam, is increasingly used to cut out specific device components for specific analysis (see Section 9.2). Moreover, the orientation of the cross-sectional plane normal is selected to be close to a low-index crystal zone axis when a single-crystal substrate is used for epitaxial layer growth, minimizing the range of sample tilting in the TEM and enabling low-C_s OL lenses to be used for very high resolution (see Section 3.4 in Chapter 6).

9.5.1 ATOMIC IMAGING OF GBS USING HREM

Direct structural imaging of crystal lattices using HREM was initially limited to minerals with relatively large lattice constants and spacing of atomic columns, e.g., the binary system of Nb_2O_5–WO_3 reported by Allpress et al. [44]. Imaging of the Si lattice and dislocation cores in Si by HREM then followed [45,46]. The spatial resolution required to image directly the atomic structure of high-angle metal boundaries, where column spacings are significantly smaller than in minerals, was first reported by Krakow and colleagues [47,48]. Krakow et al. revealed the atomic structure of several interfaces in Au using medium voltages (200–400 keV) to take advantage of improved C_s and information limit. However, very careful beam and sample tilt alignment was essential and the images were cleaned by applying both low and high band-pass computerized Fourier processing methods [47]. The integrity of the images was supported by computer-generated images of theoretical interfacial structures. In particular, the structure of a 32° [110] high-angle tilt boundary, close to the symmetric (3,−3,1)/[110] tilt boundary, was reported to contain capped trigonal prisms, repeated every 2.5 nm along the GB, with the prisms corresponding to the cores of primary lattice dislocations [49]. A 68° (1,−1,2)/[110] tilt GB, shown in Figure 9.11, was reported to contain capped trigonal prisms linked by octahedra. The structure of the experimental image (Figure 9.11a) agrees very closely with the computed image (Figure 9.11b) in showing interfacial steps, ~0.6 nm in height. What is not easy to see in the experimental image is that there is also a rigid body translation of $a/2[-1,1,1]$ which is spread over a distance of ca. 1 nm on each side of the GB. Such experiments were crucial in verifying the structural unit model of GBs and computational results. The structures of lower angle (10° and 16°) boundaries in Au were also reported by Krakow and Smith [49].

An example of HREM imaging of a heterogeneous interface between a metal, Al, and a ceramic, $MgAl_2O_4$, viewed along the [110] direction is shown in Figure 9.12 [50]. The image was recorded

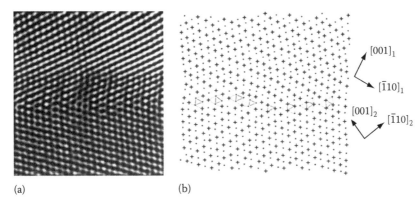

(a)   (b)

FIGURE 9.11 (a) Experimental image of a 68° (1,−1,2)/[110] tilt boundary in Au viewed along the tilt axis, showing capped trigonal prisms linked by octahedra, and (b) the corresponding computed GB structure. (Reproduced from Krakow, W. and Smith, D.A., *J. Mater. Res.*, 1, 47, 1986. With permission.)

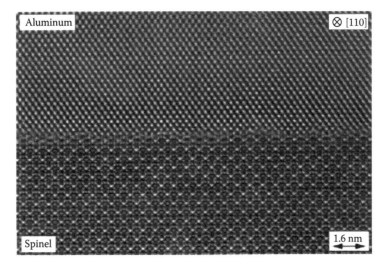

FIGURE 9.12 HREM image of a heterogeneous interface, viewed in cross section, recorded with a 1.25 MeV microscope. (Reprinted from Schweinfest, R., Kostelmeier, S., Ernst, F., Elsasser, C., Wagner, T., and Finnis, M.W., *Phil. Mag. A*, 81, 927, 2001. With permission.)

using a 1.25 MeV microscope. The interface is coherent because both phases are f.c.c. with a lattice parameter difference of <0.25% and there are no visible interfacial dislocations.

9.5.2 An Example of Interfacial Characterization by HREM

The point has been made previously that characterization of interfaces by HREM is restricted to special orientations such as tilt boundaries, where the sample can be oriented such that the beam direction is simultaneously parallel to a low-index zone axis in both grains and also to the GB/IF plane. One of an increasing number of examples in the literature that demonstrate the current power of HREM for the characterization of such GB/IFs is used to illustrate this.

The perovskite, $YBa_2Cu_3O_{7-\delta}$, where δ implies an unknown deficiency in the O sublattice, is an important superconducting material that has the orthorhombic structure shown in Figure 9.13. Its superconducting efficiency is reduced by the presence of structural disorder such as oxygen deficiency and grain boundaries. The unit cell dimensions, where $a = 0.3818$ nm; $b = 0.3887$ nm; $c = 1.1683$ nm, are such that $c/b \approx 3$ and a 90°-tilt boundary, formed by rotation around the <100> axis, has very low mismatch and strain energy. A sophisticated example of the use of HREM for the characterization of such a tilt boundary in an $YBa_2Cu_3O_{7-\delta}$ layer grown by CVD, was recently described by Houben and coworkers [51]. The experiment required the use of many of the principles described in Chapter 6, in addition to careful quantitative analysis and statistical treatment of the data. In particular, it was necessary to use the highest available spatial resolution in order to image the positions of the oxygen atoms. It is still extremely difficult to image oxygen atom columns by TEM in such compounds due to the close proximity of much heavier cation columns less than 0.2 nm away. Sufficient resolution was achieved using a 200 keV aberration-corrected TEM [52] giving an information limit of 7.7 nm^{-1}, and the residual aberrations were carefully determined. A negative value of OL C_3 was used to optimize the contrast.

A through-focal series of 20 images of the same sample area containing a [100] 90°-tilt boundary was recorded with the beam parallel to the [100] direction and processed [53,54] in order to retrieve the phase and amplitude distributions of the exit-surface wave function. Post-acquisition correction of the residual aberrations was made to the recorded images to improve

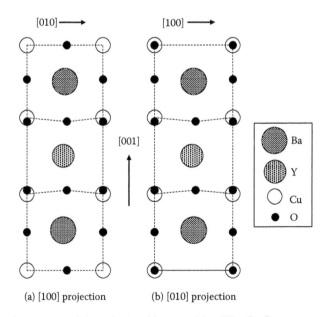

(a) [100] projection   (b) [010] projection

FIGURE 9.13 Atomic structure of the orthorhombic perovskite, $YBa_2Cu_3O_{7-\delta}$.

FIGURE 9.14 One image from a recorded through-focal series showing a 90° tilt GB in $YBa_2Cu_3O_{7-\delta}$, viewed along the [110] direction. A small area of the corresponding simulated image is inset. (Reprinted from Houben, L., Thust, A., and Urban, K., *Ultramicroscopy*, 106, 200, 2006. With permission.)

further the resolution. One image from the through-focal series is reproduced in Figure 9.14, inset within which is a small area of the corresponding computed image by the multislice method [55,56], using a commercially available software package [57]. In this image the defocus is such that the brightest dots form a roughly square grid corresponding to the projected columns of Ba and Y atoms, with Cu columns at the centers of the Ba/Y array of squares. The exit wave phase reconstruction, reproduced in Figure 9.15, greatly enhances these features and

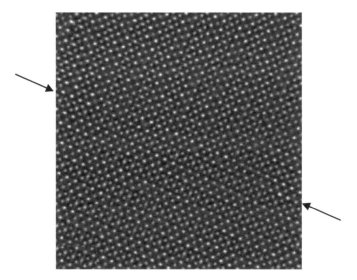

FIGURE 9.15 The phase image derived from a through-focal series of the same GB (arrowed) shown in Figure 9.14. (Reprinted from Houben, L., Thust, A., and Urban, K., *Ultramicroscopy*, 106, 200, 2006. With permission.)

shows some intensity from the O columns between adjacent Ba and Y columns. Further enhancement was achieved by averaging repeat structural image units along the GB. Finally, quantitative analysis of the pixel intensities in the phase image was performed, using a least-squares regression method, to fit ideal Gaussian intensity profiles to the image features, to determine the most likely atom positions to an accuracy as low as 2 pm. A summary of the atomic displacements in the vicinity of the GB is shown in Figure 9.16. The high accuracy of this procedure enabled detection of the displacement from the nominal positions of the Cu atoms lying in the GB plane in directions within the GB plane itself, and the displacement of the columns of Ba and Y atoms nearest to the GB plane. The O atoms can occupy four different sites in the orthorhombic structure (Figure 9.13). Analysis of the phase image intensity at the positions of the O2 site suggested that it had 0% occupancy, while O1, known to be a stable site, was assumed to be 100% occupied. This allowed estimation of the partial occupancy of the O4 sites, assuming a linear relationship between intensity and occupancy that was suggested by image calculations. A limit on the oxygen deficiency of $\delta < 0.45$ was found. Because the intensity fluctuations in the phase image at the O4 atom sites were random within the Cu-O4 chains, it was further deduced that there was no ordering of the O within the Cu-O4 chains.

This highly detailed characterization of the atomic locations and site occupancy in the $YBa_2Cu_3O_{7-\delta}$ GB illustrates what can now be achieved by combining experimental expertize with the most modern HREM instrumentation, image processing, and modeling.

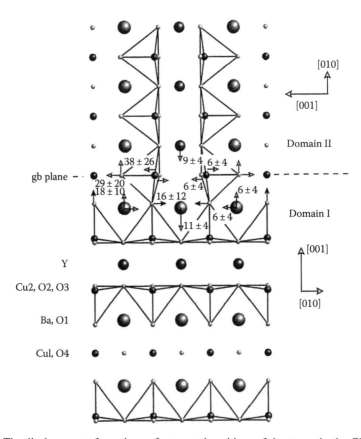

FIGURE 9.16   The displacements from the perfect crystal positions of the atoms in the GB structural unit of the 90° tilt GB in $YBa_2Cu_3O_{7-\delta}$. (Reprinted from Houben, L., Thust, A., and Urban, K., *Ultramicroscopy*, 106, 200, 2006. With permission.)

9.5.3 EXAMPLE OF ATOMIC IMAGING OF SEGREGATION USING HAADF-STEM

In addition to HREM imaging, many instruments are also now able to provide HAADF-STEM imaging with sufficient resolution for imaging individual atomic columns in most materials. The main advantage of HAADF-STEM over HREM is that the image features can be directly correlated with an atomic column because there is no "phase problem" (see Section 6.3.4.5 in Chapter 6). An example of the application of HAADF-STEM to the characterization of interfaces is described below for an important ceramic material, Si_3N_4. Polycrystalline Si_3N_4 is oxidation resistant and has excellent mechanical properties, such as good wear, creep and fatigue resistance, and high strength at high temperatures. Because its density is much lower than in metals, it is increasingly replacing steel components, e.g., in the automotive industry, and has potential for much more widespread use in moving engine parts and in high-temperature turbine components, providing further improvements can be made to its high-temperature wear and fracture resistance [58].

The hexagonal structure of β-Si_3N_4, with $c/a = 0.382$, is shown in Figure 9.17. Si_3N_4 is formed by sintering at high temperature, during which elongation is generally observed in the [001] direction relative to the perpendicular smooth <100> prism facets, with an aspect ratio of ca. 1.4:1 [59]. The addition of sintering aids, particularly of rare-earth (RE) oxides, to improve densification has been found to increase the aspect ratio and to affect the mechanical properties such as the fracture toughness. The elongated grains are separated by thin (~1 nm) amorphous GB films of oxynitride, and by wider amorphous oxynitride pockets at triple points. The grain elongation is believed to be governed by the segregation of the RE atoms to sites on the flat <100> facets, which then inhibits significant growth in the <100> and <110> directions [60,61]. The toughness has been found to increase with the average aspect ratio of the Si_3N_4 crystals which, in turn, depends on the specific RE addition, with increasing efficacy in the order: Lu, Yb, Sc, Y, and La.

The most effective method for revealing the locations of the RE atoms employs STEM-HAADF imaging. Provided the inner detector angle is made sufficiently large the elastically scattered signal is proportional to Z^n, where n has a maximum value of 2 but generally lies in the range 1.5–2. The signal from a RE atom is, therefore, much greater than from an Si, O, or N atom, so columns of RE atoms are brighter than the matrix atom columns. The Si_3N_4 samples are nonconducting so must be prepared using low-energy atom beams with shallow incidence angles to prevent damage during thinning. It is also necessary to address specific experimental problems when undertaking the examination of ceramics by high-resolution STEM-HAADF, especially when using aberration-corrected probes as in Refs. [62–65]. Samples tend to charge during examination and a very thin coating of evaporated carbon is deposited to suppress this. Aberration-corrected probes, while improving spatial resolution, are formed using a large convergence angle giving a high current that can easily

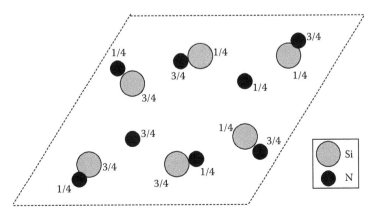

FIGURE 9.17   Structure of β-Si_3N_4 in [001] projection.

damage the sample and destroy the integrity of the atom positions of interest. It is, therefore, necessary first to optimize the imaging conditions on an area adjacent to the area used for detailed analysis, in order to minimize the risk of damage. Image distortion due to sample drift and scanning nonlinearity can be partially removed by postacquisition processing [65]. Low signal-to-noise ratio (SNR) in HAADF images can be enhanced by averaging repeating structural units in the image.

TEM investigations [62,63,66] have confirmed the segregation of RE species to the <100> prism facet interface with the amorphous intergranular film, for various RE additions using atomic imaging when viewing down the [001] direction. Segregation of both La and Lu was reported at intergranular/amorphous layers and grain/amorphous pocket interfaces [63,64]. Recently, the 3D coordinates of the RE atoms have been measured relative to the surface Si sites at the interface for La and Lu, by combining the information from atomic images viewed down both [001] and [100] directions [64,65]. In the above-referenced experiments the RE contrast was enhanced because the RE atoms occupied specific site coordinates at the Si_3N_4 surface and were, hence, viewed along a RE atom column direction. (Although the observation of single atoms of Sb in a Si matrix has recently been reported [8,67] by HAADF, it is still generally difficult to detect individual impurity atoms.)

In Figure 9.18, eight HAADF images are compared from alloy samples of Si_3N_4–2MgO–7.2 La_2O_3 and Si_3N_4–2MgO–8.7Lu_2O_3 made by sintering and hot isostatic pressing [64]. The HAADF images have been processed to remove scan irregularities, and were constructed by averaging five adjacent, identical, structural units in each case in order to improve the SNR ratio so that interfaces between facets and either (Figure 9.18a through d) glassy intergranular films, or (Figure 9.18e through h) glassy pockets in samples with either La (Figure 9.18a, b, e, and f) or Lu (Figure 9.18c, d, g, and h) sintering additions, could be shown. In Figure 9.18a, c, e, and g the interfaces are viewed down the [001] direction, while those in Figure 9.18b, d, f, and h are viewed down the [100] direction. In all eight images the crystal-amorphous interface plane is parallel to {010}. As expected from the crystal structure, the arrangement of the Si atom columns in the Si_3N_4, is completely different when viewed in the two projections.

The crystal-amorphous interfaces are marked by lines of bright dots corresponding to the locations of columns of either La or Lu atoms that have segregated. The columns of segregated atoms occur as pairs of dots when projected on [001] and equally spaced dots when projected on [100]. (Note that the N atoms have too low an atomic number to scatter sufficient intensity to be detected.) In all eight images in Figure 9.18 there is periodic contrast inside the glassy phase. This is an interesting observation which indicates that the atoms in the glass are not arranged in a random order right up to and adjacent to the interface but are partially ordered by the adjacent periodic crystal field. It is possible that such contrast arose from signal nonlocalization, analogous to observations in coherent HREM images of grain boundaries. As explained in Chapter 6, even with incoherent HAADF imaging, it is possible for scattering of a probe when it is placed close to, but not at, an atom column, to induce enhanced channeling down that column and generate intensity from the apparent probe position. However, the fact that the structure was more apparent when viewed along [001], as compared with [100], suggests that the observation of periodic variations is real, confirming the prediction of density modulations in the intergranular films of such materials using molecular dynamics simulations [68]. In the case of intergranular films, where a typical width of <1 nm is measured, it is likely that some influence of the crystal fields from the two adjacent grains is retained throughout the full thickness of the film. There is evidence to suggest that the periodic structure of the glassy phase extends further into the pockets at triple points than into the intergranular films [64,65].

When viewed in the [001] orientation (Figure 9.18a, c, e, and g), the Si_3N_4 hexagonal rings of six bright dots with a spot separation of ca. 0.24 nm can be seen corresponding to columns of Si atoms in both alloys. The centers of the hexagons are noticeably dark and correspond to the open channels in the structure running parallel to the c-axis (Figure 9.17). The Si_3N_4 facets in the four [001] images in Figure 9.18 terminate with "half-hexagons" for both the La and Lu alloys when the facet is adjacent to either an intergranular film or a glassy pocket. However, it is very obvious, when these [001] images are compared, that the locations of the La atom columns (Figure 9.18a and e) are

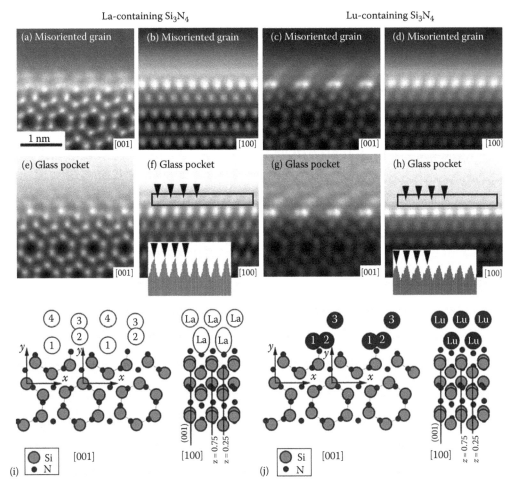

FIGURE 9.18 HAADF STEM images showing segregation of La and Lu at interfaces between {010} β-Si$_3$N$_4$ facets and amorphous intergranular films and pockets (a–h) and corresponding structural models (i) and (j). (From Winkleman, G.B., Dwyer, C., Hudson, T.S., Nguyen-Manh, D., Doblinger, M., Satet, R.L., Hoffmann, M.J., and Cockayne, D.J.H., *Appl. Phys. Lett.*, 87, 061911, 2005. With permission.)

different from those of the Lu atom columns (Figure 9.18c and g). The former have positions labeled 1, 2, 3, and 4 (Figure 9.18i), where 1 is closest to the Si$_3$N$_4$ and is symmetrically placed with respect to the Si columns of the terminating half-hexagons. Columns 2, 3, and 4 are further from the Si$_3$N$_4$, with 3 and 4 weaker than 1 and 2. Within the same distance from the Si$_3$N$_4$ in the Lu alloy there are only three distinct columns of Lu atoms (Figure 9.18j). Columns 1 and 2 are very close and located near to one of the three Si atom columns in the terminating half-hexagon, while column 2 appears brighter than the others. The corresponding images for the [100] orientation (Figure 9.18b, d, f, and h) are similar for both the La and Lu alloys, with a single uniform row of bright columns equally spaced along the interface, as though continuing the Si atom column arrangement. As indicated in Figure 9.18i and j, the Si column images are generated from two columns of Si atoms that are separated by only 0.038 nm in projection and are, therefore, unresolved. Detailed inspection reveals that the segregated atoms in both alloys are closer to the outermost line of Si columns than the separation of the lines of Si columns within the Si$_3$N$_4$.

Winkleman et al. [64,65] were able to combine the information from the two projections in Figure 9.18 in order to calculate the 3D coordinates of the RE atom positions relative to the atomic coordinates of the Si$_3$N$_4$ lattice, with an accuracy in the range 0.02–0.05 nm. The observations showed clearly that the interfacial structures are very dependent on the chemical nature of the sintering aid. The ability to provide 3D information of atomic positions at interfaces is vital to modelers trying to determine atomic positions using *ab initio* calculations, which will eventually enable an understanding and prediction of mechanical properties for future materials fabrication. What has yet to be demonstrated by calculation is that the interfacial energy variations for the observed atomic coordinates are consistent with the measured variations in mechanical properties for the various sintering agents.

9.6 CHEMICAL CHARACTERIZATION OF INTERFACES

9.6.1 MEASUREMENT OF GRAIN BOUNDARY SEGREGATION BY EDX

For polycrystalline materials, both single and multiphase, the grain boundaries are regions of high atomic disorder. The mechanical and physical properties of such materials can be strongly influenced by processes that occur on the atomic scale at these boundaries. With nanomaterials, the area of interface per unit volume is very large and the properties might turn out to be very different from materials with the same chemical composition that has a much larger grain size. Grain boundaries are readily revealed in the TEM by BF DC imaging because the two adjacent grains or phases diffract differently for any arbitrary incident beam direction. As shown in Section 9.3 it is always possible to determine (p, θ) for two grains using electron diffraction and sometimes to characterize GB dislocation structure. In special cases, when the interface plane is flat and there is structural coherence across the interface, then it may be possible using HRTEM and HAADF STEM to image the atomic structure simultaneously on both sides of the boundary and to see the structural units which compose the GB. This is very often the case with semiconductor devices, magnetic multilayered thin-film structures, and some ceramics. However, in other materials, such as metal alloys and ceramic composites, the typical grain boundaries are high angle, high energy and, in general, curved. Such boundaries are sites for preferential segregation of impurity atoms and nucleation of precipitate phases during the fabrication and service of artifacts made from these materials. The control of segregation and precipitation is often crucial in producing serviceable components in metal and ceramic alloys and so it is sometimes important to measure segregation. In this section we shall describe how EDX analysis can be used to quantify segregation.

In Section 6.5.1 of Chapter 6 the principle of EDX analysis was described, together with some of its associated strengths and limitations. A major advantage in using EDX over EELS is the ability to detect and quantify elements over most of the periodic table. When a material contains several alloying elements, and has the potential for containing traces of many other impurity elements, then EDX is especially effective.

Segregation can be broadly classified in two categories: (1) equilibrium segregation (ES) and (2) nonequilibrium segregation (NES). With ES, the segregated atoms are considered to be located at sites in the interfacial plane or immediately adjacent to the interfacial plane. The "concentration" of segregants can be defined in various ways: by a GB enrichment factor (EF) that compares the boundary and matrix concentrations, by a fractional monolayer coverage (FMC), or by the number of impurity atoms per unit of interface area. Only the last of these is immediately meaningful because it overcomes the problem that the GB atomic concentration varies with curvature and is usually less close-packed than the matrix. Although the EF is still sometimes reported, it is a rather crude parameter that should be regarded as a qualitative measurement restricted to a particular experiment on a specific microscope. Detailed examples of measurements of the number of impurity atoms per unit area of interface for ES of Bi in Cu have been reported by Rühle and coworkers [69,70] and Williams et al. [71,72]. The FMC has been used by Titchmarsh and coworkers [73–75] because it lends itself readily to a simple quantification method when the closest-packed crystal plane is used as a reference. FMC overcomes the problem of variable atomic density with boundary curvature, which is tedious to determine in the general case, and is the only method applicable to the measurement of NES where concentrations have complex profiles adjacent to interfaces. An alternative method, claimed to give improved statistical accuracy of ES measurement, has been proposed by Walther [76], but this requires successive measurements while varying the microscope operating parameters. Such a method is more time-consuming and complicated than those described here.

The use of a high-resolution probe, of order 0.1 nm in diameter, for imaging in STEM, was described in Section 9.4.3. While such a probe contains enough current (ca. 0.1 nA) to generate HAADF signals for imaging, a current of ~1 nA is necessary to generate useful EDX spectra in an acceptable time, say 100 s, in order to measure trace element concentrations. Hence, a probe of about 0.5 nm in diameter, or greater, from a FEG source in a STEM is usually required for meaningful analysis. This probe diameter is still considerably larger than the interatomic spacing within materials. Because the x-ray signal is proportional to the sample thickness, statistical accuracy is improved by using thick samples. However, high-angle Rutherford scattering will broaden the electron probe as it passes through the sample, reducing the fraction of electrons passing through the narrow volume of segregation at the boundary. Hence, for many materials, a compromise sample thickness of 50–120 nm is suitable for electron energies of 100–200 keV. Even over such a small distance, considerable curvature of the interface can still occur and dilute the contribution from the segregation.

The FMC method requires that a concentration profile be measured by recording a series of spectra along a line perpendicular to the boundary. Each spectrum is processed to yield a value for impurity concentration, c_i. The boundary position is assessed by the operator using an image that might contain phase contrast, which prevents direct identification of the precise boundary position. Hence, the plot of c_i with distance shows a maximum close to, but not necessarily right at, the apparent boundary location.

Using a simple model in which the probe current is assumed to have a Gaussian distribution $i(r)$ with 2D cylindrical symmetry, and the segregation is distributed homogeneously with a fractional concentration, f, in a layer of material of width δ, equal to the spacing of the matrix close-packed planes, the convolution of δ with $i(r)$ yields a simple estimate of the concentration profile. From Figure 9.19 it can be seen that the movement of the 2D cylindrical probe across a linear boundary image is equivalent to projecting the current into the line AB to generate a 1D Gaussian current distribution of the form:

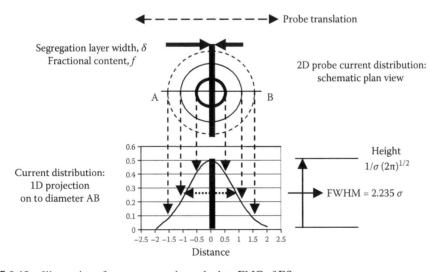

FIGURE 9.19 Illustration of geometry used to calculate FMC of ES.

$$i(x,\sigma) = (1/\sigma\,(2\pi)^{1/2})\,\exp\,(-x^2/2\sigma^2) \tag{9.5}$$

When integrated, the expression for $i(x, \sigma)$ yields a total magnitude of unity and the second moment of the distribution equates to the statistical variance, σ^2. The FWHM equals 2.235σ. When this 1D current is translated across a boundary where $\delta \ll$ probe diameter, then the measured concentration profile will also be a 1D Gaussian, with the same variance σ and a height, $h = 1/\sigma(2\pi)^{1/2}$.

The FMC, f, is calculated from the height, h, of the Gaussian in the following way. The segregation signal, S, is proportional to $f \times \delta \times h$, while the matrix signal, M, is

$$M = 1 - S\ (= \text{area under the Gaussian} - f \times \delta \times h) \sim 1$$

Hence

$$S/M = f\delta h$$

or

$$f = S\sigma(2\pi)^{1/2}/(M\delta) \tag{9.6}$$

Experimentally, the ratio, S/M is measured at the peak of the fitted line profile, σ is derived from fitting the Gaussian to the line profile, and δ is calculated from the matrix crystal structure. The method is almost insensitive to foil thickness and beam broadening because the major effect of broadening is simply to widen the Gaussian probe that is effective in generating the ES profile. Hence, the profile peak height falls but the width increases as the foil thickness increases and these changes are automatically incorporated into the measured S/M ratio. Some tailing is added to the Gaussian but any inaccuracy introduced by this is probably small compared to statistical errors when analyzing the weak signals from ES. An alternative method is to use Monte Carlo beam-broadening results to generate a theoretical line profile to which the experimental profile is fitted [74].

Many studies reported in the literature simply "quantify" ES by measuring the EF, i.e., the concentration of impurity with the electron probe positioned on the boundary, c_b, to give the maximum x-ray signal, and dividing this value by the concentration measured in the matrix, c_m, far from the boundary. It should now be clear why measurement of such an enhancement factor is meaningless. The value of c_b varies with foil thickness, boundary misalignment, and probe size, whereas c_m is almost independent of these parameters. An additional problem for ES measurement of trace impurities is that the sensitivity limits for many elements using EDX are typically 0.1 at.% in a matrix, giving rise to large uncertainties in c_b/c_m ratios.

An illustration of the FMC method is described from the work of Vatter and Titchmarsh [74,75] on the quantification of the ES of P to prior austenite grain boundaries in a 9% Cr ferritic steel. The segregation of P in ferritic steels during fabrication and service at elevated temperature can result in dramatic reduction in fracture toughness and catastrophic failure when power plant components are cooled to room temperature [77]. Four steels of identical composition, except for additions of P of 560, 300, 120, and 25 ppm by weight, were austenitized at 1150°C, quenched, tempered at 750°C, and finally aged at 500°C for 500 h to induce ES. Electropolished foil samples were analyzed using a FEG-STEM at 100 keV with a probe size measured as approximately Gaussian in shape and with a FWHM measured as 2.4 nm in one dimension. The prior austenite GBs were partially decorated with Cr-rich carbides, requiring precipitate-free sections of the GBs to be chosen for analysis. Spectral line profiles were recorded by stepping the probe along a line perpendicular to the projection of the GB and recording spectra at various distances from the GB. Great care was taken to ensure that sample drift was minimal (<~1 nm) during the 60 s of each acquisition. Figure 9.20 [74] compares two EDX spectra recorded with the probe located (a) on, and (b) 5 nm from, a GB. Each spectrum contains an inset of a magnified section to highlight the increased P (and also Mo and Cr) at position (a).

The concentration line profiles of Fe, Cr, and P, drawn for one of the GBs, are shown in Figure 9.21, and they reveal clear enhancement of P and Cr at the GB, compensated by depletion of Fe.

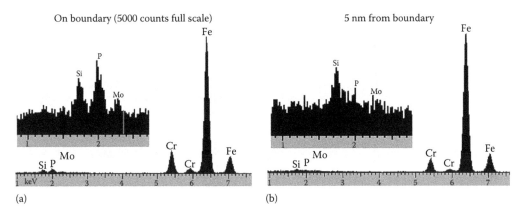

FIGURE 9.20   Comparison of two EDX spectra recorded with the probe located on, (a), and 5 nm from, (b), a GB. Increased P and Cr peaks are observed in Figure 9.20a. (Reproduced from Vatter, I.A. and Titchmarsh, J.M., *Surf. Interface Anal.*, 25, 760, 1997. With permission.)

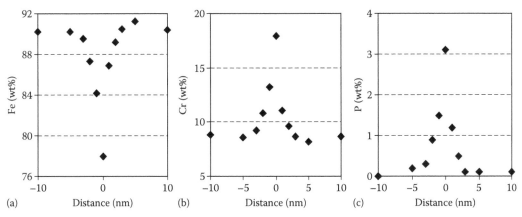

FIGURE 9.21   (a) Fe, (b) Cr, and (c) P concentration profiles across a GB in a ferritic steel. (Drawn from data in Vatter, I.A., PhD thesis, University of Bath, Bath, U.K., 1993.)

Maximum likelihood (ML) fitting of a Gaussian to the profiles yielded values for σ (Equation 9.5) from which the FMC, f, was derived. Alternative fitting using Monte Carlo profiles to account for beam broadening was also made. Analyses were made from several GBs in each of the four steels for the Monte Carlo fitting, and the results are shown in Figure 9.22 [74]. The average FMC was found to increase with P addition but, apart from alloy S4 (25 ppm P), where limited data prevented a more detailed analysis, the scatter increased as the P content fell. The low scatter and average value in alloy S4 reflect low ES due to the small P content. The scatter in the other alloys is likely to be a true reflection of the variation of ES along and between GBs, together with the increased potential for complete GB saturation with P as the content increases.

The EDX results were compared with an Auger spectroscopy (AES) study of the same alloys in which a much larger number of GBs were analyzed. Results from the two methods were compared by plotting the averaged FMC values and their errors for both Gaussian ML and Monte Carlo approaches to the EDX analysis (Figure 9.23 [74]). Good agreement was observed which supports the integrity of both methods for ES analysis. Although STEM-EDX is slower and cannot measure segregation of low Z elements, all GBs can, in principle, be analyzed. With AES, only GBs revealed by intergranular fracture can be analyzed, and this could lead to systematic overestimation of segregation

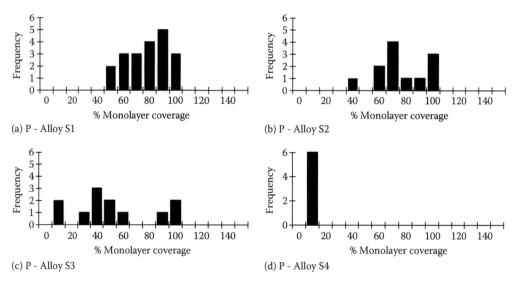

FIGURE 9.22 Phosphorus ES FMC distributions measured by STEM-EDX from four samples containing different bulk P concentrations. (Reproduced from Vatter, I.A. and Titchmarsh, J.M., *Surf. Interface Anal.*, 25, 760, 1997. With permission.)

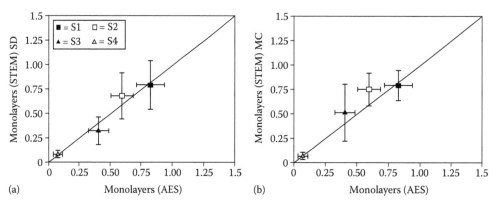

FIGURE 9.23 Comparison of EDX results of averaged FMC values and their errors for both Gaussian (a) and Monte Carlo (b) fitting methods, with corresponding measurements from the same alloys using AES. (Reproduced from Vatter, I.A. and Titchmarsh, J.M., *Surf. Interface Anal.*, 25, 760, 1997. With permission.)

in some materials if less segregation at some GBs were not revealed for analysis. A similar comparison has also been reported for the austenitic alloy PE16 [78] for which equivalence between the two methods was also found.

An alternative EDX method for segregation analysis, suitable only for ES, measures the number of segregated atoms per unit area of GB. The plane of the GB is oriented parallel to the incident probe direction, as with the FMC method, and the probe is scanned over the area $W \times L$ containing the GB in a direction parallel to the line image, of length L, of the GB. The GB is positioned centrally in the field and, unlike the FMC method, where the probe is stationary, an image of the GB is available while the EDX spectrum is acquired. Sample drift, therefore, can be continuously corrected during acquisition. The major experimental requirements are (i) that the sample has a uniform

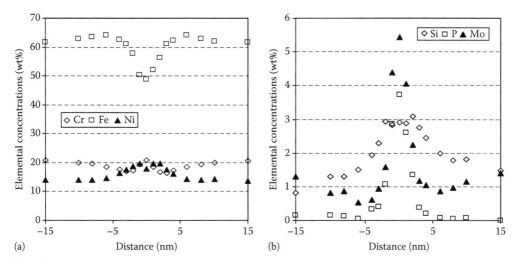

FIGURE 9.24 Complex concentration profiles for (a) Fe, Cr, and Ni and (b) Mo, Si, and P, measured by EDX across a GB in a neutron-irradiated stainless steel. (Reprinted from Dumbill, S. and Titchmarsh, J.M., *J. Microsc.*, 188, 224, 1997. With permission.)

thickness, t, over the scanned area and (ii) that dimensions of L and W are significantly larger than the (broadened) probe diameter. In this case (i) requires greater care in sample preparation than for the FMC method, and (ii) necessarily implies that the signal from the segregation is considerably diluted by a large matrix signal.

If the matrix atom density is n_m then the number of irradiated matrix atoms is $n_m WLt$. If the number of segregated atoms per unit area of GB is n_s, the total number of segregated atoms irradiated is $n_s Lt$. A single spectrum yields the atomic ratio, S/M, of irradiated segregation/matrix atoms, provided a suitable calibration is available. Hence

$$n_s = (S/M)n_m W \tag{9.7}$$

The simplicity of this method makes it extremely useful for samples where ES is known to be present. However, in complex materials, especially in metal alloys where thermal equilibrium might not have been reached and diffusion profiles are present near GBs, or when alloys have been subjected to neutron irradiation, NES might be present. In such circumstances it is essential first to record a line profile across the GB, as described above for the FMC method, in order to establish what type of segregation is present and how rapidly it is varying. An example of the complexity of segregation profiles in a neutron-irradiated alloy is shown in Figure 9.24 [79] where line profile concentrations for major alloying elements, Fe, Cr, and Ni are shown together with those from minor elements Si, P, and Mo. Close examination reveals that the detailed shapes and widths of the profile features are different for all six elements and the simple analytical methods described above for ES are impractical. The data in Figure 9.24 have been analyzed in greater detail using multivariate statistical analysis (MSA).

9.6.2 Interfacial Characterization Using EELS

The application of EELS to interfacial characterization is very exciting. Not only can the chemical composition be measured from a spectrum but also the fine structure at the ionization edges (ELNES) contains information about the electron density of states (or, rather, the vacant density of states) that can be directly related to bonding through comparisons with theoretical models of interfacial structure. 3D "data cubes" are now generated in the TEM by recording a series of energy-filtered images (EFTEM), each image containing only electrons that have lost a specific narrow energy range. The equivalent STEM method, spectral imaging (EELSI), is to record an EELS spectrum at every pixel in a field of view. From such data, maps can be extracted showing elemental distributions and even bonding variations of individual elements. The spatial resolution within such data is approaching atomic dimensions, while the spectrum energy resolution can be a few tenths of an electronvolt. Clearly, the energy and spatial resolution in a single spectrum can be significantly better than that in EFTEM or EELSI data because of the need to balance the rate of data acquisition in mapping against the degradation of resolution.

The spatial resolution of EELS measurements is generally superior to the corresponding EDX experiment because the EDX data are affected by beam broadening in all but the thinnest samples. Figure 9.25 [80] compares the simultaneously measured EDX and EELS compositional profiles across a GB in an irradiated austenitic steel at which a Cr depletion profile had been established by radiation-induced diffusion. The Cr/Fe concentration ratios from the two methods were normalized in the matrix away from the boundary. In a thin region of the foil (37 nm), the Cr/Fe ratios from EDX and EELS were similar at the bottom of the trough centered at the GB. However, in a thicker region of the foil (130 nm), the Cr depletion was shallower for the EDX profile because it was strongly affected by beam broadening. In this experiment the probe diameter containing 80% of the electrons was approximately 2.5 nm.

Except for low-energy losses (ca. <100 eV) spatial resolution in STEM is largely determined by the probe size. Until the recent development of aberration-corrected imaging filters for TEM, the characterization of IFs by EELS was performed by FEG-STEMs. An example of a STEM EELS spectrum line series across an IF between SiO_2 and TiO_2 [81,82] is shown in Figure 9.26, where successive spectra have been offset to enhance the visibility of the changes in the O–K edge with distance from the IF. The spatial separation of adjacent spectra was about 1.25 nm. The spectra from the two oxides were constant at the extremes of the profile but, when the probe partially sampled the IF, the resulting spectrum was not a simple combination of the spectra from the two oxides but contained additional fine structure. The fine structure right at the interface indicated that the environment of the O atoms at the IF was different from that of the O atoms in the pure oxides. As no impurity segregation was detected at the IF, the changes were interpreted due to bonding and coordination at the IF.

When the electron probe is larger than the interatomic spacing, as was the case for recording the data in Figure 9.26, the EELS spectrum always contains contributions from the adjacent phases, which makes isolation of the contribution from the IF spectrum very difficult. If IF spectral features can be extracted, then further understanding requires comparison with theoretical spectra calculated from possible structural models of the IF.

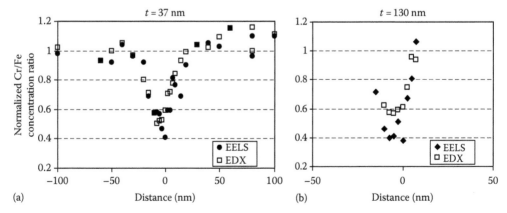

FIGURE 9.25 Comparison of Cr/Fe concentration profiles across a GB using EDX and EELS. Pairs of EDX and EELS spectra were recorded simultaneously at each point of the profile. The lower minimum in the EELS profile confirms that EELS is less affected by beam broadening and has superior spatial resolution. (Redrawn from data used in Titchmarsh, J.M., *Ultramicroscopy* 28, 347, 1989.)

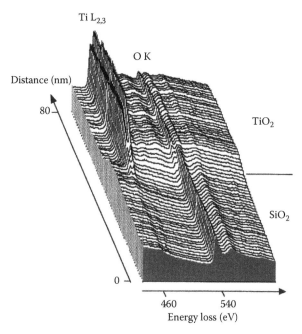

FIGURE 9.26 A series of EELS line profiles of 64 spectra across an interface between SiO_2 and TiO_2. The energy range includes the Ti-L_{23} and O–K edges. (Reprinted from Bonnet, N., Brun, N., and Colliex, C., *Ultramicroscopy*, 77, 97, 1999. With permission.)

A comparison of EELS spectra from GBs and the matrix, acquired with a FEG-STEM probe diameter of ca. 0.5 nm, was used by Muller et al. [83] to explain the improvement in the ductility of Ni_3Al by adding B. It was shown that B segregated nonuniformly along GBs when added in trace quantities to the alloy. The shape of the Ni-L_3 EELS edge at the GB, at the places of B segregation, was similar in appearance to the Ni-L_3 edge in bulk Ni_3Al whereas the same edge from B-free regions on the same GB was closer in shape to that in Ni metal. A smaller white line at the onset of the edge is consistent with an increase in s, p–d bond hybridization, and an increase in the cohesive energy of the GB, in agreement with Friedel's predictions of transition metal cohesive strength [84]. A comparison of the Ni-L_3 edge shape from GB locations with and without segregated B is shown in Figure 9.27 [83].

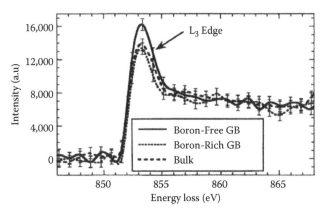

FIGURE 9.27 Comparison with bulk Ni_3Al of the Ni-L_3 ionization edges recorded at GBs with and without boron segregation. (From Muller, D.A., Subramanian, S., Batson, P.E., Sass, S.A., and Silcox, J., *Phys. Rev. Lett.*, 75, 4744, 1995. With permission.)

It was seen in Section 9.3 how the local atomic environment varies within structural units in a GB, and that the structural units are constructed from combinations of various polyhedra. The spectrum from a probe of atomic dimensions will, therefore, vary sensitively with probe position, while a broader probe will sample a larger range of unoccupied electron states and variations will be smaller. The advantage of STEM over TEM for spectroscopy on the atomic scale, for understanding real technical problems, is illustrated in Figure 9.28, where HAADF images are shown of two Si/poly-Si interfaces which have high (Figure 9.28a) and low (Figure 9.28b) electrical resistance, respectively [85]. A probe size of 0.2 nm was used and in each image the projected <110> atomic structure of the Si crystal is visible. Inset in the images are EELS spectra, acquired using the same probe, showing the Si-L_{23} ionization edge recorded at the indicated positions. In both samples there is a distinct peak at the onset of the edge at locations away from the interfaces in both the Si and the poly-Si. The peak is still present, although slightly reduced, at the low resistance interface in Figure 9.28b, while the edge at the interface in Figure 9.28a is rounded without a peak. Explanations for the spectral difference include the possible presence of a substoichiometric SiO_2 layer or H-terminated Si bonds at the high-resistance interface [85].

The EFTEM method generates elemental maps much faster and with better signal-to-noise than corresponding EDX mapping, provided samples are thin, the elements of interest have a large cross-section, and are not overlapped by adjacent peaks from other elements. However, in thicker foils, chemical analysis by EDX might still be preferable for some elemental combinations [86].

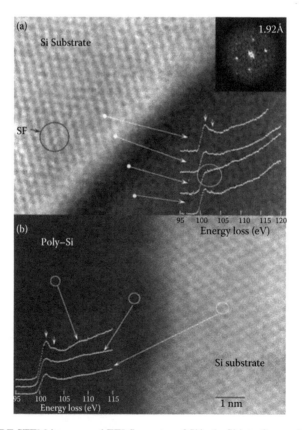

FIGURE 9.28 HAADF STEM images and EELS spectra of Si/poly-Si interfaces with (a) high and (b) low electrical resistance, showing different ELNES signatures in the Si-L_{23} edge. (Reprinted from Batson, P.E., *Ultramicroscopy*, 78, 33, 1999. With permission.)

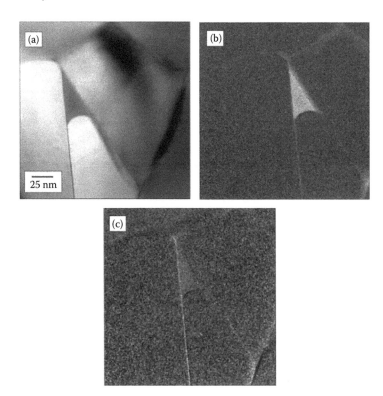

FIGURE 9.29 EFTEM images using an in-column spectrometer of a triple junction in Ca-doped Si_3N_4. (a) BF image showing triple point and several GBs; (b) O–K map; (c) Ca-L_{23} map. (From Plitzko, J.M. and Mayer, J., *Ultramicroscopy*, 78, 207, 1999. With permission.)

The potential of EFTEM for revealing GB segregation is illustrated in Figure 9.29 for the case of Ca-doped Si_3N_4 [87]. Figure 9.29a shows a BF image of a triple point with an amorphous oxide region and a straight GB between two Si_3N_4 grains running almost vertically. The images were acquired with an in-column filter using a TEM equipped with an LaB_6 electron source. The O–K image (Figure 9.29b) shows the amorphous oxygen-rich pocket, and also suggests enhanced O at the boundaries, while the Ca-L_{23} image (Figure 9.29c) shows segregation of Ca along the straight GB. It is possible for artifacts to occur in such EFTEM images but in this case it was shown by further spectroscopy that the Ca signal was genuine and the Ca segregation was quantified as 0.27 ± 0.15 monolayers, consistent with earlier measurements of an estimated 1/3 monolayer using FEG STEM [88].

REFERENCES

1. P. B. Hirsch, A. Howie, R. B. Nicholson, D. W. Pashley, and M. J. Whelan, *Electron Microscopy of Thin Crystals*, 2nd Edition, Kreiger Huntingdon, New York (1977).
2. K. C. Thompson Russell and J. W. Edington, *Electron Microscope Specimen Preparation Techniques in Materials Science*, Macmillan, Philips Technical Library, Eindhoven, the Netherlands (1977).
3. P. J. Goodhew, *Specimen Preparation for Transmission Electron Microscopy of Materials*, Oxford University Press, Oxford (1984).
4. J. C. Bravman, R. M. Anderson, and M. L. McDonald, Eds., *Specimen Preparation for Transmission Electron Microscopy of Materials*, MRS Symposium Proceedings, MRS, Pittsburgh, PA (1988), p. 115.
5. T. F. Malis, in: *Microbeam Analysis* (Ed. P. E. Russell), San Francisco Press, San Francisco, CA (1989), p. 487.
6. F. G. J. Cuisinier, R. W. Glaisher, J.-C. Voegel, J. L. Hutchison, E. F. Brès, and R. M. Frank, *Ultramicroscopy* **36** (1991) 297.

7. J. C. Bravman and R. Sinclair, *J. Electron Microsc. Technol.* **1** (1984) 53.
8. P. M. Voyles, J. L. Grazul, and D. A. Muller, *Ultramicroscopy* **96** (2003) 251.
9. K. H. Park, *Mater. Res. Symp. Proc.* **199** (1990) 271.
10. M. H. F. Overwijk, F. C. van den Heuvel, and C. W. T. Bulle-Lieuwma, *J. Vac. Sci. Technol. B* **11** (1993) 2021.
11. Y. Z. Huang, S. Lozano-Perez, R. M. Langford, J. M. Titchmarsh, and M. L. Jenkins, *J. Microsc.* **207** (2002) 129.
12. S. Lozano-Perez, unpublished work (2003).
13. A. K. Head, P. Humble, L. M. Clarebrough, A. J. Morton, and C. T. Forwood, *Computed Electron Micrographs and Defect Identification*, North Holland, Amsterdam (1973), p. 72.
14. M. H. Loretto and R. E. Smallman, *Defect Analysis in Electron Microscopy*, Chapman & Hall, London (1975).
15. A. P. Sutton and R. W. Balluffi, *Interfaces in Crystalline Materials*. Oxford University Press, Oxford (1995).
16. S. Ranganathan, *Acta Crystallograph.* **21** (1966) 197.
17. H. Grimmer, *Acta Crystallograph. A* **32** (1976) 783.
18. W. Bollmann, *Crystal Defects and Crystalline Interfaces*, Springer, Berlin (1970).
19. R. C. Pond and W. Bollmann, *Phil. Trans. R. Soc. Lond. A* **292** (1979) 449.
20. H. Grimmer, *Scripta Metall.* **8** (1974) 1221.
21. C. T. Forwood and L. M. Clarebrough, *Electron Microscopy of Interfaces in Metals and Alloys*, Hilger, Bristol, Philadelphia and New York (1991).
22. V. Randle, *The Measurement of Grain Boundary Geometry*, IoP Publishing, Bristol, 1993.
23. H. Mykura, *Grain Boundary Structure and Kinetics*, American Society for Metals, Metals Park (1980), pp. 445–456.
24. J. K. Mackenzie, *Acta Crystallograph.* **10** (1957) 61.
25. F. C. Frank, Symposium on *The Plastic Deformation of Crystalline Solids*, Office of Naval Research, Pittsburg, PA (1950), p. 150.
26. S. E. Babcock and R. W. Balluffi, *Phil. Mag. A* **55** (1987) 643.
27. T. Schober and R. W. Balluffi, *Phil. Mag.* **21** (1970) 109.
28. T. Y. Tan, S. L. Sass, and R. W. Balluffi, *Phil. Mag.* **31** (1975) 575.
29. E. P. Kvam and R. W. Balluffi, *Phil. Mag. A* **56** (1987) 137.
30. M. F. Ashby, F. Spaepen, and S. Williams, *Acta Metall.* **26** (1978) 1647.
31. R. C. Pond, D. A. Smith, and V. Vitek, *Scripta Metall.* **12** (1978) 699.
32. J. D. Bernal, *Proc. R. Soc. Lond. A* **280** (1964) 299.
33. G. H. Bishop and B. Chalmers, *Scripta Metall.* **2** (1968) 133.
34. G. H. Bishop and B. Chalmers, *Phil. Mag.* **24** (1971) 515.
35. M. J. Weins, B. Chalmers, H. Gleiter, and M. F. Ashby, *Scripta Metall.* **3** (1969) 601.
36. M. J. Weins, H. Gleiter, and B. Chalmers, *Scripta Metall.* **4** (1970) 235.
37. M. J. Weins, H. Gleiter, and B. Chalmers, *J. Appl. Phys.* **42** (1971) 2639.
38. A. P. Sutton and V. Vitek, *Phil. Trans. R. Soc. Lond. A* **309** (1983) 1.
39. A. P. Sutton and V. Vitek, *Phil. Trans. R. Soc. Lond. A* **309** (1983) 37.
40. A. P. Sutton and V. Vitek, *Phil. Trans. R. Soc. Lond. A* **309** (1983) 55.
41. H. Hashimoto, A. Howie, and M. J. Whelan, *Phil. Mag.* **5** (1960) 967.
42. H. Hashimoto, A. Howie, and M. J. Whelan, *Proc. R. Soc. A* **269** (1962) 80.
43. C. T. Forwood and L. M. Clarebrough, *Phys. Status Solidi A* **105** (1988) 365.
44. J. G. Allpress, J. V. Sanders, and A. D. Wadsley, *Acta Crystallograph. B* **23** (1969) 1156.
45. J. C. H. Spence, M. A. O'Keefe, and S. Iijima, *Phil. Mag. A* **38** (1978) 463.
46. A. Olsen and J. C. H. Spence, *Phil. Mag. A* **43** (1981) 945.
47. W. Krakow, J. T. Wenzel, and D. A. Smith, *Phil. Mag. A* **5** (1986) 739.
48. W. Krakow and D. A. Smith, *Ultramicroscopy* **22** (1987) 47.
49. W. Krakow and D. A. Smith, *J. Mater. Res.* **1** (1986) 47.
50. R. Schweinfest, S. Kostelmeier, F. Ernst, C. Elsasser, T. Wagner, and M. W. Finnis, *Phil. Mag. A* **81** (2001) 927.
51. L. Houben, A. Thust, and K. Urban, *Ultramicroscopy* **106** (2006) 200.
52. M. Lentzen, B. Jahnen, C. L. Jia, A. Thust, K. Tillman, and K. Urban, *Ultramicroscopy* **92** (2002) 233.
53. W. Coene, A. Thust, M. Op de Beeck, and D. Van Dyck, *Ultramicroscopy* **64** (1996) 109.
54. A. Thust, W. Coene, M. Op de Beeck, and D. Van Dyck, *Ultramicroscopy* **64** (1996) 221.
55. J. M. Cowley and A. F. Moodie, *Acta Crystallograph.* **12** (1959) 353.

56. J. M. Cowley and A. F. Moodie, *Acta Crystallograph.* **12** (1959) 360.
57. P. Stadelmann, *Ultramicroscopy* **21** (1987) 131.
58. M. J. Hofmann, in: *Tailoring of Mechanical Properties of Si₃N₄ Ceramics* (Eds. M. J. Hoffmann and G. Petzow), NATO ASI Series E, Applied Sciences, Vol. 276, Kluwer Academic, Dordrecht, the Netherlands (1994).
59. D. R. Clarke, *J. Am. Ceram. Soc.* **70** (1987) 15.
60. M. Kramer, D. Witmuss, H. Kuppers, M. J. Hoffmann, and G. Tetzow, *J. Cryst. Growth* **140** (1994) 157.
61. R. L. and M. J. Hoffmann, *J. Eur. Ceram. Soc.* **24** (2004) 3437.
62. N. Shibata, S. J. Pennycook, T. R. Gosnell, G. S. Painter, W. A. Shelton, and P. F. Becher, *Nature* **428** (2004) 730.
63. G. B. Winkleman, C. Dwyer, T. S. Hudson, D. Nguyen-Manh, M. Doblinger, R. L. Satae, M. J. Hoffmann, and D. J. H. Cockayne. *Phil. Mag. Lett.* **84** (2004) 755.
64. G. B. Winkleman, C. Dwyer, T. S. Hudson, D. Nguyen-Manh, M. Doblinger, R. L. Satae, M. J. Hoffmann, and D. J. H. Cockayne, *Appl. Phys. Lett.* **87** (2005) 061911.
65. G. B. Winkleman, C. Dwyer, C. Marsh, T. S. Hudson, D. Nguyen-Manh, M. Doblinger, and D. J. H. Cockayne, Mat. Sci. Eng. A 422 (2006) 77.
66. A. Ziegler, J. C. Idrobo, M. K. Cinibulk, C. Kieslowski, N. D. Browning, and R. O. Ritchie, *Science* **306** (2004) 1768.
67. P. M. Voyles, D. A. Muller, J. L. Grazul, P. H. Citrin, and H.-J. L. Gossmann, *Nature* **416** (2002) 826.
68. X. Su and S. J. Garofalini, *J. Mater. Res.* **19** (2004) 752.
69. U. Alber, H. Müllejans, and M. Rühle, *Ultramicroscopy* **69** (1997) 105.
70. V. J. Keast and D. B. Williams, *J. Microsc.* **199** (2000) 45.
71. J. R. Michael and D. B. Williams, *Met. Trans.* **15A** (1984) 99.
72. V. J. Keast and D. B. Williams, *Acta Mater.* **47** (1999) 3999.
73. I. A. Vatter and J. M. Titchmarsh, *Ultramicroscopy* **28** (1989) 236.
74. I. A. Vatter and J. M. Titchmarsh, *Surf. Interface Anal.* **25** (1997) 760.
75. I. A. Vatter, PhD thesis, University of Bath, Bath, U.K. (1993).
76. T. Walther, *J. Microsc.* **215** (2004) 191.
77. I. A. Vatter, C. A. Hippsley, and S. G. Druce, *Int. J. Pres. Ves. Piping* **54** (1993) 31.
78. A. Partridge and G. J. Tatlock, *Surf. Interface Anal.* **18** (1992) 713.
79. S. Dumbill and J. M. Titchmarsh, *J. Microsc.* **188** (1997) 224.
80. J. M. Titchmarsh, *Ultramicroscopy* **28** (1989) 347.
81. N. Brun, C. Colliex, J. Rivory, and K. Yu-Zhang, *Microsc. Microanal. Microstruct.* **7** (1996) 161.
82. N. Bonnet, N. Brun, and C. Colliex, *Ultramicroscopy* **77** (1999) 97.
83. D. A. Muller, S. Subramanian, P. E. Batson, S. A. Sass, and J. Silcox, *Phys. Rev. Lett.* **75** (1995) 4744.
84. J. Friedel, in: The Physics of Metals (Ed. J. M. Ziman), Cambridge University Press, Cambridge (1969).
85. P. E. Batson, *Ultramicroscopy* **78** (1999) 33.
86. M. Watanabe, D. B. Williams, and Y. Tomokiyo, *Micron* **34** (2003) 173.
87. J. M. Plitzko and J. Mayer, *Ultramicroscopy* 78 (1999) 207.
88. H. Gu, X. Q. Pan, I. Tanaka, R. M. Cannon, M. J. Hoffmann, H. Muellejans, and M. Ruhle, *Mater. Sci. Forum* **207–209** (1996) 729.

10 In-Depth Analysis/Profiling

François Reniers and Craig R. Tewell

CONTENTS

10.1 INTRODUCTION

In-depth analysis of materials has become of tremendous importance in academic and industrial laboratories during the last 20 years. Thermodynamics, kinetics, sample processing, sample utilization, and sample impurities can introduce compositional inhomogeneities between the surface and the bulk that affect the overall material properties. The development of multilayered materials is another driving force behind the development of powerful in-depth analysis tools. Finally, even though the interface between a coating and a substrate is often the critical region that determines the final properties, it is often poorly characterized.

The questions often put to the analyst are:

- What is the bulk composition of the sample?
- Is the bulk composition the same at every depth?
- What is the chemical state of element X at a given depth?
- Is there a film/substrate interface and where is it?
- What is the composition at the interface?

The goal of the analysis is therefore to determine the in-depth distribution of the various chemical species in a sample. The solution to this simple problem should be a profile such as shown in Figure 10.1. Depending on the nature of the sample studied, the in-depth analysis scale may vary from a few nanometers to many micrometers. Due to the progress of materials manufacturing, analysis is needed on more than one length scale: multilayered samples with an overall thickness of many micrometers may be produced, with each layer only a few nanometers (or even less than a nanometer) thick.

Many techniques can be used to analyze a sample: Auger electron spectroscopy (AES), secondary ion mass spectrometry (SIMS), secondary neutral mass spectrometry (SNMS), x-ray photoelectron spectroscopy (XPS), x-ray diffraction (XRD), Rutherford backscattering spectrometry (RBS), elastic recoil detection (ERD), enhanced-cross-section backscattering spectrometry (EBS), etc. Most of these physical techniques need the excitation of a sample and the collection of a signal which may consist of ions, electrons, neutrals, or photons. This signal is usually characterized by its intensity (I) and by the kinetic energy (KE) of the emitted particles. In-depth information is usually available

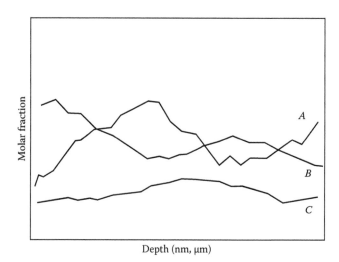

FIGURE 10.1 Absolute depth profile resulting from the in-depth analysis of a sample. *A*, *B*, and *C* are concentrations of species being sought.

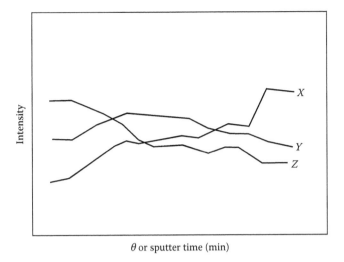

θ or sputter time (min)

FIGURE 10.2 Typical experimental depth profile of a sample. Θ is the angle of analysis; X, Y, and Z are the elements being detected by the method.

by a variation of the angle of analysis or by etching or cross-sectioning the sample. The results of the investigations are often profiles like that shown in Figure 10.2.

The general decisions confronting the analyst can be summarized as follows:

1. Which experimental technique(s) is (are) best suited to the given problem?
2. How can a profile of composition vs. depth be extracted from the experimental data?

More specifically:

- How can either the angle θ, sputter time, or particle KE be translated into depth?
- What is the relationship between the nature of X, Y, Z in Figure 10.2 and A, B, C in Figure 10.1?
- What is the depth resolution?
- How can the signal intensities be translated into the true composition of the sample?
- What are the effects of etching/erosion on the specimen, and hence on the signal from it?

Many techniques are available today and this chapter will discuss only a few of them. This subjective selection is based on the following observations:

- Only those techniques that are widely available and in common use will be considered, which means, in practice, that they must be available to both academic and industrial applications/users.
- Only "mature" techniques, from which quantitative information can be obtained routinely, will be included.

On the basis of this, the discussion will be limited to the following techniques/methodologies:

AES depth profiling, XPS depth profiling, angle resolved AES and XPS (ARAES, ARXPS), dynamic SIMS, SNMS, elastic ion scattering spectrometry (RBS, ERD, EBS), and glow discharge optical emission spectrometry (GDOES).

Compared to the description in the first edition of this volume [1], there have been some major technological advances in a few of the techniques. First, GDOES is now mostly used in the rf-mode (rf-GDOES), allowing the analysis of nonconducting samples. Second, the increase in the lateral resolution of AES, due both to the use of field emission electron guns and to the development of the *in situ* focussed ion beam (FIB) technique to prepare precise sample cross-sections, followed by line-scans with a resolution of a few nanometers. Finally, SIMS–SNMS instruments now tend to have two ion guns, one dedicated specifically to the removal of layers (the "in-depth" profiler), and the other to the SIMS–SNMS process (see Chapter 4).

The general structure of this chapter is as follows. After a general introduction, each technique will be described separately. Each subsection will contain the basic principles, the equations for quantitative analysis, the advantages and disadvantages, the applications, and the particular demerits. Finally, there is a general discussion of in-depth analysis. Table 10.1 presents a list of the principal techniques in general use for in-depth analysis. The characteristics of the incident and detected beams are listed as well as the main technical requirements. Each method, due to its individual characteristics, has advantages and disadvantages. For more detailed information, the reader is referred to Chapters 4 through 6. Many books have been published in which these techniques are described, e.g., [2–7].

The researcher who has to perform an in-depth study of a given sample must choose the appropriate technique(s) according to the information that is being sought. Table 10.2 lists the nature of the information that can, or cannot, be extracted directly from each technique. A full characterization of an unknown sample will, however, require the use of a combination of different methods. A well-designed multitechnique approach will provide a more complete analysis by offsetting the disadvantages of any one particular method against the advantages of other methods.

TABLE 10.1
List of the Main Methods Used for In-Depth Analysis (Not Exhaustive)

Method	Incident Beam	Analyzed Particle	Atmosphere	In-Depth Tool	Spectrum Obtained
AES (SDP)	Electrons[a]	Auger electrons	UHV	Ion gun	Intensity vs. sputter time
AES (FIB)	Electrons[a]	Auger electrons	UHV	FIB (Ga+ usually)	Intensity vs. electron beam position
ARAES	Electrons[a]	Auger electrons	UHV	Rotating sample	Intensity vs. angle
XPS	X-rays	Photoelectrons	UHV	Ion gun	Intensity vs. sputter time
ARXPS	X-rays	Photoelectrons	UHV	Rotating sample	Intensity vs. angle
SIMS (dynamic)	Ions	Ions	UHV	Ion gun	Intensity vs. sputter time
SNMS	Ions	Neutrals	UHV	Ion gun	Intensity vs. sputter time
XRD	X-rays	X-rays	Air	Rotating sample	Intensity vs. angle
GDOES	Ar+ ions	Photons	Vacuum	Plasma	Intensity vs. sputter time
RBS	Ions	Ions	UHV	Ion source	Intensity vs. energy of ion

Note: (a) X-ray-induced AES is no longer used in most applications.
SDP, sputter depth profiling; FIB, focused ion beam.

TABLE 10.2

Principal Information that Can Be Extracted from Each Method of Depth Profiling

Method	Elements Detected	Isotopic Analysis	Sensitivity	Chemical Information	Structure	Sampling Depth (Monolayers)
AES	$Z \geq 3$	No	0.1 at.%	Yes[b]	No	3
XPS	$Z \geq 3$	No	0.1 at.%	Yes	No	3
SIMS	All	Yes	<1 ppm	Yes[c]/No	No	1
XRD	N/A	No	1 at.%	No	Yes	
GDOES	All	No	1 ppm	No	No	10
RBS, ERD, EBS	All	No	1 at.%	No	No	100
SNMS	All	Yes	<1 ppm	No	No	10
ISS[a]	All	Yes	1 at.%	No	No	1
SSIMS[a]	All	Yes	0.01 at.%	Yes	No	1–2

[a] The last two methods are not generally used for in-depth analysis or profiling.

[b] At high energy resolution, with lineshape analysis software.

[c] When low energy cluster ions or heavy metal ions are used to generate more molecular fragments.

10.2 SAMPLE PREPARATION

Most of the techniques described in this chapter involve ultra-high vacuum (UHV)-based surface analysis methods (i.e., AES, XPS, SIMS). With them, it is imperative to handle the samples correctly, in that a sample should not be handled or treated without precautions; if washed, it must be carefully rinsed; if transported, it must be kept in a protective atmosphere (see also Refs. [3,4,7–9]).

The aim of in-depth analysis of a sample is to obtain analytical information as a function of depth. There are three methods for sampling in depth using AES or XPS: (1) varying the detection angle (ARXPS or ARAES), (2) as in SIMS, sputtering away the surface, or (3) using a combination of AES (or XPS, scanning electron microscopy (SEM), and SIMS) with FIB. Although FIB also erodes the surface, it involves a different approach, as explained later in this chapter. Classical cross-sectioning, using traditional cutting tools, combined with field emission gun (FEG)-Auger or FEG-SEM is also possible. For the least ambiguous depth information, there is one overriding requirement and that is that the surface of the sample should be as flat and smooth as possible in the analyzed area. All the theories of ion etching and angle-resolved spectroscopies assume that the roughness is minimized. Real samples, however, have surfaces which depend on their "history." Sample preparation before analysis is therefore often necessary. Descriptions of preparation methods can be found in the general references [2–6] and in Ref. [9].

The following procedures are those normally used for solid metal or inorganic samples:

1. Mechanical polishing on metallographic paper of decreasing grit size is used in order to eliminate macroscopic inhomogeneities. However, such a procedure often introduces SiC or WC particles into the sample which might interfere with the analysis.
2. Chemical, or electrochemical, polishing of the sample in an appropriate bath can be used after mechanical polishing or as the sole surface modification and/or preparation step. For metals and alloys, there are many published recipes for polishing baths [10,11].
3. Washing, rinsing, and drying the sample are typically performed to remove contaminants that remain from the surface modification steps.
4. SEM (usually available with scanning Auger microscopy (SAM) in most surface analysis systems) and/or optical microscopy is/are used to examine the surface roughness and homogeneity as well as to select a flat area suitable for analysis.

When using polymers, delicate coatings, and other sensitive samples, it is best to limit the pretreatment. For polymers, the pretreatment usually consists of washing and rinsing in an appropriate solvent, followed by degassing in the vacuum transfer chamber of the analyzer.

10.3 NONDESTRUCTIVE IN-DEPTH ANALYSIS

In nondestructive analysis, information is obtained without etching of the sample. Although grazing-incidence XRD allows structural information to be obtained as a function of depth, it will not be considered here. Recent applications can be found in the literature [12–15].

The three main physical techniques described here are angle-resolved Auger electron spectroscopy (ARAES), angle-resolved x-ray photoelectron spectroscopy (ARXPS), and elastic ion scattering (RBS, ERD, and EBS). Whereas the two first techniques are merely variations on AES and XPS, already described in Chapter 3, elastic ion scattering requires a brief description and discussion.

10.3.1 ELASTIC ION SCATTERING

10.3.1.1 Basic Principles

In elastic ion scattering techniques a beam of energetic ions (e.g., He, H, or Si) strikes a sample. Due to nuclear interactions (elastic collisions between the nuclei), the incident ion is backscattered and some lighter elements (e.g., ^1H, ^2H, and O) from the sample may be ejected. Analysis of the number and energies of the backscattered ions is called RBS. A schematic of the principle of RBS is presented in Figure 10.3. Analysis of the number and energies of ejected atoms is called ERD. By careful choice of the incident ion and ion energy, depth profiles can be obtained for elements from sodium to uranium. Depth profiles for some of the lighter elements can be obtained by EBS. Table 10.3 is a summary of the ion energies, elements that can be analyzed, analysis depth range, and depth resolution for common RBS, ERD, and EBS techniques. More extensive descriptions of these techniques can be found in the literature [16–18].

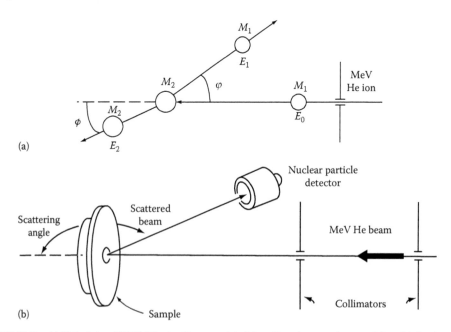

FIGURE 10.3 (a) Principle of RBS. The incident particle M_1 strikes the sample atom M_2 and is backscattered with a given angle and energy; (b) drawing of an experimental setup. See Equations 10.1 and 10.2 for definitions of the symbols.

TABLE 10.3

Summary of Elastic Scattering Methods, Including Incident Ion Energy, Range of Elements Analyzed, Sensitivity, Approximate Depth Probing Range, and Depth Resolution

	Energy (MeV)	Elements Analyzed	Sensitivity (at.%)	Analysis Range	Depth Resolution	Comments
He$^+$ RBS	1–3.5	Na to U		≈1.5 μm	0.03 μm	Depth profile
H$^+$ RBS	1–2	Na to U		≈3 μm	0.08 μm	Depth profile
He$^+$ ERD	2–3	^{1}H and ^{2}H	0.1	≈1 μm polymers	0.1 μm	Non-Rutherford cross sections
Si^{4+} ERD	12–16	H to Li	0.05	≈0.4 μm	0.02 μm	Range for H
	18–28	H to N	0.1	≈0.4 μm	0.03 μm	Range for N
	28–32	H to N and O	0.2	≈0.2 μm	0.05 μm	Range for O
He$^+$ EBS	3.5	C		≈0.5 μm	0.03 μm	C depth profile
	3.5	N		≈0.5 μm	0.03 μm	N depth profile
	8.7	O		≈7 μm	0.05 μm	O depth profile
H$^+$ EBS	2.5	O		≈4 μm	0.2 μm	O depth profile
	4.4	S		≈1 μm		S areal density

Source: Barbour, J.C., *Elastic Ion Scattering for Composition Analysis: Rutherford Backscattering Spectrometry (RBS), Elastic Recoil Detection (ERD), and Enhanced-Cross-Section Backscattering Spectrometry (EBS)*, Sandia National Laboratories, Albuquerque, NM, 1998. With permission.

10.3.1.2 Quantitative Analysis in Ion Backscattering Spectrometry

Due to the principle of the conservation of energy and momentum, a kinematic factor K can be calculated from the ratio of the energy of the detected particle (E_1) and that of the incident particle (E_0). This ratio can be shown to relate to the masses of the incident (M_1) and sample (M_2) atoms as follows:

$$K = \frac{E_1}{E_0} = \left[\frac{\left(M_2^2 - M_1^2 \sin^2 \theta\right)^{1/2} + M_1 \cos\theta}{M_2 + M_1} \right]^2 \tag{10.1}$$

where θ is the scattering angle (see Figure 10.3). By analysis of the energy of the backscattered particles, the element involved in the scattering event can be identified.

The number of backscattered particles, or the signal strength H_M, is

$$H_M = \frac{C_M Q \Delta\Omega \xi \sigma}{[\varepsilon]\cos\theta} \tag{10.2}$$

where

Q is the flux of the incident particles

C_M is the concentration of element of mass M in the sample

$\Delta\Omega$ is the aperture of the detector

ξ is the detector efficiency

ε is the stopping cross-section

σ is the scattering cross-section which can be calculated exactly from the repulsive Coulomb forces between the unscreened nuclei of the incident and target atoms

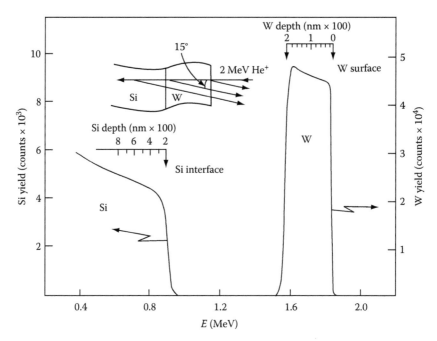

FIGURE 10.4 Simulated RBS energy spectrum from a 200 nm thick W film on Si obtained with 2 MeV He⁺, and at a scattering angle of 15°. (From Baglin, J.E.E. and Williams, J.S., *Ion Beams for Materials Analysis*, Academic Press, New York, 1989. With permission.)

However, actual scattering cross-sections deviate from such calculated values at both low and high incident ion energies. The energy loss $\Delta\varepsilon$ of the incident particle is related directly to the depth ΔX at which the elastic collision occurred by $\Delta\varepsilon = \varepsilon N \Delta X$, where ε is the stopping cross section (in eV $\cdot$ cm²/atom) and N is the particle volumetric density (atoms/cm³). A compositional depth profile ($C_M/\Delta X$) can be obtained by measuring the number of backscattered particles (H_M) if the other parameters are known (Q, $\Delta\Omega$, ζ, σ, $\Delta\varepsilon$, N, θ). In practice, a good compositional depth profile can be obtained by fitting the measured spectrum, $H_M(E_1)$, taking into account other energy loss mechanisms (i.e., straggling) and multiple scattering.

Thus in RBS, the energy loss provides the depth resolution and each element has its own depth scale. Figure 10.4 shows a simulated RBS energy spectrum from a 200 nm thick W film on Si recorded with 2 MeV He⁺ [20].

10.3.1.3 Applications of RBS

RBS is particularly suitable for following the course of interdiffusion processes. Figure 10.5 shows the RBS spectra taken at various stages in the diffusion and solid-state reaction of Er metal with a Kovar alloy during 20 min annealing cycles at the temperatures indicated [21].

Summary of the strengths and weaknesses of elastic ion scattering (RBS, ERD, and EBS).

Strengths

1. Quantitative analysis (1–2 at.% precision) without standards.
2. Depth information without sputtering.
3. Easy observation of interaction and migration of species.
4. Good near-surface depth resolution (5–20 nm depending on backscattering angle, material and total layer thickness).
5. With microbeam, a full three-dimensional profile determination is possible with 1 μm spatial resolution.

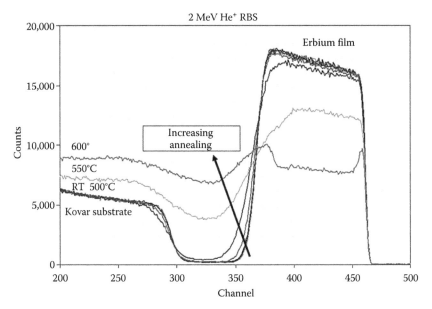

FIGURE 10.5 RBS spectra taken at various stages in the diffusion and solid-state reaction of Er metal with a Kovar alloy substrate during 20 min annealing cycles at the temperatures indicated. The results show that the reaction takes place above 500°C. (From Kammler, D.R., Wampler, W.R., Van Deusen, S.B., King, S.H., Tissot, R.G., Espada, L.I., and Goeke, R.S., *Mechanisms Governing D Loss in ErD₂ Films on Kovar with and without a Mo Diffusion Bauier*, Sandia National Laboratories, Albuquerque, NM, 1998. With permission.)

6. Analysis possible on conducting and nonconducting samples.
7. All elements can be detected.

Weaknesses

1. Large analysis area (except with microbeam)
2. Limited resolution for thick layers
3. Sampling thickness limited to 1 μm

10.3.2 ANGLE RESOLVED AUGER ELECTRON AND X-RAY PHOTOELECTRON SPECTROSCOPIES

10.3.2.1 Basic Principles

Angle Resolved Auger Electron and X-Ray Photoelectron Spectroscopies provide in-depth information close to the outer surface of a sample. In normal AES or XPS, the analyzer entrance aperture is positioned normal to the surface. In the AR mode, either the analyzer is tilted or the sample is rotated. The probing depth will depend on the exit (takeoff) angle of the emitted electrons according to the relation.

$$d = \lambda \sin \theta \tag{10.3}$$

where
d is the depth analyzed
λ is the attenuation length
θ is the exit (takeoff) angle of the ejected electrons

Conversion from signal intensity to atomic concentration is given by the usual AES or XPS relationships (see Equations 10.7 and 10.8). The maximum depth analyzed by this method is ca. 3d, which

TABLE 10.4

Analyzed Depth (d) as a Function of Analyzer Angle (θ) for Etched GaAs, Studied by ARXPS

θ (°)	d (nm)	θ (°)	d (nm)
5	0.31	35	2.06
10	0.63	40	2.31
15	0.93	45	2.55
20	1.23	50	2.76
25	1.52	55	2.95
30	1.80	60	3.12

Source: van de Walle, R., Van Heirhaeghe, R.L., Laflere, W.H., and Cardon, F., *J. Appl. Phys.*, 74, 1885, 1993. With permission.

corresponds to 95% of the electrons emitted. More information, as well as theoretical developments of the method can be found in the literature [22–25] and in Chapters 3 and 8. Table 10.4 shows the analyzed depth (d) as a function of the analyzer angle (θ) for etched GaAs studied by XPS [26] for the depth profile shown in Figure 10.6. The photoelectrons used were those from the Ga $3d$ ($E_b = 20\,eV$) and the As $3d$ ($E_b = 40\,eV$) levels. Both photoelectron peaks have an escape depth (λ) of 3.6 nm.

ARXPS is particularly suitable for the analysis of polymers. Indeed, in polymer characterization, XPS provides useful information about the chemical bonds (through the chemical shift of the photoelectron peaks). Conventional depth profiling using high energy argon ions results in the loss of this information, whereas the angle-resolved approach does not.

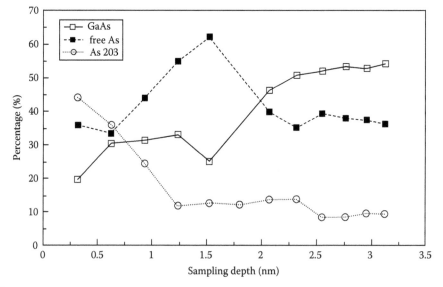

FIGURE 10.6 ARXPS depth profile using the As 3d spectrum, for a GaAs single crystal etched in H_2SO_4–H_2O_2 solution. The conversion from angle to depth is given in Table 10.5. (From van de Walle, R., Van Heirhaeghe, R.L., Laflere, W.H., and Cardon, F., *J. Appl. Phys.*, 74, 1885, 1993. With permission.)

TABLE 10.5

Sampling Depth (3d) for a Polymethylmethacrylate Sample as a Function of the Energy Source (Al or Mg Anode), the Element Analyzed, and the Take-Off Angle

Core Level	Anode	Kinetic Energy (eV)	3d (nm), 10°	3d (nm), 45°	3d (nm), 90°
O 1S	Mg	723	1.0	4.1	5.8
	Al	956	1.3	5.2	7.3
C 1s	Mg	967	1.3	5.2	7.3
	Al	1200	1.5	6.2	8.7

Source: Beamson, G. and Alexander, M.R., *Surf. Interface Anal.*, 36, 323, 2004. With permission.

10.3.2.2 Applications

These two methods are particularly useful when the region of interest is close to the surface of the sample, e.g., for ultrathin films on substrates and for surface interdiffusion processes. ARXPS is probably the best tool for the study of surface oxides, as no reduction due to ion sputtering occurs. ARXPS is also the method of choice for the analysis of polymers as illustrated by a significant change in sampling depth possible (Table 10.5) for polymethylmethacrylate. An example for 10 polymers is presented in Ref. [27].

10.3.2.3 Summary of ARXPS and ARAES (Table 10.6)

Main advantages:

1. Very good depth resolution
2. Nondestructive
3. Particularly suitable for oxides (no sputter-induced reduction) and polymers (access to chemical bond information)

Main weaknesses:

1. Limited to the depth corresponding to the escape depth of the electrons.
2. Very sensitive to surface roughness; the sample surface must be sufficiently smooth.
3. Surface contamination is overemphasized.

TABLE 10.6
Summary

	ARXPS	ARAES
Maximum depth analyzed	3d	3d
Sensitivity	0.1 at.%	0.1 at.%
Elements analyzed	$Z \geq 3$	$Z \geq 3$
Chemical information	Yes	Not easily
Sample requirements	Flat surface	Flat surface conducting

Source: Briggs, D. and Grant, J.T. (Eds.), *Surface Analysis by Auger and X-Ray Spectroscopy*, Surface Spectra, IM Publications, Chichester, U.K., 2003. With permission.

10.4 DESTRUCTIVE DEPTH PROFILING

Today, there are two main approaches to destructive depth profiling using surface spectroscopic techniques (AES, XPS, SIMS, and SNMS): the removal of surface layers by ion sputtering with either simultaneous or consecutive cycles of surface analysis, or the cross-sectioning of the sample followed by a line scan using traditional surface analysis.

10.4.1 IN-DEPTH INFORMATION BY FIB–AES

Cross-sectioning techniques are widely used for materials characterization, as in microstructural studies using transmission electron microscopy, and cross-sectioning combined with electron probe microanalysis to give in-depth chemical information. Classical cross-sectioning is often used when the layers to be analyzed are rather thick (in the micrometer range). In such cases (for metallurgy and coatings), conventional ion sputtering depth profiling is not practical. The required sputtering time is far too long to maintain an acceptable depth resolution. The cross-sectioning of a sample generates a new surface and an additional coordinate, the depth, z. A line analysis along this z-direction then gives the depth distribution of the elements directly. For a long time, the use of cross-sectioning for advanced in-depth studies was limited. The conventional methods for cross-sectioning and the limited lateral resolution of SEM and AES instruments were the greatest impediments to this approach. The depth resolution of conventional sputtering and surface analysis was far superior. However, two technological developments make the cross-sectioning approach an important alternative today:

- FIB "cuts" samples precisely with excellent lateral resolution, and without significant roughening, using liquid metal ion sources.
- Field emission guns in AES (and SEM) can achieve lateral resolutions in the range of 4 nm (SEM) to 10 nm (AES) [28]. Combined with increasing energy resolution in AES (0.05% today), exceptional spatial and spectroscopic resolution in AES "line scans" can be obtained.

The geometry of modern surface analysis instruments includes a secondary ion gun for FIB. Figure 10.7 shows in (a) such a configuration, and in (b) the cross-sectional analysis of a 0.2 mm particle [29].

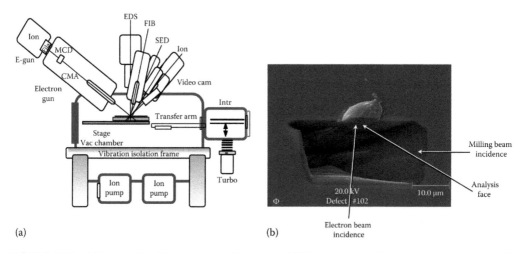

FIGURE 10.7   (a) Schematic of the geometry of a modern AES instrument with a secondary ion gun for FIB sputtering. The electron and ion guns are mounted concentric with the cylindrical mirror analyzer (CMA). (b) SEM image obtained at 20 keV of a 0.2 mm crater after cross-sectioning by a Ga$^+$ ion beam (25 keV, 10 min exposure, 100 nm spot size; the sectioned area was 10 um wide and 5 um deep). Detectors for energy dispersive x-ray spectroscopy and secondary electron imaging are shown. (From PHI-ULVAC. Smart 200 Application Notes. With permission.)

10.4.2 SPUTTER DEPTH PROFILING

Two principal applications of the sputtering of materials are of interest here: those that analyze the sputtered material (SIMS, SNMS, and GDOES), and those that analyze the remaining surface (AES, XPS, and ISS). If there is no preferential sputtering of a particular component, these methods are complementary and give the same result. Sputtered-material-based methods provide higher sensitivity than XPS and AES, making them the preferred choice for the semiconductor industry, where analysis of dopants or trace contaminants is important. On the other hand, the signal intensities in XPS and AES are less influenced by matrix effects and chemical bonding than in SIMS, which often makes these techniques more suitable for interfacial analysis.

When preferential sputtering exists, in the steady state regime, SIMS and SNMS give the correct results, whereas in XPS and AES the measured composition of the exposed surface is dependent on the relative sputtering yield [30].

With the development of multilayered materials, with the thickness of each layer of the order of 1 nm, the concept of depth resolution became crucial. The depth resolution is defined as the distance over which the initial signal decreases from 84% to 16% at a monoatomic interface (ASTM 1992).

To reconstruct the original in-depth distribution of the elements from the experimental depth profile, the concept of the depth resolution function (DRF), $g(z - z')$, was introduced. The basic relationship between the experimental depth profile and the DRF is

$$\frac{I(z)}{I(0)} = \int_{-\infty}^{\infty} X(z')g(z-z')dz' \tag{10.4}$$

Mathematical reconstruction of the DRF is nowadays successfully achieved using, e.g., the mixing-roughness-information-depth (MRI) model [31]. The three parameters involved in MRI are introduced sequentially into the DRF and lead in many cases to a successful reconstruction of the depth profile. For more information about the underlining theory in this approach, refer to Refs. [30,31]. This section presents the practical information needed for the experimentalist:

1. Atomic mixing can be described by an exponential function with a characteristic mixing zone w that can be extracted from TRIM (transport of ions in matter) computer program giving values of the projected ion range.
2. Surface roughness is modeled using a Gaussian function with a standard deviation (σ) corresponding to RMS roughness. The parameter σ is assigned either an arbitrary value, e.g., 1 nm, and later optimized by fitting, or is experimentally determined. As an example, Figure 10.8 shows the atomic force microscopy (AFM) measurement of the depth-distribution (the "roughness") of a sputter-depth-profiled Al film with texture on a Si substrate [32,33].
3. Information depth is given by an exponential function with a characteristic length (λ). λ is, e.g., for AES and XPS, taken from the National Institute of Standards and Technology tables, or can be assigned a value of 1–2 monolayers (in nm) for SIMS.

Figure 10.9 shows AES and SIMS depth profiles of a double layer of AlAs in GaAs [34]. The open circles represent the experimental data, whereas the solid curve is the MRI fitting; the parameters used for fitting are shown in the figure.

10.4.2.1 Ion Guns

The essential tool for depth profiling is invariably an ion gun. In Table 10.7, the most widely used ion sources are listed along with their typical operating characteristics. For any particular application, it is advisable to choose the ion source carefully.

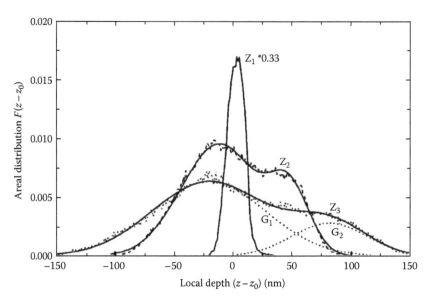

FIGURE 10.8 Areal distribution functions, measured by AFM, of Al films deposited on Si, after erosion to the three mean sputter depths, 50, 170, and 280 nm. (From Woehner, T., Ecke, G., Rossler, H., and Hofmann, S., *Surf. Interface Anal.*, 26, 1, 1998. With permission.)

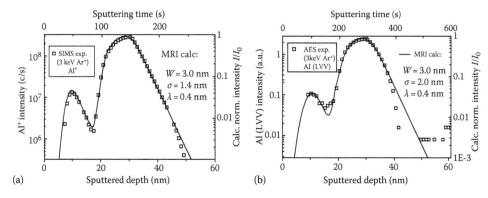

FIGURE 10.9 (a) SIMS Al⁺ intensity and (b) AES Al (LVV) intensity depth profiles of a double layer structure of AlAs (1 and 20 monolayers) in GaAs obtained with 3 keV Ar⁺ ions at (a) 52° and (b) 58° incidence angles (open circles), fitted with an MRI calculation (solid line) using the parameters shown in the figure. (From Hofmann, S., Rar, A., Moon, D., and Yoshihara, K., *J. Vac. Sci. Technol. A*, 19, 1111, 2001. With permission.)

Here, R is the erosion rate, d is the diameter of the beam, I_p is the bombarding current. The current density can be calculated from the beam current and the beam diameter. Data are extracted from the literature and the manufacturers' databases, and are strongly dependent on the application and on the sample under investigation.

10.4.2.2 Depth Determination—Conversion

Conversion of the "sputtering time" into a depth scale (in nm) is probably one of the most difficult tasks, and is a problem that has been investigated by many authors [35–37]. Basically, the depth z is related to the sputtering time t by the sputtering rate r by

TABLE 10.7

Characteristics of the Most Popular Ion Sources Available

	Beam Voltage	I_p	d	R	Application
Electron impact sources	1 kV	10^{-11} A	1 mm	0.1 nm/h	Static SIMS
	250 V to 2 kV	10–50 nA	10 μm	0.1–10 nm/min	Dynamic SIMS
	0.5–5 kV	3–10 μA	2 × 2 mm crater	1–40 nm/min	AES/XPS
Duoplasmatron sources	250 V to 8 kV	500 nA to 2 μA	2 μm	From a few nm/h to 400 μm/h	SIMS
Liquid metal ion sources	20 kV	>20 nA	60–100 nm	0.1 nm/h	Static SIMS
C_{60}^+ ion sources	20–40 kV	2 nA	2 μm	0.1 nm/h	Static SIMS

$$z = \int_0^t r\, dt \tag{10.5}$$

and

$$r = \frac{M}{\rho N_A e} Y j_p \tag{10.6}$$

where
  M is the atomic mass (kg/mol)
  ρ is density (kg/m³)
  N_A is Avogadro's number (6.02×10^{26} kg/mol)
  e is the electron charge (1.6×10^{-19} C)
  Y is the sputtering yield (atoms/ion)
  j_p is the primary ion current density (A/m²)

If r is constant, then $z = rt$.

From the literature, the most reliable method is to measure the crater depth, z_0, after a sputtering time, t_0, by means either of a mechanical profilometer, of optical techniques (interferometry) measurements [38]), or of SEM imaging. Each of these approaches permits the direct calculation of the sputter rate, $r = z_0/t_0$. Although, in some cases, it gives satisfactory results, errors of 10%–20% may occur [35]. When multilayered samples are studied, the individual sputter rate of each layer must be determined. The use of standard samples (certified reference materials) to calibrate sputter rates can be useful in this specific case.

10.4.2.3 Depth Resolution

10.4.2.3.1 Optimized Conditions for Depth Profiling
One of the major problems in sputter depth profiling (SDP) is the progressive decrease of the depth resolution during sputtering. Table 10.8 describes the major instrumental factors, sample characteristics, and radiation-induced effects that contribute to the degradation of depth resolution. To determine the depth resolution for a given experimental arrangement, measurements can be made on model samples where sharp interfaces are present, such as Ta_2O_5 and Ni/Cr multilayers.

TABLE 10.8

Factors Known to Limit the Depth Resolution and the Solutions Proposed

Instrumental Factors	Problem	Solution
Residual atmosphere	Contamination of the sample surface and a change in the sputtering rate	Use a noble gas with very low sticking coefficients
Ion mass and energy	Roughness, atomic mixing greater with high energy, low mass ions	Use low energy ions (<1 keV) and ions of high mass (Xe, Kr)
Ion beam and electron beam position (in AES)	If e-beam too wide and/or not well centered, crater wall effects	Diameter of e-beam should be at least 100 times lower than diameter of ion beam, and the centers of the ion beam and e-beam should coincide exactly
Sample characteristics		
Original roughness of the surface, and presence of microplanes	Critical effect on the depth resolution [39–46]	If sample is rough, then decrease ion incidence angle [45], or polish sample If sample is perfectly smooth, use a glancing incidence ion beam
Crystalline structure and defects	Sputtering yield depends on the crystalline orientation [47,48]	Best samples for depth profiling are single crystals or perfectly amorphous materials Otherwise use reactive ions [49] which result in a smoother surface than with argon due to the formation of an amorphous oxide layer [50]
Insulators	Positive ion implantation in insulators can lead to positive surface charging and to a gradual decrease in the effective incident ion energy and hence the sputtering yield	In XPS and SIMS flood the sample with low energy electrons [51]. Not a problem in AES
Radiation-induced effect		
Microtopography induced by sputtering	Crystalline orientation and the dependence of the sputtering yield on the ion incidence angle [47]	If sample is smooth, the effect can be minimized with lower ion energy, lower sputtering yields, and high incidence angle
Atomic mixing	Target atoms are relocated into the sample leading to atomic mixing and to the broadening of the interface. At steady state, the atomic mixing contribution to Δz is constant with z and is proportional to $E_1^{1/2}$, the primary ion energy, and is less for higher sputtering yields	Use heavy ions (Xe$^+$), low ion energies and/or glancing incidence. Liquid heavy ion sources (Ga) or even cluster ions (for SIMS) significantly reduce the atomic mixing
Preferential sputtering	Surface composition after sputtering not the same as the bulk	Use the ratio of the sputtering yields when determining the atomic concentration
Radiation-enhanced segregation and diffusion	Sputter-induced segregation is well known [52–55] and is frequent in alloys with high diffusion coefficients of the elements	Use sputtering rates greater than the diffusion process and reduce the diffusion by decreasing the temperature

TABLE 10.9
Sputtering-Induced Reduction of Oxides

No Reduction	Refs.	Reduction	Refs.
Al_2O_3	[64,66–68]	Ag_2O	[66]
BaO	[64]	Au_2O_3	[66]
Cr_2O_3	[66]	Co_3O_4	[61,64,68]
Cu_2O	[64]	CdO	[66]
FeO	[61]	Cr_2O_3	[61,64]
Ga_2O_3	[64]	$CuO\ Cu_2O$	[66]
HfO_2	[61]	Fe_2O_3, Fe_3O_4	[61,64,66,68,69]
MnO	[61]	HfO_2	[70]
MgO	[64]	IrO_2	[66]
MoO_2	[61]	MoO_3	[64,71]
SiO_2	[66]	Mn_2O_3	[64]
SnO, SnO_2	[61]	Nb_2O_5	[70,71]
Ta_2O_5	[66]	NiO	[64,66]
Ti_2O_3	[66]	$Ni(OH)_2$	[66]
V_2O_3	[61]	PbO	[61,64]
ZnO	[61,64]	PdO	[64,66]
ZrO_2	[61]	SiO_2	[71]
		Ta_2O_5	[64,70,71]
		TiO_2	[61,64,70,71]
		WO_2	[66]
		WO_3	[71]
		ZrO_2	[64,70]

Another major radiation-induced effect is the decomposition of compounds. For instance, many oxides have been observed to undergo reduction upon sputtering [56–64]. Theoretical investigations have also been conducted [65]. Table 10.9 provides examples of oxides where reduction either has or has not been observed. Usually, reduction is not observed for low Z metal oxides. Some oxides have been observed to undergo reduction under certain sputtering conditions but not others. No simple trend has been established.

10.4.2.4   Improvement of the Depth Resolution by Sample Rotation

Because the sputtering process strongly affects the achievable depth resolution, many attempts have been made to minimize the limitations. One successful strategy is to suppress the crystal orientation effects and the increase of roughness by the rotation of the sample during sputtering, as suggested by Zalar et al. This facilitates the formation of a "flat-bottom" crater where the center is the analyzed area [72,73].

Figure 10.10 presents a schematic drawing of the rotational principle and of the shape of the sputtering crater, and Figure 10.11 presents sputter profiles of a Au/Al/SiO$_2$/Si multilayer sample without (a) and without (b) sample rotation [74].

It has been shown that increases in both the rotation speed and the Ar$^+$ ion incident angle lead to a better depth resolution of the profile [75]. This rotational device, originally designed for AES, has been adapted to SIMS and XPS instruments. Figure 10.12 shows the SIMS depth profile through a GaAlAs Be-spiked sample without (a) and with (b) sample rotation. Without rotation, the depth resolution progressively decreases during sputtering, whereas it remains constant when the sample is rotated at 3 rpm [83]. The so-called "Zalar rotation" is probably one of the most important technological improvements in depth profiling in the last two decades.

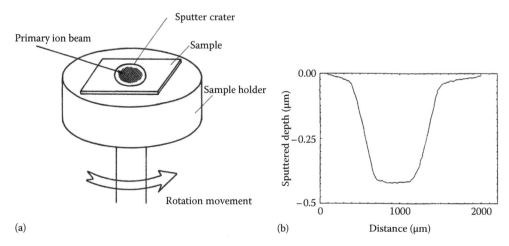

(a)

(b)

Distance (μm)

FIGURE 10.10 Principle of "Zalar rotation" as applied to AES. (a) Schematic of process; (b) example of crater profile obtained. (From Ostwald, S. and Baunack, S., *Thin Solid Films*, 425, 9, 2003. With permission.)

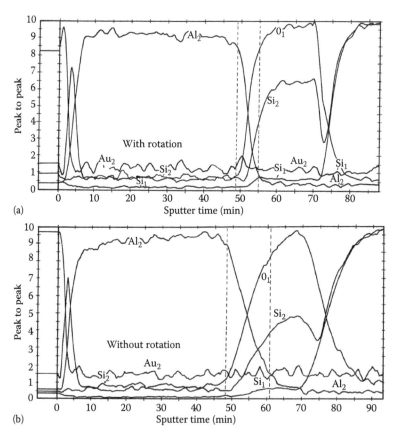

(a)

(b)

FIGURE 10.11 Effect of the rotation of the sample during sputtering on a typical Auger depth profile, showing improved depth resolution with rotation. (From Zalar, A., *Thin Solid Films*, 124, 233, 1985. With permission.)

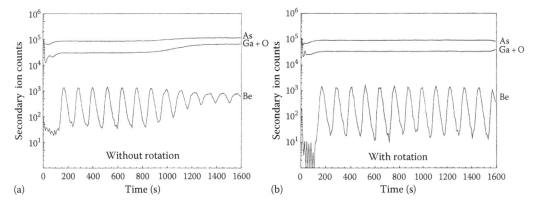

FIGURE 10.12 SIMS depth profile through a GaAlAs Be-spiked sample with and without sample rotation, showing elimination of progressive loss in depth resolution, when using rotation. (From Stickle, W.F. and Stobol, P.E., *Surf. Interface Anal.*, 19, 165, 1992. With permission.)

10.4.2.5 "Chemical" Depth Profiles

The traditional goal of a depth profile has been to provide the atomic concentration(s) as a function of depth, but until recently it was not possible to collect chemical information as a function of depth. However, the oxidation states of the elements, their chemical environment, and their chemical bonds can now be extracted from most AES, XPS, or SIMS depth profiles. In the latter case, the information is available from the specific molecular fragments that are collected. The development of heavy ion and cluster sources that reduce the fragmentation and enhance the sputtering yield was, in this respect, a significant step forward. Technological developments in AES and XPS now allow very high energy resolution (0.05% in modern AES instruments and 0.28 eV for monochromatized XPS), that reveal part of the chemical environment through the chemical changes in the AES lineshapes or through the chemical shifts of the photoelectron peaks [28]. The parallel development of powerful computerized mathematical procedures dedicated to deconvolution, and peak fitting, has also been spectacular. In XPS, peak fitting (using Gaussian–Lorentzian functions) is now routinely used. It must be stressed that background correction (i.e., as developed by Tougaard [76] or Shirley [77]) should be used correctly before performing any peak fitting (see Chapter 8). Criticism of the unthinking use of these software applications can be found in Refs. [78–81].

When a large data set of photoelectron spectra is to be studied (i.e., surface images or depth profiling), multivariate statistical methods can be used. Fulghum et al. [82] conducted a comparative study of such methods on polymer blends. They investigated different routines to find the pure chemical components after having performed factor analysis. Programs such as target factor analysis (TFA), parallel factor analysis (PARAFAC), simple-to-use interactive self-modeling mixture analysis, multivariate self-modeling curve resolution, and evolving factor analysis were tested.

PARAFAC has been used to study XPS peaks during the oxidation of aluminum [83,84]. Stickle and Stobol [85] used linear least square (LLS) and TFA to study XPS depth profiles of Ni/Cr layers deposited on Si. Oswald and Baunack applied factor analysis methods to XPS spectra of superconducting layers and to XPS depth profiles of Cr-implanted Si_3N_4 [86].

The deconvolution of Auger atomic concentration depth profiles by LLS, NLLS, and PCA (NLLS, nonlinear least squares; PCA, principal component analysis) has been extensively used [87–90] (see Chapter 15 for applications of PCA). Due to the more complex shapes of the Auger peaks, simple deconvolution of large data sets to obtain chemical information is not yet possible. Factor analysis methods have been specifically developed for this application.

To obtain a depth profile with chemical information, the following procedure is generally used:

1. Acquisition of spectra of the desired elements in defined energy windows during sputter profiling
2. Storage of these spectra in a file
3. Calculation of the peak-to-peak height in each energy window in order to obtain the usual AES depth profile
4. Deconvolution of the peak shapes for chemical information by one of the above-mentioned methods

A typical example of the power of this method can be illustrated by the WN system. In the case of the interaction between tungsten and nitrogen, the effect of nitrogen on the metal valence electrons is much weaker than in oxides. The chemical effect on the metal peak shape or shift is considerably smaller and often neglected. Figure 10.13a shows the AES depth profile of a WN film deposited on W by DC reactive sputtering [89], while (b) and (c) show the factor analysis of the W and N peaks, respectively. The left-hand diagrams show the spectra of the pure materials and the right-hand ones show the reconstructed concentration matrix for each element. Two distinct nitrogen components can be detected, surface nitrogen and nitride nitrogen. Also, three tungsten components can be identified: tungsten nitride, metallic tungsten, and interfacial tungsten corresponding to tungsten carbide, as suggested by the carbidic shape of the carbon peak at the interface (see insert in Figure 10.13a). This last chemical state of tungsten could not have been detected by the usual AES depth profile.

10.4.3　Auger Electron Spectroscopy and X-Ray Photoelectron Spectroscopy

Historically, depth profiling and 3-D analysis have been associated with AES. It is easier to focus an electron beam onto the center of a sputter crater than it is to combine a broad x-ray beam with an ion gun. For this reason, most depth profiling studies have employed AES. Today, however, the technological improvements in XPS (e.g., imaging, fine focus, high intensity sources, and parallel detection, see Chapter 3) allow mapping and depth profiling to be performed in ways similar to those in AES.

10.4.3.1　Basic Principles

In XPS, x-ray excitation of an atom leads to the direct emission of a photoelectron of characteristic energy; in AES, excitation leads to the emission of Auger electrons, also at characteristic energies, produced during the relaxation process. The theory behind the excitation processes is given in Chapter 3. AES or XPS depth profiling combines the analysis of electron energies with simultaneous ion sputtering of the sample.

The profiles obtained in each case are plotted as functions of intensity, I, vs. sputtering time, t (Figure 10.2). The challenge in SDP is the transformation of I into atomic fraction, X, and of t into depth, z. The DRF and MRI models developed in the last 10 years have helped greatly in reconstructing the original depth profile by taking into account the roughness, the mixing, and the information depth (see Section 10.4.2). However, the signal intensities must still be transformed into concentration.

10.4.3.2　Main Equations for Quantitative Analysis

It has been shown in Refs. [2–4], and in Chapter 3, that the atomic fraction can be linked to the Auger signal intensity by

$$X_i = \frac{I_i/I_i^\infty}{\sum_j F_{ji}^i (I_j/I_i^\infty)} \tag{10.7}$$

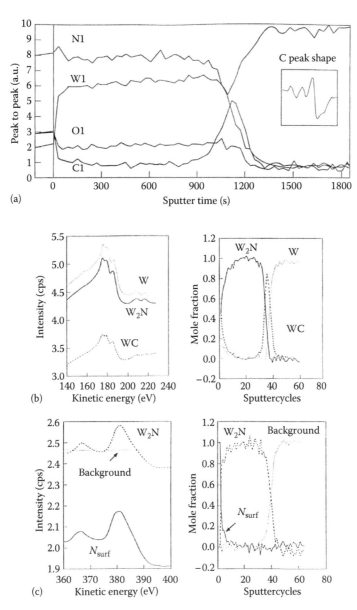

FIGURE 10.13 Factor analysis applied to the Auger sputter depth profile of a W_2N film deposited on tungsten. (a) Usual AES depth profile; (b) factor analysis of the tungsten signal; (c) factor analysis of the nitrogen signal. (From Reniers, F., Hubin, A., Terryn, H., and Vereecken, J., *Surf. Interface Anal.*, 21, 483, 1994. With permission.)

where

F_{ji}^i is a matrix factor
X_i is the atomic fraction of element i
I_i is the signal strength of element i
I_i^∞ is the signal strength of pure element i

The ratio of signal strengths for pure elements to that of a standard gives sensitivity factors. Relative sensitivity factors (RSF) are nowadays either included in surface analytical software or taken from one of several handbooks [91–94]. However, the best method is to measure the RSF on well-known standard samples or to use reference "absolute" spectra, which are now available.

An equation similar to Equation 10.7 can be used for quantification in XPS. For a diatomic sample:

$$\frac{X_A}{X_B} = F_{AB}^x \frac{I_A/I_A^\infty}{I_B/I_B^\infty} \tag{10.8}$$

where F_{AB}^x is a matrix factor for XPS. Here, the matrix effects will be smaller than in AES depth profiling because the backscattering factor for primary electrons is absent.

The correct quantification of the peak area intensities (I_i in Equations 10.7 and 10.8) in AES and XPS requires mathematical subtraction of the inelastic background. Tougaard [95] and Shirley [96] methods are included in all Auger and XPS software nowadays. Properties of the Tougaard method have been discussed in Ref. [97] and in Chapter 8. A procedure that uses subtraction of a reflected electron energy loss spectroscopy background for quantitative AES and XPS has also been reported, and seems to produce more accurate absolute intensities [98].

Equation 10.7 hides several conceptual difficulties: the sensitivity coefficient contains the inelastic mean-free path (IMFP), the electron backscattering factor, the atomic density, as well as instrumental terms. It is necessary to take into account matrix corrections, F, in most cases. Extensive discussions of each of these problems can be found in the literature [99–101] and are not the subject of this chapter. A few years ago, a new procedure was introduced that used average matrix factors, RSFs. This approach seems to reduce the systematic errors in conventional quantitative analysis performed with elemental RSFs [102]. Nevertheless, the analyst should keep in mind that all these factors affecting the linear relationship between the signal strength and the composition of a sample will modify the final results.

The work of Ostwald and Baunack [35] illustrates this well. They compared concentrations obtained by depth profiling using AES and XPS on a NiCrSi/CrSi multilayer system. The calculations carried out to determine the concentrations (in at.%) from the intensities extracted from the depth profiles were performed using the usual formulae (Equations 10.7 and 10.8), but they used different standards, and different corrections (matrix, density). Their results are presented in Table 10.10.

Their results show significant departures from the nominal concentration, especially for the CrSi layer. Indeed, these calculations neglect the effects of the depth profiling process itself, including any preferential sputtering which, they stress, is known to occur with chromium silicide.

Surprisingly, although these matrix effects have been known since the beginning of surface analysis in the 1970s, no matrix corrections have been applied to the raw data contained in most published studies except those arising from experts in metrology.

In addition, these formulae assume that the sample is homogeneous over the depth analyzed, and this is not always the case in depth profiling, especially at interfaces. A mathematical model must then be used to decompose the Auger signal into layer-by-layer contribution [103–105].

A general equation for an arbitrary AES or XPS depth profile is

$$X_i(z) = \left(\frac{I_i}{I_i^0}\right)_z - \frac{\mathrm{d}(I_i/I_i^0)}{\mathrm{d}z} \lambda_i \tag{10.9}$$

In this equation, the composition at a given depth ($X(z)$) is linked to the signal intensity I originating at depth z corrected by the effective escape depth of the electrons λ.

The effect of the sputtering process itself on the true concentration is usually neglected. Since the elements in a sample will have differing sputtering yields, some will be sputtered preferentially. Analysis of the resulting surface will thus give a composition different from the true one. If the sputtering yields Y of the components are independent of their bulk concentrations, X_b, the surface composition X_s is inversely proportional to the respective sputtering yields, i.e.,

$$\frac{X_{sA}}{X_{sB}} = \frac{Y_B}{Y_A} \frac{X_{bA}}{X_{bB}} \tag{10.10}$$

TABLE 10.10

Concentration Arising from Different Quantification Procedures for AES and XPS

	NiCrSi Layer			CrSi Layer	
	Ni (at.%)	Cr (at.%)	Si (at.%)	Cr (at.%)	Si (at.%)
Nominal concentration	37	57	6	16	84
AES					
Handbook values	32	57	11	38	62
Single element st.	30	57	13	35	65
Handbook value + density correction	35	58	7	50	50
Single element st. + matrix correction	35	58	7	48	52
$CrSi_2$ as silicide st.	32	62	6	30	70
XPS					
Handbook values	30	63	7	44	56
Single element st.	34	59	7	42	58
Single element st. + matrix correction	40	56	4	55	45
$CrSi_2$ as silicide st.				30	70

Source: Ostwald, S. and Baunack, S., *Thin Solid Films*, 425, 9, 2003. With permission.
Note: Ni (LVV), Cr (LMV), and Si (KLL) lines were used for AES.

where the subscripts A and B denote the two components (at steady state). Equation 10.10 indicates that the composition of the first layer will be modified; the element with the lower sputtering yield will become enriched at the surface.

The resulting Auger or XPS sputter profile must then be corrected by a sputtering yield ratio in order to give the true composition of the sample. Such a correction has already been developed for AgPd alloys [106].

If a steady state is not reached, a transient in surface composition is found [36], where

$$X_{sA}(t) = (X_{bA} - X_{sA})\exp\left(-\frac{t}{\tau}\right) + X_{sA} \qquad (10.11)$$

where $X_{sA}(t)$ is the instantaneous surface composition of A between the start of sputtering ($t = 0$) and the time at which secular equilibrium is reached ($t = \infty$).

As the composition of only the first (and second) atomic layers is modified by the sputtering process and since in AES and XPS signals come from a few atomic layers, the higher energy peaks (i.e., those corresponding to greater IMFP), which sample more of the bulk material, will be less affected by this problem.

10.4.4 GLOW DISCHARGE OPTICAL EMISSION SPECTROSCOPY

10.4.4.1 Basic Principles

GDOES was used in the field of metallurgy for many years before the appearance of AES, XPS, and SIMS. Recently, with the development of radiofrequency GDOES (rf-GDOES), the technique has become useful for the depth profiling of thin films (<10 nm). The basic principle remains the same.

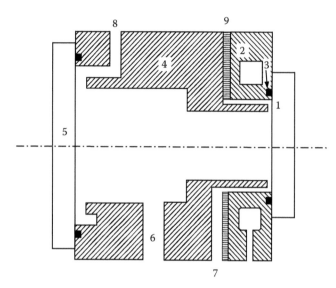

FIGURE 10.14 The Grimm lamp used in GDOES. (From Ohashi, Y., Yamamoto, Y., Tsuyonama, K., and Kishidaka, H., *Surf. Interface Anal.*, 1, 53, 1979. With permission.)

A sample is bombarded by Ar^+ ions generated in a glow discharge. Some of the sample atoms ejected into the glow discharge become electronically excited. The light emission from these atomically excited states is analyzed. The energy of the emitted photon is characteristic of the atomic species and the intensity of the photons can be used to determine the relative concentrations. The use of a high frequency potential (typically 13.56 MHz [107]) permits the analysis of electrically insulating materials. At first, a negative potential is applied to the sample. If the potential is sufficiently negative, a discharge is produced and Ar^+ ions are accelerated toward the sample surface. As the polarity of the voltage changes during the next part of the rf-cycle, electrons from the discharge are accelerated toward the sample neutralizing the positive charge accumulated on the surface. Due to the difference in mobility of the ions and electrons, a net negative bias voltage is eventually generated which maintains the discharge. A more complete description of the basic principles of rf-GDOES can be found elsewhere [107,108].

The original GDOES device was the glow discharge lamp designed by Grimm and a drawing is given in Figure 10.14 [109]. The sample (1) located at the cathode (2) is bombarded by Ar^+ ions. The gas is introduced via (8) to a pressure between 1 and 5 mbar. The glow discharge is established by application of a DC voltage of about 1000 V between the anode (4) and the cathode. The light emitted from the glow discharge is collected and analyzed by a spectrometer located behind the quartz window (5). The sputtering rate is typically from 30 nm/min to 1 μm/min, much faster than in AES or SIMS with lower energy ions. Commercially available instruments employ an rf-voltage rather than a DC voltage. The Jobin Yvon RF-5000 GDOES instrument design is based on a Marcus-type discharge lamp (see Figure 10.15 [108]). The Markus design is similar in many respects to the Grimm design. The sample is used as one of the electrodes and for this reason is still referred to as the cathode even though the voltage applied reverses polarity. The most significant change in the Markus design is the hard grounding of both the anode and the shutter above the sample. This confines sputtering to the sample surface.

10.4.4.2 Main Equations for Quantitative Analysis

For exact quantitative analysis of a sample M in GDOES, analysis of a reference material R is required. The intensity of an emission line is related to the atomic concentration of the emitting element in the sample through complex parameters such as the applied voltage, the matrix of the

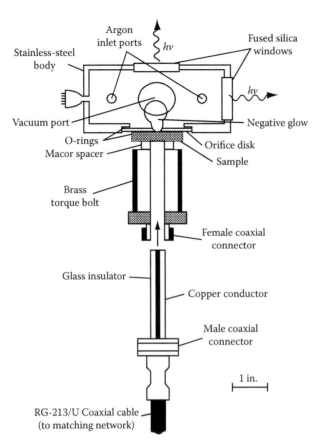

FIGURE 10.15   The Markus lamp used in rf-GDOES; the coaxial feed-through is shown withdrawn for clarity. (From Marcus, R.K., *J. Anal. Atom. Spectrom.*, 8, 935, 1993. With permission.)

sample, spectral line constants, and diffusion processes. All of these must be normalized, hence the requirement for a reference material for quantitative analysis. A good review of all these parameters was presented by Rivière [4]. If the discharge current for the analysis of the reference and the unknown sample is kept the same, the concentration is given by

$$c_{AM} = c_{AR} \frac{C_R}{C_M} \frac{I_{\lambda M}}{I_{\lambda R}} \left(\frac{V_R - V_{0R}}{V_M - V_{0M}} \right)^{X(\lambda, A)} \qquad (10.12)$$

where

c_{AM} is the unknown concentration of A in the matrix M

c_{AR} is the concentration of A in the reference material R

$C_{(M \text{ or } R)}$ is a constant dependent on the sample and the sputtering ion

$(V - V_0)$ is the voltage above the threshold for the sample M and for the sample R (V_0 being the threshold voltage)

$I_{\lambda(M \text{ or } R)}$ are the intensities of the emission lines from the unknown sample and from the reference material

$X_{\lambda, A}$ is a spectral constant

Figure 10.16 shows quantified GDOES depth profiles from three hot-dip zinc-coated steel samples [110]. Only the Zn/Fe interface region is shown; the zinc and iron concentrations vary from 0 to 100

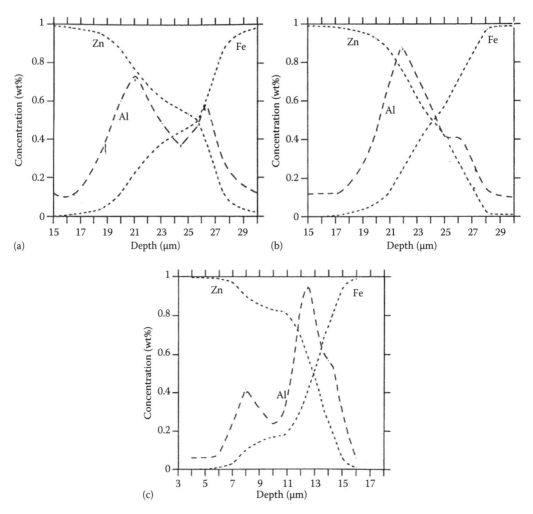

FIGURE 10.16   Quantified GDOES depth profiles from three hot-dip zinc-coated steel samples. (From Karlsson, J., Horonstrom, S.E., Klang, H., and Nilsson, J.O., *Surf. Interface Anal.*, 21, 365, 1994. With permission.)

at.% while the aluminum concentration varies from 0 to 1 at.%. This example shows two advantages of GDOES over, e.g., AES: (1) the deep location of the interface (more than 10 μm depth) is quickly reached, (2) the wide range of concentrations recorded simultaneously (0–100 at.% and 0–1 at.%) is not available in AES. However, as the authors stated, the depth resolution is very much degraded due to the sputtering process and to the grain orientation. The conversion from intensities to concentration and from sputtering time to depth was performed by software developed by Bengtson [111–113].

Summary of the strengths and weaknesses of GDOES

Main advantages

- Samples from <10 nm to 15 μm thick can be analyzed
- Quick analysis (a few min)
- All elements detected (including H)
- Cheap (in comparison with AES, SIMS, and XPS)
- No vacuum chamber required (except to detect O, N and H lines in extreme UV)

- Elemental information
- High sensitivity (ppm)

Main disadvantages

- Poor depth resolution (in comparison with AES, XPS, or SIMS)
- Not a surface analysis technique (currently in the ca. 4 nm range)
- No lateral resolution
- No chemical information

10.4.4.3 Recent Improvements in GDOES

There are many reports of increases in the surface sensitivity using rf-GDOES to <10 nm [114–118]. With such high sputtering rates, one would not expect the technique to be practical in this depth range. Improvements in data acquisition, instrumentation, and software can be used to collect data at very short time intervals (<10 ms), but these improvements alone are inadequate to achieve greater surface sensitivity. The most significant improvement, achieved with rf-GDOES, may be the uniformity of sputtering from the onset of the glow discharge. For a given power, the peak-to-peak voltage in rf-GDOES is higher than in dc-GDOES, reducing the likelihood of sparking, which can ablate the top few surface layers. Furthermore, the sputtering process in a glow discharge is significantly different from that using traditional ion gun sources. The ion energy is much lower (<50 eV), the current density is much higher (100 mA/cm^2), and there is a much wider range of incident angles. The resultant reduction in the preferential sputtering that leads to roughening and interlayer mixing induced by higher energy ion bombardment leads to an improved depth resolution.

10.4.5 Secondary Ion Mass Spectrometry and Secondary Neutral Mass Spectrometry Depth Profiling

10.4.5.1 SIMS

10.4.5.1.1 Basic Principles
SIMS is based on the emission of atomic and molecular particles (secondary ions) from the surface of a solid under bombardment with primary particles (ions).

Under DSIMS conditions, the erosion rate is of the order of monolayer(s) per second and concentration depth profiling can be performed with high spectrometric sensitivity, in favorable cases down to the ppb level. The sensitivity of SIMS is element dependent and varies from 1 ppb to 100 ppm and the technique is therefore used mainly for trace and ultratrace analysis. Additional detail about SIMS and SNMS can be found in Chapter 4.

Today, SIMS is a routinely used technique available in many laboratories, both in academia and in industry. The main problem in SIMS is quantification, because of the dependence of relative and absolute secondary-ion yields on matrix effects, on surface coverage by reactive elements (e.g., oxygen), on the background pressure in the sample chamber, on the effect of crystal orientation with respect to primary and secondary ion beam, and on singularities, etc. Great progress has been made in this area during the past 10 years, due, in part, to the use of certified reference samples, and of reactive ion beams (Cs, O_2) to increase the useful yield.

10.4.5.1.2 Main Equations for Quantitative Analysis
Some basic considerations are summarized here briefly (for more details, see Chapter 4, [3], or [6]). For both SIMS and SNMS, the ion signal of an isotope *i* (at a given mass) in a sample *j* is related to the isotope concentration X_i in the sample by

$$I_i = (AX_iY_j\varepsilon_{i,j}\Gamma_{i,j}\beta_{i,j}T_{i,j}D_i)I_p \qquad (10.13)$$

where

I_p is the primary ion flux (ions/s)

I_i is the ion signal

A is the fractional abundance of the isotope sampled (usually calibrated and neglected in SIMS)

i,j are combined indices that indicate that a parameter may depend on both the nature of the sputtered ion and on the sample matrix

X_i is the average concentration of the element taken over the sputtered atom escape depth (usually most of the sputtered atoms come from the first layer)

Y is the global sputtering yield of the sample

ε is the fraction of i, which is sputtered in the monitored form

Γ is the preferential sputtering term (relative sputtering yield efficiency of i in the matrix j)

β is the ionization probability of the sputtered entity

This last parameter causes the greatest difficulty in SIMS analysis, because β depends strongly not only on the nature of the sputtered entity, but also on the chemical state of the sample and on the nature of the primary ions. In SNMS, neutrals are post-ionized (i.e., after having been sputtered from the surface) and there is no hidden dependence of β in the sample. The other factors are

$Y^{\pm} = Y\beta$ = ion yield

$\tau^* =$ useful ion yield $= \beta T$

$T =$ mass spectrometer transmission for species i

Equation 10.13 is usually simplified to

$$I_i = K_i X_i I_p \qquad (10.14)$$

where K_i is an elemental sensitivity factor that contains all the other variables in (13), including β, which may vary over several orders of magnitude from element to element but also, for any one element, from sample to sample. The latter is called the "matrix effect." Thus, if truly quantitative SIMS analysis is required, the use of external standards, as identical as possible to the sample under study is absolutely essential. Such standards are rare and expensive since they must be prepared by ion implantation. Quantitative analysis at interfaces is even more difficult, since the matrix changes abruptly. However, due to the high sensitivities of mass spectrometers and the increased sputter yield from improved ion sources, detection limits in SIMS are very low (in the range 1 ppm and in favorable cases 1 ppb), which allows trace and ultratrace detection. This explains the success of SIMS in the microelectronics industry.

10.4.5.1.3 Applications

One of the main applications of SIMS is in the semiconductor industry where it is used for the quantification of dopants and the detection of unexpected impurities, down to the ppb level. Other applications are found in the domain of polymers and membranes (e.g., ion exchange membranes for fuel cells, electro-osmosis or electrodialysis), where the in-depth homogeneity of active grafted groups is crucial. Since the use of cluster ions increases the yield and reduces the fragmentation, in-depth information about chemical bonds in polymers becomes possible. In metallurgy, SIMS-imaging is used for applications in corrosion, while depth profiling is useful for the analysis of coating-metal interfaces. SIMS is also widely used in the glass and ceramics industry.

10.4.5.2 Secondary Neutral Mass Spectrometry

The principal rationale for SNMS is the elimination of the matrix effects present in SIMS, by decoupling the emission and the ionization processes. The ionization probability no longer depends on the sample characteristics and composition but becomes an equipment constant. Thus, quantitative analysis may be performed even at nondilute concentrations. Moreover, SNMS can be operated with low energy incident ions (a few hundred eV), which eliminates atomic mixing. It is also worth remembering that in the sputtering process only about 1% of the ejected particles are ionized, the remaining 99% being neutrals. Finally, the artifacts due to preferential sputtering, which are present in depth profiling using the electron spectroscopies, are avoided in SNMS.

10.4.5.2.1 Basic Principles

When bombarded by an incident ion beam, most of the particles ejected from a solid are neutrals. However, the analysis of neutrals requires post-ionization. Many techniques have been developed and they are summarized in Table 10.11. The most common are those using electron beam and electron gas post-ionization, although laser post-ionization (LPI) is gaining popularity. Typical operating conditions for an electron gas plasma are given in Table 10.12.

TABLE 10.11
Post-ionization Techniques

Technique	Thermal Ionization	Electron Beam	Electron Gas	Penning Ionization	Pulsed Laser
Elements ionized	Low ionization potential	All	All	All, not noble gases	All[a]
Ionizing process	Surface ionization	Electron impact	Electron impact	Collisions with noble gas atoms and electron impact	Photoionization
Ionizing medium	Heated chamber	Electron beam plasma	Low pressure plasma	Medium pressure	Laser light
Residual gas	Mass resolution	Energy discrimination	Energy discrimination	Mass resolution	Mass resolution
Efficiency	Poor	Satisfactory	Satisfactory	Satisfactory	Satisfactory
Mass spectrometer	Quadrupole	Quadrupole	Quadrupole	Quadrupole	ToF

Sources: Adapted from Jede, R., Ganschow, O., and Kaiser, U., in *Practical Surface Analysis*, Wiley, Chichester, 1992; Muller, U., Schittenhelm, M., Schmittgen, R., and Helm, H., *Surf. Interface Anal.*, 27, 904, 1999.

[a] Using nonresonant multiphoton ionization (NRMPI) [120].

TABLE 10.12
Typical Operating Conditions for SNMS, Using Electron Gas Post-ionization

Type of gas	Argon
Pressure	Around 10^{-3} mbar
Ion current density	0.8–3 mA/cm^2
Magnetic field	2–3 mT
rf-Frequency	27.12 MHz
rf-Power	80–200 W
Plasma potential	45–55 V
Dimensions	150 mm diameter 150 mm length

Typically, LPI uses an ArF excimer laser operating at 193 nm with 70–80 mJ pulses of 10–20 ns duration with a repetition rate of 300 kHz [119].

10.4.5.2.2 Quantification in SNMS

The basic relationship between the ion current $I^0(X)$ of a particular mass X collected and the concentration of the element of that mass in the sample is the same as in SIMS (see Equation 10.13) and can be written

$$I^0(X) = Y_{tot}C(X)D(X)I_p \qquad (10.15)$$

where
   I_p is the primary current
   Y_{tot} is the total sputtering yield of the sample per primary ion
   $C(X)$ is the atomic concentration of element or isotope X
   $D(X)$ is the useful yield (the ratio of the number of detected particles to the number of sputtered particles) (Table 10.13)

Equation 10.15 assumes, contrary to SIMS, that $D(X)$ does not depend on (a) the concentration of X and (b) the concentration of the other constituents. Equation 10.15 also allows the use of RSFs (as in AES or XPS).

For two elements i and j, one can write:

$$\frac{I^0(X_i)}{I^0(X_j)} D_j(X_i) = \frac{C(X_i)}{C(X_j)} \qquad (10.16)$$

where $D_j(X_i)$ is the RSF between i and j. This RSF must be determined from a standard sample, which may be rather different from the sample under investigation. In electron impact SNMS, RSFs are usually referred to Fe. The general equation is then

$$C(X_i) = \frac{I^0(X_i^{mi})\eta^{mi}D_{Fe}(X_i)}{\sum_j I^0(X_j^{mj})/\eta^{mj}D_{Fe}(X_j)} \qquad (10.17)$$

where
   $I^0(X_i^{mi})$ is the SNMS signal of isotope with mass mi of element Xi
   η^{mi} is the abundancy of the isotope

The RSFs are relatively independent of the matrix of the sample but a dependence, especially for light elements, has been shown for the ion energy and the ejection angle. Tables of RSFs can be found in the literature [121].

The best way to perform quantitative SNMS with reliable standards is as follows:

1. Choose the standard sample with microstructure and texture similar to the sample to be analyzed in order to avoid topographical and angular effects (use a microscope for comparison if necessary). Similar composition is necessary of course.
2. Calibrate the instrument using the standard and fix the instrumental parameters (i.e., ion energy, plasma or e-beam characteristics, spectrometer position, etc.).
3. Measure the elemental RSF.
4. Analyze the unknown sample under the same instrumental conditions.

5. Steels are often used as standards because they have good bulk homogeneity and they contain a large number of elements with a range of concentrations of width usually sought in SNMS (from 10 at.% to 1 ppm) [122–127]; brass and glass substrates have also been used [125].

10.4.5.2.3 Applications

SNMS is particularly suitable for the analysis of alloys in which a wide range of concentrations must be measured. Semiconductors can also be analyzed, because the detection limits of SNMS are of the same order of magnitude as SIMS. Finally, because the depth resolution is better in SNMS than in SIMS or AES, SNMS is particularly suitable for the study of interfaces [128].

10.4.5.3 Optimum Conditions for Performing SIMS and SNMS Depth Profiling

Due to the similarity of the two techniques, most of the operating conditions of SIMS and SNMS are similar, and are summarized here. As the reader will notice, many different operating conditions are possible, depending on the nature of the sample to be analyzed (metal, inorganic, polymer, and semiconductor), the nature of information required, and the available instrumentation. Currently, SIMS–SNMS instruments tend to have two ion guns: one dedicated specifically to the removal of layers (the "in-depth" profiler) and the other to the SIMS–SNMS analysis. The separation of these functions achieves the best conditions for both processes. A low energy ion source is oriented at a grazing angle for smooth etching and therefore good depth resolution, while a higher energy ion source is oriented normal to the surface for maximum secondary ion or neutral yield.

In addition to the unique problem in the quantification of SIMS–SNMS, the two techniques share the problem common to all sputter depth profile techniques, namely, the decrease of depth resolution with sputtering time. This degradation is due to the progressive atomic mixing and roughening of the surface under ion bombardment. As noted previously (Section 4.2.5), great efforts have been made to reconstruct mathematically the DRF and, from the experimental point of view, to reduce the effects. One experimental approach, as shown in Figure 10.10, is sample rotation.

Another improvement is the development of sources producing low energy ions, known to limit roughening and sputter-induced atomic mixing. Current research in SIMS is devoted to the use of increasingly heavier projectiles and clusters. Today, Ga^+, Au_3^+, SF_5^+, Bi_3^+, and C_{60}^+ ion sources are commercially available. The effect of heavier ions has the following consequences, as pointed out by Winograd [129,130]:

- Decrease of the damage in the layer (less atomic mixing)
- Limitation of the sputter-induced surface roughness (for specific samples)
- Decrease of the secondary ion escape depth by the use of heavy ions or clusters as projectile in some cases [130]
- Increase in the sputtering yield (higher sensitivity)

Table 10.13 illustrates a significant change in neutral species yield for water-ice using different primary ions.

TABLE 10.13
Number of Water Molecule Equivalents for Different Primary Ions [129]

	Au^+	Au_2^+	Au_3^+	C_{60}^+
Number of H_2O equivalents removed	94	570	1200	1830

10.4.5.3.1 Bombarding Conditions: Ions
Ar^+, O_2^+, O^+, O^- at energies from 1 to 10 keV (reactive gases mainly for SIMS).
Cs^+ at energies from 15 to 30 keV.
Ga^+, In^+ at energies from 15 to 40–60 keV (for small high brightness ion probes <100 nm in diameter).
C_{60}^+ from 10 to 20 keV.
For depth profiling, the beam diameter is typically between 10 and 1000 nm with an ion beam current density of $1–10\,\mu A/mm^2$.

10.4.5.3.2 Angle of Incidence
There are two mutually exclusive requirements in SIMS. On the one hand, there is a need for high sensitivity, which is optimum at normal incidence, and on the other, as in AES and XPS, a need for good depth resolution which is optimum at glancing incidence. As mentioned previously, there are two approaches to the solution of this apparent conflict. Two ion sources can be used—one for depth profiling and the other for obtaining a high yield of secondary ions or neutrals. Alternatively, a single ion source can be used at normal incidence, and then the sample is rotated as in sputtering using Zalar rotation [73]. This produces a depth resolution relatively independent of the ion incidence angle while achieving a high sensitivity.

10.4.5.3.3 Effect of the Choice of the Gas in SIMS and SNMS
SIMS depth profiling is normally performed using O_2^+ and Cs^+ ions. Bombardment with these species increases the secondary ion yield by several orders of magnitude, notably at normal and near normal ion beam incidence [33,131,132]. Cluster ions also increase the secondary yield [129]. The introduction of reactive gases is nevertheless risky because there are also some adverse effects. Additional comments include:

1. O_2^+ bombardment at normal incidence prevents roughening of the surface of polycrystalline metals such as Ti, Cr, Mn, Fe, and Mo by formation of an amorphous surface oxide layer [133,134]. This increases the depth resolution.
2. N_2^+ and O_2^+ have been found to increase the depth resolution in the study of $^{28}Si/^{30}Si$ multilayers [135].
3. O_2^+ is not recommended when noble metals and alkali metal impurities are to be studied [136–141].
4. Reactive gases are not required for SNMS, since the ionization of the ejected particles does not depend on the sputtering process itself. Noble gas or cluster ions may be used instead.
5. Cluster ions are recommended in order to limit fragmentation in the SIMS depth profiling of polymers.
6. Heavy ions reduce atomic mixing and therefore increase depth resolution.

10.4.5.3.4 Choice of Ion Beam Energy
SIMS sputtering yield has an hourglass-shaped dependence on ion energy with the maximum between 1 and 10 keV for Ar^+, N_2^+, O_2^+ ions [135,142,143].

Because the only function of the ion source in SNMS is to eject particles from the solid and not to provide the extra energy necessary for ionization, very low energy ion beams can be used (<1–2 keV).

For heavier ions, typical energies used today are in the range 10–20 keV, but values up to 40–60 keV for finely focussed probes (<100 nm) can be found in the literature.

Tables of sputtering yields of metals as a function of ion energy have been published, and the analyst must choose the ion type and energy taking into account the following requirements:

- The sensitivity must be as high as possible, which requires a high energy ion beam.
- The depth resolution must be sufficient, which is best achieved with low energy ions (<2 keV) [144,145].
- The sputtering yields of the different elements must be as similar as possible.
- If high lateral resolution is also required, low energy ion beams are not practicable due to the difficulty in focusing them. Therefore, high lateral resolution, which requires higher energy ions, often induces a loss of depth resolution. High energy C_{60}^+ ion sources present an alternative. As already mentioned, clusters create less damage than atomic ions and maintain a higher depth resolution for a given energy. Winograd et al. [130] have shown that the use of C_{60}^+ in SNMS depth profiling of NiCr layers gives a much better depth resolution than Ga^+ ions (see Figure 10.17).
- A detailed discussion of all these parameters can be found in the literature [146].

10.4.5.3.5 Basic Interferences in SIMS and SNMS Depth Profiling
The mass spectra often contain peaks of adventitious ions such as Na^+, K^+, Al^+, C^-, O^-, OH^-, and Cl^- [147,148].

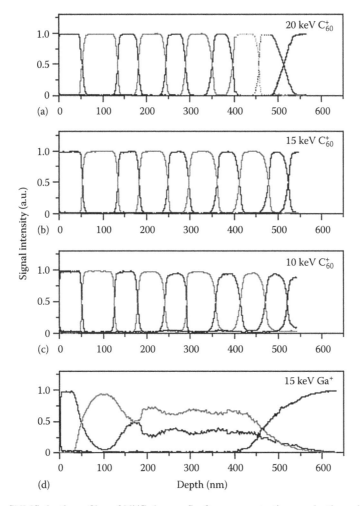

FIGURE 10.17 SNMS depth profiles of Ni/Cr layers. Surface concentration vs. depth, under bombardment with (a) 20 keV C_{60}^+, (b) 15 keV C_{60}^+, (c) 10 keV C_{60}^+, and (d) 15 keV Ga^+. It is apparent that C_{60}^+ bombardment provides better depth resolution, and less interlayer mixing than Ga bombardment. (From Sun, S., Szakai, C., Roll, T., Mazarov, P., Wucher, A., and Winogard, N., *Surf. Interface Anal.*, 36, 1367, 2004. With permission.)

10.5 DISCUSSION AND GENERAL CONCLUSIONS

Every analytical technique discussed in this chapter has its own advantages and disadvantages, arising from both the physical processes involved and technical requirements. The theory behind each of the techniques would in itself fill a book, whereas the present purpose was to set out briefly the depth profiling approach as it related to each one. Indeed, depth profiling, sputtering, and depth resolution have themselves been the subjects of individual investigation. The analysis of specific materials, and processes, such as biomaterials, nanomaterials, composites, catalysts, corrosion, adhesion, etc., are the subject of individual chapters in this book.

To complete the brief overview, a selection is given below of some of the typical problems encountered in depth profiling, with proposed solutions. Finally, some basic questions are posed to help the analyst choose the appropriate analytical method.

10.5.1 Some Typical Problems Encountered in the Sputter Profiling of Certain Types of Sample and Their Solutions

1. The surface of the sample is rough. The effects of roughness can be eliminated by mechanical and electrochemical polishing (if possible), and normal ion beam incidence, and/or noble gas ions.
2. The surface of the sample is already smooth. The depth resolution is enhanced by glancing beam incidence, low ion energy, noble gas ions, and/or cluster ions.
3. If the sample is polycrystalline and/or metallic, use a reactive gas ion beam.
4. If heavy element oxide sample, be aware of metal oxidation state, use ARAES or (better) ARXPS (see Table 10.6).
5. If the sample is an insulator, that means that there will probably be positive surface charging of the sample. If XPS or SIMS analysis is to be carried out, use an electron flood gun.
6. If the sample is a thin film and/or has poor thermal conductivity, use XPS (best), SIMS, or AES with a low-current electron beam.
7. If the sample is a metallic alloy with high diffusion coefficients, use a high sputtering rate and a cooled sample holder (if available).
8. If the sample is an organic compound, use XPS or SIMS.
   Note that if C–F bonds are present bond breaking may occur under x-ray irradiation, resulting in F contamination of the vacuum system. When using SIMS, sputtering with clusters is preferred, to minimize fragmentation and maximize yield.
9. If boron analysis is required, use AES, but do not use Ar^+ for sputtering. If there might be spectral interferences with the B 179 eV peak, use Xe^+ instead.
10. If Cr, O, or I are to be analyzed using AES, do not use Xe^+ due to spectral interference.
11. If the sample is a superconductor where decomposition to lower oxidation states is possible, use ARAES or, better ARXPS.

10.5.2 Key Parameters/Considerations for Choice of the Appropriate Analysis Method

1. Is sample damage acceptable?
   Yes: AES, XPS, SIMS, SNMS, GDOES, and FIB-AES are the preferred techniques.
   No: ARAES, ARXPS, and RBS/ERD/EBS should be used.

2. Is the sample conducting?
   Yes: all the techniques are available.
   No: do not use (AR)AES, or any other electron beam technique.

3. Type of information required:
   Quick, quantitative depth profile: GDOES.
   Elemental information: GDOES, XPS, AES, SIMS, SNMS, and RBS/ERD/EBS are suitable.

Chemical and elemental information: XPS is best, SIMS is suitable in some individual cases, and AES if careful lineshape analysis is carried out.
Trace impurities: SIMS, SNMS.
Interdiffusion reactions (in situ): RBS/ERD/EBS.
Interface to be studied: AES, XPS, SIMS (RBS).
Near-surface in-depth analysis: ARAES, ARXPS.
Quantitative depth profile analysis: AES, XPS, GDOES, SNMS.
Good lateral resolution: AES, microspot RBS, small-area XPS, imaging SIMS.
Nanometer range lateral resolution: FEG-AES.
Nanometer range depth resolution: SDP–AES, SIMS, FIB–AES, GDOES.
Organic sample: SIMS, SNMS, XPS.
Isotopic detection: SIMS, SNMS.
Surface analysis: SSIMS, ISS, AES, XPS.

4. Best lateral and depth resolution:

	AES	XPS	SIMS	GDOES	RBS
Lateral	10 nm	3 μm	100 nm to 1 μm	Several mm	Typically several mm
Depth	2–3 monolayers	2–3 monolayers	1 monolayer	<10 nm	<0.1 μm

Finally, without prejudice to the "catalogue" of techniques described in this chapter, the solution of an analytical problem today requires a multitechnique approach that combines complementary information needed to solve complex questions about increasingly advanced materials.

ACKNOWLEDGMENTS

F.R. wishes to thank Profs. Hubin, Terryn, and Vereecken from the Vrije Universiteit Brussel (VUB) for their continuous collaboration in surface analysis and in AES depth profiling. F.R. also thanks F. Elyoussofi for her help in the preparation of this manuscript.

REFERENCES

1. Rivière, J. C. and Myhra, S. (Eds.), *Handbook of Surface and Interface Analysis*, 1st edn. Marcel Dekker, New York, 1998.
2. Briggs, D. and Seah, M. P. (Eds.), *Practical Surface Analysis*, Vol. 1, 2nd edn. Wiley, Chichester, 1990.
3. Briggs, D. and Seah, M. P. (Eds.), *Practical Surface Analysis*, Vol. 2, 2nd edn. Wiley, Chichester, 1992.
4. Rivière, J. C., *Surface Analytical Techniques*, Oxford Science Publications, Oxford, 1990.
5. Briggs, D. and Grant, J. T. (Eds.), *Surface Analysis by Auger and X-Ray Spectroscopy*, Surface Spectra, Ed., IM Publications, Chichester, U.K., 2003.
6. Vickerman, J. (Ed.), *Surface Analysis, the Principal Techniques*, Wiley, New York, 1997.
7. Mathiew, H. J. Thin film and depth profile analysis. In *Topics in Current Physics*, Vol. 37, Oechsner, H. (Ed.), Springer-Verlag, Berlin, 1984.
8. Czanderna, A. W. (Ed.), *Surface Analytical Techniques*, Vol. 1, Elsevier, Amsterdam, 1975.
9. Yates, J. T. (Ed.), *Experimental Innovations in Surface Science*, AIP-Press, Melville, 1997.
10. Dettner, P., *Electrolytic and Chemical Polishing of Metals*, Ordentlich, Tel Aviv, 1988.
11. MacTegart, W. F., *The Electrolytic and Chemical Polishing of Metals in Research and Industry*, Pergamon Press, London, 1959.
12. Bolognesi, A., Botta, C., Mercogliano, C., Porzio, W., Jukes, P. C., Geoghegan, M., Grell, M., Durell, M., Trolley, D., Das, A., and Macdonald, J. E., *Polymer* 2004, *45*, 4133.
13. Tse, Y. Y., Babonnea, D., Michel, A., and Abadias, G., *Surf. Coat. Technol.* 2004, *180–181*, 470.
14. Ueda, M., Gomes, G. F., Abramof, E., and Reuther, H., *Surf. Coat. Technol.* 2004, *186*, 291.

15. Zaumseil, P. and Schroeder, T., *J. Phys. D* 2005, *38*, A179.
16. Chu, W. K., Mayer, J. W., and Nicolet, M. A., *Backscattering Spectrometry*, Academic Press, New York, 1978.
17. Feldman, L. C., Mayer, J. W., and Picraux, S. T., *Materials Analysis by Ion Channeling*, Academic Press, New York, 1982.
18. Tesmer, J. R., Natasi, M., Barbour, J. C., Maggiore, C. J., and Mayer, J. W. (Eds.), *Handbook of Modern Ion Beam Materials Analysis*, Materials Research Society, Pittsburgh, 1995.
19. Barbour, J. C., *Elastic Ion Scattering for Composition Analysis: Rutherford Backscattering Spectrometry (RBS), Elastic Recoil Detection (ERD), and Enhanced-Cross-Section Backscattering Spectrometry (EBS)*, Sandia National Laboratories, Albuquerque, NM, 1998.
20. Baglin, J. E. E. and Williams, J. S., In *Ion Beams for Materials Analysis*, Bird, J. R. and Williams, J. S. (Eds.), Academic Press, New York, 1989.
21. Kammler, D. R., Wampler, W. R., Van Deusen, S. B., King, S. H., Tissot, R. G., Espada, L. I., and Goeke, R. S., *Mechanisms Governing D Loss in ErD$_2$ Films on Kovar with and without a Mo Diffusion Barrier*, Sandia National Laboratories, Albuquerque, NM, 2006.
22. Bonzel, H. P., Breuer, U., and Knauff, O. *Surf. Sci. Lett.* 1990, *237*, L398.
23. Jablonski, A., *Surf. Interface Anal.* 1990, *15*, 559.
24. Jablonski, A., *Surf. Interface Anal.* 1994, *21*, 758.
25. Werner, W. S. M., Gries, W. H., and Stori, H., *Surf. Interface Anal.* 1991, *17*, 693.
26. van de Walle, R., Van Heirhaeghe, R. L., Laflere, W. H., and Cardon, F., *J. Appl. Phys.* 1993, *74*, 1885.
27. Beamson, G. and Alexander, M. R., *Surf. Interface Anal.* 2004, *36*, 323.
28. Reniers, F. and Tewell, C. R., *J. Electron Spectrosc.* 2005, *142*, 1.
29. PHI-ULVAC. Smart 200 Application Notes.
30. Hofmann, S., *Phil. Trans. R. Soc. Lond. A* 2004, *362*, 55.
31. Hofmann, S., *Thin Solid Films* 2001, *398*, 336.
32. Woehner, T., Ecke, G., Rossler, H., and Hofmann, S., *Surf. Interface Anal.* 1998, *26*, 1.
33. Wittmaack, K., *Int. J. Mass. Spectrom. Ion Phys.* 1975, *17*, 39.
34. Hofmann, S., Rar, A., Moon, D., and Yoshihara, K., *J. Vac. Sci. Technol. A* 2001, *19*, 1111.
35. Ostwald, S. and Baunack, S., *Thin Solid Films* 2003, *425*, 9.
36. Werner, H. W. and Boudewijn, P. R., In *Beam Effects, Surface Topography and Depth Profiling in Surface Analysis*, Czanderna, A. W., Madey, T. E., and Powell, C. J. (Eds.), Plenum Press, New York, 1998, p. 355.
37. Hofmann, S., *Reports Progr. Phys.* 1998, *61*, 827.
38. Kempf, J. E. and Wagner, H. H., Chapter 5, Thin film and depth profile analysis. In *Topics in Current Physics*, Vol. 37, Oechsner, H. (Ed.), Springer-Verlag, Berlin, 1984.
39. Hofmann, S. and Zalar, A., *Surf. Interface Anal.* 1987, *10*, 7.
40. Hofmann, S., Erlewin, J., and Zalar, A., *Thin Solid Films* 1977, *43*, 275.
41. Zalar, A. and Hofmann, S., *Nucl. Instr. Meth. Phys. Res. B* 1987, *18*, 655.
42. Zalar, A. and Hofmann, S., *Vacuum* 1987, *37*, 169.
43. Zalar, A. and Hofmann, S., *J. Vac. Sci. Technol. A* 1987, *5*, 1209.
44. Mathieu, H. J., McLure, D. E., and Landolt, D., *Thin Solid Films* 1976, *38*, 281.
45. Seah, M. P. and Lea, C., *Thin Solid Films* 1981, *81*, 257.
46. Hofmann, S., *Surf. Interface Anal.* 1986, *8*, 87.
47. Oechsner, H., *Appl. Phys.* 1975, *8*, 185.
48. Rosendaal, H. E., Sputtering by ion bombardment. In *Topics in Applied Physics*, Vol. 1, Behrisch, R. (Ed.), Springer-Verlag, Heidelberg, 1981, p. 219.
49. Seah, M. P. and Kuhlein, M., *Surf. Sci.* 1985, *150*, 273.
50. Tsunoyama, K., Suzuki, T., Okahasbi, Y., and Kishikada, H., *Surf. Interface Anal.* 1980, *2*, 212.
51. Hunt, C. P., Hoddard, C. T., and Seah, M. P., *Surf. Interface Anal.* 1981, *3*, 157.
52. Lam, N. Q., *Surf. Interface Anal.* 1988, *12*, 65.
53. Shimizu, R., *Nucl. Instr. Meth. Phys. Res. B* 1987, *18*, 486.
54. Hofmann, S., *Mater. Sci. Eng.* 1980, *42*, 55.
55. Kelly, R., *Surf. Interface Anal.* 1985, *7*, 1.
56. Hofmann, S. and Sanz, J. M., *Frez. Z. Anal. Chem.* 1983, *314*, 215.
57. Hofmann, S. and Sanz, J. M., *J. Trace Microprobe Technol.* 1982–1983, *1*, 213.
58. Betz, G. and Wehner, G. K., Sputtering by ion bombardment. In *Topics in Applied Physics*, Vol. II, Behrisch, R. (Ed.), Springer-Verlag, Heidelberg, 1987, p. 11.
59. Mathieu, H. J. and Landolt, D., *Appl Surf. Sci.* 1982, *10*, 100.
60. Hofmann, S. and Sanz, J. M., *Surf. Interface Anal.* 1984, *6*, 78.

61. Kelly, R., *Nucl. Instr. Meth.* 1978, *149*, 553.
62. Storp, S. and Holm, R., *J. Electron Spectr. Rel. Phenom.* 1979, *16*, 183.
63. Malherbe, J. B., Sanz, J. M., and Hofmann, S., *Appl. Surf. Sci.* 1986, *27*, 355.
64. Mitchell, D. F., Sproule, G. I., and Graham, M. J., *Surf. Interface Anal.* 1990, *15*, 487.
65. Kelly, R., *Surf. Sci.* 1980, *100*, 85.
66. Kim, K. S., Baitinger, W. E., Amy, J. W., and Winograd, N., *J. Electron Spectr. Rel. Phenom.* 1978, *5*, 19.
67. Sun, T. S., McNamara, D. K., Ahearn, J. S., Chen, J. M., Ditchek, D., and Venables, J. D., *Appl. Surf. Sci.* 1980, *5*, 406.
68. Chuang, T. J., Brundle, C. R., and Wandelt, K., *Thin Solid Films* 1978, *53*, 19.
69. Konno, K. and Nagayanna, M., In *Passivity of Metals*, Frankenthal, R. P. and Kruger, J. (Eds.), The Electrochemical Society, Princeton, NJ, 1978.
70. Sanz, J. M., Dissertation, University of Stuttgart, Stuttgart, Germany, 1982.
71. Holm, R. and Storp, S., *Appl. Phys.* 1977, *12*, 101.
72. Hofmann, S., Zalar, A., Cirlin, E.-H., Vajo, J. J., Mathieu, H. J., and Panjan, P., *Surf. Interface Anal.* 1993, *20*, 621.
73. Zalar, A., *Thin Solid Films* 1985, *124*, 223.
74. De Boeck, K., Thesis, Vrije Universiteit Brussel, Belgium, 1982.
75. Hofmann, S. and Zalar, A., *Surf. Interface Anal.* 1994, *21*, 304.
76. Tougaard, S., *Surf. Interface Anal.* 1988, *11*, 453.
77. Shirley, D. A., *Phys. Rev. B* 1972, *5*, 4709.
78. Leclercq, G. and Pireaux, J. J., *J. Electron Spectr. Rel. Phenom.* 1995, *71*, 179.
79. Leclercq, G. and Pireaux, J. J., *J. Electron Spectr. Rel. Phenom.* 1995, *71*, 165.
80. Leclercq, G. and Pireaux, J. J., *J. Electron Spectr. Rel. Phenom.* 1995, *71*, 141.
81. Seah, M. P. and Brown, M. T., *J. Electron Spectr. Rel. Phenom.* 1998, *95*, 71.
82. Artyushkova, K. and Fulghum, J., *J. Electron Spectr. Rel. Phenom.* 2001, *121*, 33.
83. Do, T. and McIntyre, N., *Surf. Interface Anal.* 1999, *27*, 1037.
84. Do, T., McIntyre, N., Harshman, R., Lundy, M., and Splinter, S., *Surf. Interface Anal.* 1999, *27*, 618.
85. Stickle, W. F. and Stobol, P. E., *Surf. Interface Anal.* 1992, *19*, 165.
86. Oswald, S. and Baunack, S., *Surf. Interface Anal.* 1997, *25*, 942.
87. Reniers, F., Hubin, A., Terryn, H., and Vereecken, J., *Surf. Interface Anal.* 1994, *21*, 483.
88. Van Lier, J., Baretzky, B., Zalar, A., and Mittemeijer, E. J., *Surf. Interface Anal.* 2000, *30*, 124.
89. Bubert, H. and Mucha, A. *Surf. Interface Anal.* 1992, *19*, 187.
90. Asteman, H., Norling, R., Svensson, J.-E., Nylund, A., and Nyborg, L., *Surf. Interface Anal.* 2002, *34*, 234.
91. Davis, L. E., MacDonald, N. C., Palmberg, P. W., Riach, G. E., and Weber, R. E., *Handbook of Auger Electron Spectroscopy*, 2nd ed. Physical Electronics Industries, Inc., Eden Prairie, Minnesota, 1976.
92. Shiokawa, Y., Isida, T., and Hayashi, Y., *Auger Electron Spectra Catalogue—A Data Collection of Elements*, Anelva Corp., Tokyo, Japan, 1979.
93. Sekine, T., Nagasawa, Y., Kudoh, M., Sakai, Y., Parkes, A. S., Geller, J. D., Mogami, A., and Hirota, H., *Handbook of Auger Electron Spectroscopy*, JEOL, Tokyo, Japan, 1982.
94. Wagner, C. D., Riggs, W. M., Davis, L. E., Moulder, J. F., and Muilenberg, G. E., *Handbook of X-Ray Photoelectron Spectroscopy*, Perkin-Elmer Corp., Eden Prairie, Minnesota, 1979.
95. Tougaard, S., *Surf. Interface Anal.* 1997, *25*, 137.
96. Shirley, D. A., *Phys. Rev. B* 1972, *5*, 4709.
97. Seah, M. P., *Surf. Sci.* 1999, *420*, 285.
98. Seah, M. P., Gilmore, I., and Spencer, S. J., *Surf. Sci.* 2000, *461*, 1.
99. Sekine, T., Hirata, K., and Mogami, A., *Surf. Sci.* 1983, *125*, 565.
100. Tanuma, S., Sekine, T., Yoshihara, K., Shimizu, R., Homma, T., Tokutaka, H., Goto, K., Uemura, M., Fujita, D., and Kurokawa, A., *Surf. Interface Anal.* 1990, *15*, 466.
101. Reniers, F., *Surf. Interface Anal.* 1995, *23*, 374.
102. Seah, M. P. and Gilmore, I., *Surf. Interface Anal.* 1998, *26*, 908.
103. Pons, F., Le Hericy, J., and Langeron, J. P., *Surf. Sci.* 1977, *69*, 565.
104. Gallon, T. E., *Surf. Sci.* 1969, *17*, 486.
105. Reniers, F., Jardinier-Offergeld, M., and Bouillon, F., *Surf. Interface Anal.* 1991, *17*, 343.
106. Mathieu, H. J. and Landolt, D., *Appl. Surf. Sci.* 1982, *10*, 455.
107. Winchester, M. R. and Payling, R., *Spectrochim. Acta B* 2004, *59*, 607.
108. Marcus, R. K., *J. Anal. Atom. Spectrom.* 1993, *8*, 935.
109. Ohashi, Y., Yamamoto, Y., Tsuyonama, K., and Kishidaka, H., *Surf. Interface Anal.* 1979, *1*, 53.

110. Karlsson, J., Horonstrom, S. E., Klang, H., and Nilsson, J. O., *Surf. Interface Anal.* 1994, *21*, 365.
111. Bengtson, A., *Spectrochim. Acta B* 1984, *40*, 631.
112. Bengtson, A. and Lundholm, M., *J. Anal. Atom. Spectrom.* 1988, *2*, 537.
113. Bengtson, A., Eklund, A., Lundholm, M., and Saric, A., *J. Anal. Atom. Spectrom.* 1990, *5*, 563.
114. Hoffmann, V., Dorka, R., Wilken, L., Hodoroaba, V. D., and Wetzig, K., *Surf. Interface Anal.* 2003, *35*, 575.
115. Shimizu, K., Habazaki, H., Skeldon, P., Thompson, G. E., and Wood, G. C., *Surf. Interface Anal.* 1999, *27*, 998.
116. Mato, S., Thompson, G. E., Skeldon, P., Shimizu, K., Habazaki, H., and Masheder, D., *Corros. Sci.* 2001, *43*, 993.
117. Shimizu, K., Payling, R., Habazaki, H., Skeldon, P., and Thompson, G. E., *J. Anal. Atom. Spectrom.* 2004, *19*, 692.
118. Shimizu, K., Habazaki, H., Skeldon, P., and Thompson, G. E., *Surf. Interface Anal.* 2002, *35*, 564.
119. Muller, U., Schittenhelm, M., Schmittgen, R., and Helm, H., *Surf. Interface Anal.* 1999, *27*, 904.
120. Becker, C. H. and Gillen, K. T., *J. Opt. Soc. Am. B* 1985, *2*, 1438.
121. Jede, R., Ganschow, O., and Kaiser, U., Sputtered neutral mass spectrometry. In *Practical Surface Analysis*, Briggs, D. and Seah, M. P. (Eds.), Wiley, Chichester, 1992.
122. Lipinski, D., Jede, R., Ganschow, O., Kaiser, U., and Seifert, K., *J. Vac. Sci. Technol. A* 1985, *3*, 2007.
123. Jede, R., Peters, H., Dunnebier, G., Ganschow, O., Kaiser, U., and Seifert, K., *J. Vac. Sci. Technol. A* 1988, *6*, 2271.
124. Mueller, K. H. and Oeschner, H., *Microchim. Acta* 1983, *10*, 51.
125. Wucher, A., Novak, F., and Reuter, W., *J. Vac. Sci. Technol. A* 1988, *6*.
126. Muller, K. H., Seifert, K., and Willmers, M., *J. Vac. Sci. Technol. A* 1985, *3*, 1967.
127. Tumpner, J., Wilsch, R., and Benninghoven, A., *J. Vac. Sci. Technol. A* 1987, *5*, 1186.
128. Schoof, H. and Oeschner, H., *Proceedings of the 4th International Conference on Solid Surfaces and 3rd ECOSS*, Degav, D. A. and Costa, M. J. (Eds.), Paris, 1981, Vol. II, 1921.
129. Winograd, N., Postawa, Z., Cheng, J., Szakal, C., Kozole, J., and Garrison, B., *Appl. Surf. Sci.* 2006, *252*, 6836.
130. Sun, S., Szakai, C., Roll, T., Mazarov, P., Wucher, A., and Winograd, N., *Surf. Interface Anal.* 2004, *36*, 1367.
131. Wittmaack, K., *Surf. Sci.* 1981, *112*, 168.
132. Wittmaack, K., *Surf. Sci.* 1983, *126*, 573.
133. Tsunoyama, K., Ohashi, Y., Suzuki, T., and Tsuruoka, K., *Jap. J. Appl. Phys.* 1974, *13*, 1683.
134. Tsunoyama, K., Suzuki, T., Ohashi, Y., and Kishidaka, H., *Surf. Interface Anal.* 1980, *2*, 212.
135. Wittmaack, K. and Poker, D. B., *Nucl. Instrum. Meth. B* 1990, *47*, 224.
136. Williams, P. and Baker, J. E., *Nucl. Instrum. Meth.* 1981, *182/183*, 15.
137. Boudewijn, P. R., Akerboom, H. W. P., and Kempeneers, M. N. C., *Spectrochim. Acta B* 1984, *39*, 1567.
138. Hues, S. M. and Williams, P., *Nucl. Instrum. Meth. B* 1986, *15*, 206.
139. Deline, V. R., Reuter, W., and Kelley, R., *Secondary Ion Mass Spectrometry, SIMS V.* In *Springer Series in Chemical Physics*, Vol. 44, Benninghoven, A., Colton, R. J., Simons, D. S., and Werner, H. W. (Eds.), Springer, Berlin, 1986, p. 299.
140. Wittmaack, K., *App. Phys. Lett.* 1986, *48*, 1400.
141. Boudewijn, P. R. and Vriezema, C. J., In *Secondary Ion Mass Spectrometry, SIMS VI*, Benninghoven, A., Huber, A. M., and Werner, H. W. (Eds.), Wiley, Chichester, 1988, p. 499.
142. Zalm, P. C., *J. Appl. Phys.* 1983, *54*, 2660.
143. Hechtl, E. and Bohdansky, J., *J. Nucl. Mater.* 1985, *133–134*, 301.
144. Dowsett, M. G., *Appl. Surf. Sci.* 2003, *203*, 5.
145. Hofmann, S., *Philos. Trans. R. Soc. Lond. A* 2004, *362*, 55.
146. Wittmaack, K., *Surf. Interface Anal.* 1994, *21*, 323.
147. Benninghoven, A., Werner, H. W., and Rudenauer, F., *Secondary Ion Mass Spectrometry*, Vol. 55, Secs. 5.1.4.1 and 5.2.6, Wiley, Chichester, 1987,
148. Clegg, J. B., In *Secondary Ion Mass Spectrometry, SIMS VI*, Benninghoven, A., Huber, A. M., and Werner, H. W. (Eds.), Wiley, Chichester, 1987, p. 689.

11 Characterization of Nanostructured Materials

Matthias Werner, Alison Crossley, and Colin Johnston

CONTENTS

11.1 INTRODUCTION

Nanotechnology is a very general term and in some of the media has come to be used as a buzzword. Currently, there is no definition of the term "nanotechnology" that is accepted worldwide. Most fundamental physical properties change if the geometrical size in at least one dimension is reduced to a critical value below 100 nm, depending on the material itself. Such changes can in fact allow tuning of the physical properties of a macroscopic material if it consists of nanoscale building blocks whose size and composition can be controlled. Each property has a critical length scale, and if a nanoscale building block is smaller than that critical length, then the fundamental physics of that property change. By altering the sizes of those building blocks, controlling their internal and surface chemistry, their atomic structure, and their assembly, it is possible to engineer properties and functionalities in completely new ways, thus opening up potential future markets.

In this chapter, the term nanotechnology will be taken to mean

- All products with a controlled geometrical size in which at least one functional component is of dimension below 100 nm, allowing physical, chemical, or biological effects to be available that could not be so achieved above the critical dimensions without loss of performance
- Equipment for analytical or manipulatory purposes that allows controlled fabrication, movement, or measurement resolution, with a precision below 100 nm

According to this definition, a nanotechnology product should contain at least one functional component fulfilling one of the above-mentioned criteria. Obviously, in only a few cases would do such a product consist of nanoscale building blocks alone, that is, without any macroscopic dimensions. Therefore, a "nanotechnology product" can be defined as one which contains at least one unit with a nanotechnological functionality that can be sold commercially in the market place. For example, the functional component of a magnetic disk drive is the read/write head that is based on the giant magnetoresistance (GMR) effect. However, the smallest unit that is commercially available is the magnetic disk drive and not the read/write head. The value of the nanotechnology contribution to such a product is difficult to estimate. It may be relevant only to the market price of the end product. Market data are most often based on the market prices of the smallest commercially available units with functional nanotechnology components. According to various sources [1–3], the nanotechnology market is currently in the range of several $100 billion and is expected to increase to more than $1 trillion by 2015 (Figure 11.1). However, it should be borne in mind that different definitions of the term "nanotechnology" have been used in order to arrive at these estimates.

These market figures do not indicate either in which nanotechnology area the most rapid changes can be expected, or the likelihood of the projected market figures being realized. Nanomaterials in the raw state account for a significant niche in the current market. The niche is known as the classical nanomaterials market consisting of materials such as carbon blacks, dyes, pigments, and zeolites. New nanomaterials include metal oxides, fullerenes, and carbon nanotubes (CNTs), which currently account for a relatively small market in the range 5%–15% of the whole nanomaterials market, but they are becoming of increasing significance. However, the integration into products of these new materials, based on good basic and applied science, is the key if the true potential of nanotechnology is to be fulfilled.

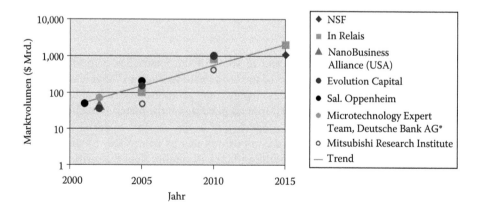

FIGURE 11.1 Market predictions for the "whole nanotechnology market." A "nanotechnology product" is defined as the smallest unit with a functional nanotechnology component that can be commercially sold in the marketplace. (From NMTC Report, *Nanotechnologie und Nanomaterialien—Applikationen und Marktpotenzial Verfahrenstechnik*, April 2005. With permission.)

The International Standards Organization (ISO) has defined "nano" as a deliberately engineered structure with at least one dimension less than 100 nm and have suggested a nomenclature [4] (note that other definitions are used elsewhere in this book):

1-D—where one dimension is less than 100 nm, for example, thin films
2-D—where two dimensions are each less than 100 nm, for example, nanofibers, nanotubes
3-D—where all three dimensions are less than 100 nm, for example, nanoparticles

Deliberately constraining the dimensions of structures to 100 nm or less, in order to introduce novel properties, also presents considerable challenges to their characterization. Many of the conventional analytical techniques employed for characterizing materials are unable to access information at that level of spatial resolution; for example, the physical dimensions of a nanoparticle might be an order of magnitude below the diffraction limit of visible light, making it impossible to image it directly with photon microscopy. Established tools such as electron and scanning probe microscopy (SPM), and x-ray photoelectron spectroscopy (XPS), however, all have a role to play in the characterization of nanostructured materials.

There are many examples of materials that are nanoscale in one dimension, for example, thin films in electronics, or surface layers for improved tribological properties. Characterization of this type of nanostructure is covered elsewhere in this book (Chapters 3, 5, and 12 in particular). However, as well as the techniques described there, there are also novel techniques such as the atom probe, which can be used to produce detailed information about the structure of extremely thin film structures such as those encountered in GMR devices.

Similarly, materials that are nanoscale in two dimensions have also been considered in detail elsewhere in this book, in particular with respect to the use of electron microscopy (Chapters 6 and 9). Sometimes, techniques that are generally not considered in a nano-context, for example, Raman spectroscopy, can be used to obtain additional structural information about such materials as CNTs.

This chapter focuses on the characterization of materials that are nanoscale in three dimensions; more specifically, nanoparticles. These can exist in single, fused, aggregated, or agglomerated forms with spherical, elongated, or irregular shapes. They can be presented for characterization as powders or dispersed in a fluid. They can also be introduced into other materials, such as a polymer as a nanocomposite. Obviously, the form of the nanomaterial has an impact on the choice of characterization method as well as on handling, safety aspects, and sample preparation.

When making a choice of characterization methods, irrespective of whether the material is on the 1-, 2-, or 3-D nanoscales, consideration must first be given to as to what information is required, and then, based on that, as to which techniques might be appropriate. In general, characterization techniques need not only to be easy to use, relatively rapid, widely available, and cost effective, but also to be able to be applied to unknown samples, which may be a mixture of materials, with a good/high degree of certainty. This is particularly challenging with nanostructured materials. Different methods of analysis may even give different results, so it is always advisable to use as many appropriate techniques as are available before choosing, say, a quality control method. For example, TEM might be the only method available to image nanoparticulates, but a more cost-effective method of particle sizing, such as dynamic light scattering (DLS), might be more convenient for routine use, with occasional cross-calibration. The proposed end use of the nanoparticulate should also be borne in mind. As an example, in oxide-supported metal catalysts, it is important to know the specific surface areas of the active metal catalyst and of the oxide support. It may be less important to know the physical size and shape of the oxide particles, but critically important to be able to image the shape of the metal particle supported on the oxide.

The physicochemical properties of nanomaterials can be determined from their chemical composition, surface structure (including any surface coatings or modifications), size and associated increased surface-to-volume ratio, shape, and aggregation behavior. Figure 11.2 shows some of the parameters that should be considered in characterizing nanoparticulate materials [5].

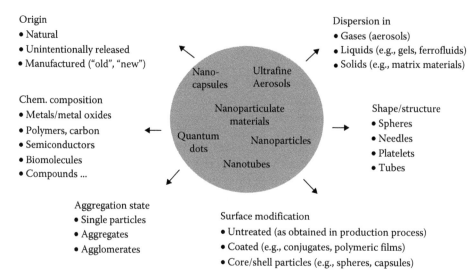

Origin
• Natural
• Unintentionally released
• Manufactured ("old", "new")

Chem. composition
• Metals/metal oxides
• Polymers, carbon
• Semiconductors
• Biomolecules
• Compounds ...

Dispersion in
• Gases (aerosols)
• Liquids (e.g., gels, ferrofluids)
• Solids (e.g., matrix materials)

Shape/structure
• Spheres
• Needles
• Platelets
• Tubes

Aggregation state
• Single particles
• Aggregates
• Agglomerates

Surface modification
• Untreated (as obtained in production process)
• Coated (e.g., conjugates, polymeric films)
• Core/shell particles (e.g., spheres, capsules)

FIGURE 11.2 Characterization parameters for nanoparticulate materials. (From Schulenburg, M., *Nanotechnologie: Innovationen für die Welt von morgen Hrsg.*, 3 aktualisierte Auflage *Koordination: VDI-Technologiezentrum*, W. Luther and G. Bachmann, Eds., BMBF, Bonn/Berlin, 2006.)

The main variables of interest with respect to nanoparticulate characterization are

- Physical properties
  ○ Size, shape, and aspect ratio
  ○ Agglomeration and aggregation behavior
  ○ Surface morphology/topography
  ○ Structure, including crystallinity and defect structure
  ○ Solubility
- Chemical properties
  ○ Bulk elemental/molecular composition
  ○ Phase identity and purity
  ○ Surface composition
  ○ Surface charge
  ○ Interfacial characteristics

Each of these variables may be of significance to properties and performance, and, while numerous techniques exist for their measurement, each technique has its own advantages and disadvantages. Measurements that are specific to a particular property may also be necessary, for example, in order to assess the explosive hazard of a nanopowder, it may be necessary to use calorimetric methods to determine ignition temperature and burn rate as a function of particle size.

11.2 MATERIALS ON THE 1-D NANOSCALE

Techniques for the characterization of thin films are given in detail elsewhere in this book together with many examples of applications (Chapters 3 through 6). However, a specialist technique, the atom probe (an evolution of field-ion microscopy), and its application to the characterization of GMR thin films will be described here.

Atom probe technology is based on the field-ion microscope (FIM), and all atom probes can also perform basic field-ion microscopy. The FIM uses a specimen in the form of a sharp needle, of

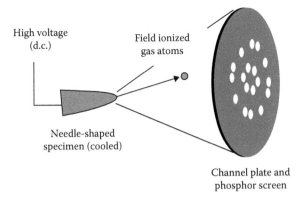

High voltage
(d.c.)

Field ionized
gas atoms

Needle-shaped
specimen (cooled)

Channel plate and
phosphor screen

FIGURE 11.3 Schematic representation of the FIM. The specimen is held in UHV and cooled to temperatures below 100 K. An inert gas, usually helium, is admitted to a pressure of about 10⁻⁹ Pa. A high positive voltage (typically 10–15 kV) is applied to the specimen needle, and because of its shape, a very intense electric field is produced at the tip. Field ionized gas atoms are then accelerated in straight paths to a detector. The operation of both the FIM and the atom probe depends on the physical phenomena occurring very close to the tip as a result of the high electric field (see text). (From Smith, G.D.W. and Cerezo, A., *Handbook of Microscopy*, VCH, Weinheim, Germany, 1997, 775–801.)

radius 25–100 nm at the apex (Figure 11.3) [6]. Specimen fabrication is typically carried out by electropolishing, but more sophisticated techniques involving focused ion beams (FIBs) are used for difficult materials.

When a low pressure (10^{-9} Pa) of an inert gas such as helium is admitted into the FIM chamber, the high field at the apex of the specimen causes ionization of the gas molecules adjacent to the apex surface. The positively charged gas ions so produced are then repelled from the positively charged specimen along straight field lines, to a phosphor screen, where they generate a highly magnified image of the tip. The magnification is typically 10^6, and thus the image on the screen represents a real-space map of the positions of individual atoms on the apex surface.

The atom probe uses time-of-flight mass spectrometry as the main added feature in order to identify single atoms removed from a sharp specimen tip by the effect of very short voltage pulses superimposed on the static large electric field. This process is known as field evaporation. The ionized atoms are repelled from the specimen, just as gas ions are repelled in the FIM. The transit time to the screen, of ions removed from the tip, depends on their atomic mass number, as well as on their charge. Thus the chemical identities can be determined and mapped laterally, as well as layer by layer, since field evaporation strips the surface progressively.

The three-dimensional atom probe (3DAP) combines

- 10^6 magnification of the FIM
- Time-of-flight identification of single ions as in the conventional atom probe
- Position sensitive detection

resulting in an instrument which is able to map the 3-D positions and chemical identities of the majority of atoms within a volume of $20 \times 20 \times 100$ nm³ of a conducting sample.

The evolution of cheap mass-digital data storage has been driven by the development of nanoscale magnetic materials. Such materials rely heavily on the ability to engineer the properties of the magnetic layered structures from which the devices are fabricated. These properties are strongly dependent on the nature of the interfaces between the individual nanoscale magnetic layers, so that knowledge of the interface chemistry is crucial. 3DAP is one of the few techniques available that can characterize these materials. As an example, 3DAP has been used to characterize

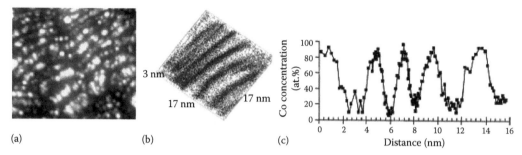

(a)   (b)   (c)

FIGURE 11.4   (a) FIM image showing Co (bright) and Cu (dark) layers; (b) a 3DAP atom map of the Co/Cu multilayer, in which only the Cu atoms are shown for clarity—note the curvature imparted to the map from the physical curvature of the sample; and (c) a selected line scan profile showing the variation in Co composition across the layered structure. (From Larson, D.J., Petford-Long, A.K., Ma, Y.Q., and Cerezo, A., *Acta Mater.*, 52, 2847, 2004. With permission.)

the interfaces in Co/Cu and CoFe/Cu multilayers that form part of the read sensor in magnetic recording heads [7]. A needle is formed from a thin film of Co/Cu, which then acts as the specimen for 3DAP. Figure 11.4 shows the FIM image and the resultant 3DAP atom map of the Co/Cu—note the curvature due to the shape of the sample needle. Such maps can then be converted into profiles, which can be used to determine the degree of intermixing between the layers.

Sample preparation techniques have been developed (e.g., annular FIB milling) that remove the curvature imparted by needle samples, and that demonstrate that true atomic resolution across interfaces can be achieved [8]. Figure 11.5 shows a 3DAP atom map and profile through a CoFe/Cu/CoFe trilayer. The atom map shows qualitatively that there is more intermixing at the CoFe/Cu (i.e., upper) interface than there is at the Cu/CoFe (i.e., lower) interface, resulting in the former being wider than the latter. The strength of the 3DAP technique lies in the fact that in addition to being able to show such effects in a qualitative manner, the extent of the interdiffusion can be measured quantitatively by taking composition profiles through localized regions of the data perpendicular to the local interface plane. Measurement of the interface widths from the compositional profile, shown in

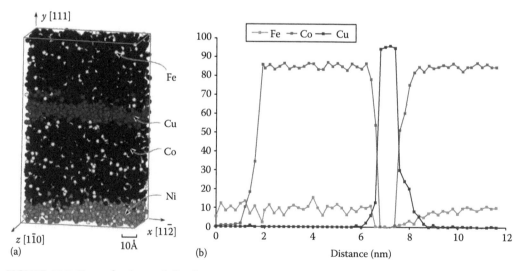

(a)   (b)   Distance (nm)

FIGURE 11.5 (See color insert following page 396.)   (a) 3DAP atom map of Co (blue), Fe (yellow), and Cu (red) atoms in a CoFe/Cu/CoFe trilayer on an NiFe substrate (Ni [green]), and (b) a compositional profile across the same region. (From Zhou, X.W., Wadley, H.N.G., Johnson, R.A., Larson, D.J., Tabat, N., and Cerezo, A., *Acta Mater.*, 49, 4005, 2001. With permission.)

Figure 11.5b, using as a measure 10%–90% of the Cu concentration, gave values of interfacial widths of 1.08 ± 0.18 nm for CoFe on Cu and 0.4 ± 0.14 nm for Cu on CoFe.

11.3 MATERIALS ON THE 2-D NANOSCALE

A good example of a currently topical area of R&D in nanomaterials is that of CNTs. Although market figures differ markedly, estimates of the compound annual growth rates of companies involved in the production of CNTs are in the range from 70% to 300% [8]. Currently, CNTs are primarily research materials, and, like others, are produced in small volumes at high cost. However, commercial products including tennis rackets with CNT reinforcement, and high-aspect-ratio probes for SPM, are beginning to emerge. CNTs could play a pivotal role in the up-and-coming nanotechnology age if their remarkable electrical and mechanical properties can be exploited commercially. Nanotubes offer significant advantages over comparable existing materials, such as carbon fiber, including an impressive tensile strength-to-weight ratio, attractive mechanical properties, and interesting field emission properties. For CNTs to be successful in mass-market applications, large-scale production of them with carefully controlled and predictable properties needs to be implemented. A potential tool for the aid of production control is Raman spectroscopy, possibly a future in-line quality control tool.

11.3.1 RAMAN SPECTROSCOPY

When intense monochromatic laser light impinges upon a material, light scattering can occur in all directions. When the scattered light has the same frequency as the original light source, the scattering is elastic and is known as Rayleigh scattering. However, inelastic scattering can also occur and is known as Raman scattering, where a frequency/wavelength shift either to higher or lower (Stokes and anti-Stokes) energy is observed, depending on the vibrational state of the molecules (Figure 11.6) [10]. Raman scattering can be used to elucidate the structure of molecules and crystals.

By far the stronger of the two inelastic processes is Stokes scattering, in which the photon is scattered to lower energy (wavelength shifted toward the red end of the spectrum). The reason for this is that at room temperature the population state of a molecule is principally in its ground vibrational state, so that the Stokes transition predominates.

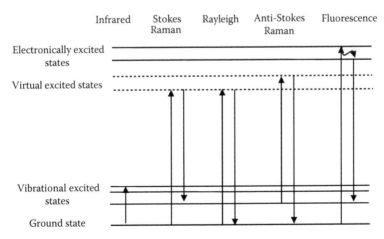

FIGURE 11.6 Energy diagram of light-induced excitation. (From Wikipedia, http://en.wikipedia.org/wiki/Raman_spectroscopy)

A small number of molecules will be in a higher vibrational level, and hence the scattered photon can be scattered to a higher energy (a gain in energy—a blue-shifted wavelength). This is the much weaker anti-Stokes Raman scattering.

The energy loss or gain experienced by a photon on interaction with a molecule is characteristic of the vibrational modes of a particular bond in that molecule. Not all modes will be observable with Raman spectroscopy (depending upon the symmetry of the molecule), but sufficient information will usually be present to enable a definitive identification and characterization of the molecular structure. Thus, the energy shift for a C–H bond is different from that found for a C–O bond, and different again from that associated with a metal–O bond. When an entire spectrum is recorded, spectral features can be identified as being associated with the various different bonds and vibrations.

The key property of a molecule with respect to the scattering of a photon is the polarizability, α, which represents the ability of an applied electric field, E, to induce a dipole moment, μ_{in}, in an atom or molecule, and is given by the expression

$$\mu_{in} = \alpha E \tag{11.1}$$

For example, large atoms such as xenon have a strong polarizability because their electron clouds distant from the nucleus are relatively easy to distort with an applied electric field. Hydrogen atoms, which are smaller and more compact, have a small polarizability. Polarizabilities for atoms are isotropic whereas polarizabilities for molecules may vary with position within the molecule, depending on the inter- and intramolecular symmetry.

A photon can act as an induced electric field, thus Equation 11.1 applies when light interacts with atoms and molecules. However, as a result of the various degrees of freedom associated both with the whole molecule and with its component parts, certain actions such as rotation, stretching, and vibration, can cause the polarizability of the molecule to change. At the equilibrium nuclear geometry of a molecule, the polarizability is some value α_0. At some perturbation, Δr, of the equilibrium geometry, the instantaneous polarizability α is given by

$$\alpha = \alpha_0 + \frac{\partial \alpha}{\partial r} \Delta r \tag{11.2}$$

where the derivative $\frac{\partial \alpha}{\partial r}$ represents the change in polarizability with the instantaneous change in position. If the molecule is vibrating or rotating in a sinusoidal fashion, Δr can be written as a sinusoidal function in terms of the frequency of the vibration, v, and the time, t. viz.

$$\Delta r = r_{max} \cos(2\pi v t) \tag{11.3}$$

where r_{max} is the maximum vibrational amplitude. Incident light that has a particular frequency, v_{in}, will induce an electric field, E, that also has sinusoidal behavior, given by

$$E = E_{max} \cos(2\pi v_{in} t) \tag{11.4}$$

where E_{max} is the maximum electric field amplitude. Thus, substituting α, Δr, and E into Equation 11.1 gives

$$\mu_{in} = \left[\alpha_0 E_{max} \cos(2\pi v_{in} t) \right] + \left[E_{max} r_{max} \frac{\partial \alpha}{\partial r} \cos(2\pi v r) \cos(2\pi v_{in} t) \right] \tag{11.5}$$

which can be reduced to

$$\mu_{in} = \left[\alpha_0 E_{max} \cos(2\pi v_{in} t) \right] + \left(\frac{E_{max} r_{max}}{2} \cdot \frac{\partial \alpha}{\partial r} \left\{ \cos\left[2\pi(v_0 + v) \right] + \cos\left[2\pi(v_0 - v) \right] \right\} \right) \tag{11.6}$$

This final expression can be used as the dipole moment operator in a transition moment integral. The first term contains the variable v_{in}, which is the frequency of the incoming light, and relates to the outgoing, scattered photon that has the same frequency as the incoming photon. That is, it describes Rayleigh scattering. The second term contains two cosines. One contains the variable $(v_0 + v)$, which relates to an outgoing, scattered photon that increases in frequency by some amount v, which is the frequency of the molecular vibrational mode. The other cosine term contains the variable $(v_0 - v)$, relating to a scattered photon negatively frequency-shifted by the same amount, v. These two terms correspond to Raman scattering, and show that incoming photons will suffer frequency shifts, up and down, by amounts corresponding to particular characteristic vibrational modes of the molecules. This is the basis of Raman spectroscopy and gives a gross selection rule for what are called Raman-active motions. The derivative, $\frac{\partial \alpha}{\partial r}$, is the change in the polarizability with nuclear position. If that derivative equals zero, the entire second term is zero and there will be no Raman scattering. Thus the gross selection rule is that a molecular motion will be Raman active only if the motion is associated with a change in polarizability.

11.3.2 RAMAN SPECTROSCOPY OF CNTs

Even though conventional Raman microprobe spectrometers cannot provide laser spots smaller than ~0.75 μm in diameter, it should be remembered that the spectra actually arise from the vibrations of individual chemical bonds. This intrinsic "nanoprobing" makes Raman spectroscopy sensitive to the short-range structure. Also, since almost no sample preparation is necessary, it is often possible and cost effective to use the technique as a basic nanoprobe.

Raman spectroscopy has been used to investigate all carbon-containing materials for many years [11]. The allotropy presented by carbon provides a rich vein for research into relating spectral features to molecular and electronic structure. CNT vibrational properties have been widely investigated and still represent a "hot" topic [12–18].

The basic characteristics of CNT, such as crystalline quality and phase purity, can be monitored using quality indicators derived from the Raman spectra. The Raman spectrum of CNT consists of bands in a low energy range, due to the radial breathing modes (RBM), dispersive or D-bands around 1300–1400 cm^{-1}, and so-called G-bands near 1600 cm^{-1} due to tangential modes. The RBM modes are sensitive to the diameter and chirality/helicity of small diameter tubes, and also to the excitation wavelength, with narrow resonance profiles (<100 meV) and a small broadening and energy shift (when the CNTs are in bundles) [19].

The wave numbers v_{RBM} of the RBM modes depend inversely on the tube diameter d, according to the relationship

$$v_{RBM} = \frac{A}{d} + B \tag{11.7}$$

However, various values have been reported for the constants A and B, the variations often being attributed to environmental effects, such as whether the SWCNTs are present as individual tubes covered with a surfactant or isolated on a substrate or in the form of bundles [20]. But for isolated, freestanding, SWCNTs, where the chiral vector and therefore the diameter have been determined independently by electron diffraction, values of A and B of 204 and 27 cm^{-1} have been established [20–22].

The tangential modes are the most intense high energy modes of SWCNTs and form the so-called G-band, observed near 1600 cm^{-1}, which is close to the position of the G mode in graphite. It consists of two bands, each of symmetric line shape, at 1590 cm^{-1} (G$^+$), and near 1572 cm^{-1} (G$^-$), for tubes with diameters of 1.4 nm. The position of the G$^+$ mode is not sensitive to the tube diameter. The precise position of the G$^-$-band depends on (C/d^2), where C is a constant depending on the semiconducting or metallic properties, and d is the diameter. Coupling with valence electrons in metallic CNTs has a strong influence on the band shape and intensity of the G-bands [23]. For multiwall carbon nanotubes (MWCNTs), the number of walls and the presence of faceted graphitic particles complicate the G-band intensity and band shape. On the higher wave number side of the G-band, a shoulder, due to the so-called D′-band, is also sometimes observed [24].

Probably the most discussed mode for the characterization of functionalized SWCNTs is the D-band, observable at 1300–1400 cm⁻¹, and related to defect-induced double-resonant scattering processes involving elastic scattering of electrons by structural defects. In their early work on Raman spectroscopy of graphite, Tuinstra and Koenig [27] showed that the intensity of this band scales linearly with the inverse of the crystallite size, and its appearance was interpreted as being due to a break down of the k-selection rule. However, for graphite and SWCNTs, the position of this mode, as well as its intensity, with respect to the G-band, depends on the laser energy. The D-band also displays a dependence on the diameters of the SWCNTs, which has been interpreted as a double-resonance process in which not only one of the direct, k-conserving electronic transitions but also the emission of the phonon is a resonant process. In contrast to single-resonant Raman scattering, in which only phonons around the center of the Brillouin zone ($q = 0$) are excited, the phonons that contribute to the D-band exhibit a nonnegligible q-vector. Overall, k-conservation for the Raman scattering process is fulfilled by elastic defect scattering. Taking the electronic as well as the phonon dispersion relation into account, this double-resonance theory is able to explain the intensity of the D-band as a function of laser energy, as well as the fine structure observed in the D-band spectra [27].

The overtone of the D-band, the G′-band (sometimes called the D*-band), observed at 2600–2800 cm⁻¹, does not require explanation by defect scattering, since two phonons with q and $-q$ are excited. This mode is therefore observed independently of defect concentration. In fact, the intensity ratio I_D/I_G is thought by some authors [28,29] to be a sensitive measure of the defect concentration in CNTs. An assignment of additional modes visible in the Raman spectra of SWCNTs can be found in the review articles by Saito et al. [30] and Dresselhaus et al. [14].

Figure 11.7 shows a typical spectrum from an SWCNT, excited by red light of wavelength 633 nm from a HeNe laser, in which the RBM, D-, and G-bands, as well as the two-phonon overtone

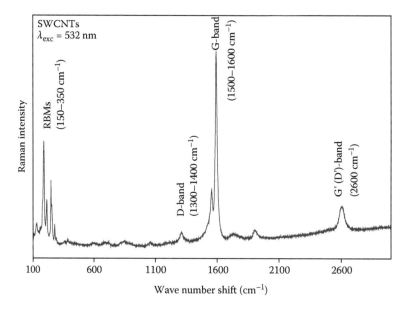

FIGURE 11.7 Raman spectrum from an SWCNT, showing the RBM, D-, and G-bands and the overtone band at ~2600 cm⁻¹. The exciting laser wavelength was 633 nm. (From Rao, A.M., Bandow, S., Ritcher, E., and Eklund, P.C., *Thin Solid Films*, 331, 141, 1998.)

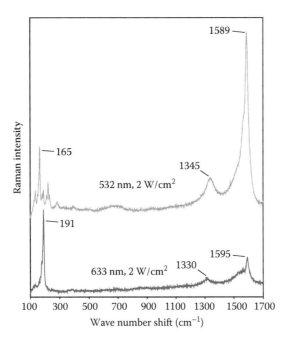

FIGURE 11.8 Room temperature Raman spectra from purified SWCNTs excited at five different laser wavelengths and power densities. (From Rao, A.M., Bandow, S., Ritcher, E., and Eklund, P.C., *Thin Solid Films*, 331, 141, 1998.)

band at ~2600 cm^{-1}, can be seen [20], while Figure 11.8 shows Raman spectra recorded from purified SWCNTs excited by five different wavelengths [16].

11.4 MATERIALS ON THE 3-D NANOSCALE

The size of a nanoparticle is a fundamental parameter to be determined. However, ascertaining the size of a three-dimensional object, whether it is nanosized or not, is a conundrum in that one number will not suffice in general. There is only one 3-D object that can be described by an unique number, and that is a perfect sphere, the number being its radius. The notion of an "equivalent" sphere has been used to describe nanoparticles, that is, a particle is assumed to approximate to a perfect sphere, and its average dimension is then taken as equivalent to the "perfect" radius, but this assumption can lead to problems. Equivalent sphere theory assumes that whatever dimension is measured can then be reduced to that of a sphere. Hence by definition, the theory does not take into account any deviation from a spherical shape, and the "radius" so derived may therefore be a meaningless parameter. For example, a CNT of radius 1 nm and length 100 nm has an equivalent sphere radius of 4.2 nm. In addition, it must be remembered that the derivation of this parameter does not take into account whether the CNT is of the single-wall or multiwall variety.

Furthermore, particulates are rarely monodisperse, and hence it is often necessary to measure the size distribution of particles. This also introduces a number of different ways of reporting similar, but different, data—such as the number, mass, and volume averages. Although these averages are interrelated and can often be converted, the accuracy of the interconversion relies heavily on the statistics of the original measurement. A reliable outcome requires that large numbers of particles must be measured before any interconversion should be attempted. An excellent treatment of the statistics relating to number, mass, and volume counting is presented

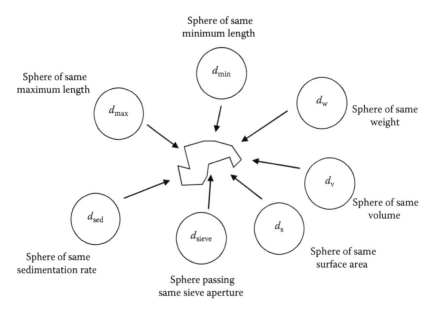

FIGURE 11.9 Equivalent measurements that may be used to describe an irregular particulate. (From Rawle, A., *Basic Principles of Particle Size Analysis*, Malvern, V.K., Downloadable application note from www.malvern. com. With permission.)

in Ref. [31]. A number of such equivalent measures are available to describe particulates, as shown in Figure 11.9.

All results, when reported, should outline the measurement method used, to aid comparison, and to allow the reader to draw appropriate conclusions. Even if the particulate is measured in an electron microscope only those particles which end up on the TEM grid will be "seen." The average particle size or distribution quoted will depend both on the how the particles arrived there, and on the number of particles that have been imaged and measured. Obtaining a statistically sound particle size distribution by this method can be both costly and time consuming, but it may be the only technique which can give the primary particle size. Bear in mind that even with the TEM what is observed is a 2-D projection of the particle. It is necessary to be aware that information about particle size obtained from different characterization methods may be related to different properties of that particle, and all the methods may therefore not give the same answer. Few techniques (with the exception of TEM) can measure the size of the smallest nanoparticles (~1 nm), so that the latter may well be overlooked by most methods. Similarly, if the nanoparticles are agglomerated the inferred size distribution may be distorted.

There are many complementary methods for particle sizing of nanoparticles, some of which will be described here.

Particle sizing techniques can be divided roughly into three categories:

1. Ensemble methods collect mixed data from all the differently sized particles in a sample at the same time, and then digest the data to extract a distribution of particle sizes for the entire population. The most common ensemble methods which collect information on nanoparticles dispersed in a liquid are photon correlation spectroscopy (PCS) (also known as DLS) and backscattering spectroscopy (e.g., Ref. [32]).
2. Separation methods all apply an outside separation force to the particles in a distribution to separate physically the particles according to size. Since particles of different sizes are actually physically separated, the problems of accurate characterization of individual particles (counting methods) and of calculating a distribution from mixed data (ensemble techniques) are reduced or eliminated. The accuracy of these methods depends upon whether the particles react to the separation force as expected, and their effective resolution depends

on how completely the particles are separated according to size. Common separation techniques include sieves, gravitational sedimentation, the disk centrifuge, capillary hydro-dynamic fractionation, sedimentation field flow fractionation, and others. The disk centrifuge is the method of choice for nanoparticles [33].

3. Counting methods all characterize the sample distribution one particle at a time, basically by accumulating counts of particles with similar sizes. Some common counting methods are the electrozone counter, the light counter, the time-of-flight counter, and electron or scanning probe microscopies. In each of these methods, particles are classified and placed in "size bins," one particle at a time. In all cases, it is necessary to ensure that multiple particles are not counted together, and thus cause errors in the reported size distribution (e.g., due to "co-incident counting"). Their accuracy and resolution depend on how accu-rately the size of each particle can be characterized during the (usually) very brief time that it is counted. Counting methods will not be described in this section since some, for exam-ple, TEM and SPM, are described elsewhere in this book (Chapters 6 and 5, respectively).

Two techniques used to gather specific information will now be discussed, that is, zeta potential—most commonly associated with PCS instrumentation—and surface area measurement from isother-mal gas adsorption—the so-called BET analysis.

11.4.1 Photon Correlation Spectroscopy or Dynamic Light Scattering

For particles or agglomerates of size less than 1000 nm, DLS is often the method of choice. DLS is based on the dependence of Brownian motion on particle size. Brownian motion is the random movement of particles in a solution due to collisions with the solvent molecules surrounding them. Larger particles move more slowly and over shorter distances, whereas smaller particles move faster and further. Autocorrelation of successive speckle patterns (see below) allows the particle size dis-tribution to be determined within the measurement window; larger particles will move slowly and hence successive speckle patterns will change slowly, whereas small particles will move more quickly and successive patterns will change rapidly. Thus a correlation function can be constructed which relates to particle size distribution. Temperature stability is an important consideration when performing DLS measurements as liquid viscosity changes with temperature.

11.4.1.1 Theory

Particle diameters can be calculated from the translational diffusion coefficient using the Stoke–Einstein equation:

$$d(H) = \frac{kT}{3\pi\eta D} \tag{11.8}$$

where
  $d(H)$ is the hydrodynamic diameter
  D is the translational diffusion coefficient
  k is the Boltzmann constant
  T is the absolute temperature
  η is the viscosity

The diameter so measured is based on the mode of diffusion of a particle within a fluid, and is referred to as the hydrodynamic diameter (see Figure 11.10) [34]. It is larger than the true particle diameter as determined by other techniques, for example, by SEM, whose consideration may need to be taken into account since the properties of biological systems are sensitive to the hydrodynamic diameters of particles.

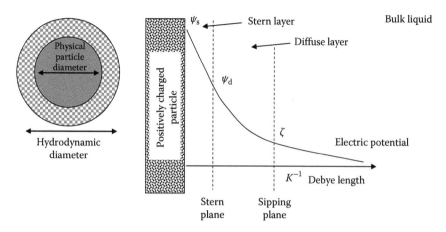

FIGURE 11.10 Schematic showing the apparent increase in a particle diameter caused by the presence of an electrical double layer, leading to measurement of the so-called hydrodynamic diameter. (From Measurement and Interpretation of Electrokinetic Phenomena, International Union of Pure and Applied Chemistry, Technical Report, *Pure Appl. Chem.*, 77, 1753, 2005.)

The hydrodynamic diameter of a particle corresponds to the diameter of a sphere having the same translational diffusion as that particle, which depends on the size of the particle core, its surface structure, the concentrations, and the types of ions in the medium. Factors that affect the measurement of the hydrodynamic diameter include ionic concentration of the medium (which changes the electric double layer, defined by the Debye length, K^{-1}), changes to the surface structure, and extent of nonsphericity.

11.4.1.2   PCS/DLS in Practice

In DLS, the speed at which particles diffuse due to Brownian motion is measured from the rate at which the intensity of the scattered light fluctuates when detected using a suitable optical arrangement. In an imaginary situation in which suspended particles are stationary, that is, with no Brownian motion, then if the cell containing them is irradiated with suitable laser light and the light scattered from the particles is projected onto a frosted glass screen, a so-called classic "speckle" pattern would be seen. The arrangement is shown schematically in Figure 11.11. Because the whole system has been assumed to be stationary, the speckle pattern would also be stationary in both speckle size and position. In the pattern, the dark spaces correspond to destructive interference, that is, where the vector additions of the scattered light are out of phase and cancel each other (Figure 11.12a). On the other hand, the bright regions correspond to constructive interference, that is, where the light scattered from the particles arrives in-phase, with consequent phase reinforcement (Figure 11.12b).

When the particles are not stationary, but undergoing Brownian motion, a dynamic speckle pattern would be observed, since the phase addition from the moving particles would then be evolving constantly and forming new patterns. The rate at which these intensity fluctuations occur will depend on the size of the particles. Figure 11.13 illustrates schematically typical intensity fluctuations arising from dispersions of large and small particles. Small particles cause the intensity to fluctuate more rapidly than large ones.

It is possible to measure directly the spectrum of frequencies contained in the intensity fluctuations arising from the Brownian motion of particles, but it is inefficient to do so. A better way is to use a device called a digital autocorrelator. A correlator is basically a signal comparator. It is designed to measure the degree of similarity between two signals, or one signal with itself at varying time intervals.

If the intensity of a signal is compared with itself at a particular point in time, and then again at a much later time, it is obvious that for a randomly fluctuating signal the intensities will not be

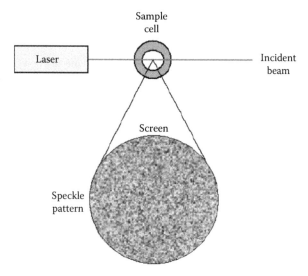

FIGURE 11.11 Schematic representation of a speckle pattern.

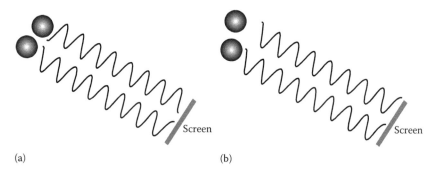

FIGURE 11.12 The signal observed by the detector depends on the result of the phase addition of the scattered light. (a) The two beams interfere destructively canceling each other, resulting in a *decreased* intensity at the detector; and (b) the two beams interfere constructively, resulting in an *increased* intensity at the detector.

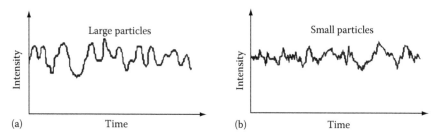

FIGURE 11.13 Typical intensity fluctuations for (a) large and (b) small particles. (From Dynamic light scattering: An introduction in 30 minutes, Downloadable application note from www.malvern.com. With permission.)

related in any way, that is, there will be no correlation between the two signals. Knowledge of the initial signal intensity does not allow the signal intensity at time $t = \infty$ to be predicted, which will be true of any random process such as diffusion.

However, if the intensity of a signal at time t is compared to the intensity of the same signal at an incremental time later $(t + \delta t)$, when δt is significantly shorter than the timescale of evolution of the random dynamics of the system under study, there will be a strong relationship or correlation between the intensities of two signals.

If the signal at t, derived from a random process such as Brownian motion, is compared to the signal at $t + 2\delta t$, there will still be a reasonable correlation between the two signals, but it will not be as good as that at t and $t + \delta t$. The strength of the correlation reduces with time. The period of time δt is usually very small, perhaps nanoseconds or microseconds, and is called the sample time of the correlator, while $t = \infty$ may be of the order of 1–10 ms.

If the signal intensity at t is compared with itself, then there is perfect correlation as the signals are identical. Perfect correlation is indicated by unity (1.00) and no correlation is indicated by zero (0.00). If the signals at $t + 2\delta t$, $t + 3\delta t$, $t + 4\delta t$, etc. are compared with the signal at t, the correlation of a signal arriving from a random source will decrease with time until at some time, effectively $t = \infty$, there will be no correlation. A plot of correlation coefficient versus time is called a correlellogram, as in Figure 11.14. If the particles are large, the signal will change slowly and the correlation will persist for a long time (Figure 11.14a), but if they are small and moving rapidly, then the correlation will reduce more quickly (Figure 11.14b).

Simple examination of a correlellogram can provide a great deal of information about the sample. The time at which the correlation starts to decay significantly is an indication of the mean size of the particles making up the sample, while greater steepness of the slope of the decay line correlates with greater monodispersivity of the particle distribution. Conversely, the more extended the decay, the greater the polydispersity of the particle distribution.

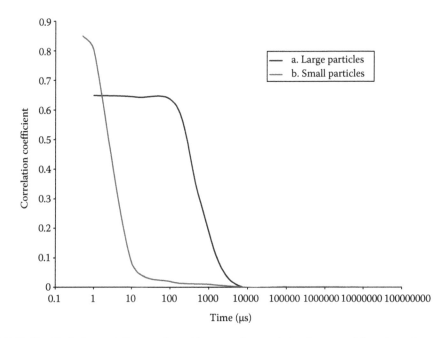

FIGURE 11.14 Typical correlellogram from (a) a sample containing large particles where the correlation signal takes a long time to decay, and (b) a sample containing small particles, where the correlation signal decays more rapidly. (From Measurement and Interpretation of Electrokinetic Phenomena, International Union of Pure and Applied Chemistry, Technical Report, *Pure Appl. Chem.*, 77, 1753, 2005. With permission.)

Particles in a dispersion are in constant random Brownian motion, causing the intensity of scattered light to fluctuate as a function of time. The correlator used in a PCS instrument constructs the correlation function $G(\tau)$ of the scattered intensity, according to

$$G(\tau) = \langle I(t) \cdot I(t+\tau) \rangle \tag{11.9}$$

where τ is the time difference (the sample time treated as a continuous variable) of the correlator.

For a large number of monodisperse particles in Brownian motion, the correlation function is an exponentially decaying function of the correlator time delay τ:

$$G(\tau) = A\left(1 + B\exp^{-2\Gamma\tau}\right) \tag{11.10}$$

where
  A is the baseline [$G(\tau)$ for $\tau \to \infty$] of the correlation function
  B is the intercept on the ordinate axis at $\tau = 0$ of the correlation function

The decay parameter is defined by

$$\Gamma = Dq^2 \tag{11.11}$$

where
  D is the translational diffusion coefficient
  q is given by

$$q = \left(\frac{4\pi n}{\lambda_0}\right)\sin\left(\frac{\theta}{2}\right) \tag{11.12}$$

where
  n is the refractive index of the dispersant
  λ_0 is the wavelength of the laser
  θ is the scattering angle

For polydisperse particle distributions, Equation 11.10 can be written as

$$G(\tau) = A\left[1 + Bg_1(\tau)^2\right]$$

where $g_1(\tau)$ is the sum of all the exponential decay terms contained in the correlation function.

Size can be obtained from the correlation function by using various algorithms. There are two approaches that can be taken, viz.

1. Fit a single exponential to the correlation function to obtain the mean size (z-average diameter) and an estimate of the width of the distribution (polydispersity index) (this is called the Cumulants analysis and is defined in ISO13321 Part 8).
2. Fit multiple exponential terms to the correlation function to obtain the distribution of particle sizes (using software such as non-negative least squares [NNLS] or CONTIN).

The size distribution so obtained is a plot of the relative intensity of the light scattered by particles in various size classes, and is therefore known as an intensity/size distribution.

If the distribution by intensity is a single fairly smooth peak, then there is little point in carrying out the conversion to a volume distribution using the procedure described above [35]. If the optical parameters are correct, that will merely provide a peak of slightly different shape. However, if the plot shows a substantial tail, or more than one peak, then, using appropriate refractive indices for the

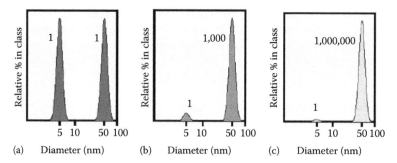

FIGURE 11.15 (a) Number, (b) volume, and (c) intensity, distributions of a bimodal mixture of 5 and 50 nm particles present in equal numbers. (From Measurement and Interpretation of Electrokinetic Phenomena, International Union of Pure and Applied Chemistry, Technical Report, *Pure Appl. Chem.*, 77, 1753, 2005. With permission.)

sample, the intensity distribution can be converted to a volume distribution, which will give a more realistic view of the importance of the tail or of the presence of a second peak. In general terms it is found that the particle size, d, derived from different methods of calculation will vary as follows:

$$d_{intesnity} > d_{volume} > d_{number} \tag{11.13}$$

It is important to note that large particles make a disproportionate contribution to volume, in comparison with small particles. Also, scattering from particles have a $1/d^6$ dependence so small particles are swamped by larger particles in terms of scattered intensity.

A simple way of illustrating the difference between intensity, volume, and number distributions is to consider two populations of spherical particles of diameters 5 and 50 nm, respectively, present in equal numbers (Figure 11.15). If a number distribution of the two particle populations is plotted, then the plot should consist of two peaks (positioned at 5 and 50 nm) in a ratio of unity. If this number distribution were to be converted into a volume distribution, then the ratio of the two peaks would change to 1:1000 (because the volume of a sphere is proportional to $(d/2)^3$. If this were converted further into an intensity distribution, a 1:1,000,000 ratio between the two peaks would be found (because the intensity of scattering is proportional to d^6 [from Rayleigh's approximation]). The significance is that in DLS, the distribution obtained from a measurement is based on intensity.

In contrast to traditional diffraction theory, when the particle size approaches the wavelength of light, the scattered intensity becomes a complex function of particle size with a series of maxima and minima. The application of Mie (or Lorenz–Mie) scattering theory [35] embraces all possible ratios of diameter to wavelength and solves the equations for the interaction of light with matter. In most cases, applications based on Mie theory provide accurate characterization over a range from ~10 nm upward. It does, however, assume that the refractive indices for the material and the medium are accurately known and that the absorption part of the refractive index is also known (an informed estimate may suffice). For the majority of cases these values are either known or can be measured. Figure 11.16 shows the theoretical plot of the log of the relative scattering intensity versus particle size at scattering angles of 173° and 90°.

11.4.1.3 Instrumentation

A typical DLS system consists of six main modules as shown in Figure 11.17. Firstly, a laser (1) provides a light source to illuminate the sample contained in a cell (2). For dilute concentrations, most of the laser beam passes through the sample, but some is scattered by the particles within the sample into all angles. A detector is used to measure the scattered light. In one of the popular instruments, the detector position can be at either 173° or 90° (3), depending on the particular mode to be used for data reduction [35].

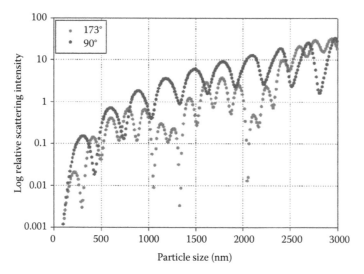

FIGURE 11.16 (See color insert following page 396.) Theoretical plot of the log of the relative intensity of scattering versus particle size at scattering angles of 173° and 90°, assuming a laser beam at a wavelength of 633 nm, real refractive index of 1.59, and an imaginary refractive index of 0.001. (From Dynamic light scattering: An introduction in 30 minutes, Downloadable application note from www.malvern.com. With permission.)

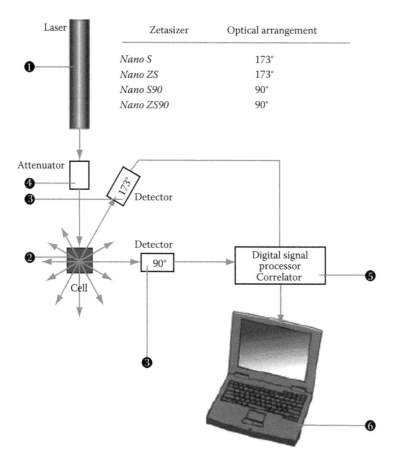

FIGURE 11.17 Optical configuration for DLS. (From Dynamic light scattering: An introduction in 30 minutes, Downloadable application note from www.malvern.com. With permission.)

The intensity of the scattered light must be within a specific range for optimal sensing by the detector. If that intensity is too high, the detector will become saturated, in which case an attenuator (4) can be used to reduce the intensity of the laser source and hence reduce the intensity of scattered light. For samples that do not scatter much light, such as those with particles that are either very small or in low concentration, the amount of scattered light must be increased. In that case, the attenuator will allow higher laser light intensity through to the sample. For some models, the appropriate attenuation is determined automatically by software and covers a transmission range of 100%–0.0003%

The scattering intensity signal from the detector is passed to a digital processing board where a correlator (5) compares the scattered intensity at successive time intervals in order to derive the rate at which the intensity is varying. The information from it is then passed to a computer (6), where software analyzes the data and derives size information.

The advantages and disadvantages of DLS can be summarized as follows:

Advantages	• A minimal amount of information about the sample is needed to run an analysis. Even mixtures of different materials can be accurately measured; only the viscosity of the medium must be known accurately.
	• Very small minimum measurable particle size, approximately less than 10 nm.
	• Only a small sample is needed, less than 10 µL or less than 0.001 mg.
	• The analysis is fast and simple.
	• Testing is nondestructive, so samples can be recovered if needed.
	• The instrumentation can be coupled to flow cells for automated measurements of the zeta potential.
Disadvantages	• Low resolution; particles must differ in size by 50% or more for DLS to detect reliably two peaks. The method does not really provide much "size distribution" data, only a mean size and estimate of standard deviation.
	• A small quantity of a small size particle can easily be "lost" in a much larger quantity of a large size particle.

11.4.2 Zeta Potential

Zeta potential refers to the electrostatic potential generated by the accumulation of anions at the surface of a colloidal particle in a fluid medium; the potential becomes organized into an electrical double-layer consisting of the Stern layer and the diffuse layer (as shown in Figure 11.10). The zeta potential of a particle can be calculated, if the electrophoretic mobility of the sample is known, by Henry's equation [36,37], which is

$$U_e = \frac{2\varepsilon\zeta f(ka)}{3\eta} \tag{11.14}$$

where
   U_e is the electrophoretic mobility
   ε is the dielectric constant of the sample
   ζ is the zeta potential
   $f(ka)$ is Henry's function (most often the Hückel and Smoluchowski [37] approximations of
      1 and 1.5, respectively, are used)
   η is the viscosity of the solvent medium

The zeta potential is easily measured using a laser Doppler velocimeter (LDV). An electrical field of known strength is applied across a colloidal sample, through which a laser beam is then passed. The electrophoretic mobility of the colloid dictates the velocity of the charged particles, which in turn induces a frequency shift in the incident laser beam. Using the dielectric constant of the sample, the viscosity of the solvent, the measured electrophoretic mobility, and either the Hückel or the Smoluchowski approximation for Henry's function, the zeta potential of the particles within the colloid can be calculated. It is obvious that PCS/DLS can also be used to measure the zeta potential from changes in the hydrodynamic particle size as a function of, for example, ionic strength, or more commonly pH.

The importance of the zeta potential of a colloid is that of a relative measure of the stability of the system. The DLVO theory [38,39] for colloidal interactions dictates that a colloidal system will remain stable, if, and only if, the Coloumbic repulsion arising from the net charge on the surface of the particles in a colloid is greater than the van der Waals force between those same particles. When the reverse is true, the colloidal particles will cluster together and form flocculates and aggregates (depending on the strength of the van der Waals attraction and the presence/absence of steric effects). The higher the absolute zeta potential, the stronger the Coloumbic repulsion between the particles within a colloid, and therefore the less the impact of the van der Waals force on the colloid. The line dividing stable and unstable suspensions is usually taken to be around ±30 mV (numerically greater than 30 mV and the suspension would normally be considered stable).

Factors that affect stability include pH, conductivity, and concentration. For example, the addition of more alkali will induce a more negative charge, and vice versa with addition of acid. Figure 11.18 shows a typical plot of zeta potential against pH.

The thickness of the double layer depends on the concentration of ions in solution, and can be calculated for any particular ionic strength. High ionic strength will compress the double layer as will the valency (e.g., Fe^{3+} vs. Na^+). Specific adsorption of ions onto the surface can have a dramatic effect on the isoelectric point, which may lead to complete reversal of the surface charge. Colloidal concentration can also affect the zeta potential. The effects described above can be used to great advantage during formulations to resist, for example, flocculation (see Chapter 15).

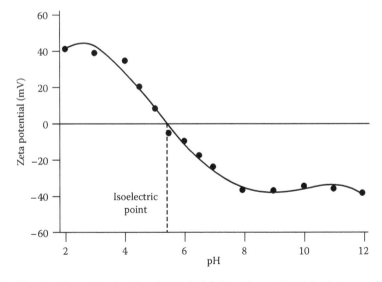

FIGURE 11.18 Plot showing an isoelectric point at pH 5.5. It can be predicted that between pH values of less than 4 and greater than 7.5, stability problems will arise since the zeta potential values are numerically greater than 30 mV in those regions. (From Zeta potential: An introduction in 30 minutes, Downloadable application note from www.malvern.com. With permission.)

11.4.3   Separation Methods

There are a number of separation methods but here the emphasis will be on differential centrifugal sedimentation (DCS), a novel and innovative, yet simple, technique, which has become "reborn" in recent years. Previous limitations and difficulties with the technique of sedimentation have been overcome with recent advances in technology, and DCS is now a powerful tool for the measurement of nanoparticle size distributions down to 3 nm. With its unique ability to resolve multimodal particle distributions of closely spaced sizes, even within 2%, and to distinguish extremely small shifts and changes in particle size, DCS is once more becoming a valuable particle characterization tool. The practical range of the technique is from about 3 nm up to about 80 μm (the exact range will be dependent on particle density). DCS is relatively user friendly, is highly accurate and reproducible, can measure up to 40 samples in the same "run," does "speed ramping" for measurement of broad distributions in a single sample, and can even measure "buoyant" or "neutral density" particles (i.e., particles having a lower density than that of the medium in which they are dispersed). Due to the high resolution achievable, DCS is ideal for resolving aggregates and agglomerates, and for observing very small relative shifts in peaks and tails of particle size distributions. It may also be used to measure absolute particle size; however, the density of the particulate material must be known. It can even be used for quantitative measurements if the refractive index of the particulate is known. Number or weight distributions can also be easily calculated and displayed. Figure 11.19 shows a typical multi-particle measurement [40].

11.4.3.1   Basics of Differential Sedimentation

Sedimentation of particles in a fluid has long been used to characterize particle size distribution. It is based on the use of Stokes' law for the determination of an unknown distribution of spherical particle sizes, by measuring the time required for the particles to settle a known distance in a fluid of known viscosity and density. Sedimentation can be either gravitational (under 1 G force), or centrifugal (many G force). For a centrifuge running at constant speed and temperature, all the parameters in the expression for Stokes' law, except time, are constant during an analysis. Their values will be either well known or can be accurately measured. Within a broad range of analysis conditions, a modified form of Stokes' law can be used to measure accurately the diameter of spherical

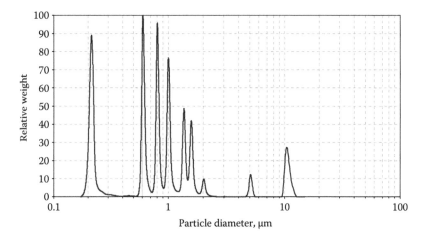

FIGURE 11.19 A trace showing a typical example of a multimodal particle distribution measured by DCS. Nine polystyrene standards, all of different sizes, were mixed and injected into the instrument. Each particle size can be seen to be resolved clearly. (From CPS application note, see www.cpsinstruments.eu/pdf/General%20Brochure.pdf. With permission.)

particles, based on their arrival time at the detector. Hence by introducing a known, traceable standard, the timescale can be calibrated against particle size, via the expression

$$V = \frac{D^2(\rho_P - \rho_F)G}{18\eta} \tag{11.15}$$

where
　　D is the particle diameter (cm)
　　ρ_P is the particle density (g/mL)
　　ρ_F is the fluid density (g/mL)
　　G is the gravitational acceleration (cm/s^2)
　　η is the fluid viscosity (poise)

11.4.3.2 DCS Instrument Design

The most common design for DCS instruments is that of a hollow, optically clear, disk mounted vertically and driven by a variable speed motor. A typical front view and disk cross section are shown in Figure 11.20. The outer reinforcing ring is made of Kevlar-reinforced aluminum to withstand the high-G-force when the disk is spinning at high speed (up to 24,000 rpm). The disk has a central closure cap, which can be opened when the contents need to be emptied. When the disk is rotating at speed, any fluid in it is forced centrifugally to the outer edge. The amount of fluid through which sedimentation is to occur corresponds to a sedimentation depth of 1–2 cm.

For the above-mentioned sedimentation fluid, rather than using pure water, a so-called gradient fluid (see Figure 11.20) is used, in which a gradated density profile is achieved by the use of, for example, solutions of sucrose in water, typically 8% and 24%. Despite difficulties in the concept of density gradients in miscible liquids, the gradient can in fact be set up in the following way. First the empty disk is set in motion to the chosen maximum rotational speed, then the sucrose/water solutions are injected along the axis of the spinning disk, with the solutions of lower concentration being injected before those of higher concentrations. As long as the disk remains spinning at high speed the solutions will not mix appreciably, and a density gradient is maintained that increases more or less continuously from inside to the outer edge. Of course, when the disk stops the gradient is destroyed. Injection can be carried out either by an automatic "gradient builder," or manually.

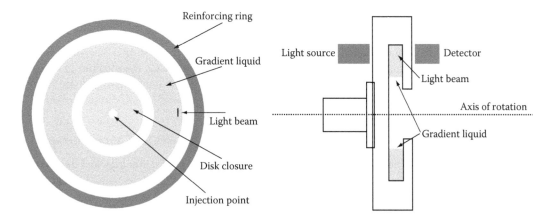

FIGURE 11.20 Typical disk from a DCS instrument. Front view and cross section, illustrating how a liquid with a density gradient would fill the hollow section during rotation. The optically clear disk is mounted vertically as shown on the right, and driven by a variable speed motor. (From CPS application note, see www.cpsinstruments.eu/pdf/General%20Brochure.pdf. With permission.)

If such a density gradient is not used, then a sample of dispersed particles in suspension would simply pass through the sedimentation fluid without appreciable particle separation. This is commonly called streaming or sedimentation instability, and results in a broad particle size distribution peak rather than in separate peaks for each band of particle size, resulting in the loss of all information about the particle size distribution. Streaming was a common problem in older instrumentation which used integral rather than differential sedimentation (i.e., no density gradient).

The disk can be of virtually any size, but manufacturers have settled on a diameter of about 125–150 mm. A light source, usually monochromatic light of relatively short wavelength (400–500 nm), and a detector, are placed on opposite sides of the moving disk so that there is continuous irradiation through the fluid near the edge of the disk. Although some instruments use a longer wavelength (650 nm), or x-rays, it has been accepted that near-UV wavelength, light gives better detector sensitivity when particles smaller than 100 nm are to be measured. At first the detector indicates maximum intensity, but as particles undergo sedimentation through the gradient fluid and reach the irradiated sector, the signal reduces. The reduction in intensity is related to the concentration of particles in the detector beam (the Mie theory of light scattering can be applied here). When all the particles have passed the detector, the signal returns to its original level. The raw data (time vs. intensity) can be used to draw a plot of particle concentration against calculated particle diameter, once a NIST traceable standard has been used to calibrate sedimentation time (plotted on the x-axis) versus particle size.

Samples are prepared for analysis by dilution in a fluid (normally water) of density slightly lower than that of the least dense fluid in the disk. When a sample (of volume normally around 100 μL) is injected at the center of the moving disk, it first strikes the back inside face of the disk, and then forms a thin film, which spreads as it accelerates radially toward the surface of the gradient fluid. When the dispersed sample solution reaches the gradient fluid surface, it quickly spreads over that surface because of its lower density. In fact, the sample solution "floats" on the gradient fluid, and the individual particles start immediately to be pulled centrifugally through the gradient to the outer edge, passing in front of the detector, on their way. The injection of a sample is rapid (typically <50 ms), so the starting time for an analysis is well defined, and the precision of sedimentation time is correspondingly good. When an analysis is complete, all the particles will have passed the detector and accumulated at the edge of the disk, where they will stay for as long as the disk is in motion. Since the dispersed particles have at that stage all passed through the gradient fluid, the instrument is then ready for the next sample. There is no need to empty and clean the disk, therefore many samples can be run in sequence without stopping it.

A calibration standard, which has a known particle size and particle density and is NIST traceable, is run before the sample, or optionally in between samples, so that the instrument can be calibrated, enabling sample analysis time to be converted to particle size. Figure 11.21 shows a centrifuge disk in an instrument during operation.

Disks have been developed that enable other types of analysis to be performed that were previously very difficult. For example, when particles have a lower density than the medium in which they are dispersed, they have a tendency to float rather than settle. Special low density disks, combined with reversal of the detector position, can now allow these types of samples to be measured.

11.4.3.3 Characterization of Coated Gold and Silver Nanoparticles for Cancer Therapy

Much research is currently being carried out in the context of cancer nanotechnology and the therapeutic benefits of using drug-coated metal nanoparticles. These coated particles, mainly gold and silver based, have multifunctional capabilities with reduced toxicity, high solubility and stability, and high efficacy for the target area. It has always been difficult to characterize such particles and to be able to distinguish the difference in particle size between the coated and uncoated core material, when using traditional light scattering techniques for particle sizing. With DCS, distribution peaks for both the coated and uncoated particles can be separated easily and resolved, making the technique an ideal characterization tool.

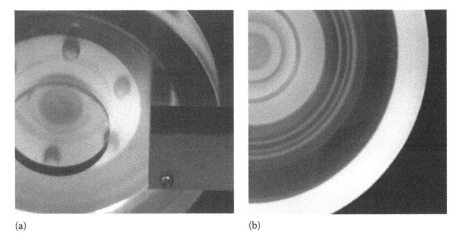

(a)   (b)

FIGURE 11.21 (a) The images show a centrifuge disk inside an instrument and the light source-detector toward the outside of the disk, and (b) the same disk in rotation during an analysis, where the separated bands of differently sized particles can be seen clearly as they approach the detector toward the outside of the disk. (From CPS application note, see www.cpsinstruments.eu/pdf/General%20Brochure.pdf. With permission.)

When a core metal particle is coated, its average bulk density is much reduced, and hence coated particles will settle more slowly than the core material itself when injected into the same solvent gradient in the rotating disk. Using the same density value as that for the core material in the instrument, software causes the distribution peak for the coated material to appear at a smaller particle size, since it has taken the coated particles longer to sediment and to pass in front of the detector. Initially, therefore, the particle size peak of the coated material appears oddly positioned, but it can be corrected simply by recalculating data using the correct density. However, the great benefit of this approach is that, due to the density difference, each peak for core and for coated metal particles can be resolved, and the positions of their mean particle size, together with the densities of the core and coating materials can be used to calculate accurately the thickness of the coating. No other commonly used particle sizing techniques are able to achieve this since they are unable to resolve the two peaks. For example, with light scattering methods only a single broad peak would be seen, with a maximum peak size indicative of the average of the mixture of core and coated particle sizes.

Figure 11.22a and b shows particle size distributions measured on a gold core with and without a polymer coating, respectively. The shell thickness of the coating was calculated to be 3.5 nm. The example shows two clearly resolved peaks, one at 13.2 nm for the gold core particles and another at 10.8 nm for the polymer-coated gold particles.

11.4.4 SURFACE AREA DETERMINATION

The surface area of a solid material is the total area of the material that is in contact with the external environment, and is generally expressed as m^2/g of dry sample. In many materials this area is not merely that of the external, visible, surface but includes that of the internal surfaces as well. Materials in the forms of both solid and powder, either of natural origin (e.g., stones, soils, minerals, etc.) or of industrial origin (e.g., catalysts, pharmaceuticals, metal oxides, ceramics, carbons, zeolites, etc.), can contain internal void volumes. These are distributed within the solid in the form of pores, cavities, and cracks, and the total sum of the volumes is called the porosity. Porosity is a strong determinant of certain important physical properties of materials, such as durability, mechanical strength, permeability, adsorption behavior, etc., and knowledge about pore structure, particularly the internal surface area, is important in characterizing materials and predicting their behavior. When significant

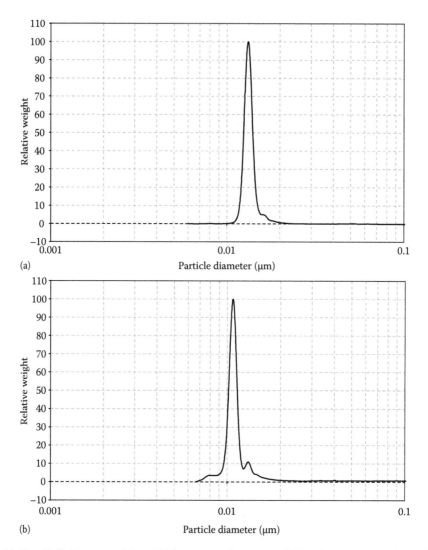

FIGURE 11.22   (a) Gold core particles of 13.2 nm mean diameter, and (b) polymer-coated gold particles with an *apparent* mean diameter at 10.8 nm due to lower overall particle bulk density. Note that in (b) a small number of uncoated gold core particles at 13.2 nm can still be seen. (From CPS application note, see www. cpsinstruments.eu/pdf/General%20Brochure.pdf. With permission.)

porosity is present, the ratio of the external surface area to that of the total surface area may well be quite small. Measurement of surface area can be made from an adsorption isotherm (see below).

There are two principal types of pores: closed and open. Closed pores are isolated completely from the external surface, and direct access by liquids and gases is not possible. However, closed pores influence mechanical and thermal properties as well as density. Open pores are by definition connected to the external surface, and are therefore accessible to fluids and gases, depending on the pore nature and size and on the particular fluid or gas. They can be subdivided further into dead-end and interconnected pores.

The characterization of porosity in solids consists of the determination of the average pore size. Pore dimensions cover a very wide range, and can be classified into three main groups:

- Micropores: diameters less than 2 nm
- Mesopores: diameters between 2 and 50 nm
- Macropores: diameters greater than 50 nm

11.4.4.1 Adsorption

Adsorption occurs when molecules from the gas phase impinge on a surface and remain attached to it. It can take two forms, depending on the strength of the bond attaching the molecules to the surface. If there is chemical reaction between a molecule and the surface, involving the transfer of electrons and the resultant formation of a chemical bond, then the process is called chemisorption. In the other form, no chemical reaction is involved, but molecules approaching a surface experience a weak attractive force due to the intrinsic surface energy. Such a force is called the van der Waals force, and the process is called physisorption. Chemisorbed layers are difficult to remove, because of the strength of the bond, whereas physisorbed layers can be desorbed easily by an increase in temperature. Physisorption is therefore a reversible phenomenon. Because of this temperature dependence, experiments in physisorption are normally carried out at a low temperature, most conveniently at that of liquid nitrogen.

In chemisorption the surface reaction is site-selective, and the course of the reaction can be complicated, with the possible formation of islands of reaction products, in which there are several successive reacted layers. On the other hand, in physisorption the coverage of the surface occurs in a consecutive layer-by-layer fashion, usually but not always with the first adsorbed layer complete before the second starts on top of it, and is not site-selective. Thus, to a good approximation, the amount of gas adsorbed in the first layer is proportional to the surface area of the solid that is available to the gas. That area will consist of both external and internal (e.g., pore) surface areas, assuming that the pore sizes allow ingress of the gas molecules. This proportionality forms the basis of a very widely used method for the measurement of surface area, called the BET (Brunauer, Emmett, and Teller) method [41].

11.4.4.2 Surface Area Measurement by the BET Method

The BET method is an extension of the Langmuir theory [42], which deals with monolayer molecular adsorption leading to multilayer adsorption based on the following hypotheses: (a) physical adsorption of gas molecules on a solid occurs indefinitely in a layer-by-layer manner; (b) there is no interaction between each adsorption layer; and (c) the Langmuir theory can be applied to each layer. The resulting BET equation is

$$\frac{1}{\upsilon\left(\dfrac{P_0}{P} - 1\right)} = \frac{c-1}{\upsilon_m c}\left(\frac{P}{P_0}\right) + \frac{1}{\upsilon_m c} \tag{11.16}$$

where
  P and P_0 are the equilibrium and the saturation pressures, respectively, of an adsorbate at the temperature of adsorption
  υ is the adsorbed gas quantity (e.g., in volume units)
  υ_m is the monolayer adsorbed gas quantity
  c is the BET constant, given by

$$c = \exp\left(\frac{E_1 - E_L}{RT}\right) \tag{11.17}$$

where
  E_1 is the heat of adsorption for the first layer
  E_L is the heat of adsorption for the second and further layers and is equal to the heat of liquefaction

Equation 11.16 represents an adsorption isotherm and can be plotted as a straight line with $1/[\upsilon[(P_0/P) - 1]]$ on the y-axis and $\varphi = P/P_0$ on the x-axis. This plot is called a BET plot, Figure 11.23.

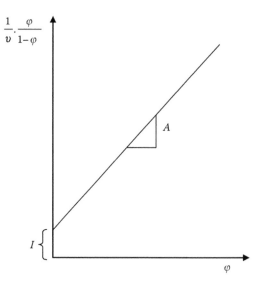

FIGURE 11.23 Schematic of BET plot.

The linear relationship of this equation is maintained only in the range of $0.05 < P/P_0 < 0.35$. The value of the slope A and the y-intercept I of the line are used to calculate the monolayer adsorbed gas quantity v_m and the BET constant c. The following expressions can be used:

$$v_\mathrm{m} = \frac{1}{A+I} \quad \text{and} \quad c = 1 + \frac{A}{I} \qquad (11.18)$$

A total surface area S_total and a specific surface area S can be evaluated from the following equations:

$$S_\mathrm{total} = \frac{v_\mathrm{m} Ns}{V} \quad \text{and} \quad S = \frac{S_\mathrm{total}}{a} \qquad (11.19)$$

where

N is Avogadro's number
s is the adsorption cross section
V is the molar volume of adsorbent gas
a is the molar weight of adsorbed species

The most common analysis based on the adsorption isotherm uses the multipoint BET method to measure total surface. A typical isotherm for a zeolitic catalyst is shown in Figure 11.24. The large uptake of N_2 at low P/P_0 indicates filling of the micropores (of size <2 nm) in the catalyst. The linear portion of the curve represents multilayer adsorption of N_2 on the surface of the catalyst, and the concave upward portion of the curve corresponds to the filling of meso- (of size 2–50 nm) and macro-pores (of size > 50 nm). An entire isotherm is needed to calculate the pore size distribution of the catalyst. However, for a surface area evaluation, data in the relative pressure range 0.05–0.30 are generally used.

Modern instruments usually measure volume and pressure and relate them to a modified BET equation, in which theory predicts a linear relationship between slope A and intercept I when the pressure of the adsorbate gas P, saturation vapor pressure P_0, and volume of gas adsorbed V is plotted against the relative pressure, as in

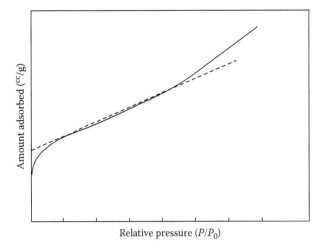

FIGURE 11.24 Typical nitrogen adsorption isotherm, obtained for a zeolite.

$$\frac{P}{V(P_O - P)} = A\frac{P}{P_O} + I \tag{11.20}$$

The linear portion of the curve is restricted to a limited portion of the isotherm, generally between 0.05 and 0.30 of normalized pressure. The slope and intercept are used to determine the quantity of nitrogen adsorbed in the monolayer and to calculate the surface area. For a single point method, the intercept is taken as zero or a small positive value, and the slope from the BET plot used to calculate the surface area. The surface area reported, usually quoted in m^2 g^{-1}, will depend upon the method used, as well as on the partial pressures at which the data were collected.

11.4.4.3 BET Particle Analysis and the Equivalent Sphere

To derive a particle size from BET measurements several assumptions must be made, that is, all the particles are spherical in shape, and have sharply defined diameters and densities. Based on these assumptions, an equivalent particle diameter can be calculated from the specific surface area and known density. In reality, this is not strictly valid, and so the calculated diameter is referred to as the "equivalent sphere" diameter, that is, an average of the particle diameters.

Additionally, further consideration should take into account the fact that the specific surface area of a particle can be divided into two parts: the outer surface and the inner surface areas. The outer surface area is correlated with the particle size and shape, whereas the inner surface area reflects the particle porosity. A distinction can be made by using gases of different molecular weight/molecular size. For example, nitrogen is a relatively small molecule and can penetrate into small pores. A full isotherm can be undertaken to determine the specific pore density as well as the specific surface area.

11.5 OVERVIEW: CHOICES OF TECHNIQUES

This chapter has addressed the question of how to characterize nanoparticles (i.e., objects on the size-scale well below that corresponding to the optical diffraction limit). The techniques fall into two broad categories, depending on what is meant by characterization.

In an industrial context, quality control becomes an important variation on the general theme of characterization. The emphasis is then essentially on ensuring that nothing has changed from one day to the next or from one batch to another, where "nothing" might refer to the functionality of the product (e.g., nanoparticles with a particular desired functionality).

In an R&D context, on the other hand, the aim is to determine the attributes of a single or a small number of nanoparticles, where attributes refer to such things as crystalline structure, composition, size, morphology, etc. This kind of characterization usually takes place during an R&D phase, where a great deal of expertise, time, energy, and cost can be invested on one-off or a small number of analyses. The techniques can then be chosen from a category that is generally not cost effective in the case of quality control. The relative freedom to choose may also be exercised for the purposes of fault finding or problem solving, when quality control says that there is a problem, and when quality control does not give sufficient information to solve the problem.

Another category of techniques, of greater relevance to quality control, is concerned with obtaining ensemble averages of attributes of nanoparticles, again with respect to structure, composition, size, etc. In many cases it is neither practical nor necessary to carry out analyses on a sufficiently large number of individual particles in order to infer average or representative values for a large population of, say, particles from a broad size range, and possibly with different compositions, surface chemistries, and so on.

The most widely used and available techniques from the two categories have been listed in Tables 11.1 and 11.2 with reference to information obtainable from single nanoparticles or ensembles of such particles.

TABLE 11.1
Characterization of Individual Nanoparticles

Attributes	Techniques	Comments
Crystal structure	HRTEM	Lattice imaging, see Chapter 6
Composition	EDS	(S)TEM/SEM, see Chapter 9
Molecular structure		Useful information from EELS, see Chapter 9
Size and shape	SEM/TEM/AFM	See Chapters 5, 6, and 9
Surface chemistry		Useful information from AFM/F-d, usually in an aqueous solution, see Chapter 5
Surface area	SEM	Inferred/estimated from imaging, see Chapter 6

TABLE 11.2
Characterization of Ensemble Averages

Attributes	Techniques	Comments
Crystal structure	XRD[a]	Down to 10 nm particle size
Composition	XPS[a]/EDS[a]/WDS[a]	See Chapters 3 and 9
Molecular structure	Raman[a]/IR[a]EXAFS[a,b]	See Chapter 15
Number density	CPC[c]	Down to 3 nm size range, dynamic range up to 10^6 particles/cm^3
Equivalent size	DLS	Monodisperse samples
	DLS	Bimodal size distributions
	DCS	Multimodal size distributions
Surface chemistry	XPS[a]/SRXPS[b]	See Chapters 3 and 15
Surface area	BET	

[a] Specimen configuration usually dense thin packed layer of particles on flat substrate with typical coverage of 1 cm^2.
[b] Synchrotron-based technique.
[c] Condensation particle counter.

REFERENCES

1. A. McWilliams, *Nanocomposites, Nanoparticles, Nanoclays, and Nanotubes*, Report ID. NAN021C, Published by BCC Inc., Norwalk, CT, 2006.
2. J. Swartz, *The Nanotech Report*, 5th Edition, Lux Research, 2007.
3. Nanotechnology Sector Report, Technology Roadmap Project for The Centre for Economic Growth, prepared by Rensselaer Lally School of Management and Technology, 2004.
4. International Standards Organisation TC 229/WG 1: Terminology and Nomenclature.
5. M. Schulenburg, *Nanotechnologie: Innovationen für die Welt von morgen Hrsg.*, 3 aktualisierte Auflage *Koordination: VDI-Technologiezentrum*, W. Luther and G. Bachmann (eds.), BMBF, Bonn/Berlin, 2006.
6. G. D. W. Smith and A. Cerezo, Field emission and field-ion microscopy (including Atom Probe). Chapter in *Handbook of Microscopy*, VCH, Weinheim, Germany (1997), Vol. 2, Section VI, Chapter 2, pp. 775–801.
7. D. J. Larson, A. K. Petford-Long, Y. Q. Ma, and A. Cerezo, *Acta Mater.* 52 (2004) 2847.
8. X. W. Zhou, H. N. G. Wadley, R. A. Johnson, D. J. Larson, N. Tabat, and A. Cerezo, *Acta Mater.*, 49 (2001) 4005.
9. J. Oliver, *Carbon Nanotubes: Technologies and Commercial Prospects*, Report ID NAN024C C2007 published by BCC Inc., Norwalk, CT, 2007.
10. For example, Wikipedia, http://en.wikipedia.org/wiki/Raman_spectroscopy
11. For example, see I. R. Lewis and H. G. M. Edwards, *Handbook of Raman Spectroscopy*, CRC Press (2001), with over 1000 references to Raman spectroscopy of elemental carbon-based materials.
12. C. Castiglioni et al., in *Carbon: The Future Material for Advanced Technology Applications*, G. Messina and S. Santangelo (eds.), Springer Series, Heidelberg, *Top. Appl. Phys.* 100 (2006) 403–426.
13. T. Shimada, T. Sugai, C. Fantini, M. Souza, and L. G. Cançado, *Carbon*, 43 (2005) 1049.
14. M. S. Dresselhaus, G. Dresselhaus, R. Saito, and A. Jorio, *Phys. Rep.*, 409 (2005) 47.
15. V. W. Brar, G. G. Samsonidze, and M. S. Dresselhaus, *Phys. Rev. B*, 66 (2002) 155418.
16. A. M. Rao, S. Bandow, E. Richer, and P. C. Eklund, *Thin Solid Films*, 331 (1998) 141.
17. M. S. Dresselhaus, G. Dresselhaus, and A. Jorio, *Carbon*, 40 (2002) 2043.
18. R. Pfeiffer, H. Kuzmany, W. Pank, and T. Pichler, *Diamond Relat. Mater.*, 11 (2002) 957.
19. M. G. Donato, G. Messina, S. Santangelo, S. Galvagno, C. Milone, and A. Pistone, *J. Phys. Conf. Ser.*, 61 (2007) 931.
20. R. Graupner, *J. Raman Spectrosc.*, 38 (2007) 673.
21. P. Puech, E. Flahaut, A. Bassil, T. Juffmann, F. Beuneu, and W. S. Basca, *J. Raman Spectrosc.*, 38 (2007) 714.
22. S. Gui, R. Canet, A. Derre, M. Couzi, and P. Delhaes, *Carbon*, 41 (2003) 41.
23. J. A. Fantini, M. A. Pimenta, R. B. Capaz, G. G. Samsonidze, G. Dresselhaus, M. S. Dresselhaus, J. Jiang, N. Kobayashi, A. Gruneis, and R. Saito, *Phys. Rev.*, B 71 (2005) 075401.
24. J. C. Meyer, M. Paillet, T. Michel, A. Moreac, A. Neumann, G. Duesberg, S. Roth, and J. L. Sauvajol, *Phys. Rev. Lett.*, 95 (2005) 217401.
25. V. N. Popov and P. Lambin, *Phys. Rev. B*, 73 (2006) 85407.
26. M. Paillet, T. Michel, J. C. Meyer, V. N. Popov, L. Henrard, S. Roth, and J. L. Sauvajol, *Phys. Rev. Lett.*, 96 (2006) 257401.
27. F. Tuinstra and J. L. Koenig, *J. Phys. Chem.*, 53 (1970) 1126.
28. J. Maultzsch, S. Reich, C. Thomsen, S. Webster, R. Czerw, D. L. Carroll, S. M. C. Vieira, P. R. Birkett, and C. A. Rego, *Appl. Phys. Lett.*, 81 (2002) 2647.
29. H. Murphy, P. Papakonstantinou, and T. I. T. Okpalugo, *J. Vac. Sci. Technol. B*, 24 (2006) 715.
30. R. Saito, A. Gruneis, G. G. Samsonidze, V. W. Brar, G. Dresselhaus, M. S. Dresselhaus, A. Jorio, L. G. Cancado, C. Fantini, M. A. Pimenta, and A. G. S. Filho. *New J. Phys.*, 5 (2003) 1571.
31. A Rawle, and V. K. Malvern, *Basic Principles of Particle Size Analysis*, downloadable application note from www.malvern.com
32. R. Pecora (ed.), *Dynamic Light Scattering—Applications of Photon Correlation Spectroscopy*, Plenum Press, New York, 1985.
33. http://www.cpsinstruments.eu/pdf/Compare%20Sizing%20Methods.pdf
34. Measurement and Interpretation of Electrokinetic Phenomena, International Union of Pure and Applied Chemistry, Technical Report, *Pure Appl. Chem.*, 77 (2005), 1753.
35. Dynamic light scattering: An introduction in 30 minutes, Downloadable application note from www.malvern.com

36. Zeta potential: An introduction in 30 minutes, Downloadable application note from www.malvern.com
37. Malvern ZetaSizer NS Operating Manual, 2007.
38. B. V. Derjaguin and E. M. Landau, *Acta Physicochim. URSS*, 14 (1941) 633.
39. E. J. W. Verwey and J. Th. G. Overbeek, *Theory of the Stability of Lyophobic Colloids*, Elsevier, 1948.
40. CPS application note, see www.cpsinstruments.eu/pdf/General%20Brochure.pdf
41. S. Brunauer, P. H. Emmett, and E. Teller, *J. Am. Chem. Soc.*, 60 (1938) 309.
42. I. Langmuir, *J. Am. Chem. Soc.*, 41 (1919) 861.
43. NMTC Report, *Nanotechnologie und Nanomaterialien—Applikationen und Marktpotenzial Verfahren-stecnik*, April 2005.

12 Problem-Solving Methods in Tribology with Surface-Specific Techniques

Christophe Donnet and Jean-Michel Martin

CONTENTS

12.1 TRIBOLOGY AND SURFACE-RELATED PHENOMENA

Tribology is involved in both industrial and basic scientific researches. In industry, tribological considerations arise in all requirements and applications relating to the control of friction and wear in mechanical components. This domain includes extension of the lifetime of components, increase in their mechanical and energy efficiency, and improvement in their safety in, for example, the transportation, chemical, textile, food, plastic, biomedical, and space industries. The widespread application of tribology in industry is evident in a wide variety of situations. Historically, the optimization of contact geometries and the choice of bulk materials and lubricants have been the main technological thrusts in the design of specific remedies for industrial applications.

From a basic research point of view, the field of tribology involves the contact of surfaces in relative motion, leading to the introduction of surface science concepts, which are coupled to those from other more traditional scientific fields, such as materials science (including metallurgy) and continuum and fluid mechanics. Even if tribology may be considered a multidisciplinary field, embracing mechanics, physics, and chemistry, it can also be described as a scientific discipline in its own right, with its own methodologies. When two solid surfaces are in dynamic contact, the material phases between them may exist in either gas or liquid forms, or as a solid thick enough for the interface media to be described adequately in terms of traditional concepts and to be amenable to the experimental techniques of bulk materials science. However, numerous situations occur in which the interface responsible for overall tribological properties can be so thin (as little as a monomolecular layer) that surface-specific analysis methods must be used to determine its nature and to quantify its intrinsic properties.

For readers not familiar with this field, the basic aspects concerning the lubrication of two contacting surfaces should be summarized briefly, in order that the surface analysis requirements and methodologies carried out in tribology can be understood [1,2]. When two surfaces undergo sliding or rolling under load, energy dissipation leads to friction in which the production of wear particles and the plastic deformation of the contacting surfaces are the main causes of serious material loss. In the case of sliding friction, the three basic phenomena of concern are adhesion, ploughing (due to wear particles or to asperities on the harder counterface), and asperity deformation. Lubrication consists of separation of the moving surfaces by an interposed film of solid, liquid, or gas, characterized by low shear resistance and minimization of surface damage. Various lubrication modes can be identified [1] from the Stribeck curve (Figure 12.1), which shows the dependence of the coefficient of friction μ (tangential force/normal force) on the Sommerfeld number S, which is given by

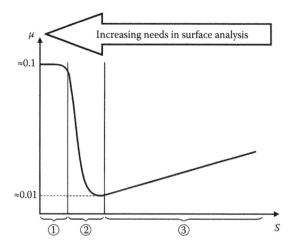

FIGURE 12.1 Three basic phenomena in sliding friction: (1) adhesion, (2) ploughing, and (3) asperity deformation.

$$S = \frac{\eta V}{W}$$

where

η is the dynamic viscosity of the lubricant
V is the relative motion of the two contacting surfaces
W is the normal load applied in the contact

From Figure 12.1, three principal lubricant regimes may be identified; as the Sommerfeld number decreases, so increasing demands are made on surface analysis:

1. Hydrodynamic lubrication (see 3 in Figure 12.1) is based on the formation of a thick lubricant film (typically of thickness from 1 to 100 μm) that inhibits contact between the moving surfaces. This lubrication mode is governed by the bulk physical properties of the lubricant, mainly the viscosity, and by the speed of the relative motion. No specific needs in surface analysis are required.
2. Elastohydrodynamic (EHD) lubrication (see 2 in Figure 12.1) occurs when the extent of surface deformation is comparable with the lubricant film thickness, so that a heavy load causes local elastic deformation of the contacting surfaces, but without any significant asperity interaction. Taking into account the low lubricant thicknesses (typically from 0.01 to 10 μm) and the high contact pressures (typically in the GPa range), the lubricant properties differ from those of a traditional bulk liquid, since a strong viscosity increase may occur, when the lubricant behaves more like a solid than a liquid. Moderate temperature rises may occur, thus inducing some thermochemical reactions between the surfaces and the lubricant additives, but no tribochemical phenomena are involved, unlike in the next lubrication mode.
3. Boundary lubrication (see 1 in Figure 12.1) is probably one of the most complex phenomena studied by tribologists, in view of the numerous experimental parameters that influence the tribological behavior. Considerable asperity deformation is usually observed, since the contacting surfaces move very close to each other. The role of the thin-film lubricant (thicknesses from 1 to 100 nm) adsorbed on the solid surfaces is that of inhibition of asperity welding, thus reducing wear, and of lowering of the coefficient of friction, as a result of the formation of a friction-induced tribochemical film with a shear strength lower than that of the bulk solid material. Generally speaking, the boundary lubricant modifies the solid–solid interactions, due to various solid-lubricant reactions (physisorption, chemisorption, and chemical reactions), which depend strongly on the environmental conditions and on the lubricant composition.

As pointed out in an early publication by Bowden and Tabor [2], a frictional interface under boundary conditions can be considered as a specific chemical reactor in which the combined effects of pressure and temperature, and of the fresh surfaces created by the friction process in the presence of gaseous, liquid, or solid phases acting as lubricants or contaminants are unique to the particular system. Lubricants contain additives that act as packages with several functions, including improvement of the pressure–velocity (rheological) properties of the lubricant, control of friction, and inhibition of oxidation of contacting surfaces and of oil at the highest operating temperatures. Understanding of the mechanisms of friction and wear requires a precise identification and quantification of the tribochemical phenomena, which in turn allows the correct selection of materials, surface treatments, surface morphology, lubricants with additives, and operating conditions, for a given engineering device.

The complex phenomena occurring in boundary lubrication goes some way to explaining why no model for tribochemical reaction exists, thus emphasizing the crucial role of surface and interface analysis in the investigation of tribosystems, and in the understanding of their behavior. Nevertheless, there are two nonexclusive schools of physicochemical thought. The first states that the tribochemical process is nonequilibrium, and that materials in nonequilibrium condition are generated in the contact [3]. This might explain the amorphous states of most tribochemical films formed in boundary lubrication [4]. The second school asserts that the effects of friction (increase of temperature, formation of fresh metal surfaces, and mechanical deformation) only enhance, accelerate, and stimulate reactions that would occur in the same but static interaction without friction [5]. With this latter assumption, it is believed that the final products are thermochemically stable, and that the film composition can be predicted from the thermochemical considerations. Thus, one of the major motivations in boundary lubrication is the identification of the chemical products and an understanding of the chemical reactions, leading to formation of the beneficial tribochemically induced lubricant films [6].

Mixed lubrication occurs when the contact behavior is governed by a mixture of EHD and boundary lubrications. Asperity contact and deformation may occur, even when the surfaces are separated by a lubricant film from 0.01 to 1 μm thick.

Solid-film lubrication, more recently developed, is used when conventional lubricants can no longer operate, as in extreme conditions, such as at very low (cryogenic) or very high temperatures (>300°C), and in vacuum. The principle consists of the interposition of a solid film (of thickness usually <10 μm) between the two sliding or rolling surfaces, prior to friction. Shear takes place within and across the film or between the film and the sliding surfaces. The film is applied directly by conventional coating deposition techniques, or it can also be formed by reaction between the triboactivated sliding surface and the environment [7,8].

From this well-established and generic classification, it appears that there is a strong need for surface analysis to investigate tribological contact and to understand interfacial and surface interactions, particularly in the boundary, mixed, and solid lubrication modes. Tribologists agree that recording of the friction force and measurement of wear rates are necessary, but not sufficient to investigate such complex tribosystems. Since the pioneering work of Buckley in the 1970s [9], the combination of tribology and surface science has allowed progress in the development of low-friction and wear-resistant contacts, using optimized liquid or solid lubricants, in various mechanisms and devices. What is now needed in particular is the identification and classification of the surface analytical methodology in the examination of tribological contacts, and these are described in the next section.

12.2 SURFACE ANALYSIS REQUIREMENTS FOR TRIBOLOGY

> If an understanding of the nature of surfaces calls for such sophisticated physical, chemical, mathematical, materials and engineering studies in both macro- and molecular terms, how much more challenging is the subject of…interacting surfaces in relative motion? [10]

12.2.1 OVERVIEW

This section discusses the surface and interface analysis methodology needed for the investigation of a tribological contact. It will not describe the potentialities, performances, and limitations of the various surface analysis techniques that have been applied in this field, since that description can be found in other chapters. Nor is it intended to be exhaustive with respect to the numerous studies that have been reported. However, it emphasizes a logical procedure leading to an integrated top-down approach, since surface analysis is considered as a means to a complete understanding of a tribological contact and to the elucidation of tribo-induced surface modifications. Several examples, entitled generic studies, are then used to illustrate, in Section 12.3, the different procedures that couple a tribological experiment to relevant surface investigation of the contacting bodies. Each generic study has been selected on the basis of how various analysis techniques have been brought together to extract crucial information from a tribological point of view. A more exhaustive and detailed presentation of surface investigations in tribology is given in Ref. [11].

Basically, surface analysis is applied in tribology in two ways:

1. *Post mortem* examination of failures in real working systems. This helps to identify the origins of the failures, with the possibility of recommendation for remedial or preventative action.
2. Systematic surface analytical studies on model or ideal systems. These are performed in order to understand the mechanisms and underlying physics and chemistry of various phenomena, such as the reactions promoted by additives, the effects of adsorption on friction and adhesion, and the degradation of a thin lubricant coating in a severe environment.

The first of these requires a classical surface analysis methodology, which is not necessarily specific to the field of tribology. The second is more specific and requires a detailed methodology, based on the following three basic criteria that must be considered when a tribologist has to follow a coherent methodology in the investigation of surface-related phenomena in a tribological contact:

1. *Dimensional criterion.* In tribological contact processes, both lateral and depth dimensions are important, the former because of the size and shape of wear debris and the latter because of the presence of thin films. It is therefore vital to match the characteristics of a technique to the dimensions of the analysis.
2. *Time-scale criterion.* In most tribological tests, friction is measured continuously, while, on the other hand, surface composition and topography are still determined only at the completion of the friction test. Moreover the surface phenomena can not only vary continuously during the friction experiment, but can also be altered by air exposure between the end of the tribological test and analysis, due to environmental effects such as oxidation and hydration. Careful thought must thus be given to the chosen surface analysis configuration in order to compensate for these experimental difficulties, as far as possible.
3. *Information criterion.* The surface analytical information necessary for elucidation of the tribological mechanisms in terms of physicochemical, structural, morphological, and rheological modifications of the contacting surfaces must be identified precisely.

These three criteria are examined in detail. Sections 12.3.1 through 12.3.10 describe 10 generic studies. They have been chosen to illustrate the complementarity of surface and interface investigations, in relation to the above three criteria, in the implementation of the top-down approach to the field of analytical tribology.

12.2.2 DIMENSIONAL CRITERION

Surface analysis applied to tribology requires the dimensions of the various surface elements and related phenomena to be borne in mind, when two surfaces are probed that have undertaken relative motion under contact pressure. This consideration has to be coupled with others concerning the nature of the probed information, as indicated in Section 12.2.4; chemical information, such as the nature of the oxidation states of the surface elements, is present in both a thin film of a few molecular monolayers (in boundary lubrication) and in a film of micrometer thickness (in thin-film lubrication by a solid lubricant coating). Here, the depth dimension is of great importance in selecting the appropriate technique and analysis procedure; x-ray photoelectron spectroscopy (XPS) on its own, with a depth resolution of <5 nm, is suitable in the first case, whereas XPS with sequential sputtering of the coating is required to analyze the entire thickness of the solid lubricant film, with the usual caveat that damage induced by ion beam sputtering must be minimized.

Figure 12.2 [11, p. 37] shows an overview of the size range of surface elements and surface-related phenomena in tribology. A continuous size distribution is observed of surface-related

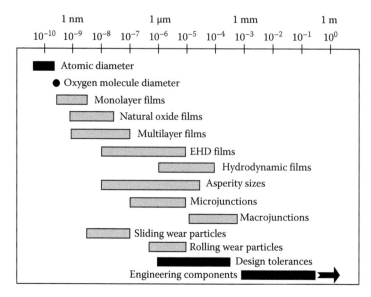

FIGURE 12.2 Magnitudes of the relevant surface related phenomena in tribology. (From Holmberg, K. and Mathews, A., in Dowson, D. (Ed.), *Coatings Tribology*, Tribology Series 28, Elsevier, Amsterdam, the Netherlands, 1994. With permission.)

phenomena characteristic of tribological behavior, since the pertinent data cover at least six orders of magnitude, ranging from 10^{-9} m to the scale of design tolerances (10^{-6} to 10^{-3} m), and then to the size of engineering components (from 10^{-3} m). Monolayer films influencing friction and wear behavior in boundary lubrication are in the 10^{-9} to 10^{-8} m range. Natural oxide films, multilayer films, and typical sliding wear particles range in size and thickness from 10^{-9} to 10^{-7} m. As stated in Section 12.1, EHD films range in thickness from about 10^{-8} to 10^{-5} m. This size range includes the typical sizes of rolling wear particles and microjunctions. Further up the scale, hydrodynamic films cover a thickness range from 10^{-6} to 10^{-4} m, whereas asperity heights cover a wide range from 10^{-8} to 10^{-4} m, depending on the surface geometric finish. Finally, macrojunctions observed in catastrophic adhesive wear can reach the 10^{-3} m range, which is of the same order of magnitude as some less restrictive design tolerances.

This dimensional criterion strongly influences the choice of analytical technique for obtaining the required surface information, taking into account also the available spatial resolution. The reader should refer to other chapters for information about the spatial resolution range of the techniques identified as being suitable for acquiring the surface information classified in Section 12.2.4.

12.2.3 TIME-SCALE CRITERION

This criterion concerns the different chronologies within which the friction experiment and the surface analysis of the contacting surfaces can be arranged. Different configurations, also called modes, can be considered on the basis of Figure 12.3. Whatever the configuration, prefriction analysis must be performed in order to have complete knowledge of the experimental system prior to the tribotest. Both qualitative and quantitative analyses are required of the surfaces that will be in contact, and of the lubricant phase, as described in Section 12.2.4. This routine step needs only moderate spatial resolution, since virgin surfaces prior to friction are usually laterally homogeneous. Prefriction analysis is performed either outside the tribometer with traditional surface analysis apparatus, or inside the tribometer if it is equipped with dedicated analytical instrumentation.

The time-scale criterion can be broken down into three main analysis modes for the investigation of a tribo-contact. Compared to the analysis carried out prior to friction, the three modes all

Ex situ Outside the tribometer	*Post mortem* After friction	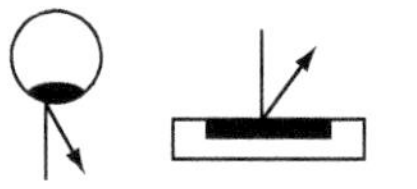	See Sections 12.3.2 and 12.3.6 through 12.3.9
In situ Inside the tribometer Outside the contact	*Pre mortem* During friction	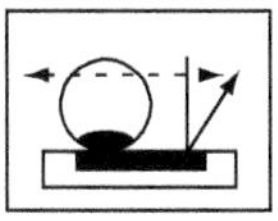	See Section 12.3.5
	Post mortem After friction	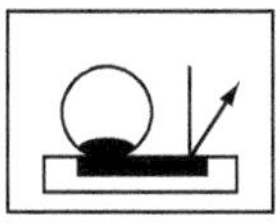	See Sections 12.3.2 through 12.3.4, 12.3.6 and 12.3.8
In vivo Inside the tribometer Inside the contact	*Pre mortem* During friction	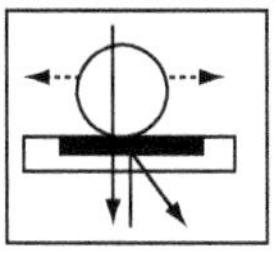	See Sections 12.3.1 and 12.3.2
	Post mortem After friction	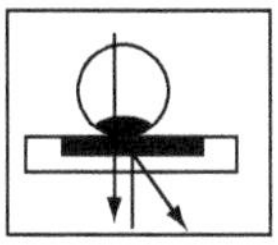	See Section 12.3.10

FIGURE 12.3 Time-scale criteria combining friction experiments and surface analysis in different modes. Letters in the right-hand column refer to the generic studies in Section 12.3 that illustrate the use of the corresponding technique [52]. (From Donnet, C., in Rivière, J.C. and Myhra, S. (Eds.), *Handbook of Surface and Interface Analysis: Methods for Problem-Solving*, Marcel Dekker, New York, 1998, 704. With permission.)

require spatial resolution consistent with the size of the area affected by friction and wear (which is often less than a few $100\,\mu m^2$), assuming that the intrinsic lateral resolution of the technique to be used is adequate. The three modes are the following:

1. *Ex situ* analysis: surface analysis performed outside the tribometer, after the friction test.
2. *In situ* analysis: surface analysis performed inside the tribometer, inside the wear scars but outside the contact. *In situ* analysis can also be performed during (*pre mortem*), or after (*post mortem*), the friction test.
3. *In vivo* analysis: surface analysis performed inside the tribometer, inside the wear scars, and inside the contact. *In vivo* analysis can also be performed during (*pre mortem*), or after (*post mortem*), the friction test.

These three modes allow the identification of friction-induced surface modifications, such as the development of transfer films, tribochemical phase formation, or the removal of material and lubricant films by wear. Correlations may be revealed that might help to elucidate the friction and wear mechanisms. There follows a description of these modes, with emphasis on their interests, potentialities, and limitations, as illustrated by the generic studies developed in Section 12.3, and summarized in Figure 12.3.

1. *Ex situ* analysis is performed outside the tribometer and obviously after the friction test, with analytical instruments independent of the tribometer. Direct observation of the surfaces may therefore be coupled with physicochemical and structural investigations performed inside the wear scars of the surfaces, which have been separated from each other at the end of the tribotest. Third-body products, such as wear particles, can be extracted from the contact for independent examination. The main advantage of *ex situ* analysis is the possibility

of using the great potential of various complementary modern surface analysis techniques, providing sensitive measurements from ever more restricted areas. The first drawback is the inevitability of air exposure between the tribometer and the analytical instruments, thus leading to adsorption or reaction of surfaces with air, water vapor, or other mineral and organic contaminants, complicating the tribological process under investigation. A second drawback is the usual size limitation of samples for analysis in conventional analytical instruments, which are often not adapted to surface investigation of larger tribotested samples or components. The generic studies described in Sections 12.3.2 and 12.3.6 through 12.3.9 illustrate the surface analysis needs and the use of various complementary *ex situ* methods.

2. *In situ* analysis is performed inside the tribometer, which will then incorporate the surface probe. This mode allows exploration of the friction and adhesion of clean solid surfaces, the effect of adsorbed species, or the nature of the lubricant phase, in order to elucidate the tribo-induced surface modifications in as direct a manner as possible. Depending on the analytical technique used, the nature of the environment during the tribotest is not necessarily the same as that during analysis; *in situ* Raman spectroscopy can be performed in ambient air, whereas *in situ* XPS and Auger electron spectroscopy (AES) require ultra high vacuum (UHV). This mode can be performed in either of two configurations according to the time-scale criterion.

   a. Analysis can be performed after the friction test (*in situ post mortem* analysis) by direct observation inside the wear scars. Possible limitations arise from the geometry of the analyzed surfaces, since some techniques (e.g., LEED for crystallinity study) require flat surfaces, thus prohibiting the analysis of transfer films on the spherical balls often used in tribological tests. XPS/AES can be performed on both the plane and the ball, the latter having curvature diameters in the range of several millimeters or more. Examples are given in the generic studies in Sections 12.3.2 through 12.3.4, 12.3.6, and 12.3.8.

   b. Analysis can also be performed during the friction test (*in situ pre mortem* analysis), but is less systematic, since it places greater demands on the coupling between the tribotest and the analysis procedure. Such a configuration requires rotational motion of the contact, with analysis carried out on the rotating disk, inside the wear scars but outside the dynamic contact. Ideally, the analysis should be carried out during rotation of the disk counterface, but there are obvious difficulties: (1) the analyzed area changes continuously, with a period related to the rotation speed, and (2) most of the analytical techniques require acquisition times longer than the normal rotation periods. That is probably why such a direct analytical configuration has never been attempted. However, in earlier work, it was approximated by a sequential procedure, that is, by stopping the rotation very so often to allow analysis by AES (see the generic study in Section 12.3.5). In that way, a succession of tribotests/AES analyses was performed in order to study continuous friction-induced surface modification.

3. *In vivo* analysis consists in performing spatially resolved surface analysis directly inside the contact. This mode imposes drastic conditions on both the tribometer configuration and the performance of the analytical technique. First, the use of electron or x-ray transparent materials is necessary in order that the probe may reach the interface with only minimal interaction with the bulk contacting bodies. Second, a dedicated tribometer must be compatible with the analytical configuration. As in the case of *in situ* analysis, two time-scale configurations are possible.

   a. Analysis can be performed after the friction stage (*in vivo post mortem* analysis), thus providing analytical data relating to the contact with minimal perturbation, since the two contacting bodies are not separated; this is illustrated in the generic study in Section 12.3.10.

b. Analysis can be performed during the friction stage (*in vivo pre mortem* analysis): this mode is the dream of all tribologists, since it corresponds to the ultimate goal in recording real-time and spatially resolved information from the friction interface, which can then be correlated with simultaneous friction forces, surface forces, or wear rate measurement. Of course, as for the *in situ pre mortem* analysis mode, a balance has to be struck between the dynamic of the friction movement on the one hand, and the spatial resolution/acquisition time of the analysis on the other. More details are given in Section 12.3.1 where ultrathin thickness measurements of a lubricant film inside a dynamic contact are described. Attempts to record physicochemical and structural data during friction are also mentioned in Section 12.3.2.

12.2.4 Information Criterion

There is now an extensive body of accumulated experimental work concerning relevant surface analytical data in tribology [11]. Three categories of information can be listed, as follows:

12.2.4.1 Physicochemical and Structural Information

This includes

- Nature of the elements and their chemical states
- Nature of the chemical bonds and orbital hybridization
- Nature of the chemical phases

Structure is related to the spatial arrangement of atoms and groups of atoms or molecules. The order ranges from long-range, characteristic of crystalline compounds, and established by conventional diffraction techniques, to friction-induced short-range (local order) established by more specialized techniques, as detailed hereunder.

The above types of information are of great importance in boundary and solid-film lubrication, since the lubricant phase and the surface composition can change continuously during the friction test, depending on the experimental conditions (contact pressure, velocity, temperature, and nature of the surrounding environment). Acquisition of such information is vital for clarification of the relationships that exist between the nature of the friction-induced surface and of the interface modifications and the interpretation of the tribological mechanisms. Only specialized techniques can provide the physicochemical and structural information, taking into account the dimensional and time-scale criteria discussed above. Table 12.1 compares qualitatively the capabilities of the analytical techniques most commonly used in tribology, with references to the generic studies described in Section 12.3. More detailed information regarding the characteristics and capabilities of each technique can be found in Chapters 3 and 4. From the literature [12], it can be seen that the spectroscopic techniques XPS and AES have been used extensively in the field of tribochemistry, as illustrated in Sections 12.3.2, 12.3.4 through 12.3.6, 12.3.8, and 12.3.9. More recently, x-ray absorption spectroscopy (XAS) [13] (see Chapter 7) and electron energy-loss spectroscopy (EELS) (see Chapter 6) [14] have revealed unique capabilities in terms of chemical, electronic, and crystallographic information on the atomic scale. The chemical bonding, oxidation state, and hybridization of selected atoms, can be extracted from a careful analysis of the near-edge fine structure of the corresponding absorption (or ionization) edge (Figure 12.4). This is called x-ray absorption near-edge spectroscopy (XANES) for x-rays and electron (energy) loss near-edge spectroscopy (ELNES) for electrons.

Local order can be studied by processing the extended fine structure found above an absorption edge (Figure 12.4), called extended x-ray absorption fine structure (EXAFS) for x-rays and extended electron-loss fine structure (EXELFS) for electrons (see Chapter 7). The EXAFS arises from the

TABLE 12.1
Capabilities of Some Analytical Techniques in Terms of Physical, Chemical, and Structural Data [52]

	Elemental Identification	Chemical Bonding	Oxidation State	Orbital Hybridization	Local Order	Crystalline Structure	See Generic Studies[a]
ToF-SIMS	•	○	○	○	○	○	12.3.9
XPS	•	•	•	○	○	○	12.3.4–12.3.8
AES	•	•	•	•	○	○	12.3.2, 12.3.4–12.3.6, 12.3.8, and 12.3.9
Raman	○	•	•	•	○	○	12.3.10
XAS	•	•	•	•	•	○	12.3.2
EELS	•	•	•	•	•	○	12.3.2, 12.3.4, and 12.3.6
FTIR	○	•	•	•	○	○	12.3.9
TEM	○	○	○	○	•	•	12.3.2, 12.3.6, and 12.3.8
XRD	○	○	○	○	○	•	12.3.7
CEMS	○	•	•	•	•	•	12.3.2

Source: Donnet, C., in Rivière, J.C. and Myhra, S. (Eds.), *Handbook of Surface and Interface Analysis: Methods for Problem-Solving*, Marcel Dekker, New York, 1998, 707. With permission.
Note: •, good and ○, poor or unadapted. ToF-SIMS = Time-of-flight secondary ion mass spectroscopy.
[a] Refers to generic studies in Section 12.3 illustrating the use of the corresponding techniques.

oscillatory variation of x-ray absorption as a function of photon energy beyond the absorption edge. When the x-ray photon energy is tuned to the binding energy of the core level of an atom of a given material, an abrupt increase in absorption (called the absorption edge) occurs. For isolated atoms (e.g., as a gas), the absorption coefficient decreases monotonically as a function of the energy beyond the edge. For atoms in a molecule or in a condensed phase (such as a lubricant), the variation of the absorption coefficient at energies ranging from 40 to 1000 eV above the absorption edge displays a complex fine structure (Figure 12.4). A single-scattering short-range order theory is adequate to explain the structure beyond 50 eV from the threshold (EXAFS region). The technique is particularly

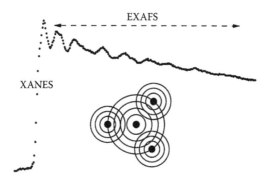

FIGURE 12.4   General shape of the ionization K-edge of an element as obtained by XAS. XANES gives chemical and electronic information; EXAFS reflects the short-range order in the material. Below: schematic view of the diffraction of the electron wave [52]. (From Donnet, C., in Rivière, J.C. and Myhra, S. (Eds.), *Handbook of Surface and Interface Analysis: Methods for Problem-Solving*, Marcel Dekker, New York, 1998, 708. With permission.)

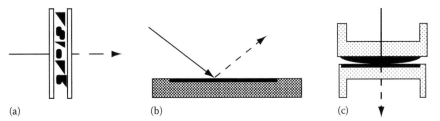

(a)   (b)   (c)

FIGURE 12.5   XAS as applied to tribology: (a) *ex situ* wear particle analysis in the transmission mode, (b) *in situ* surface analysis (total electron yield, x-ray fluorescence, or photoelectrons), and (c) *in vivo* frictional interface analysis in the transmission mode [52]. (From Donnet, C., in Rivière, J.C. and Myhra, S. (Eds.), *Handbook of Surface and Interface Analysis: Methods for Problem-Solving*, Marcel Dekker, New York, 1998, 710. With permission.)

successful in deriving the precise local order around selected atoms; the result of the EXAFS processing is the so-called radial distribution function (RDF), which is a function of the local environment of the selected atoms or, in other words, the number of neighboring atoms versus the radial distance. In a crystalline structure, two or three coordination spheres are typically observed, whereas in an amorphous structure only the first atomic shell is usually detected in the RDF.

XAS requires synchrotron radiation and a relatively large amount of material but no vacuum condition. On the other hand, EELS can be performed directly using an electron spectrometer fitted to a scanning transmission electron microscope (STEM). Here, the main advantage is the high spatial resolution attainable. (The incident electron beam can be as small as 1 nm in diameter, see Chapter 6.) EELS can also be coupled with conventional transmission electron microscope (TEM) facilities, and particularly with high-resolution transmission electron microscopy (HRTEM) and energy dispersive x-ray spectroscopy.

A general strategy for applying the above analytical techniques to a tribological test has been developed [15], as described in the generic study in Section 12.3.2. Bulk lubricants can be analyzed using a liquid cell, at different temperatures, and in various environments, since no vacuum is required. Local analysis of wear debris collected at the end of the test (*ex situ*) is always possible and can be carried out by XAS in the transmission mode (Figure 12.5a), preparation in that case being very easy. Alternatively, when high spatial resolution is needed, selected wear fragments can be analyzed in the STEM. For this purpose, they are deposited on a holey carbon film supported by a copper grid.

The analysis of surfaces is also possible by either x-rays or electrons (Figure 12.5b). If photoelectrons are used (as in XPS), their mean free paths are very small (~2 nm) and the outer surface only is probed, thus requiring UHV conditions. The techniques are referred to as SEXAFS (surface EXAFS) and SEELS (surface EELS). If high-energy secondary electrons or Auger electrons are detected instead, then the corresponding mean free paths indicate that the depth being analyzed is a few tenths of a nanometer, which is very suitable for tribochemical films. Finally, if x-rays are detected from fluorescence x-ray fluorescence (XRF) in the material, the information comes from much deeper regions.

The analysis of friction interfaces in motion is feasible (*in vivo* configuration, see Section 12.2.2) due to the high transparency of the matter to x-rays (Figure 12.5c). Martin and Belin have developed a tribometer that can be coupled to both XAS and Raman spectroscopy (Figure 12.6) [15], illustrating the desirable coupling of the time-scale criterion mentioned in Section 12.2.3 with the information from the analysis. For this purpose, x-ray-transparent supporting materials were machined, which allow the x-ray beam to probe the friction interface even during sliding, thus providing *in vivo pre mortem* experimental conditions, as depicted in the previous section. The test rig consists of a compact reciprocating pin-on-flat tester with the x-ray beam crossing the contact area. The motion is actuated by a vibrator, with a maximum amplitude of 5.0 mm and a frequency range from 0.2 to 20 Hz. The normal load is applied by a spring system. The beam dimension at the exit of the synchrotron monochromator has a rectangular section that can be collimated by shutters in both directions.

From a triboanalytical point of view, the EXAFS data, in terms of the RDF, allow identification of the location of the shearing process in the friction interface. For example, if the friction process

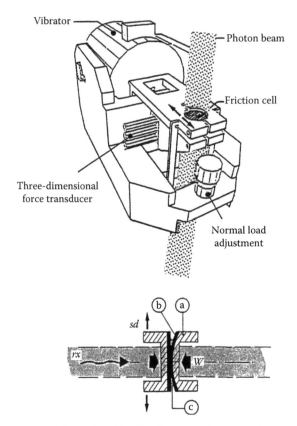

FIGURE 12.6 Schematic view of the *in vivo* EXAFS tribometer. Top: general arrangement. Bottom: the friction cell, consisting of (a) low x-ray absorbent carbon supports, (b) thin iron foils (2 × 7 μm²), (c) the lubricant or coating suffering friction; *sd* is the sliding direction, and *W* the applied normal load. The photon beam, represented by the gray line (*rx*), is transmitted through the friction cell [52]. (From Donnet, C., in Rivière, J.C. and Myhra, S. (Eds.), *Handbook of Surface and Interface Analysis: Methods for Problem-Solving*, Marcel Dekker, New York, 1998, 711. With permission.)

involves only a few easily sheared planes, as in lamellar compounds, the EXAFS parameter of the whole interface material (wear debris plus surface films) will be practically unchanged. This is because the EXAFS signal is averaged over all identical atoms while shear planes involve only a few. Alternatively, if the velocity accommodation occurs on the atomic scale, implying that all atoms are continuously but perhaps only slightly displaced from their initial crystal sites during friction, then the EXAFS parameter will be drastically modified. Thus, EXAFS is able to give information about the localization of the shearing process and to specify the velocity accommodation mechanisms.

XANES (or ELNES) reflects the electronic structure of selected atoms and can give vital information on electron transfer processes, due to the well-known chemical shift of the ionization edge energy. In addition, the hybridization of light elements (carbon, nitrogen, and oxygen) can be deduced from the near-edge structure because that structure is a function of the distribution of antibonding orbitals (or unoccupied levels). As both structural and chemical modification often occur in many tribochemical processes, this type of analysis is unique in that all the information is contained in the XAS (or EELS) spectrum and can be processed accurately. The generic study in Section 12.3.2 describes investigations of tribochemical films in boundary lubrication performed by this methodology.

12.2.4.2 Surface Morphology

Surface roughness strongly influences the tribological behavior of contacting bodies in engineering devices and occurs over a wide dimensional range. At one extreme, the microscopic surface roughness (up to 10^{-5} m) depends on the process by which the surface was produced. At the other extreme, investigations by the atomic force microscope (AFM) and the scanning tunneling microscope (STM) show surface roughness on an atomic scale. Surface topography anisotropy can also appear over a wide size range and can influence the friction and wear process. At the grossest level, some surface preparatory processes can induce anisotropy (e.g., turning), whereas others do not (e.g., lapping). At the finest level, the surface crystallographic structure can also induce anisotropy of the crystalline orientation. Whatever the scale of anisotropy, surface topographic investigations are crucial, since not only can surface morphology influence tribological behavior, but also conversely the tribological process can modify surface morphology, due to local stresses and heating of the contacting asperities during friction.

The techniques used extensively (but not exclusively) for obtaining surface topographical information are stylus profilometry, stereomicroscopy, optical interferometry, and scanning probe microscopies (SPMs), including STM and AFM (see Chapter 5). Chung [16] has made a comparison of the principles, strengths, and weaknesses of these techniques. As mentioned by Bhushan et al. [17], recent use of the probe microscopies has allowed a systematic investigation of surface and interface phenomena at the atomic and molecular scales in tribology. STM, which provides images of electrically conducting surfaces with atomic resolution, is used for the imaging of clean surfaces and lubricant molecules. AFM provides topographical measurements of electrically conducting or insulating surfaces on the nanoscale. Moreover, when coupled with computational techniques for simulating tip-surface interactions and interfacial properties, the SPMs provide a better fundamental understanding (in terms of adhesion and friction) of interacting surfaces in relative motion, as well as a guide and a methodology for rational design in the lubrication of microdevices (see the next subsection). Since direct contact exists between the probed surface and an AFM tip, the latter technique also allows the measurement of ultra-small surface forces. Specific modifications of the AFM have led to the development of the friction (or lateral) force microscope, which measures forces transverse to the surface, allowing atomic and microscale studies of friction. Reviews of recent developments in the use of tip-based microscopies in the field of tribology can be found in Ref. [17] and in Chapter 5.

12.2.4.3   Physical, Mechanical, and Frictional Surface and Interface Properties

New fundamental insights into boundary and thin-film lubrication have been provided recently, thanks to increased interest in the measurement of the properties of molecular thin liquid films sandwiched between two contacting surfaces [17,18]. Since the scope of such studies does not match exactly classical surface analysis, it is not proposed to go far in that direction here. Nevertheless, the study of confined thin films is a traditional area that appeared first in colloid and adhesion science prior to tribology. In tribology, it has considerable practical importance since the interaction between loaded engineering counterfaces occurs through the close proximity of high spots or asperities, with locally high contact pressures and temperatures. Homola [18] has reviewed the different experimental approaches to the direct measurement of the physical properties of thin liquid films with thicknesses approaching the dimensions of the liquid molecules themselves. Bhushan et al. [17] have provided an overview dealing with the development of experimental and computer simulation techniques for the study of these phenomena on the atomic scale, thus providing an emerging global understanding of the molecular mechanisms of tribology in thin films and at surfaces.

Experimental methods included the surface force apparatus (SFA) for measuring very weak forces, submicroscopic surface geometries, and surface separation on the nanoscale. The SFA, originally developed to study static or equilibrium interfacial forces, has been modified recently to measure the dynamic shear response of liquid confined between sliding surfaces [19–23]. The objectives were to ascertain (1) how the physical properties of a liquid localized in small spaces (pores, crevices, and thin films) differ from those of the bulk material and (2) how these properties vary with the size or the thickness of the confined molecules, in order to provide a description of the transitions from continuum to molecular behavior of very thin layers. When films are globally thicker than 10 molecular diameters, both static and dynamic properties can be described in terms of their bulk properties. However, for thinner films, the shear behavior becomes progressively more solid-like and their bulk properties are altered drastically, due to relationships between the molecular ordering (or glassing) and the structural (molecular architecture), physical, and mechanical properties of the interface. For example, SFA experiments carried out with complex fluid or polymer films indicate that branched-chain molecules lubricate better than straight-chain molecules, even though the former have much higher bulk viscosities; the symmetrically shaped straight-chain molecules are prone to ordering and freezing, which dramatically increases their shearing resistance, whereas the irregularly shaped, branched molecules remain in the liquid state even under high loads [17]. This explains why straight-chain molecules are particularly appropriate when high friction is required, such as in clutch mechanisms.

From a technological point of view, the interest in this kind of study is due to the possibility of controlling friction and wear in real conditions approaching boundary lubrication, by judicious choice of the lubricant fluid and by chemically grafting chain-like molecules such as surfactants or polymers to surfaces. In the latter case, the absence of traditional fluid lubricant owing to the self-lubricating grafted surfaces indicates that surface investigations on the nanometric scale in molecular tribology open up the possibilities of new types of lubricant systems for many applications, including microdevices.

12.3   GENERIC STUDIES

Ten generic studies have been chosen to illustrate how a variety of surface and other analytical techniques can be combined intelligently to investigate tribological systems, taking into account the dimensional and time-scale criteria along with the nature of the required surface information, as discussed in the previous section. The common objective is to arrive at the tribological mechanisms responsible for the observed friction and wear behavior. Table 12.2 sets out the subjects of the selected studies, whose relevant results are summarized in the following.

12.3.1   Ultrathin Boundary Lubricant Films

This section addresses the following question: How does one measure the minimum lubricant film thickness in boundary lubrication? [24,25].

In boundary lubrication, several complex and interactive phenomena occur (adsorption and tribochemical reaction) at film thicknesses below the minimum at which quantitative measurements of most relevant surface properties can be made. One of the most influential tribological parameters in the lubrication mode of an engineering device is the thickness of the lubricant film. Historically, there has been considerable controversy as to whether boundary films could be considered thin (a few molecular thicknesses) or thick (>10–100 molecular thicknesses). Since the 1980s, Johnston et al. [25] have developed the technique of ultrathin film interferometry (UFI), consisting of an extension of conventional optical interferometry, allowing lubricant film thickness measurements in the hydrodynamic mode to be made. UFI has been refined recently so that it can be used to measure film thicknesses down to about 1 nm, that is, of molecular dimensions, thus being a powerful means of measuring the thicknesses of boundary lubricating films in actual rubbing contacts. It thus illustrates the *in vivo pre mortem* analysis configuration performed in tribology.

The principle of UFI can be described on the basis of Figure 12.7 [24]. A steel ball is loaded hydraulically against the underside flat surface of a rotating glass disk to form a circular concentrated contact. The ball is contained in a lubricant bath and is driven by the disk in nominally pure rolling motion. The test rig can be heated to temperatures up to 200°C, that is, consistent with conditions relating to very low lubricant thickness due to low viscosities. The contact area is illuminated by white light. Compared to conventional optical interferometry, this apparatus is original in that it has, first, a spacing layer of transparent silica about 400 nm thick coated on the underside of the

TABLE 12.2

Overview of the Generic Studies in Section 12.3, Illustrating the Analytical Methodologies Used to Investigate Surfaces and Interfaces in Tribology [52]

Study	System	Methodology	Information Sought	Technique	References
12.3.1	Ultrathin boundary lubricating films	*In vivo pre mortem*	Film thickness	Thin-film interferometry	[24,25]
12.3.2	Tribochemistry of antiwear additives	*Ex situ*	Composition/bonding	TEM/EELS	[4,15,26–28]
		In situ post mortem	Hybridization/phases	XAS/CEMS	[2,29]
		In vivo pre mortem	Local structure	AES	
12.3.3	Tribochemical activity of nascent surfaces	*In situ post mortem*	Chemical activity	Mass spectrometry	[31]
12.3.4	Nature of surface versus tribochemsitry	*In situ post mortem*	Composition/bonding	XPS/AES HREELS/STM	[32]
12.3.5	Monolayers—dry friction	*In situ pre mortem*	Composition/bonding	AES	[33]
12.3.6	Tribochemistry of SiC versus O_2 partial pressure	*Ex situ*	Composition/bonding	AES TEM/EELS	[34]
		In situ post mortem	Phase/local structure	EXELFS	—
12.3.7	Durability versus microstructure MoS_2 coatings	*Ex situ*	Structure	XRD	[35]
12.3.8	MoS_2 sliding in UHV	*Ex situ*	Composition/bonding	XPS/AES	[36–38]
		In situ post mortem	Nanocrystallinity	HRTEM	
12.3.9	Tribology of DLC films	*Ex situ*	Composition/bonding and phase	FTIR/SPM	[41–43]
		In situ post mortem		ToF-SIMS AES	
12.3.10	Tribochemistry of C_{60}	*In vivo post mortem*	Hybridization/phase	Raman	[29]

Source: Donnet, C., in Rivière, J.C. and Myhra, S. (Eds.), *Handbook of Surface and Interface Analysis: Methods for Problem-Solving*, Marcel Dekker, New York, 1998, 707. With permission.

chromium-plated glass disk. Acting as a supplementary oil film, the layer enables optical interference to be observed even when the actual oil film present is very thin. Second, it has a spectrometer dispersing the reflected interfered light, allowing a significant increase in the precision of the thickness measurement. The dispersed spectrum is computer-analyzed to identify the wavelength for which there has been the most constructive interference. When measurements of the coefficient of friction are needed, the rig is the same, except that the glass disk is replaced by a smooth, hardened, polished stellite disk, with a surface finish similar to that of the glass disk.

Experimental investigations carried out with various kinds of mineral- and organic-based lubricants have shown that some of them can form boundary layers up to 20 nm thick, with viscosities different from the bulk fluid. These layers may be formed either by adsorption of monolayers of polymers on surfaces or by the accumulation of more polar, low-molecular-weight species near the solid surface due to van der Waals attraction. Deviation from classical EHD film-forming behavior is introduced by these tribo-induced layers, resulting in the observed variations in film thickness. The boundary layers can also shift the point of onset of mixed lubrication, as observed by changes in the coefficient of friction. Thus, the UFI technique leads to considerable improvement in clarifying the chemical and rheological contributions of the different additives and base oils to the surface

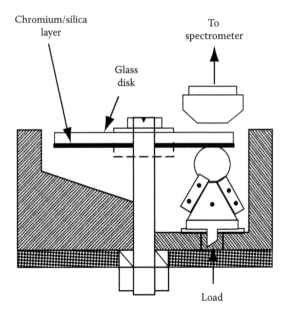

FIGURE 12.7 Test rig for ultrathin layers used for *in vivo pre mortem* thickness measurements in boundary-mixed lubrication. (From Spikes, H.A., in *Proceedings of the International Tribology Conference, Satellite Forum on Tribochemistry*, Yokohama, Japan, 1995, 49. With permission.)

separation of contacts under realistic conditions, including the highest temperatures of use consistent with the lowest viscosities and therefore thicknesses.

12.3.2 Tribochemistry of Antiwear Additives in Boundary Lubrication

Here the questions asked are as follows: What kinds of physicochemical interactions occur between antiwear additives and the contacting surfaces, and how do they depend on the nature of the environment during friction? [4,15,26–29].

The role of antiwear additives in boundary lubrication has been mentioned briefly in Section 12.1. The way in which zinc dithiophosphate (ZDDP) acts as an antiwear additive is principally that of the elimination of the abrasive wear contribution of crystallized iron oxide species, resulting from their tribochemical reactions with the additive molecules and various degradation products. The use of XAS and TEM/EELS together with complementary conversation electron mössbauer spectroscopy (CEMS) and AES represents a powerful analytical combination in elucidating the tribochemical reactions responsible for antiwear mechanisms. The first part of this section summarizes analytical results obtained with friction tests carried out in the ambient atmosphere. The second part describes some original results obtained with friction tests and *in situ post mortem* AES analysis in UHV. In this way, the intrinsic frictional properties of the solid reaction film itself, without the effect of the environment, and the ability of friction to modify the nature and the structure of the surface films have been investigated.

12.3.2.1 *Ex Situ* Surface Analytical Investigations

The solid wear particles from ZDDP films produced in lubricated contacts contain phosphorus, sulfur, and zinc from the ZDDP molecule and oxygen mainly from the surrounding air environment, and also a low iron content, depending on the severity of the test [4,15,26,29]. *Ex situ* examination (see Figure 12.5a by XAS, EELS, and CEMS) of these types of particle has been carried out [4] to provide local analysis of the iron atoms, since their atomic environment in the surface film is of great

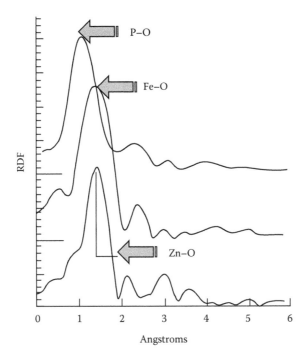

FIGURE 12.8 RDFs of iron, phosphorus, and zinc obtained by XAS recorded from wear debris, with dodecane + 1% of ZDDP antiwear additive, as the lubricant.

interest, not only because of its direct relationship to the wear of the steel surface, but also because it can play a key role in the adhesion mechanism of the film. XAS (EXAFS plus XANES) measurements were carried out on a collection of wear debris from a lubricated test with ZDDP in the lubricant base. The fully processed EXAFS data in Figure 12.8 show the RDFs of iron, phosphorus, and zinc atoms (uncorrected phase shifts) in the presence of ZDDP. From these data, some important results can be deduced:

- Iron atoms have become bound to oxygen in the wear debris. The main peak in the RDF is attributed to an Fe–O separation of 0.19 nm.
- In the presence of the additive, and during the steady-state friction regime, iron atoms in the wear debris are found to be isolated from each other, because the second Fe–Fe peak in the RDF has decreased considerably, indicating a high level of crystallographic disorder.
- Phosphorus and zinc are also bound to oxygen in an amorphous manner.

To complete these studies, XANES was used [29] to record the Fe–K near-edge structure, and ELNES the O–K edge, in a TEM. Results indicated the presence of a fully oxidized Fe^{3+} state in octahedral symmetry and the presence of some residual iron oxide Fe_3O_4 [4]. Additional CEMS measurements (Figure 12.9) confirmed the presence of octahedrally coordinated ferric cations.

From these results, the transition from the abrasive wear to the low-wear regime could be explained in terms of velocity accommodation mechanisms in the interface film. The most important phenomenon is a complex polyphasic tribochemical reaction, responsible for the formation of a solid transition-metal phosphate glass material as an adherent thin film. This acts as a protective layer against wear due to its superplastic behavior when it is formed in the contact area. Crystalline iron oxides have very rigid atomic skeletons with selective plastic deformation processes and, as particles, are abrasive. By comparison, the tribo-induced phosphate glass has chemical bonding

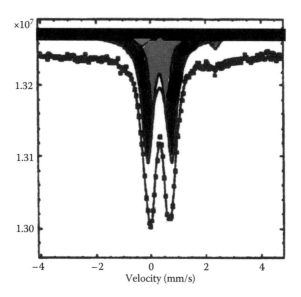

FIGURE 12.9   Mössbauer spectrum recorded on wear debris from a lubricated test with ZDDP (recorded at 295 K, Gbq ^{57}Co /Rh source). From the processing of the spectrum, it was found that the material is composed of 95% of Fe^{3+} in octahedral sites and that most of these sites are distorted.

angles, which are flexible in all directions, and which can therefore undergo slight deformation; in addition, ionic iron atoms can diffuse easily in the multiple-connected channels. Consequently, the velocity is accommodated during the shearing process by distribution throughout the film thickness due to the accumulation of atomic-scale displacements, somewhat similar to the hydrodynamic regime. The abrasive wear at the early stage of friction was thus almost eliminated since the crystalline iron oxides were progressively digested in the glassy phosphate material, which was able to support much greater deformation because of its superplastic behavior.

The glassy structure of the tribofilm can be demonstrated well by focused ion beam-transmission electron microscopy (FIB-TEM) [30]. The wear fragment material is fixed, the rapid relaxation of shear and pressure being equivalent to thermal treatment, and analysis of it should therefore be able to reveal what was on the surface before and during friction. The validity of this assumption can be checked by studying the possible differences between the microstructure of wear debris (Figure 12.5a) and of the surface film (Figure 12.5b), formed in a single friction experiment. This has been achieved by comparing the environments of zinc atoms present in both ZDDP wear fragments and surface films, with the advantage that the same analytical tool can be used. Wear debris is normally analyzed by XAS in the transmission mode while worn surfaces are analyzed in the reflection mode, using the total electron yield, including high-energy Auger electrons (in this case the thickness analyzed is ~20 nm). EXAFS was recorded at the Zn–K edge (9659 eV).

The analysis showed that the zinc atoms had similar environments both in wear fragments and on the top of the surface film, suggesting that the wear mechanism was dominated by nonequilibrium states, the tribochemical reaction products being quenched, and the microstructure being fixed at the end of contact, due to the rapid relaxation of shear and pressure [4].

12.3.2.2 *In Vivo Pre Mortem* Surface Analytical Investigations

Localized analysis of zinc atoms in a dynamic interface (*in vivo pre mortem* analysis) has been carried out successfully in the tribometer whose design is shown in Figure 12.6, although the lubricant quantity analyzed in the operating interface was very small (e.g., a few microliters) [27]. Despite lack of quality due to poor signal/noise ratio, the zinc signal was strong enough to be processed. The frictional motion does not disturb unduly the *in vivo* analysis (the acquisition time was 5 min in the sequential mode, but could probably be reduced to a few seconds with a parallel detection system). It could be claimed that the analysis of interfaces in motion was now feasible, but in fact any tribo-chemical reaction of zinc atoms could not be observed in this, the first experiment on the Orsay synchrotron line, since the initial Zn–S bonding of the molecule remained Zn–S after a friction experiment of 7 h duration. Unfortunately, no reaction film was formed on the surfaces, possibly due to the relatively gentle friction conditions. Future studies involving real-time EXAFS on the frictional interface in motion will be possible with a parallel detection system, which will allow much shorter spectral recording times.

12.3.2.3   *In Situ Post Mortem* Surface Analytical Investigations in UHV

Some fundamental aspects of the tribochemistry of ZDDP have been investigated by performing tribotests in UHV on selected chemisorbed films previously formed on steel bearing surfaces [28]. In order to isolate the effect on friction of the outer surface contamination layers, their tribological behavior was studied by coupling friction tests with *in situ post mortem* AES microanalysis before and after ion etching. It was found that the native oxide film, of thickness on the nanometer scale on the steel surface, had high wear resistance but very high friction in UHV. However, at the beginning of the test, friction was reduced by the carbon-rich contamination layer, which was removed mechanically by a few reciprocating friction passes without reaction with the surfaces. Friction between the pure metals (after ion etching) was lower than unity but adhesive wear and considerable plastic deformation took place.

In further experiments, *in vacuo* friction tests were performed using a steel pin rubbing against a previously formed ZDDP tribofilm. Two experiments were then carried out; the first with only three-cycles of friction using the reciprocating pin-on-flat tribometer, the second for seven cycles of the steel pin on the ZDDP tribofilm. A transfer film composed of tribofilm elements (sulfur, zinc, oxygen, and occasionally phosphorus) was always observed after rubbing the steel pin on the tribo-film. The transfer film was localized in the center of the track, where the pressure was at its maximum, and did not accumulate at the periphery. The radius of the wear scar was near the theoretical Hertzian calculated value.

In the AES depth profile on the pin wear scar (after seven cycles), phosphorus appeared after etching for 200 s, although zinc and sulfur had been detected from the start. A depth profile outside the track (not shown here) was performed in exactly the same etching conditions. It took 1500 s of etching to remove the native oxide layer, estimated to be 4 nm thick. If the etching speed in the transfer film is assumed to be similar to that of iron oxide, the thickness of the transfer film could be calculated to be 10 nm. Moreover, it was found that the top surface layer was composed of zinc and sulfur (probably ZnS) with a thickness of about 0.5 nm.

The magnitude of the O_{KLL} Auger peak during depth profiling is shown in Figure 12.10 for depth profiles of the pin transfer film after three and seven cycles of friction (Figure 12.10a and b, respectively). After three cycles, from the start to around 2000 s of etching, the O_{KLL} peak could be deconvoluted using the two reference spectra of the oxide and phosphate forms, respectively, of iron. The results showed that oxygen as both oxide and phosphate formed at the same time. After the three-cycle test, when the substrate was reached by ion etching, a thin layer with oxygen in oxide form was found. Thus, a significant part of the native iron oxide was still intact between the transfer film and the substrate. However, after seven cycles, it was clear that the oxide form had practically disappeared, in favor of the phosphate form.

The disappearance of the native oxide layer can be explained by a chemical reaction between the transfer film on the flat and the native oxide layer on the pin, following Pearson's HSAB (hard and soft acids and bases) approach [48]: a hard base prefers to react with a hard acid, and a soft base with a soft acid. Here, the hard base is the phosphate ion, and the hard acid is the ferric ion. In other words, the phosphate glass has digested the iron oxide after seven cycles, according to the reaction

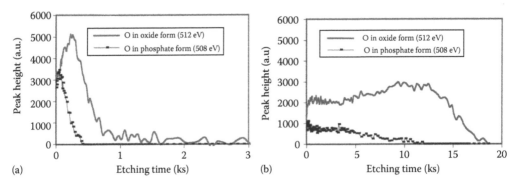

FIGURE 12.10 AES profiles performed on the transfer film on the pin. Friction test in UHV after (a) three cycles and (b) seven cycles. During the friction process, it was shown that the iron oxide film was digested progressively by the phosphate film. (From Minfray, C., Martin, J.M., Lubrecht, T., Belin, M., and Le Mogne, T., in *Proceedings of the 30th Leeds-Lyon Symposium*, September 02–05, 2003, Lyon, France, *Transient Process in Tribology*, Dalmaz, E. and Lubrecht, T. (Eds.), Elsevier, Amsterdam, the Netherlands, 2004, 43. With permission.)

$$5Zn(PO_3)_2 + Fe_2O_3 \rightarrow Fe_2Zn_3P_{10}O_{31} + 2ZnO \tag{12.1}$$

$$(ZnO, P_2O_5)(Fe_2O_3, 3ZnO, 5P_2O_5)$$

The formation of ZnS is also predicted by this reaction because residual sulfur in the polyphosphate chain can promote the formation of zinc sulfide (ZnS) in the chain-shortening polyphosphate process:

$$4Zn_6(P_{10}O_{29}S_2) + 10Fe_2O_3 \rightarrow 10Fe_2Zn(P_2O_7)_2 + 8ZnS + 6ZnO \tag{12.2}$$

Thus for friction experiments under UHV, a transfer film on the pin was formed after only three cycles of friction, consisted of a tribofilm material containing phosphorus, oxygen, sulfur, and zinc. After three cycles, the native oxide layer was still present between the transfer film and the substrate, but after seven cycles, it had mostly disappeared, due to a digestion by the phosphate glass (acid–base reaction). A very thin ZnS film covered the surface and is presumed to have controlled the overall friction value.

Since the effect of the environment was eliminated, the overall results demonstrate that tribochemical reactions can take place during friction in an UHV environment. A judicious combination of analysis by *ex situ* and *in situ* methods has thus allowed the complex tribochemical mechanisms responsible for friction and wear behavior in boundary lubrication to be elucidated.

12.3.3 TRIBOCHEMICAL ACTIVITY OF NASCENT SURFACES

Here the question is as follows: What is the role of friction-induced fresh surfaces in boundary lubrication? [31].

As mentioned above, tribological behavior in boundary lubrication is strongly affected by surface reactions as well as by chemisorption of additives on contacting surfaces. The molecular structure of the additives is not the only parameter responsible for any particular tribochemical behavior. During the friction process, contamination layers on the outer surface are removed mechanically, leading to the formation of fresh, nascent surfaces characterized by enhanced chemical activity, and it is the chemical nature of these nascent surfaces that is one of the most important parameters, since such surfaces are active sites for tribochemical reactions. *In situ* quantification of the chemisorptive activity

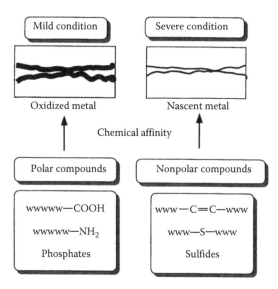

FIGURE 12.11 Tribological conditions and surface chemistry in boundary lubrication; the symbol "w" refers to a chemical bond. (From Mori, S., in *Proceedings of the International Tribology Conference*, Satellite Forum on Tribochemistry, Yokohama, Japan, 1995. With permission.)

of fresh surfaces created in controlled environments (i.e., nature and pressure of ambient gases), following a standardized scratch test of the studied surfaces, has been undertaken. The chemisorptive activity of a given gas on a given scratched surface can be estimated from the time dependence of the pressure decrease during the adsorption process, since the slope of the pressure–time curve is proportional to the sticking coefficient. During a systematic investigation of the chemical activities of various metals and alloys with respect to gaseous organic compounds, systematic-enhanced activities were observed on the nascent surfaces. A comparison of the chemical reactivities of various organic compounds as a function of the nature of the surfaces is shown in Figure 12.11.

The gas/surface chemical reactivity depends on the polarity of the organic compound and on the chemical nature of the solid surface, that is, either nascent or oxidized; nonpolar compounds chemisorb on nascent metallic surfaces more easily than on oxidized metal surfaces, whereas polar compounds can chemisorb readily on the latter. From Pearson's HSAB principle [43], nonpolar compounds are classified as soft bases, and interact more strongly with soft acids such as clean metal surfaces rather than with hard acids, that is, oxidized metals. Metal surfaces are covered with oxide layers, which under severe conditions can be removed by friction. Thus, the chemical nature of metal surfaces is closely dependent on the lubricating conditions, that is, oxidized under mild conditions and nascent under severe conditions. Consequently, the additives that should be effective under mild conditions will be polar compounds such as fatty acids and phosphates, whereas those likely to be effective under severe conditions will be nonpolar compounds, such as organic sulfides.

This confirms previous tribological investigations, which showed that phosphate additives were effective under low load conditions (mild wear), whereas sulfides were effective under high load conditions (severe wear).

12.3.4 INFLUENCE OF THE NATURE OF THE SURFACE ON THE TRIBOCHEMISTRY OF VARIOUS TRIBOMATERIALS

The questions to be answered here are as follows: What kinds of tribochemical reactions occur at the friction interface, and how do they depend on the nature of the contacting bodies and of the lubricant? [32].

In order to understand the influence of the surface chemistry on the friction of two contacting solids, DeKoven [32] has carried out friction measurements in UHV and controlled gaseous environments. They were coupled with *in situ post mortem* XPS/AES and *ex situ* STM analyses on various pairs of contacting materials, that is, metal/metal (Fe/Fe), metal/ceramic (Fe/Al$_2$O$_3$), ceramic/ceramic (B$_4$C/B$_4$C), and steel/steel, with organic lubricants. Friction was measured in well-controlled environments, ranging from UHV to pure gases and even liquids. With a purpose-built pin-on-flat device that allowed contact interchange in UHV, *in situ post mortem* XPS/AES analyses were performed inside the tribometer. The clean and lubricant-covered steel surfaces were also examined by HREELS to study the evolution of molecular bonds in the metal/organic compound interface, after various thermal treatments were carried out independently of the friction test, in order to separate thermochemical and tribochemical phenomena.

The friction of Fe/Fe contacts in UHV decreases as the superficial oxide thickness increases within the controlled range from zero in the clean metallic state to about 4 nm in the oxidized. O/Fe ratios corresponding to different oxide thicknesses were established by AES linescans outside and on each side of a wear track. The nature of the thin oxide interfaces in metal/metal contacts was found to govern the friction process, as shown already in the two generic studies described in Sections 12.3.2 and 12.3.3. Attempts to understand the surface chemical and mechanical interactions of Fe, with films formed from the reaction of a model perfluorodiethylether, were also made. The fluorine films were found to be able to sustain loads much better than thin Fe oxide films, as demonstrated by lower friction values and by STM profiles of the wear scars.

Between Fe and Al$_2$O$_3$ in UHV, the friction depends both on the presence of an oxidized outer layer on the iron pin surface and of a carbonaceous contamination layer on the alumina plate. AES analyses carried out on the transfer film formed on the iron pin showed that the film contained relatively pure carbon in some areas and iron oxide plus carbon in others. The amount of carbon transferred from the alumina outer surface to the iron therefore had considerable influence on the level of friction.

B$_4$C is one of the hardest ceramic materials in the temperature range 20°C–1800°C. Its friction behavior was investigated in UHV and in ambient air, for clean surfaces and for those exposed to air prior to friction. Friction coefficient values ranged from 0.2–0.3 to 0.8–1.8, depending on the environment during friction (i.e., low in air, high in UHV) and on the surface cleanliness (i.e., either air-exposed or cleaned prior to UHV friction). AES linescan analyses of the wear scars revealed that the compositions of both the pin and the flat wear track were quite different from that of the noncontacted area. For the contact whose surface had been cleaned prior to friction, an increase in the boron signal and a decrease in the carbon signal were observed. For the surface exposed to oxygen prior to friction, there was evidence of a tearing of the oxide layer near the center of the scar.

Answers to the following questions in both tribology and surface science were also sought: (1) What is the nature of the chemical interactions between thin layers of adsorbed high-temperature lubricants and both clean and oxidized surfaces? (2) Do enhanced surface chemical reactions occur at liquid/metal interfaces under the influence of friction? (3) If so, is the formation of the reaction products due to momentary thermal excursions or to other consequences of friction, such as mechanically induced bond breaking, to be regarded as a tribochemical phenomenon?

In order to separate thermochemical effects unambiguously from tribochemical effects, surface chemical and friction studies were carried out in parallel on steel (M50) surfaces, either oxidized or in an almost atomically clean condition after preparation in UHV. The thermal surface chemistry was studied at different temperatures (from 90 to 750 K) using HREELS and XPS, while the tribochemical interactions were investigated by *in situ post mortem* XPS after friction tests in UHV. Three fluids, strong candidates for high-temperature lubrication under high-temperature conditions, were used; a classical polyphenyl ether (PPE) and two aryloxicyclotriphosphazenes (HMPCP and HTFMPCP), with the same molecular structure except for the nature of the group (–CH$_3$ or =CF$_3$) on the phenoxy ring (Figure 12.12). Despite the similarity of the two phosphazene molecules, substantial differences in the thermochemistry and tribochemistry were found from the surface analytical investigations. Thermally induced chemical changes were observed on the clean and oxide-covered

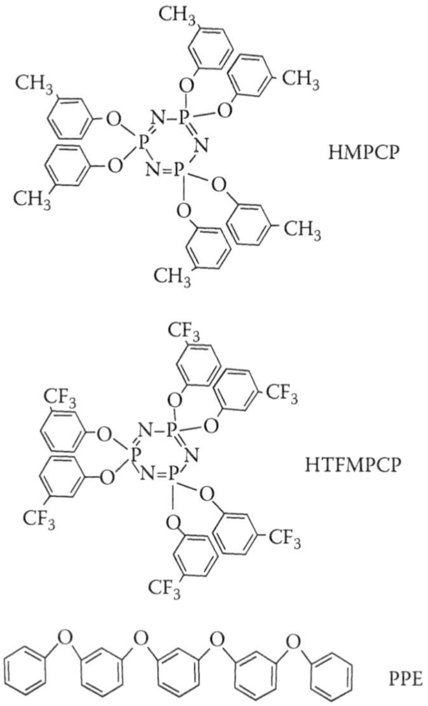

FIGURE 12.12 Molecular structures of the three organic lubricant fluids used for thermochemical and tribochemical investigations. (From DeKoven, B.M., *Surface Diagnostics in Tribology*, World Scientific, Singapore, 1993. With permission.)

surfaces for both phosphazenes after thermal treatment above 470 K. The CF_3-containing molecule reacted both at the aromatic ring (as did PPE) and at the CF_3 group. Surprisingly, the CH_3-containing molecule did not show any reactivity at the aromatic ring. Breaking of the aryl-O bond was observed for both phosphazenes. In tribological behavior, the CF_3-containing molecule and the PPE both exhibited tribochemical phenomena, unlike the CH_3-containing molecule.

The tribochemical changes were identified by *in situ post mortem* XPS that could analyze the interface directly, because of the low thickness of the lubricant film (<2 nm). It was found that metal fluorides and oxides were formed after the use of the CF_3-containing molecule and the PPE, respectively. On the other hand, no chemical change associated with the CH_3-containing molecule was found. Since the tribochemistry was not believed to be thermally induced, because of the low loads and sliding speeds, the most likely mechanism was one in which the shear stress due to sliding caused mechanical bond breaking (i.e., a mechanochemical effect), for both the CF_3-containing molecule and the PPE. If only one layer of fluid molecules separated the two sliding surfaces, all the shear stress (except that dissipated by plastic deformation of the solid surfaces) must have been concentrated in this single monolayer. Since the PPE and the CF_3-containing molecules were anchored strongly to the surface, some of the energy dissipation must have induced bond breaking in the fluid, leading to metal-O or metal-F tribo-induced bondings. By contrast, the stress applied to the more weakly bonded CH_3-containing molecules might have been relieved by ploughing through the lubricant film, which would have broken the surface-fluid bonds without inducing any tribochemical changes in the molecular structure.

Studies of this kind emphasize the role of surface chemistry and the reactivity of organic compounds on solid surfaces in being able to separate the thermochemical from the tribochemical effects in boundary lubrication.

12.3.5 EFFECT OF ADSORBATE MONOLAYERS ON DRY FRICTION

The question here is as follows: What is the effect of adsorbed monolayers on friction? [33].

As mentioned in Section 12.2.1 in connection with the time-scale criterion, the use of surface analysis techniques separately from a friction rig normally requires the exposure of the specimen to air in transit between the friction test and the analysis procedure. That is why tribologists have developed *in situ* analytical tribometers that allow direct surface chemical investigations. The triboanalytical coupling may combine the friction device and the analytical technique in the same chamber, ideally in a *pre mortem* configuration. An example of this mode is shown in Figure 12.13 [33]. Direct *in vivo* AES analysis inside the dynamic contact is not possible in this case, unlike in the configuration described above in Figure 12.6. Thus, the choice of coupling mode linking the tribological

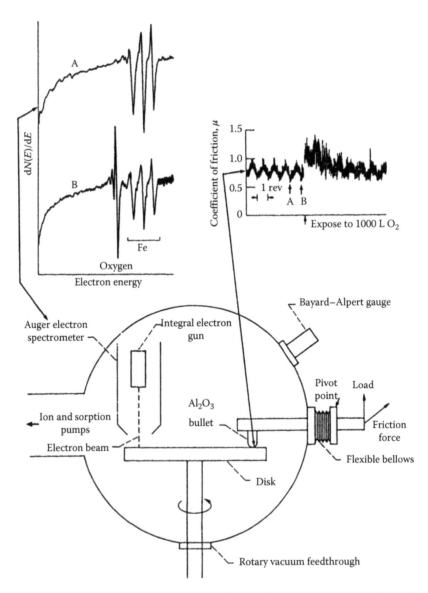

FIGURE 12.13 Example of *in situ pre mortem* analysis in a tribological experiment. (From Pepper, S.V., *J. Appl. Phys.*, 47, 2579, 1976. With permission.)

test and the analytical procedure depends strongly on the surface analytical technique chosen: *in vivo* AES is not possible, although XAS performed with x-ray transparent substrates might be. In a study by Pepper [33], *in situ pre mortem* AES was used to demonstrate the effect of monolayer thicknesses of adsorbates on friction in UHV and in various pure gases, for aluminum oxide sliding against iron, nickel, and copper. Because of the acquisition time required for AES, the surface wear scar analysis was performed sequentially by stopping the rotational motion just long enough to record an analytical signal. Clean metal surfaces were exposed to O_2, Cl_2, C_2H_4, and C_2H_3Cl, and the changes in friction due to the adsorbed species were measured. Figure 12.13 shows that an oxide-free iron surface (AES spectrum A) has a friction coefficient of about 0.7. Exposure to an oxygen partial pressure of about 10^{-7} hPa allowed oxygen to adsorb on the clean iron (AES spectrum B) and resulted in an increase in friction, demonstrating the significant effect of adsorbate monolayers on friction. It was also found that systems exposed to Cl_2 exhibited low friction, interpreted as being due to van der Waals interactions between the alumina and the metal chloride. The increase in friction resulting from the formation of metal oxides by oxygen exposure can be interpreted as due to strong interfacial bonds formed by the reaction of the metal oxide with alumina to form a complex spinel. The only effect of the C_2H_4 was to increase the friction of the Fe system, but C_2H_3Cl exposures decreased friction in both the Ni and Fe systems, indicating the dominance of the chlorine over the ethylene radical on the surface.

12.3.6 Tribochemistry of SiC/SiC under a Partial Pressure of Oxygen

An interesting question is "Is it possible to lubricate ceramic materials by a reactive gas phase?" [34].

The surface-chemistry-related tribological properties of α-silicon carbide (α-SiC) have been studied by Martin et al. [34] using an analytical UHV tribometer equipped with XPS/AES for *in situ post mortem* surface investigations. The experimental configuration was the same as that described in Section 12.3.2.3. The studies were carried out at room temperature in order to isolate the role of tribochemistry from that of any static interaction of the SiC surface with oxygen prior to friction. Spatially resolved *in situ post mortem* AES analyses were coupled with *ex situ* TEM/EELS and HRTEM analysis of the wear fragments. Figure 12.14 shows the friction and associated AES results, leading to the following observations:

1. Under high vacuum (10^{-10} hPa), SiC friction was dominated by high adhesion and by fracture, compaction, and attrition of the spheroidal grains in the interfacial region. No preferential shear plane appeared to form, but amorphization of SiC took place mainly at the edges of the grains. Consequently, the friction was high (0.8).
2. Under 0.5 hPa O_2 partial pressure, SiC quickly oxidized at the interface, producing silicon oxide, as identified by AES in Figure 12.14. *Ex situ* TEM/EELS and HRTEM examination (not shown here) identified the presence of rolling pins inside the wear track at the end of the friction test. These wear fragments consisted of bulk amorphous silicon oxide with a pre-graphitic carbon outer layer. This pre-graphitic component, having planes with low shear strength in the presence of oxygen, seemed to be responsible for the low shear strength of the interface film. Consequently, friction was lowered (<0.1) and corrosive wear was predominant on the SiC pin surface.

Thus, friction of α-SiC appears to be tribochemically controlled and very dependent on the reactive gases present in the surrounding environment. The simple tribochemical reaction involved in this experiment was identified by a judicious combination of *in situ* and *ex situ* surface analysis.

12.3.7 Relationship of Durability to Microstructure of IBAD MoS₂ Coatings

This section addresses the following question: What are the correlations between the microstructure of a solid lubricant and its durability? [35].

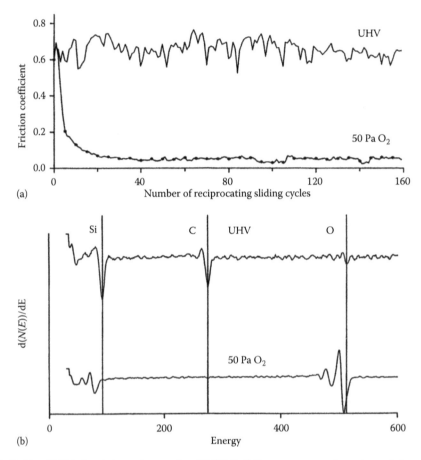

FIGURE 12.14 (a) Frictional dependence of an SiC/SiC sliding contact on the nature of the surrounding environment (UHV or pure oxygen) during 160 cycles of sliding friction. (b) Corresponding *in situ post mortem* AES analyses inside the flat wear scars. Similar AES analyses were performed outside the wear scars and showed persistence of the Si and C signals from both areas.

MoS_2 is one of the most widely studied solid lubricants, since it is in general use in conditions of high vacuum or extreme temperature, when conventional liquid lubricants fail. One of the determining factors in choosing a solid lubricant coating is the durability; once applied to the contacting surfaces, the coating should either lubricate throughout the system lifetime (especially in space mechanisms) or until possible renewal. The lubricating properties of MoS_2 are related to its layer structure (Figure 12.15). Covalent bonds join sulfur and molybdenum atoms in planar arrays of hexagonal S–Mo–S sandwiches, in which weak van der Waals interactions between adjacent and superimposed sulfur planes allow easy shear parallel to the sliding direction. The factors determining the durability of solid lubricants are complex and include the nature of the substrate, the coating thickness, the adhesion, the coating morphology, and the impurities dispersed in the lubricant layer.

Ion beam assisted deposition (IBAD) of various MoS_2 coatings was carried out by Seitzman et al. [35] on steel substrates in such a way that careful control of the crystallite orientation could be maintained by variation of the deposition conditions. Friction tests (rotational motion) in dry air were performed with standardized and reproducible contact pressure and speed conditions, and friction values in the 0.02 range within the steady-state regime were measured. *Ex situ* x-ray diffraction (XRD) established the precise crystalline orientation before friction, which provided very good correlation between the crystalline structure and the lifetime as defined by the duration of friction to

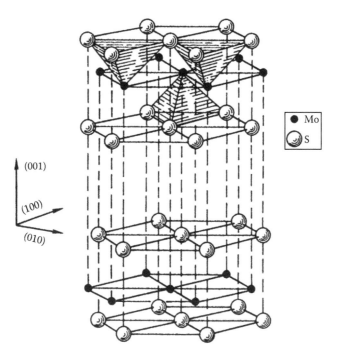

Mo
S

(001)

(100)

(010)

FIGURE 12.15 Crystalline structure of the 2H-MoS$_2$ solid lubricant showing the layered-sandwich atomic arrangement and the crystallographic axes. The (0001) basal surfaces are perpendicular to the (001) axis.

coating failure. The grain size of the IBAD deposited coatings was below 10 nm, based on the width of the characteristic MoS$_2$ diffraction peaks. From the XRD experiments, the (002) basal and the (100) edge intensity ratios were determined, taking into account the thickness and substrate variations from one coating to another and using an intense steel substrate peak as a reference. These intensity ratios were then used to define orientation, since they were a measure of the relative number of crystals with either a single element or both elements exposed in the outer surface. The MoS$_2$ crystal structure is ordered hexagonal with each basal plane containing only one elemental species. A high basal intensity ratio meant that the (002) basal planes were preferentially oriented parallel to the surface, and therefore to the sliding direction.

Coatings with high basal intensity ratios generally had high durability (>100,000 cycles), while those with low basal intensity ratios tended to have lower lifetimes. Thus, coatings with no edge-orientated crystallites exhibited the best durability. In addition, no relationship between durability and the coating thickness, or the nature of the substrate, or the presence of a TiN interlayer was found. This work illustrates the analytical needs required to study the optimization of the lifetime of a typical solid lubricant as a function of its preparative procedure.

12.3.8 Frictionless Sliding of Pure MoS$_2$ in UHV

This section addresses the following question: What is the origin of the superlow frictional behavior in solid lubrication by a pure MoS$_2$ coating? [36–38].

Whereas the above generic study emphasized the durability of MoS$_2$ as a lubricant, this one focuses on its low friction properties. Many tribological and analytical investigations have been devoted to this subject, in particular by Didziulis and Fleischauer [39] and by Roberts [40]. In this example, the vital contribution made by surface analysis in providing complementary information is well illustrated; it has helped in the understanding of why a particular MoS$_2$ structure should have an unusually low coefficient of friction under UHV conditions. In fact, it was found to be in the 10^{-3}

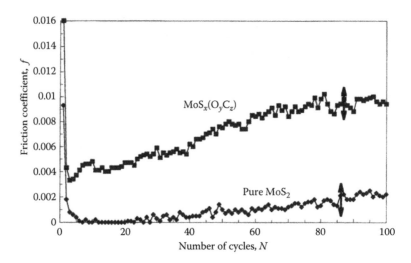

FIGURE 12.16   Average coefficient of friction versus the number of reciprocating sliding cycles of pure and contaminated molybdenum disulfide coatings, in UHV (10^{-7} Pa).

range, the lowest value ever observed in solid lubrication. This frictional behavior was compared to that of a conventional MoS_2 coating, as shown in Figure 12.16. The coefficient of friction of the conventional coating was generally about 10 times greater than that of the superlubricant. In both cases, a transfer film buildup was observed on the pin counterface. Analysis of the coating composition prior to friction, and of the wear scar and wear particle structure after friction (*in situ post mortem* and *ex situ* examinations), revealed the origin of the differences in frictional behavior. From a theoretical point of view, frictionless sliding between two atomic planes needs three conditions to be satisfied:

1. Weak interaction forces between sliding atoms
2. Atomically clean surfaces
3. Incommensurate sliding atomic lattices; that is, the friction force is minimized when a misfit angle shifts two superimposed sliding crystal planes with respect to each other, as visualized in Figure 12.17

Condition 1 is satisfied in the MoS_2 structure, which accounts for the ultralow coefficients of friction reported in the literature (10^{-2} range), and confirmed here with the conventional coating. Condition 2 requires UHV during friction together with the absence of oxygen contaminant. The oxygen content of both coatings prior to friction was measured by nuclear reaction analysis (NRA) using the resonant nuclear reaction $^{16}O(\alpha,\alpha')^{16}O$ stimulated with α particles at 7.5 MeV. The oxygen concentration versus depth for both coatings was also deduced from the NRA spectra, as shown in Figure 12.18. The conventional MoS_2 coating contained about 13 at% oxygen, whereas the nearly pure MoS_2 coating contained only 4 at% oxygen at its outer surface (probably due to surface oxidation subsequent to coating synthesis). The nuclear method did not of course indicate the site of the oxygen species in the conventional coating. The oxygen might have been present either as substitutes for sulfur atoms, or in small molybdenum oxide precipitates, or in water molecules dispersed in the coating. The first hypothesis is contradicted by the fact that oxygen substitution on sulfur lattice sites expands the c-axis, thus lowering the van der Waals interaction between chalcogen atoms, leading to a decrease in the frictional force. Since the opposite phenomenon (higher friction) was observed, the second hypothesis is more plausible. EXAFS investigations would be necessary to take further the correlation between frictional behavior and the nature of the coatings.

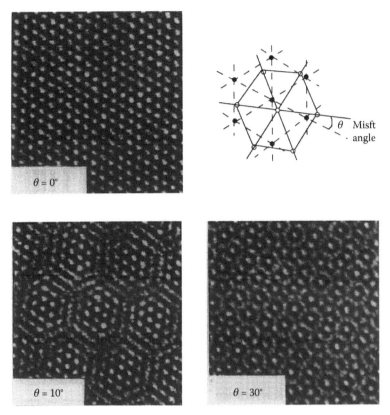

FIGURE 12.17 Visualization of two superimposed hexagonal lattices (representing adjacent sulfur planes in MoS$_2$), with no misfit angle ($\theta = 0°$) and with misfit angles of $\theta = 10°$ and 30°. The frictional force is known to be minimum at $\theta = 30°$, because of the hexagonal symmetry of the crystalline structure.

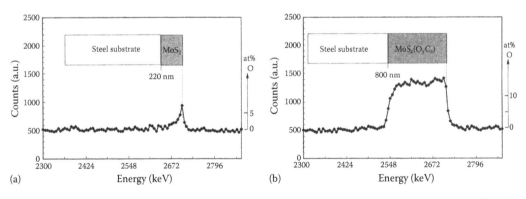

FIGURE 12.18 Quantitative oxygen depth profiles in (a) pure and (b) contaminated molybdenum disulfide coatings, as recorded by NRA.

The role of any oxygen contaminant was also established by *in situ post mortem* AES micro-analysis performed inside the wear tracks of both coatings in the same UHV conditions as during friction. Figure 12.19 compares the two AES spectra, which show clearly the absence of oxygen and carbon on the outer surfaces of the wear scars of the superlubricant MoS$_2$, in contrast to the commercial coating that contained both carbon and oxygen.

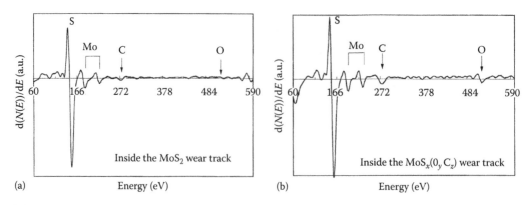

FIGURE 12.19 *In situ post mortem* AES microanalysis performed inside the wear scars of the superlubricant pure MoS_2 coating (a) and the contaminated coating (b) exhibiting higher friction.

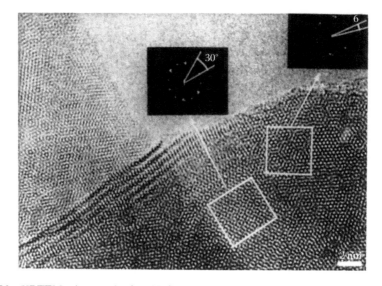

FIGURE 12.20 HRTEM micrograph of an MoS_2 wear particle, with atomic resolution. The electron beam was parallel to the c-axis. The calculated diffractograms correspond to the image frames showing Moiré patterns. Areas where atomic coincidence has disappeared are characterized by a misfit angle of 30° between two superimposed crystals.

Condition 3 was checked by a combination of *ex situ* AFM and *ex situ* HRTEM carried out respectively inside the wear scars and on the wear particles extracted from the contact area. AFM images (not reproduced here) confirmed the cleanliness of the external surface inside the wear scar, with the (002) easy-shear basal planes oriented parallel to the sliding direction. HRTEM micrographs from the wear fragments (Figure 12.20) showed the existence of Moiré patterns characteristic of superimposed crystals with a misfit angle between them, as simulated in Figure 12.17. Such information on the atomic scale thus helped to confirm the atomistic origin of the superlow frictional behavior and showed how it is possible for basic research to advance knowledge about solid lubrication.

12.3.9 TRIBOLOGY OF DIAMOND-LIKE CARBON COATINGS

The question to be answered here is as follows: What is the influence of the composition and thickness of carbonaceous films on their lubrication properties? [41–43].

12.3.9.1   Tribological Performance of DLC Films

Amorphous DLC (diamond-like carbon) coatings, whose primary use is as hard coatings, possess some of the lowest friction coefficients ever measured in a wide range of environments from high vacuum to ambient air and dry air [44]. For the last 15 years, a great deal of work has been devoted to tribo-investigations of DLC coatings synthesized by physical and chemical vapor deposition processes, revealing a considerable spread in the friction and wear results. This may be explained by the diversity of structures and compositions, arising from the preparative procedures and parameters.

Figure 12.21 shows the locations of the various DLCs and other carbon films in a ternary phase diagram. This diagram was proposed by Ferrari and Robertson, who have performed perhaps the most comprehensive structural and chemical studies on these films using spectroscopic techniques [45]. In this ternary diagram, the regions of various DLC films are clearly identified, and based on the respective fractions of sp^3 bonds and of hydrogen content. The films are classified into several kinds, ranging from hydrogenated amorphous carbons (or a-C:H) to tetrahedral amorphous carbon (or ta-C). Dopant or alloying elements (such as N, Si, F, and various metals) can be added to enhance adhesion, to decrease the surface energy, and to modify some mechanical properties, such as hardness and residual stress [46]. Of particular interest in the use of DLC is the deposition temperature, which can be lower than 200°C, thus making it possible to deposit coatings on most types of engineering materials, including polymers. Adhesion properties are generally improved by the use of intermediate layers, such as TiC, SiC, Ti, or a TiN/TiC composite, the choice depending on the nature of the substrate. The strong atmospheric dependence of the friction and wear of DLC coatings has been the subject of many studies [44]. One of the principal motivations in the search for DLC coatings with low friction and wear under the widest range of environmental conditions is the need for reliable solid lubricants in technological fields that require the very highest reliability.

12.3.9.2   *Ex Situ* FTIR Microanalysis inside the Wear Tracks of DLC Films

In the first study considered here [41], tribochemical reactions have been identified by *ex situ* FTIR microanalysis of the contacting surfaces, transfer film, and particles, thus allowing a better understanding of the tribological mechanisms as a function of the nature of the environment (air or vacuum) during friction. Analytical measurements in the reflection mode were performed using

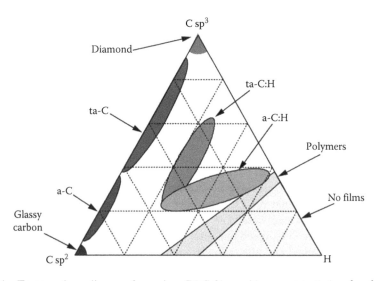

FIGURE 12.21   Ternary phase diagram for various DLC films with respect to their sp^2, sp^3, and hydrogen contents. (From Donnet, C., *Surf. Coat. Technol.*, 100/101, 180, 1998. With permission.)

a micro-FTIR spectroscope equipped with a wire-grid polarizer. Two kinds of DLC coatings were studied using this methodology: a-C:F coatings produced by radio frequency (RF) fluoropolymer sputtering, and a-C:H and a-C:H:Si coatings obtained from the decomposition of gaseous precursors. In the case of the a-C:F coatings, *ex situ* infra red (IR) measurements at the end of the friction process provided correlation of the frictional fluctuations with molecular transformations of the C:F structure.

It was found that the halogenated carbon framework decomposed into amorphous carbon and activated carbon, which then reacted with fresh metal through bond cleavage between halogen and carbon. The C–C bond cleavage also reduced the molecular size. The low-friction state occurred during the decomposition process, due to the presence of molecules containing fluorinated C=C moieties, such as polydifluorinated acetylene, which were oriented in the sliding direction (as seen by the polarized IR probe). In the case of the a-C:H and a-C:H:Si coatings, micro-IR investigations were performed at the point corresponding to the lowest coefficient of friction. This temporary high-lubrication performance was found to be attributable to the formation and adherence of sp^3 hydrocarbons produced on the ball surface from the rubbed film and oriented along the sliding direction.

12.3.9.3 *Ex Situ* ToF-SIMS Microanalysis inside the Wear Tracks of DLC Films

Other more recent studies have attributed the origin of the superlow friction (coefficients in the 10^{-3} range) of some hydrogenated DLC films in UHV to the weak van der Waals interactions between hydrogen atoms belonging to the a-C:H network [47]. Such a behavior has been observed only at the highest hydrogen content incorporated in the films [48]. Indeed, low hydrogenated films are known to exhibit very high friction in UHV conditions. However, the introduction of pure hydrogen gas during the friction process induces tribochemical reactions at the sliding interfaces, thus decreasing friction to superlow values, as with films containing a high concentration of hydrogen. This hydrogen gas effect on the superlow friction behavior of DLC has been confirmed by *ex situ* ToF-SIMS investigations inside and outside the wear tracks performed in deuterium gas at a pressure of 1000 Pa [42]. Being isotopes, hydrogen and deuterium are chemically equivalent. Their quantification inside and outside the wear tracks is possible since the probe diameter of the SIMS apparatus is lower than the lowest dimension of the wear track (a few tens of micrometers).

Outside the wear track, deuterium could be found, but was removed after a short etching, suggesting a weak bonding to the surface. On the other hand, large amounts of deuterium were found inside the wear track, and etching could not remove it completely. Thus, deuterium has a tendency to physisorb on the a-C:H surface, but is strongly bonded to the surface inside the wear track. This suggests that tribochemical reactions between the DLC film and the hydrogen gaseous environment occur during sliding, thus lowering friction to superlow values in vacuum.

12.3.9.4 *In Situ Post Mortem* AES Elemental Depth Profiling inside the Wear Tracks

The achievement of low friction values with DLC films is reported to be related to the buildup of a transfer film, consisting mainly of carbon [42]. When performing a friction experiment under UHV on a highly hydrogenated film, such a transfer film can indeed be observed on the steel pin (Figure 12.22). The growth of such a film has been related to the decrease of the friction coefficient, from about 0.2 to less than 0.02 [49]. To elucidate further the mechanisms of transfer film buildup, friction experiments on highly hydrogenated DLC films have been performed in UHV with different pre-treatments of the steel pins. These were (1) a raw pin, as polished and cleaned, with adventitious carbon and oxide layers; (2) an oxidized pin, after a short etching with Ar ions to remove adventitious carbon; and (3) an etched pin, after a longer etching to remove the oxide layers as well. The respective friction curves are plotted in Figure 12.23a. For the raw and oxidized pins, the initial coefficient of friction was the same, at 0.22, but the decrease was much faster for the oxidized pin, with less than 10 cycles to reach a friction value of 0.01, instead of more than 35 cycles for the raw pin.

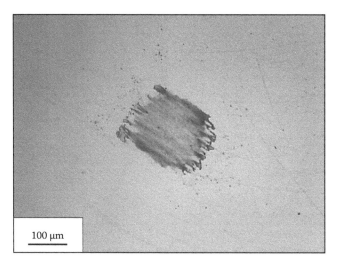

FIGURE 12.22 Optical micrograph of a transfer film on a steel pin after a friction experiment under UHV for sample AC5. (From Fontaine, J., Le Mogne, T., Loubet, J.L., and Belin, M., *Thin Solid Films*, 482, 99, 2005. With permission.)

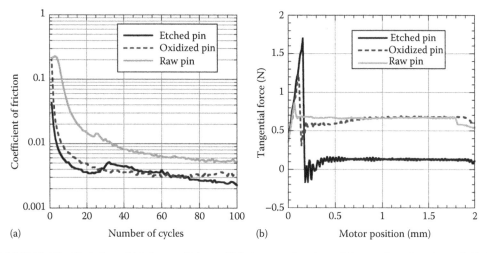

FIGURE 12.23 Variation of (a) the coefficient of friction as a function of the number of cycles and (b) the tangential force as a function of motor position during the first cycle, for a friction experiment on an AC5 flat sample with three different steel pin surfaces, under UHV. (From Fontaine, J., Le Mogne, T., Loubet, J.L., and Belin, M., *Thin Solid Films*, 482, 99, 2005. With permission.)

For the etched pin, the friction coefficient started immediately at about 0.05 and the subsequent decrease was very fast.

However, if the variation of the tangential force during the first cycle of sliding, for each of these three experiments (Figure 12.23b), as a function of the motor displacement, is considered, it can be seen that there was a linear increase of the force followed by a sudden decrease and after a short transient period, a constant value. This was due to the relatively low stiffness of this particular tribometer; an initial linear friction increase corresponded to a bending of the tribometer, that is, a motion of the motor without motion of the contact. Then, when the tangential force became high enough, relative motion of the contacting surfaces could actually start. The maximum force necessary before the start of real sliding can be attributed to sticking, and it is thus possible to extract a static coefficient of friction, while the constant value reached during sliding can be related to a dynamic

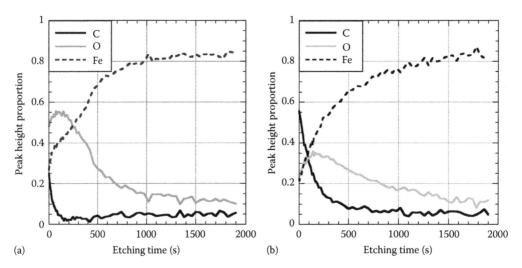

FIGURE 12.24 Variation of the proportion of carbon, oxygen, and iron peak heights in Auger spectra as a function of etching time: (a) outside the pin wear scar and (b) inside the pin wear scar. (From Fontaine, J., Le Mogne, T., Loubet, J.L., and Belin, M., *Thin Solid Films*, 482, 99, 2005. With permission.)

coefficient of friction. The more the pin was etched, the higher was the static friction, increasing from less than 0.3 to more than 0.5. For the etched pin, without oxide layers, the value of the dynamic friction coefficient was small, below 0.05, compared to the high static friction coefficient. These results suggest that a strong sticking phenomenon occurs before or at the very beginning of sliding. The stronger is this phenomenon, the faster is the friction decrease, implying a faster transfer film buildup. Adventitious carbon and particularly iron oxide seem to slow the growth of the transfer film.

Furthermore, an *in situ post mortem* Auger depth profiling performed concomitantly outside (Figure 12.24a) and inside (Figure 12.24b) the pin wear scar on the raw pin revealed that the amount of oxygen underneath the transfer film had been significantly reduced, apparently being replaced by carbon. This confirms that iron oxide has to be removed to favor transfer film growth. The HSAB concept, introduced by Pearson [49], and already mentioned in Section 12.3.2.3, could account for this phenomenon. It attempts to explain the chemical reactions between Lewis acids (electron acceptors) and bases (electron donors) by stating that "hard acids bind strongly to hard bases and soft acids bind strongly to soft bases." Hard elements will take part in ionic bonding, while soft elements will participate in covalent or metallic bonding. Within this classification, iron is a soft acid in the metallic state and a hard acid in the oxide state, while the hydrogenated amorphous carbon network can be considered as a soft base. Thus, reactions of hydrogenated amorphous carbon should be less favorable with iron oxide than with iron as metal. By removing the oxide layers, either by etching or by sliding, reactions between carbon and iron, leading to the buildup of the transfer film, are allowed.

Such studies show how powerful the spectroscopies and microscopies are in determining both the initial structures and the friction-induced modified structures of carbonaceous films, which are representative of typical thin coatings requiring specialized analytical methods, due to their composition (light elements) and their amorphous structure.

12.3.10 TRIBOCHEMISTRY OF C_{60} COATINGS

Another interesting question is "Do any of the new carbonaceous compounds exhibit lubrication properties?" [29].

Carbon has always been an element of interest in tribological applications [50]. C_{60} (fullerene) molecules have unique properties in terms of stability, high load capacity, low surface energy, weak intermolecular bonding, and spherical shape. Moreover, thin films of fullerenes can be formed easily either from evaporation of a solvent solution or directly by sublimation of the powder in high vacuum. Fullerenes are therefore good candidates for solid lubrication, but little is known of their tribochemistry. Tribochemical modifications and frictional behavior of C_{60} coatings have been investigated by an analytical tribometer coupled to a laser Raman optical microprobe (Figure 12.6). *In vivo post mortem* Raman spectroscopy was carried out inside a static contact in order to study the structural modifications of C_{60} solid lubricant films in the contact area between two sapphire substrates transparent to the laser probe at the wavelength used.

Fullerene thin films were deposited by sublimation of C_{60} powder at 450°C. Raman tribometry analysis was performed with a micro-Raman spectrometer, on C_{60} deposited on sapphire sliding against C_{60} deposited on steel in air at room temperature. Friction tests were carried out with an average contact pressure of 45 MPa, a radius of contact of 100 μm, and a normal force of 1.6 N. The contact zone was recorded by a video device, which showed that the argon ion laser beam (of power <1 μW at 541 nm) was focused to a spot of 1 μm diameter. Because of the sensitivity characteristics of the detector and of the nature of the coatings, *in vivo pre mortem* analysis was not possible, since, with an acquisition time of 20 min, the sliding velocity would have been too low. Raman spectra were therefore acquired while the sliding motion was halted temporarily.

Figure 12.25 shows a series of Raman spectra acquired (1) before friction, and after friction tests (2) of 15 min (coefficient of friction 0.82 after 15 min), and (3) of a total of 110 min (coefficient of friction 0.40 after 110 min). Contact was maintained during analysis. Before the beginning of the friction test, the Raman spectrum revealed the presence of the characteristic fullerene peaks, located at 1425, 1459, 1469, and 1570 cm^{-1}, in agreement with published results [51] measured from C_{60} in powder or solution form. After 15 min (470 cycles) of friction (2), the same Raman peaks could still

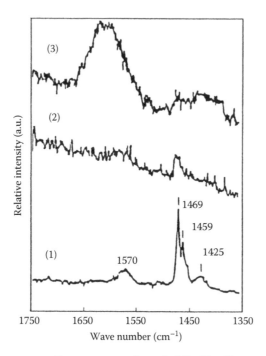

FIGURE 12.25 *In vivo post mortem* Raman spectra of a typical C_{60} film (1) as-received, (2) after 15 min friction exposure, and (3) after 110 min of friction exposure. (From Erdemir, A. and Donnet, C., *J. Phys. D: Appl. Phys.*, 39, R311, 2006. With permission.)

be distinguished but their intensities were much lower. From optical observations, the presence of rolling pins in the interface was noted, evidence for C_{60} film degradation. In spectrum (3), the C_{60} characteristic peaks can be seen to have disappeared and to have been replaced by two broadbands centered around 1430 and 1600 cm^{-1}, characteristic of a disordered sp^2 carbon structure. Simultaneously, a large number of rolling pins were observed inside the contact.

A correlation between the coefficient of friction and the C_{60} lubricant film tribochemistry can be deduced from these experimental results; rolling pin formation produces a progressive decrease in the coefficient of friction, and the lubricant film modification corresponds to the amorphization of the C_{60} molecules.

12.4 SYNTHESIS AND CONCLUSION

Analytical tribology consists basically of the judicious coupling of surface analysis with tribological experiments. There are several essential types of information that are needed, such as surface chemistry, structure, morphology, and other physical properties, on different dimensional scales in order to elucidate the tribological mechanisms and thus to be able to recommend specific mechanical designs in various technological fields. Analytical tribology can take full advantage not only of the substantial advances in the well-established surface analytical techniques but also of the emergent SPMs. However, the said coupling between tribology and surface analysis requires a top-down approach as the current best practice procedure to guide the problem-solving methodology into the most rational, direct, and valid analytical paths.

This chapter has given an overview that can be summarized by the global generic scheme illustrated in Figure 12.26. Prefriction analysis is often required in order to have precise knowledge of the contacting surfaces and the lubricant phase prior to any dynamic contact between the tribopartners. The dynamic friction process must be perfectly controlled, in terms of mechanical parameters (contact geometry, contact pressure, and vibrations), physical parameters (temperature and radiation), and chemical parameters (nature and pressure of the ambient atmosphere). The lack of standardization in tribological investigations is a real obstacle in reviewing studies in this field. However, the well-standardized surface analytical procedures allow the many generic results obtained by the numerous research laboratories working in this field to be correlated, thus providing significant progress in the control of friction and wear in various technological configurations. Surface analysis

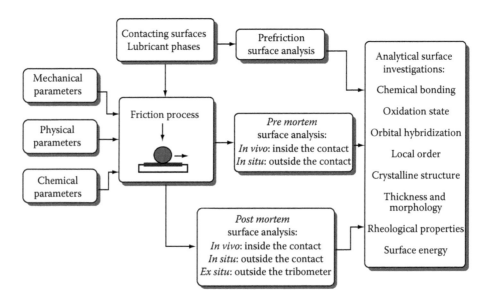

FIGURE 12.26 Synopsis of a typical top-down approach for surface analytical investigations in tribology.

of contacting surfaces is performed more commonly after the end of the friction experiment (*post mortem* examination), requiring spatially resolved analysis inside the wear scars. Ideally, the *pre mortem* examination would allow a direct real-time surface investigation of the contact, but specific triboanalytical combinations in terms of the nature of the materials or the tribometer configuration would be required. In both cases, the analysis can be performed inside the tribometer either outside (*in situ*) or inside (*in vivo*) the contact. In addition, *ex situ* examinations can be carried out systematically, which allows advantage to be taken of the full capabilities of the various surface analysis techniques, in spite of indirect examination due to air exposure of the tribosurfaces, which separates the friction experiment and the analysis step. Whatever the configuration, the analytical tribologist has to define precisely the nature of the pertinent surface analytical information required to clarify the tribological mechanisms, and thus to propose significant improvements for controlling friction and minimizing wear in engineering devices.

REFERENCES

1. B. Bhushan and B. K. Gupta (Eds.), *Handbook of Tribology*, McGraw-Hill, New York, 1991, p. 229.
2. F. P. Bowden and D. Tabor, *The Friction and Lubrication of Solids*, Part 2, Clarendon Press, Oxford, 1964, p. 210.
3. G. Heinicke, *Tribochemistry*, Carl Hanser Verlag, München, Germany, 1984.
4. J. M. Martin, M. Belin, J. L. Mansot, H. Dexpert, and P. Lagarde, *ASLE Trans. 49(4)*, 523, 1986.
5. I. L. Singer, *Surf. Coat. Technol. 49*, 474, 1991.
6. S. M. Hsu and R. S. Gates, Tribochemistry and lubrication, *Proceedings of the International Tribology Conference*, Japanese Society of Tribologists (Ed.), Satellite Forum on Tribochemistry, Yokohama, Japan, October 28, 2005, p. 43.
7. K. Holmberg and A. Mathews, *Coatings Tribology*, D. Dowson (Ed.), Tribology Series 28, Elsevier, Amsterdam, the Netherlands, 1994.
8. I. L. Singer, Solid lubrication process, in *Fundamentals of Friction: Macroscopic and Microscopic Processes*, I. L. Singer and H. M. Pollock (Eds.), Kluwer, Boston, MA, 1991, p. 237.
9. K. Miyoshi and D. H. Buckley, *ASLE Trans. 22*, 245, 1979.
10. D. Dowson, *History of Tribology*, Longman, 1979, p. 3.
11. K. Miyoshi and Y. W. Chung (Eds.), *Surface Diagnostics in Tribology*, World Scientific, Singapore, 1993.
12. K. Miyoshi, in *Surface Diagnostics in Tribology*, K. Miyoshi and Y. W. Chung (Eds.), World Scientific, Singapore, 1993, p. 93.
13. B. K. Teo and D. C. Joy, *EXAFS Spectroscopy: Techniques and Applications*, Plenum Press, New York, 1979.
14. R. F. Egerton, *EELS in the Electron Microscope*, Plenum Press, New York, 1986.
15. J. M. Martin and M. Belin, *Thin Solid Films 236*, 173, 1993.
16. Y. W. Chung, in *Surface Diagnostics in Tribology*, K. Miyoshi and Y. W. Chung (Eds.), World Scientific, Singapore, 1993, p. 33.
17. B. Bhushan, J. N. Israelachvili, and U. Landman, *Nature 374*, 609, 1995.
18. A. M. Homola, in *Surface Diagnostics in Tribology*, K. Miyoshi and Y. W. Chung (Eds.), World Scientific, Singapore, 1993, p. 271.
19. J. N. Israelachvili, P. M. McGuiggan, and A. M. Homola, *Science 240*, 189, 1988.
20. J. M. Georges, D. Mazuyer, J. L. Loubet, and A. Tonck, in *Fundamentals of Friction: Macroscopic and Microscopic Processes*, I. L. Singer and H. M. Pollock (Eds.), Kluwer, Boston, MA, 1991, p. 263.
21. S. J. Hirz, A. M. Homola, G. Hadziioannou, and C. W. Franck, *Langmuir 8*, 328, 1992.
22. S. Granick, *Science 253*, 1374, 1991.
23. J. Klein, D, Perahia, and S. Warberg, *Nature 352*, 143, 1991.
24. H. A. Spikes, Boundary lubrication films, *Proceedings of the International Tribology Conference*, Japanese Society of Tribologists (Ed.), Satellite Forum on Tribochemistry, Yokohama, Japan, October 28, 2005, p. 49.
25. G. J. Johnston, R. Wayte, and H. A. Spikes, *STLE Tribal. Trans. 34*, 187, 1991.
26. M. Belin and J. M. Martin, *STLE Tribol. Trans. 32*, 410, 1989.
27. M. Belin and J. M. Martin, In situ structural changes of lubricated surfaces, as studied by EXAFS, *Proceedings of the Leeds-Lyon Symposium on Tribology—Wear Particles*, D. Dowson et al. (Eds.), Elsevier, 1992, p. 413.

28. C. Minfray, J. M. Martin, T. Lubrecht, M. Belin, and T. Le Mogne, The role of mechanical and chemical processes in antiwear properties of ZDDP tribofilms, *Proceedings of the 30th Leeds-Lyon Symposium (September 02–05, 2003, Lyon, France)—Transient Process in Tribology*, E. Dalmaz and T. Lubrecht (Eds.), Elsevier, Amsterdam, the Netherlands, 2004, p. 43.

29. J. M. Martin, Antiwear mechanisms of zinc dithiophosphate: A chemical hardness approach, *Trib. Lett. 6*, 1, 1999.

30. C. Minfray, J. M. Martin, C. Esnouf, T. Le Mogne, R. Kersting, and B Hagenhoff, *Thin Solid Films 447–448*, 272, 2004.

31. S. Mori, Tribochemical activity of nascent metal surfaces, *Proceedings of the International Tribology Conference*, Japanese Society of Tribologists (Ed.), Satellite Forum on Tribochemistry, Yokohama, Japan, October 28, 2005, p. 37.

32. B. M. DeKoven, in *Surface Diagnostics in Tribology*, K. Miyoshi and Y. W. Chung (Eds.), World Scientific, Singapore, 1993, p. 299.

33. S. V. Pepper, *J. Appl. Phys. 47*, 2579, 1976.

34. J. M. Martin, T. Le Mogne, and M. N. Gardos, Friction of alpha silicon carbide under oxygen partial pressure: High resolution analysis of interface films, *Proceedings of the Japan International Tribology Conference*, Nagoya, Japan, 1990, p. 1407.

35. L. E. Seitzman, R. N. Bolster, I. L. Singer, and J. C. Wegand, *Tribol. Trans. 38*, 445, 1995.

36. T. Le Mogne, C. Donnet, J. M. Martin, N. Millard-Pinard, A. Tonck, S. Fayeulle, and N. Moncoffre, *J. Vac. Sci. Technol. Al 2(4)*, 1998, 1994.

37. J. M. Martin, C. Donnet, T. Le Mogne, and T. Épicier, *Phys. Rev. B48(14)*, 10583, 1993.

38. J. M. Martin, H. Pascal, C. Donnet, T. Le Mogne, J. L. Loubet, and T. Épicier, *Surf. Coat. Technol. 68/69*, 427, 1994.

39. S. V. Didziulis and P. D. Fleischauer, in *Surface Diagnostics in Tribology*, K. Miyoshi and Y. W. Chung (Eds.), World Scientific, Singapore, 1993, p. 135.

40. E. W. Roberts, *Trib. Int. 23(2)*, 95, 1990.

41. S. Miyake, in *Surface Diagnostics in Tribology*, K. Miyoshi and Y. W. Chung (Eds.), World Scientific, Singapore, 1993, p. 183.

42. J. Fontaine, T. Le Mogne, J. L. Loubet, and M. Belin, *Thin Solid Films 482*, 99, 2005.

43. A. Erdemir and C. Donnet, *J. Phys. D: Appl. Phys. 39*, R311, 2006.

44. J. Robertson, *Mater. Sci. Eng. R37*, 129, 2002.

45. C. Donnet, *Surf. Coat. Technol. 100/101*, 180, 1998.

46. C. Donnet, J. Fontaine, A. Grill, and T. Le Mogne, *Trib. Lett. 9(3/4)*, 137, 2001.

47. C. Donnet and A. Grill, *Surf. Coat. Technol. 94/95*, 456, 1997.

48. C. Donnet, *Surf. Coat. Technol. 80*, 151, 1996.

49. R. G. Pearson, *Hard and Soft Acids and Bases*, Dowden, Hutchinson & Ross Inc., Stroudsburg, PA, 1973.

50. H. O. Pierson, *Handbook of Carbon, Graphite, Diamond and Fullerenes*, Noyes, Park Ridge, NJ, 1993, p. 356.

51. B. Chase, N. Herron, and E. Holler, *J. Phys. Chem. 96*, 4262, 1992.

52. C. Donnet, Problem-solving methods in tribology with surface-specific techniques, in J.C. Rivière and S. Myhra (Eds.), *Handbook of Surface and Interface Analysis: Methods for Problem-Solving*, Marcel Dekker, New York, 1998.

13 Problem-Solving Methods in Metallurgy with Surface Analysis

R. K. Wild

CONTENTS

13.1 INTRODUCTION

The failure of a relatively small component in a plant can have catastrophic results. In 1969, a blade failed in one of the turbines at Hinkley Point A Power Station during a routine overspeed test [1]. Figure 13.1 shows the resultant devastation, with components having passed through all the outer

FIGURE 13.1 Damage to Hinkley turbine caused by temper embrittlement failure of a rotor blade. (From Akhurst, K.N., *CEGB Res.*, 21, 5, 1998.)

casings and the roof of the turbine hall and some parts ending up several hundred meters distant. Fortunately, there were no injuries to personnel but the cost of repair and prevention ran into millions of dollars. It is therefore important to be able to predict how a metal or alloy will behave in service. Despite all the knowledge concerning embrittlement, examples continue to occur where components have failed as the result of segregation of trace elements in alloys. There is also the important problem of intergranular stress corrosion cracking where components under load in an aggressive environment can fail because small cracks are opened up by the combined effects of stress and oxidation. Knowledge of the grain boundary chemistry can assist in reducing or eliminating both these and other problems. There are many ways in which a metal or alloy can be characterized and the structure and composition of the grain boundaries determined. Some techniques analyze the internal surface that has first been exposed, others analyze the internal surface in thin sections of the material, while others section slowly through the material until a boundary is encountered. This chapter outlines some techniques and methods for determining the grain boundary chemistry and hence the properties of metals and alloys.

13.1.1 STRENGTH OF MATERIALS

The strength of a metal or alloy will be determined by the strengths of the weakest bonds between atoms. In some cases, such as single crystal whiskers with few defects, the theoretical strength can be approached, but in most practical alloys the ultimate strength is several orders of magnitude lower than that and is usually determined by the strength of the interface between grain boundaries. The grain boundary strength can be altered dramatically by segregation to the grain boundaries of trace elements present in the bulk, by the enrichment or depletion of bulk elements in the vicinity of the grain boundary, or by the formation of particles, such as carbides, at the grain boundary positions. Thermal treatments and aging in service will modify the grain boundary composition, and hence the mechanical properties will vary with time; the grain boundary composition therefore requires monitoring at various stages throughout the lifetime of the component. In nuclear power plants, the radiation flux will modify the diffusion rates of the various alloying components, further influencing the mechanical properties as a function of time in service.

13.1.2 Failure Mechanisms

Most metals and alloys are made up of small crystals or grains oriented in a random manner to one another. These grains can vary in size from less than a micron to several hundreds of microns. The grain size is frequently determined by the heat treatment and mechanical work that the alloy has received. Aging at high temperature allows some grains to grow at the expense of others, resulting in large average grain sizes, while mechanical work breaks up and reorients grains, producing much smaller grains. Within the grains may be found small particles such as carbides or sulfides, which can have the effect of increasing the hardness and reducing the ductility of the alloy by pinning dislocations and preventing plastic flow. However, the interface between the grains has the greatest effect on the mechanical properties. Grain boundaries may have particles formed on them, or elements segregated to the interface. Frequently, more than one element will be found segregated to the grain boundary surface. Some segregating elements may increase the strength of the boundary, usually small atoms such as boron, while others, such as phosphorus, tin, antimony or arsenic, reduce the strength. How a metal or alloy will fail is determined by many factors, but if the grain boundaries have high cohesive strength the failure will be either ductile, when the metal tears apart leaving a surface looking like the surface of toffee that has been broken on a warm day, or by cleavage, when failure is across sheets of atoms, leaving relatively flat surfaces similar to toffee that has been broken after removal from a refrigerator. If the grain boundary is the weak point, the material will fail by the fracture path following the boundaries between individual grains. Grain boundaries will be randomly oriented and some will be oriented favorably for failure while others will not. A grain boundary oriented parallel to the direction of fracture has a high probability of failure, whereas a grain boundary oriented normally to the direction of fracture is unlikely to fail. In practice, a fracture surface will often contain examples of all types of failure with varying fractions of intergranular, cleavage, and ductile fracture.

13.1.3 Segregation

13.1.3.1 Thermal

Elements present in metals and alloys in trace quantities can, under certain heat treatments, segregate to the grain boundaries, and there are many examples in the published literature. A review by Lejcek and Hofmann [2] details many of the segregating species and systems, and outlines the theories describing grain boundary segregation. Many materials will fracture in an intergranular manner by impact at low temperature, and for these materials Auger electron spectroscopy (AES) is an ideal analytical tool. The systems studied most extensively by this means are iron and ferritic steels where segregants such as P, S, Sn, Sb, and As cause severe embrittlement [3–7]. However, other systems have also been studied. For example, Powell and Woodruff [8] used AES to study the segregation of bismuth in copper and showed that the bismuth segregated as a single atom layer at the grain boundary, while Chuang et al. [9] studied antimony in copper in a similar manner. Smiti et al. [10] showed that carbon could increase grain boundary cohesion in tungsten while oxygen decreased it. Other systems can be induced to fracture in an intergranular manner by first charging the metal or alloy with hydrogen. In some cases, the hydrogen-charged alloy will fracture intergranularly by impact [11], but in other cases it is necessary to use a slow tensile pull [12–14]. Austenitic stainless steels [15] and nickel-based superalloys [16] have been studied extensively using this technique. In addition, Ni_3Al containing B [17], Co-50%Ni containing Sb [18], and Ni containing In [19] will all fracture in an intergranular manner following hydrogen charging.

Not all metals and alloys can be induced to fracture along grain boundaries. Some of the pure stainless steels will not fail along the boundaries even after hydrogen charging followed by slow tensile fracture. In these cases, other techniques must be used. Transmission electron microscopy (TEM) can reveal grain boundaries in suitably thinned specimens. Field emission gun scanning (FEG)STEM, combined with energy dispersive x-ray (EDX) [20,21] analysis or parallel electron loss spectroscopy (PEELS) [22], has proved to be very successful in determining the grain boundary composition in these difficult alloys. Indeed the technique is now used to complement the data from AES in specimens that can be induced to fracture in an intergranular manner. It does, however, suffer from the drawback that it is laborious to produce thin specimens and the thinned region may contain only a few grain boundaries.

Other techniques can be used to detect and determine the levels of segregants at grain boundaries in bulk specimens, and there are examples of the use of autoradiography [23], radioactive tracer for tellurium in silver [24], and field ion microscopy (FIM) for detection of niobium in cobalt [25]. FIM has also been used in combination with a position sensitive detector [26].

Grain boundary segregation in metallurgy has been comprehensively described in a number of papers by Seah [27–30]. McLean [31] analyzed the kinetic behavior of segregation under equilibrium conditions from diffusion theory. He visualized a grain boundary as a bubble raft and considered a lattice with N undistorted sites with P solute atoms amongst them and n distorted sites with p solute atoms. If the energy of the solute atom in the lattice is E and on segregation is e, the free energy of the solute atoms is given by

$$G = pe + PE - kT[\ln n!N! - \ln(n - p)!p!(N - P)!P!]$$ (13.1)

By considering the minimum in G, Equation 13.1 may be developed into the familiar McLean equation:

$$\frac{X_b}{(1 - X_b)} = \left[\frac{X_c}{(1 - X_c)}\right]\exp\left[\frac{E_1}{RT}\right]$$ (13.2)

where
  X_b is the adsorption level of the segregant as a molar fraction of a monolayer
  X_c is the solute molar fraction
  E_1 is the molar heat of adsorption of the segregant at the grain boundary

Gibbs [32] related adsorption to the grain boundary energy (γ_b) with the following:

$$\frac{d\gamma_b}{dX_c} = -\left(\frac{RT}{X_c}\right)\Gamma_b$$ (13.3)

where Γ_b is the excess solute at the grain boundary in mol m^{-2}.

Carr et al. [33] have produced a time–temperature diagram (Figure 13.2) for the embrittlement of the steel SAE3140. The diagram indicates that heat treatment to temperatures between 500°C and 550°C for times between 1 and 100 h produces the greatest embrittlement. Seah [34] has used the McLean equation, together with the segregation levels of phosphorus and the equation for the volume diffusivity of phosphorus in iron [35],

$$D = 1.58\exp(-52,300/RT)$$ (13.4)

to determine the time–temperature diagram for the segregation of phosphorus for the same SAE 3140 steel used by Carr et al. [33], as shown in Figure 13.3. This plot shows that the grain boundary segregation of phosphorus follows the same time–temperature relationship as that which Carr et al. had determined for embrittlement. It can be demonstrated that the addition of other elements such as molybdenum and chromium reduces the diffusivity of phosphorus in iron while manganese increases it, so that these equations cannot be applied to all steels without using a modified D.

When an element segregates to a grain boundary, it occupies a vacancy site between two adjacent grains. If this element were to cause the material to be weakened or embrittled, then there would have to be considerable enrichment of it compared with its level within the bulk. Seah and Hondros [27] have correlated grain boundary enrichment with atomic solubility for a number of alloy systems (Figure 13.4). It can be seen that enrichment ratios up to 10^4 are possible. The extent

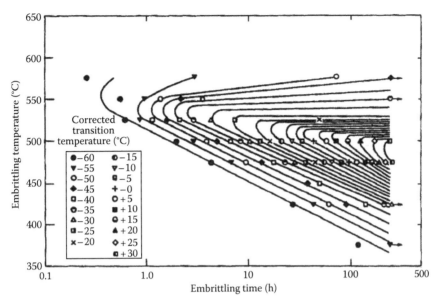

FIGURE 13.2   Time vs. temperature diagram for the embrittlement of steel SAE 3140. (From Carr, F.L., Goldman, M., Jaffe, L.D., and Buffum, D.C., *Trans. AIME*, 197, 998, 1953. With permission.)

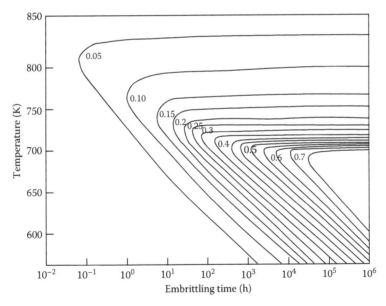

FIGURE 13.3   Predicted phosphorus segregation levels at grain boundaries as a function of embrittlement time and temperature for steel SAE 3140 containing a bulk level of phosphorus of 0.015 wt.%. The numbers on the curves indicate the fraction of monolayer coverage of phosphorus for the plotted times and temperatures. (From Seah, M.P., *Acta Met.*, 25, 345, 1977. With permission.)

of the segregation away from the grain boundary has been determined experimentally by Palmberg and Marcus [36], who exposed grain boundaries in materials that had high levels of grain boundary segregation. The segregant was then removed in a controlled manner by bombarding the grain boundary with inert gas ions while the level of segregant was determined. The solid line on this figure shows that good agreement was observed between the theoretical prediction for a layer of segregant 0.45 nm thick and the experimental points indicating that segregation is confined to a

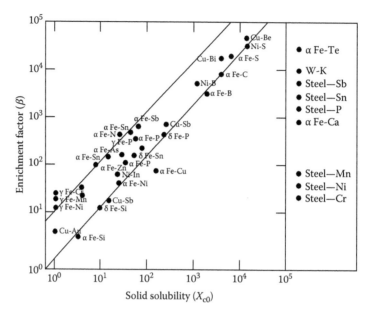

FIGURE 13.4 Grain boundary enrichment as a function of atomic solubility for various alloys. (From Seah, M.P. and Hondros, E.D., *Proc. Roy. Soc. Lond. A*, 335, 191, 1973. With permission.)

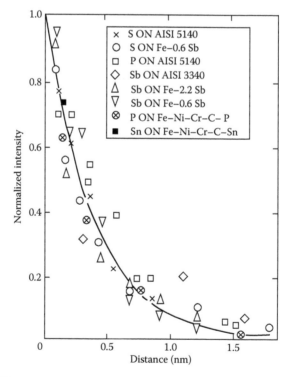

FIGURE 13.5 FEG-STEM EDX determination of phosphorus distribution across a grain boundary in nickel-based alloy Inconel 600.

single atom layer at the grain boundary. This was further demonstrated by EDX analysis across a grain boundary in a nickel-based alloy Inconel 600. Figure 13.5 shows the profile for phosphorus segregated to the boundary with the phosphorus confined to a layer a few nanometers thick.

Atomic size must be included when considering whether an element will embrittle a metal or alloy or not. In general, large atoms decrease the grain boundary strength while small atoms increase it [30]. Thus, in the case of iron and steel, atoms such as S, P, Sn, Sb, and As which are all larger than Fe will tend to make a steel more brittle (i.e., lower its brittle-to-ductile transition temperature), while atoms such as B, Be, and C, which are smaller than Fe will tend to make the steel less brittle.

13.1.3.2 Irradiation Assisted

Irradiation can alter dramatically the level of segregants at grain boundaries. In a nickel-based alloy such as PE16, in the absence of grain boundary carbide formation, the boundaries can be depleted in chromium and iron and enriched in nickel. A mechanism for the observed behavior of the major alloying elements is the inverse Kirkendall effect [37,38]. Here, fast neutron irradiation of the material produces numerous point defects (interstitials and vacancies). These move through the lattice in a random manner, but on encountering a sink such as a grain boundary, will be annihilated, which produces a net flux of point defects toward grain boundaries. A flow of vacancies toward grain boundaries implies a flow of atoms in the opposite direction; however, the rate of flow depends upon the jump frequency J of the particular atoms [39]. For the alloying elements the jump frequency magnitudes are in the order Cr > Fe > Ni [40], so that Cr and Fe will flow away from the boundaries at a greater rate than Ni, and the boundaries thus become enriched in Ni by default.

On the other hand, studies of the Cr depletion in type 316 stainless steel as a result of neutron irradiation, using a high spatial resolution instrument, showed segregation as well as depletion. Here the benefits of high spatial resolution instruments using analyzing electron probes of diameter ~1 nm were evident. Cr profiles were determined in relation to the positions of grain boundaries. The results (Figure 13.6) show a Cr-depleted region adjacent to the grain boundary, but an increase in concentration at the grain boundary itself. However, in the case of these irradiated materials, unlike similar depletions in thermally sensitized steels [41], there were no carbide precipitates formed at the grain boundaries [42]. The measurements, which were made using the high resolution capability of the STEM-EDS x-ray microanalysis system, need to be extended if the true composition profile is to be derived, since the latter is limited by the spatial resolution even for such a small, focused electron probe.

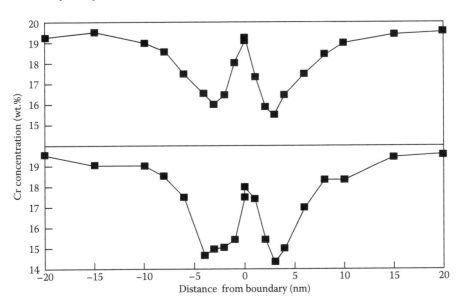

FIGURE 13.6 Experimental measurements of Cr concentrations at two grain boundaries in neutron-irradiated stainless steel. Each profile demonstrates clearly the presence of a depletion profile adjacent to but not at the boundary, in addition to segregation at the boundary. (From Titchmarsh, J.M. and Dunbill, S., *J. Nucl. Mat.*, 227, 203, 1996. With permission.)

In any attempt to characterize the internal interfaces in a metal or alloy, there are a number of steps that need to be considered. In the following sections, these will be considered, starting with the simple and nondestructive, and finishing with the destructive or complicated.

13.2 ANALYTICAL METHODS FOR DETERMINING GRAIN BOUNDARY SEGREGATION

13.2.1 INTRODUCTION

In this section, a suggested route is described for the determination of the type and amount of grain boundary segregation. A flowchart, Figure 13.7, can be used to illustrate the sequence of steps that should be taken with any metallurgical investigation that involves surfaces. It starts with techniques that do not require the metal or alloy to be damaged by fracture or thinning, but that can probe a metallographically polished specimen. Following that, surface analytical techniques will be considered, and they will require a means of fracturing in an intergranular manner to expose the grain boundary surface. Finally TEM methods of determining grain boundary structure, as well as more sophisticated techniques such as (time-of-flight) TOF-FIM, are described.

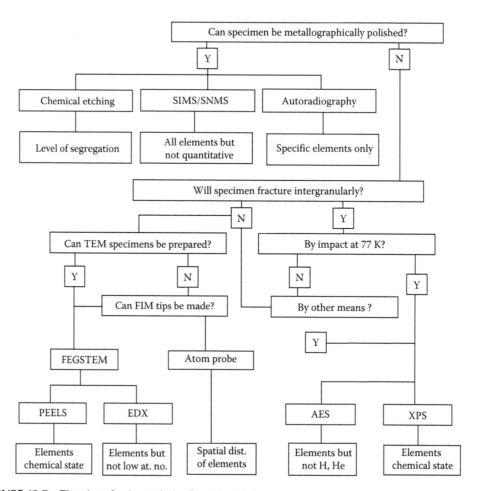

FIGURE 13.7 Flowchart for the analysis of metals and alloys.

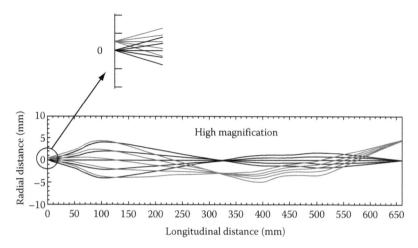

FIGURE 3.32 Trajectories of 1000 eV electrons through a transport lens set at $M = 10$ with retardation of the electrons to a final energy of 100 eV. Black lines indicate trajectories from a point on axis at angles of $0°$, $\pm 1°$, and $\pm 2°$ to the axis, and blue lines the same from a point 0.5 mm off axis. (From SPECS literature: www.specs.com.)

FIGURE 4.12 Positive ion mass resolved ToF-SIMS images of Ba halide crystals covered by Sr halide. The Ba signal is in red, and the Sr signal in green.

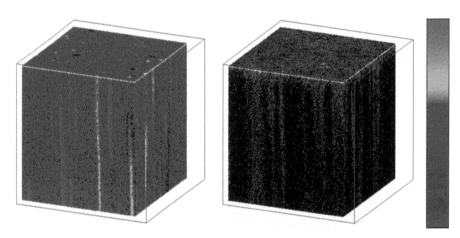

FIGURE 4.14 Three-dimensional ToF-SIMS study of diffusion. The sample consisted of a polycrystalline metal oxide $O_{2.8} Mg_{0.2} Ga_{0.8} Sr_{0.2} La_{0.8}$ with a $Cr+Fe+Y$ overlayer. Left, the distribution of Mg^+, and, right, that of Y^+; a color coded map is used to represent ion intensities with blue representing high intensity and red representing low intensity, respectively. Experimental conditions: sputtering; O_2^+, 1 keV on 150×150 μm²; analysis; Ga^+, 25 keV on 60×60 μm². The sample was kindly provided by Professor Martin, RWTH Aachen.

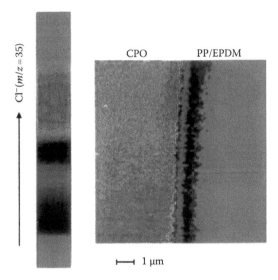

FIGURE 4.16 Cl-ToF-SIMS image of a cryosection of a CPO layer deposited on a PP substrate (diffusion time: 57 days). A color code is used to represent image intensities (see left color bar; pink and red correspond to low and high intensities, respectively). (From Rulle, H., PhD thesis, Münster, Germany, 1996.)

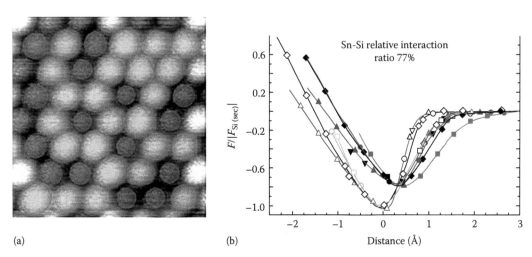

(a)   (b)

FIGURE 5.34 (a) False color image of surface alloy composed of Si (red), Pb (green,) and Sn (blue) atoms deposited on a Si(111) surface. The field of view is $4.3 \times 4.3 \, \text{nm}^2$. (b) Normalized $F–d$ curves show the distinct difference in interaction between the tip and Si and Sn surface atoms. Curves of different color refer to data obtained with different probes. The curves demonstrate that the assignments are independent of the tip material. The $F–d$ data have been normalized to those obtained for the Si species ($F/F_{\text{Si-set}} = 1$ at distance = 0). (From Sugimoto, Y., Pou, P., Abe, M., Jelinek, P., Pérez, R., Morita, S., and Custance, Ó., *Nature*, 446, on-line05530, 2007.)

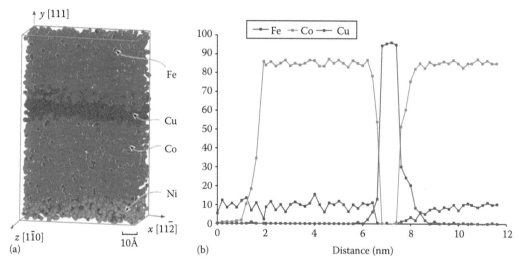

FIGURE 11.5 (a) 3DAP atom map of Co (blue), Fe (yellow), and Cu (red) atoms in a CoFe/Cu/CoFe trilayer on an NiFe substrate (Ni [green]), and (b) a compositional profile across the same region. (From Zhou, X.W., Wadley, H.N.G., Johnson, R.A., Larson, D.J., Tabat, N., and Cerezo, A., *Acta Mater.*, 49, 4005, 2001.)

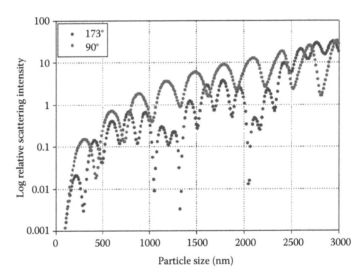

FIGURE 11.16 Theoretical plot of the log of the relative intensity of scattering versus particle size at scattering angles of 173° and 90°, assuming a laser beam at a wavelength of 633 nm, real refractive index of 1.59, and an imaginary refractive index of 0.001. (From Dynamic light scattering: An introduction in 30 minutes, Downloadable application note from www.malvern.com.)

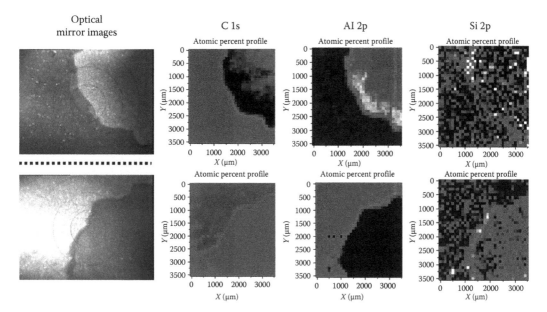

FIGURE 18.9 SAX images (field of view 3.5 mm) recorded from the failure surfaces of adhesively bonded aluminum using an epoxy resin to which had been added 1% of an organosilane. The optical images (LHS) show failure surfaces of the joint opened out as a book along the dotted line. Adhesive and metal regions are identified by the intense C 1s and Al 2p images, respectively, in the middle. The Si 2p images (RHS) show how the silane has segregated to one failure surface (toward the RHS of these images), but not the other.

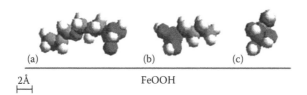

FIGURE 18.19 Molecular dynamics simulations of the positional arrangement of fully hydrolyzed organosilane molecules on an FeOOH substrate, obtained using the Cerius^2 Sorption package: (a) GPS, in which the silane head is at the right-hand end of the molecule, (b) γ-APS, in which the silane head is at the left-hand end of the molecule, and (c) vinyl triethoxysilane, in which the silane head of the molecule is adjacent to the FeOOH surface. The scale bar is 0.2 nm. The diagram was produced using the original (color) computer graphics files obtained by Davis. (From Tatoulian, M., Arefi-Khonsari, F., Shahidzadeh-Ahmadi, N., and Amoroux, J., *Int. J. Adhes. Adhes.*, 15, 177, 1995.)

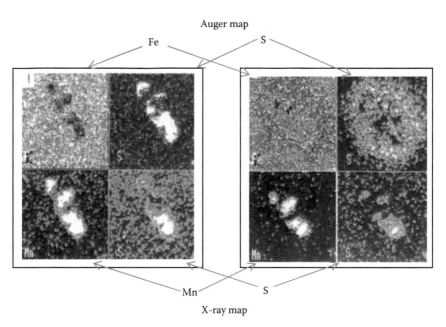

Auger map

Fe S

Mn S

X-ray map

FIGURE 19.19 Study by SAM and EDX of corrosion at an inclusion, showing the separation of anodic and cathodic sites. The left-hand maps show the surface before exposure, and the right-hand ones after exposure to a corrosive environment for 35 min. Changes in surface chemistry can be recognized in the Auger maps but not in the x-ray maps. (From Castle, J.E. and Ke, R., *Corros. Sci.*, 30, 409, 1990.)

13.2.2 Metallographically Polished Specimens

In this section, it is assumed that sufficient material is available for a specimen to be mounted in a conducting mount and polished using various grit sizes ending finally with a micron or submicron diamond paste.

13.2.2.1 Chemical Etching

It is useful to start by polishing a metallographic specimen to a mirror finish. This may then be etched to allow the identification of grain boundaries and inclusions and to give an indication of segregation levels. Figure 13.8 is an example of a steel plate containing oxide particles and manganese sulfide inclusions. It has been polished to a 1 µm grade diamond finish and then etched in 2% Nital (a solution of alcohol and nitric acid). The grain boundaries are clearly visible and the two types of oxide inclusions can be differentiated easily with optical microscopy due to their different light reflection and contrast. The MnS inclusions appear gray colored and the oxide inclusions are shiny and bright. The method for detecting segregation on steels has been described by Dryer et al. [23]. The etchant is a saturated picric acid solution containing 1 g of sodium didecylbenzene sulphurate/100 mL. Etching times of approximately 1 h are normally sufficient to expose grain boundaries without significant pitting. Once the ideal etching time for the specimens has been determined, those with different segregation levels at grain boundaries should then be etched for equal times. The depth of the grain boundary grooving, which can be measured optically, will then be proportional to the level of segregant.

13.2.2.2 SIMS

A detailed description of the secondary-ion mass spectrometry (SIMS) technique can be found in Chapter 4. When applying SIMS to metallurgical analysis, the specimen should first be polished metallographically as described above. In SIMS, spatial resolution can be achieved either by rastering a fine focus ion beam over the surface while detecting the secondary ions, or by irradiating the area of interest with a broad ion beam and detecting the secondary ions by focusing to equivalent points in the image plane. An example of the latter mapping method is shown in Figure 13.9, in which two carbon maps from a region of diameter 120 µm on a metallographically polished specimen of steel are displayed [43]. The instrument employed in this study was a CAMECA 3f in which an oxygen bleed is introduced to increase the secondary ion yield. The method is very useful for detecting and identifying the segregating species. Detection levels in SIMS for some elements can

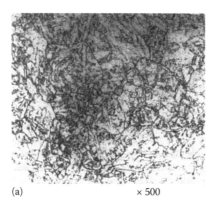 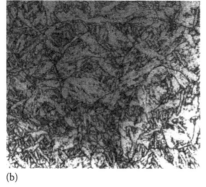

(a)   × 500   (b)

FIGURE 13.8 Area of mechanically polished steel plate with grain boundaries and sulfides present within the ferrite phase exposed by chemical etching.

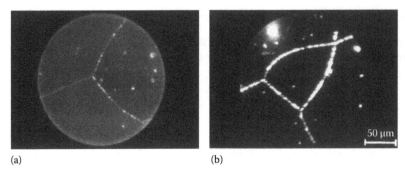

(a)   (b)

FIGURE 13.9   SIMS carbon maps from two regions of a metallographically polished steel specimen. (From Karlsson, L., Norden, H., and Odelius, H., *Acta Metal.*, 36, 1, 1988. With permission.)

be a low as parts per billion in certain circumstances. However, the technique has the drawback that its range of sensitivities for different elements is very wide, ranging over four orders of magnitude, and matrix effects are significant, with the result that it cannot easily be made quantitative (for further details see Chapters 4 and 10).

Another feature of SIMS is that ionized mass clusters appear in the mass spectrum, and it is common for such clusters, from fragments with different elemental composition to have the same mass, making positive identification of the cluster difficult. This is particularly true for low mass ions and fragments in which carbon–hydrogen combinations are common. In such cases, it is often possible to use different isotopes of the element being sought in the analysis, to arrive at a positive identification. For example, the method has been used to identify the presence of B in a C-Mn steel, where the B was present in the bulk at a concentration less than 5 ppm [44]. Boron has two isotopes with masses at 10 and 11 Da in the ratio of approximately 1:4, respectively. By recording a SIMS spectrum in the low mass region, the presence of B could be inferred from both the positions and the ratios of the peaks, which could then be used to map the location of the B, as seen in Figure 13.10. Figure 13.10a shows the ion-induced secondary electron image for a Si-killed (i.e., where Si is added to deoxidize or "kill" the steel) C-Mn steel plate containing 4 ppm of B, while in Figure 13.10b the mass spectrum recorded in the B region confirms that the two peaks were at the expected positions and relative intensities. The peaks at 10 and 11 Da were then used to map the surface; an exact match was found between the images in (c) for 10 Da and in (d) for 11 Da, in contrast to the maps for Al and Si shown in (e) and (f).

13.2.2.3 Autoradiography

Autoradiography is a technique that can be used to identify certain elements in metals and alloys. It relies on the formation of a radioactive isotope of the element whose identification is required, following exposure to an intense radioactive source. The decay of the isotope is then used to form an image on a photographic plate placed in contact with the surface. A metallographic sample of the material is prepared, exposed to the radioactive source for a given time, and the photographic plate placed in contact with the polished surface. Autoradiography is particularly useful for detecting the presence of boron in steel, as demonstrated in a study on casts of stainless steel [45]. Figure 13.11 shows autoradiographs from two casts (A and B) of stainless steel, with A containing only 3 ppm of boron, distributed mostly within the bulk, while cast B had a boron content of 90 ppm. There is strong evidence for segregation of the boron to the grain boundaries. Cast A fractured readily in an intergranular manner while cast B was ductile and was extremely difficult to fracture intergranularly. Boron is an element with an atomic diameter that is small compared with the matrix atoms and would be expected to increase the grain boundary energy on segregation.

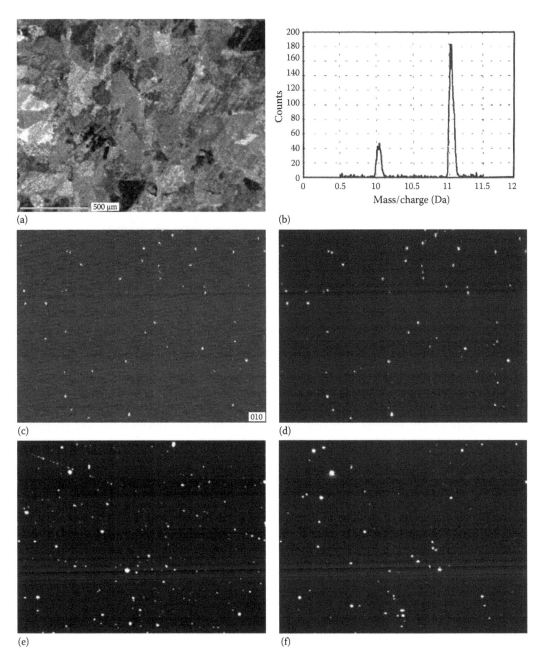

FIGURE 13.10   (a) Ion-induced secondary electron image of a steel plate containing 4 ppm boron. (b) Mass spectrum recorded in the boron mass region, showing peaks due to ^{10}B and ^{11}B in the correct ratio. (c–f) SIMS elemental distribution maps obtained at the same magnification, (c) B at 10 Da, and (d) at 11 Da. (e) Al, and (f) Si. (From Jones, R.B., Younes, C.M., Heard, P.J., Wild, R.K., and Flewitt, P.E.J., *Acta Mater.*, 50, 4395, 2002. With permission.)

13.2.2.4   Auger Electron Spectroscopy

The technique of AES is described in detail in Chapter 3. In addition to elemental identification, AES can be used to provide chemical state information with good spatial resolution from a polished surface, if the surface is first given a light ion etch. In the example shown here, a superalloy coating

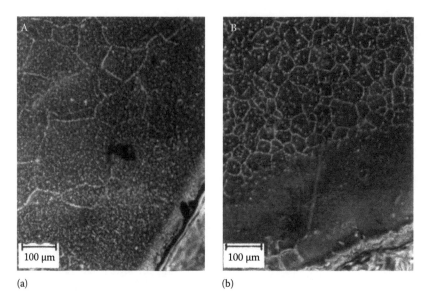

(a)   (b)

FIGURE 13.11 Autoradiographs for two casts of stainless steel containing (a) 3 ppm B and (b) 90 ppm B showing enhanced B segregation to grain boundaries in (b). (From Wild, R.K., *Mater. Sci. Eng.*, 42, 265, 1980. With permission.)

was first metallographically polished, then argon ion etched to remove any surface oxide or contamination [46]. AES spectra were then recorded from various points on the sample, and revealed the presence of Al, Cr, Ti, Ni, Co, O, and N. A secondary electron image of a region of the sample is shown in Figure 13.12a, together with Ti in (b), Al in (c), and N in (d) AES maps. The structure consisted of phases with bright and dark contrast aligned along crystallographic planes, confirming the distribution of elements found in SIMS maps. Spectra from the dark phases showed that there were two types of phase, one containing N and Al, and the other N and Ti, with some C and Ni being present. The N KLL peak was found in the kinetic energy range 380–390 eV. It can be seen in Figure 13.13 that the exact location of the nitrogen peak was dependent on the position of the electron beam, with that in the Al phase being at 383 eV, and that in the Ti phase being at 388 eV. This chemical shift can be used to distinguish between N bound to Al and that to Ti, as shown in Figure 13.12d and e. The data in Figure 13.12d revealed the N peak at 388 eV, while in Figure 13.12e the N peak position was at 383 eV. The energy resolution of the electron detector of this instrument is $E/\Delta E = 200$, so that a peak at a kinetic energy of 400 eV would have an full width half maximum (FWHM) peak width of 2 eV. This was sufficient to resolve the shift in the N peak between aluminum nitride and titanium nitride. When taken in association with the aluminum and titanium maps, they show very clearly the presence of separate aluminum nitride and titanium nitride phases.

13.2.3 INTERGRANULAR FRACTURE

Ferritic steels have a body-centered cubic structure, and in their most basic form contain carbon additions, but Cr may be added in quantities up to 12 wt.% to improve corrosion characteristics, while other additions of up to a few wt.% may include Mo, V, and Ni. The steels have a ductile-to-brittle transition temperature that is normally below room temperature but above 80 K. The transition temperature tends to rise from this minimum in pure materials, to room temperature and above, as the grain boundary segregation levels increase. Frequently, these steels contain carbides, usually chromium and molybdenum carbides, on the grain boundaries and failure may be through or around them. Cu will also fail in an intergranular manner if impurities such as Bi [8] and Sb [9] have segregated to the grain boundaries. Mo and W normally contain impurities such as C, O, and S present at

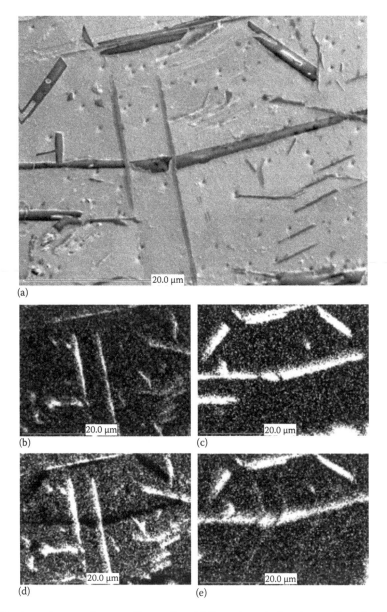

FIGURE 13.12 Auger images of a superalloy coating. (a) Secondary electron image. (b–e) Auger maps: (b) Ti, (c) Al, (d) N using peak at 388 eV, indicating N associated with Ti, and (e) N using peak at 383 eV, indicating N associated with Al. (From Heard, J., Day, J.C.C., and Wild, R.K., *Microsc. Anal.*, 77, 9, 2000. With permission.)

grain boundaries and will also fracture in an intergranular manner [10]. These and other similar metals and alloys are therefore suitable for analysis by AES following impact fracture at liquid nitrogen temperature.

13.2.3.1 Impact at Low Temperature

A specimen of the material to be fractured is prepared and cleaned by ultrasonically washing in isopropanol. The specimens may be machined to standard forms, usually rods of diameter 3–5 mm and length 25–30 mm, with a notch 1 mm deep at the mid-point. Small specimens, of dimensions down to $1 \times 1 \times 10 \text{ mm}^3$, can be fractured by mounting in prepared sleeves. The specimens are

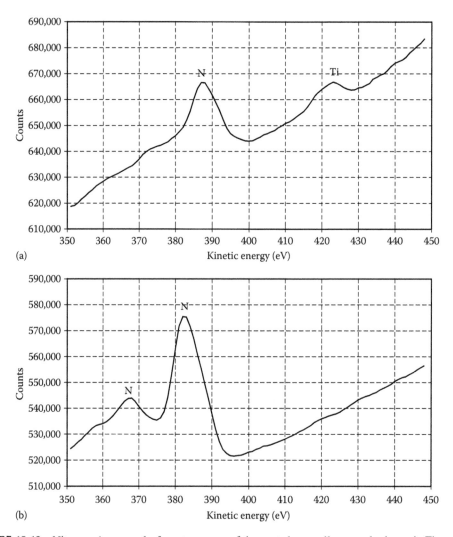

FIGURE 13.13   Nitrogen Auger peaks from two areas of the coated superalloy sample shown in Figure 13.2. (a) From the Ti-rich region, and (b) From the Al-rich region. (From Heard, J., Day, J.C.C., and Wild, R.K., *Microsc. Anal.*, 77, 9, 2000. With permission.)

introduced into the ultra high vacuum (UHV) system through an air lock and positioned for fracture. Alternatively, several may be loaded directly into the fracture stage before evacuation. An example of a low temperature impact stage is shown schematically in Figure 13.14 [47]. It is important that the vacuum in the system be as low as possible before fracture is attempted in order that the freshly exposed surfaces do not become contaminated. Seah [29] has demonstrated the importance of fracturing in UHV by depositing a monolayer of Sn onto a clean surface of Fe in UHV, which was then characterized using AES; the spectrum contained intense peaks from Sn in the region of 430 eV. Following exposure to air for a short period, the Sn peaks were barely detectable. To achieve the all-important UHV, it is good practice to load the specimens into the fracture stage at the end of a day and leave the system to pump overnight before starting to cool the next morning. Cooling is normally continued for approximately 1 h before fracture is attempted. In most systems, fracture is achieved by transmitting a sharp impact to the specimen via a bellows arrangement. A typical fracture surface from an Fe steel with 3 wt.% Ni which contained 530 ppm of P and 110 ppm of Sn and had been heat treated at 853 K for 48 h is shown in Figure 13.15. The fracture path is almost 100% intergranular

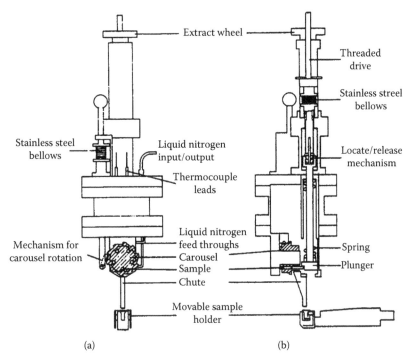

FIGURE 13.14 Low temperature impact fracture stage, showing (a) front view and (b) side view. (From Coad, J.P., Rivière, J.C., Guttmann, M., and Krahe, P.R., *Acta Met.*, 25, 161, 1977. With permission.)

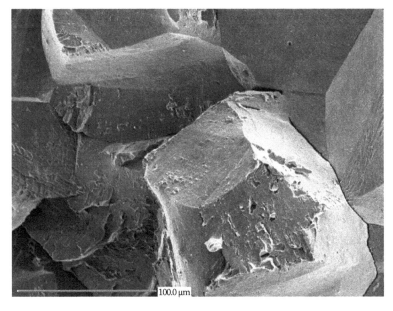

FIGURE 13.15 SEM image of an impact fracture surface of an Fe/3%Ni steel with P and Sn segregated to the grain boundaries.

with very few ductile or cleavage faces. This is an extreme example; most materials fracture with many cleavage and ductile facets. Cleavage can be identified from the appearance of a very flat surface containing many "river" lines running across it. Ductile failure has an appearance of warm toffee after it has been pulled apart, with cavities and peaks.

13.2.3.1.1 Grain Boundary Analysis by AES

Analysis of grain boundaries exposed in UHV is normally carried out using AES. The Auger electrons detected are in the energy range 0–2000 eV and since electrons of such energies have mean free paths in most materials of only a few nanometers, the resulting spectrum arises from the top few atom layers of the surface. AES can detect all elements in the periodic table with the exception of hydrogen and helium. Chemical state effects can be detected as shifts in peak position and/or changes in peak shape (see Chapter 3 Section 3.3.7). Most elements have similar electron yields for the most intense Auger transitions, and as a result the technique can give a quantitative elemental analysis of the surface. However, care must be exercised when attempting to quantify the levels of a segregant at a grain boundary. The standard method for determining the surface composition assumes a surface homogeneous in depth, whereas grain boundary segregants almost always consist of a single atom layer sitting on top of a matrix. A typical Auger spectrum, recorded from the grain boundary surface of the material pictured in Figure 13.15 is shown in Figure 13.16. The spectrum is displayed as the differential of the number $N(E)$ of Auger electrons multiplied by the electron energy E, that is, $(\mathrm{d}N(E)/\mathrm{d}E) \times E$, vs. the energy E and shows peaks resulting from Auger transitions within the surface atoms. In this example, peaks can be observed at approximately 600, 650, and 700 eV from Fe, at 800 and 850 eV from Ni, at 120 eV from P, and at 430 and 435 eV from Sn.

The extent of segregation at a grain boundary will be determined by the heat treatment that the material has received and by the relative orientations of the adjacent grains. P segregation has been found to increase with increasing tilt angle of grain boundary misorientation [48] while others have found that high index planes have a high level, and low index planes a low level, of segregation [49]. In practice, when a material is fractured intergranularly there will be a distribution of levels of segregation from grain boundary to grain boundary. It is therefore essential that a sufficient number of grains be analyzed on each fracture surface in order to obtain statistics of significance. A recommended number of grains is between 16 and 25. The time required for each analysis determines in practice the upper limit to the number of grains that can be analyzed, while in many cases it is often difficult to find 16 grains on a fracture face that contains only a small percentage of intergranular fracture.

Quantification of the composition of the grain boundary from the Auger spectrum is not trivial. For a homogeneous material, the composition is given by

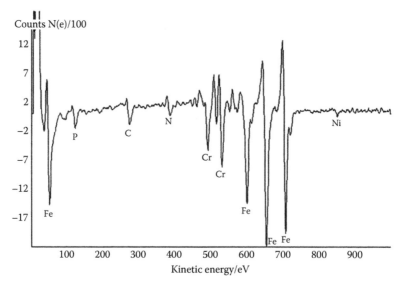

FIGURE 13.16 Auger spectrum from the grain boundary surface of the Fe/3%Ni steel imaged in Figure 13.15, showing segregation of P and Sn to the grain boundary.

$$X_i\% = (X_i/S_i)/\sum X_j/S_j \tag{13.5}$$

where

X_i and X_j are the ith and jth element peak heights
S_i and S_j are the sensitivity factors for the ith and jth elements

In most cases of segregation to a grain boundary, the segregating species is present as a single atom layer on the matrix surface. It is also assumed that half the segregating species remains on each exposed surface following fracture. To calculate the amount of segregant (A) on the surface of a matrix (B), it is necessary to know the inelastic mean free paths of electrons in the segregating species (λ_A) and in the matrix (λ_B), as well as the energy of the incident electron beam, and to determine the degree of electron backscattering (Γ). This is treated in some detail by Briggs and Seah [50] where an equation is derived which allows the amount of segregant to be calculated from:

$$X_i = Q_{AB}\frac{I_A/I_A^{Inf}}{I_B/I_B^{Inf}} \tag{13.6}$$

where

$$Q_{AB} = \left[\frac{\lambda_A(E_A)\cos\Theta}{a_A}\right]\left[\frac{1+\Gamma_A(E_A)}{1+\Gamma_B(E_A)}\right] \tag{13.7}$$

where

a_A is the size of atom A
Γ is the electron backscattering factor
Θ is the angle of the incident electron beam relative to the surface normal

For specific segregants such as P in standard matrices such as Fe, using known incident beam energies, it is possible to derive Q_{AB} once and for all and this may then be applied to the standard equation to give the segregation in monolayers.

Elements may not always segregate to the grain boundaries in a uniform manner. Certain elements may induce a grain boundary to form cavities or the element may segregate preferentially to surface cavities. Other elements diffusing to the grain boundary may combine to form particles on the boundary. It is therefore important that the boundary be analyzed at high spatial resolution. The advent of FEGs in both SEM and TEM has increased dramatically the spatial resolution for analysis. The inelastic mean free path for Auger electrons is only a few atom layers, which ensures that the lateral spatial resolution is determined essentially by the size of the incident probe and not, as is the case with EDX, the size of the incident electron scattering volume. Thus, analysis with a lateral resolution of a few tens of nanometers is now possible. The distribution of a segregating species across a fracture surface can be obtained by first recording a secondary electron image of the surface, then setting the analyzer to detect a major peak from the element of interest together with two backgrounds, one on either side of the peak, while the electron beam is rastered over the field of view. The peak minus background divided by the background is then displayed as a digital map in which the brightness at any point is proportional to the elemental concentration. Dividing the peak height by the background reduces the effect of surface topography on electron yields and hence on elemental distribution.

An example of the mapping procedure in scanning AES scanning Auger microscopy (SAM) is shown in Figure 13.17 [51], which demonstrates element mapping with a spatial resolution of 100 nm. Figure 13.17a is the secondary electron image from the surface of an Fe-3 wt.% Ni alloy fractured in UHV. Grain boundaries have been exposed which contain cavities varying in size from 0.1 to 1 μm. The heat treatment given to this material has caused Sn to segregate to the cavity surface, as shown by the Sn elemental map in Figure 13.17b and P to segregate to the grain boundary (Figure 13.17c). Such maps demonstrate that SAM can yield information with high spatial resolution.

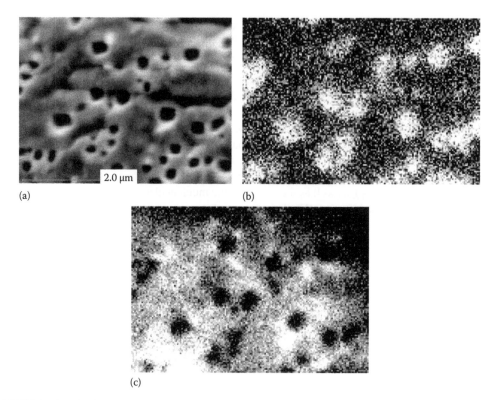

(a)

(b)

(c)

FIGURE 13.17   Grain boundary fracture surface of an Fe/3%Ni steel analyzed in a SAM spectrometer with an FEG, showing (a) the secondary electron image, (b) the Sn elemental map, and (c) the corresponding P elemental map.

Ferritic steels can become embrittled by P, or other trace elements, segregating to grain boundaries, and by using SAM to determine the amount of P segregation the degree of embrittlement can be determined. Figure 13.18 shows a secondary electron image (a) of the fracture surface of a CrMoV steel which had been fractured by impact at 77 K, together with the P Auger map (b) from the surface. Although intergranular fracture was not 100%, large areas of grain boundary are clearly visible, both by observation of the secondary electron image and from the intensity of the P Auger map.

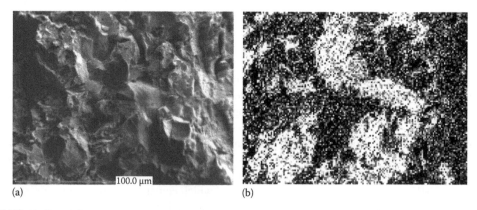

(a)

(b)

FIGURE 13.18   (a) Secondary electron image of the fracture surface of a CrMoV steel and (b) the corresponding P Auger map. (From Bulloch, J. and Wild, R.K., *Adv. Eng. Mater.*, 99-1, 169, 1995. With permission.)

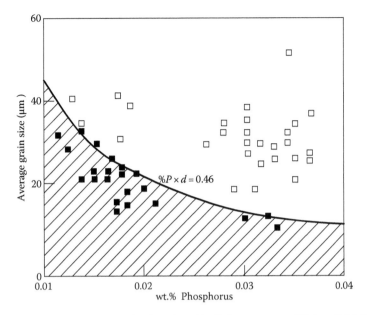

FIGURE 13.19 Relationship between grain size, d, and bulk P content, wt.%P, in a CrMoV steel heated to 713 K for 10^5 h. Embrittled specimens are indicated by solid points and lie below the curve. (From Bulloch, J. and Wild, R.K., *Adv. Eng. Mater.*, 99-1, 169, 1995. With permission.)

By determining the degrees of segregation of P in CrMoV steels containing different bulk levels of P, which had been heated at 713 K for 10^5 h, it was possible to identify embrittled and nonembrittled specimens [52]. The average grain size varied in these steels from 5 to >60 μm, and hence the area of grain boundary surface available for P segregation varied accordingly. Plotting grain size against the percentage of P in the bulk, and identifying the embrittled and nonembrittled specimens, gave the result in Figure 13.19, from which a relationship between grain size, d, and bulk P content, %P, could be derived, viz.:

$$d \times \%P = 0.28 \tag{13.8}$$

13.2.3.1.2 XPS

Although AES has good spatial resolution and can determine quantitatively the presence of most elements, it is not usually able to give much information on the chemical state of atoms at the surface. X-ray photoelectron spectroscopy (XPS), on the other hand, can give chemical state information but until recently did not have the spatial resolution necessary to distinguish grain boundaries. Since the first edition of this book, when XPS instruments could image with a spatial resolution of 10 μm and analyze from areas less than 30 μm in diameter, further improvements have been achieved. Currently, a spatial resolution of 1 μm with analysis from an area a few micrometer across is possible (see Chapter 3 for details). They therefore have the capability of analyzing grain boundary surfaces, which are frequently greater than 30 μm in size. Recently this capability has been demonstrated by fracturing and analyzing the grain boundaries of a ferritic steel [53]. Figure 13.20 is an example of XPS analysis of the Fe/Sn specimen described above. It shows the secondary electron image of the surface together with a Sn $3d_{5/2}$ map of the surface. Although the spatial resolution was still less than desired, it was possible to identify and analyze individual grains. The XPS spectrum from the Sn region is shown in Figure 13.21 and indicates that in this instance the Sn was present in its elemental form and not chemically combined with any other element.

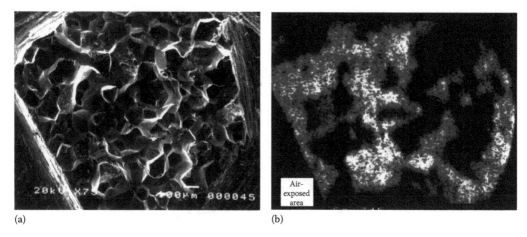

(a)   (b)

FIGURE 13.20   (a) Secondary electron image of the fracture surface of an Fe/3%Ni steel in which P and Sn have segregated to the grain boundaries, together with (b) the imaging XPS Sn $3d_{5/2}$ map. (From Hallam, K. and Wild, R.K., *Surf. Interface Anal.*, 23, 133, 1995. With permission.)

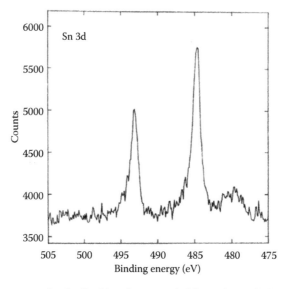

FIGURE 13.21   XPS spectrum for the Sn 3d region recorded from the grain boundaries of the Fe/3%Ni steel. (From Hallam, K. and Wild, R.K., *Surf. Interface Anal.*, 23, 133, 1995. With permission.)

13.2.4   TRANSMISSION ELECTRON MICROSCOPY

Not all metals and alloys can be induced to fracture intergranularly by the methods outlined above. For example, very pure austenitic 316 stainless steel is very difficult to fracture with any significant percentage of intergranular failure. It is also desirable to compare Auger and XPS results with independent methods to determine if the fracture path is indeed along the grain boundary and to confirm the levels of segregants. TEM combined with other analytical methods offers this possibility although it can be time-consuming.

13.2.4.1   Production of a Thin Foil

Sample preparation is often the most time-consuming aspect of TEM analysis. It is relatively straightforward if the sample is a homogeneous metal or alloy and if a thin foil can be produced from any part of it. Problems arise if areas near the surface or the boundary between two dissimilar materials are required to be analyzed. Normally, a thin section of material 3 mm in diameter is produced mechanically, which is then polished to the minimum practical thickness before being finally thinned to electron transparency. This final stage tends to be carried out by either electrolytic or ion thinning methods. When thinning electrolytically, it is so arranged that the center is thinned at a rate faster than that of the outer edges, and thinning is halted when a hole is observed near the center of the foil. Ion etching is carried out by bombarding the foil with a beam of inert gas ions, usually argon. The ion beam can be stationary and the foil moved, or the ion beam can be rastered to thin uniformly a large area. If the thinnest section is required from an area near the foil surface, then it is necessary to build a layer of supporting material onto that surface. Thus a specimen of stainless steel with an oxide on the surface might have a layer of nickel electrodeposited on it. The whole sandwich would then be used to produce a thin foil, with the thinned area containing the interface between the stainless steel and the nickel. There are specific instances where special thinning procedures need to be adopted. To examine radioactive materials, it is desirable to keep the volume of active material in the microscope to a minimum to reduce operator dose. This is accomplished by making a 3 mm disc in which the outer 1 mm perimeter consists of unirradiated material, while the irradiated material is restricted to a disc 1 mm in diameter, which is fitted into a hole of corresponding diameter in the unirradiated material. Since the central region is the thinned area, the dose is reduced considerably by more than a factor of 10.

A number of automatic electrolytic thinning systems are available commercially in which a light detection system is used to indicate when penetration of the foil has occurred. These techniques and the electrolytes appropriate for use with a range of materials have been extensively reviewed [54,55].

A major development in the preparation of TEM specimens has been the introduction of the focused ion beam (FIB) technique [56]. Field emission ion guns have been developed with intense beams capable of being focused to a diameter of less than 20 nm [57,58]. These FIBs can be used to machine specimens from all types of material. They are fast, since a TEM specimen can be produced in less than 2 h, and they are precise; not only can an exact location be specified, but they can also be used in awkward positions, e.g., specimens can be cut from surfaces and edges. The technique is illustrated in Figure 13.22, where a TEM specimen has been cut from a semiconductor device and is ready for transfer to the copper grid and then to the TEM. Initially, a thin metallic layer is laid down and this is visible as the light band at the top of the TEM specimen. The metallic layer is formed by passing over the surface a gaseous compound of the metal to be deposited, while the ion beam is rastered over the area to be covered. Two slots are then cut, defining the sides of the specimen which is typically 100 nm wide. Cuts are then made down the sides and along the base to produce a specimen approximately $20 \times 10 \times 0.1\,\mu m^3$, suitable for TEM. The whole specimen is then removed from the FIB system and transferred to a TEM.

A typical application has been described by Lozano-Perez et al. [59] who used the above technique to prepare specimens through cracks in an austenitic stainless steel. In their approach, it was possible to retain a thicker frame around the specimen and thereby provide support at all stages of preparation and observation. The advantages of the above procedure are that the TEM specimen is obtained from the exact desired location, many specimens may be produced from a single sample, and from a small volume, and the specimen may contain oxides, metals, glasses, and ceramics with little or no variation in TEM specimen thickness. Figure 13.23 shows a TEM specimen cut from a Ti/TiO_2 sample. In Figure 13.23a, a low magnification image shows the protective metal laid down before the specimen was cut, the thin layer of TiO_2 of thickness 1 μm, and the metal substrate. The high resolution image (b) and the selected area diffraction patterns (c)–(f) correspond to zones (1)–(4), respectively [60].

13.2.4.2 Field Emission Gun STEM

The advent of the field emission gun scanning transmission microscope (FEGSTEM) has given the electron microscopist a very bright fine focus electron source (see Chapters 6 and 9). With this tool (Figure 13.24), an analysis can be obtained from a thin foil using a beam that is only 1 or 2 nm in

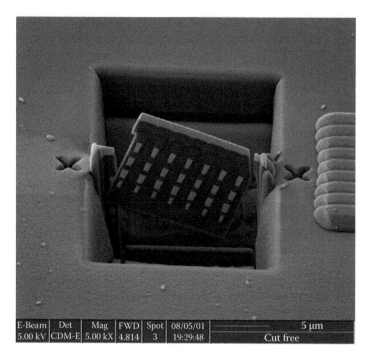

| E-Beam | Det | Mag | FWD | Spot | 08/05/01 | 5 µm |
| 5.00 kV | CDM-E | 5.00 kX | 4.814 | 3 | 19:29:48 | Cut free |

FIGURE 13.22 Production of a TEM specimen from a semiconductor device using a FIB. A metallic layer is deposited first and is visible at the top of the specimen. The ion beam first cuts a slice approximately 100 nm thick before making cuts to the side and base to allow the sample to be removed to the TEM. (Courtesy of FEI.)

diameter, and without significant spread on passing through the foil. If an electron-transparent specimen is prepared in which a grain boundary can be oriented so that it is parallel to the incident electron beam, then it is possible to use either electron energy loss spectroscopy (EELS) or EDX to determine the composition of a cylinder approximately 2 nm in diameter times the thickness of the foil. By scanning the beam across the grain boundary, a composition profile can be obtained.

13.2.4.2.1 Parallel EELS

As an incident electron beam passes through a specimen some of the electrons lose energy by interaction with the electron cloud surrounding an atom and with the nucleus. One of the types of loss is due to plasmon excitation, and they extend over a region from 0 to 50 eV. Losses greater than 50 eV arise from inelastic collisions with the inner atomic shells of atoms, while the general background results from valence shell excitations. Electrons, of a given energy, emerging from the thin foil, are focused using a magnetic prism to an image point on the detector. Electrons with displaced energy ΔE are brought to a focus position displaced by Δx from the image point. The electrons focused on the detection slit are detected by a scintillator-multiplier or an array of solid-state diodes, and are collected either in series or parallel. The parallel version, PEELS, based on solid-state detectors, is faster by some two orders of magnitude than serial detection, and its use increases detection limits, reduces damage to the foil, and improves the collection of extended energy loss fine structure. With these systems, information concerning the chemical state of atoms at the grain boundary can be obtained (see Chapter 9).

13.2.4.2.2 Energy Dispersive X-Ray Analysis

When an incident electron beam passes through a thin foil, it ionizes atoms within the foil. The ionized atoms then decay with the emission of x-rays, and the energy of the emitted x-ray is

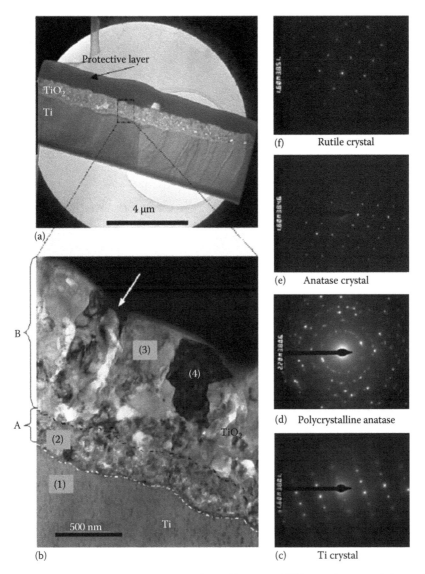

FIGURE 13.23   Cross section and TEM images of the lifted-out Ti/TiO$_2$ specimen at (a) low and (b) high resolution. (c), (d), (e), and (f) are selected area diffraction patterns corresponding to zones (1), (2), (3), and (4), respectively, from Figure 13.23b. (From Gueneau de Mussy, J.P., Langelaan, G., Decerf, J., and Delplancke, J.L., *Scr. Mater.*, 48, 23, 2003. With permission.)

characteristic of the atom from which it came. Detection of these x-rays with a solid-state detector allows the composition of the volume of the thin foil through which the beam has passed, to be determined.

Figure 13.25 shows the concentrations of Cr, Fe, Ni, and P as the incident beam was rastered across a grain boundary in an Inconel 690 alloy [61]. Inconel 690 contains approximately 60 wt.% Ni, 30 wt.% Cr, and 10 wt.% Fe. The profile shows that at the grain boundary, the Ni concentration could be enhanced to 70 wt.%, while that of the Cr could be depleted to 20 wt.%. In this case, the depleted region extended approximately 10–20 nm to either side of the boundary. Depletion of Cr is thought to be a major factor in stress corrosion cracking. Cr, when present in steels to greater than 12 wt.%, will form a protective surface oxide. Microcracks can form at or near a grain boundary at

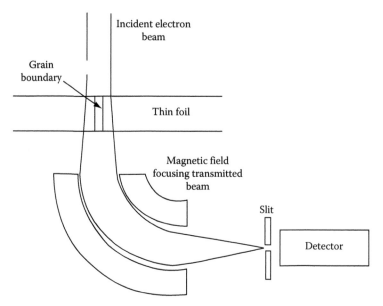

FIGURE 13.24 Schematic showing method for EDX and PEELS analysis of grain boundaries in a FEGSTEM.

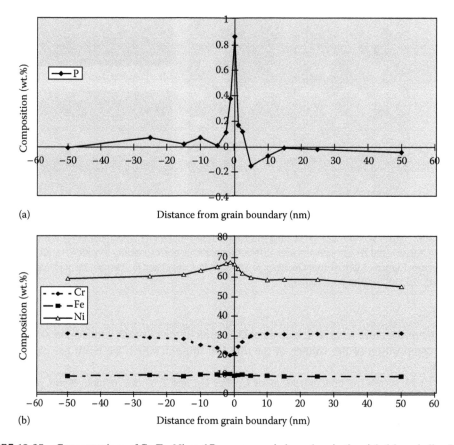

FIGURE 13.25 Concentrations of Cr, Fe, Ni, and P across a grain boundary in the nickel-based alloy Inconel 690 using EDX, (a) phosphorus and (b) chromium, iron, and nickel. (From Wild, R.K., *Micros. Anal.*, July 5, Wiley, New York, 1997. With permission.)

the surface. Depletion in the region of grain boundaries allows the surface to be attacked and, with no protective oxide able to form, corrosion extends down the grain boundary surface, the crack continuing to open up aided by the stress and the growing oxide. P is also shown to have been enriched at the grain boundary, but here the profile was much narrower, extending only 1 or 2 nm to either side of the grain boundary. In this case, P has segregated to the grain boundary as a single atom layer. The incident beam extended over a diameter of 2 nm and therefore included matrix material as well as grain boundary which was approximately 0.5 nm wide. FEGSTEM EDX will therefore have underestimated the level of P segregated to the boundary. P is a well-known embrittling agent and plays a significant role in stress corrosion cracking. The action of P at the grain boundary aids the opening of the crack both by applied stress and by oxide formation.

13.2.4.2.3 Comparison of AES and FEGSTEM

AES and FEGSTEM sample a material in different ways. AES analyzes directly one side of a grain boundary, sampling mainly the top atom layer but with contributions included from a few layers below the surface. FEGSTEM combined with EDX analyzes a volume that is 2 nm in diameter and thus includes material either side of the 0.3 nm wide grain boundary. Each method attempts to determine the actual level of the segregant at the grain boundary, but because of the many assumptions made there may be errors. It is instructive to compare results from the two methods. Walmsley et al. [62] have analyzed several systems with SAM and EDX FEGSTEM, plotting the concentrations obtained from one against those obtained from the other. The results of this exercise are shown in Figure 13.26. Converting each of the signals from Auger and EDX to wt.% of that element in the matrix leads to a relationship in which SAM gives approximately twice the concentration of P as EDX.

13.2.4.3 Time-of-Flight Atom Probe

The detailed internal composition of a metal can be determined with atomic resolution using an atom probe combined with a time-of-flight mass spectrometer (see Chapter 11). Atoms can be stripped from the probe tip layer by layer by pulsing the field in the vicinity of the tip, and become ionized by the high field. The ions then travel through the flight tube with the lighter ones arriving first and the heavier ones last. The ion mass is determined from the time of flight from desorption to detection. The spatial distribution, laterally across the tip, is determined by using a multi-channnel detector to sense the position of the ion at the detector, which is directly related to the position on the tip. Depth information is obtained by stripping individual layers of atoms in a known and controlled manner. The information concerning time and position of arrival is stored by computer, and used subsequently to map the distribution of elements in three-dimensional space. Clearly, this technique is time consuming and the number of specimens that can be analyzed is limited. It can however give some unique information. In one example, a tip was prepared from a Ni base superalloy (Astroloy) with composition approximately 52% Ni, 16% Cr, 16% Co, 8% Al, and 4% Ti, and which contained a grain boundary. The alloy contained 0.11% B as a trace impurity. The boundary position was first located using a TEM. The tip was then analyzed as described above, and it was possible to identify the position of the various elements around the grain boundary. Figure 13.27 shows the distributions for Al + Ti, Cr, Mo, and B + C. It can be seen that the boundary is enriched in Cr, Mo, and B + C but depleted in Al and Ti. It was also possible to determine the width of segregation at the grain boundary and in this case the width of the segregated B layer was 0.5 nm [63,64].

13.3 CRACKS IN METALS AND ALLOYS

When a material fails in service, it is necessary to analyze the failed component to determine the causes of failure. While the metallographic and analytical methods already described in this

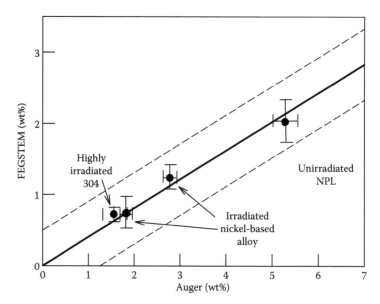

FIGURE 13.26 Comparison of a grain boundary composition determined by AES and FEGSTEM EDX showing an approximately 2:1 ratio of the respective quantifications by the two techniques. (From Walmsley, J., Spellwood, P., Fisher, S.B., and Jenssen, A., *Proceedings of the 7th International Symposium on Environmental Degradation of Materials in Nuclear Power Systems—Water Reactions*, Breckenridge, CO, p. 985, 1985. With permission.)

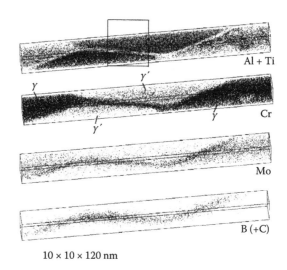

$10 \times 10 \times 120$ nm

FIGURE 13.27 Position-sensitive atom probe images across a grain boundary in the nickel-based alloy Astroloy for Al + Ti, Cr, Mo, and B + C. (From Blavette, D., Deconihout, B., Bostel, A., Sarrau, J.M., Bouet, M., and Menand, A., *Rev. Sci. Instrum.*, 64, 2911, 1933; Letellier, L., Guttmann, M., and Blavette, D., *Phil. Mag. Lett.*, 70, 189, 1994.)

chapter should be undertaken, the failed surface itself can often yield valuable information. Failure frequently occurs by the gradual opening of a crack for some time before catastrophic failure actually occurs. Often a crack may be detected in the component during routine nondestructive testing and before failure has occurred. In these cases, it is important to know when the crack started, and in the case of the crack detected before failure, if the crack is growing and at what rate.

The oxide thickness can be used to determine the age of different parts of the crack and hence the rate of growth. A freshly exposed metal surface will, in general, oxidize in such a way that the thickness of the oxide increases as the square root of the time of exposure at temperature [65,66], namely:

$$x = A + kt^2 \tag{13.9}$$

where
  A is a constant
  k is the parabolic rate constant

Frequently, the oxide may be many micrometers thick as in the case of a ferritic steel operating at several hundred degrees centigrade, in which a crack had opened over a period of months or years before final failure. In such cases, the oxide thickness may be determined readily from optical examination of a metallurgically prepared specimen. The thickness may then be used, together with the time/temperature history, to build up a picture of the crack history.

However, if a crack has grown in a material which oxidizes only slowly, then the oxide thickness may be too thin to determine by metallographic means. In this case, surface analytical techniques may be used to determine the oxide thickness and, in the case of the crack which has yet to cause failure, may also yield some information regarding the causes of crack initiation and growth. Ideally, the crack should be opened inside the spectrometer so that any contaminant or segregation can be identified without being first contaminated by the ambient air. Sometimes the component may have been cleaned with solvents prior to receipt, or it may be necessary to cut it in order to produce a specimen suitable for fracture in the spectrometer. Both these operations may themselves cause contamination, which may add to the difficulties of identifying the cause of the crack. However, it is still preferable to open the crack under UHV conditions. Figure 13.28 shows a secondary electron image of a crack tip in an Fe/Cr/Mo/V steel together with an Auger elemental map for O which allows the crack tip to be identified. The oxide thickness can be determined by depth profiling through the oxide at regular distances from the crack tip. These profiles are obtained by determining the peak heights of selected elements, in this case O and Fe, at each point on the oxide surface, removing a fixed amount of oxide by bombarding with energetic argon ions, and then repeating the process. The operation is controlled by computer, allowing an elemental depth distribution for O and Fe to be built up at each point. The time taken to reach the oxide/metal interface is then converted into oxide thickness using calibration data obtained from standard oxides. The depth profiles at two

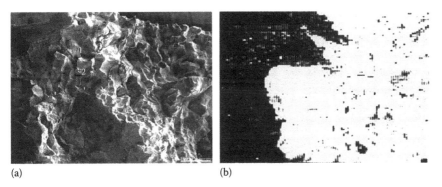

(a)   (b)

FIGURE 13.28   (a) Secondary electron image of a crack tip opened by fracture in the UHV environment of the spectrometer, together with (b) the O elemental Auger map.

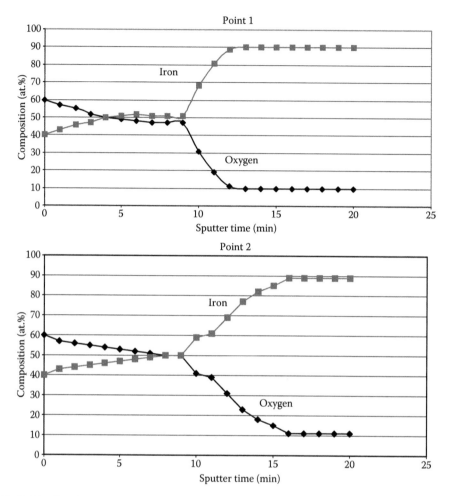

FIGURE 13.29 Depth profiles through the oxide at two points on the crack shown in Figure 13.28. The oxide/metal interface is broad, and also varies in width from point to point.

points on the oxide are reproduced in Figure 13.29. It can be seen that the oxide/metal interface was very broad, indicating that the grain boundary surface was not completely flat, and that the interface width varied from point to point. The interface was reached at point 1 in about 10 min, indicating an oxide thickness of 400 nm, whereas the interface at point 2 was reached in approximately 13 min, suggesting an oxide thickness of nearly half a micrometer. It can also be seen that the Fe/O ratio varied from 40:60 at the outer surface to 50:50 within the oxide, which may indicate that the oxide composition was changing, but which may also have been due to ion beam reduction. From a series of such profiles the crack history can be built up, from which it is possible to state whether the crack is continuing to grow or whether it has stopped.

13.4 CONCLUSION

Progress has been made in our ability to analyze and characterize metallurgical surfaces and interfaces in a number of areas. The advances in fast ion bombardment to produce transmission electron microscope specimens from metals and oxides have opened the way for TEM to analyze the interfaces between different materials and to characterize the grain boundaries. Improvements have been made in both spatial resolution and sensitivity of both Auger and x-ray photoelectron spectrometers. XPS instruments can image at around 1 μm making grain boundary analysis possible for many more specimens, while Auger spectrometers can record from areas only a few nanometers in diameter.

REFERENCES

1. K. N. Akhurst, *CEGB Res.*, **21**, 5 (1988).
2. P. Lejcek and S. Hofmann, *Crit. Rev. Solid State*, **20**, 1 (1995).
3. H. Erhart and H. J. Grabke, *Met. Sci.*, **15**, 401 (1981).
4. C. L. Briant, *Acta Metall.*, **33**, 1241 (1985).
5. M. P. Seah and C. Lea, *Phil. Mag.*, **31**, 627 (1975).
6. M. Guttmann, *Surf. Sci.*, **53**, 213 (1975).
7. H. J. Grabke, *Segregation at Interfaces in Chemistry and Physics of Fracture*, R. M. Latanision and R. H. James (eds.), Martinus Nijhoff, Dordrecht, the Netherlands, 1987, 388.
8. B. D. Powell and D. P. Woodruff, *Phil. Mag.*, **A34**, 169 (1976).
9. T. H. Chuang, W. Gust, L. A. Heldt, M. B. Hintz, S. Hofmann, R. Lucic, and B. Predel, *Scripta Met.*, **16**, 1437 (1982).
10. E. Smiti, P. Joufrey, and A. Kobylanski, *Scripta Met.*, **18**, 673 (1984).
11. C. L. Briant, *Met. Trans. A*, **18A**, 691 (1987).
12. G. P. Airey, *Corrosion NACE*, **41**, 2 (1985).
13. A. Elkholy, J. Galland, P. Azou, and P. Bastien, *C. R. Acad. Sci. Paris*, **284C**, 363 (1977).
14. G. C. Allen and R. K. Wild, *Phil. Mag. A*, **54**, L37 (1986).
15. C. L. Briant, *Surf. Interface Anal.*, **13**, 209 (1988).
16. D. J. Nettleship and R. K. Wild, *Surf. Interface Anal.*, **16**, 552 (1990).
17. C. T. Liu and E. P. George, in *Structure and Property Relationships for Interfaces*, J. L. Walter, A. H. King, and K. Tangri (eds.), ASM, Metals Park, OH, 1991, 281.
18. Y. Ishida, S. Yokoyama, and T. Nishizawa, *Acta Met.*, **33**, 255 (1985).
19. T. Muschik, W. Gust, S. Hofmann, and B. Predel, *Acta Met.*, **37**, 2917 (1989).
20. D. I. R. Norris, C. Baker, and J. M. Titchmarsh, *Proceedings of Radiation Induced Sensitisation of Stainless Steels*, Berkeley Nuclear Laboratories, Berkeley, UK, 1986, 86.
21. D. I. R. Norris, C. Baker, C. Taylor, and J. M. Titchmarsh, *Effects of Radiation on Materials: 15th International Symposium. ASTM STP 1125*, R. E. Stoller, A. S. Kumar, and D S. Gelles (eds.), ASTM, Philadelphia, 1992.
22. O. L. Krivanek, *Ultramicroscopy*, **28**, 118 (1989).
23. G. A. Dryer, D. E. Austin, and W. D. Smith, *Metal Progr.*, **86,** 116 (1964).
24. C. Herzig, J. Geise, and Y. M. Mishin, *Acta Met. Mater.*, **41**, 1683 (1993).
25. R. Herschitz and D. N. Seidman, *Scripta Met.*, **16**, 849 (1982).
26. A. Cerezo, T. J. Godfrey, and G. D. W. Smith, *Rev. Sci. Instrum.*, **59**, 862 (1988).
27. M. P. Seah and E. D. Hondros, *Proc. Roy. Soc. Lond. A*, **335**, 191 (1973).
28. M. P. Seah, *Surface Sci.*, **53**, 168 (1975).
29. M. P. Seah, *Proc. Roy. Soc. Lond. A*, **349**, 535 (1976).
30. E. D. Hondros and M. P. Seah, *Intern. Met. Rev.*, **22**, 867 (1977).
31. D. McLean, *Grain Boundaries in Metals*, Oxford University Press, London, 1957.
32. J. W. Gibbs, *Collected Works*, Vol. 1, p. 219, Yale University Press, 1948.
33. F. L. Carr, M. Goldman, L. D. Jaffe, and D. C. Buffum, *Trans. AIME*, **197**, 998 (1953).
34. M. P. Seah, *Acta Met.*, **25**, 345 (1977).
35. P. L. Gruzin and V. V. Mural, *Fiz. Metal. Metalloved.*, **17**, 62 (1964).
36. P. W. Palmberg and H. L. Marcus, *Trans. Amer. Soc. Metals*, **62**, 1016 (1969).
37. D. A. Smith and A. H. King, *Phil. Mag.*, **A44,** 333 (1981).
38. R. W. Balluffi and J. W. Cahn *Acta Metall.*, **29**, 493 (1981).
39. J. M. Perks, A. D. Marwick, and C. A. English, *Proceedings of Radiation Induced Sensitisation of Stainless Steels*, Berkeley Nuclear Laboratories, Berkeley, UK, 1986, 15–34.
40. A. D. Marwick, R. C. Piller, and M. E. Horton, *AERE-R10895*, Atomic Energy Research Establishment, Harwell, UK, 1983.
41. P. Doig and P. E. J. Flewitt, *Analytical Electron Microscopy*, Institute of Metals, London, 1988.
42. J. M. Titchmarsh and S. Dunbill, *J. Nucl. Mat.*, **227**, 203 (1996).
43. L. Karlsson, H. Norden, and H. Odelius, *Acta Metal.*, 36, 1 (1988).

44. R. B. Jones, C. M. Younes, P. J. Heard, R. K. Wild, and P. E. J. Flewitt, *Acta Mater.*, **50,** 4395 (2002).

45. R. K. Wild, *Mat. Sci. Eng.*, **42,** 265 (1980).

46. J. Heard, J. C. C. Day, and R. K. Wild, *Microsc. Anal.*, **May** 9, 2000.

47. J. P. Coad, J. C. Rivière, M. Guttmann, and P. R. Krahe, *Acta Met.*, **25,** 161 (1977).

48. K. Tatsumi, N. Okumura, and S. Funaki, in *Grain boundary Structure and Related Phenomena*, Y. Ishida (ed.), *Proc 4th JIM Intl. Symp. Trans. Jap. Inst. Met. Suppl.*, **27,** 427 (1986).

49. S. Suzuki, K. Abiko, and H. Kimura, *Scripta Metall.*, **15,** 1139 (1981).

50. D. Briggs and M. P. Seah, *Practical Surface Analysis*, Vol. I, J. Wiley and Sons, Chichester, UK, 1990.

51. R. K. Wild, *Mater. World*, **May** 389 (1997).

52. J. Bulloch and R. K. Wild, *Adv. Eng. Mater.*, **99-1,** 169 (1995).

53. K. Hallam and R. K. Wild, *Surf. Interface Anal.*, **23,** 133 (1995).

54. I. S. Brammar and M. A. P. Dewey, *Specimen Preparation for Electron Metallography*, American Elsevier Publishing Co., New York, 1966.

55. P. J. Goodhew, *Specimen Preparation in Materials Science*, North Holland, Amsterdam, 1972.

56. D. J. Barber, *Microsc. Anal.*, **May** 5 (1995).

57. E. Kirk, *Microsc. Semicond. Mater. Proc. Royal Microsc. Soc.*, **15,** 831 (1989).

58. M. Overwijk, *J. Vac. Sci. Technol. B*, **21,** 2021 (1993).

59. S. Lozano-Perez, Y. Huang, R. Lanford, and J. M. Titchmarsh, *Electron Microsc. Anal., IOP Conf. Ser.*, **168,** 191 (2001).

60. J. P. Gueneau de Mussy, G. Langelaan, J. Decerf, and J. L. Delplancke, *Scripta Mater.*, **48,** 23 (2003).

61. R. K. Wild, *Micros. Anal.*, July 5, Wiley, New York (1997).

62. J. Walmsley, P. Spellwood, S. B. Fisher, and A. Jenssen, *Proceedings of 7th International Symposium On Environmental Degradation of Materials in Nuclear Power Systems—Water Reactions*, Breckenridge, Colorado, 1995, p. 985.

63. D. Blavette, B. Deconihout, A. Bostel, J. M. Sarrau, M. Bouet, and A. Menand, *Rev. Sci. Instrum.*, **64,** 2911 (1993).

64. L. Letellier, M. Guttmann, and D. Blavette, *Phil. Mag. Lett.*, **70,** 189 (1994).

65. C. Wagner, *J. Electrochem. Soc.*, **99,** 369 (1952).

66. P. Kofstad, *High Temperature Oxidation of Metals*, Wiley, New York, 1966.

APPENDIX

Documentary Standards in Surface Analysis

A.1   INTRODUCTION

Commercial surface analysis systems have been available since around 1970. Most of the early instruments were dedicated to longer-term fundamental research, even if they were located at industrial research centers. However, since the latter part of the 1980s, the most popular surface analytical techniques such as AES, SIMS, and XPS have gained a greater level of acceptance in industry due to their improved reliability. Surface analysis is now routinely used to solve complex industrial problems in both research and quality assurance environments. It has been specifically the move toward the use of these techniques in quality-assurance-type applications that has started to force the development of national/international documentary standards in order to formalize the methods of application of the techniques.

In 1991, the International Standards Organisation (ISO) set up ISO technical committee 201 on surface chemical analysis (ISO/TC 201) specifically to develop documentary standards for the most industrially developed surface analytical techniques.

A.2   ISO TECHNICAL COMMITTEE 201 ON SURFACE CHEMICAL ANALYSIS

There are currently 30 countries that are members of ISO/TC 201, with 11 of those participating in the development of the standards (Australia, Austria, China, France, Hungary, Japan, Korea, Russia, Switzerland, United Kingdom, and United States).

A.2.1 SCOPE

Standardization in the field of surface chemical analysis in which beams of electrons, ions, neutral atoms or molecules, or photons are incident on the specimen material and scattered or emitted electrons, ions, neutral atoms or molecules, or photons are detected.

A.2.2 STRUCTURE OF ISO TECHNICAL COMMITTEE 201

The main ISO committee on surface analysis has nine subcommittees (SCs) and one working group (WG). Also, each SC may have a number of WGs. The SCs are as follows:

SC1	Terminology	Secretariat USA (NIST)
SC2	General procedures	Secretariat USA (NIST)
SC3	Data management and treatment	Secretariat UK (BSI)
SC4	Depth profiling	Secretariat Japan (JISC)
SC5	AES	Secretariat USA (NIST)
SC6	Secondary ion mass spectroscopy	Secretariat Japan (JISC)
SC7	XPS	Secretariat UK (BSI)
SC8	Glow discharge spectroscopy	Secretariat to be decided
SC9	Scanning probe microscopy	Secretariat Korea

Current list of standards from ISO/TC 201 (for most standards, a summary is available in *Surface and Interface Analysis* in the volume given in square brackets).

(I) ISO 14976:1998—*Surface chemical analysis—Data transfer format* [*SIA* 1999; **27**: 693].

(II) ISO 14237:2000—*Surface chemical analysis—Secondary-ion mass spectrometry—Determination of boron atomic concentration in silicon using uniformly doped materials* [*SIA* 2002; **33**: 361].

(III) ISO 14707:2000—*Surface chemical analysis—Glow discharge optical emission spectrometry—Introduction to use* [*SIA* 2002; **33**: 363].

(IV) ISO 14606:2000—*Surface chemical analysis—Sputter depth profiling—Optimisation using layered systems as reference materials.* [*SIA* 2002; **33**: 365].

(V) ISO 14975:2000—*Surface chemical analysis—Information formats* [*SIA* 2002; **33**: 367].

(VI) ISO 14706:2000—*Surface chemical analysis—Test method of surface elemental contamination on silicon wafers by total reflection X-ray fluorescence spectrometry* [*SIA* 2002; **33**: 369].

(VII) ISO 15472:2001—*Surface chemical analysis—X-ray photoelectron spectrometers—Calibration of energy scales* [*SIA* 2001; **31**: 721].

(VIII) ISO 18115:2001—*Surface chemical analysis—Vocabulary* [*SIA* 2001; **31**: 1048].

(IX) ISO TR 15969:2000—*Surface chemical analysis—Depth profiling—Measurement of sputtered depth* [*SIA* 2002; **33**: 453].

(X) ISO 17560:2002—*Surface chemical analysis—Secondary-ion mass spectrometry—Method for depth profiling of boron in silicon* [*SIA* 2005; **37**: 90].

(XI) ISO 17974:2002—*Surface chemical analysis—High resolution Auger electron spectrometers—Calibration of energy scales for elemental and chemical state analysis* [*SIA* 2003; **35**: 327].

(XII) ISO 17973:2002—*Surface chemical analysis—Medium resolution Auger electron spectrometers—Calibration of energy scales for elemental analysis* [*SIA* 2003; **35**: 329].

(XIII) ISO 18114:2003—*Surface chemical analysis—Secondary-ion mass spectrometry—Determination of relative sensitivity factors from ion-implanted reference materials* [*SIA* 2006; **38**: 171].

(XIV) ISO TR 19319:2003—*Surface chemical analysis—Auger electron spectroscopy and X-ray photoelectron spectroscopy—Determination of lateral resolution, analysis area and sample area viewed by the analyser* [*SIA* 2004; **36:** 666].

(XV) ISO 20341:2003—*Surface chemical analysis—Secondary-ion mass spectrometry—Method for estimating depth resolution parameters with multiple delta-layer reference materials* [*SIA* 2005; **37:** 646].

(XVI) ISO 15470:2004—*Surface chemical analysis—X-ray photoelectron spectroscopy—Description of selected instrumental performance parameters.*

(XVII) ISO 15471:2004—*Surface chemical analysis—Auger electron spectroscopy—Description of selected instrumental performance parameters.*

(XVIII) ISO 19318:2004—*Surface chemical analysis—X-ray photoelectron spectroscopy—Reporting of methods used for charge control and charge correction* [*SIA* 2005; **37:** 524].

(XIX) ISO 17331:2004—*Surface chemical analysis—Chemical methods for the collection of elements from the surface of silicon-wafer working reference materials and their determination by total-reflection X-ray fluorescence (TXRF) spectroscopy* [*SIA* 2005; **37:** 522].

(XX) ISO 18118:2004—*Surface chemical analysis—Auger electron spectroscopy and X-ray photoelectron spectroscopy—Guide to the use of experimentally determined relative sensitivity factors for the quantitative analysis of homogeneous materials* [*SIA* 2006; **38:** 178].

(XXI) ISO 21270:2004—*Surface chemical analysis—X-ray photoelectron and Auger electron spectrometers—Linearity of intensity scale* [*SIA* 2004; **36:** 645].

(XXII) ISO 22048:2004—*Surface chemical analysis—Information format for static secondary-ion mass spectrometry* [*SIA* 2004; **36:** 642].

(XXIII) ISO 24236:2005—*Surface chemical analysis—Auger electron spectroscopy—Repeatability and constancy of intensity scale* [*SIA* 2007; **39:** 86].

(XXIV) ISO 24237:2005—*Surface chemical analysis—X-ray photoelectron spectroscopy—Repeatability and constancy of intensity scale.*

(XXV) ISO 18116:2005—*Surface chemical analysis—Guidelines for preparation and mounting of specimens for analysis.*

(XXVI) ISO 16962:2005—*Surface chemical analysis—Analysis of zinc- and/or aluminium-based metallic coatings by glow-discharge optical-emission spectrometry.*

(XXVII) ISO/TR 18392:2005—*Surface chemical analysis—X-ray photoelectron spectroscopy—Procedures for determining backgrounds* [*SIA* 2006; **38:** 1173].

(XXVIII) ISO 18115:2001/Amd. 1:2006—*Surface Chemical Analysis—Vocabulary—Amendment 1.*

(XXIX) ISO 20903:2006—*Surface chemical analysis—Auger electron spectroscopy and X-ray photoelectron spectroscopy—Methods used to determine peak intensities and information required when reporting results.*

(XXX) ISO/TR 18394:2006—*Surface chemical analysis—Auger electron spectroscopy—Derivation of chemical information.*

ISO standards may be purchased from national standards bodies, directly from the ISO Central Secretariat, Case Postale 56, CH-1211 Geneva 20, Switzerland, or through the Internet at http://www.iso.ch. More information about ISO/TC 201 on Surface Chemical Analysis may be obtained from this Internet site or from Mr. Yukio Hirose, Secretariat of ISO/TC 201, Japanese Standards Association, Toraya Bldg 7F, 4–9–22 Akasaka, Minato-ku, Tokyo 107–0052, Japan.

14 Composites

Peter M. A. Sherwood

CONTENTS

14.1   INTRODUCTION

Surface analytical techniques have been especially useful for the study of composite materials because the properties of such materials are so strongly influenced by the nature of the internal and external interfaces. Composite materials consist of a low modulus continuous phase matrix strengthened by an embedded high modulus and strong discontinuous reinforcement phase. This combination allows for the development of a large number of new materials with properties tied to different applications. In composite materials, the stress is carried in proportion to the moduli of the constituent phases, weighted by their respective volume fractions. The reinforcement phase is generally a fiber, which thus bears the principal load, while the matrix not only binds the composite together but also deforms under load thus distributing the majority of the stress to the fibers. Isolation of the fibers in the reinforcement phase reduces the possibility that the failure of an individual fiber will lead to catastrophic failure. These mechanical properties, combined with light weight, corrosion resistance, an ability to be fabricated into complex geometries, and other desirable properties, have meant that composites have found many important applications.

The interface between the reinforcement phase and the matrix phase plays a major role in the mechanical properties of the composite, and studies of this interface are essential in order to provide the necessary understanding to develop a composite with the desired properties. The detailed topography (i.e., microstructure) of, and the chemistry at, this interface are needed for that understanding. The interface is, of necessity, buried beneath the surface of the composite, providing challenges to the analyst wishing to use surface analytical techniques to investigate it.

The surface of the composite is also important because, as with other materials, it is this surface that meets the environment. In the case of composites, the lack of homogeneity presents some special problems because the penetration of an active agent (such as oxygen, water, etc.) through the surface region allows that agent to attack preferentially the interface region between the reinforcement phase and the matrix phase. Carbon–carbon composites are a good example of an oxidation sensitive system, since they are generally employed at high temperatures. Oxidation protective films may be applied to the surfaces of high temperature composites, but such films may crack and allow composite oxidation if the coefficient of thermal expansion is different for the protective coating and the composite, which is the case for many oxidation protective films on carbon–carbon composites. Thus the interface between the composite and any oxidation protective film is also an important one for study.

The reinforcement phase in composites is usually in the form of fibers. They may be natural fibers, but are usually glass fibers, carbon fibers, carbon nanofibers, carbon nanotubes, fibers of inert materials such as boron carbide-coated boron fibers, silicon carbide (Nicalon), and others. The matrix material may be an epoxy resin, a plastic, a ceramic material such as carbon or a glass or a metal.

Fibers used in composites present some special practical considerations for the surface scientist, and this chapter focuses upon these considerations. In particular, sample presentation and sample analysis with a view to obtaining chemical and topographical information will be addressed. The chapter does not attempt to review the large body of work in this area (e.g., Refs. [1–7] for some earlier reviews), but will rather focus upon the relevant problem-solving methods required for composite studies. Since the first edition of this book, there has been a very substantial increase in the number of papers that use surface analysis for fibers. For example, Figure 14.1 shows the numbers of papers using different surface analytical probes that have been published in this area, while in Figure 14.2 it can be seen that the number of papers describing the application of XPS to fiber surfaces has grown substantially over the past decade. The use of time-of-flight secondary ion mass spectrometry (ToF-SIMS) has also grown very rapidly in the past 5 years, reflecting the general availability of such instrumentation in industrial and university laboratories. The past decade has

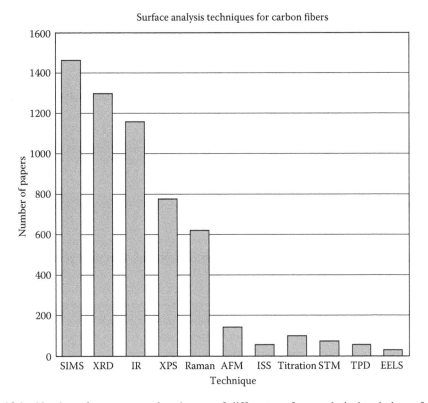

FIGURE 14.1 Number of papers reporting the use of different surface analytical techniques for carbon fibers.

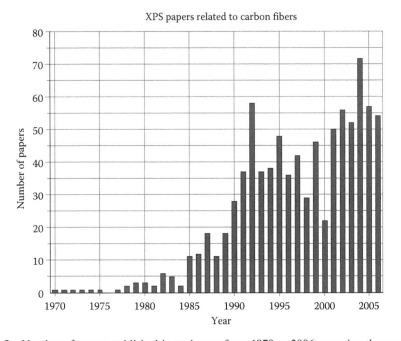

FIGURE 14.2 Number of papers published in each year from 1970 to 2006 reporting the use of XPS for the study of carbon fiber surfaces. The number of papers during the first half of 2007 suggests a comparable number of papers in 2007 as in 2006.

seen many surface analytical techniques reaching maturity in the investigation of fiber and composite surfaces and the composite–fiber interface. Many papers report the use of a variety of surface analytical probes. In evaluating and selecting a particular surface analytical probe, one needs to be aware of the possibility that the probe itself may change the fiber surface as a result of the analysis. This is particularly true of oxidized fibers, where electron and ion beams usually remove the surface oxidized groups as well as introducing considerable disorder into the fiber itself.

14.2 PRESENTING FIBERS FOR SURFACE ANALYSIS

The analysis of fibers by surface analysis presents some special problems and considerations. These will be discussed in Sections 14.2.1 through 14.2.3.

14.2.1 PRESENTATION OF MULTIPLE FIBERS FOR ANALYSIS

Fibers are generally small. Most commercial fibers are of diameter greater than 1 μm, though sub-micrometer fibers are of interest, and may present special problems. Nanotubes are extremely small. Typical carbon fiber diameters are 5–10 μm, while others may be as large as 50 μm. In all cases, however, it is not normally practical to examine single fibers. Surface analytical techniques that lack high lateral spatial resolution, such as XPS, generally require a sample of adequate size (e.g., 5 mm²) to produce a reasonable signal. Surface analytical techniques with high lateral spatial resolution, such as scanning Auger microscopy (SAM), can examine small samples, but the narrower the probe beam (and hence the greater the current density) the greater the possibility of sample damage. In practice, then it is normal to group fibers together. The most convenient unit is a "tow" of fibers. Fibers are manufactured in tows in which a large number of fibers are bundled together. Typical carbon fiber tows contain 3,000 fibers, though tows containing as many as 12,000 fibers have been used in applications requiring cheaper fibers. Samples may thus be presented as a tow attached to the spectrometer sample holder. A good arrangement is to use a "brush" of fibers, where one end of the fiber tow is attached to the holder; the end to be attached is wrapped in aluminum foil and then clamped in the sample holder (Figure 14.3a). The problem with this approach, however, is that the fiber tow often "splays out" at the end (which may also lead to some loss of sample), and it is advisable to wrap the other end of the sample as well with aluminum foil to prevent splaying (Figure 14.3b). When this is done it is important to ensure that the surface analytical probe does not "see" either the aluminum foil or the sample holder.

Sample orientation is also a consideration. The easiest arrangement is that of the sample aligned normal to the direction of the surface analytical probe, allowing attachment to a rotatable shaft, as illustrated in Figure 14.3c. Such an arrangement is compatible with most sample insertion systems such as those of the rod in an insertion lock or of the end of an x,y,z manipulator. Where angle-resolved experiments are to be attempted, the takeoff angle from the sample surface must be varied. In this case, often used in XPS, the sample should be mounted so that the fibers are aligned normal to the direction of rotation, as shown in Figure 14.3d. The only problem that arises then is that of holding the sample in such a way that no signal is derived from the sample holder; clearly, for angle-resolved experiments the sample length must be greater.

The study of nanotubes presents special challenges. An effective way to study such materials is to spread the nanotubes on carbon tape ensuring complete coverage of the tape.

14.2.2 PROBLEMS IN THE STUDY OF CONDUCTING FIBERS

Conducting fibers such as those of carbon have the advantage that surface charge does not accumulate on them in techniques such as XPS. On the other hand, the small size of the fibers coupled with their conducting nature presents a very serious practical problem. If sample mounting is not carried out so that the fibers are very firmly anchored, then individual fibers can be lost into the instrument. Once in

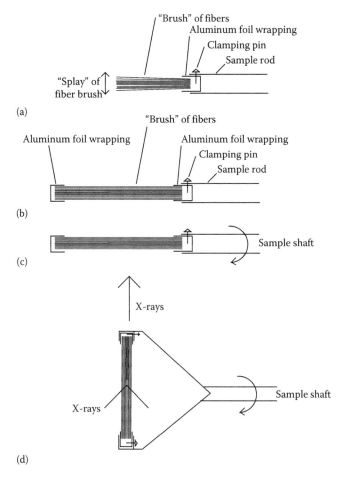

FIGURE 14.3 Mounting methods for fibers for surface analysis. (a) "Brush" of fibers, clamped at one end after wrapping in Al foil. Unfortunately in this method, the fibers "splay" out at the end of the brush. (b) "Brush" clamped at one foil-wrapped end, with the free end also wrapped in Al foil. (c) Shows a typical arrangement for XPS studies where the fibers are arranged with the fiber axis along the sample shaft axis. (d) In this arrangement, the fibers are mounted so that their axes are normal to the sample shaft axis.

the instrument, the fibers can cause havoc by short-circuiting electron lenses, multipliers, insulators, feed-throughs, etc., and they are very difficult to remove. The first problem is to establish exactly where the lost fibers are without having to dismantle the instrument. If they can be located and then connected to an external power source, they can be burnt away by passage of a large current. Destruction of the offending fibers in this way is, however, not always possible. Other methods involve breaking the vacuum in the system. Sometimes the fibers can be blown out of a critical location by a sudden increase in pressure from vacuum in that part of the system. When all else fails, as can happen unfortunately, then there is nothing for it but to dismantle the instrument completely and examine each dismantled part for the offending fibers. Such action may sound excessive to some readers, but the author is aware of a number of laboratories in which that approach has had to be taken.

14.2.3 QUESTION OF FIBER DECOMPOSITION

Any surface analytical approach must be evaluated critically with respect to the possibility that surface decomposition might occur during analysis; if it does, then the surface analyst is actually

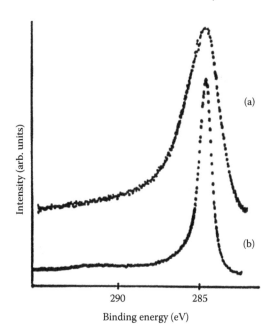

FIGURE 14.4 Damage caused by ion beams on fibers. The C 1s XPS spectrum from a PAN-based carbon fiber is shown (b) before etching, and (a) after a 10 min argon ion etch. (From Sherwood, P.M.A., *Mat. Res. Soc. Proc.*, 270, 79, 1992.)

recording data from a surface whose chemistry has been influenced strongly by the surface analytical conditions themselves. Many fibers used in composites, and the composites themselves, are relatively inert. Unfortunately, the surface chemical groups formed at the important interfaces described above may be highly sensitive to surface decomposition. For example, in carbon fibers surface oxidation leads to −C/O functionality, which can be lost by the combination of heat, high vacuum, and high-energy radiation. Electron beams, especially those focused for high spatial resolution, can cause substantial damage. Such damage may consist not only of the loss of surface functionality but also of changes in lattice arrangement. It is this problem that seriously limits the number of techniques that can be applied to composites and their component parts. The least-damaging approach is to use XPS, but even the heat and the Bremsstrahlung radiation from an x-ray gun can cause significant surface decomposition. By far the best solution is to record XPS data using an x-ray monochromator in which the white-hot x-ray target is about 2 m from the sample, rather than the 1–2 cm common to most achromatic x-ray sources. Monochromatic x-ray sources also remove Bremsstrahlung radiation that can cause decomposition.

Figure 14.4 indicates how sensitive carbon fibers are to decomposition. Figure 14.4 (curve a) shows the C 1s XPS spectrum [2] from a fiber that had been exposed to 10 min of argon ion etching, compared to the spectrum (curve b in Figure 14.4) of the same fiber before etching. The considerable broadening is a result of damage to the lattice caused by the etching process. The relationship between the C 1s line-width and lattice order is discussed below.

14.3 PRESENTING COMPOSITES FOR SURFACE ANALYSIS

The external surfaces of composites can be examined in the normal way by surface analysis, but usually there is greater interest in their internal properties. To gain analytical access to the interior, composites are therefore fractured, either outside the surface analytical instrument or within it. Clearly the latter approach would be preferable since any atmospheric changes could then be eliminated. Some surface analytical instruments have facilities for internal fracture, but a note of caution must be sounded, since such fracture will almost certainly lead to the loss of fibers into the instrument with the possible consequences discussed in Section 14.2.2.

14.4 SURFACE ANALYSIS TECHNIQUES FOR COMPOSITES AND FIBERS

A range of surface analytical techniques can be applied to composites. Some of these are inherently surface-specific, such as XPS, UV photoelectron spectroscopy (UPS), electron energy loss spectroscopy (EELS), SIMS, and ion scattering spectroscopy (ISS), while others, such as x-ray diffraction (XRD), titration, fluorescent labeling of surface species (FLOSS), XANES, Raman, and IR, may provide surface analytical information in appropriate cases.

A good illustration of how current studies deal with issues arising in the development of desirable composite materials can be found in a series of papers by Zielke et al. [8–11]. In Ref. [8], different types of surface-oxidized carbon fibers were examined by scanning electron microscopy (SEM), Fourier transform infrared spectroscopy (FTIR), scanning tunneling microscopy (STM), contact angle measurement, temperature-programmed desorption (TPD), XRD, BET (surface area measurement by nitrogen adsorption experiments), and XPS. Fiber oxidation was conducted by ozone treatment (a method often used commercially), but in fact a whole range of different surface treatment methods for fibers is available. In Ref. [9], the surface functional groups on the fibers were modified by chemical reaction, while in Ref. [10] the acidity and basicity of the functionalized fibers were determined from the work of adhesion at different pH values. In Ref. [11], the interaction of the surface-oxidized carbon fibers with high temperature thermoplastics to produce a composite material was examined.

14.4.1 X-Ray Diffraction

XRD has been used extensively for the examination of thin films. For the crystalline fibers used in composites, XRD can provide useful information on the extent to which surface treatment of the fiber has affected the bulk of the material. XRD is being used increasingly in a thin-film mode, employing very small takeoff angles, to derive surface information, but generally speaking, it must be regarded as a bulk structural technique. Bulk, or even near-surface, structural information is valuable because there are several surface treatment modes that can cause changes affecting the interior of the fiber. Examples of the application of XRD to the study of carbon fibers can be found in author's own work [12–15]. Figure 14.5 shows that the XRD data showed little change when surface oxidation was taking place (a–d), but when substantial oxidation occurred (e–g), changes had clearly penetrated into the bulk of the fiber; the main peak broadened and a new peak at about 15° (2θ) appeared, which might have been associated with lattice disorder.

14.4.2 FTIR and Raman Spectroscopies

FTIR and Raman spectroscopies can in principle provide information about surface functionality on fibers and composites. In both techniques, however, the sampling depth is considerable, which means that if there is any chemical functionality in the interior of the fiber or composite, then the information is ambiguous. Nevertheless, it can sometimes be a useful complement to the information from surface analytical techniques [3,4], as, for example, when investigating carbon functionality. FTIR can distinguish between different forms of oxidized carbon (e.g., by distinguishing between −OH, >C=O, and >COO⁻, see Ref. [16]). There have been more FTIR than Raman studies, and that includes studies with the Raman microprobe, which allows individual fibers to be

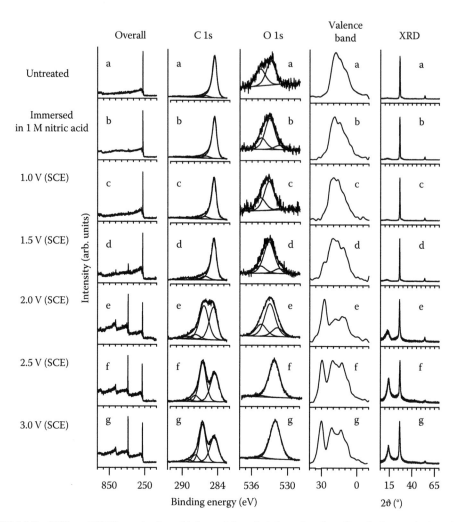

FIGURE 14.5 XPS and XRD results for a high-modulus pitch-based carbon fiber before and after electrochemical treatment in 1 M nitric acid. Each treatment consisted of 20 min at a series of increasing polarization potentials. The spectra labeled "overall" are also known as "survey" or "wide scan." (From Xie, Y. and Sherwood, P.M.A., *Appl. Spectrosc.*, 45, 1158, 1991.)

examined. Raman spectroscopy can also be used to determine the extent of lattice order in fibers. For example, it has been found that surface-treated carbon fibers have a more disordered nature than untreated fibers [17].

The variation of Raman spectroscopy called surface-enhanced Raman scattering (SERS) has proved useful for the analysis of surface features on carbon fibers. In a recent study [18], silver colloids were chemically deposited on different carbon fiber samples. The commercial carbon fiber (curve a in Figure 14.6) showed bands at 1350 and 1596 cm^{-1}, typical of graphitic carbon. When the fiber was oxidized by heating in nitric acid and then exposed to maleic anhydride, new features (curve b in Figure 14.6) appeared, corresponding to $-C=C-$, $-CH_2-$, and $>C-O$ vibrations. Further reaction of the maleic-anhydride-treated fibers with bismaleimide in N-methyl-2-pyrrolidinone (NMP) then caused further spectral changes (curve c in Figure 14.6), corresponding to a number of overlapping features including those arising from $>C-N$ and the imide ring, as well as other vibrations.

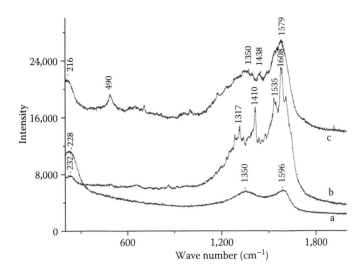

FIGURE 14.6 Surface-enhanced Raman scattering (SERS) spectra of carbon fibers, which have had silver colloids deposited onto them. (a) "as received" fiber; (b) after oxidation by heating in nitric acid followed by exposure to maleic anhydride; and (c) after treatment in (b) and then further treatment with bismaleimide in *N*-methyl-2-pyrrolidinone (NMP). (From Xu, B., Wang, X., and Lu, Y., *Appl. Surf. Sci.*, 253, 2695, 2006. With permission.)

14.4.3 SEM

SEM has been used extensively for the study of composites, and is the principal method used to determine sample topography. It has been applied to both whole composites and fibers. SEM can provide good evidence for the nature of the overall interaction between fiber and matrix. Thus, good wetting between fiber and matrix—an essential requirement for a composite with desirable qualities—can be studied easily by SEM. Composites prepared with surface-treated fibers have shown generally much better wetting and adhesion than those made from untreated fibers [2]. The examples are so numerous that no attempt will be made here to review them. A good example of the application of SEM to microstructural changes is given by a study of fiber–matrix interaction in carbon–carbon composites [10]. Surface treatment of fibers can lead to significant topographical changes on the fiber surface, as, for example, when different electrochemical or plasma treatments are used [13,20,21].

14.4.4 TEM

It is not easy to apply TEM to the study of carbon fibers, but some useful studies have nevertheless been made. For example, the technique was able to show that in fluorine-intercalated carbon fibers [22], the graphitic planes in a pitch-based carbon fiber were altered by fluorination. TEM has also proved valuable in the case of nanofiber composites, which often have a complex and very random fiber configuration (e.g., Ref. [23]).

14.4.5 STM and AFM

STM and AFM are useful complements to SEM, since they can provide microstructural information at the sub-micrometer level of lateral resolution, where SEM becomes resolution limited. With their use, important information has been obtained about fiber micro-topography and the effect of surface treatment (e.g., Ref. [24]). The topography of protective surface coatings can also be observed; in Figure 14.7, a $5 \times 5\,\mu m^2$ area of a silicon carbide coating on a carbon fiber, recorded by AFM, is shown [25].

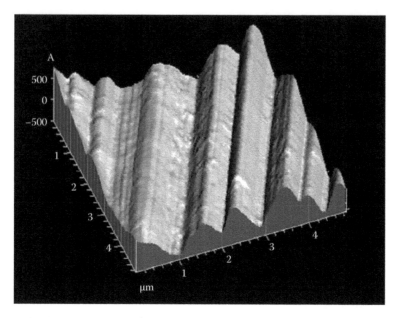

FIGURE 14.7   AFM image of a $5 \times 5\,\mu m^2$ area of a high-modulus pitch-based carbon fiber coated with a SiC protective film. (From Rooke, M.A. and Sherwood, P.M.A., *Carbon* 33, 375, 1995.)

14.4.6   Wavelength Dispersive X-Ray Emission in an Electron Microprobe

The distribution of elements on a fiber sample can be obtained by using wavelength dispersive x-ray emission in an electron microprobe. This approach has the advantage that it is possible to investigate the locations of carbon, oxygen, and other light elements that cannot be studied with the normal energy dispersive x-ray emission (EDAX) analytical attachment found on most SEM instruments, and has been used for carbon fiber studies [26].

14.4.7   Surface Energy

The surface free energy can be obtained by measuring the contact angle for a series of liquids (e.g., Refs. [27–30]), by inverse gas chromatography (i.e., by placing fibers in a column and passing probe molecules through the column [30,31]), by surface tension, or by calorimetry [32]. The surface energy can be considered to be composed of a dispersive component and a polar component. Liquids chosen for the contact-angle studies have known values for these two components. Results show typically an increase in the polar component of the surface-free energy on fiber surface treatment, although sometimes the dispersive component increases. The changes in the polar and dispersive components have been related [27,31,32] to different surface functionalities.

Watts and coworkers [30] have provided an excellent example of how surface energy measurements can be combined with other techniques, including XPS and ToF-SIMS, to provide complementary information.

14.4.8   Titrimetric Methods

Functional groups on fiber surfaces can be analyzed by carrying out titration reactions with appropriate reagents. For example, this approach has been used to analyze functional groups on carbon fiber surfaces [16,33]. However, the method is generally not very sensitive.

14.4.9 Mass Spectrometry

An important probe of carbon surface functionality is mass spectrometry, in monitoring the nature of the gaseous phase produced from heated carbon fibers by TPD. In the author's group, TPD has been applied by heating the fibers electrically, while simultaneously monitoring the temperature by optical pyrometry and the gaseous phase by mass spectrometry; XPS data were recorded at the same time [14].

TPD continues to be a popular method for the determination of oxygen surface functionality on carbon fibers and other materials, by monitoring the products of thermal decomposition of the functionality groups (e.g., H_2O, CO, CO_2, and H_2) [34]. For example, the loss of such gases from untreated fibers, and from fibers oxidized in a plasma, can give useful indications of the surface functionality [35].

Laser-desorption mass spectrometry has also proved valuable for the analysis of surface functionalized carbon nanofibers. For example [36], a mass peak corresponding to the formic acid ion ($HCOOH^+$), seen in such an experiment, could have arisen from the presence of carboxylic acid surface functional groups, and higher mass number peaks from surface amine-containing groups.

14.4.10 Secondary Ion Mass Spectrometry

SIMS has considerable potential for the study of composites, and is particularly effective for polymers [37,38], which often form the matrix material of composites (see Chapter 4). The advantages of the technique include the ability to identify hydrogen-containing fragments, to distinguish between different isotopes and hence to use isotopic labeling, and to provide spatial information by the use of scanning SIMS. For example, Briggs [38,39] examined a carbon fiber/thermoplastic fracture surface and found an intense negative ion peak at 26 amu (due to $C_2H_2^-$), characteristic of the carbon fiber surface. A peak was also found at 24 amu (due to C_2^-), which could not be used to discriminate between the fiber and the thermoplastic polymer used in a fiber composite. SIMS can be especially valuable when studying fibers produced in less conventional ways, such as those that are vapor grown. The latter process may use supported catalysts, leading to characteristic features in the SIMS spectra, e.g., iron from vapor-grown fibers using an iron carbonyl catalyst on a graphite support [40]. ToF-SIMS studies are often combined with other surface analytical probes such as XPS, providing complementary information, and different depth sensitivities [41]. SIMS is now the most widely used technique for the analysis of carbon fibers. However, as can be seen from Figure 14.4, ion beams can be destructive, particularly when using a liquid metal ion gun; in that example a beam diameter of 50 nm was used.

14.4.11 Ion Scattering Spectroscopy

ISS can also be used for studies of composites and their component fibers, e.g., carbon fiber surfaces [42,43]. ISS is an especially surface-specific technique, and is good at detecting and quantifying light elements both as surface impurities (e.g., sodium, potassium, and calcium), and as outer monolayers of a matrix (e.g., carbon, oxygen, and nitrogen).

14.4.12 Electron Energy Loss Spectroscopy

EELS is another technique that has proved valuable in the analysis of carbon fiber surfaces. It is often used with electron microscopy, in which the electron energies are so high that it is bulk information that is normally provided. However, if all the functionality were at the surface, then it could provide surface information. Such measurements are often conducted in conjunction with inherently surface-specific techniques such as XPS. A good illustration of this approach is the study [44] of the changes that occur when polyacrylonitrile (PAN) is converted to carbon by heating (a common method for forming carbon fibers).

14.4.13 SMALL ANGLE X-RAY SCATTERING

Small angle x-ray scattering has been used as an *in situ* approach to investigate the presence of adsorbed water molecules on carbon fibers. The approach can be combined with methods such as XPS to monitor the water adsorption characteristics of surface-treated carbon fibers [45].

14.4.14 SOLID-STATE NMR

Solid-state NMR has also been used as a means of investigating fiber surface composition. Although not an inherently surface-sensitive technique, it can provide useful information, and can be complemented by surface-specific techniques such as XPS. For example, ^{13}C and ^{31}P NMR has been used to study carbon fibers treated in phosphoric acid [46].

14.4.15 SURFACE AREA STUDIES

Surface area measurement by nitrogen adsorption has played a role in the surface studies of carbon fibers since the earliest days of such studies. Surface areas were first obtained in this way using an approach published in 1938 [47] and now called the BET method (see Chapter 11). Such measurements provide an insight into the topographical changes which always accompany surface treatment. The approach continues to be used, for example, in studies of fibers [48] and nanofibers [49].

14.4.16 FLUORESCENT LABELING OF SURFACE SPECIES

Fluorescent labeling has become a valuable analytical tool. In FLOSS, relatively low concentrations of surface functionality can be identified. The approach has been applied to carbon fibers [50], where it was combined with other surface analytical probes such as XPS.

14.5 X-RAY PHOTOELECTRON SPECTROSCOPIC STUDIES OF COMPOSITES AND FIBERS

14.5.1 INTRODUCTION

Much of the rest of this chapter is devoted to the use of XPS in the study of composites and fibers. XPS has been able to provide a great deal of valuable chemical information about these materials. There are, however, a number of very important practical considerations that need to be considered if the quantity and quality of this information is to be optimized.

Both the core levels and the valence-band regions in XPS can be used to extract the desired information about the surface region (to depths of no more than 10 nm) of fibers and composites. The core region (electron binding energies >30 eV) is the easier to interpret since each type of atom has core electrons with a set of binding energies characteristic of that atom; thus atomic identification is immediate. Information about the chemical environment of an atom is provided by the chemical shift in the binding energy. The valence-band region (0–30 eV), on the other hand, will contain features arising from all species present on the surface. Since all energy levels contributing to the valence band are involved in chemical interaction, and since that interaction will be specific to the species present at or near the surface, it follows that valence-band spectra in XPS will be highly sensitive to surface chemical states. Core electron chemical shifts are sometimes insufficient to be able to distinguish between subtle chemical differences, and the valence-band region can thus play an important role. In order to understand the valence-band region, reliable spectra from model compounds are needed, as well as some means of reliable prediction of the spectra. Spectral prediction is especially important because many compounds have surface composition and structure unrepresentative of the bulk (e.g., most lithium compounds have a surface consisting of lithium carbonate), and misleading information can result from a reliance on model compounds alone.

XPS core electron chemical shifts can be interpreted using a potential model which relies upon point charge values for the atoms present. In its most effective form, the relaxation potential model, the relaxation energy is included. The author has used such calculations to predict shifts associated with carbon surface functionality [51]. Figure 14.8 provides an example of such a calculated spectrum compared with the experimental spectrum [52] of NMP. The agreement is excellent, the calculated spectrum showing where the chemically shifted peaks would be expected for each of the chemically different carbon atoms.

Since the C 1s photoelectron peak exhibits significant shifts with changes in the carbon oxidation state, changes occurring in the core region spectra can provide considerable chemical information. The shifts can be measured accurately in carbon fiber spectra since there is always an intense C 1s feature at the lowest binding energy, associated with the graphitic backbone, and the separations between this peak and the other C 1s features in the same spectrum can be identified easily. The measurements are not therefore subject to the normal calibration problems associated with XPS.

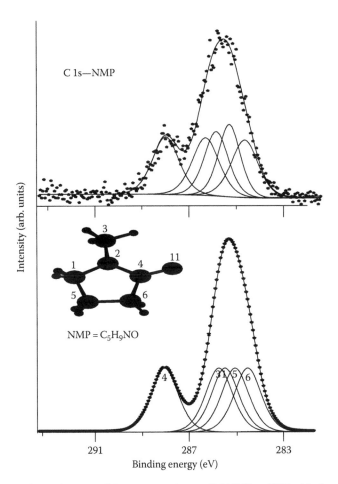

FIGURE 14.8 Comparison of (upper) C 1s spectrum from solid NMP at 77 K with (lower) calculated XPS spectrum generated with the relaxation potential model using CNDO calculations. (From Viswanathan, H., Wang, Y.-Q., Audi, A.A., Allen, P.J., and Sherwood, P.M.A., *Chem. Mater.*, 13, 1647, 2001.)

XPS valence-band spectra require more sophisticated methods of calculation. The author has found that scattered-wave Xα, *ab initio* molecular orbital, and band structure calculations, are good approaches, which he has developed so that they may be used to generate a predicted valence-band spectrum simultaneously with the experimental observations [53]. Agreement with experiment has been found to be excellent. In the valence band, the separations of features, and their relative intensities, which can be compared with calculations and model compounds, provide the chemical information. As in the core region such separations can be measured accurately and are not subject to the uncertainties of calibration.

Some 43 papers have now appeared that use valence-band XPS to study carbon fibers and composites. All but four of these [54–57] have come from the author's laboratory, and it is hoped that this valuable technique will find wider application in future.

UPS has been used for the study of carbon fiber surfaces, but the very high surface-specificity of UPS can present some practical problems. This is because any contaminant on the surface will dominate the spectrum. Further, the theoretical interpretation of UPS spectra is more complex than that of valence-band XPS because of the need to consider the joint density of states (ground and excited states) in the photoelectron process. The differences in surface specificity between UPS and valence-band XPS (the latter is less surface- specific because in standard XPS the kinetic energies of the photoelectrons are at their greatest when ejected from the valence band) can be used to probe different depths into the surface, when examining a buried interface [58]. A paper by Cazorla-Amorós and coworkers [59] provides an illustration of the use of He(I) and He(II) UPS for the study of various carbon fiber surfaces.

Figure 14.5 gives an example of how a comparison of core and valence-band spectra in XPS with each other, and with changes in XRD data (to be discussed later), can provide useful information. The spectral and diffraction data are from a high-modulus pitch-based fiber (Dupont E-120) first immersed in 1 M nitric acid, and then oxidized electrochemically for 20 min at different voltages (vs. the saturated calomel electrode [SCE]). SCE is used as a reference electrode in many electrochemical studies, and has a potential of 0.242 V vs. the normal hydrogen electrode (NHE) whose internationally accepted reference potential is 0.0 V. The overall (often called survey) spectra show an intense peak in the C 1s core region, with an increasing contribution from the O 1s feature as the extent of oxidation increased. Detailed analysis of the C 1s and O 1s regions by spectrum fitting, as shown in Figure 14.5, revealed a number of overlapping features corresponding to different chemical functionalities. The valence-band spectra (which have had nonlinear backgrounds subtracted) also showed significant differences. The surface chemical information that can be extracted from the data in Figure 14.5 will be discussed at greater length later.

14.5.2 QUESTION OF SURFACE CHARGING

Sample charging arises because the positive charges that result from the photoelectron process are not neutralized by electrons flowing from the earth. Differential sample charging occurs when different parts of a sample experience different degrees of charging. Any application of XPS must be conducted carefully in order to avoid differential surface charging. When grounded conducting samples such as carbon and carbon fibers are examined, the surface charging is not a problem as the surface charge is instantly eliminated by the flow of electrons from ground. On the other hand, when insulating materials such as the polymers and fibers found in some composites, or in coatings of carbon composites and fibers, are studied, then differential surface charging can be a problem. Calibration methods such as the use of an external calibrant, or of an internal core peak as a calibrant, can be effective [60]. Differential sample charging can cause what is actually a single peak to appear as though it were a mixture of overlapping peaks [61].

Figure 14.9 illustrates a demonstration of the problem by devising an experiment in which the differential sample charging situation is deliberately induced. Spectrum 1 in Figure 14.9 shows the

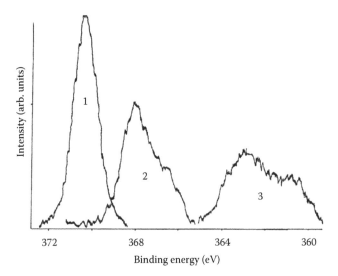

FIGURE 14.9 Ag $3d_{5/2}$ XPS spectra from a silver iodide sample placed on a fine copper mesh. 1, no bias voltage; 2, −4 V DC bias voltage; and 3, −10 V DC bias. (From Seah, M.P., in Briggs, D. and Seah, M.P. (Eds.), *Practical Surface Analysis by Auger and X-Ray Photoelectron Spectroscopy*, John Wiley, Chichester, 1990.)

Ag $3d_{5/2}$ region for a silver iodide sample mounted on a fine copper grid. Spectrum 2 results from the generation of a differential sample charge by biasing the grid at −4 V. It can be seen that part of the peak has shifted the full 4 V, but another part has shifted by a lesser amount, giving the impression of an asymmetric peak corresponding to the overlap of at least two component peaks. Spectrum 3 shows the result of exaggerating the differential sample charging by the application of a bias of −10 V. It might be thought that the spectrum could now be resolved into three or more peaks. Of course, in all cases only one Ag $3d_{3/2}$ peak should be seen.

It is possible to exploit the lateral differences in a differentially charged sample in order to separate different surface components, by using large bias potentials to amplify the effect of differential charging. This approach has been used to distinguish between different carbon environments (disordered graphitic carbon and lubricant hydrocarbon) in alloy systems [62].

An example of a situation in which differential sample charging can occur in fibers and composites is that of an oxidation protective film on a fiber or composite. Such films are often thick (a few micrometers) and may consist of insulating materials. In checking as to whether or not differential charging is present on the surface of a sample, and its effects, it is advisable to apply both positive and negative bias voltages. This is illustrated in Figure 14.10, in which the C 1s, O 1s, N 1s, and Si 2p spectra from a carbon fiber coated with a film of silicon nitride about 2 μm thick are shown [63]. In Figure 14.10, spectra (d) and (a) were recorded before and after biasing, respectively, while in (b) and (c) the specimen was biased +15 and −15 V, respectively. The appearance of single peaks in the C 1s and N 1s regions indicates that differential charging was largely eliminated when the specimen was given a positive bias. On application of a negative bias, there seems to have been sufficient stress built up in the silicon nitride coating to cause the coating to crack and expose the underlying fiber. It is suggested that the different behavior with respect to positive and negative biasing arises from the electrical properties of silicon nitride, the electrical current being able to flow from coating to fiber during positive bias, but unable to do so on application of a negative bias (i.e., a type of diode effect). The resultant electric field across the silicon nitride coating may have been high enough to cause cracking. The differences observed in this experiment should alert the experimentalist to the potential problems of differential sample charging.

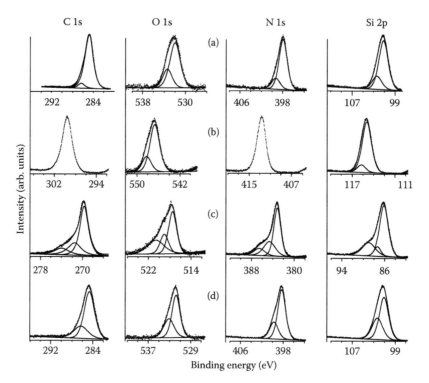

FIGURE 14.10 XPS core spectra from argon-ion-etched silicon nitride coated carbon fibers. (a) The final spectra at the conclusion of the biasing experiment, (b) +15 V DC bias voltage, (c) −15 V DC bias voltage, and (d) initial spectra before biasing. (From Dickinson, T., Povey, A.F., and Sherwood, P.M.A., *J. Electron Spectrosc. Relat. Phenom.*, 2, 441, 1973.)

14.5.3 DEPTH PROFILING OF CARBON COMPOSITES AND FIBERS

Depth profiling by the usual Ar^+ bombardment cannot be used because of the damage caused to the fibers or the composite (see discussion above and illustration in Figure 14.4). There is information about the variation with depth of the surface functionality available in XPS spectra, however, based on the relationship between photoelectron kinetic energy and the average electron escape depth (see also Chapters 3 and 10). Thus the more energetic the photoelectrons, the greater the depth from which spectral features will arise, which means that the valence-band region will always correspond to greater average depths than the core regions (except when UV light is used, as discussed above). The same relationship can also be exploited by using either a range of x-ray sources of different photon energies, so that any one characteristic core line then appears at various kinetic energies, corresponding again to a range of escape depths, or to use the variable photon energy available from a synchrotron.

Angle-resolved photoemission has also been used to study differences with depth of fiber surface functionality (see Chapters 3 and 10). The method needs flat surfaces to work properly, since photoemission from the surface layers is favored at preferentially low takeoff angles, and a rough surface would therefore blur the dependence of spectral information on depth. When fibers are to be studied, the approach is often based upon using the sample orientation shown in Figure 14.3d. The principle is that the fiber looks to the energy analyzer more like a flat surface when turned at right angles to the x-ray beam than it does when normal to the direction of the beam. However, any fiber orientation (either Figure 14.3c or d) is only an approximation to a flat surface, and so the results of angle-resolved experiments are less informative than they would be for a truly flat surface. In fact, both orientations (Figure 14.3c and d) will provide some information since mutual geometrical "shadowing" by neighboring fibers will promote the angular selectivity. Pittman [56,57,64] has provided examples of this approach.

14.5.4 DECOMPOSITION OF CHEMICALLY SHIFTED FEATURES DURING SPECTRAL COLLECTION

An important point is that surface functionality is usually sensitive to heat, and may therefore be lost by decomposition caused by heat from the x-ray source (see Section 2.3). Such decomposition is illustrated in Figure 14.11, which shows how the C 1s spectrum from a PAN-based carbon fiber, oxidized in 0.22 M nitric acid at 2.0 V (vs. SCE), changed after spectral recording for 10 min and for 90 min. The data were fitted using the same peak positions as those in Table 14.2 (discussed later), and a nonlinear background (also discussed later). The amount of oxidized carbon fell from about 50% to about 40% during the period of data acquisition [20].

Figure 14.12 shows similar data for a pitch-based carbon fiber oxidized in 1 M nitric acid at 3.0 V (vs. SCE) [13]. In this case, the C 1s, O 1s, and valence-band regions of the photoelectron spectra are shown. The kinetic energies of the photoelectrons in these regions are, in descending order, VB > C 1s > O 1s, indicating that the surface sensitivity should be in the order O 1s > C 1s > VB. Thus, if it is assumed that any x-radiation damage would have had the greatest effect on the outer surface, then decomposition should have occurred from the outer surface inward and its effects should have been more noticeable on the O 1s and C 1s spectra than on the valence-band spectra. Figure 14.12 shows that that is exactly what was observed. For instance,

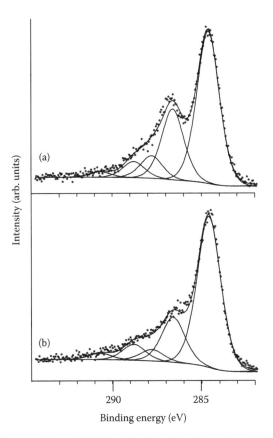

FIGURE 14.11 C 1s XPS spectra from a PAN-based carbon fiber electrochemically oxidized in 0.22 M nitric acid at 2.0 V (vs. SCE) for 20 min. (a) After 10 min x-ray exposure, and (b) after 90 min exposure. (From Yue, Z.R., Jiang, W., Wang, L., Toghiani, H., Gardner, S.D., and Pittman, C.U. Jr., *Carbon*, 37, 1607, 1999.)

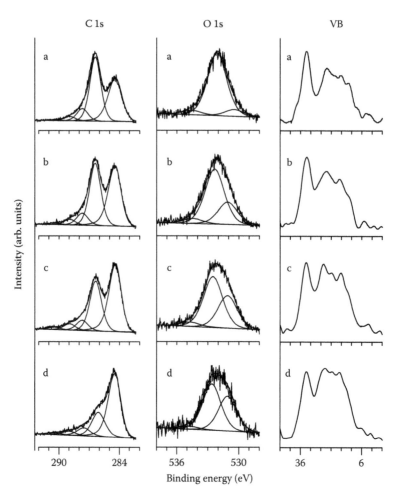

FIGURE 14.12   C 1s, O 1s, and valence band XPS spectra from an E120 pitch-based carbon fiber electrochemically oxidized in 1 M nitric acid at 3.0 V (vs. SCE) for 20 min. X-ray exposure times for the C 1s spectra were (a) 12 min, (b) 1.3 h, (c) 3 h, and (d) 24 h; for the O 1s spectra, (a) 6 h, (b) 20 h, (c) 44 h, and (d) 67 h; and for the valence-band (VB) spectra, (a) 1.5 h, (b) 7 h, (c) 28 h, and (d) 48 h. (From Xie, Y. and Sherwood, P.M.A., *Appl. Spectrosc.*, 44, 1621, 1990.)

compare the C 1s spectrum after 1.3 h exposure (Figure 14.12b) and the valence-band spectrum after 1.5 h exposure (Figure 14.12a), with the C 1s spectrum after 24 h exposure (Figure 14.12d) and the valence-band spectrum after 28 h exposure (Figure 14.12c). Whereas the C 1s spectrum had altered dramatically after 24 h exposure, there were only minor changes in the valence-band spectrum over approximately the same time.

Detailed discussion of the features in the core and valence-band regions will be left until later. However, some simple observations can be made now. The peak at highest binding energy in the valence band is largely due to the O 2s contribution, while that at the next highest binding energy arises mostly from the C 2s band; in the C 1s core level spectrum the component peak at the highest binding energy is due to oxidized carbon. It is thus clear that the fall in the intensity of the oxidized carbon contribution to the C 1s region is much greater than in the valence-band region, supporting the suggestion that decomposition occurs from the outer surface inward into the bulk. The data have been fitted to the same peak positions as those in Table 14.1 (discussed later).

TABLE 14.1

Curve-Fitting of C 1s and O 1s Spectra of Oxidized Carbon Fibers Exposed to X-Radiation for Different Lengths of Time

Exposure Time (h)	O 1s/C 1s Area Ratio	Area ("Bridged"/–C=O) in C 1s Spectra	Area ("Bridged"/–C=O) in O 1s Spectra
0.5	0.76	3.6	All bridged
3.0	—	2.7	
6.0	0.43	—	17.57
20	0.43	—	2.83
24	—	2.00	—

In the O 1s region two principal peaks can be identified, that at higher binding energy being due to –C bridged functionality and that at lower binding energy to –C=O functionality (from –C=O or –CO$_2$X). In the C 1s region the corresponding features in Figure 14.12a are that at the highest intensity, –C bridged, and those at the next two highest binding energies, –C=O functionality. The relative changes in the ratios of these peak areas are shown in Table 14.1. Although the spectra were recorded for different lengths of time, it is clear that the number of –C bridged species appeared to fall more rapidly with respect to that of the >C=O species, based on the changes in the O 1s spectrum than those in the C 1s spectrum. This difference presumably reflects both the greater surface-specificity of photoemission in the O 1s region, and the greater tendency of the –C bridged functionality to decompose under the influence of heat or x-radiation.

The data in Figures 14.11 and 14.12 were collected with achromatic x-radiation. This decomposition would have been avoided if monochromatized x-rays had been used.

14.5.5 XPS Data Analysis and Interpretation of Core Chemical Shifts

Careful data analysis can yield important chemical information that might otherwise be missed [65–67], and the extraction of such information is essential for a number of systems, especially fibers and composites. Their core electron spectra contain a number of overlapping features associated with different surface functionalities. In many papers (e.g., Refs. [68–71]), the author and coworkers have shown how such overlapping features can be resolved by combining data analysis methods with the requirement that, where the surface chemistry of a sample is changing in a continuous fashion (e.g., by electrochemical oxidation), then the fit of individual spectral components to the resultant series of core spectra must be entirely self-consistent.

One of the principal motivations for surface science studies of carbon fibers and other fibers used in composites is to evaluate the effect of surface treatment. Commercially available carbon fibers have nearly always undergone commercial surface and sizing treatments (usually confidential). Surface treatment can also be applied in the laboratory, and many of the papers referenced in this chapter describe such treatment. It is important to note that, although commercially surface-treated fibers can be surface treated subsequently in the laboratory, the resulting surface chemistry is nearly always different from that which would have resulted if the same fibers without the commercial treatment had been laboratory treated. Removal of size is discussed below. Surface treatment can be removed by heating in vacuo to high temperatures, and this approach has been used to investigate the nature of the surface treatment of carbon materials by monitoring the emitted gases via mass spectrometry (see Section 14.4.9). A detailed study of IM7 carbon fibers has shown the effect of the removal of surface treatment, and also the effect of prior commercial treatment on subsequent laboratory treatment [72].

14.5.5.1 Fitting C 1s Spectra

For many fibers and composite materials, the C 1s spectrum is the most important. Care must be taken, however, in the interpretation of the C 1s region because fibers and composites are subject to hydrocarbon contamination. In experiments using single-crystal metal samples such contamination can be removed by Ar^+ etching and heating, and by other cleaning methods, but such methods are quite inappropriate for fibers and composites because of the resulting sample decomposition. In carbon fibers and composites, analysis of the as-received surface can give very important information about surface treatment and interfacial chemistry. Fortunately, carbon fibers have a rather low affinity for hydrocarbon contamination, so that the original surface can be analyzed reliably in most cases. Some manufacturers treat carbon fibers with size, which is very difficult to remove [73]. If the size is not removed, then the C 1s spectrum will be indicative of the size and not the fiber.

Depending on their origin, untreated and unsized carbon fibers can show significant differences in their C 1s spectra. In general, high-modulus fibers have the narrowest C 1s spectra. Typical data using achromatic XPS are shown in Figure 14.13 [74], in which the C 1s spectra from five different types of fiber are compared with that from pyrolytic graphite. It can be seen that the spectra from the untreated fibers and from the graphite can be curve-fitted by including components due to oxidized carbon. The main difference between the various spectra in Figure 14.13 is that the C 1s line-width varies considerably, though in all cases the principal peak has an exponential line-shape. The latter arises from conduction band interaction, typical of conducting materials. The peaks also have considerable Lorentzian character. When carbon fibers are subjected to considerable oxidation, the exponential line-shape is lost since the fiber surface is then no longer conducting. Examples of this

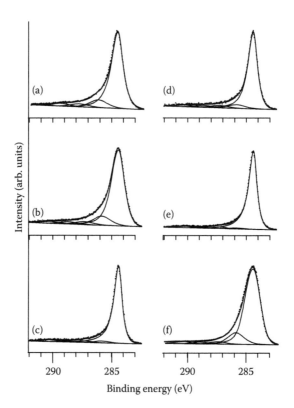

FIGURE 14.13 C 1s XPS spectra from various forms of untreated and unsized carbon fiber and from carbon; (a) a type II PAN-based fiber stored for 18 months; (b) an AU4 PAN-based fiber; (c) a P55X pitched-based fiber; (d) an E-35 pitch-based fiber; (e) an E-120 pitch-based fiber; and (f) pyrolytic graphite. (From Weitzsacker, C.L., Bellamy, M., and Sherwood, P.M.A., *J. Vac. Sci. Technol. A*, 12, 2392, 1994.)

condition will be discussed below. The differences in peak width in Figure 14.13 can be related [74] to differences in the degree of graphitic character and in the degree of order in the lattice, and may be compared with measurements of bulk order by XRD [74]. Ar+ etching would be expected to disrupt the graphitic order, and cause a considerable broadening of the C 1s peak, as seen in Figure 14.4.

Figure 14.5 illustrates how the C 1s and O 1s regions can be fitted by a number of component peaks, each peak corresponding to a different surface functionality. The curve-resolved spectra in Figure 14.5 have been derived by self-consistent fits to a sequence of spectra from progressively chemically modified carbon fiber surfaces, i.e., surfaces subjected to electrochemical oxidation at increasingly positive potentials. Note how the component peaks change in relative intensity with oxidation—a low intensity peak at low potential grows into a major peak at high (oxidizing) potentials. In the C 1s example, the feature at lowest binding energy corresponds to the fiber carbon, and those at the higher binding energies to various different oxidized carbon species. It is tempting to correlate the C 1s spectra with the same features at the same intensities as those resolved in the O 1s region since correspondence might be expected. However, it should be remembered that the depths sampled by photoemission from these two core levels are different and differing information would therefore be given by each if the surface were not homogeneous.

The common approach to the interpretation of core chemical shifts is to compare the shifts with those corresponding to known functionality, normally by using compounds of known composition (i.e., "model" compounds). This approach has potential difficulties as model compounds may sometimes have a surface composition that is different from that of the bulk. Calculated spectra, such as that shown in Figure 14.8, provide an important way of verifying the spectra of model compounds. Figure 14.14 provides an example of how C 1s chemical shifts in carbon fibers can be compared

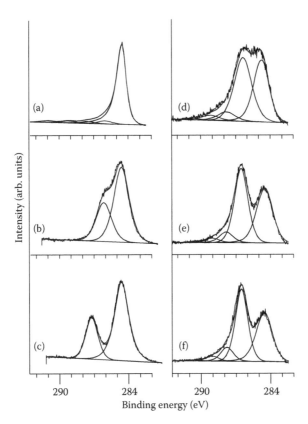

FIGURE 14.14   C 1s XPS spectra from (a) an untreated pitch-based carbon fiber, compared with (b) solid hydroquinone, (c) solid benzoquinone, and (d–f) fibers oxidized at 2.0, 2.5, and 3.0 V, respectively. (From Xie, Y. and Sherwood, P.M.A., *Chem. Mater.*, 3, 164, 1991.)

with those of compounds of known structure. The spectra from solid hydroquinone at 77 K (an aromatic molecule with −C−OH functionality), and of solid benzoquinone (an aromatic molecule with −C=O functionality), provide useful model compounds. It is clear that the separation between the C 1s peaks arising from the aromatic ring (at lower binding energy), and from the −COH functionality, is less than that between the former and that due to −C=O functionality. A high-modulus pitch-based carbon fiber oxidized to 2.0 V (Figure 14.14d) shows a C 1s peak shift comparable to that seen in hydroquinone, so it is reasonable to assume that the surface functionality on that surface is due to −C−OH groups. The same fiber oxidized to 2.5 (Figure 14.14e) or 3.0 V (Figure 14.14f) produces a shift intermediate between those due to −C−OH in hydroquinone and to −C=O in benzoquinone. It has been suggested [53,75] that such an intermediate shift corresponds to an intermediate functionality involving a "bridged" structure, discussed below. Typical shifts between the C 1s peak from the oxidized sample and the so-called graphitic C 1s peak are 1.5 eV for −OH functionality, 2.0 eV for the "bridged" structure, 3.2 eV for a C=O type functionality, and about 4.2 eV for carboxyl (COOH/ CO_2^-) or ester (COOR) type functionality. Features due to π–π^* transitions occur at shifts of around 6.1 eV.

14.5.5.2 Detailed Fitting Considerations

Curve-fitting of the type discussed above has the advantage that an observer can see clearly how the summation of the fitting components compares with the original experimental data. The key point to note about curve-fitting is that there is never an unique solution to fitting the data, and it is essential that as much chemical and physical reality as possible is incorporated into the overall procedure. It is also important that as much additional supporting information as possible is included.

An important starting point is the selection of an appropriate fitting function. In XPS, the basic peak shape is Lorentzian, modified by instrumental and other factors (such as phonon broadening) whose combined effects are equivalent to a Gaussian contribution. In some cases, especially for electrically conducting materials, the final peak shape is asymmetric due to various loss processes, such as the exponential tail on the C 1s peak discussed above. Various functions have been used in XPS over the years, and the author has reviewed this matter in more detail elsewhere [65,66]. A function that combines Gaussian with Lorentzian character, and has the ability to represent asymmetric peak features is essential if overlapping peaks are to be correctly identified. Such a function has been introduced by the author [65,66,76] and is used throughout this chapter.

Most XPS data are recorded using achromatic x-radiation, for which the x-ray line-widths are ~0.8 and 1.0 eV for the popular Mg and Al x-ray sources, respectively. If achromatic x-radiation is used, then there will appear associated with each XPS peak a series of weak peaks resulting from the weaker x-ray satellite radiation in the achromatic x-ray source. In this chapter all the spectra generated by achromatic x-radiation have been fitted in such a way that these satellite features are included, using the function previously described [76] with a 50:50 Gaussian/Lorentzian character.

In core-level XPS data from fibers and composites, the background of the spectrum represents a significant unknown. Where the slope of the background is small, the background can be fitted effectively by a line, and such a simple background has been used in Figures 14.5, 14.12 through 14.14. In other cases the spectra of a mixture of surface components may contain substantial background contributions either due to the inhomogeneous distribution of the various components in the mixed surface layer region, or to the presence of surface contaminants, or to the particular surface topography. Fortunately there have been considerable advances in the understanding of such backgrounds, in particular, the ideas and formulations of Tougaard and coworkers [77,78] and others (see Chapter 8, references therein, and Ref. [66]). Background analysis is generally performed over a large energy range, but curve-fitting over a small energy range. The reason for this is that the range of chemical shifts is generally no more than 10 eV and often much less. The spectral range chosen for curve-fitting needs to be large enough to include the main features, especially any arising from the K$\alpha_{3,4}$ satellite peaks from achromatic x-radiation, but also small enough to allow data with good statistics to be collected in a reasonable length of time. The reality is that the energy range chosen for curve-fitting, being significantly smaller than that needed for background analysis, makes subtraction of a correct background more difficult. The author has found that a nonlinear iterative background [79] gives the best results for spectra from composites and fibers [80]. The background should be included in the curve-fitting process, as has been carried out in all the examples of curve-fitting in this chapter, thus allowing simultaneous display of the fit and of the original experimental data.

Spectra from composites and fibers are often in the form of a series changing progressively with surface treatment. An example of this is given in Figure 14.3 in which the set of spectra corresponds to increasing oxidation of carbon fibers. The challenge in curve-fitting is then to achieve self-consistency throughout the series of changing spectra. Such consistency might be achievable, for example, by fixing the energetic positions of the component peaks but varying their intensities. Although few chemical systems produce photoelectron spectra in which the component peaks show no variation in position, the approximation is often a good one in electrochemical oxidation processes. If a suitable system is chosen, electrochemical treatment, such as that leading to oxidation, for example, produces changes in electrode surface chemistry that are highly controllable and reproducible, so that such treatment is very useful in the study of surface chemistry. In Figure 14.15 are shown the

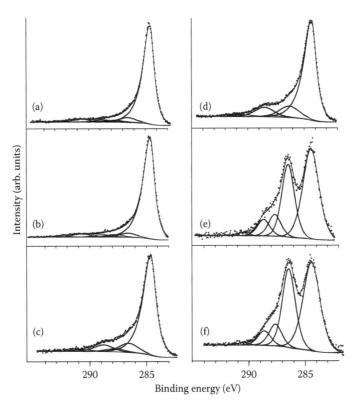

FIGURE 14.15   C 1s XPS spectra from a PAN-based carbon fiber after potentiostatic polarization at different potentials (vs. SCE) for 20 min in 2.7 M nitric acid solution. The spectra have each been fitted with five peaks. Four peaks (G/L mix = 0.5) can be seen separated from the principal fiber peak (at 284.6 eV with G/L mix = 0.8) by 2.0 eV (carbon attached to "bridged" oxygen), 3.2 eV ($>C=O$ carbon), 4.2 eV ($-CO_2H/R$), and 6.1 eV (carbonate/$\pi \rightarrow \pi^*$ shake-up processes), respectively. The FWHM of the peaks arising from oxidation were at about 2 eV (for polarization potentials 0.5–2.0 V) and 1.4 eV (for polarization potentials 2.5 and 3.0 V) (e) and (f). A nonlinear background has been included in the fit. Results of the curve-fitting are shown in Table 14.2. (From Proctor, A. and Sherwood, P.M.A., *Anal. Chem.*, 54, 13, 1982.)

TABLE 14.2
Curve-Fitting a Series of C 1s XPS Spectra: Changes in Surface Chemistry

Peak Shift	Percent of Total Area due to Peak after Various Electrochemical Treatments					
	Potential (Volts vs. SCE)					
	0.5	1.0	1.5	2.0	2.5	3.0
0	86.8	86.9	78.8	74.8	50.4	49.5
2.0	5.4	4.4	9.3	12.1	31.7	33.7
3.2	2.2	2.9	3.4	1.2	9.5	9.5
4.2	2.5	2.6	6.7	9.9	7.1	6.3
6.1	3.1	3.2	1.8	2.0	1.3	1.0

results of the electrochemical oxidation of carbon fibers at increasing oxidizing potentials. The C 1s spectra in Figure 14.15 were recorded with achromatic radiation from PAN-based carbon fibers that were oxidized electrochemically in 2.7 M nitric acid for 20 min at various potentials. The spectra have been fitted with five component peaks, four of which correspond to different surface functionalities on the carbon fiber. Clearly any attempts to fit the data at 0.5 and 1.0 V (Figure 14.15a and b, respectively) would have been difficult without additional information. On the other hand, the spectra at 2.5 and 3.0 V (Figure 14.15e and f, respectively) could be resolved into at least three features and (using additional information) one or two weaker features can be added as well. The peak positions established in Figure 14.15e and f were then used to fit the whole series giving the results shown (the peak positions were fixed at those values, and all the features arising from oxidized carbon were placed in the same "group" with the same peak widths). Table 14.2 shows that the steady decrease in the intensity of the peak at 284.6 eV, corresponding to carbon in the fiber, with the simultaneous growth in the intensity of the two most intense features due to oxidized carbon, is what would have been expected from the electrochemistry. The shifts are the same as those listed above in the preceding subsection. Although there is no unique fit to the data, the chosen fit is consistent with the chemistry of the process. Note how the inelastic tail on the "graphitic" carbon peak was lost following oxidation levels at 2.5 and 3.0 V, corresponding to the loss of graphitic (and therefore conducting) character in the surface region as the graphite sheets became highly oxidized.

Figure 14.16 shows how the fit to the C 1s spectrum from an epoxy resin on a carbon fiber, where the background has been fully incorporated into the fit, varies with the choice of background type. Although the fits are of comparable quality, Table 14.3 shows significant variation in the relative amounts of the different components depending on the choice of background. Clearly there is no way of assessing which is the most chemically realistic fit in this particular case, but in a review of other cases [80], the author found that the iterative nonlinear background [79] gave the best result.

14.5.5.3 Use of Monochromatic X-Radiation

Monochromatic x-radiation has a number of advantages, already discussed. In the case of carbon fibers, x-ray-source-induced decomposition is virtually eliminated, and when used with the best instrument resolution a spectacular improvement in line-width can be obtained. Figure 14.17 compares the C 1s spectra from an untreated and unsized Toray M40 PAN-based carbon fiber recorded with achromatic (Figure 14.17a) and with monochromatic (Figure 14.17b) radiation [81]. The spectrum in Figure 14.17a has been fitted to the component peaks listed in Table 14.2, plus an additional peak shifted by 1.0 eV from the principal "graphitic" carbon peak at 284.6 eV. This additional peak is

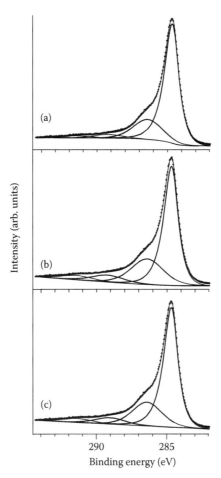

FIGURE 14.16 C 1s XPS spectra from a type II HT PAN-based carbon fiber coated with a thin film of epoxy resin after potentiostatic oxidation for 20 min at 2 V (vs. SCE) in 2 M ammonium bicarbonate solution, fitted with various backgrounds: (a) a nonlinear background, (b) a Tougaard background, and (c) a linear sloping background. X-ray satellite features have been included in the curve-fitting. Area ratios and the quality of the fitting are given in Table 14.3. (From Proctor, A. and Sherwood, P.M.A., *Anal. Chem.*, 54, 13, 1982.)

TABLE 14.3
Curve-Fitting of the Same C 1s Spectrum with Different Backgrounds

Background Used	Percent of Total Area due to Different Types of Peak			
	Carbon	Oxide I	Oxide II	Oxide III
Nonlinear (χ^2 =1140)	65.6	24.0	6.5	3.9
Tougaard (χ^2 = 1117)	61.8	26.6	7.2	4.4
Linear sloping (χ^2 = 925)	72.4	21.4	4.0	2.2

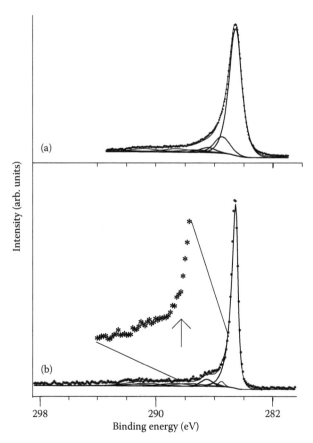

FIGURE 14.17 C 1s XPS spectra from a PAN-based untreated and unsized carbon fiber, recorded with (a) achromatic Mg x-radiation, and (b) monochromatic x-radiation using an instrumental resolution of 0.2 eV. (From Sherwood, P.M.A., *J. Vac. Sci. Technol. A*, 14, 1424, 1996.)

in fact a non-graphitic contribution to the principal peak. Of course, it would have been possible to fit the spectrum in Figure 14.17a without the extra peak simply by increasing the slope of the exponential tail of the principal graphitic peak. It is believed, however, that there is a genuine peak in that position, based on examination of the spectrum recorded with monochromatic radiation. The use of monochromatic radiation reduces the FWHM of the C 1s graphitic peak from about 1.00 to 0.32 eV. This reduction allows further spectral features to be identified. There is a clear shoulder marked by an arrow in the expanded part of the spectrum in Figure 14.17b, believed to be due to β-carbons. The latter are carbon atoms bonded to other carbon atoms to which in turn oxygen atoms are attached. The author suggested earlier that such peaks might be present [82], but as is clear from Figure 14.17a they could not be identified unambiguously when achromatic radiation was used. Obviously the way in which the β-carbon peak has been included in Figure 14.17a overestimates its contribution. Since that contribution could be adjusted easily by merely increasing the amount of exponential tail, it is clear that the fit in Figure 14.17a is ambiguous, illustrating how difficult it is to fit correctly a β-carbon peak when using achromatic radiation. Figure 14.17b also indicates that chemically shifted features, especially the peak shifted by 2 eV, are much more readily identifiable than in the data recorded with achromatic radiation. In fact, the only evidence in Figure 14.17a of oxidation is based on subsequent information from which it is known that continued oxidation causes an increase in the contributions of peaks at the positions used in the fit.

When monochromatic radiation is used to examine the effects of electrochemical oxidation of fibers, the spectra obtained are of better quality than with achromatic radiation, allowing more and

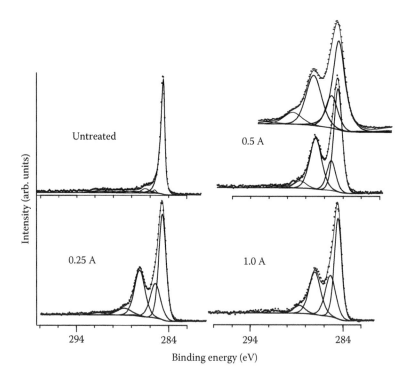

FIGURE 14.18 C 1s monochromatic XPS spectra from a PAN-based carbon fiber galvanostatically oxidized in a pilot plant for 40 s in 1 M nitric acid at various currents. For the sample oxidized at 0.5 A the achromatic XPS spectrum is shown above the monochromatic spectrum. (From Sherwood, P.M.A., *J. Vac. Sci. Technol. A*, 14, 1424, 1996.)

more precise information to be extracted. For example, in Figure 14.18 [81] are shown the C 1s spectra recorded using monochromatic radiation, from the same fibers as in Figure 14.17, following galvanostatic oxidation in 1 M nitric acid in a pilot plant [14] after exposures for 40 s at various currents [81]. Spectra from the oxidized fibers have the same general shape as those in Figure 14.16, for which achromatic radiation was used, but a much improved spectral quality. Three features are clearly resolved in addition to the principal peak, with the β-carbon peak being included effectively in the fit. The components were fitted according to the positions given in Table 14.2, together with a plasmon feature at a shift of about 6.9 eV. A nonlinear background has also been included. The result at 0.5 A shows both the monochromatic (lower) and achromatic (upper) data. The valley between the principal peak and that shifted by about 2 eV from it in the monochromatic spectrum is clearly deeper than when the data were collected with achromatic radiation. Oxidation caused an increase in the width of the principal "graphitic" peak (to around 0.9 eV), although the peaks arising from oxidized carbon were narrower than with achromatic radiation. The increase in width of the graphitic peak was expected since oxidation increases the disorder in the lattice, as discussed in Section 14.5.5.1.

14.5.5.4 Fitting O 1s Spectra

The O 1s region of the photoelectron spectrum from organic materials is generally rather broad, but can be resolved into three principal features as seen in Figures 14.5 and 14.12. These features correspond to $-C=O$ at about 531.8 eV, to $-C-OH/C-$bridged at about 533.3 eV, and to adsorbed water and some chemisorbed oxygen at about 536.1 eV. Analysis of O 1s spectra can complement the information provided by analysis of C 1s spectra, although it should be remembered that since the

O 1s photoelectron kinetic energies are lower than those of the C 1s, the O 1s sampling depth is smaller, and therefore the O 1s spectra are slightly more surface specific.

14.5.5.5 Fitting N 1s Spectra

Although the N 1s spectrum from an organic material is often of low intensity and therefore difficult to resolve accurately, it is capable of giving some distinctive information [83]. On the other hand, inorganic nitrogen, which might, for example, be deposited on a surface from adsorbed nitrate ions as a result of surface treatment, can be distinguished readily from organic nitrogen by the much higher binding energies of inorganically combined nitrogen (e.g., 407.3 eV for nitrate ions).

14.5.6 Interpreting the Valence-Band Spectrum

Valence-band spectra in XPS can be especially valuable in the study of fibers and composites because the analysis of such spectra may be able to assist in the interpretation of otherwise ambiguous spectral features in the core regions. The damaging and spectral broadening effects of achromatic x-radiation are also much less obvious in valence-band spectra for the reasons discussed in Section 14.5.4. With the help of the valence-band spectrum, it is possible to distinguish between unfunctionalized hydrocarbons [84], while useful information about polymers can also be extracted [67]. When comparing valence-band data with calculation, background and x-ray satellite removal from the experimental data is usually helpful, and can be performed effectively by data processing methods [65–67].

14.5.6.1 Using Calculations to Predict Valence-Band Spectra

The valence band can be modeled by performing Xα, *ab initio*, or band structure calculations on appropriate structural units. For Xα, *or ab initio* calculations, the unit is the polymeric repeat unit for a polymer or the substituted coronene structure for a carbon fiber. Comparison between the calculated spectrum of a polymer repeat unit and a full band structure calculation can assist in identifying features such as hydrogen bonding between polymer chains [15]. The calculations allow the energy level positions within the band to be calculated, and the intensities of the component peaks in the spectrum to be estimated; the latter are found by taking the molecular orbital coefficients (or the partial density of states in the case of band structure calculations) from the calculation and multiplying them by their photoelectric cross-sections. The component peaks are then set to have equal widths, comparable to the experimental resolution. The calculations thus do not include any experimental "adjustment" factors, and can be used for predictive purposes. Calculations of this type allow the experimental and theoretical spectra to be compared, especially the separations and relative intensities of the peaks. Good agreement between experiment and theory gives reasonable confidence in assigning a particular compound to an experimental spectrum.

Polymers have rich valence-band spectra, and in Ref. [85] there are examples of polymer valence-band spectra of high performance polymer fibers interpreted by corresponding calculated spectra. Carbon nanotubes also have valence-band spectra that can be very useful in the determination of their composition, especially when the spectra are compared with calculation [86]. Figure 14.19 illustrates how the valence-band region interpreted by calculated spectra can be used to identify a Si–B–N–C ceramic on a fiber surface (an oxidation protection coating), showing that the ceramic has been polymerized, since the valence band is different from that expected for the polymer precursor [87]. Figure 14.20 shows how the valence-band region can be used to determine the degradation product of a PEKK-coated carbon fiber exposed to sodium chloride solution at elevated temperatures. The calculated spectrum for a degradation product with a –COOH group agrees well with the experimental spectrum [88]. In Figure 14.21, the spectra predicted for oxidized carbon fibers are shown, in which multiple scattered-wave Xα calculations on a set of substituted coronene units were used [53]. The use of the band structure approach in polymers has been demonstrated in a study of polystyrenes [89].

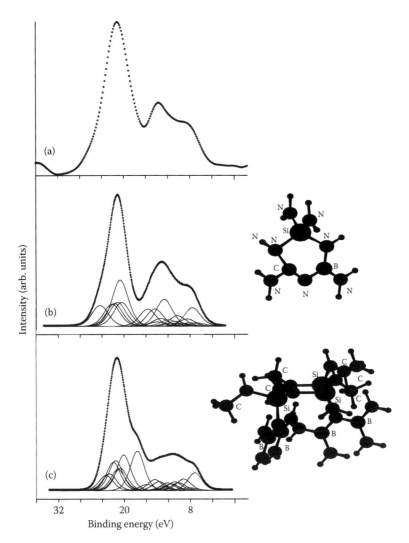

FIGURE 14.19 XPS valence-band spectrum from a Si–B–N–C ceramic on a fiber surface: (a) smoothed difference spectrum obtained by subtracting the valence-band spectrum of SiO_2 from that of a coated fiber; (b) spectrum obtained from an *ab initio* calculation on a fragment of the Si–B–N–C ceramic lattice shown; and (c) spectrum obtained from an *ab initio* calculation on a fragment of the polymer precursor molecule. (From Wang, Y.-Q. and Sherwood, P.M.A., *Chem. Mater.*, 16, 5427, 2004.)

14.5.6.2 Understanding the Valence-Band Spectra of Carbon Fibers

The calculated spectra of Figure 14.21 indicate that two important pieces of information can be obtained from the valence-band spectra of carbon fibers:

1. The energy region around 20–30 eV corresponds to C 2s/O 2s mixing, and the separation between the two major peaks, one at the highest binding energy (O 2s) and the other distanced from it by some 8–11 eV to lower binding energy (C 2s), are sensitive to surface functionality. Thus the −O− and the −OH functionalities may be differentiated by their different separations, 10.58 (Figure 14.21b) and 8.77 eV (Figure 14.21e), from the O 2s peak. Such differentiation is not possible from core spectra in the C 1s and O 1s regions.

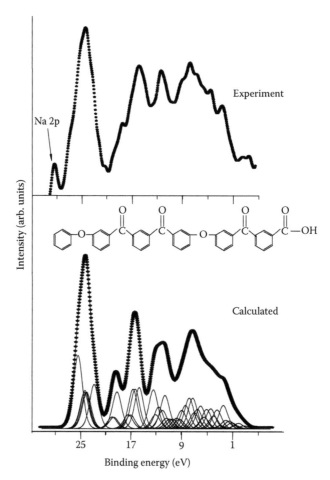

FIGURE 14.20 XPS valence-band spectrum from PEKK-coated carbon fibers after exposure to sodium chloride solution at 365 K. The experimental spectrum is compared with a spectrum calculated for a model compound with a −COOH group, which is a likely candidate for the surface degradation product. (From Rooke, M.A. and Sherwood, P.M.A., *Chem. Mater.*, 9, 285, 1997.)

Note how the spectrum (Figure 14.21a) of carboxyl (CO_2^-) has two features in the O 2s region (arising from the ability of the O 2s orbital to form bonding and anti-bonding interactions with C 2s orbitals).

2. The region below a binding energy of 18 eV serves as a "fingerprint" region. Calculations show that it contains a number of "peaks" resulting from the overlap of mixed O 2p and C 2p components. Note how that region changes substantially with changes in surface functionality. In order to be able to compare the calculated valence band spectra with the experimental spectra from oxidized carbon fibers, the photoelectron cross-sections have been adjusted to account for variable amounts of carbon and oxygen [53].

The C-bridged structure referred to earlier is shown in Figure 14.21d with a hydrogen atom placed symmetrically between two oxygen atoms. The C−O bond length used (0.12825 nm) was intermediate between that of −C=O (0.1215 nm) and −C−OH (0.135 nm). The real situation may be more complex, probably with a double potential well, but the "bridged" structure is regarded as a reasonable model for this type of oxidation.

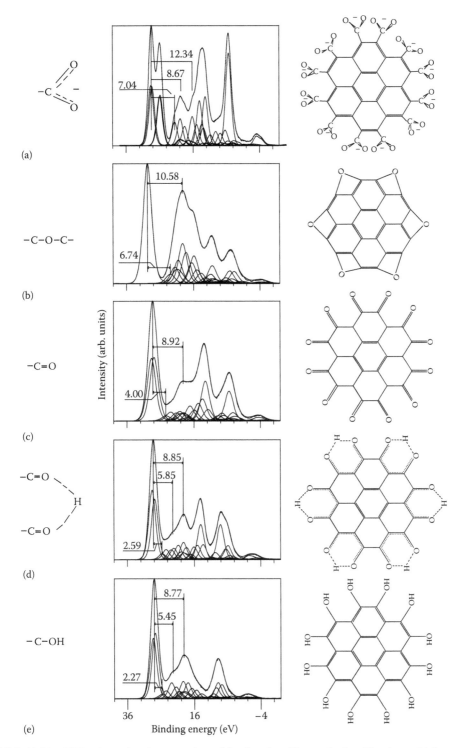

FIGURE 14.21 XPS valence-band spectra for oxidized carbon fibers calculated by the multiple scattered-wave Xα method. The functional groups shown in the left-hand column, and attached to the coronene nucleus in the right-hand column are (a) carboxyl, (b) epoxide, (c) carbonyl, (d) bridged, and (e) hydroxide. (From Weitzsacker, C.L., Bellamy, M., and Sherwood, P.M.A., *J. Vac. Sci. Technol. A*, 12, 2392, 1994.)

The additional chemical information that can be obtained from the valence-band region, in comparison with information from the core region, is demonstrated by the author's work on the electrochemical oxidation of pitch-based fibers in ammonium carbonate solution [90]. Both potentiostatic (holding the potential constant during electrochemical treatment) and galvanostatic (holding the current constant during electrochemical treatment) oxidation led to the appearance of two peaks in the O 2s region, as in Figure 14.21a. The two peaks are believed to correspond to epoxide or −C−O−C− (near 30 eV) and −C−OH (near 25 eV).

14.5.7 INTERFACIAL STUDIES

Interfaces are very important in the study of composites, especially the buried interfaces mentioned in the introduction. Such an interface is that between a fiber surface and the surrounding matrix in a composite material, and its study is important because it is susceptible to oxidation in composites exposed to high temperatures under atmospheric conditions. The author [58,89,91,92] has developed a method that allows a buried interface to be examined without damage, by arranging for the interface to be close to the surface and within the sampling depth of valence-band photoemission (see Section 5.6). The method involves preparation of an interface by coating the surface of a fiber with a very thin film of the matrix material, which is carried out by immersing the fiber in a solution of the matrix material, and allowing the solvent to evaporate. The spectrum of the interface can then be derived from the following spectra:

1. Spectrum from matrix alone = M.
2. Spectrum from fiber surface alone = F.
3. Spectrum from the fiber coated with a very thin layer of matrix material = S. It is essential that the spectrum from the underlying fiber can be seen in the XPS data, to make sure that the buried interface is within the sampling depth.

The situation can be illustrated schematically:

- Vacuum
- Matrix
- Buried Interface
- Fiber

If there is no chemical interaction at the fiber/matrix interface

$$S = M + F$$

If there is chemical interaction at the fiber/matrix interface

$$S \neq M + F$$

In the latter case, the spectrum from the interface region can be found by extracting a difference spectrum (65), i.e., by subtracting M and F from S.

An example of this approach is illustrated in Figure 14.22. The difference spectrum (Figure 14.22d) is the result of subtraction of the spectra (Figure 14.22a) of an oxidized carbon fiber, and (Figure 14.22b) of a phenolic matrix, from the total spectrum (Figure 14.22c) of the fiber and matrix coupled with a titanium alkoxide coupling agent (TOT = tetrakis(2-ethylhexyl)titanate) [91]. The difference spectrum is characteristic of the −O−TiR$_2$−O− group that might be formed by reaction of the titanium alkoxide with both the oxidized fiber surface and the phenolic resin according to the scheme.

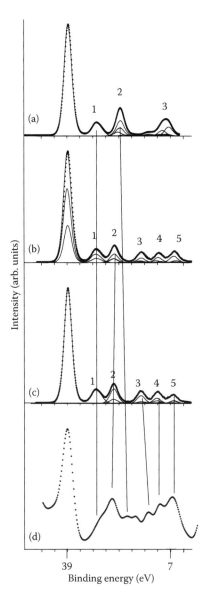

FIGURE 14.22 Illustration of the method of extracting the valence-band spectrum from the buried interface between an oxidized carbon fiber and a phenolic resin, with an intermediate titanium alkoxide agent. (a) Spectrum from the oxidized carbon fiber alone, (b) spectrum from phenolic matrix alone, (c) spectrum from the composite, and (d) difference spectrum—the result of subtracting spectra (a) and (b) from (c). Spectrum (d) represents the desired valence-band spectrum of the interface. (From Xie, Y., Wang, T., Franklin, O., and Sherwood, P.M.A., *Appl. Spectrosc.*, 46, 645, 1992.)

14.6 CONCLUDING COMMENTS

Surface analysis has an important role to play in the study of composites and fibers, as evidenced by a very substantial growth in the number of papers in this area in the past 10 years. Many surface analytical methods are suitable, and new methods of analysis are being introduced. ToF-SIMS is now the most commonly used method, though XPS continues to be the most valuable for obtaining surface chemical information. The two most important practical points are that samples be mounted and handled very carefully in the spectrometer, and that the data be analyzed thoroughly. Complete problem solving cannot be achieved unless the surface analyst uses the full extent of information available (e.g., by careful curve-fitting, and by making use of the valence band). This chapter has given many examples of the types of information obtainable, and has discussed factors that need to be considered in conducting an analysis in the most effective manner.

ACKNOWLEDGMENTS

The material quoted in this chapter is based upon work supported by the National Science Foundation, the Air Force Office of Scientific Research, and NASA. The author acknowledges his coworkers in studies of carbon fibers, Dr. Andrew Proctor, Dr. Carol Jones, Dr. Yaoming Xie, Dr. Cara Weitzsacker, Dr. Tiejun Wang, Dr. Michael Bellamy, Dr. Tom Schuerlein, Dr. Michael Rooke, Dr. Nathan Havercroft, Dr. Hema Viswanathan, Dr. Yuqing Wang, Feng-Qiu Zhang, and Oliver Franklin.

REFERENCES

1. P. M. A. Sherwood, *J. Electron Spectrosc. 81*, 319 (1996).
2. P. M. A. Sherwood, *Mat. Res. Soc. Proc. 270 (Novel Forms of Carbon),* 79 (1992).
3. J. Matsui, *Crit. Rev. Surf. Chem. 1*, 71 (1990).
4. M. S. Dresselhaus, G. Dresselhaus, K. Sugihara, I. L. Spain, and H. A. Goldberg, *Graphite Fibers and Filaments, Springer Series in Materials Science*, Vol. 5, Springer-Verlag, Berlin, 1988.
5. J. E. Castle and J. F. Watts, The study of interfaces in composite materials by surface analytical techniques, in *Interfaces in Polymer, Ceramic, and Metal Matrix Composites* (H. Ishida, Ed.), Elsevier, Amsterdam, 1988.
6. J.-B. Donnet, T. K. Wang, J. C. M. Peng, and S. Rebouilat (Eds.), *Carbon Fibers*, 3rd edition, Marcel Dekker, Inc, New York, 1998.
7. M. S. Dresselhaus, G. Dresselhaus, and P. Avouris (Eds.), *Carbon Nanotubes: Synthesis, Structure, Properties and Applications*, Springer, New York, *Top. Appl. Phys. 80*, (2001).
8. U. Zielke, K. J. Hüttinger, and W. P. Hoffman, *Carbon 34*, 983 (1996).
9. U. Zielke, K. J. Hüttinger, and W. P. Hoffman, *Carbon 34*, 999 (1996).
10. U. Zielke, K. J. Hüttinger, and W. P. Hoffman, *Carbon 34*, 1007 (1996).
11. U. Zielke, K. J. Hüttinger, and W. P. Hoffman, *Carbon 34*, 1015 (1996).
12. Y. Xie and P. M. A. Sherwood, *Appl. Spectrosc. 44*, 797 (1990).
13. Y. Xie and P. M. A. Sherwood, *Appl. Spectrosc. 44*, 1621 (1990).
14. Y. Xie and P. M. A. Sherwood, *Appl. Spectrosc. 45*, 1158 (1991).
15. Y.-Q. Wang, F.-Q. Zhang, and P. M. A. Sherwood, *Chem. Mater. 11*, 2573 (1999).
16. E. Pamula and P. G. Rouxhet, *Carbon 41*, 1905 (2003).
17. G. Katagiri, H. Ishida, and A. Ishitani, *Carbon 26*, 565 (1988).
18. B. Xu, X. Wang, and Y. Lu, *Appl. Surf. Sci. 253*, 2695 (2006).
19. C. Ahearn and B. Rand, *Carbon 34*, 239 (1996).
20. C. Kozlowski and P. M. A. Sherwood, *J. Chem. Soc., Faraday Trans. I 80*, 2099 (1984).
21. C. Kozlowski and P. M. A. Sherwood, *J. Chem. Soc., Faraday Trans. I 81*, 2745 (1985).
22. A. Tressaud, M. Chambon, V. Gupta, S. Flandrois, and O.P. Bahl, *Carbon 33*, 1339 (1995).
23. S. M. Rhodes, B. Higgins, Y. Xu, and W. J. Brittain, *Polymer 48*, 1500 (2007).
24. W. P. Hoffman, W. C. Hurley, T. W. Owens, and H. T. Phan, *J. Mater. Sci. 26*, 4545 (1991).
25 M. A. Rooke and P. M. A. Sherwood, *Carbon 33*, 375 (1995).
26 Y. Xie and P. M. A. Sherwood, *Chem. Mater. 6*, 650 (1994).
27. L. T. Drazal, J. A. Meschen, and D. L. Hall, *Carbon 17*, 375 (1979).
28. K. K. C. Ho, A. F. Lee, and A. Bismark, *Carbon 45*, 775 (2007).
29. X. Zhang, Y. Huang, and T. Wang, *Appl. Surf. Sci. 253*, 2885 (2006).
30. P. E. Vickers, J. F. Watts, C. Perruchot, and M. M. Chehimi, *Carbon 38*, 675 (2000).
31. J. Schultz, L. Lavielle, and H. Simon, Surface and adhesion properties of carbon fibres, *Proceedings of the International Symposium on the Science and New Applications of Carbon Fibers '84*, Toyohashi University of Technology, Japan, 1984, p. 125.

32. B. Rand and R. Robinson, *Carbon 15*, 311 (1977).
33. J.-B. Donnet, H. Dauksch, J. Escard, and C. Winter, *C. R. Acad. Sci. Paris 275*, 1219 (1972).
34. J. A. Turner and K. M. Thomas, *Langmuir 15*, 6416 (1999).
35. J. P. Boudou, J. I. Paredes, A. Cuesta, A. Martinez-Alonso, and J. M. D. Tascón, *Carbon 41*, 41 (2003).
36. J. Li, M. J. Vergne, E. D. Mowles, W.-H. Zhong, D. M. Hercules, and C. M. Lukehart, *Carbon 43*, 2883 (2005).
37. D. Briggs, *Surface Analysis, in Encyclopedia of Polymer Science and Engineering*, Vol. 16, Wiley, New York, 1989, pp. 399–442.
38. D. Briggs, *Surf. Interface Anal. 9*, 391 (1986).
39. M. J. Hern and D. Briggs, *Surf. Interface Anal. 17*, 421 (1991).
40. Ph. Serp, J. L. Figueiredo, P. Bertrand, and J. P. Issi, *Carbon 36*, 1791 (1998).
41. M. R. Alexander and F. R. Jones, *Carbon 33*, 569 (1995).
42. L. T. Drzal, *Carbon 15*, 129 (1977).
43. D. J. D. Moyer and J. P. Wightman, *Surf. Interface Anal. 14*, 496 (1989).
44. L. Laffont, M. Monthious, V. Serin, R. B. Mathur, C. Guimon, and M. F. Guimon, *Carbon 42*, 2485 (2004).
45. T. Ohba and K. Kaneko, *J. Phys. Chem. C 111*, 6207 (2007).
46. R. Fu, L. Liu, W. Huang, and P. Sun, *J. Appl. Polymer Sci. 87*, 2253 (2003).
47. S. Brunauer, P. H. Emmett, and E. Teller, *J. Am. Chem. Soc. 60*, 309 (1938).
48. S. Tang, N. Lu, J. K. Wang, S.-K. Ryu, and H.-S. Choi, *J. Phys. Chem. C 111*, 1820 (2007).
49. P. V. Lakshminarayanan, H. Toghiani, and C.U. Pittman, Jr., *Carbon 42*, 2433 (2004).
50. X. Feng, N. Dementev, W. Feng, R. Vidic, and E. Borguet, *Carbon 44*, 1203 (2006).
51. A. Proctor and P. M. A. Sherwood, *J. Electron Spectrosc. Relat. Phenom. 27*, 39 (1982).
52. H. Viswanathan, Y.-Q. Wang, A. A. Audi, P. J. Allen, and P. M. A. Sherwood, *Chem. Mater. 13*, 1647 (2001).
53. Y. Xie and P. M. A. Sherwood, *Chem. Mater. 3*, 164 (1991).
54. H. H. Madden and R. E. Allred, *J. Vac. Sci. Technol. A 4*, 1705 (1986).
55. Z. R. Yue, W. Jiang, L. Wang, S. D. Gardner, and C. U. Pittman, Jr., *Carbon 37*, 1785 (1999).
56. S. D. Gardner, C. S. K. Singamsetty, G. L. Booth, G.-R. He, and C. U. Pittman, Jr., *Carbon 33*, 587 (1995).
57. S. D. Gardner, G. He, and C. U. Pittman, Jr., *Carbon 34*, 1221 (1996).
58. Y.-Q. Wang and P. M. A. Sherwood, *J. Vac. Sci. Technol. A 21*, 1120 (2003).
59. E. Raymundo-Piñero, D. Cazorla-Amorós, A. Linares-Solano, J. Find, U. Wild, and R. Schlögl, *Carbon 40*, 597 (2002).
60. M. P. Seah, in *Practical Surface Analysis by Auger and X-Ray Photoelectron Spectroscopy*, 2nd edition (D. Briggs and M. P. Seah, Eds.), John Wiley, Chichester, 1990, Appendix 2, pp. 541–554.
61. T. Dickinson, A. F. Povey, and P. M. A. Sherwood, *J. Electron Spectrosc. Relat. Phenom. 2*, 441 (1973).
62. N. J. Havercroft and P. M. A. Sherwood, *Surf. Interface Anal. 29*, 232 (2000).
63. M. A. Rooke and P. M. A. Sherwood, *Surf. Interface Anal. 21*, 681 (1994).
64. Z. R. Yue, W. Jiang, L. Wang, H. Toghiani, S. D. Gardner, and C. U. Pittman, Jr., *Carbon 37*, 1607 (1999).
65. P. M. A. Sherwood, in *Practical Surface Analysis by Auger and X-Ray Photoelectron Spectroscopy* (D. Briggs and M. P. Seah, Eds.), John Wiley, Chichester, 1983, Appendix 3, pp. 445–475.
66. P. M. A. Sherwood, in *Practical Surface Analysis by Auger and X-Ray Photoelectron Spectroscopy*, 2nd edition (D. Briggs and M. P. Seah, Eds.), John Wiley, Chichester, 1990, Appendix 3, pp. 555–586.
67. P. M. A. Sherwood, in *Surface Analysis of Advanced Polymers* (L. Sabbatini and P. G. Zambonin, Eds.), VCH Press, Weinheim, 1993, Chapter 7, pp. 257–298.
68. P. M. A. Sherwood, *J. Vac. Sci. Technol. A 9*, 1493 (1991).
69. P. M. A. Sherwood, *J. Vac. Sci. Technol. A 11*, 2280 (1993).
70. S. Thomas and P. M. A. Sherwood, *Anal. Chem. 64*, 2488 (1992).
71. S. Thomas and P. M. A. Sherwood, *Surf. Interface Anal. 20*, 595 (1993).
72. Y.-Q. Wang, H. Viswanathan, A. A. Audi, and P.M.A. Sherwood, *Chem. Mater. 12*, 1100 (2000).
73. C. L. Weitzsacker, M. Bellamy, and P. M. A. Sherwood, *J. Vac. Sci. Technol. A 12*, 2392 (1994).
74. Y. Xie and P. M. A. Sherwood, *Chem. Mater. 2*, 293 (1990).
75. C. Kozlowski and P. M. A. Sherwood, *Carbon 25*, 751 (1987).
76. R. O. Ansell, T. Dickinson, A. F. Povey, and P. M. A. Sherwood, *J. Electroanal. Chem. 98*, 79 (1979).
77. S. Tougaard and P. Sigmund, *Phys. Rev. B 25*, 4452 (1982).

78. H. S. Hansen and S. Tougaard, *Surf. Interface Anal. 17*, 593 (1991).
79. A. Proctor and P. M. A. Sherwood, *Anal. Chem. 54*, 13 (1982).
80. P. M. A. Sherwood, *J. Vac. Sci. Technol. A 14*, 1424 (1996).
81. H. Viswanathan, M. A. Rooke, and P. M. A. Sherwood, *Surf. Interface Anal. 25*, 409 (1997).
82. A. Proctor and P. M. A. Sherwood, *Carbon 21*, 53 (1983).
83 Y. Xie and P. M. A. Sherwood, *Chem. Mater. 1*, 427 (1989).
84. P. M. A. Sherwood, *J. Vac. Sci. Technol. A 10*, 2783 (1992).
85. L. E. Hamilton, P. M. A. Sherwood, and B. M. Reagan, *Appl. Spectrosc. 47*, 139 (1993).
86. Y.-Q. Wang and P. M. A. Sherwood, *Chem. Mater. 16*, 5427 (2004).
87. M. A. Rooke and P. M. A. Sherwood, *Chem. Mater. 9*, 285 (1997).
88. Y.-Q. Wang, F.-Q. Zhang, and P. M. A. Sherwood, *Chem. Mater. 13*, 832 (2001).
89. E. Orti, J. L. Bredas, J. J. Pireaux, and N. Ishihara, *J. Electron Spectrosc. Relat. Phenom. 52*, 551 (1990).
90. Y. Xie, T. Wang, O. Franklin, and P. M. A. Sherwood, *Appl. Spectrosc. 46*, 645 (1992).
91. T. Wang and P. M. A. Sherwood, *Chem. Mater. 7*, 1031 (1995).
92. A. L. Asunskis and P. M. A. Sherwood, *J. Vac. Sci. Technol. A 25*, 872 (2007).

15 Minerals, Ceramics, and Glasses

Roger St. C. Smart and Zhaoming Zhang

CONTENTS

15.1 INTRODUCTION

The surfaces of minerals and ceramics have many features in common. This statement can be justified by comparison of the nature of the information required from their respective surface analyses [1–3]. Indeed, tailored ceramics often result from judicious design involving combinations of mineral structures [4,5]. Similarities in phase structure and composition, surface structure and surface sites, microstructure, and surface reactivity have been demonstrated in numerous studies. Surface reactions involving oxidation, leaching, dissolution, weathering, precipitation, and phase transformation, are now well documented. Surface modification of minerals as a result of adsorption, reaction, processing (e.g., plasma spraying), and surface coatings (e.g., sol-gel deposition), has been found to be equally applicable to ceramic materials. Hence the methodologies for determining the surface properties that control the mechanisms of reaction and transformation of surfaces of minerals and ceramics will in general require similar surface analytical techniques.

Glass surfaces can also be characterized in terms of composition, adsorption, reaction, and surface modification, all of which require the same analytical techniques as are applied to minerals and ceramics. Structural information is of course less easily obtainable by diffraction methods and related imaging techniques, but the scanning probe microscopies (SPM) are useful for the definition of surface structure and surface sites. There will be less concern with grain boundaries and intergranular films but, instead, rather greater interest in *in situ* reaction, reprecipitation, and phase transformation. Changes in composition and chemistry with depth, relating particularly to segregation and reaction, are important parameters for characterizing the reactivity of glass surfaces, and for defining mechanisms of adsorption, reaction, and surface modification.

These three broad groups of materials therefore require essentially the same approach to their characterization.

In this chapter, the approach to problem solving by surface analytical methods will focus firstly on the nature of the information required and on the surface analytical techniques able to provide it. In structural terms, phase identification (e.g., lattices, grain size, and size distribution), as well as information about surface structure and surface sites (from atomic to millimeter scales), grain boundaries, intergranular films, and defects (point, linear, and planar) may be required to understand the mechanisms controlling the behavior of the material. In chemical terms, surface sites and their reactivity, elemental segregation, composition of the surface layers, depth profiles, grain boundaries, and intergranular films, may all be important. The ability to study adsorption (including surface distribution, molecular form, bonding, and layer formation) and surface reaction (e.g., oxidation, dissolution, precipitation, and phase transformation), and the mechanisms of these reactions, is central to the design and modification of mineral, ceramic, and glass surfaces. Surface modification for controlled interfacial behavior in diverse applications now requires the use of advanced surface analytical techniques, not only during the research and development phases, but also for quality control and problem solving during fabrication and manufacture.

Section 15.2 will therefore be concerned with the choice of technique(s) likely to deliver the required information for a particular material. Such a choice is usually made on the basis of a strategy. A strategy of analysis often begins with the most readily available and user-friendly instrumentation providing general and/or averaged information over the surface. More detailed and/or specific information then requires higher resolution (in each of the spatial, temporal, or spectral domains), which implies the deployment of more specialized instrumentation. In Section 15.3 of this chapter, an analytical strategy identifying different techniques appropriate to the provision of different kinds and levels of information is discussed.

Finally, in Sections 15.4 through 15.6, relevant case studies relating to mineral, ceramic, and glass surfaces, will be presented. The intention behind these sections is to use specific examples from the literature to demonstrate the utility of the different surface analytical techniques and the type of information that is obtainable in each case.

15.2 INFORMATION REQUIRED: ANALYTICAL TECHNIQUES

Table 15.1 summarizes the types of information normally required from surface analysis of minerals, ceramics, and glasses. Accompanying each category of information is a suggested list of analytical techniques most applicable to the extraction of that information. It can be seen that, for any

TABLE 15.1

Types of Information Required and Relevant Techniques for Mineral, Ceramics, and Glass Surfaces

Information	Minerals	Ceramics	Glasses	Techniques
1. Phases				
Structure	√	√	X	XRD: GAXRD: STM: AFM: LEIS: TEM(ED)
Composition (uniformity)	√	√	√	EDS: EPMA: XRF: XPS: STS: LEIS: EELS
Distribution (size)	√	√	X	OM: SEM(BSE, EDS): AFM: STM: TEM
2. Surface structure				
Macroscopic (1 μm to 1 mm)	√	√	√	OM: SEM: profilometry
Microscopic (10 nm to 1 μm)	√	√	√	FESEM: AFM: TEM
Nanoscopic (<10 nm)	√	√	√	STM: AFM: TEM
3. Surface sites				
Structure (defects)	√	√	√	LEED: GAXRD: EXAFS: XSW: STM: AFM: LEIS
Chemistry (reactivity, defects)	√	√	√	XPS: SAM: FTIR: Raman: NMR(SS): SIMS: STS: LEIS: EELS
4. Grain boundaries				
Structure	√	√	X	TEM: STM: AFM
Segregation	√	√	X	EELS(TEM): EDS(TEM): STS: XPS: SAM: SIMS
5. Intergranular films				
Structure	√	√	X	TEM: STM: AFM
Composition	√	√	X	EELS(TEM): EDS: STS: XPS: SAM: SIMS
6. Depth profiles				
Structure	√	√	X	GAXRD: XSW: RBS: EXAFS: LEED: RHEED: TEM(X-section)
Chemistry (composition)	√	√	√	XPS(AR): SAM: SIMS: PIXE
7. Adsorption				
Molecular form (bonding)	√	√	√	FTIR: Raman: XPS: NMR(SS): SIMS(TOF): EXAFS(AR): NEXAFS
Coverage (layers)	√	√	√	FTIR: Raman: XPS(AR): STM: AFM
Distribution	√	√	√	SAM: SIMS(TOF): STM: AFM
8. Surface reactions				
Oxidation (hydrolysis)	√	√	√	SEM: FTIR: Raman: XPS: SAM: SIMS: STM(STS): AFM: LEIS
Dissolution (leaching)	√	√	√	XPS: SAM: SIMS: STM: AFM: PIXE: RBS: FTIR
Phase transformation	√	√	√	GAXRD: LEED: EXAFS: AFM: STM
9. Surface modification				
Calcination	√	√	X	XRD: XPS: SEM(BSE): TEM(ED): SIMS: FTIR: AFM
Plasma	√	√	√	XPS (Auger): FTIR: XRD: TEM
Surface layers (sizing)	√	√	√	FTIR: XPS: SIMS: SEM
Flotation separation	√	X	X	XPS: FTIR Raman: SAM: STM: AFM: SIMS(TOF): SEM(BSE)

particular category, more than one technique can be used, and in practice, it is almost always useful to study surfaces with at least four analytical techniques. The combination of analytical scanning electron microscopy (SEM) with energy or wavelength disperse spectrometer (EDS) or WDS, atomic force microscopy (AFM) x-ray photoelectron spectroscopy (XPS), and fourier transform infrared spectroscopy (FTIR) (or Raman), has been gaining some acceptance as a basis set for the characterization of these materials. This combination, and other analytical techniques providing more detailed information, is discussed in Section 15.3.

In the problem-solving mode, it may be sufficient to use only one technique or a subset of techniques to obtain the required information. For instance, accelerated corrosion of ceramic or glass surfaces may be the result of an activating agent introduced by contact with a solution or a gaseous ambient. It is often possible to determine the primary cause of the accelerated corrosion using XPS analysis alone [6]. In other cases, it is the molecular structure of the adsorbed layer and its alteration on reaction that is of direct interest; FTIR and Raman spectroscopies may then be required to produce the necessary additional information [7,8].

Table 15.1 constitutes a starting point for the choice of those analytical techniques that will provide the information required for research, development, or problem solving. Rigid separation between structural/chemical and compositional information is not intended to be inferred from the list. There are many techniques that provide information in both categories (e.g., scanning Auger microscopy (SAM), low-energy ion scattering (LEIS); the latter is also called ion scattering spectroscopy (ISS). The separate sections on minerals, ceramics, and glasses will illustrate the application of the techniques and their complementary information content.

15.3   ANALYSIS STRATEGY

In the "top-down" approach taken in this book, it is useful to consider the most efficient use of the surface-specific and other analytical techniques for obtaining the information necessary to address a particular problem. With the focus on the material rather than on the technique(s), experience suggests that information on structure and composition is most likely to be obtained by proceeding through the sequence in Table 15.2.

The usefulness of information obtainable from optical microscopy (OM) in reflection mode is underestimated in most approaches to the examination of these materials [9]. In many cases, phases can be identified directly from OM in combination with x-ray diffraction (XRD). The distribution of phases, and their grain size (when larger than 0.5 µm), can also readily be obtained. In glasses, devitrified (crystalline) regions can be identified together with structural defects and surface topography. Thin sections, when investigated in transmission mode with polarized light, can reveal strain fields, macroinclusions, and triple points, and provide evidence for nonuniformity of composition [9]. OM is an excellent starting point for the examination of the surfaces of minerals, ceramics and glasses, and gives the analyst a clear idea of the overall structure of the material at low resolution, for subsequent comparison with information from higher spatial resolution techniques further down the sequence. The merits of OM are strengthened by the fact that surface roughness and uniformity on the micrometer to millimeter scale are parameters of considerable importance in surface reactions, modification, bonding, adhesion, and incorporation into other matrices.

The next higher level of spatial resolution (Table 15.2) is that of routine SEM. Imaging in secondary electron (SE), backscattered electron (BSE), and topographic modes extends that resolution to better than 100 nm. The complementary information available on phase identification and distribution, from comparisons of SE and BSE imaging, will be illustrated in the following sections. Similar images can be obtained by SAM in combination with identification of species and thus composition. Field emission SEM (FESEM) instruments have improved the resolution further to 2–5 nm on reasonably conductive surfaces with high SE contrast. The high brightness of the field emission sources has an additional advantage in that it can allow examination of insulating surfaces without conductive coatings. The accelerating voltage of the electron beam can be reduced, with

TABLE 15.2

Analysis Strategy for the Application of Techniques to Give Information in a Top-Down Approach, i.e., Macroscopic to Microscopic to Atomic Scales

Structural Information

Imaging	Lattices:Sites
OM	XRD
↓	↓
SEM: SAM	GAXRD: LEED
↓	↓
FESEM	EXAFS: XSW
↓	↓
AFM: STM: TEM	LEIS (ISS)

Chemical Information

Near Surface	Surface	Molecular
EDS: EPMA: XRF	XPS	FTIR: Raman
↓	↓	↓
EXAFS: RBS: PIXE	SAM	NMR (solid state)
↓	↓	
EELS (TEM)	SIMS (ToF)	
	↓	
	STS(STM): LEIS	
	↓	
	SRXPS	

some loss of resolution, to allow images to be obtained from materials as insulating as mullite ($3Al_2O_3 \cdot 2SiO_2$), alumina (Al_2O_3), and quartz (SiO_2), without serious surface charging.

To obtain resolution at the unit cell level, one or more of the techniques of high-resolution transmission electron microscopy (HRTEM, requires thin sections), scanning tunneling microscopy (STM, requires conducting samples), or AFM (able to deal with nonconductors), can be used. Lattice imaging by TEM can reveal defect structures in microdomains as linear or planar defects, and can also provide direct information about phase relationships, grain boundary mismatch, and intergranular films, at the nanometer level. STM and AFM can identify individual reactive sites, impurity atoms, and defect structures.

The full range of imaging techniques is therefore capable of resolving structure on a dimensional scale from atomic to millimeter. Which information is to be sought can often be decided only after an initial survey of the material at the lower spatial resolutions of OM and SEM in combination with general information on surface and near-surface compositions from XPS and EDS/WDS.

The second group of techniques relating to structure, in this case based on diffraction or scattering, can also be used in the top-down approach. A routine examination of the mineral or ceramic sample with XRD is almost always worthwhile. The presence of unexpected phases, and possible variations of unit cell size with impurity incorporation, can be revealed from a few hours work. If more specific information is required, e.g., the structure of a surface layer, then glancing angle x-ray diffraction (GAXRD) from large cleaved or polished surfaces and/or low energy electron diffraction (LEED), can reveal any surface restructuring resulting from reaction, relaxation, "rumpling", or reconstruction. The synchrotron-based technique of x-ray standing wave (XSW) studies can probe surface structure at the next level of detail. The XSW method is particularly sensitive to the z-position of the atom. Standing waves generated from different diffraction planes allow the positions of surface atoms to be determined to high accuracy. The method can be applied not only to the structure of the native surface, but also to that of an adsorbate on a surface [10,11] and to the distribution of ions in the aqueous electrical double layer near the solid surface [12]. Low energy ion scattering (LEIS, also called ISS) can, at the atomic level, determine structure and composition of the top atomic layer of a surface [13,14]. The relative positions of the surface atoms can be determined layer by layer in the first few nanometers of the surface using this extremely surface-specific technique.

Turning now to chemical and compositional information, an approach similar to that described above can be taken to determine chemical states (e.g., oxidation state, bonding, compound, or species identification) and composition (e.g., for quantification, for determination of stoichiometry, and for impurity identification).

It is necessary to distinguish between near-surface (i.e., 10 nm to 1 μm) and surface-specific (i.e., less than 5 nm) techniques [15] (Table 15.2). The most readily available technique for establishing composition in the near-surface region is SEM with EDS or WDS capabilities, which parallels x-ray fluorescence (XRF) in its ability to determine the approximate composition to depths of a few micrometers. The electron probe microanalysis (EPMA) technique, i.e., SEM with WDS, can provide more accurate quantification in the near-surface region. Ultrathin window and windowless detector technology extend the analytical range to include the low-Z elements C, O, N, and F, thus enhancing the method for the purpose of obtaining initial information. It is always useful to carry out EDS/WDS analysis for comparison with surface analysis so that the surface/bulk ratio of individual elemental species can be correlated. This ratio is often important in studies of segregation, leaching, dissolution and reprecipitation, and of surface reactions in general.

In some industrial processes involving minerals, e.g., flotation, it is of vital importance to be able to distinguish the changes in surface composition from those in the bulk, while in ceramics and glasses surface/bulk compositional ratios are often altered by processing at high temperature or by subsequent weathering reactions. Extended x-ray absorption fine structure (EXAFS) is a synchrotron-based technique capable of revealing the chemical environment (i.e., nearest neighbor atoms, coordination, and bond distances) of specific atoms in minerals, ceramics, and glasses [10,16–18]. Adsorption processes and the formation of new chemical species and phases can be followed directly with EXAFS [16]. Near edge x-ray absorption fine structure near edge x-ray absorption fine structure (NEXAFS), also called x-ray absorption near edge structure (XANES), is another technique for obtaining information about the oxidation states of specific elements. If higher spatial resolution of the near-surface structure is required, then the combination of TEM with electron energy loss spectroscopy (EELS) can be particularly valuable. The related, but not TEM-based, technique of high spectral resolution EELS has poor spatial resolution, but is useful for providing information on the molecular vibrational structure of adsorbed species, complementary to that obtainable from FTIR and Raman spectroscopies.

The most commonly used technique for obtaining chemical and compositional information from the surfaces of minerals, ceramics, and glasses is XPS. Assignment of chemical states from shifts in binding energy (BE), and quantitative analysis (to a precision of ca. ±10% for most elements), can be obtained by routine analyses from regions of diameter 0.1–10 mm. High-resolution (spatial and energy) imaging XPS instruments (see Chapter 3) are capable of mapping the lateral distributions of specific chemical species with spatial resolutions at the 5–50 μm level, depending on the type of instrument and the material under investigation. There is real promise of improvement in the lateral spatial resolution to 0.1 μm in the most recent instruments. XPS can be applied to both conductive and nonconductive materials, making it ideally suitable for minerals, ceramics, and glasses. If even higher lateral spatial resolution is required, then compositions may be mapped in the 25–50 nm range by SAM with LaB_6 electron sources or further to a few nanometers with field emission sources. Correlation of point analyses and maps of elemental distribution, with physical images in either the SE or the BSE modes, can help elucidate the behavior of minerals, ceramics, and some glasses. A major disadvantage of SAM is that it is not in general suitable for insulating surfaces because of charging effects, but many minerals, ceramics, and reacted glasses do have sufficient surface conductivity to allow analysis by it.

The next category of detailed information on surface chemistry and composition is that available from time-of-flight secondary ion mass spectrometry (ToF-SIMS). In that technique, the positive and negative secondary ion mass spectra correspond to the fragmentation patterns from adsorbed molecules and reaction products, with contributions from sputtered substrate atoms. Although the mass spectra can be complex, their analysis has allowed the mechanisms of surface bonding, reaction between adsorbed molecules, and substrate reactions, to be studied [19]. The scanning imaging mode, using liquid metal ion guns (e.g., Ga, Au, and Bi), and very low ion beam current densities (with beam diameters less than 100 nm), can provide direct information on the distribution and surface coverage of adsorbates and reaction products [19–22] (see also Chapter 4).

Detail at the atomic level of spatial resolution concerning chemical states and composition is available from STM operated in different modes—e.g., scanning Kelvin probe and scanning tunneling spectroscopy (STS) (Table 15.2). Single atom resolution and site identification with these techniques are applicable only to reasonably conductive samples. The alternative is LEIS from which a combination of local atomic structure and quantification of surface composition, especially for light elements, can be obtained [14].

Very surface-specific (1–2 molecular layers) chemical information can now be obtained by the more specialized technique of synchrotron radiation XPS (SRXPS). There, the energies of the exciting x-rays are chosen to be 40–50 eV higher in energy than that of the specific photoelectron, to minimize the escape depth of the photoelectrons. The technique has provided new insight on surface states and their reactivity not available in conventional XPS studies (Section 15.4.3).

For information on molecular structure, in particular as it relates to adsorbed molecules and their reactions on mineral, ceramic, and glass surfaces, the most commonly used techniques are FTIR and Raman spectroscopies [23]. FTIR is used increasingly in the diffuse reflectance (DRIFT) mode for samples in air, and in the attenuated total reflectance (ATR) mode, particularly as circle cell attachments, for *in situ* adsorption, and for reactions in solution [24,25]. These techniques can reveal the functional groups of adsorbed molecules bonded to the surface, and can also track changes in molecular structure during kinetically controlled reactions at the surface. Quantitative estimates of adsorption, displacement by competitive adsorbates, and the roles of acid and base surface sites on the substrate can also be followed. Another useful technique is solid-state nuclear magnetic resonance (NMR), from which detailed information has been derived about the behavior at specific sites (e.g., ^{29}Si and ^{27}Al) of hydroxyl groups and organic adsorbates [26,27].

In summary, it is clear that the normal starting point for an analytical strategy would be that of the readily accessible techniques, i.e., OM, SEM/EDS, XRD, XPS, and FTIR or Raman spectroscopy. Based on the outcome of the initial survey, the researcher/analyst should then be able to make a more informed choice as to the detail and nature of information required subsequently, in terms of spatial resolution, atomic imaging, molecular structure, chemical states, and lateral distribution. The requirements will also dictate whether the analysis should go in the directions of either or both improved spatial and better spectral resolutions.

In the next three sections, the strategy sketched out above will be illustrated with examples demonstrating the merits of different techniques and methodologies.

15.4 MINERALS

In this section, examples will be chosen of the information that can be generated by the techniques listed in Table 15.1. Rather than following in strict order the categories of information in that table, some typical problems will be addressed, and the contributions made by different techniques to the understanding required to resolve the questions will be discussed. In general, the approach will follow the scheme in the table, but it will become apparent that any one technique may be able to contribute to more than one category. Reference to more comprehensive reviews, book chapters, and books will be provided, so that the reader may have access to additional details on experimental and instrumental conditions. Two recent reviews in particular [28,29] have illustrated the multitechnique approach to obtaining this information for mineral systems including those obtained by single phase synthetic and multiphase processing routes.

The process of selective separation of valuable minerals from ores by froth flotation probably has the highest throughput (by mass) of any process in which control of the chemistry, physics, and engineering at the solid/liquid interface decides the economic outcome. Worldwide, this industry generates more than US$B120 p.a. with throughput in excess of 10^9 ton p.a. Effectiveness of the separation, in both grade (selectivity) and total mineral recovery, is heavily dependent on several processing steps: grinding to liberate separate particles of the valuable minerals; control of hydrophilic surface reaction products on the mineral surfaces; and adsorption of "collector" molecules designed to induce hydrophobic surfaces on the valuable minerals for bubble–particle attachment [28]. Bubbles usually much larger than the ground particles are generated in the flotation cell, rising through the pulp (or slurry) to collect the hydrophobic particles into the froth concentrate. The unwanted minerals are collected into the tails stream. The primary objective is to achieve complete liberation and a hydrophobic solid/liquid interface only on the valuable mineral phase to be concentrated in this step but these processes, reviewed in Ref. [28], are complex and inter-dependent. Some examples of the information necessary for effective control will be illustrated here.

In this and following sections, mineral compositions will be specified where first used, but more detailed information on structures, substitution, etc. can be obtained from the excellent web site webmineral.com.

15.4.1 Phase Structures

The use of OM as a first step in understanding the structure of minerals and their surfaces has been discussed in the previous section and reviewed in the literature [9,30]. In summary, it is possible to obtain the following information:

- Particle size range (from image analysis) and shape
- Phase identification (combination of transmitted, reflected, and polarized light)
- Phase distribution (volume fractions and spatial distribution)
- Mineral associations/locking (down to an average dimension of ca. 0.5 μm)
- Surface texture, morphology, and porosity
- In mineral separation, identification of losses in grade or recovery due to incomplete liberation [30]

Quantitative analysis of multiphase specimens can determine the proportion of different phases present in an ore, a mineral product or, indeed, a fabricated ceramic. Routines for Rietveld analysis [31] of x-ray (and neutron) diffraction patterns are available in software packages for PC platforms [32]. A least-squares fit of a diffraction pattern, calculated from the crystal structure parameters of the known phases, is made to the observed pattern of the multiphase mixture. The intensities of all the peaks from one phase are proportional to a sample scale factor Z and to the mass of that phase, from which the Rietveld program [32] can determine the scale factors and masses of the crystalline phases present. For instance, the wt.% of each of six minerals (rutile TiO_2, galena PbS, pyrite FeS_2, sphalerite ZnS, chalcopyrite $CuFeS_2$, and quartz SiO_2), in a prepared mixture, has been matched by neutron diffraction analysis to the respective known percentages with an error of less than 5% in the final measure of fit [33]. The method is particularly useful for the determination of the abundances of phases with the same nominal chemical composition but with different crystal structures. As an example, the relative proportions of the cubic, tetragonal, and monoclinic phases of zirconia in partially stabilized zirconia ceramics can be determined by this method [33]. The method is unreliable when either preferred orientation, due to directional grain growth or disposition, or high concentrations of amorphous material are present. The results from Rietveld analysis are sometimes difficult to interpret when major changes in the intensities of particular diffraction peaks are the result of incorporation of impurities, or when there are variations in stoichiometry between grains of the same phase within a single sample or between different samples. Nevertheless, Rietveld analysis is now a well-established tool for determining the crystalline phases in minerals and ceramics. The proportion of different phases likely to be present in surface analysis can be determined directly so that surface/bulk phase ratios can be measured.

Identification of different phases and their distributions can be made in mineral mixtures, ores, calcined products, and ceramics, using SEM/EDS. A comparison of the types of information available from SE, BSE, and topographic images is given in Figure 15.1. The SE image reveals pores, cracks, grain edges, and other topographical features. In the BSE image, most of the contrast is due to compositional variations arising from the atomic number dependence of the yield of BSEs. In many cases, it is possible to distinguish different phases by the difference in contrast and thus to map the distribution of each phase. Also, it is apparent that the BSE image does not reveal the structure of pores, cracks, and other topographical features, which are seen more easily in the SE image. The third image was obtained by taking the difference signal from the split BSE detector, thereby suppressing the compositional dependence. This mode is sensitive predominantly to topography. The image shows the broad features of surface roughness at the cost of increased noise level [34].

A "locked", or composite, mineral particle, comprising two or more phases, can be analyzed by use of the sequence: SE imaging, BSE imaging, and EDS analysis. In Figure 15.2, e.g., the BSE image from a muscovite ($K_2Al_4[Si_6Al_2O_{20}](OH,F)_4$, a common mica) and pyrite (FeS_2)-locked particle shows that the higher average-atomic-number pyrite phase (light contrast) can be clearly

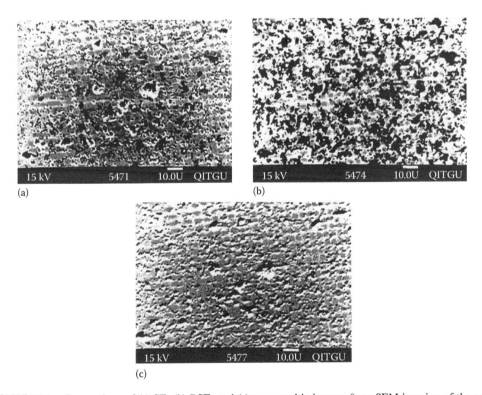

(a)

(b)

(c)

FIGURE 15.1 Comparison of (a) SE, (b) BSE, and (c) topographic images from SEM imaging of the same area of a polished ceramic surface showing grain pull-out and porosity. The scale bar is 10 μm. Note that the BSE image indicates phase regions of low (dark) and high (light) average atomic number. (From Klauber, C. and Smart, R.St.C., in O'Connor, D.J., Sexton, B.A., and Smart, R.St.C. (Eds.), *Surface Analysis Methods in Materials Science*, Springer-Verlag, Berlin, 2003, 3–70. With permission.)

FIGURE 15.2   Comparison of SE (left) and BSE images (right) from the same area of pyrite and muscovite (mica) grains showing composites of the two minerals. The central muscovite particles have locked regions of pyrite (light contrast in BSE image) not liberated during grinding of the ore. (From Smart, R. and Skinner, W.M., unpublished report.)

distinguished from the lower average-atomic-number muscovite (dark contrast). EDS point analysis confirmed this differentiation by recording signals from Fe, S, and O in the partially oxidized pyrite (light region) and from K, Si, Al, and O in the region with dark contrast. Versions of this analysis are now used extensively in minerals processing under the acronyms QEM*SEM (quantitative evaluation of minerals by SEM) [35], recently renamed QEMSCAN [36], and MLA (mineral liberation analyser) [37]. The technique measures pixellated BSE images and EDS spectra stereoscopically and refers the analytical information to a database of compositions to identify the mineral phases. It can provide quantitative estimates of the percentage of each phase in different particle size fractions, and the proportion of each phase in locked composite particles. It is used widely for the assessment of ores, and for the evaluation of mineral separation processes which include flotation, electrostatic and magnetic separation, and separation by gravity.

TEM of sections (usually <300 nm thick) can be used to study the internal microstructure, crystal structure, and surface features, of mineral crystals. Electron diffraction from areas of diameter down to 300 nm, as selected by an aperture, can be observed by imaging at the back focal plane of the objective lens, which gives specific identification of the mineral structure from the diffraction pattern. Amorphous films or regions give rise to broad rings in the diffraction pattern, with radii which correspond to the most probable interatomic spacings either in multiphase samples, or in samples with intergranular films, triple point regions, microdomains, and nonuniform grains. Information can be obtained about the extent of crystallinity, and about the orientation and crystal structure of individual grains. Weak diffraction spots arising from surface structures can, in some cases, be used to form dark field images which enhance surface detail. For instance, phase contrast imaging of MgO crystals after exposure to water vapor can reveal small (1–2 nm) rectangular depressions and protrusions which had formed across the original smooth {100} surfaces [34]. Lattice imaging of very thin sections or nanoparticles with the electron beam orientated along a zone axis can be related directly to the crystal structure, with resolution corresponding to the spacing of the planes giving rise to the Bragg reflections. Analysis of these TEM images can reveal extraordinary detail of size, shape perfection (point defects and dislocations), orientation, composition (with EDS or EELS), and surface structure. Profile imaging with a TEM [38] can reveal detail on the atomic scale of the surface structures of facets at the thin edges of crystals. Matching the experimental images to those computed from dynamical electron scattering models [39] has established this technique as

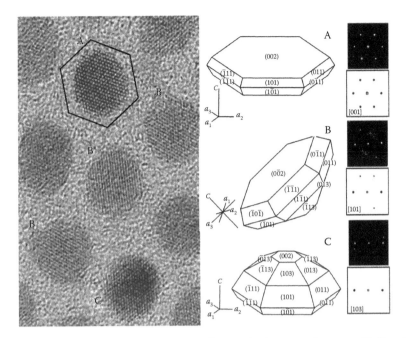

FIGURE 15.3 $Zn_xCd_{1-x}Se$ nanocrystals deposited onto Formvar/carbon-coated copper grids from hexane solution. Fourier transform patterns derived from each selected particle are shown in the black boxes. The calculated diffraction patterns matched to the orientation and faceting in the particle models are shown in the light boxes. (From Zhong, X., Han, M., Dang, Z., White, T.J., and Knoll, W., *J. Am. Chem. Soc.*, 125, 8589, 2003; White, T.J. Nanyang Technological University, Singapore, unpublished personal communication. With permission.)

complementary to SPM on the atomic scale. A recent example of some of this analysis for $Zn_xCd_{1-x}Se$ nanocrystals [40] is shown in Figure 15.3. The lattice images can reveal directly point defect and dislocation structures. Size distribution histograms were obtained from over 200 particles. For each of the nanocrystals selected (A, B, and C), a Fourier transform of the lattices gives the diffraction patterns from which the crystal shape, orientation, and surface faceting can be calculated. Similar analysis can be done for thin sections of minerals, ceramics, and glass/ceramics.

The compositions of individual phases in the near-surface region are normally obtained by EDS or by the more accurate WDS, while surface composition can be determined by XPS or SAM. Examples of the latter will be discussed later. Other techniques which can be applied to the determination of composition and/or chemical environment are EXAFS and NEXAFS. Examples of their applications can be found in Chapter 7, and in reviews by Greaves [10] and Garrett and Foran [18]. For instance, the ratio of monovalent to divalent copper in tetrahedrites (e.g., $(Cu,Fe)_{12}(Sb,As)_4S_{13}$) can be determined from NEXAFS peaks recorded at different x-ray absorption thresholds (which are dependent on the oxidation state) [41]. Chemical state imaging of regions having the same element in different oxidation states by collecting microfocus XRF data above and below the Fe K-edge can reveal hematite and magnetite inclusions in sintered iron ore [42]. Impurity sites (e.g., incorporating Ti and Mn in staurolite, $(Fe,Mg)_4Al_{17.3}(Si_{7.6}Al_{0.4})O_{48}H_3$) have been identified by refinement of the near-edge structure, and by comparison with model compounds. It was shown that Ti was associated with octahedral Al sites and Mn with tetrahedral Fe^{2+} sites; this was in direct contrast to the conclusions drawn from XRD refinement of the structure [10]. The implications of studies such as these, which determine the expected site occupancy in the surfaces by impurity atoms, are of considerable importance for studies using SPM.

Where surface analytical techniques are to be used on multiphase samples, such as minerals involved in separation by flotation, the very surface sensitivity of those techniques can be a problem in phase identification, since their information is derived from the outermost molecular layer or the top few layers. These signals can be dominated by adsorbed or reacted molecular species obscuring the underlying bulk phases, making it difficult to associate reliably the specific surface chemistry with a particular mineral, which is often the primary reason for the analysis. A second issue is that, in real mineral samples, different particles or grains are likely to have different surface concentrations of reagents or reaction products. Hence, the data reduction and interpretation, where it is related to the surface chemistry, are not only mineral-specific, but must also be based on a statistical average of the mineral particles in that phase. An example of these problems and a method for their resolution can be found in ToF-SIMS information from selected mineral phases in separation by flotation [43]. The methods can, however, be applied equally to any complex mineral mixture and processing sequence.

Diagnosis of the surface chemical factors playing a part in separation by flotation of a valuable metal sulfide phase (e.g., chalcopyrite $CuFeS_2$, sphalerite ZnS) requires measurement of the adsorbed hydrophobic (e.g., collectors such as xanthate $ROCS_2^-$ ions) and hydrophilic (e.g., hydroxide) species statistically distributed between the concentrate (hydrophobic) and tail (hydrophilic) streams. Statistical methods, based on the monolayer-sensitive ToF-SIMS technique with principal component analysis (PCA), have been developed for measuring mass signals from hydrophobic species (e.g., collector ions) as well as hydrophilic metal ions, hydroxides, and other flotation-depressant species. Reliable identification of specific mineral particles is central to this statistical analysis. PCA identifies combinations of factors strongly correlated (positively or negatively) from sets of data. In ToF-SIMS, PCA selects these correlations (as principal components) from the mass spectra recorded at each of 256×256 pixels ($>6 \times 10^7$ data points) in a selected area of particles, and in the image mode has proved to be a much better method of selecting particles by mineral phase with clearer definition of particle boundaries due to multivariable recognition.

For example, a chalcopyrite/pyrite/sphalerite mineral mixture was stirred at pH 9 for 20 min to study the transfer of Cu from chalcopyrite via solution to the other two mineral surfaces. Collector molecules will form complexes with copper ions causing unwanted flotation of the sphalerite and pyrite minerals during chalcopyrite recovery, lowering the concentrate grade. The uptake of Cu on sphalerite and pyrite will govern this loss of selectivity. The PCA results for the first two principal components are illustrated in Figure 15.4. The first principal component PC1, with positive factor loadings (statistical correlations) at each mass, and associated image, is representative of the largest variance in the data set, i.e., topography and matrix (ion yield intensity) fluctuations. The second and subsequent PCs have this variance removed and are topography- and matrix-corrected, a major advantage in the processing of ToF-SIMS data. PC2 shows positive correlation values for zinc mass peaks and negative correlation values for iron and copper mass peaks. Thus, bright areas on the image score are indicative of zinc-rich (sphalerite) phases. The dark areas are rich not only in iron and copper as in chalcopyrite but also in pyrite (with Cu transferred from chalcopyrite via solution). Further separation of pyrite and chalcopyrite phases is found in other PCs (e.g., PC4 here) where specific Cu-containing particles (bright areas) are separated from Fe-containing particles (dark areas). After the mineral phase areas are identified from the PC sets of factor loadings, the other mass peaks in each set are statistically correlated with these mineral phases. These mass fragments can include metal ions, oxidation products such as hydroxides, hydrocarbons and, in real ore systems, the flotation control reagents (e.g., collectors) providing the essential information on adsorption selectivity between minerals in the processing. Without PCA processing, no statistical differences in the Cu intensities between pyrite and sphalerite surfaces, selected from Fe and Zn images, could be found after conditioning. On the other hand, PCA processing clearly separated a statistical difference in Cu intensities between the sphalerite and pyrite phases [43] illustrating the improved phase selection for surface chemical differences.

15.4.2 SURFACE STRUCTURES

The use of SPM in studies of mineral surfaces has been reviewed by Hochella [44]. At an atomic level of detail, a combination of STM and STS has been used to reveal the positions of surface atoms

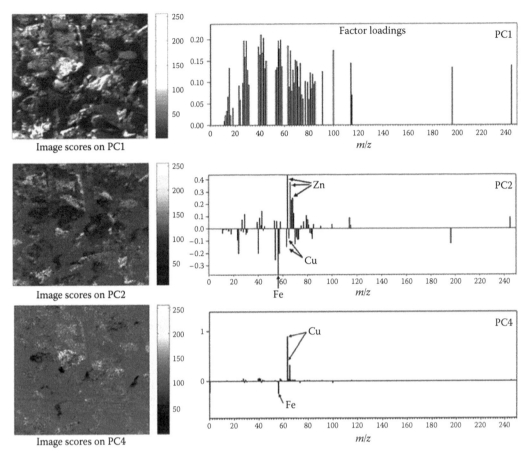

FIGURE 15.4 Principal component image scores and factor loadings for the chalcopyrite/pyrite/sphalerite mixture. Images are of area ca. 100 × 100 μm. (From Hart, B., Biesinger, M.C., and Smart, R.St.C., *Min. Eng.*, 19, 790, 2006. With permission.)

and the electronic structure of rutile (TiO_2) single crystal surfaces [45,46]. The sites of surface titanium cations, but not those of the anions, were revealed clearly. STM studies of sulfide mineral surfaces have also been carried out by Eggleston and Hochella [47–53]. Figure 15.5 illustrates the superb spatial resolution of the technique in the case of the {100} surface of galena (PbS); the atomically resolved image was recorded in the constant height mode (see Chapter 5). In Figure 15.5a, a surface unit cell is outlined with an edge dimension of approximately 0.6 nm. With a sample bias of −599.1 mV (Figure 15.5a), the high intensity maxima in the image arise from tunneling contributions from the 3p electron states of individual sulfur atoms, whereas with a positive sample bias of +200 mV (Figure 15.5b) the main maxima still have sulfur character, but the subsidiary maxima now show the locations of electron states associated with lead cations [47].

The same authors obtained an image from the {100} surface of pyrite (FeS_2) with unit cell resolution, with the main maxima arising from the Fe 3p states [49]. Steps on the surface with dimensions of half a unit cell were imaged. In addition, imaging of hematite (Fe_2O_3) at atomic resolution [50] revealed similar information about monatomic steps and the arrangement of oxygen atoms. Atomic positions in the STM map could be matched to theoretical models of a stepped (001) hematite surface, showing rows of oxygen atoms being offset as a step is crossed. The atomic structure and morphology of the more complex, nonconductive albite ($NaAlSi_3O_8$) {010} surface was also studied with a combination of AFM and electron diffraction [54].

SPM can also be used to study directly the reactivity of mineral surfaces. The oxidation of sulfide mineral surfaces, weathering (e.g., leaching and dissolution) of nonsulfide minerals, sorption of

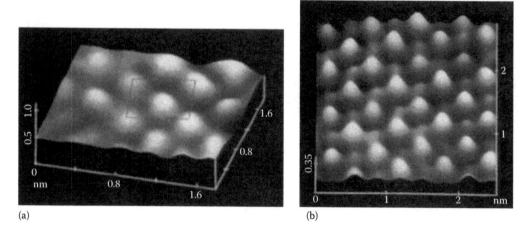

(a)   (b)

FIGURE 15.5   (a) STM image of the {100} surface of galena (PbS) taken in the constant height mode, at −599.1 mV sample bias and 2.9 nA tunneling current. The peaks in the image are the tunneling contribution for 3p electrons from individual S atoms on the surface. The box outlines the surface unit cell, ca. 0.6 nm on edge. (b) STM image of the same surface as in (a) but taken at +200 mV sample bias and 1.8 nA tunneling current. The major peaks still have S character, but there are now smaller peaks in between with Pb character. All scales are in nanometers. (From Eggleston, C.M. and Hochella, M.F. Jr., *Geochim. Cosmochin. Acta*, 54, 1511, 1990. With permission.)

cations and anions, and reactions taking place during surface modification, can be studied on the atomic scale. Eggleston and Hochella [47,48,51] have proposed models for oxidation of the galena surface on the basis of information from images in which individual atoms were observed to react on contact with air or water. Figure 15.6 shows the {100} surface of galena after exposure to water for 1 min. On oxidation, the tunneling intensity maxima from a surface sulfur atom disappeared due to alteration of the local chemical structure. In the image (Figure 15.6), the apparent vacancies are in fact the sites of oxidized sulfur atoms. From their work, and from another study described below,

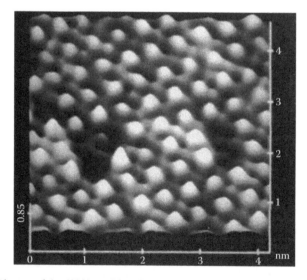

FIGURE 15.6   STM image of the {100} surface of galena taken after exposure to water for 1 min. The conditions of image collection were the same as in Figure 15.5b. The peaks represent surface S, and the lower maxima in between correspond to Pb atoms. The apparent vacancies are *oxidized sulfur sites; the tunneling* current to these sites is very low. (From Hochella, F.M. Jr., in Vaughan, D.J. and Pattrick, R.A.D. (Eds.), *Mineral Surfaces*, Minerological Society Series No. 5, Chapman & Hall, London, 1995, 17. With permission.)

it is clear that, when a surface S atom is oxidized, S atoms in adjacent sites are activated, thus promoting further reaction rather than initiating new reaction at S sites further away. A trend for [110] directionality was observed at the boundaries of the oxidized patches. Eggleston and Hochella [51] suggested that this was due to the association of only one nearest-neighbor oxidized S atom with an adjacent unoxidized S across the [110] front, whereas across a [100] front there are two nearest-neighbor oxidized S atoms adjacent to an unoxidized S. This suggests that the [100] fronts move faster and ultimately become lost.

Sorption of gold, (Au(III)), from solution onto PbS surfaces was followed by the same authors over time scales ranging from seconds to several days [48,51]. The metallic nature of the small islands (of dimensions <10 nm) was confirmed by STS measurements on gold foil standards, and by an XPS study by Bancroft and Hyland [55]. Island growth on unoxidized PbS surfaces again showed [110] directionality and was faster by a factor of three than growth on the preoxidized surfaces. In similar studies combining STM, XPS, FTIR, and SAM, the sorption of uranium on pyrite and galena surfaces [55,56] was examined. Reduction of U(VI), coincident with oxidation of the sulfide surface to polysulfides (on PbS), or to Fe(III) oxides/hydroxides (on FeS_2), at inhomogeneously distributed reaction sites, could be followed and was thought to be associated with surface roughness and defect sites.

Precipitation reactions involving the formation, growth, and coalescence of surface nucleation sites are important in studies of the processes of dissolution, solution speciation, and saturation followed by precipitation. AFM has been used to follow the appearance and development of the nucleated growth of calcite ($CaCO_3$) on a calcite crystal substrate in solution [57]. Growth was observed to take place along steps 1–3 monolayers in height. The effect of inhibitors such as phosphate on the growth mechanism suggested preferential adsorption at steps, edges, and corners, which blocked sites for incipient nucleation and crystalline growth. As a result, the transition from surface nucleation to spiral growth was apparently delayed until after about 2 h in solution.

At an intermediate level of spatial resolution, studies of oxidation and reaction processes using SPM can also be of fundamental importance for understanding the mechanism(s) and progress of alteration of a surface. Continuing with the model PbS {100} surfaces, a systematic combination of STM and XPS investigations of the oxidation in air of an area of size 70×70 nm^2 illustrates the usefulness of information at this level of detail. Figure 15.7 shows the progressive growth

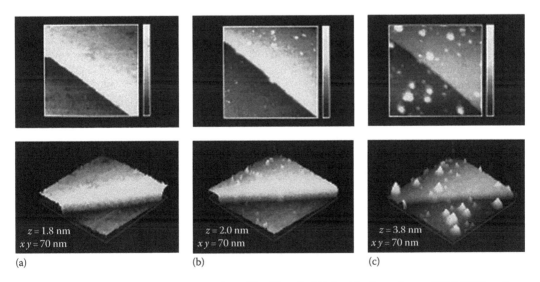

(a)							(b)							(c)

FIGURE 15.7 STM images from an area of size 70×70 nm^2 of (a) freshly cleaved natural PbS {100} surface; (b) the same surface after 70 min standing in air; and (c) after 270 min in air. The upper row contains gray scale images, while the lower row shows three-dimensional (rotated) images with vertical scales of 1.8, 2.0, and 3.8 nm, respectively. (From Laajalehto, K., Smart, R.St.C., Ralston, J., and Suoninen, E., *Appl. Surf. Sci.*, 64, 29, 1993. With permission.)

of oxidation products; the original surface features with lateral dimensions of less than 0.6 nm eventually evolved into overlapping regions >10 nm in diameter [58]. The final stage was that of a nearly continuous oxidized surface with "holes" of dimensions of a few nanometers, through which continued reaction could apparently proceed [59]. XPS information, obtained in parallel, could be correlated with the initial appearance of hydroxide (and possibly peroxide (O^-)) species followed by growth of the predominant product lead hydroxycarbonate from reaction with CO_2. It is interesting that, even after reaction with air for 120 min, the S 2p spectra still showed only sulfide character, i.e., sulfur-oxygen species were not formed despite oxygen and CO_2 interaction with the lead ions. This may have been due either to attenuation of the signal from any sulfur-oxygen species by the overlying lead hydroxycarbonate or, as suggested by Buckley and Woods [60], to diffusion of lead ions into the metal-deficient surface. There was also an apparent absence of preference for initiation of reaction at low-coordination sites (i.e., corners and steps), but instead an apparently random initiation at sites on the {100} surfaces. In a subsequent investigation [61], comparison of the oxidation of natural galena and high-purity synthetic galena revealed that the difference was likely to have been due to the presence of impurities in the surface of natural galena. Growth of oxidation products on the synthetic sample occurred preferentially at step edges and dislocation sites instead of on {100} faces as on the natural PbS surface. The rate of surface oxidation of the synthetic PbS, as inferred from both STM images and XPS O/Pb ratios, was lower than that for natural galena by factors >10. The XPS spectra showed in both cases that lead hydroxide was the main oxidation product. Exposure beyond 6h in air generated, for the natural PbS but not for the synthetic one, small signals from sulfate and carbonate species.

It is also possible to use SPM techniques for *in situ* studies of the mechanisms of oxidation and reaction in solution. A different mechanism for the oxidation of PbS occurs in aqueous solution when compared with that in air [62]. STM imaging over 500×500 nm² fields of view is shown in Figure 15.8, illustrating the development with time of pits of subnanometer dimensions. The z-dimensions of these pits corresponded to half or full unit cell parameters of PbS [62]. AFM imaging under the same conditions revealed a closely similar development of pits. The main surface process occurring was that of congruent dissolution; this was confirmed by XPS, which showed unaltered Pb 4f and S 2p spectra during the initial stages. The rate of formation of dissolution pits was strongly dependent on the concentration of impurities (as above), on pH, and on the gas used to purge the aqueous solution. High-purity synthetic PbS reacted much more slowly. Analysis of the STM images allowed quantification of the rate of development of the dissolution pits in terms of area (x-, y-dimensions) and depth (z-dimension). Figures 15.9 and 15.10 illustrate the dependence of the reaction rate on pH and on the reactive gas, respectively.

These examples may be related specifically to the information categories in Table 15.1 under the headings of phase structures, surface structure, surface sites (e.g., structure, chemistry, reactivity,

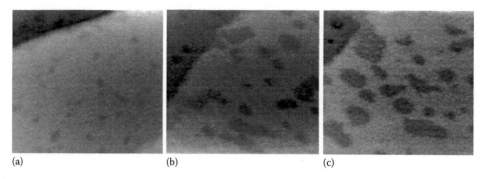

(a)   (b)   (c)

FIGURE 15.8 STM topview images of galena in air purged water at pH 7 (xy scale = 500 nm): (a) 20 min (z scale = 1.7 nm), (b) 40 min (z scale = 2.1 nm), (c) 60 min (z scale = 1.8 nm) and (d) 80 min (z scale = 2.5 nm).

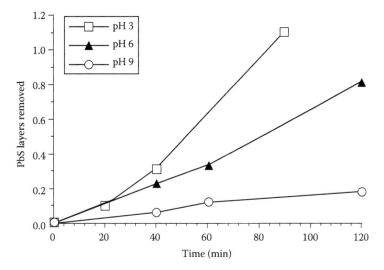

FIGURE 15.9 Number of equivalent PbS monolayers removed from the galena surface as a function of time in air-purged water, at different pH values. (From Kim, B.S., Hayes, R.A., Prestidge, C.A., Ralston, J., and Smart, R.St.C., *Langmuir*, 11, 2554, 1995. With permission.)

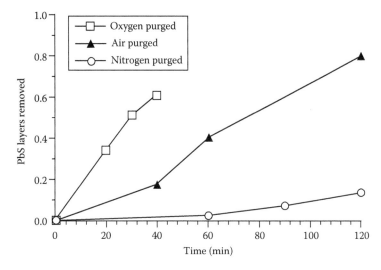

FIGURE 15.10 Number of equivalent PbS monolayers removed from the galena surface as a function of time in water at pH 7, with different purging gases. (From Kim, B.S., Hayes, R.A., Prestidge, C.A., Ralston, J., and Smart, R.St.C., *Langmuir*, 11, 2554, 1995. With permission.)

and defects), adsorption (e.g., distribution, coverage, monolayer versus multilayer, colloidal, molecular form, and bonding), and surface reactions (e.g., oxidation, dissolution, precipitation, and phase transformation).

15.4.3 SURFACE SITES

Structural investigation of surface layers, which aims to define the positions of atoms in the surface phase and the structure of surface layers, can be carried out by LEED [63], LEIS [13], GAXRD [10], and EXAFS [18,64]. LEED and LEIS methodologies have been dealt with in the review literature [13,63]. The applications of GAXRD and EXAFS methodologies and analyses to minerals have not been reviewed so thoroughly, but have received some attention [10,64, Chapter 7]. Here, it is relevant to note that an overlayer of α-Fe_2O_3 of thickness only $9.0 \pm 1.5\,nm$ on a γ-Fe_2O_3 substrate has been identified by GAXRD [65]. XRD and EXAFS techniques have a major advantage in that experiments can be performed in air or in liquids (provided that the liquid is less dense than the solid). They can therefore be used to study reactions *in situ*.

A good example of the information on surface sites available from the combined use of XPS and SAM is the series of studies of iron sulfide mineral surfaces carried out by Pratt et al. [66,67], who have published the first spectroscopic evidence for the existence of Fe(III) in pyrrhotite (Fe_7S_8) surfaces. Careful calculation and curve-fitting of the multiplet and satellite structures associated with the Fe(II) and Fe(III) atoms bonded to sulfur indicated that 32% of the surface sites exhibited Fe(III) character, in close agreement with stoichiometry (which suggests 29% Fe(III) in the particular pyrrhotite studied). The fitted data are illustrated in Figure 15.11 for the Fe $2p_{3/2}$ and S 2p spectra. Sulfur was present primarily as monosulfide (S^{2-}) with minor contributions from disulfide (S_2^{2-}) and

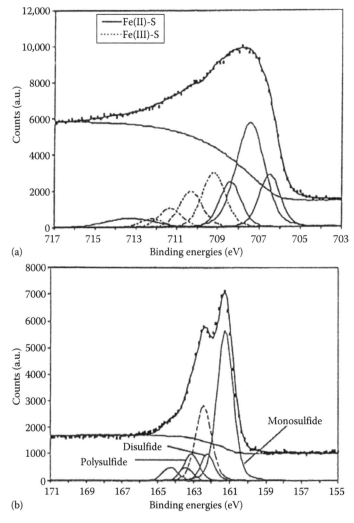

(a)

(b)

FIGURE 15.11 XPS spectra of a pyrrhotite (Fe_7S_8) sample fractured under high vacuum, 10^{-7} Pa: (a) Fe $2p_{3/2}$ spectrum; the series of peaks with solid lines represent Fe(II) (multiplets and satellite) bonded to S, and those with dotted lines represent Fe(III) (multiplet structure) bonded to S; (b) S 2p spectrum; doublets of S $2p_{3/2}$ and S $2p_{1/2}$ are seen for the monosulfide, disulfide, and polysulfide. For the monosulfide, the S $2p_{3/2}$ peak is plotted as a solid line and the S $2p_{1/2}$ peak as a dotted line. (From Pratt, A.R., Muir, I.J., and Nesbitt, H.W., *Geochim. Cosmochim. Acta*, 58, 827, 1994. With permission.)

polysulfide (S_n^{2-}). After 6.5 h oxidation in air, 58% of the iron was in the Fe(III) state, and bonded to oxygen, while most of the Fe(II) remained bonded to sulfur. The sulfur spectra showed a range of oxidation states from sulfide to sulfate. Angle-resolved XPS (ARXPS) and Auger (AES) depth profiles revealed three compositional zones: an outer zone composed of Fe-oxyhydroxides, an underlying zone of Fe-deficient, S-rich, reacted pyrrhotite, and then a continuous compositional grading to bulk pyrrhotite. The outer zone was 0.5–1.0 nm in thickness, and was separated by a sharp interface from the reacted layer, which had a thickness of ca. 3 nm. The mechanism of an Fe diffusion reaction with oxide, hydroxide, and water, and the production of disulfide species in the S-rich zone beneath the Fe-oxyhydroxide layer could all be demonstrated using the combination of the two techniques.

It is also noteworthy [68] that the adsorption of water alone on the pyrrhotite or pyrite surfaces did not result in oxidation. The O 1s spectra were composed of contributions from hydroxide and water species, but there was no evidence for the formation of Fe-oxide, or oxidized sulfur products. Careful analysis of the multiplet and satellite contributions to the Fe $2p_{3/2}$ structure revealed that the interaction of water with the two minerals occurred as a result of fundamentally different processes. Pyrrhotite reaction involves the transfer of electronic charge through iron vacancies, whereas on pyrite water interacts with the Fe 3p (e_g) molecular orbital, suggesting that hydrogen bonding with the disulfide sites may be important.

The role of the various surface S species, i.e., monosulfide, disulfide, and polysulfides, defined by XPS spectra, in controlling acid dissolution kinetics, has been explained in studies of pyrrhotite by Thomas et al. [69–72] and of sphalerite by Weisener et al. [73,74]. The sequence of formation of these species in the oxidation of galena has also been followed by correlating their S 2p doublets in XPS spectra with the appearance of S_n^{2-} ($n = 1$–6) mass fragments in ToF-SIMS spectra [75].

The surface specificity (95% of signal) in conventional XPS with Al K_α or Mg K_α sources is usually in the range 4–8 nm or roughly 10–20 molecular layers, depending on the kinetic energy of the photoelectron emission being studied. This dilutes the information available from sites in the top molecular layer which are often the critical ones in reaction or adsorption. ARXPS can yield nondestructive, more nearly surface-specific information, but requires surfaces that are close to atomically-flat, while the depth accuracy requires knowledge of the photoelectron inelastic mean free paths in overlying reaction products. In the more recently developed technique of SRXPS, the analysis depth can be reduced by controlling the kinetic energy and, hence, the escape depth, of the emitted photoelectrons of interest. In the universal curve of photoelectron escape depth as a function of kinetic energy, the maximum surface specificity is attained when the kinetic energy of the emitted photoelectron is in the range 40–50 eV, which implies the use of a source of energy 40–50 eV above the BE of the photoelectron. For example, to achieve this for the S 2p core-line (BE ~160 eV), an x-ray energy of approximately 210 eV should be used. This condition yields an almost fivefold enhancement in surface specificity over that using a conventional Al K_α source (1487 eV) for the S 2p signal. An example from pyrite mineral spectra is shown in Figure 15.12.

Similarly detailed information on the chemistry of atoms at surface sites, and on the relationships of such sites to surface reactivity, may be obtained for oxide, silicate, aluminosilicate, and other minerals using XPS in association with EXAFS and adsorption studies. The reviews by Bancroft and Hyland [55], Brown et al. [16], and Casey [17] provide many examples of the types of information obtainable for these surfaces.

Although not an experimental surface analytical technique, recent advances in the application of quantum mechanical molecular modeling to the determination of location, electronic structure, and chemistry of atoms at surface sites on mineral and ceramic surfaces should be noted here. Modeling has now reached a stage of maturity from which, in some cases, it is possible to predict experimental results ahead of actual measurement. An early example concerned the prediction that sites on non-defective MgO {100} surfaces would not cause dissociation of small molecules such as CO and H_2O [78]; this was later verified experimentally using atomically flat MgO smoke particles [79]. Dissociation and reaction at, and activation of, adjacent sites were predicted and found to occur at defects, and at low-coordination sites at corners, steps, and edges [80]. Recently, the mechanism of reaction

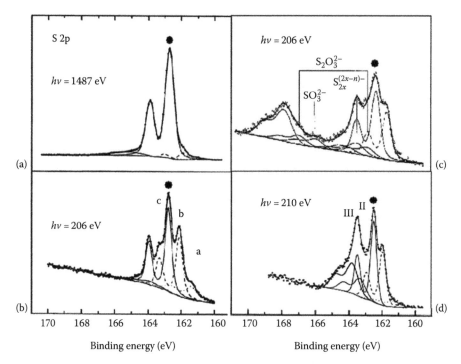

FIGURE 15.12 S 2p spectra from fractured pyrite (FeS$_2$) surfaces using (a) a conventional monochromated Al K$_\alpha$ source [76] and (b) a 206 eV synchrotron source [76]. Spectra from pyrite surfaces after subsequent reaction are shown for (c) air oxidation for 14 h [76] and (d) adsorption of mercaptobenzothiazole from solution [77]. The spectral contributions from bulk sulfur dimers (S$_2^{2-}$) are indicated by the solid circles above the primary S 2p$_{3/2}$ components. (From Szargan, R., Schaufuss, A.G., and Rossbach, P., *J. Electron. Spectrosc. Relat. Phenom.*, 100, 357, 1999. With permission.)

at these sites to produce [100] directed pits and protrusions a few unit cells in dimensions, as observed experimentally [79], has been explained using more advanced calculations [81].

The SRXPS interpretation by Harmer and Nesbitt [82] of sulfide mineral surface reconstruction and polymerized surface species, as a result of fracture, has been confirmed in recent *ab initio* calculations for pyrite [83] and for chalcopyrite and molybdenite (MoS$_2$) [84]. Such calculations provide a high level of support for the interpretation of synchrotron XPS of other important sulfide mineral fracture surfaces.

Tossell [85] has provided a useful review of the various quantum-mechanical methods for calculation of the free energy of mineral surfaces. Examples dealing with surface reconstruction of the (111) face of sphalerite (ZnS), the (001) face of Al$_2$O$_3$, and the electronic structure and spectra of the reduced oxide TiO have been given by him. Major reviews of oxide surface models have also been provided by Henrich [86] and Kunz [87].

15.4.4   GRAIN BOUNDARIES AND INTERGRANULAR FILMS

Segregation to grain boundaries and surfaces is of particular importance in the behavior of ceramic materials during fabrication processes, and some examples of relevant studies of grain boundaries and intergranular films in ceramics will be given in Section 15.5.4. Solid-state reactions, sintering, grain growth, and resulting mechanical (e.g., fracture strength, toughness, high temperature creep, and wear resistance) and electrical properties (e.g., conductivity and dielectric loss) are all affected directly by structural and compositional factors. In principle, information on segregation, structure, and composition in grain boundaries and intergranular films is equally important in the study and behavior of minerals, particularly where multiphase composites or ores are involved.

Methods for studying grain boundaries have been reviewed by Burggraaf and Winnubst [88]. The most direct method is that of HRTEM in combination with EELS or EDS analyses, which allows a grain boundary to be imaged so that any lattice mismatch of adjacent grains, and the presence of segregated elements or impurities, can be identified. Grain boundaries and intergranular films can also be studied by exposure of fracture faces, if the predominant mechanism of fracture is intergranular rather than transgranular. For that, a variety of surface analytical techniques, including XPS, SAM, static SIMS (SSIMS), and LEIS, may be used to determine the composition, distribution, and chemical states at the fracture interface. Fracture in the vacuum environment of the spectrometer, or carried out externally under liquid nitrogen with the use of a sample transfer vessel, may be required to avoid oxidation or other reactions of the fracture face with the ambient.

It is important to realize that segregation factors (i.e., surface/bulk elemental ratios) can range from 2 to more than 10^3, and that equilibrium segregation at grain boundaries and surfaces is normally limited to depths of a few nanometers, but in cases of nonequilibrium segregation much thicker layers can be found. Studies have revealed that the driving force of segregation can arise either from strain relaxation, when there is a large misfit of ionic radii of the elements within the mineral phases, or from space-charge effects with impurities of differing valency, or from a combination of both mechanisms. Hence, in naturally formed minerals, the structure and composition of fracture faces may be significantly different from the bulk mineral, thereby affecting surface chemistry and reactivity.

Grain boundaries can also be important in the formation and stabilization of microinclusions. A specific example can be found in the work of Pring and Williams [89] in which TEM studies revealed inclusions as small as a few nanometers of galena (PbS) in pyrite (FeS_2). Structures of this type have potential implications in the selective separation of minerals.

Properties of intergranular films with depths of a few nanometers, and with compositions and structures distinctly different from those of the bulk mineral phases, have been studied by XPS, SIMS, SAM, and related techniques. A specific example [90] concerns the presence of relatively thick (i.e., >25 nm) surface coatings of hydrophobic graphitic carbon on iron sulfide fracture faces after grinding a Cu/Pb-Zn sulfide ore. Some of the iron sulfide particles with this type of surface layer float naturally, which is an undesirable effect during the concentration of the Cu-, Pb-, and Zn-sulfides by mineral flotation separation. The carbonaceous layers are formed during ore genesis and may vary considerably from one deposit to another.

15.4.5 Depth Profiles

Depth profiling using ARXPS (a nondestructive method), or by ion beam sputtering in combination with XPS, SAM or SIMS, is often used to study the variation of composition with depth. Examples of the multitechnique approach to the study of the composition of surface layers can be found in a series of papers describing the analyses of weathered or dissolved feldspar mineral surfaces [91–95]. The plagioclase feldspar ($NaAlSi_3O_8$-$CaAl_2Si_2O_8$) mineral series proceeds from albite ($NaAlSi_3O_8$) through oligoclase, andesine, labradorite, and bytownite to anorthite ($CaAl_2Si_2O_8$) with an increasing percentage of Ca-containing (or decreasing percentage of Na-containing) end members. A combination of TEM, XPS, SIMS, and SEM was used. Structural and compositional profiles through the surface layers provided strong evidence for the formation of leached layers from which Ca and Al had been removed, leaving a residual porous structure enriched in Si. The thicknesses of the altered layers were dependent on the pH of the reactant solution, and on the composition of the particular plagioclase feldspar. The dependence on pH of the depth profiles obtained by SIMS for labradorite ($2NaAlSi_3O_8 \cdot 4CaAl_2Si_2O_8$) in pH 3.5 and 5.7 is illustrated in Figure 15.13 [92]. The altered layers increased in thickness from a few tens to over a 100 nm in the order: albite < oligoclase < labradorite < bytownite. SEM and TEM micrographs of labradorite specimens showed lamellae of calcium- and sodium-rich phases of widths ca. 70 nm. The calcium-rich phase was weathered preferentially to a greater average depth (135 nm) than that of the sodium-rich phase, thus producing a corrugated surface.

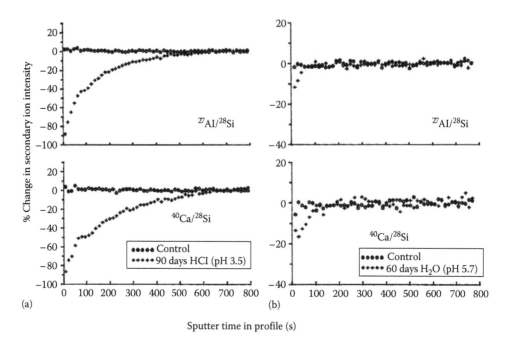

FIGURE 15.13 SIMS depth profiles from labradorite after dissolution (a) for 90 days in an aqueous solution of HCl (pH 3.5) and (b) for 60 days in doubly distilled and deionized water (pH 5.7). (From Muir, I.J., Bancroft, G.M., Shotyk, W., and Nesbitt, H.W., *Geochim. Cosmochim. Acta*, 54, 2247, 1990. With permission.)

XPS analysis provided support for a mechanism of hydrolysis of the Si-O-Al bonds, which allowed ion exchange during the initial leaching process. In more recent work [93], it was shown that the release of Ca and Al from labradorite could not be modeled on the basis of diffusion through homogeneous media, but required a "hopping" model with ion exchange between vacated (reacted) structural sites of the feldspar. This process has a well-established analog in the weathering of glass surfaces [4]. The surface reaction process could be followed in the spatial, structural, and chemical domains by using the multitechnique approach with angle-resolved analysis and depth profiling by ion etching.

15.4.6 ADSORPTION

According to Table 15.1, information concerning adsorption may be obtained as follows. Spatial distribution and surface coverage of adsorbed molecules can be measured from the atomic to the submicrometer level of resolution by using STM (with STS) or, for nonconductive surfaces, with AFM, as shown in Section 15.4.2. For imaging at lower spatial resolution, ToF-SIMS, particularly when used in the SSIMS mode, is a valuable tool. For instance, the spatial distribution of the molecular ion of the collector molecule diisobutyl dithiophosphinate on PbS surfaces has been observed directly [20] by comparison of a positive ion SIMS image of Pb abundance with the negative ion image of the abundance of the molecular ion of the adsorbed dithiophosphinate. The observed uneven spatial distribution suggested face-specific adsorption.

ToF-SIMS has also been used to determine directly the surface compositions of mineral particles from concentrator plant operation [22]. The particles ranged in dimension from 20 to 100 μm and were probed to obtain surface composition. The initial ion pulse is claimed to provide submonolayer depth resolution. Spot analyses (from areas of diameter 250 nm) showed the presence of flotation activating species (i.e., Cu and Pb ions), and oxidation products, on sphalerite, pyrrhotite, pyrite, and quartz. Spatial distributions across the particle surfaces were mapped, and the distributions of adsorbed collector molecules, namely, amyl xanthate and diisoamyl dithiophosphate, were also mapped on surfaces of laboratory-prepared PbS. As above, uneven distributions of both the adsorbed cations and the adsorbed xanthate or dithiophosphate were again observed. Specificity, in terms of crystal facet, of the adsorption of xanthate and dithiophosphate was confirmed in images of the molecular ion and ion fragments. Correlations of the sites of the adsorbed molecules with surface concentrations of hydroxyl groups could be made in both sets of experiments.

Information about surface coverage (e.g., monolayer, multilayer, or colloidal), molecular form, and bonding can be obtained from a combination of XPS and FTIR analyses and electrochemical experiments. Some of the methods used to model the surface of PbS with xanthate adsorption have been summarized [96]. Careful analysis of the results from several studies has shown that there is good agreement between estimates of coverage by XPS and FTIR (used in the ATR mode), when adsorption is measured under controlled potentials applied to the mineral in solution. The techniques have established that there is a submonolayer coverage of adsorbed lead ethyl xanthate at potentials below those at which lead ethyl xanthate is expected to form in bulk solution. It was also established that, even at relatively high concentrations of xanthate in solution, there was still significant coverage of the surface by adsorbed oxide/hydroxide species. Orientational effects, evident from the ARXPS spectra, as well as small differences in the positions of IR absorption bands corresponding to adsorbed product and crystalline lead ethyl xanthate, suggested that the adsorbed xanthate molecules had perpendicular orientation with respect to the surface. In this case, multilayers of adsorbed molecules were apparently not formed. Examples of multilayer formation have been discussed in Section 15.6.4 on glass surfaces.

Infrared spectroscopy is particularly well suited to the determination of the molecular structure of minerals with adsorbed molecules and of small changes in the molecular properties at the interfaces. Importantly, measurements can be made *in situ* at both gas–mineral and aqueous solution–mineral interfaces. An IR external reflection technique has been developed [97–100] that can provide reliable data for the monitoring and understanding of molecular structural changes at any mineral interface. IR internal reflection spectroscopy, and Raman and sum-frequency generation (SFG) techniques have been applied in recent studies of adsorption at mineral surfaces. SFG is a nonlinear optical process, in which the signal is generated at a frequency which is the sum of the frequencies of two incident optical fields due to the nonlinear interaction of infrared and visible lasers. SFG stimulation is forbidden in a medium with inversion symmetry, but this nonlinear optical process occurs at surfaces where the inversion symmetry is broken. Most bulk materials have inversion symmetry, thus do not generate SFG signals. This unique feature makes SFG a sensitive and powerful tool for the study of various interfaces and surfaces [101]. For instance, the SFG spectrum of adsorbed molecules with long hydrocarbon chains is very sensitive to hydrocarbon chain order; loosely packed chains and disordered monolayers will, in general, have more random orientations of the methyl and methylene groups, and the intensity of the SFG signal will be much smaller than in the case of an all-trans state. When an alkyl chain is in an all-trans conformation, it is locally centrosymmetric around the C–C bond, and the CH_2 symmetric stretching mode at ca. 2850 cm^{-1} is SFG inactive and does not appear in the spectra. When the chains are disordered because of gauche defects (i.e., with left-hand chirality), this local inversion symmetry is lifted, and a peak at 2850 cm^{-1} will appear in the spectra at an intensity that depends on the degree of disorder. The ratio of intensities of the CH_3 symmetric stretching to CH_2 symmetric stretching modes can be used to provide a relative measurement of hydrocarbon chain ordering.

The recent reviews by Smart et al. [28,29] describe the principles of these techniques and provide examples of their application to mineral systems.

15.4.7 SURFACE REACTIONS

Some of the entries under this heading in Table 15.1 will be illustrated here with particular examples. The variety of surface reactions on minerals, and the techniques used for their study, is far too broad for all aspects to be included in this chapter. Several monographs [1–3,15,86] and a review [4] provide many relevant case studies.

The surface oxidation of sulfide minerals has been reviewed recently by Smart et al. [28]. Studies of the physical and chemical forms of oxidation products by SAM, XPS, STM, AFM, SEM, and ToF-SIMS have revealed several different processes of oxidation. The seminal work of Buckley et al. [102,103] was the first to identify the process of formation of oxyhydroxide products on underlying metal-deficient, sulfur-rich layers of similar forms to those described in Section 15.4.3. Other oxidation products have been observed directly, such as polysulfides, elemental sulfur, oxidized fine sulfide particles, colloidal hydroxide particles, and flocculated aggregates, as well as continuous surface layers of oxyhydroxide species (of variable depth), and sulfate and carbonate species. Different spatial distributions (i.e., isolated, patchwise or face-specific products) have already been described in this chapter. The above review also identified the actions of adsorbed molecules in several different modes, such as adsorption at specific surface sites, adsorption or precipitation of colloidal species from solution, detachment by competitive adsorption of small particles from surfaces of larger particles, detachment of small oxyhydroxide particles, removal of adsorbed amorphous oxidized surface layers, inhibition of surface oxidation, disaggregation of larger particles, and patchwise or face-specific coverage.

Several reviews have described leaching, dissolution, and precipitation reactions for oxides [104,105], silicates, aluminosilicates [16,17,106], and titanates [4,107], and point in particular to the necessity for studying both solution and surface phases to understand fully the mechanisms governing each reaction. For instance, while solution analysis may suggest that simple ion exchange is occurring at the surface, surface analysis has shown that bond-breaking reactions in many cases precede ion exchange and leaching, with the products of the reaction being retained *in situ* at the surface. The kinetics of leaching may therefore be governed by that of the hydrolysis reaction (often base-catalyzed), and not by ion exchange (i.e., due to diffusion limitations). *In situ* phase transformation and precipitate formation are also observed to take place without significant alterations to the solution phase, as in the transformation of $CaTiO_3$ to TiO_2 as brookite and anatase [107]. Precipitation on the same site of layers with different compositions has been observed in SAM depth profiles, which show that sequential saturation of the solution will occur with nucleated growth at crystallites [107].

An example of surface phase transformation can be found in the work of Jones et al. [108]. Pyrrhotite ($Fe_{1-x}S$) surfaces in acid solution restructured to a crystalline, defective, tetragonal Fe_2S_3 surface phase due to loss of Fe to solution. The metal-deficient, S-rich product was identified by XRD in combination with XPS. Linear chains of sulfur with a nearest neighbor distance similar to that of elemental S were observed, but with an S 2p BE 0.2–0.6 eV less than that of elemental sulfur (S_8). As a result of the oxidation process, hydrophilic iron hydroxides were found at the surface.

15.4.8 SURFACE MODIFICATION

An example of the surface modification of minerals will be discussed to illustrate the merits of a combined-technique approach. The breadth of the subject constrains the discussion to that dealing with strategies for gathering information and for choosing the most relevant surface analytical techniques.

Surface modification of the mineral kaolinite ($Al_2Si_2O_5(OH)_4$) arising from low temperature water vapor plasma reactions, followed by adsorption, can produce tailored interfaces ranging from fully hydrophilic to fully hydrophobic [109]. The surfaces of the basal planes of kaolinite, containing siloxane (Si-O-Si-) and aluminum hydroxide (Al-OH) structures, are relatively inert, with the edge sites of the platelets being most reactive for the unmodified mineral. Reaction in a water vapor plasma (for a duration of 3 h) renders the surfaces reactive due to an increase in the number of sites now apparently located on the basal planes, as well as at the edges. The reaction could be monitored by FTIR analysis based on the absorbance of a new peak at 1400 cm^{-1}. The reacted surface had a contact angle of 0°, and the kaolinite then underwent total dispersion in aqueous solution. The activated surface could be rendered irreversibly hydrophobic by reaction with conventional adsorbates, such as monochlorosilanes. The uptake of the latter could be monitored by measuring the two resolvable contributions to the Si 2p XPS spectrum. SEM and TEM images failed to reveal any observable changes to the morphology of the kaolinite platelets as a result of plasma reaction or adsorption. Molecular modeling studies [110] suggested that the plasma reaction led to the attachment of -OH species at the Si sites in the siloxane basal plane, rather than to the breaking of Si-O-Si bonds. The energetics of this attachment route are more favorable than those of the bond-breaking reaction, in contradiction to earlier interpretations. These studies illustrate clearly the power of molecular modeling for arriving at alternative interpretations of experimental results.

15.5 CERAMICS

15.5.1 PHASE STRUCTURES

Multiphase titanate ceramics designed to immobilize high level nuclear waste have been characterized by SEM, TEM, electron diffraction, XRD, XPS, SIMS, SAM, and dissolution studies [4,111–113]. SEM characterization at low magnification in SE and BSE imaging modes to reveal phase distribution has been illustrated earlier in Figure 15.1. Phase distribution at higher spatial resolution can be revealed by TEM analysis. An example is given in Figure 15.14 for synroc-C (a multiphase titanate waste form ceramic) [114]. Identification of the phase structure of individual grains by EDS has been described [111–113]. In the same publications, other examples involving twinned structures and phase transformations, and the use of special techniques such as selected area diffraction (SAD) and lattice imaging, have been described. Figure 15.15 reveals the structural intergrowth of pyrochlore [$(Ca,Gd,U,Pu)_2(Ti,Hf)_2O_7$] and zirconolite [$Ca(Hf,Gd,U,Pu)Ti_2O_7$] in another titanate waste form containing plutonium using high-resolution TEM and SAD [114].

Resolution of phase structures within individual grains at the level of the unit cell is similar to that described in Section 15.4.2 using STM and AFM techniques. The application of SPM to ceramic materials has expanded greatly in recent years. SPM-based techniques are employed not only for surface morphology measurements, but also for studying the properties of materials on the nanometer and atomic scales. A recent paper by Kalinin et al. [115] provides an excellent review of the use

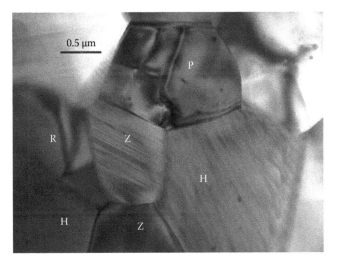

FIGURE 15.14   TEM image showing the microstructure of synroc-C prepared by hot pressing under reducing conditions. Note the lack of porosity, the lamellar structure in zirconolite (Z), and the twin domains in perovskite (P). Other phases are hollandite (H) and rutile (R). Excluding waste species, the formulations are nominally $CaZrTi_2O_7$ for zirconolite, $CaTiO_3$ for perovskite, $BaAl_2Ti_6O_{16}$ for hollandite, and TiO_2 for rutile. (From Lumpkin, G.R., *Elements*, 2, 365, 2006. With permission.)

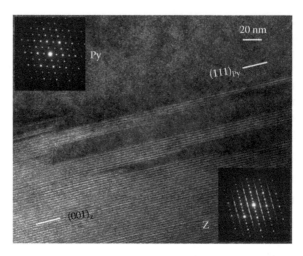

FIGURE 15.15 High-resolution TEM image of a pyrochlore-based titanate waste form containing pluto-nium, showing the structural intergrowth of pyrochlore (Py) and zirconolite (Z). The corresponding SAD patterns are also shown. (From Lumpkin, G.R., *Elements*, 2, 365, 2006. With permission.)

of "advanced" SPM to study local transport, and electronic, mechanical, and optical properties not just in "model" systems (i.e., single or bi-crystals) but in "real" polycrystalline materials as well.

The application of nuclear methods to studies of ceramic surfaces and interfaces is well developed. Relevant applications of Rutherford backscattering spectroscopy (RBS), particle induced x-ray emis-sion (PIXE), neutron reaction analysis (NRA), elastic recoil detection analysis (ERDA), and high-resolution α-spectroscopy have been described in the review by Matzke [116]. Analyses by RBS, in the channeling mode, and NRA, have revealed structural defects in U_4O_{9-y} [117]. RBS is sensitive mainly to the metal atom sublattice, whereas NRA can be used selectively to study the nonmetal sublattice. It was shown that the U sublattice is nearly identical to that of UO_2, whereas significant cluster formation occurs in the O sublattice, these observations having implications for surface sites in the material. Investigations of surface reactions (e.g., leaching and dissolution) of titanates and UO_2 have also been carried out by nuclear techniques [118–120]. Studies by ERDA, with He^+, of the hydra-tion of titanates in water at 150°C [118,119], showed uptake of hydrogen in the first ca. 80 nm, in agreement with SIMS analysis [121]. RBS channeling experiments also revealed the incorporation of additional oxygen atoms into the UO_2 structure due to leaching, resulting in a structural transformation of the surface layer from cubic UO_2 to a monoclinic $UO_{2.28\pm0.05}$ phase [122].

The effects of ion beam damage in ceramics, particularly the formation and annealing of defects introduced during ion implantation, can be studied by RBS channeling techniques as well. High-dose ion implantation may cause ceramic surfaces to become metamict (i.e., radiation-damaged) or amorphous. In the case of $CaTiO_3$, after Pb-implantation, loss of crystallinity was observed to a depth of ca. 0.2 μm [123]. After annealing at 425°C the amorphous/crystalline interface, which was sharp and planar over lateral dimensions of 100 nm, moved some 0.1 μm toward the surface. Chan-neling analysis [123] was also able to identify the movement of two interfaces in the surface region of implanted Al_2O_3 crystals during annealing; an amorphous phase crystallizing into γ-Al_2O_3, and the transformation of the γ- to the α-phase at a well defined interface.

15.5.2 SURFACE STRUCTURES

The information that can be derived for ceramic surface structures, and the necessary methodologies, are similar to those described in Section 15.4.2 for surface structures on minerals. Practical applica-tions of AFM on ceramic surfaces have been described by Yamana et al. [124]. Filtering by the Fast Fourier Transform (FFT) method for enhancement of atomic structure has been demonstrated in the case of a fluoride-containing mica $(KMg_3(Si_3AlO_{10})F_2)$ surface of a mica-glass ceramic, as

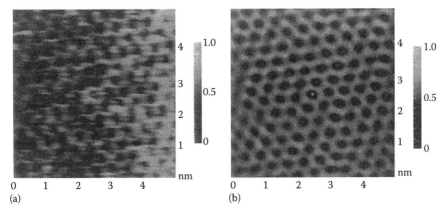

(a)   (b)

FIGURE 15.16 (a) AFM image of fluoro-mica cleavage. Note that noise in the image recording made atomic detail difficult to analyze. (b) The same image after removal of noise by FFT filtering. (From Yamana, K., Miyamoto, M., Nakamura, S., and Kihara, K., in Lutze, W. and Ewing, R.C. (Eds.), *Scientific Basis for Nuclear Waster Management XII*, Materials Research Society Symposium Proceedings, Vol. 127, Pittsburgh, PA, 1989, 57. With permission.)

seen in Figure 15.16. A zirconia-dispersed mica/glass ceramic, formed from zircon ($ZrSiO_4$) added to the mixture of starting materials (resulting in both mica and zirconia grains in the fluorine-containing glass matrix), was also examined [124]. The topography of a surface resulting from cutting, grinding, and polishing can reveal patterns characteristic of the fracture or abrasion mechanisms. It was demonstrated that fracture and cutting occur not only at grain boundaries, but also within the mica grains. Networks of cracks were formed by compressive pressure during grinding, and the loss of zirconia grains appeared to be due to polishing. A recent study of polycrystalline $CaTiO_3$ ceramic surfaces reported that thermal annealing is an effective method of removing surface damage caused by mechanical polishing [125]. The extent of damage was examined by cross-sectional TEM, revealing that the damage extended to a depth of up to ca. $0.1\,\mu m$ for an as-polished surface. For a subsequently annealed surface, high-resolution TEM showed lattice fringes extending all the way to the surface. However, the annealed surface was no longer flat, as a result of microfaceting. Figure 15.17 shows the AFM and TEM images of a polished $CaTiO_3$ surface following subsequent annealing, showing terraces and steps on the surface of each grain and their crystallographic orientations.

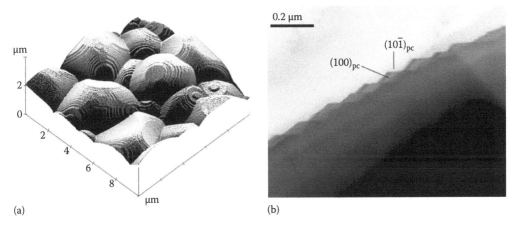

(a)   (b)

FIGURE 15.17 (a) AFM and (b) TEM images of a mechanically polished polycrystalline perovskite ($CaTiO_3$) surface following subsequent annealing at 1320°C for 2 h. The crystallographic orientation of the terraces and steps on the surface was determined by TEM/SAD (not shown here). The subscript "pc" indicates that the assignations were set in terms of pseudo-cubic perovskite symmetry. (From Zhang, Z., Blackford, M.G., Lumpkin, G.R., Smith, K.E., and Vance, E.R., *J. Mater. Res.*, 20, 2462, 2005. With permission.)

15.5.3 SURFACE SITES

The examples concerning minerals discussed in Section 15.4.3 are also illustrative of the information required about surface sites in ceramic materials. The techniques applied on a routine basis to obtain this information are essentially the same.

It is worthwhile emphasizing the extensive work that has been carried out with LEIS on ceramic surface sites [13,14]. The review by Brongersma et al. [14] summarized studies of composition, structure, diffusion, segregation, growth, adsorption, desorption, depth profiling, sputtering, neutralization, annealing, oxidation/reduction, and dispersion, for materials such as SiO_2, Al_2O_3, MgO, ZrO_2, VO_x, SnO_2, FeO_x, and high temperature superconductors, and for other ceramics and oxides. In addition, cation and anion sites on MgO {100} surfaces, surface-segregated Ca impurity sites, and surface defects were investigated. Site labeling in SnO_2 with oxygen isotopes allowed identification of the positions of those O atoms bridging Sn sites, that were selectively removed by heating for 3 min in vacuum at 700 K. Amorphous and polycrystalline ceramic materials were also studied. In the case of spinels ($MgAl_2O_4$), cations in tetrahedral and octahedral sites could be identified. Surface segregation taking place during oxidation, e.g., Fe in Al_2O_3, and as a result of ion bombardment (e.g., alkali metal diffusion), can cause profound changes to surface sites, which will affect subsequent reactions.

The phenomena of growth and wetting during oxidation of metals and semiconductors, or during growth of oxides on oxide supports, have also been considered in the literature dealing with the use of LEIS [14]. Due to its unique capability of selectively probing the outermost surface layer (i.e., with much higher surface sensitivity than other techniques such as XPS and AES), LEIS is particularly valuable in elucidating the segregation-induced surface phenomena. Based on their recent LEIS work, de Ridder et al. have proposed a new four-layered surface model for 10 mol% yttria-stabilized zirconia (YSZ), which is a common electrolyte material used in solid oxide fuel cells (SOFC) [126]. In their model, the top monolayer consists entirely of impurity oxides segregated from the bulk during sintering, the second monolayer is enriched in yttria, the next sublayer of ca. 6 nm in thickness is depleted in yttria (possibly as monoclinic YSZ), followed by the cubic bulk phase. They also performed oxygen isotope exchange experiments, and suggested that the operating temperature of an SOFC can be significantly reduced if the bulk impurities are prevented from segregating to the surface. The presence of an yttria-depleted sublayer is, however, inconsistent with previously reported results. This discrepancy has been discussed by Nowotny et al. and attributed to nonequilibrium segregation associated with sample cleaning procedures used in the LEIS study [127].

As with mineral surfaces, the matching of theoretical models to experimental data by molecular modeling methods is a powerful tool for the prediction of adsorption and surface reactivity on ceramics. The reviews by Henrich [86], Kunz [87], Stoneham and Tasker [128], and Colbourn and Mackrodt [78,129], all discuss the applications of theoretical techniques and their correlations with experimental results.

15.5.4 GRAIN BOUNDARIES AND INTERGRANULAR FILMS

Studies of interfaces in ceramic materials have a long and distinguished history. The merit of using TEM to observe boundaries between grains of the same phase can be illustrated by Figure 15.18a, in which the lattice spacings can be seen in adjacent Magnéli phases (Ti_nO_{2n-1}) in a titanate ceramic. The boundaries between the phases were found to be coherent in this case, without the presence of intergranular films; similar coherent boundaries were also reported by Ball et al. [111]. Intergranular glassy films, however, can be present between grains of the same phase, such as the thin (0.5–0.9 nm) disordered layers found in doped alumina incorporating Si and Y impurities/dopants [130]. Other examples include the studies of grain boundaries in ceramics by Ruhle [131]. A series of systematic studies of intergranular films by Clarke [132–134] established the formation of interphase regions differing in structure and composition from the adjacent crystalline grains. The equilibrium thickness of intergranular films is typically of the order of 1 nm in ceramic materials [135]. An example of the imaging of thin glassy films is shown in Figure 15.18b in which the apparently amorphous

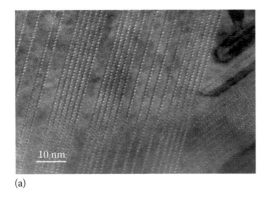
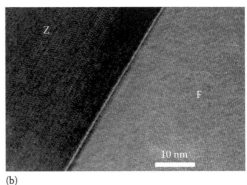

(a)   (b)

FIGURE 15.18 HRTEM images of a titanate ceramic waste form, showing (a) a coherent grain boundary between Magnéli (Ti_nO_{2n-1}) phases and (b) a disordered grain boundary between zirconolite (Z) and defect fluorite (F) phases. The formulations were nominally $(Ca,Pu)(Zr,Hf)Ti_2O_7$ for zirconolite and $(Ca,Zr,Hf,Ti)_4O_7$ for defect fluorite. (Courtesy of Lumpkin, G.R. and Blackford, M.G., Australian Nuclear Science and Technology Organisation, unpublished data.)

film, ca. 1 nm wide, between zirconolite and defect fluorite grains, can be seen in a titanate ceramic waste form. Intergranular films often promote the concentration of impurities and some minor constituents by segregation. The subject of equilibrium segregation at such interfaces has been reviewed by Cabane and Cabane [136].

The application of surface analytical techniques to the study of fracture faces, at which grain boundaries and intergranular films are preferentially exposed, also has a well-established literature (e.g., Refs. [137,138]). Observations by SSIMS of alkali metal segregation into intergranular regions of ceramics and glasses [139] have demonstrated enhancement of Cs and Na, but not of Ca or Ti, as shown by the depth profile in Figure 15.19. SAM elemental maps of the distribution of Cs around microvoids in fracture faces showed high concentrations of the volatile Cs species due to trapping in the voids and segregation into the intergranular film regions [137]. XPS can also be used to determine the composition, chemical states, and bonding of elements exposed in the fracture faces. If intergranular films are very thin (i.e., less than 2 nm), the XPS signal is likely to include contributions from the underlying crystalline grains, but consistency between SIMS, XPS, and SAM analyses has been demonstrated in these studies.

Many ceramics contain a variety of dopants, the concentration of which in the starting material is calculated according to the desired bulk chemical formula. However, due to dopant segregation to

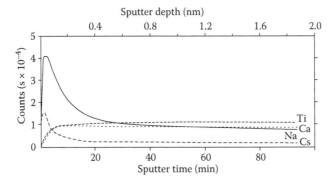

FIGURE 15.19 SSIMS depth profile of a fresh fracture face of the ceramic synroc-C, demonstrating enhancement of Cs and Na, but not Ca and Ti, in the intergranular region. (From Cooper, J., Cousens, D.R., Hanna, J.A., Lewis, R.A., Myhra, S., Segall, R.L., Smart, R.St.C., Turner, P.S., and White, T.J., *J. Am. Ceram. Soc.*, 69, 347, 1986; Smart, R.St.C., *Appl. Surf. Sci.*, 22/23, 90, 1985.)

grain boundaries (and external surfaces) during the high temperature fabrication process, the actual dopant concentration in the bulk may be lower than its target value. The extent of deviation could be significant when the dopant concentration is low, as demonstrated in a recent study of Ca/U-doped thorutite $(Ca_xU_{2x}Th_{1-3x})Ti_2O_6$, where $x = 0.015$, 0.05, and 0.15 [140]. Substitution of Ca^{2+} on the Th^{4+}/U^{4+} sites in the lattice would be expected to cause oxidation of all the uranium from U^{4+} to U^{5+} if there were no Ca^{2+} segregation. However, it was discovered by both XPS and diffuse reflectance spectroscopy that oxidation from U^{4+} to U^{5+} was incomplete, suggesting that the amount of Ca^{2+} inside each grain was less than the nominal value. This incomplete conversion of U^{4+} to U^{5+} was attributed to the segregation of Ca^{2+} to the grain boundaries, since the Ca concentration on the fractured surface, which contained more grain boundary regions, was approximately 50% higher than that on the polished surface. The residual amount of U^{4+} in the bulk was quantified by XPS analysis of the polished surfaces, and found to decrease with increasing x: $U^{4+}/(U^{4+} + U^{5+}) \sim 81\%$, 64%, and 30% for $x = 0.015$, 0.05, and 0.15, respectively. This trend is consistent with Ca-segregation, since a higher percentage of the Ca dopant is expected to segregate to the grain boundaries with decreasing dopant concentration in the bulk (x) (i.e., the enrichment factor, defined as $C_{grain\ boundary}/C_{bulk}$, is expected to increase with decreasing x) [141]. Therefore, when the dopant level is low, the actual concentration within each grain is often lower than its nominal value. This segregation-induced effect is illustrated schematically in Figure 15.20 for the cases of both high and low dopant concentrations [142].

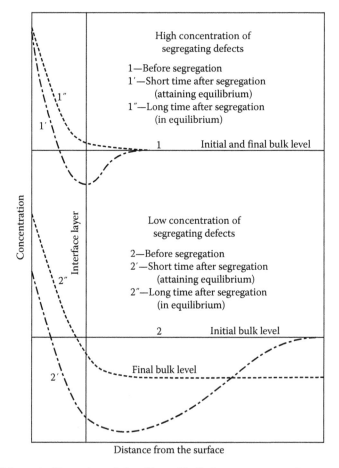

FIGURE 15.20 Schematic illustration of the effect of bulk dopant concentration on segregation-induced concentration gradients within the interface layer and the bulk during different stages of segregation. (From Nowotny, J., Bak, T., Nowotny, M.K., and Sarrell, C.C., *Adv. Appl. Ceram.*, 104, 165, 2005. With permission.)

15.5.5 Depth Profiles

Examples of SIMS depth profiles obtained from minerals and ceramics have already been given, e.g., Figures 15.13 and 15.19 [92,137,139]. Depth profiling, with nuclear techniques such as RBS-channeling, and with ARXPS, on large flat single crystal surfaces can be carried out without suffering the uncertainties introduced by ion beam sputtering. These techniques were used by Morais et al. [143] to investigate the effect of postdeposition annealing on the atomic transport and chemical stability of zirconium silicate films on a Si (001) substrate. The depth distributions of Si, O, and Zr in the thin films (ca. 9 nm thick) were determined by RBS, nuclear resonance profiling, and LEIS, while the chemical environments of these elements in the near-surface and near-interface regions were revealed by ARXPS. It was shown that, although the interface region was stable, Si had segregated to the surface region after thermal annealing.

15.5.6 Adsorption

Studies of the adsorption of long chain alcohols and surfactant molecules on silica surfaces, using a combination of XPS, FTIR, SEM, SAM, and contact angle measurements, have provided information on the quantity, distribution, structure, and chemistry of adsorbed layers [109,144–146]. Adsorption of primary alcohols on silica powder surfaces can be measured quantitatively using DRIFT in the absorbance mode of operation. For instance, quantitative measurements [144–146] on stearyl alcohol ($C_{18}H_{37}OH$) have given densities of between 0.9 and 1.4 molecules nm^{-2}, depending on the absorbance band chosen for quantification (e.g., 3740, 2930, or 1460 cm^{-1}). Pyrolysis experiments [144] give an average value of ca. 0.8 molecules nm^{-2}. Previous studies [147] of silica powder surfaces have suggested that there should be 4.6 reactive silanol groups nm^{-2} accessible for reaction. DRIFT and pyrolysis results therefore suggest that less than 25% of the surface silanol groups had reacted with adsorbed stearyl alcohol molecules. Estimations of film thickness and surface coverage can be derived from overlayer equations applied to XPS signal intensities, from both adsorbed long chain alcohols and the silica substrate. A series of long chain alcohols of differing lengths was adsorbed on silica plates [145,146]. XPS at a fixed angle combined with variation of the film thickness, and in the angle-resolved mode at constant film thickness, was used to determine the thickness of the adsorbed layer. The combined results from XPS, FTIR, pyrolysis, and contact angles showed that adsorption was in the form of patchy, disordered, and multilayered structures, rather than as an ordered-oriented monolayer. For instance, at an exit angle of only 5°, with a sampling depth of less than 1 nm (using $C_{22}H_{45}OH$), ARXPS revealed 3 at.% Si and 11 at.% O, indicating that exposed areas of silica had not reacted with the alcohol molecules.

The interaction of water vapor with metal oxide surfaces has been studied widely due to its technological importance. Many of the investigations have been conducted on single crystal samples using surface techniques operating under UHV conditions. Such studies have the advantage of probing the reaction at the molecular level on clean and well-ordered surfaces; many examples have been provided in a review by Brown et al. [148], which also described adsorption studies on polycrystalline and amorphous samples under ambient conditions using solid-state NMR.

15.5.7 Surface Reactions

Reactions involving hydrolysis, leaching, dissolution, precipitation, and *in-situ* phase transformation have been discussed in reviews by Myhra et al. [4,107], Blesa et al. [105], Casey [17], Brown et al. [16], and Schoonheydt [106]. The strategies, methodologies, and techniques are essentially the same as those described in Section 15.4.7. SEM, AFM, TEM, XPS, and solution analysis were employed to investigate the dissolution behavior of polycrystalline perovskite $CaTiO_3$ surfaces in deionized water [125]. Anatase crystals were found to form readily on both polished and annealed $CaTiO_3$ (i.e., defect-free) surfaces after dissolution testing at 150°C. Although surface damage did

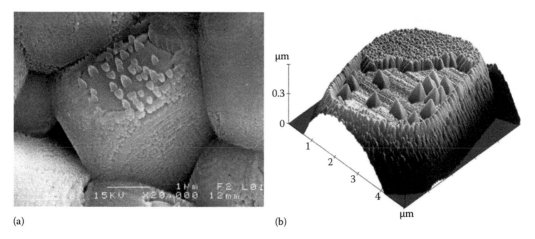

(a)   (b)

FIGURE 15.21   (a) SEM and (b) AFM images of a thermally annealed polycrystalline CaTiO$_3$ surface after dissolution testing in deionized water at 150°C for 1 week, showing individual anatase crystals grown in preferred orientations with respect to the underlying perovskite grains. Note that the shapes of the anatase crystals are not depicted accurately in the AFM image as no attempt was made to deconvolute tip shape (which was of comparable size to the anatase crystals). (From Zhang, Z., Blackford, M.G., Lumpkin, G.R., Smith, K.E., and Vance, E.R., *J. Mater. Res.*, 20, 2462, 2005. With permission.)

not result in a notable increase in the total amount of Ca released from perovskite in the course of a week, the removal of surface damage by annealing did lead to the spatial ordering of the alteration products on perovskite surfaces. The alteration layer on the annealed surface was composed of well-aligned anatase crystals grown in preferred directions with respect to the underlying perovskite, as shown in Figure 15.21a and b, in contrast to the inhomogeneous layer observed on the mechanically polished surface (not shown here). The preferred orientation of anatase crystals on perovskite is attributed to the similarity between their anionic sublattices, despite their overall crystallographic dissimilarities. This pseudo-epitaxial relationship between anatase and perovskite provides direct experimental evidence for "congruent dissolution" (Ca and Ti released at equal rates, followed by precipitation of anatase) as the dissolution mechanism for perovskite at 150°C.

15.5.8   SURFACE MODIFICATION

Examples of surface modification of ceramic materials resulting from adsorption, plasma, and sol-gel reactions have been reviewed by Smart et al. [109]. There are many other examples [2,3,15,86,149].

The so-called modified Auger Parameter α^* is often more useful in identifying chemical state than shifts in the BEs as measured in XPS, or in Auger KEs. It is defined as

$$\alpha^* = KE \text{ (Auger)} + BE \text{ (XPS)} \tag{15.1}$$

where KE is the kinetic energy of the most intense Auger transition, and BE is the binding energy of (normally) the most intense core photoemission peak, from the same element, measured in the same spectrum. The Parameter is independent of surface charging and does not depend on the reference energy (i.e., Fermi or vacuum) used in the spectral analysis. It reflects directly changes in the screening of atoms in different chemical environments, and is particularly useful for silicon. An example of its use is that of the study of silicate structures produced in thin oxide layers on metal surfaces, by low temperature plasma reactions [109,150]. Oxide layers of predetermined thickness can be produced with an H$_2$O vapor plasma followed by reaction with tetraethoxysilane TEOS (Si(OC$_2$H$_5$)$_4$):H$_2$O plasma products. Mass spectra [150] of the plasma have demonstrated the presence of reactive SiO species that produce a sequence of silicate structures ranging from orthosilicate ($SiO_4^{[2-]}$), pyrosilicate ($SiO_7^{[6-]}$), and other oligomeric structures, to layer silicates and bulk silica, depending on the duration of reaction. In addition to identifying the changes in chemical states with progressive reaction (i.e., increasing Si concentration in the surface layers), α^* could also be correlated with the change in concentration of oxygen in the film. There was an increase in the O/Si ratio as the plasma reaction proceeded from $SiO_{1.5}$ to SiO_2 (Figure 15.22).

The plasma reaction produces a reaction layer functionally graded in composition, structure, hardness, and reactivity from the nickel metal to its oxide with silicate species and to bulk silica. These ceramic oxide/silicate/silica surface layers are strongly resistant to hydrolysis and acid attack. The dissolution rate of the plasma-treated nickel in pH 2 acid at 60°C is reduced by factors of more than 300 in comparison with the uncoated oxidized metal. Similar functionally graded layers, reviewed in Ref. [152], can be formed on Ti, stainless steel, and brass. Earlier studies, correlating

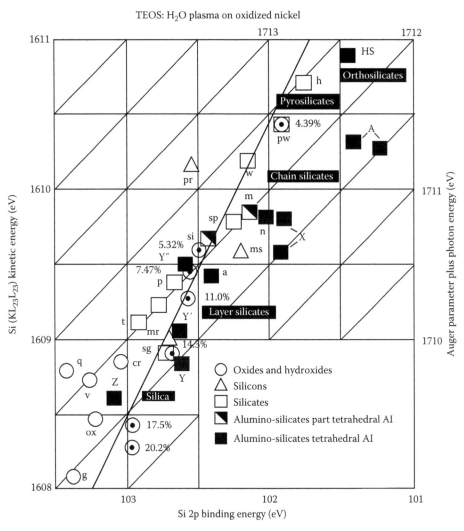

FIGURE 15.22 Modified Auger Parameter plot for a number of oxy-silicon species specified in the insert (symbols as in Ref. [151]), and plasma-reacted (TEOS:H₂O) nickel surfaces; % values on bulls-eye symbols show the Si atomic concentrations of a sequence of surfaces plasma reacted for increasing times (1–20 min) measured by XPS. Note the transition from pyrosilicate (SiO_7^{6-}) structures to bulk silica (SiO_2) species as the reaction progresses. (From Arora, P. and Smart, R.St.C., *Surf. Interface Anal.*, 24, 539, 1996. With permission.)

XPS, FTIR, XRD, and TEM observations, established that the silicate reaction also takes place on bulk nickel oxide surfaces [153]. However, the use of α^* provided evidence more directly for the change in chemical form of the silicate species in the surface layer [109,150].

15.6 GLASSES

Many of the examples concerning mineral and ceramic surfaces can be used equally well to illustrate information, measurements, and processes for glass surfaces. The reader will find it useful to consider examples under the same headings in Sections 15.4 and 15.5 in conjunction with those presented here, which have been selected to illustrate processes that have some generality in glass surface reactions and modifications.

15.6.1 SURFACE COMPOSITION

The bulk structural constituents of glasses consist of bridging and nonbridging oxygen atoms (e.g., -Si-O-Si- and -Si-O), other network forming elements replacing Si (e.g., Al and B), network modifying cations (e.g., Fe and Ca), and ion exchangeable cations (e.g., Na and K). The structure of glasses is relatively complex, and yet flexible, in comparison with crystalline structures; the locations of alkali and alkaline earth ions in holes in the Si–O structure exhibit an irregular pattern of repetition, and there is therefore no characteristic chemical composition or short-range order. Discussion of the relationship between structure and composition can be found in reviews, e.g., Henderson et al. [154], Galoisy [155], Paul [156], and Kruger [23]. Recent research by Dalby et al. [157,158] has found links between the detailed analyses of FTIR and XPS spectra from silicate glasses, giving strong evidence for the changing structure of bonding with dopant content.

Surface composition and the nature of surface sites are governed by the fact that planar termination of the bulk structure, even for flat defect-free surfaces, does not produce an equilibrium structure or composition similar to that of the bulk material. The surface and subsurface layers exhibit gradual changes down to relatively large depths, which may correspond to thousands of atoms, from the surface into the bulk, thus causing even a rigid glass to allow ingress and hence reaction with small molecules at those depths. Reacted surface layers with thicknesses in the micrometer range will result from the "microporosity" of the subsurface layer. Another consequence of the structural changes with depth is that small diameter glass fibers differ in structure and composition from the glass material used in their manufacture; their chemistries may not therefore resemble that of the bulk glass.

The observations above can be illustrated by a compositional analysis by XPS of E-glass surfaces of fibers and plates; the results, and a comparison with nominal bulk composition, can be found in Table 15.3 [159]. The surface compositions of the fibers and plates were significantly different from the nominal composition of the glass; there was depletion of B, Mg, Al, Na, and, particularly for the plate, of Ca species. The fibers showed higher surface concentrations of Si and O in comparison with bulk composition, while the plates showed surface enhancement of O (as hydroxyl groups) without Si enhancement. The surface elements were apparently lost, or depleted, by segregation or dissolution, during manufacture of the fiber and plate specimens. The glasses included all three types of species, namely, network formers, network modifiers, and ion exchangeable cations. FTIR studies of surface functionality, as related to surface composition and surface sites, have been covered in the literature, much of which has been reviewed by Kruger [23]. Assignments of bridging and nonbridging oxygens to hydroxyl groups associated with different cations can now be made with reasonable confidence using FTIR. Dehydroxylation during heating to higher temperatures has also been followed by FTIR. SIMS [139], NMR [160], LEIS [14], and nuclear techniques [116] complement XPS and FTIR studies of the surface compositions of glasses. In view of the discussion above, it is clear that processing and pretreatment, as well as exposure to air, will result in the compositions of surfaces of glass being different from those of the bulk.

TABLE 15.3
Composition of E-Glass as Determined from the Constituents Used in the Manufacturing Process and by XPS[a] [159]

	Composition (at.%)		
		XPS	
Element	Nominal	Fiber	Plate
O	61.7	66.2	72.7
Si	18.8	21.9	17.9
Ca	7.9	7.1	2.6
Al	5.9	4.3	4.7
Na/K	1.1	0.5[b]	0.4[b]
B	4.4	<0.05	1.7
Mg	0.3	<0.05	<0.05

[a] XPS percentages were adjusted after the carbon contamination was taken as zero.

[b] Atomic percentages of sodium only; no potassium was found on the glass fiber or plate surface.

15.6.2 SURFACE SITES

Much of the information from studies of surface composition can be related to the chemical environments of different surface sites. AFM with nanometer resolution can give information on the distribution of surface sites, and on surface topography and surface reactions, including those leading to the formation of hydrolyzed, leached surface layers (e.g., [124]). Because glasses are insulating, it is difficult to assign sites unequivocally to specific atoms, or to map the distribution of a particular elemental species on a glass surface. Thus STS and Auger techniques cannot be used. Much of the information on the chemistry and reactivity of specific surface sites has to be inferred from techniques which average over relatively large surface regions, e.g., XPS, FTIR, and ToF-SIMS. LEIS offers the most atom-specific information about surface composition and individual sites.

15.6.3 DEPTH PROFILES

Depth profiling glasses with ion beam techniques is also difficult due to surface charging problems. Pulsed ion beams in ToF-SIMS with relatively long inter-pulse periods have been used successfully, however. ARXPS can also be used to study variations in composition and chemical states over the outer 10 nm. Nuclear techniques, such as PIXE, NRA, RBS, and ERDA, are particularly useful in studying the composition of deeper-lying altered layers to depths of micrometer. The latter techniques generally have depth resolutions of the order of 10 nm which is sufficient in most cases [116].

15.6.4 ADSORPTION

Adsorption on glass surfaces is an important area of investigation. A good example is that of composites consisting of glass fibers in inorganic or polymer matrices, where there is concern about the coupling agents designed to reduce the effects of water on the mechanical properties of the composites. Section 15.6.1 noted that glass surfaces may undergo hydrolysis during manufacture, and as a result of exposure to air; the result is the production of a surface layer with a chemical composition different from that of the bulk [159]. Diffusion and reaction at the glass–matrix interface may lead to delamination and to loss of strength of glass-reinforced composites [161,162] unless a suitable adsorbed layer is incorporated at the interface. Silanes are the most common coupling agents used to provide a water resistant interfacial bond between glass fibers and organic polymer matrices. Surface modification using siloxane alternatives, instead of silanes, have been discussed in Section 15.6.6.

For many applications, however, it is sufficient to render the glass surface hydrophobic by adsorption of simpler and cheaper reagents. Long-chain alcohols adsorb strongly on silica surfaces (see Section 15.5.6); adsorption of these alcohols on E-glass plates, fibers, and powders has been studied [159,163]. Alcohols with chain lengths C_{12}, C_{14}, C_{18} were reacted at 130°C with the glass surfaces. Analysis by FTIR of the methylene absorption peak at 2950 cm^{-1} showed that the C_{14} and C_{18} alcohols adsorbed strongly on the glass surface, but the C_{12} alcohol adsorption resulted in lower surface coverage. Even after seven extraction cycles in warm cyclohexane, it was found that all three alcohols retained an absorbance value of about 0.01, and high contact angles (85°–97°) were measured. These angles were those expected for methylene groups rather than those (112°) characteristic of highly oriented monolayers exposing predominantly methyl groups. Quantitative DRIFT analysis gave results agreeing with the 0.01 methylene absorbance, indicating a relatively high surface coverage of 1–5 monolayer equivalents, and suggesting disordered multilayer adsorption of the alcohols. ARXPS analyses of C 1s and Si 2p signals also suggested that complete coverage of the glass surface was unlikely. ARXPS measurements were also carried out on E-glass plates and fibers before and after seven cyclohexane washing cycles. The clean surfaces of the plates and fibers showed the "normal" hydrocarbon contamination acquired as a result of exposure to air (ca. 20–25 at.% C at 45°), but the concentrations of hydrocarbons on the alcohol-reacted surfaces were considerably greater. There was a systematic increase in carbon concentration before washing, with increasing alcohol chain length, at all take-off angles. Washing had the effect of producing similar thicknesses of hydrocarbon for each of the alcohols, suggesting that the stable residual adsorbed layers, or regions, had similar structures with respect to thickness and molecular orientation. Analyses of the Si 2p spectra showed lower silica signals at all take-off angles on the alcohol-treated samples in comparison with clean samples but there was still a considerable presence of silica at the 10° angle, i.e., 25%–55% of the signal obtained from clean surfaces, even though the depth of analysis was of the order of the molecular size of the alcohol molecules.

The combined results from DRIFT, contact angle, and ARXPS analyses showed no evidence for a high degree of orientation of the adsorbed alcohols; instead disordered multilayers in separate regions or islands could be inferred. Nevertheless, major changes in surface composition and hydrophobicity were measured despite apparently incomplete monolayer coverage.

A previous study combining XPS, streaming zeta potential, and Wilhelmy plate wetting [164] confirmed that substantial adsorption and alteration of surface properties took place with only fractional surface coverage at very low surfactant concentrations (i.e., 0.1 wt.%). These results again support an interpretation in terms of patchwise adsorption on relatively thick multilayered regions.

15.6.5 SURFACE REACTIONS

Most studies of surface reactions on glasses have been concerned with the three processes of leaching or ion exchange, base-catalyzed hydrolysis of the glass network, and *in situ* phase transformation or reprecipitation. Reviews of the techniques applicable to physical characterization of glass surfaces can be found in the work of Hench and Clark [165,166], Myhra et al. [4], Kruger [23], and references therein.

TABLE 15.4

Nominal Composition (wt.%) of Elements in PNL 76-68 Glass (at.% Values in Brackets) with Leach Rates (g m⁻² day⁻¹) of Selected Elements in Water at 90°C for 7 days

	Composition	Leach Rate
B	2.9 (4.2)	1.74
O	41.1 (59.0)	
Na	9.6 (10.8)	1.94
Si	18.6 (17.0)	1.47
P	0.2 (0.1)	1.06
Ca	1.4 (1.4)	0.43
Ti	1.8 (1.0)	0.002
Cr	0.3 (0.1)	
Fe	7.3 (2.0)	0.008
Ni	0.2 (0.1)	
Zn	4.0 (2.4)	0.003
Rb	0.1 (<0.05)	
Sr	0.3 (0.2)	0.53
Y	0.2 (<0.05)	
Zr	1.4 (0.4)	
Mo	1.6 (0.3)	1.79
Cs	1.0 (0.2)	2.1
Ba	0.4 (0.1)	0.15
La	0.5 (0.1)	
Ce	1.3 (0.2)	
Pr	0.6 (<0.05)	
Nd	4.7 (0.4)	
Sm	0.4 (<0.05)	
Eu	0.1 (<0.05)	
Mass	100.0	1.01

Source: Myhra, S., Smart, R.St.C., and Turner, P.S., *Scanning Microsc.*, 2, 715, 1988.

The reaction processes can be described by a combination of results from leach tests and SIMS [4]. The nominal composition and leach rates (g m⁻² d⁻¹) of a glass formulation, PNL 76–68 (Pacific North-West Laboratories), used as the standard high-level nuclear waste glass matrix, are shown in Table 15.4. The definition of leach rate used normalizes the loss (in g) to the proportion of that element in the nominal bulk composition. It can be seen that the leach rates for Cs, Na, Mo, Si, and B exceeded the overall rate of mass loss from the glass surface, whereas the elements Fe, Zn, and Ti showed no significant loss to solution under the conditions used. There were also several elements, e.g., Ca, Ba, Sr, and P, that were neither rapidly leached nor apparently retained in the surface layers. Studies described in the above-mentioned reviews, using XPS, FTIR, SIMS, SEM, and dissolution rates, established clearly that the primary reaction occurring in solution is the bond-breaking attack by OH⁻ at Si (or Al and B) atoms of the glass network (i.e., -Si-O-Si-). The resulting -Si-OH and -Si-O⁻ species then react further with water to produce two -SiOH groups and an accompanying release of an OH$^-$ for still further reaction. This bond-opening mechanism allows ion exchange of mobile ions, particularly alkali metal cations. Hence the initial rates of loss to solution reflect release of the exchangeable cations, while the rate of loss of Si is closer to that of the overall loss rate of mass. The surface layer thus becomes depleted in alkali metal cations and hydrolyzed to a silicaceous hydrogen-bonded structure. Those elements which exhibit very low leach rates are retained in the surface as new and relatively insoluble products.

SIMS profiles from a disk of waste glass, recorded before and after hydrothermal attack in doubly distilled, deionized water for 7 d at 300°C, showed the loss of Na, and the partial loss of Si and B [4]. On the other hand, Fe, Ti, and Zn were retained in the surface layer with maximum abundances in the near-surface region. SIMS and FTIR showed the formation of SiOH groups, and SIMS also revealed hydroxylated ions such as CaOH$^+$ and SiOH$^+$. While XRD did not reveal evidence of crystallinity in the reacted surface layer, TEM imaging in combination with SAD found crystalline products such as β-cristoballite (SiO$_2$) and two pyroxene structures similar to wollastonite (CaSiO$_3$) and rhodonite ((Mn,Ca,Fe)SiO$_3$). Several studies of this and other glasses [167–169] have found crystalline products containing Fe and Zn [169]. SEM images of a Zn-containing glass attacked hydrothermally at 170°C for 60 h showed small crystallites of micrometer dimensions, which were shown by TEM and electron diffraction to be zinc hemimorphite (Zn$_4$Si$_2$O$_7$(OH)$_2 \cdot$(H$_2$O)) and zinc orthosilicate (Zn$_2$SiO$_4$). Not all the retained elements were in crystalline form, however, since crystalline structures retaining Mg, Ni, Nd, and Ce have not been identified despite their accumulation in the near-surface, as shown by depth profiles from SIMS and XPS [4].

15.6.6 Surface Modification

Surface modification of glasses can be the result of adsorption and surface reactions, and may be studied by methods similar to those covered in the previous two sections. The focus here will be on those forms of surface modification in which the lateral interactions are at least as important as the bonding to the glass surface. The characterization of these "sheathing" layers again requires a combination of surface analytical techniques. Separation of the mechanisms of lateral bonding from those involving interaction with the glass surface can be difficult even with information from several different techniques. For instance, silanes are believed to adsorb initially through hydrogen bonding interactions, with subsequent condensation and lateral reactions generating siloxane structures. There are examples in which lateral polymerization occurs, apparently without the formation of bonds to the substrate [170], and the siloxane film has been shown to consist of multiple layers [171–174].

An example of this type of surface modification is that of "sizing" of glass fiber surfaces, which involves both adsorption of coupling agents and formation of relatively thick adhering overlayers. XPS can be used to monitor the thickness of the overlayer by providing the ratio of abundances of substrate elements before and after application of the coupling agents or overlayers. For instance, use of the less expensive siloxanes, instead of silanes, as the interacting species, has been studied with E-glass surfaces [175,176]. Because polymerized siloxane is the final product of the lateral reaction of adsorbed silanes, the application of the adsorbate to the surface in this form may be more direct, less expensive, and of equal value in surface modification. There is also potentially greater control and reproducibility. Adsorption of six types of siloxane (i.e., hydrido, methacryl, epoxy, unsaturated amino, pendant amino, and terminal amino) was studied [175,176]. The siloxanes bearing amino, hydrido, and methacryl functional groups bonded more strongly to the E-glass and resisted removal more effectively by a variety of organic solvents, than the silane coupling agent used extensively in industrial applications (i.e., vinyltrimethoxyethoxy silane). It is noteworthy that the functional groups on these three reagents do not undergo the same hydrolysis reaction that silane coupling agents are believed to require when bonding to glass surfaces. Nevertheless, they are either equally or more strongly bound which suggests that other mechanisms may be just as important for the formation of the final polymerized siloxane structures resulting from silane adsorption.

The thicknesses of the surface modifying layers could be inferred relatively easily from the calcium concentration in the underlying E-glass substrate and with the aid of the formula [177]:

$$t = -\lambda_{Ca} \ln[I_{Ca}/I_{Ca}^*] \, 2/\pi \qquad (15.2)$$

where

I_{Ca} is the measured concentration (proportional to the area under the Ca 2p envelope) after formation of the modifying surface layer

I_{Ca}^* is the corresponding concentration for a clean E-glass surface

λ_{Ca} is the attenuation length for Ca 2p photoelectrons

The factor $2/\pi$ accounts for the curvature of the fiber, which increases the effective path length of photoelectrons in the surface layer. The attenuation length can be estimated from the empirical determinations by Seah and Dench [178] to be 2.6 nm for Ca 2p electrons (KE 907 eV). There was an inverse correlation, as expected, between the abundances derived from the C 1s and Ca 2p intensities as a function of increasing surface coverage. Ion etching of the sample systematically increased the Ca concentration as the overlayer was removed, re-exposing the E-glass surface. Layer thicknesses between 1.2 and 4.1 nm were measured for different functionalized siloxanes at different surface coverages.

The coupling agents and overlayers applied to the glass surface must be able to interact with both the glass surface and the polymer matrix into which the glass fibers are to be incorporated. It is now well established that interactions require multiple layers of silane coupling reagents to generate an interpenetrating network within the polymer matrix, thus optimizing the strength of the resulting composite [172]. Aminohydroxysiloxanes have many hydrolyzable alkoxy groups along the polymer backbone, and have been shown [176] to produce multiple-site attachment and close-packed structures analogous to those resulting from adsorbed silane reagents. DRIFT and XPS techniques were also used in this study of the reagents. The work demonstrated that, when the mechanisms of surface attachment and lateral polymerization were understood, it was possible to design alternative reagents to perform the same surface modification.

ACKNOWLEDGMENTS

The work described in this chapter has been carried out by a large number of scientists with whom we have collaborated over many years and we express our gratitude for their many and varied contributions.

For RStCS, they include John Ralston, Tim White, Slavek Sobieraj, Stephen Grano, Bill Skinner, Clive Prestidge, Rob Hayes, Pawittar Arora, Tom Hörr, Angus Netting, Andrea Gerson, and Darren Simpson at UniSA as well as Robert Segall, Peter Turner, Sverre Myhra, Roger Lewis, David Cousens, and Colin Jones at Griffith University (Brisbane, Australia). The collaboration with the Åbo Akademi (Finland) which involved Jarl Rosenholm and Heidi Fagerholm has contributed to the work on glass surfaces. The recent work with PCA methods in ToF-SIMS of minerals owes much to Brian Hart, Mark Biesinger, and Stewart McIntyre at Surface Science Western, University of Western Ontario. For ZZ, her collaborators include Mark Blackford, Kim Finnie, Greg Lumpkin, Kath Smith, and Lou Vance at the Australian Nuclear Science and Technology Organisation (ANSTO), and Janusz Nowotny at the University of New South Wales.

The award of a Senior Research Fellowship by the Australian Research Council to RStCS (1992–1996) was of central importance for much of the work described in this review.

REFERENCES

1. D.J. Vaughan and R.A.D. Pattrick (eds.), *Mineral Surfaces*, *Mineralogical Society Series No. 5*, Chapman and Hall, London, 1995.
2. J. Nowotny (ed.), *Science of Ceramic Interfaces*, *Material Science Monographs No. 75*, Elsevier, Amsterdam, 1991.

3. J. Nowotny (ed.), *Science of Ceramic Interfaces II, Material Science. Monographs No. 81*, Elsevier, Amsterdam, 1994.

4. S. Myhra, R.St.C. Smart, and P.S. Turner, *Scanning Microsc. 2*, 715 (1988).

5. T.J. White and I.A. Toor, *J. Mater. 48*, 54 (1996).

6. P.S. Arora, A.K.O. Netting, and R.St.C. Smart, *Mater. Forum 17*, 293 (1993).

7. C.A. Prestidge, J. Ralston, and R.St.C. Smart, *Int. J. Min. Proc. 38*, 205 (1993).

8. J.R. Mycroft, G.M. Bancroft, N.S. McIntyre, J.W. Lorimer, and I.R. Hill, *J. Electroanal. Chem. 292*, 139 (1990).

9. C.D. Gribble and A.J. Hall, *Optical Microscopy: Principles and Practice*, UCL Publ., London, 1992, pp. 1–34.

10. G.N. Greaves, New X-ray techniques and approaches to surface mineralogy, in Ref. 1, p. 87.

11. J.C. Woicik, T. Kendelewicz, K.E. Miyano, P.L. Cowan, C.E. Bouldin, B.A. Karlin, and W.E. Spicer, *Phys. Rev. Lett. 68*, 341 (1992).

12. H.D. Abruna, G.N. Bommarito, and D. Acevedo, *Science 280*, 69 (1990).

13. D.J. O'Connor, Low energy ion scattering, in *Surface Analysis Methods in Materials Science*, D.J. O'Connor, B.A. Sexton, and R.St.C. Smart (eds.), *Springer Series in Surface Science No. 23*, 2nd edn., Springer-Verlag, Berlin, 2003, pp. 287–306.

14. H.H. Brongersma, P.A.C. Groenen, and J.-P. Jacobs, Application of low energy ion scattering to oxide surfaces, in Ref. 3, p. 113.

15. C. Klauber and R.St.C. Smart, Solid surfaces, their structure and composition, in Ref. 13, pp. 3–70.

16. G.E. Brown Jr., G.A. Parks, and P.A. O'Day, Sorption at mineral-water interfaces: macroscopic and microscopic perspectives, in Ref. 1, p. 129.

17. W.H. Casey, Surface chemistry during the dissolution of oxide and silicate materials, in Ref. 1, p. 185.

18. R.F. Garrett and G.J. Foran, EXAFS, in Ref. 13, pp. 347–373.

19. J.C. Vickerman, Static secondary ion mass spectroscopy, in *Methods of Surface Analysis*, J.M. Walls (ed.), Cambridge University Press, Cambridge, 1989, p. 169.

20. J.S. Brinen and F. Reich, *Surf. Interface Anal. 18*, 448 (1992).

21. J.S. Brinen, S. Greenhouse, D.R. Nagaraj, and J. Lee, *Int. J. Min. Proc. 38*, 93 (1993).

22. K.G. Stowe, S.L. Chryssoulis, and J.Y. Kim, *Min. Eng. 8*, 421 (1995).

23. A.A. Kruger, The role of the surface on bulk physical properties of glasses, in *Surface and Near-Surface Chemistry of Oxide Materials*, J. Nowotny and L.-C. Dufour (eds.), *Material Science Monographs No. 47*, Elsevier, Amsterdam, 1988, p. 413.

24. J.M. Cases, M. Kongolo, P. deDonato, L. Michot, and R. Erre, *Int. J. Min. Proc. 30*, 35 (1990).

25. J.O. Leppinen, *Int. J. Min. Proc. 30*, 245 (1990); J.W. Stojek and J. Mielcarzski, *Adv. Coll. Interf. Sci. 19*, 309 (1983).

26. P.F. Barron, R.L. Frost, J.O. Skjemstad, and A.J. Koppi, *Nature 302*, 49 (1982); R.L. Frost and P.F. Barron, *J. Phys. Chem. 88*, 6206 (1984).

27. J.O. Skjemstad, R.L. Frost, and P.F. Barron, *Aust. J. Soil Res. 21*, 539 (1983).

28. R.St.C. Smart, J. Amarantidis, W.M. Skinner, C.A. Prestidge, L. LaVanier, and S. Grano, Surface analytical studies of oxidation and collector adsorption in sulfide mineral flotation, in *Solid–Liquid Interfaces: Macroscopic Phenomena—Microscopic Understanding*, K. Wandelt and S. Thurgate (eds.), *Topics in Applied Physics 85*, Springer-Verlag, Berlin, 2002, pp. 3–60.

29. R.St.C. Smart, W.M. Skinner, A.R Gerson, J. Mielczarski, S. Chryssoulis, A.R. Pratt, R. Lastra, G.A. Hope, X. Wang, K. Fa, and J.D. Miller, Surface characterisation and new tools for research, in *Froth Flotation: Centenary Volume*, M. Fuerstenau and R-H Yoon (eds.), SME Publ., USA, 2007, pp. 283–338.

30. J.R. Craig and D.J. Vaughan, *Ore Microscopy and Ore Petrography*, Wiley Interscience, New York, 1981.

31. H.M. Rietveld, *J. Appl. Cryst. 2*, 65 (1969).

32. R.J. Hill and C.J. Howard, *A Computer Program for Rietveldt Analysis of Fixed Wavelength X-ray and Neutron Powder Diffraction Patterns*, Australian Atomic Energy Commission Report AAEC/M112, July, 1986; J.C. Taylor, *Powder Diffr. 6*, 2 (1991); Siroquant Version 3, www.sietronics.com.au (2006).

33. C.J. Howard, R.J. Hill, and M.A.M. Sufi, *Chem. Aust. 55*, 367 (1988).

34. P.S. Turner, C.E. Nockolds, and S. Bulock, Electron microscope techniques for surface characterisation, in Ref. 13, pp. 85–106.

35. D. Sutherland, R. Creelman, P. Gottlieb, R. Jackson, V. Quittner, G. Wilkie, M. Zuiderwyck, N. Allen, and T. Maclean, in *CHEMECA 87* (*Proceedings of the Chemical Engineering Conference*), Vol. 2, Aust. Inst. Min. Met. Publ., 1987, pp. 106.1–106.6; P. Gottlieb, B.J.I. Adair, and G.J. Wilkie, *Proceedings of the Mill Operators Conference*, Aust. Inst. Min. Met. Publ., Melbourne, 1994, pp. 5–13.

36. P. Gottlieb, G. Wilkie, D. Sutherland, and E. Ho-Tun, Using quantitative electron microscopy for process mineralogy applications. *Journal of Minerology 52*(4), 24–27, 2000.
37. Y. Gu, in *Applied Mineralogy: Developments in Science and Technology*, M. Pecchio, F.R.D Andrade, L.Z.D'Agostino, H. Kahn, L.M. Sant'Agostino, and M.M.M.L. Tassinari (eds.), *Proceedings of the 8th International Conference on Applied Mineralogy*, Sao Paulo, Brazil, September, 2004, Publications of the International Council for Applied Mineralogy of Brasil, pp. 119–122.
38. L.D. Marks and D.J. Smith, *Nature 303*, 316 (1983); D.J. Smith, *Surf. Sci. 178*, 462 (1986).
39. L.D. Marks, *Surf. Sci. 139*, 281 (1984).
40. X. Zhong, M. Han, Z. Dong, T.J. White, and W. Knoll, *J. Am. Chem. Soc. 125*, 8589 (2003).
41. G. van der Laan, R.A.D. Pattrick, C.M.B. Henderson, and D.J. Vaughan, *J. Phys. Chem. Solids 53*, 1185 (1992).
42. A. Iida, T. Noma, S. Hayakawa, M. Takahashi, and Y. Gohshi, *Jpn. J. Appl. Phys. 32*, 160 (1993).
43. B. Hart, M.C. Biesinger, and R.St.C. Smart, *Min. Eng. 19*, 790 (2006).
44. M.F. Hochella Jr., Mineral surfaces: Their characterisation and their chemical, physical and reactive nature, in Ref. 1, p. 17.
45. F. Fan and A.J. Bard, *J. Phys. Chem. 94*, 3761 (1990).
46. S.E. Gilbert and J.H. Kennedy, *Langmuir 5*, 1412 (1989).
47. C.M. Eggleston and M.F. Hochella Jr., *Geochim. Cosmochim. Acta 54*, 1511 (1990).
48. C.M. Eggleston and M.F. Hochella Jr., *Science 254*, 983 (1991).
49. C.M. Eggleston and M.F. Hochella Jr., *Am. Mineral. 77*, 221 (1992).
50. C.M. Eggleston and M.F. Hochella Jr., *Am. Mineral. 77*, 911 (1992).
51. C.M. Eggleston and M.F. Hochella Jr., *Am. Mineral. 78*, 877 (1993).
52. C.M. Eggleston and M.F. Hochella Jr., in *The Environmental Chemistry of Sulphide Oxidation, ACS Symposium Series, 550*, C.N. Alpers and D.W. Blowes (eds.), American Chemical Society, Washington DC, 1994, p. 201.
53. M.F. Hochella Jr., C.M. Eggleston, V.B. Elings, G.A. Parks, G.E. Brown Jr., C.M. Wu, and K. Kjoller, *Am. Mineral. 74*, 1235 (1989).
54. M.F. Hochella Jr., C.M. Eggleston, V.B. Elings, and M.S. Thompson, *Am. Mineral. 75*, 723 (1990).
55. G.M. Bancroft and M.M. Hyland, in *Mineral-Water Interface Geochemistry*, M.F. Hochella Jr. and A.F. White (eds.), *Reviews in Mineralogy*, Vol. 23, Mineralogical Society of America, Washington DC, 1990, p. 511.
56. P. Wersin, M.F. Hochella Jr., P. Persson, G. Redden, J.O. Leckie, and D.W. Harris, *Geochim. Cosmochim. Acta 58*, 2829 (1994).
57. P.M. Dove, M.F. Hochella Jr., and R.J. Reeder, in *Water-Rock Interaction*, Y. Kharaka and A. Maest (eds.), Balkema, Rotterdam, 1992, p. 141.
58. K. Laajalehto, R.St.C. Smart, J. Ralston, and E. Suoninen, *Appl. Surf. Sci. 64*, 29 (1993).
59. G.F. Cotterill, R. Bartlett, A.E. Hughes, and B.A. Sexton, *Surf. Sci. Lett. 232*, L211 (1990).
60. A.N. Buckley and R. Woods, *Appl. Surf. Sci. 17*, 401 (1984).
61. B.S. Kim, R.A. Hayes, C.A. Prestidge, J. Ralston, and R.St.C. Smart, *Appl. Surf. Sci. 78*, 385 (1994).
62. B.S. Kim, R.A. Hayes, C. A. Prestidge, J. Ralston, and R.St.C. Smart, *Langmuir 11*, 2554 (1995).
63. P.J. Jennings and C.Q. Sun, Low energy electron diffraction, in Ref. 13, pp. 319–336.
64. G.N. Greaves, *Adv. X-ray Anal. 34*, 13 (1991).
65. T.C. Huang, *Adv. X-ray Anal. 33*, 91 (1991).
66. A.R. Pratt, I.J. Muir, and H.W. Nesbitt, *Geochim. Cosmochim. Acta 58*, 827 (1994).
67. J.R. Mycroft, H.W. Nesbitt, and A.R. Pratt, *Geochim. Cosmochim. Acta 59*, 721 (1995).
68. S.W. Knipe, J.R. Mycroft, A.R. Pratt, H.W. Nesbitt, and G.M. Bancroft, *Geochim. Cosmochim. Acta 59*, 1079 (1995).
69. J.E. Thomas, W.M. Skinner, and R.St.C. Smart, *Geochim. Cosmochim. Acta 62*, 1555 (1998).
70. J.E. Thomas, R. St.C. Smart, and W.M. Skinner, *Min. Eng. 13*, 1149 (2000).
71. J.E. Thomas, W.M. Skinner, and R.St.C. Smart, *Geochim. Cosmochim. Acta 65*, 1 (2001).
72. J.E. Thomas, W.M. Skinner, and R.St.C. Smart, *Geochim. Cosmochim. Acta 67*, 831 (2003).
73. C.G. Weisener, R.St.C. Smart, and A.R. Gerson, *Geochim. Cosmochim. Acta 67*, 823 (2003).
74. C.G. Weisener, R.St.C. Smart, and A.R. Gerson, *Int. J. Min. Proc. 74*, 239 (2004).
75. R.St.C. Smart, M. Jasieniak, K.E. Prince, and W.M. Skinner, *Min. Eng. 13*, 857 (2000).
76. A.G. Schaufuss, H.W. Nesbitt, I. Kartio, K. Laajalehto, G.M. Bancroft, and R. Szargan, *J. Electron Spectrosc. Relat. Phenom. 96*, 69 (1998).
77. R. Szargan, A.G. Schaufuss, and P. Rossbach, *J. Electron Spectrosc. Relat. Phenom. 100*, 357 (1999).
78. E.A. Colbourn and W.C. Mackrodt. *Surf. Sci. 117*, 571 (1982); *Solid State Ionics 8*, 221 (1983); E.A. Coulbourn, J. Kendrick, and W. Mackrodt, *Surf. Sci. 126*, 550 (1983).

79. C.F. Jones, R.A. Reeve, R. Rigg, R.L. Segall, R.St.C. Smart, and P.S. Turner, *J. Chem. Soc. Faraday I* *80*, 2609 (1984).
80. C.F. Jones, R.L. Segall, R.St.C. Smart, and P.S. Turner, *Proc. Roy. Soc., A374*, 141 (1981); *J. Mater. Sci. Lett. 3*, 810 (1984); *J. Chem. Soc. Faraday I 74*, 1615 (1978).
81. D.J. Simpson, T. Bredow, R.St.C. Smart, and A.R. Gerson, *Surf. Sci. 516*, 134 (2002).
82. S.L. Harmer and H.W. Nesbitt, *Surf. Sci. 564*, 38 (2004).
83. G.U. von Oertzen, W.M. Skinner, and H.W. Nesbitt. *Phys. Rev. B 72*, 235427 (2005).
84. G.U. von Oertzen, S.L. Harmer, and W.M. Skinner, *Mol. Simulat. 32*, 1207 (2006).
85. J.A. Tossell, Mineral surfaces: theoretical approaches, in Ref. 1, p. 61.
86. V.E. Henrich, Electronic and geometric structure of defects in oxides and their role in chemisorption, in Ref. 23, p. 23.
87. A.B. Kunz, Theoretical study of defects and chemisorption by oxide surfaces, in *External and Internal Surfaces in Metal Oxides*, L.-C. Dufour and J. Nowotny (eds.), *Material Science Forum*, Vol. 29, Trans. Tech. Publ., Switzerland, 1988, p. 1.
88. A.J. Burggraaf and A.J.A. Winnubst, Segregation in oxide surfaces, in Ref. 23, p. 449.
89. A. Pring and T.B. Williams, *Min. Mag. 58*, 453 (1994).
90. S. Grano, J. Ralston, and R.St.C. Smart, *Int. J. Min. Proc. 30*, 69 (1990).
91. I.J. Muir, G.M. Bancroft, and H.W. Nesbitt, *Geochim. Cosmochim. Acta 53*, 1235 (1989).
92. I.J. Muir, G.M. Bancroft, W. Shotyk, and H.W. Nesbitt, *Geochim. Cosmochim. Acta 54*, 2247 (1990).
93. H.W. Nesbitt and W.M. Skinner, *Geochim. Cosmochim. Acta 65*, 715 (2001).
94. W.H. Casey, M.J. Carr, and R.A. Graham, *Geochim. Cosmochim. Acta 52*, 1545 (1988).
95. W.P. Inskeep, E.A. Nater, P.R. Bloom, D.S. Vandervoort, and M.S. Erich, *Geochim. Cosmochim. Acta 55*, 787 (1991).
96. K. Laajalehto, P. Nowak, A. Pomianowski, and E. Suoninen, *Coll. Surf. 57*, 319 (1991).
97. J.A. Mielczarski and R.-H. Yoon, *J. Phys. Chem. 93*, 2034 (1989).
98. J.A. Mielczarski, *J. Phys. Chem. 97*, 2649 (1993).
99. J.A. Mielczarski and E. Mielczarski, *J. Phys. Chem. 99*, 3206 (1995).
100. J.A. Mielczarski and E. Mielczarski, *J. Phys. Chem. 103*, 5852 (1999).
101. Z.S. Nickolov, X. Wang, and J.D. Miller, *Spectrochim. Acta A 60*, 2711 (2004).
102. A.N. Buckley, *Coll. Surf. A 93*, 159 (1994).
103. A.N. Buckley, I.C. Hamilton, and R. Woods, in *Flotation of Sulfide Minerals*, K.S.E. Forssberg (ed.), Elsevier, Amsterdam, 1985, p. 41.
104. R.L. Segall, R.St.C. Smart, and P.S. Turner, Oxide surfaces in solution, in Ref. 23, p. 527.
105. M.A. Blesa, A.E. Regazzoni, and A.J.G. Marito, Reactions of metal oxides with aqueous solutions, in Ref. 99, p. 31.
106. R.A. Schoonheydt, Clay mineral surfaces, in Ref. 1, pp. 303–332.
107. S. Myhra, D.K. Pham, R.St.C. Smart, and P.S. Turner, Surface reactions and dissolution of ceramics and high temperature superconductors, in Ref. 2, p. 569.
108. C.F. Jones, S. LeCount, R.St.C. Smart, and T.J. White, *App. Surf. Sci. 55*, 65 (1992).
109. R.St.C. Smart, P. Arora, B. Braggs, H.M. Fagerholm, T.J. Horr, D.C. Kehoe, J.G. Matisons, J. Ralston, and J.B. Rosenholm, in *Interfaces of Ceramic Materials: Impact on Properties and Applications*, K. Uematsu, Y. Moriyoshi, Y. Saito, and J. Nowotny (eds.), *Key Engineering Materials*, Vols. 111–112, Trans. Tech. Publ., Switzerland, 1995, p. 361.
110. A.R. Gerson, in *Proceedings of the PacRim2 (2nd International Meeting of Pacific Rim Ceramic Societies)*, P. Walls, C. Sorrell, and A Ruys (eds.), published as CD, Australian Ceramic Society, ANSTO, Sydney, 1996.
111. C.J. Ball, W.J. Buykx, F.J. Dickson, K. Hawkins, D.M. Levins, R.St.C. Smart, K.L. Smith, G.T. Stevens, K.G. Watson, D. Weedon, and T.J. White, *J. Am. Ceram. Soc. 72*, 404 (1989).
112. W.J. Buykx, K. Hawkins, H. Mitamura, R.St.C. Smart, G.T. Stevens, K.J. Watson, D. Weedon, and T.J. White, *J. Am. Ceram. Soc. 71*, 678 (1988).
113. J.A. Cooper, D.R. Cousens, R.A. Lewis, S. Myhra, R.L. Segall, R.St.C. Smart, P.S. Turner, and T.J. White, *J. Am. Ceram. Soc. 68*, 64 (1985).
114. G.R. Lumpkin, *Elements 2*, 365 (2006).
115. S.V. Kalinin, R. Shao, and D.A. Bonnell, *J. Am. Ceram. Soc. 88*, 1077 (2005).
116. Hj. Matzke, Nuclear methods in studies of ceramic surfaces and interfaces, in Ref. 2, p. 457.
117. Hj. Matzke, J.A. Davies, and N.G.E. Johansson, *Can. J. Phys. 49*, 2215 (1971).
118. A.G. Solomah and Hj. Matzke, in *Scientific Basis for Nuclear Waste Management XII*, W. Lutze and R.C. Ewing (eds.), *Materials Research Society Symposium Proceedings*, Vol. 127, Pittsburgh, PA, 1989, p. 241.

119. Hj. Matzke, G. Della Mea, F.L. Freire Jr., and V. Rigato, *Nucl. Instrum. Meth. Phys. Res. B 45*, 194 (1990).
120. Hj. Matzke and A. Turos, *Solid State Ionics 49*, 189 (1991).
121. D.K. Pham, F.B. Neall, S. Myhra, R.St.C. Smart, and P.S. Turner, in *Scientific Basis for Nuclear Waste Management XII*, W. Lutze and R.C. Ewing (eds.), *Materials Research Society Symposium Proceedings*, Vol. 127, Pittsburgh, PA, 1989, p. 231.
122. L. Nowicki, A. Turos, C. Choffel, F. Garrido, L. Thomé, J. Gaca, M. Wójcik, and Hj. Matzke, *Phys. Rev. B 56*, 534 (1997).
123. C.W. White, L.A. Boatner, P.S Sklad, C.J. McHargue, J. Rankin, G.C. Farlow, and M.J. Aziz, *Nucl. Instrum. Meth. Phys. Res. B 32*, 11 (1988).
124. K. Yamana, M. Miyamoto, S. Nakamura, and K. Kihara, in Ref. 109, p. 57.
125. Z. Zhang, M.G. Blackford, G.R. Lumpkin, K.E. Smith, and E.R. Vance, *J. Mater. Res. 20*, 2462 (2005).
126. M. de Ridder, R.G. van Welzenis, H.H. Brongersma, and U. Kreissig, *Solid State Ionics 158*, 67 (2003).
127. J. Nowotny, C.C. Sorrell, and T. Bak, *Surf. Interface Anal. 37*, 316 (2005).
128. A.M. Stoneham and P.W. Tasker, The theory of ceramic surfaces, in Ref. 23, p. 1.
129. E.A. Colbourn, *Surf. Sci. Rep. 15*, 281 (1992).
130. I. MacLaren, R.M. Cannon, M.A. Gülgün, R. Voytovych, N. Popescu-Pogrion, C. Scheu, U. Täffner, and M. Rühle, *J. Am. Ceram. Soc. 86*, 650 (2003).
131. M. Ruhle, *J. Phys., Colloq. (Orsay, Fr.) 43*, 115 (1982).
132. D.R. Clarke, *J. Am. Ceram. Soc. 63*, 104 (1980).
133. D.R. Clarke, *Ultramicroscopy 8*, 95 (1982).
134. D.R. Clarke, *Ultramicroscopy 4*, 33 (1979).
135. D.R. Clarke, *J. Am. Ceram. Soc. 70*, 15 (1987).
136. J. Cabane and F. Cabane, Equilibrium segregation in interfaces, in *Interface Segregation and Related Processes in Materials*, J. Nowotny (ed.), Trans. Tech. Publ., Switzerland, 1991, p. 1.
137. J. Cooper, D.R. Cousens, J.A. Hanna, R.A. Lewis, S. Myhra, R.L. Segall, R.St.C. Smart, P.S. Turner, and T.J. White, *J. Am. Ceram. Soc. 69*, 347 (1986).
138. A.B. Harker, D.R. Clarke, and C.M. Jantzen, in *Surfaces and Interfaces in Ceramic and Ceramic-Metal Systems*, J. Pask and A. Evans (eds.), Material Science Research, Vol. 14, Plenum Press, New York, 1981, p. 207.
139. R.St.C. Smart, *Appl. Surf. Sci. 22/23*, 90 (1985).
140. K.S. Finnie, Z. Zhang, E.R. Vance, and M.L. Carter, *J. Nucl. Mater. 317*, 46 (2003).
141. G.S.A.M. Theunissen, A.J.A. Winnubst, and A.J. Burggraaf, *J. Mater. Sci. 27*, 5057 (1992).
142. J. Nowotny, T. Bak, M.K. Nowotny, and C.C. Sorrell, *Adv. Appl. Ceram. 104*, 165 (2005). www.maney. co.uk/journals/aac <http://www.maney.co.uk/journals/aac> and www.ingentaconnect.com/cantent/ maney/aac <http://www.ingentaconnect.com/content/maney/acc>
143. J. Morais, E.B.O. da Rosa, L. Miotti, R.P. Pezzi, I.J.R. Baumvol, A.L.P. Rotondaro, M.J. Bevan, and L. Colombo, *Appl. Phys. Lett. 78*, 2446 (2001).
144. T.J. Horr, J. Ralston, and R.St.C. Smart, *Coll. Surf. 64*, 67 (1992).
145. T.J. Horr, J. Ralston, and R.St.C. Smart, *Coll. Surf. 92*, 277 (1994).
146. T.J. Horr, J. Ralston, and R.St.C. Smart, *Coll. Surf. A 97*, 183 (1995).
147. T.J. Horr, P.S. Arora, J. Ralston, and R.St.C. Smart, *Coll. Surf. A 102*, 181 (1995).
148. G.E. Brown Jr., V.E. Henrich, W.H. Casey, D.L. Clark, C. Eggleston, A. Felmy, D.W. Goodman, M. Grätzel, G. Maciel, M.I. McCarthy, K.H. Nealson, D.A. Sverjensky, M.F. Toney, and J.M. Zachara, *Chem. Rev. 99*, 77 (1999).
149. K.N. Strafford, R.St.C Smart, I. Sare, and C. Subramanian (eds.), *Surface Engineering: Processes and Application*, Technomic Publ. Co., Pennsylvania, 1995.
150. P. Arora and R.St.C. Smart, *Surf. Interface Anal. 24*, 539 (1996).
151. C.D. Wagner, D.E. Passoja, H.F. Hillery, T.G. Kinisky, H.A. Six, W.T. Jansen, and J.A. Taylor, *J. Vac. Soc. Technol. 21*, 933 (1982).
152. R.St.C. Smart, P.S. Arora, M. Steveson, N. Kawashima, G.P. Cavallaro, H. Ming, and W.M. Skinner, New approaches to metal-ceramic and bioceramic interfacial bonding, in *Ceramic Interfaces: Properties and Applications II*, H-I Yoo and S-J Kang (eds.), Institute of Materials Press, UK, 2001, pp. 293–326.
153. W.R. Pease, R.L. Segall, R.St.C. Smart, and P.S. Turner, *J. Chem. Soc. Faraday Trans. I 76*, 1510 (1980).
154. G.S. Henderson, G. Calas, and J.F. Stebbins, *Elements 2*, 269 (2006).

155. L. Galoisy, *Elements 2*, 293 (2006).
156. A. Paul, *J. Mater. Sci. 12*, 2246 (1977).
157. H.W. Nesbitt and K.N. Dalby, *Canad. J. Chem*, 85, 782 (2007).
158. K.N. Dalby, H.W. Nesbitt, V.P. Zakaznova-Herzog, and P.L. King, *Geochim. Cosmochim. Acta*, 71, 4292 (2007).
159. H.M. Fagerholm, J.B. Rosenholm, T.J. Horr, and R.St.C. Smart, *Coll. Surf. A 110*, 11 (1996).
160. D.W. Sindorf and G.E. Maciel, *J. Am. Chem. Soc. 105*, 1489 (1983).
161. A.G. Atkins, *J. Mater. Sci. 10*, 819 (1975).
162. J.O. Outwater, *J. Adhesion 2*, 242 (1970).
163. J.M. Olinger and P.R. Griffiths, *Anal. Chem. 60*, 2427 (1988).
164. H.M. Fagerholm, C. Lindsjö, J.B. Rosenholm, and K. Rokman, *Coll. Surf. 69*, 79 (1992).
165. L.L. Hench, D.E. Clark, and E. Lue Yen-Bower, *Nucl. Chem. Waste Manage. 59*, 1 (1980).
166. D.E. Clark and L.L. Hench, *Nucl. Chem. Waste Manage. 2*, 93 (1981).
167. C.Q. Buckwalter and L.R. Pederson, N.W. Labs Report PNL-SA-9940, Hanford, USA, 1983.
168. G.L. McVay and C.Q. Buckwalter, N.W. Labs Report PNL-SA-10474, Hanford, USA, 1984.
169. R.A. Lewis, S. Myhra, R.L. Segall, R.St.C. Smart, and P.S. Turner, *J. Non-Cryst. Sol. 53*, 299 (1982).
170. C.P. Tripp and M.L. Hair, *Langmuir 8*, 1120 (1991).
171. E.P. Pleuddemann, *Silane Coupling Agents*, Plenum Press, New York, 1982.
172. J.M. Park and R.V. Subramanian, *J. Adhesion Sci. Technol. 5*, 459 (1991).
173. K.W. Allen, *J. Adhesion Sci. Technol. 6*, 23 (1992).
174. E.K. Drown, H. Al Moussawi, and L.T. Drzal, *J. Adhesion Sci. Technol. 5*, 865 (1991).
175. D.R. Bennett, J.G. Matisons, A.K.O. Netting, R.St.C. Smart, and A.G. Swincer, *Polymer Int. 27*, 147 (1992).
176. L. Britcher, D. Kehoe, J. Matisons, R.St.C. Smart, and A.G. Swincer, *Langmuir 9*, 1609 (1993).
177. C.G. Pantano and T.N. Wittberg, *Surf. Interface Anal. 15*, 498 (1990).
178. M.P. Seah and W.A. Dench, *Surf. Interface Anal. 1*, 2 (1979).

16 Catalyst Characterization

Wolfgang E. S. Unger and Thomas Gross

CONTENTS

16.1 INTRODUCTION

The analytical characterization of heterogeneous catalysts occupies a central position in catalysis research, whether fundamental, applied, or industrial. The various analytical activities are directed toward the overall objective of gaining a detailed chemical and structural picture of the catalysts themselves. The information obtained can be used to develop a better understanding of the function of a catalyst, leading to the formulation of new catalysts and the improvement of existing ones. In industry the results of characterization are often useful for quality management and for control of process economy. Consequently, there are two different approaches to catalyst characterization. One includes all those efforts intended to establish empirical relations between catalyst performance and such parameters as elemental composition and chemical state, particle size and dispersion, etc. The other, more fundamental and academic approach is to determine the catalyst's surface composition and structure on the atomic scale, preferably under reaction conditions but with a simplified catalyst model. The latter studies are often carried out on well-defined surfaces. To match the overall aim of this book, the emphasis here will be on the characterization of real catalysts and on appropriate models.

Nowadays it is common practice to bring multiple analytical techniques to bear on the characterization of catalysts. Very often bulk and surface analysis methods are combined to yield detailed complementary information. A typical scheme of catalyst parameters to be determined,

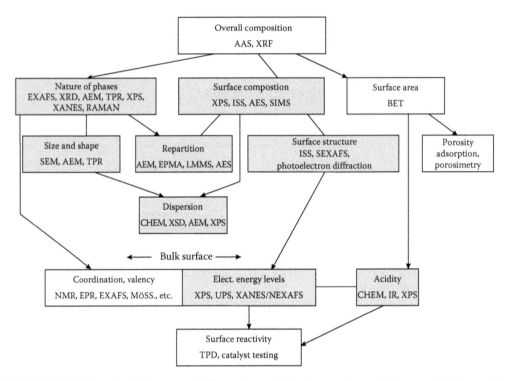

FIGURE 16.1 Schematic of the types of information available from various techniques applied in catalyst bulk and surface characterization. Fields in which surface spectroscopies can contribute are highlighted. (From Delmon, B., *Surf. Interface Anal.*, 9, 195, 1986. With permission.)

and of appropriate analytical methods, has been presented by Delmon [1], who reviewed the potential of surface analysis in the characterization of hydrodesulfurization (HDS) catalysts. This scheme, redrawn and amended, is shown in Figure 16.1. Unfortunately, due to certain inherent problems, the whole range of the well-established surface spectroscopic methods cannot be applied in every case. Another disadvantage is that the application of the surface-sensitive spectroscopies requires in almost all cases the transfer of catalyst samples into vacuum, thus restricting the possibilities for genuine *in situ* surface analyses of working catalysts. However, strenuous efforts have been made during the last few years, with reasonable success, to enable real *in situ* x-ray photoelectron and absorption spectroscopic experiments on model catalysts at pressures in the millibar range to be performed [2–4]. Figure 16.2 presents an example in which the oxidation of methanol over copper was investigated for different methanol–oxygen feed gas mixtures, by use of the O 1s signal in x-ray photoelectron spectroscopy (XPS) [3]. In that experiment the incident x-rays irradiated not only the catalyst surface but also the gas phase over it. That gave rise to photoemission signals originating not only from oxygen species at and in the catalyst surface, but from gaseous oxygen species as well. In the example, O 1s peaks could be detected in the x-ray photoelectron (XP) spectrum, arising from the reaction partners O_2 and CH_3OH; the reaction products CH_2O, CO_2, and H_2O; the catalyst components surface and subsurface oxygen; and Cu_2O as well. The relative magnitudes of the O 1s peaks in the spectra, which were measured at 400°C, depended on the methanol–oxygen proportion in the feed gas mixture. The activity of the catalyst for each of the mixtures was evaluated using XP O 1s and C 1s peak areas and, as usual, mass spectrometric data. Oxidizing reaction conditions resulting from a feed gas proportion of $CH_3OH:O_2 = 1:2$ could be distinguished from reducing ones at proportions of $CH_3OH:O_2 = 3:1$ and 6:1. It was concluded that the catalyst surface active for the oxidation of methanol is characterized

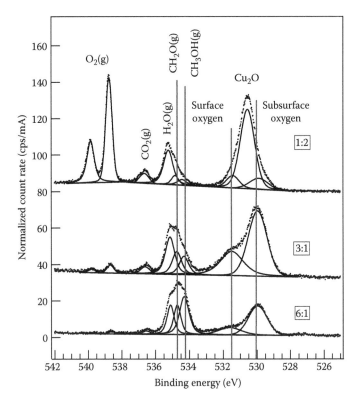

FIGURE 16.2 O 1s spectra from a working copper catalyst for methanol oxidation at 400°C and 0.6 mbar at different $CH_3OH:O_2$ feed gas mixtures, by *in situ* XPS. Reaction partners O_2 and CH_3OH; reaction products CH_2O, CO_2, and H_2O; catalyst constituents such as surface and subsurface oxygen species; and Cu_2O can be differentiated. (Courtesy of Knop-Gericke, A., Fritz-Haber-Institute, Berlin; Bluhm, H., Hävecker, M., Knop-Gericke, A., Kleimenov, E., Schlogl, R., Teschner, D., Bukhtiyarov, V., Ogletree, D., and Salmeron, M., *J. Phys. Chem. B.*, 108, 14340, 2004.)

by metallic copper (as deduced from the Cu 2p XP spectra) and subsurface oxygen species, and is established under reducing conditions. The CH_2O reaction yield correlates with the amount of subsurface oxygen, both estimated from *in situ* XPS O 1s spectra.

A statistical analysis of papers dealing with catalyst characterization has revealed that XPS is the most frequently applied surface analysis technique in the field. For that reason this chapter will deal preferentially with applications of XPS. Where appropriate, combinations of XPS with other characterization methods will be described. The temperature-programmed and sorption techniques, which are important methods in the surface characterization of catalysts, are not considered here in much detail because to do so would be beyond the scope of this book. Easily accessible information on strategies and problems in those fields can be found in Refs. [5,6]. It should be mentioned that calibration and validation, necessary for successful application of the sorption methods, can be accomplished using surface spectroscopic tools.

For the "best practice" requirements in the application of surface analytical methods frequent reference is made below to the most recent standards and technical reports published by ISO TC 201. The ISO TC 201 committee on surface chemical analysis was established in November 1991 to address the need for international standards following a "Strategy Policy Statement for ISO/TC 201," which is itself in continual revision. In 2008, 34 international standards and technical reports were published. The committee E-42 on surface analysis of the ASTM in the United States has also published standards for surface analysis over the past 30 years.

16.2 APPLICABILITY OF SURFACE SPECTROSCOPIES TO CATALYST CHARACTERIZATION

This section considers the applicability of the five most commonly used surface analytical techniques to catalyst surface characterization. They are XPS (also known as ESCA), which also includes x-ray excited Auger electron spectroscopy (XAES), electron excited Auger electron spectroscopy (AES), x-ray absorption spectroscopy (XAS, x-ray absorption near-edge structure [XANES], near-edge x-ray absorption fine structure [NEXAFS], and extended x-ray absorption fine structure [EXAFS]); secondary ion mass spectrometry (SIMS), including its nondamaging "static SIMS" (SSIMS) mode; and, last but not least, ion-scattering spectroscopy (ISS).

XPS is the most commonly used, and most useful, surface analysis method in catalyst characterization. It can be used for both qualitative and quantitative analyses for almost all kinds of catalysts used in heterogeneously catalyzed reactions. XPS studies of oxide, sulfide, fluoride, halide, etc., catalysts; supported metal catalysts; Raney or metal gauze catalysts; and zeolite catalysts are all possible. The samples can be studied in the precursor, calcined, reduced, activated, deactivated, aged, or poisoned states. Real industrial catalysts can be analyzed as well as academic model systems. Quantitative analysis is possible either with the help of empirical sensitivity factors or in a "standard-free" way. In the latter case appropriate theoretical models are used, along with photoionization cross-section tables, effective attenuation length (EAL) data, and individual spectrometer transmission functions. However, there are also some features that restrict the application of XPS in catalyst characterization:

1. The analysis must be carried out under high vacuum conditions that must be good, and preferably in the ultra-high vacuum (UHV) region.
2. X-ray damage must be kept under control.
3. The analysis can suffer from photoelectron and Auger line interferences, which may occur easily with multicomponent catalyst samples.
4. Speciation for minority constituents (<1 at.%) can be impossible due to lack of sensitivity.
5. Lateral resolution is limited at present to a few micrometer, using the most advanced commercial spectrometers.

The information depth in XPS is defined as the maximum depth from which useful information can be obtained, normally regarded as the sample thickness from which 95% of the detected signal originates. In that case the information depth is approximately three times the EAL [7], i.e., of the order of a few nanometers. A consequence of the magnitude of the information depth in XPS is that, with high BET surface area catalysts, inner and outer surfaces as well as the material in between (bulk) contribute to the signals detected. Obviously, in that situation XPS is no longer a true surface-specific analytical technique.

For more detailed information about the application of XPS the reader is referred to Ref. [5], which is a book on catalyst characterization by surface spectroscopies.

Finally, it should be noted that a worldwide effort, under the heading of spectromicroscopy [8], is currently about to achieve significantly enhanced lateral resolution in XPS (see Chapter 3). The overall objective is a lateral resolution in the range of a few tens of nanometers, which would satisfy effectively the demands of catalyst characterization.

The applicability of AES to the characterization of real surfaces is limited. This is due to the fact that, in many cases, nonconducting samples cannot be analyzed because of surface charging problems. Thus only certain types of samples from the real world, e.g., Raney or metal gauze catalysts, are suitable for AES analysis. Recent state-of-the-art spectrometers have sophisticated charge compensation facilities, and it is expected that nonconductive catalysts will also be able to be analyzed in the future. The applicability of AES, however, is wide in the case of the study of academic models in catalytically motivated surface science. Many aspects of gas and temperature treatments of metal or semiconductor surfaces, including segregation phenomena in multicomponent systems, have been studied extensively and successfully with the technique. AES may, in principle, provide chemical state information from information depths very similar to those characteristic of XPS. However, the damage induced by the intense beam of energetic primary electrons must be carefully controlled. Another feature of AES, the possibility of applying it in the scanning mode (SAM) with a lateral resolution as good as some tens of nanometers, makes it attractive in special cases. Niemantsverdriet [5] has considered the potential of AES in his book on spectroscopies applied to catalyst characterization.

XAS [9] is based on electronic transitions from a deep core-level orbital to (1) unoccupied antibonding states in the valence region of the emitter atoms or (2) the continuum (see Chapter 7). These transitions are excited by x-ray photons with tunable energy from a synchrotron radiation source. The decay of core holes is accompanied by the emission of electrons initiated by Auger transitions, known as "electron yield," which provides a surface-sensitive information channel with an information depth similar to that in XPS or AES. Another information channel that provides bulk characteristic data for a sample is given by the radiative decay of the core hole, i.e., by photon emission in the fluorescence mode. This mode is known as fluorescence yield XAS. In principle, both yields can be measured in one and the same XAS experiment. Recently many synchrotron radiation sources have offered beamline end-stations with the opportunity of applying XAS to a variety of analytical problems.

The analytical information that can be obtained by XAS on samples relevant to catalysis is

- Identification of bonding situations and valence states of emitter atoms by the evaluation of preabsorption-edge spectral features in the NEXAFS, representing resonant transitions into unoccupied orbitals
- Determination of nearest-neighbor distances and coordination numbers for emitter atoms by evaluation of the EXAFS above the absorption edge, resulting from interference of the photoelectron wave leaving the absorber atom and the wave backscattered onto neighboring atoms

SIMS is acceptable in catalyst characterization but has been used less often than XPS. It can, in principle, be used for studying real catalysts including zeolites. Again there is a severe charging problem, which must be solved for each individual case with the appropriate means, e.g., by an electron flood gun. Recent SIMS instruments have powerful charge neutralizers and the problem of charging can be kept under control. The most impressive SIMS studies have been made on model catalysts simulating selected features of real catalysts. In such systems the samples can be "designed" in a way that allows optimal SIMS analysis. Finally, the SIMS method has contributed successfully to surface science studies of chemisorption of molecules on active single crystal surfaces motivated by catalysis science.

There are important advantages of SIMS. The first is the enhanced surface specificity with respect to XPS. Detailed simulations have revealed that the secondary ions detected in SIMS originate exclusively from the first two monolayers (roughly 85% stem from the first layer [10]). The second advantage of SIMS is its unique elemental sensitivity, down to the parts per million level for certain elements, e.g., the alkali metals. This feature is useful when traces of promoters or poisons are of interest. The third advantage is the possibility of elemental or chemical mapping of a catalyst surface, with lateral resolution down to about 50 nm (imaging SIMS), by using secondary element or cluster ions. Cluster ions may carry chemical state information provided SIMS is applied in its so-called static mode (SSIMS).

The main disadvantage of SIMS is that the quantification of raw data is extremely difficult. Sputter yields and ionization probabilities are the main parameters defining secondary ion yields in SIMS. Sputter yield data for multicomponent materials such as real catalysts scarcely exist. The ionization probability itself varies over five decades across the periodic table of elements. Furthermore, there is a huge matrix effect when, for example, the same element or cluster ion is sputtered from chemically different surface sites. On the other hand, once understood, this same extreme matrix effect can be exploited for purposes of qualitative analysis on model catalysts. The use of SIMS in catalyst characterization has been reviewed in Refs. [11,12].

A thorough overview of the advantages and disadvantages of the applicability of low-energy ion-scattering (LEIS) spectroscopy, usually referred to as ISS, in catalysis research has been given elsewhere [13]. ISS can be applied to real as well as to academic catalyst samples. Moreover, it can make a valuable contribution in catalytically motivated surface science studies. When the surfaces of nonconductive catalyst samples are to be analyzed, charge compensation by flood gun is recommended by most workers in the field. The ISS method is extremely surface-specific because its analytical data are defined almost exclusively by the topmost atomic layer. The main application of the method in catalyst characterization is that of following the location or dispersion of elemental constituents at a catalyst's surface. This can be achieved, in the most favorable cases, in a semiquantitative manner by utilizing elemental intensity ratios. A shortcoming of ISS is that it cannot, in general, provide chemical state information. ISS is therefore used in catalyst studies most often with other surface analysis techniques, for instance with XPS, providing complementary information. The ISS/XPS combination is favored because in standard photoelectron spectrometers the energy analyzer, normally used for the measurement of the kinetic energies of electrons, can be switched in polarity so that the energies of the scattered positive ions are measured.

16.3 SAMPLE DAMAGE

Without any doubt analysis must be invalid when the sample is damaged by the probe itself. X-ray-, electron-, and ion-induced sample damage is common during XPS, AES, XAS, SIMS, and ISS analyses of catalysts and can cause the spectrum to change as a function of time of exposure. The most prominent damage effect is the chemical reduction of sensitive species either by heat or by energetic particle or photon irradiation in vacuum. In particular, catalyst precursors such as precipitates, impregnated supports, metal salts, and organometallic compounds must be analyzed with

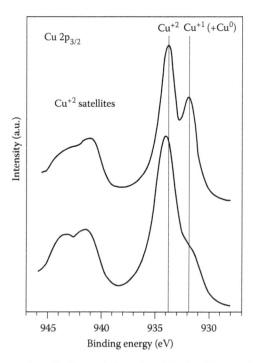

FIGURE 16.3 Cu $2p_{3/2}$ spectra from $CuCl_2$ partially reduced during 3 h x-ray irradiation in different spectrometers (upper spectrum, high thermal load; lower spectrum, reduced thermal load due to water-cooled x-ray housing) at equal x-ray doses of $\approx 8 \times 10^{17}$ photons/mm². (From Klein, J., Li, Ch.P., Hercules, D., and Black, J., *Appl. Spectrosc.*, 38, 729, 1984. With permission.)

utmost caution. There are also examples in which catalysts in calcined states are sensitive to irradiation. A well-known one is the reduction of Cu^{+2} species in calcined copper catalysts during XPS analysis. A detailed study of the damage effects observed with several Cu^{+2} compounds has been made in Ref. [14]; in Figure 16.3 are shown the results in an extreme case. It is of course always possible to minimize the irradiation time itself, especially when a sensitive state-of-the-art spectrometer and appropriate sample preparation are used. Compromises with respect to the optimum signal-to-noise ratio then have to be considered.

In an XPS database on polymers [15], another class of radiation-sensitive materials, the authors sought an approximate guide to the damage rate that would be useful in the optimization of measurement conditions, including acquisition times. They defined a degradation index that gives the percentage damage after $t = 500\,min$ at a fixed x-ray source power. The x-ray degradation index of a sample is obtained from a graph of X_t/X_0 plotted vs. time, t, where X is some characteristic parameter obtained from sequentially recorded XPS spectra (e.g., an atomic ratio or the intensity of some selected spectral feature). This procedure can be applied easily to the XPS analysis of catalysts and also to the other surface spectroscopies. In an analysis of calcined Cu^{+2} catalysts Unger et al. [18] used successfully the 2p satellite: main peak intensity ratio as the decreasing X_t parameter in order to determine optimal times for the acquisition of spectra from a nearly nondamaged sample.

The conclusion is that, because sample damage cannot be fully ruled out, quotation of a damage index as defined above may improve significantly the traceable quality of analytical data. The requirements of the ASTM standard practice E996-94 [16] could be met in this way.

16.4   SAMPLE PREPARATION

Industrial catalysts consist, in most cases, of pellets pressed from active component powders, often mixed with certain binders, for example, technical grade graphite or kaolinite. In their normal form the pellets are often unsuitable for surface analysis because of several experimental problems, amongst them surface charging and high rates of out-gassing into the spectrometer vacuum system. Despite this they can in principle be analyzed by XPS. However, it is recommended that catalyst pellets be carefully ground to powder before analysis. Where binders have been used, the samples are extremely heterogeneous and there is a risk of spectral interference between the active component features and those originating from the binder. Powders are also the usual form of laboratory-made catalysts, but the addition of binders should be avoided whenever possible. A very simple way of mounting a catalyst powder for analysis is by sticking it onto a sample holder with the help of a double-sided adhesive film. Here the analyst must be aware of spectral interferences (typically C and Si signals) arising from the adhesive material itself. Improved signal-to-noise-ratios, together with a reduced risk of sample damage (cf. Section 16.3), may be achieved when the powders are pressed into self-supporting thin wafers. For example, alumina-supported catalysts can be prepared in this way. However, it often occurs that binder-free catalyst powders do not give stable wafers, in which case the powders can be pressed as thin films into soft metal foils, e.g., Au, In, or Pb [11,17]. When a study of heat treatment effects is intended, it has been found useful to press the powder into a thin but integral film on a low-volatility metal foil that assists the lateral homogeneity of the temperature distribution. Al or Au foils are highly suitable [18]. Another useful approach for the fixing of catalyst powders on a sample holder is the use of catalyst particle dispersion in a simple, high-purity organic solvent [19]. The sample holder must be completely covered by the dispersion. A stable layer of powder particles is obtained after evaporation of the solvent, provided that the right combination of solvent, particle concentration, and metal substrate has been found. Good results have also been obtained with zeolite dispersions in isopropyl alcohol or high-purity water deposited on Ni surfaces.

Finally, reference must be made to the international standard ISO 18116:2005 [20] and the international standard under development [21], which covers common recommendations and requirements for sample handling prior to, during, and following surface analysis and are of basic importance for catalyst characterization, too.

Often a catalyst sample has to be analyzed in a condition that is sensitive to atmospheric exposure. For instance, samples taken from pilot or plant reactors can be stored in containers under a protective atmosphere and then inserted into a spectrometer with the help of a glove box [17]. This sounds simple, but a fully controlled realization of this experimental concept is a great challenge for the analyst. Another approach relies on the preparation of a selected catalyst state within a special reaction chamber containing heat and gas treatment options, with facilities for transferring the sample into the analysis chamber without exposure to the ambient. Some surface analysis instrumentation manufacturers offer this optional chamber on their standard spectrometers. There are many papers on this topic, covering the whole range from academically motivated analysis of intermediates in catalyzed reactions on single crystal faces [22] to the activation of, for instance, commercial Cu–Zn–Al oxide methanol synthesis catalysts [18].

16.5   MODELING A CATALYST

Nowadays it is known that, for instance, thin alumina, silica, titania, niobia, ceria, or iron oxide films can be deposited on conductive substrates, and can then act as model supports [23–27]. These supports model the real counterparts by providing the required defects and active sites, thus offering excellent possibilities for a stepwise study of catalyst preparation by all surface analysis techniques currently available. The advantage of thin oxide supports is that they are so sufficiently conducting that the effects of sample charging (cf. Section 16.6) occurring in XPS, AES, and SIMS can be minimized. Catalytically active components can be anchored on these supports with exactly the same wet chemical impregnation techniques and heat/gas formation treatments as are applied normally in technical catalyst production. Sophisticated procedures have also allowed the preparation

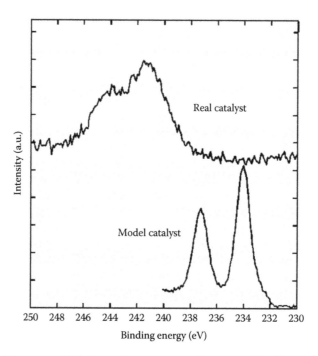

FIGURE 16.4   Mo 3d spectra of MoO_3 on (upper spectrum) commercial silica and (lower spectrum) a thin-film SiO_2 model support, recorded under the same conditions. The spectra are presented as recorded. The positive binding energy shift and the line broadening in the upper spectrum are due to charging. Clearly the signal-to-noise-ratio and the (chemical) resolution of the method are significantly improved in the lower spectrum. (From Niemantsverdriet, J.W., *Spectroscopy in Catalysis*, VCH Verlagsgesellschaft mbH, Weinheim, 1993. With permission.)

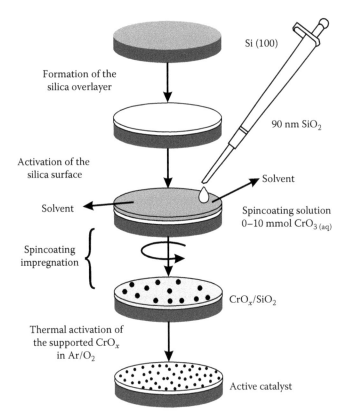

FIGURE 16.5 Preparation of a CrO$_x$/SiO$_2$ ethylene polymerization model catalyst ideally accessible to spectroscopic characterization tools such as XPS and SIMS. (Based on Thüne, P., Loos, J., de Jong, A., Lemstra, P., and Niemantsverdriet, J., *Top. Catal.*, 13, 67, 2000. With permission.)

of well-defined γ-alumina grained layers on Al substrates [29]. The particular advantage of the modeling approach is the strongly enhanced spectral, and therefore also chemical, resolution obtained, for instance, in XPS analysis, as illustrated by Figure 16.4.

In this approach, the surface probed in the spectrometer is identical to the catalytically active surface in a catalytic reactor, and a correlation between catalytic activity and surface characterization results can in principle be established. On the other hand, model supports usually possess only a small catalytically active surface of area, typically 1 cm^2. This small area of catalytically active material, corresponding only to nanomoles, results in difficulties in the measurement of catalytic properties, e.g., conversion, etc., of the working model catalyst.

An instructive example of the approach of using flat model supports is given in Figure 16.5, where an SiO$_2$-on-Si model support is used to prepare a CrO$_x$/SiO$_2$ ethylene polymerization catalyst, similar to its function in the chemical industry [28]. On a silicon wafer a thin film of silica comprising surface silanol groups is established. The precursor for the active phase is deposited onto this silica surface by spin-coating impregnation using well-defined CrO$_{3(aq)}$ solutions. After evaporation of the solvent a controlled and evenly dispersed layer of CrO$_x$ precursors is produced and, finally, after activation, a working ethylene polymerization catalyst is obtained. Its surface is ideally accessible to spectroscopic characterization tools such as XPS and SIMS.

16.6 CHARGING OF INSULATOR SURFACES BY THE PROBE

Irradiation of an insulator by photon, electron, or ion beams results in charging of the specimen surface. Catalysts are normally insulating. Real catalysts are usually multiphase systems with complex morphologies. In the course of analysis the surface of this type of specimen can become differentially charged. The phenomenon of lateral surface potential distributions on irradiated insulator surfaces has been verified for XPS by a spectromicroscopic approach [30,31]. Individual spectra originating from the differentially charged regions of the same sample can be shifted mutually in energy. Merging them together results in a broadened "total spectrum," as usually recorded in an XPS analysis without active charge compensation. As a consequence, not only is the chemical resolution of the technique reduced, but any attempt to analyze the peak shape in terms of physical parameters would be irrelevant.

The charging problem is invariably critical in chemical state analysis of an insulating catalyst surface because reliable and reproducible results can be obtained only when the charging is properly under control. That is to say, stabilization, homogenization, and compensation of the surface charge must be established during analysis. In the literature a broad range of individual approaches to the handling of the charging problem can be found. Solutions are based on the use of electron flood guns. For XPS there is an international standard, ISO 19318:2004 [32], that describes the state of the art. State-of-the-art monochromatic XPS instrumentation comes with substantially enhanced photon flux density, as well as improved charge neutralization strategies. Redesigned electron flood guns with high output that meet the requirements of an enlarged effective emission area and a small energy spread of thermionic electrons are available now [31]. Successful approaches use either a well-stabilized flood gun built coaxially into the analyzer lens assembly, which is combined with magnetic fields to guide the electrons to the right place [33,34], or a combination of low-energy ions and electrons [35]. In this respect, the performance of recent spectrometers has been confirmed by the observation of multiplet splitting features in Cr 2p XP spectra of a chromia powder catalyst mounted simply on the sample holder with double-sided adhesive tape [36].

16.7 CHEMICAL STATE ANALYSIS WITH XPS BY FINGERPRINTING AND REFERENCE TO DATABASES OR TO CHEMICAL STATE PLOTS

Deriving chemical state information about the elements present at a catalyst surface is the most important objective of qualitative catalyst characterization. It can be achieved, more or less successfully, by employing XPS to record specific chemical shifts in the binding energies of photopeaks. These energies depend on the respective nearest-neighbor coordination spheres of the emitter atoms. One approach is that of simple fingerprinting, which relies on reference binding energy data taken from the literature, from databases (the most prominent is the NIST Standard Reference Database [37]), or from the analysis of well-defined reference samples with properly calibrated instruments. Energy calibration should follow the international standard ISO 15472:2001 [38]. An ideal reference is a single-phase sample with no differences between its bulk properties (which can be determined by standard chemical and structure analytical methods) and those of its surface. Unfortunately, in most cases relevant to catalyst characterization, such ideal reference samples do not exist. Often a specimen surface will undergo reaction with atmospheric gases including water vapor; thus, for instance, oxide surfaces may become hydroxylated. An expensive way out of this problem is to prepare reference samples *in vacuo* when possible. Once reliable binding energy reference data are available, an equally reliable method for comparing them with data obtained from the catalyst surface under consideration must be found. But first the binding energy scale of the photoelectron spectrum must be corrected for charging. The earliest, and still the most commonly used method of static charge correction of the binding energy scale of insulating catalyst specimens, is to employ the C 1s peak binding energy of the adventitious carbon, which is present on all samples of interest. Usually it is fixed at a value, BE_{ref}, between 284.6 and 285.0 eV (there are different conventions), and the whole photoelectron spectrum is then shifted by the binding energy difference $BE_{ref} - BE_{measured}$ [32]. This calibration method possesses several uncertainties, as has been pointed out [31,39]. In particular, during XPS analysis of real catalysts it is often found to be impossible to apply the correction

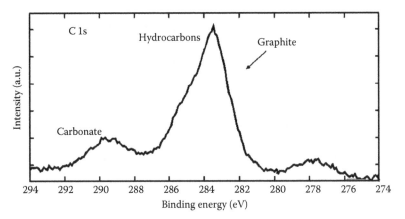

FIGURE 16.6 Complex XP C 1s spectrum recorded from a calcined Cu–Zn–Al oxide/graphite binder methanol synthesis catalyst; the spectrum contains carbonate, hydrocarbon, and graphite contributions. Obviously, determination of the aliphatic hydrocarbon reference C 1s binding energy by fitting this spectrum would not yield an unambiguous result. (From Gross, Th., Lippitz, A., Unger, W.E.S., Lehnert, A., and Schmid, G., *Appl. Surf. Sci.*, 78, 345, 1994. With permission.)

procedure adequately because large systematic errors may intrude due to complexities in the C 1s spectrum itself. The problem can be regarded as a special effect arising from the fact that most of the recent XPS instruments use oil-free pump systems.

Because of that the internal source of the uniform external reference material "adventitious carbon" no longer exists. Under these conditions the nature of the contamination carbon, which is still observed, can be very different depending on the individual history of a sample. It can no longer be taken for granted that the aliphatic hydrocarbon species unequivocally dominates the C 1s spectrum. Figure 16.6 gives a typical example taken from Ref. [40]. In such a situation it is necessary to look for other photopeaks that could be used for static charge referencing. In catalyst characterization support-related photopeaks are interesting candidates provided they can be proved to be sufficiently stable. For instance, Moretti [41] has reported results for an impregnated alumina catalyst where binding energies used for qualitative analysis were obtained by charge referencing with the Al 2p and Al 2s lines (at 74.5 and 119.3 eV, respectively, for the oxide). In silica-supported catalysts the corresponding Si photopeaks could also be used for charge referencing. Another example is the use of the Zn $2p_{3/2}$ photopeak as a static charge reference in the study of different CuO–ZnO impregnated and coprecipitated catalysts [42]. The possibilities and limitations of these approaches have been discussed to some extent in Ref. [40], where it was concluded that the errors in the resulting binding and kinetic energies obtained after referencing could not be smaller than ±0.2 eV.

An alternative, but also not universally used, method for binding energy referencing is to use the photoemission from well-defined noble metal particles deposited on a sample surface to provide an external static charge reference [40,43,44]. Suitable Au and Pd particles can be obtained from a colloidal dispersion. An interlaboratory comparison under VAMAS TWA 2 has been undertaken in order to investigate the performance of charge referencing approaches based on adventitious carbon as well as Au nanoparticles [45].

The simple fingerprint approach outlined above for qualitative analysis of a real-world catalyst sample usually allows differentiation between coexisting chemical species of any one constituent element when the characteristic binding energy differences are ≥0.5 eV, at a rough estimate.

A more sophisticated approach to qualitative analysis is to use the photopeak binding energies plotted against the kinetic energies of x-ray excited Auger peaks and the modified Auger Parameter, in the so-called chemical state plot. Characteristic photoemission and Auger transitions for an element must be recorded in the same spectrum. Details of the principles of the chemical state plots have been reviewed in Ref. [46]. Briefly, the binding energy of the most intense photopeak of the element of interest is plotted on the abscissa, while the kinetic energy of a prominent Auger transition is plotted on the ordinate. The modified Auger Parameter α^* (see also Chapter 15) is defined as the sum of the kinetic energy of the Auger electron and the binding energy of the photoelectron. For α^*, charge corrections cancel as a result of the summation, in contrast to the simple fingerprinting method above, in which binding and kinetic data have to be corrected for charging. Moreover, the basic principles underlying the concept of the Auger Parameter imply that more detailed considerations of the physical chemistry of a sample are possible. For instance, differences in the extra-atomic relaxation energy of the final state can be measured for an element in different chemical bonding states. This energy is related to core hole screening and, consequently, also to polarizability and the local geometry of the nearest neighbors surrounding a photoionized atom.

An α^* grid in chemical state plots is usually drawn as a family of parallel lines with unity slope. All points on any of these lines have the same α^*. One advantage of presenting data in this way is that each well-defined chemical state occupies a unique position on the two-dimensional grid. Fingerprinting based on more than one parameter can thus be achieved. Another advantage is that, where chemical shifts in core-binding energies are too small for unambiguous simple fingerprinting, α^* often has greater chemical sensitivity because additional information from the x-ray excited Auger transitions is embedded. There are many examples of the successful application of the chemical state plotting method in both catalyst characterization and catalysis related studies.

Some instructive examples include a study of precipitation, calcination, and reduction steps for Cu–Zn–Al oxide methanol synthesis catalysts [18,40] and the determination of the coordination of Al and Si in aluminosilicates including zeolites [47]. Figure 16.7 presents the results of another study in which γ-alumina was analyzed before and after activation for catalyzed halogen exchange reactions [48]. Activation of γ-alumina with $CHClF_2$ at 525 K produced a single Al surface species whose position on the Al chemical state plot (Figure 16.7a) was shifted toward the aluminum hydroxyfluoride position, but whose modified Auger Parameter remained much the same as before activation. In the F chemical state plot (Figure 16.7b) the position of the activated γ-alumina was a long way from those of the Al-fluoride or Al-hydroxyfluoride reference samples. The conclusion was that this Al-F surface could be regarded as a precursor of an Al-hydroxyfluoride; i.e., at a certain number of sites on the γ-alumina surface OH^- had exchanged with F^- during activation. Activation in $CHClF_2$ at 675 K, on the other hand, resulted in two coexisting species, indicated by the two identical symbols in Figure 16.7. For one of these species the positions on the two plots were very similar to those observed for the single species found after activation at 525 K. For the other species the positions on the plots were at α^* values similar to those of Al-fluorides or hydroxyfluorides, but with binding energies substantially higher than in AlF_3. The conclusion reached in Ref. [48] was that the latter sample consisted probably of a disordered high F/OH ratio Al-hydroxyfluoride species coexisting with a slightly fluorinated γ-alumina surface, as well as chlorinated and fluorinated coke. The Al-hydroxyfluoride species might have been present as nanoparticles, either with departures from octahedral coordination or under the strong influence of the fluorinated γ-alumina support.

Activation in a different gas atmosphere, CH_3–$CClF_2$, at 525 K resulted in single Al and F species, combined to form an amorphous Al-hydroxyfluoride phase undetectable by XRD analysis. Only this activation procedure provided the full catalytic activity obtained with β-AlF_3, the "reference catalyst"; this correlates well with the fact that γ-alumina activated in this way can be seen to occupy positions on the two chemical state plots very close to those of β-AlF_3.

Another method of chemical state analysis by fingerprinting, which relies on the evaluation of satellite features on the high binding energy sides of core-level peaks in XP spectra, can also be applied to catalyst characterization. These spectral features can be observed with high intensities in the XP spectra of compounds (oxides, halides, fluorides, etc.) of catalytically active transition and rare-earth metals, such as Fe, Co, Ni, Cr, Cu, and La. Detailed discussions of the physical origins of

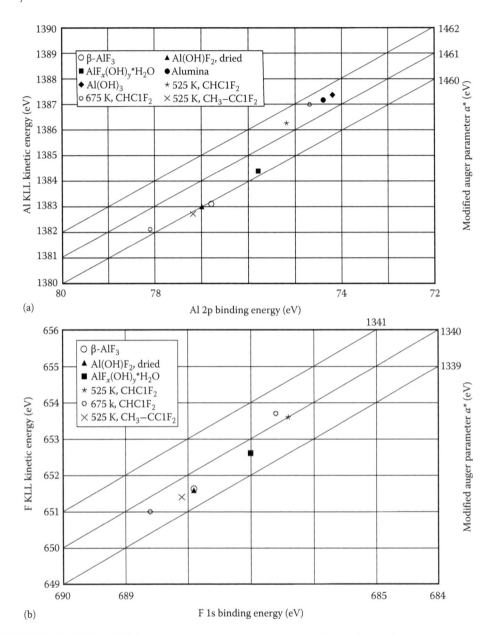

FIGURE 16.7 Al(a) and F(b) chemical state plots for γ-alumina activated for catalyzed halogen exchange reactions with $CHClF_2$ and CH_3–$CClF_2$, at different temperatures. Data recorded from reference samples (β-AlF_3, $Al(OH)F_2$, $AlF_{2.3}(OH)_{0.7}$ *H_2O, and $Al(OH)_3$) are included for comparison. (Redrawn from Hess, A., Kemnitz, E., Lippitz, A., Unger, W.E.S., and Menz, D.-H., *J. Catal.*, 148, 270, 1994.)

the satellite features have been published [49]. For Cu especially these satellites have often been exploited successfully to differentiate quantitatively between Cu^+ and Cu^{+2} species (for a typical Cu^{+2} 2p spectrum, from $CuCl_2$, see Figure 16.3; cf. also Ref. [18]). However, transition metal XP spectra may contain multiplet splitting and satellite features that hamper their straightforward spectral interpretation. Recently, features in Cr^{+3} 2p spectra recorded from Cr_2O_3 and other Cr^{+3} compounds have been analyzed in detail stimulated by investigations of chromia halogen exchange catalysts. Rather complex multiplet structures were revealed [50]. The Cr $2p_{3/2}$ and $2p_{1/2}$ "main" peaks are usually reported as being associated with a screened final state in which charge is transferred from adjacent polarizable ligands to the Cr core hole. There is a coexisting unscreened final state that is represented by the satellite doublet in the spectrum. The Cr $2p_{3/2}$ and $2p_{1/2}$ "main" peaks are themselves split by strong multiplet interactions into many components resulting from the coupling of the unpaired 3d valence electrons in an unfilled d shell with the 2p core hole. Additionally, the multiplet splitting of Cr^{+3} in a solid Cr compound is influenced by adjacent ligands. It is a challenge to resolve all these spectral features by peak fitting approaches, and no simple characteristic binding energy and relative peak area results can be extracted from these Cr^{+3} 2p spectra.

Finally, it must be emphasized that the usefulness of the fingerprint approaches discussed above is limited to extended solid phases. Final state effects in XP spectra, often called particle-size effects, have to be taken into account when nanodispersed active components are present, as in the case of the important supported metal catalysts. Metal ions or metal clusters on a zeolite framework represent other difficult examples relevant to catalysis. The problem is that, in a fingerprint analysis relying only on chemical shifts in core-level and Auger electron energies, it is impossible to differentiate between an extended phase of an oxidized metal and highly dispersed zero-valent nanoparticles (clusters) of the same metal, because in both cases there are usually positive binding energy shifts with respect to the extended metallic phase. This is very important when the reduction states of supported metal catalysts and other systems forming nanophases during formation, e.g., Cu–Zn–Al oxide methanol synthesis catalysts, need to be studied. Binding energy shifts in XPS are generally discussed by considering initial and final state effects [49]. In the case of nanoparticles or clusters, the size and nuclearity, respectively, determine the electron structure, possibly giving rise to initial state alterations. The reduced core hole screening in metal clusters supported on insulating supports definitely affects the final state.

Returning to the problem of the surface chemical state analysis of supported metal catalysts, it should be emphasized that the particle size dependence of the binding energy should be known, or at least reveal a reliable trend. There are different multitechnique ways of measuring it. The metal of interest can be evaporated or sputtered onto a flat substrate, for instance an oxidized Si wafer, a sapphire single crystal, or pyrolytic graphite. Mean particle size-related submonolayer coverages can then be estimated by quantitative XPS [51,52]. More directly, STM has been used to measure a mean particle size for Pd particles supported on highly oriented pyrolytic graphite [55]. Alternatively, it has been found possible with a mass separation technique to collect single-sized particles from a liquid metal ion source on an amorphous carbon substrate, which could then be analyzed by XPS [56]. However, the preferred approach is to prepare the dispersed metal in the same way as in catalyst preparation on typical supports, where the mean particle size can be estimated carefully by titration experiments or, when possible, directly by high-resolution electron microscopy. In this way a direct correlation between XPS binding energy data and mean supported metal particle size can be obtained (cf., for instance, Ref. [53] and Figure 16.8). Additional checks on the structural properties of the supported metal particles could be provided by XAS (EXAFS).

In the case of small clusters entrapped in zeolites, or of zeolites with exchanged metal cations, the complexity of the XPS data interpretation increases. Here the effects of the zeolite lattice field and of the coordination to ligands on a photoionized metal atom must be considered. Ligands have strong effects on the relaxation of the core hole at the emitter atom and on its correlated final state energy. The unusual chemical situation is demonstrated, for example, by the fact that intermediate Me^+ states of transition metals can be stabilized by a zeolite lattice field. However, the results from dedicated academic research confirm clearly the diagnostic potential of XPS here [57]. This includes redox and dehydration treatments, chemisorption, and migration effects considered to occur during temperature treatments of zeolites. Binding energy shift data from transition metals along with their valence state changes in zeolites, useful for fingerprint analysis, have been collected in Ref. [58]. Moreover, it has been shown for oxidized and reduced Cu-exchanged A-, X-, and Y-zeolites that the Auger Parameter as employed in a chemical state plot is also very useful in chemical state analysis of Cu in a zeolite [54,59].

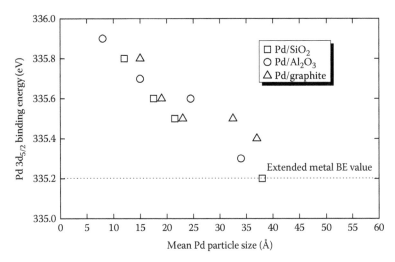

FIGURE 16.8 The particle size dependence of the Pd $3d_{5/2}$ binding energy, measured after Pd dispersion by reduction of precursors on γ-alumina (200 m²/g), silica (250 m²/g), and wide-pore graphite-like carbon. The mean particle sizes were estimated by H_2-titration and electron microscopy. The dotted line represents the binding energy, measured from a metal foil. (Redrawn from Nosova, L., Stenin, M., Nogin, Y., and Ryndin, Y., *Appl. Surf. Sci.*, 55, 43, 1992.)

16.8 CHEMICAL STATE ANALYSIS WITH SIMS BY FINGERPRINTING AND REFERENCE TO DATABASES

An alternative, but more limited, way of obtaining chemical state information about the elements present at a catalyst surface is the application of SIMS in the static mode, SSIMS. Such information can be extracted only when it can be proved that the secondary ion cluster emission is a direct result of fragmentation of the surface itself, i.e., that direct cluster emission processes determine the spectra. Assuming that this requirement is fulfilled, an identified cluster ion can yield unique information about which elements were nearest neighbors on the surface of a catalyst. SSIMS applications in catalyst characterization are usually based on this assumption. The alternative process of cluster emission, that is, atomic recombination in the proximity of the sputtered surface, cannot lead to such a straightforward interpretation of SSIMS data. Moreover, the unequivocal identification of the elemental composition of a cluster ion requires, in the case of complex multicomponent samples such as catalysts, high-resolution mass spectrometry with sufficient sensitivity, e.g., sector field or, a better choice since sensitivity can never be too high, time-of-flight (TOF) instruments. The most important approach, similar to that in XPS, is fingerprinting that relies on reference secondary ion spectra from the analysis of well-defined reference samples or from databases. The most comprehensive set of data can be found in the *Handbook of Secondary Ion Mass Spectrometry* or the CD-ROM version of the Static SIMS Library [60]. Unfortunately, the number of inorganic compounds relevant to catalysis, whose spectra are given in the above database, is rather limited. One of the problems is that these compounds often possess surfaces that are reactive to the environment, and they therefore suffer from surface corrosion. In addition, standard reagent grade quality chemical compound surfaces sufficiently free of organic contamination are difficult to find. Bearing in mind the high surface specificity of SSIMS, spectra may become distorted by both organic contaminants and corrosion by atmospheric components. When reporting the SSIMS spectra of reference samples for inclusion in the common fingerprint database, the recommendations of the relevant ASTM standard practice [61] should be considered carefully. As well as direct fingerprinting, logical deduction of the surface structure from the form of the fragmentation pattern and knowledge of the principles determining fragmentation processes can be helpful. In actual fact, such applications depend on the state of the art in the interpretation of inorganic mass spectra, which must be extracted from the current literature.

A comprehensive review of the application of SIMS techniques to the analysis of the surface composition and structure of catalysts to the study of catalyst preparation and the effects of promoters and poisons, as well as to the study of reactivity, has been given in Ref. [11]. Various kinds of samples have been investigated, amongst them metal oxide catalysts, supported metal catalysts, and zeolites in a variety of states. In most of the studies the occurrence of certain molecular ions or fragmentation patterns has been exploited successfully for surface chemical analysis. One SSIMS

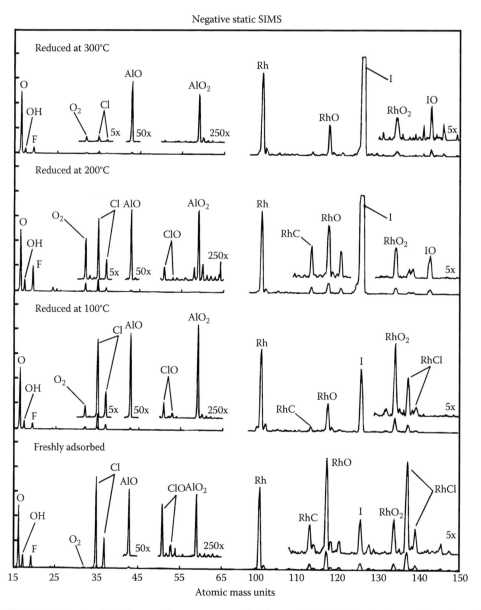

FIGURE 16.9 Negative SSIMS spectral sequences recorded from an Rh/thin-film alumina model catalyst. The freshly adsorbed state and the states after reduction in 1 atm of H_2 at different temperatures were analyzed. Key secondary ions are Cl^-; ClO^-, indicating chlorine-support contact; and $RhCl^-$, indicating rhodium-chlorine contact. (From Borg, H., van den Oetelaar, L., and Niemantsverdriet, J., *Catal. Lett.*, 17, 81, 1993. With permission.)

study [23] of catalyst preparation, the formation of metallic Rh finely dispersed on alumina, which is an NO reduction component in automotive exhaust catalysis, is worth reporting here. It shows again the potential in the use of a thin-film alumina model support in providing optimal conditions for the application of surface spectroscopies. The catalyst was prepared in the usual way by electrostatic adsorption of negatively charged $[RhCl_x(OH)_y(H_2O)_z]^{3(-x-y)+}$ complexes on the surface of an alumina film support that was positively charged in an acid Rh chloride solution. Rh^{3+} precursors were decomposed by subsequent reduction in hydrogen, thus providing dispersed and catalytically active metallic Rh. Analysis of molecular secondary ions of the type $RhCl^-$ showed that a precursor-related Rh–Cl species existed at the alumina surface after adsorption, which then decomposed at reduction temperatures below 200°C, as verified by a decrease in $RhCl^-$ yields. Higher reduction temperatures were needed to remove chlorine from the support, to which chloride ions are known to bind strongly. ClO^- secondary ions are regarded as diagnostic of this chlorine-support binding. Residual chlorine is considered to have adverse effects on the catalytic properties of the activated sample whereas high reduction temperatures may lead to Rh sintering, which is also undesirable. It may be concluded that measurement of the Cl^-, ClO^-, and $RhCl^-$ secondary ion yields and their correlation with activation process parameters (cf. Figure 16.9) may help, when taken together with additional data obtained from other methods, to optimize the catalyst activation procedure itself.

In another study static TOF-SIMS was used to study phase formation on CCl_2F_2-activated chromia catalysts used for Cl/F exchange reactions [62]. Characteristic key fragments and fragment patterns in negative TOF-SIMS spectra were obtained from well-characterized CrF_3 and $CrCl_3$ references (Figure 16.10a and b). These fragment patterns were not found for the activated catalysts. The high-abundance $CrClOF^-$ secondary fragment ion in the spectra of activated chromia was identified as being the "key SIMS fragment" characteristic of chromium oxide halide species (Figure 16.10c). It points to mixed $CrO_xCl_yF_z$ species instead of separated CrX_3 (X = F, Cl) phases. As a result, the $CrClOF^-$ yield increased with increasing halogenation of chromia. These TOF-SIMS results strongly underpin earlier conclusions from XPS, XANES, and tracer studies, that the accumulation of fluorine and chlorine at the surface of chromia by a heterogeneous reaction with a chlorofluorocarbon compound results in the formation of mixed chromium oxide halide species and not in the nucleation of CrF_3 and $CrCl_3$ phases.

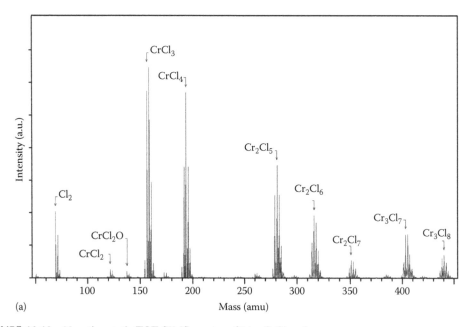

(a)

FIGURE 16.10 Negative static TOF-SIMS spectra of (a) a $CrCl_3$ reference.

(continued)

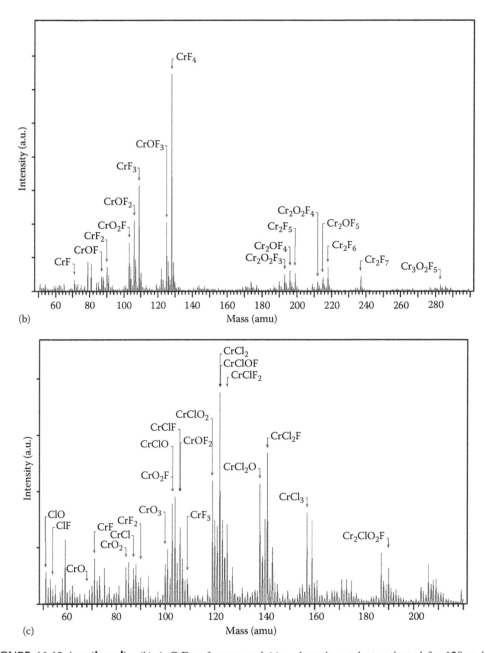

FIGURE 16.10 (continued) (b) A CrF_3 reference and (c) a chromia catalyst activated for 120 m in a CCl_2F_2 atmosphere at 390°C. (Reprinted from Ünveren, E., Kemnitz, E., Oran, U., and Unger, W., *J. Phys. Chem. B.*, 108, 154, 2004. With permission.)

16.9　MISCELLANEOUS

16.9.1　Molecular Probe Approach: Assessment of Acid–Base Properties

The qualitative and (semi)quantitative determination of acid–base properties is of major interest in catalyst characterization. Measurements of these properties are normally made with infrared (IR) spectroscopy, calorimetry, or gravimetry using appropriate molecular probes such as H-bond acceptors, nitriles, and pyridine. However, in certain situations, such as when dealing with solids of low surface area, or when acid site concentrations are of interest, application of these techniques becomes difficult. XPS has proved to be an alternative approach that can also be used for a validation of the former, more common, methods. Chemical shifts exhibited by key atoms in an adsorbed molecule can be used in such a way as to provide qualitative information about the character of the acid sites at the sample surface. The approach can be made quantitative by evaluation of the XPS intensities of the key elements, provided that an appropriate model has been developed for the analysis of the respective XPS intensity data [64,65]. As an example, the binding energy shifts in the N 1s spectra from adsorbed pyridine can be used semiquantitatively to differentiate between Lewis and weak and strong Brønstedt acid sites on various zeolites (H-ZSM-5, H-Y) [64,66]. The strength of the Lewis basicity of basic sites on alkali cation zeolites has also been estimated by measuring the N 1s binding energy shifts in chemisorbed pyrrole molecules. The shifts were correlated with the electron donor capability of lattice oxygen atoms adjacent to the alkali cations [67]. The approach is not restricted to zeolite analysis. Acid site concentrations normalized per aluminum atom have also been obtained with silica–alumina gels [65]. It is interesting that, more recently, this molecular probe XPS approach has been extended successfully to the assessment of the acid–base properties of a completely different class of material, i.e., polymers. Workers in this field have used various molecular probes, e.g., trichloromethane, dimethyl sulfoxide, and hexafluoroisopropanol, in which the key atoms whose core-level shifts were determined following adsorption were Cl, S, and F, respectively [68].

It is clear that there is further potential for the molecular probe approach. For instance, it is attractive to apply another type of molecular probe, developed for the quantitative determination of hydroxyl groups on polymer surfaces by XPS [69], to the characterization of catalyst supports such as silica-gel or alumina. Derivatization of surface OH groups by trifluoroacetic anhydride (TFAA) was shown to be specific and quantitative, thus allowing the total number of such groups on the surface to be estimated. This approach has been applied successfully to inorganic samples, e.g., to the determination of surface concentrations of OH groups on various lithium niobate films formed from aqueous solutions [70]. However, it should be emphasized that such XPS analysis is rather sophisticated and should not be regarded as routine.

16.9.2 ALLOYING AT BIMETALLIC SUPPORTED CATALYSTS

Bimetallic systems have long been of interest in catalysis. Adding a metallic modifier to the primary metal may provide ligand and ensemble effects or multifunctionality, resulting in enhanced catalyst performance. The characterization of bimetal catalysts has been found to be a considerable challenge to surface analysis techniques, and multitechnique approaches are usually required. The contribution of XPS has been that of the confirmation of the presence of zero-valent metals at the surface, which is a prerequisite for alloy particle formation on the catalyst support. Of course, particle size effects on the binding energies must be considered carefully (cf. Section 16.7). Moreover, there have been XPS studies on Pt/sp-metal alloys, which have revealed that alloying may also have characteristic effects on binding energies (shifts of up to ≈ 0.5 eV in Sn 3d and Pt 4f in Pt–Sn alloys), and on widths and asymmetries of the core-level lines of alloy constituents [71]. SSIMS can be used too [63,72], to look for $A_x B_y^+$ cluster ions in the secondary spectra; they should appear when metal A atoms are adjacent to metal B atoms forming an alloy cluster. The association of Pt and Sn on γ-alumina with a high surface area, as measured by the BET method, studied by monitoring PtSn$^+$ and PtSnO$^+$ secondary cluster ions by FAB-MS, has been described in detail in Ref. [63]. The interpretation, supported by dedicated temperature programmed reduction (TPR) studies, was that Sn, Pt, and O are indeed the nearest neighbors at the surface of the catalyst after reduction in hydrogen. The other interesting result was that there were many different support-related $Al_xO_yH_z^{\pm}$ cluster ions in the spectra, but no MeAl$^+$ or MeAl$_xO_yH_z^+$ ions. The latter species would be expected if cluster ion formation by recombination determined the mass spectrum.

A powerful tool for tackling the problem of alloy formation in highly dispersed metals on high-surface area supports is the EXAFS technique (cf. Section 16.2, and Chapters 7 and 15), but it requires that the constituents have high scattering cross-sections [73]. From an analysis of EXAFS data the number of atoms of a particular kind at a particular distance from a chosen absorber atom, i.e., those in coordination spheres, can be evaluated. An EXAFS determination for each alloy constituent can provide information on the individual environment of atoms of that constituent on the support, indicating whether or not alloy formation has occurred. It should be mentioned that EXAFS is also a very powerful method for investigating promoter–support interactions [73].

The interpretation of XPS, SSIMS, and EXAFS data obtained from complex systems such as bimetallic catalysts should be supported by other more direct methods, e.g., XRD, Mössbauer spectroscopy, or analytical electron microscopy (whenever possible), or by the indirect methods usually applied in catalyst characterization, e.g., chemisorption, TPR, etc. As an illustration there have been some successful multitechnique studies of Pt–Re and Pt–Sn reforming catalysts [74]. Employing XPS, EXAFS, and L-edge x-ray absorption spectroscopy (XANES), it was found that instead of substantial Pt–Re alloying, coexisting Pt metal and rhenium(IV) oxide species were formed. It was suggested that the latter provide enhanced stability of the catalyst against the sintering of Pt particles. Study of the Pt–Sn reforming catalyst by XPS, EXAFS, TPR, XRD, and reaction measurements revealed that the formation of bimetallic Pt–Sn entities depended on the conditions used in the preparation of the catalysts and on the nature of the support. Thus Sn is present primarily as an Sn(II) oxide species with a high affinity for alumina after high-temperature reduction of Pt–Sn/γ-Al$_2$O$_3$, whereas on silica alloy formation was found to occur under the same conditions.

16.9.3 IN-DEPTH ANALYSIS

Depth profiling of a catalyst surface by ion etching combined with surface analysis would seem to be a very attractive tool in catalyst characterization. However, the range of application of this technique is rather restricted. It has indeed been used successfully, in the analysis of Raney nickel catalysts [75]. The effects of segregation phenomena at Al/Ni alloy surfaces, and the degree of reactivity of the aluminum on the development of the surface during leaching with NaOH to produce active hydrogenation catalysts, were investigated by Auger depth profiling and XPS. Such in-depth analyses are in principle not very different from those usually undertaken in the field of metallurgy in which corrosion, passivation, and segregation phenomena are studied.

In general, however, an in-depth analysis based on ion etching performed on standard multiphase/multicomponent catalysts is full of pitfalls and should be avoided at all costs (see Chapter 10). The notion that it should be possible to derive structural models for catalysts (e.g., layered promoter or "cherry" models, diffusion of promoter components into the bulk, etc.) from such profile data must be firmly rejected. It is also dangerous to believe that surface oxides, contaminant carbon or coke, as examples, can be removed from a catalyst surface by sputter cleaning in order to reveal the "real" sample surface. An instructive example has been given in Ref. [76] in which a study of Ni–Mo/Al$_2$O$_3$ catalysts by ISS sputter depth profiling was reported. The result was that the shape of the curve that related Ni/Mo ISS intensity ratios to increasing bombardment time was influenced strongly by the incident ion energy. The reason is that multiphase/multicomponent samples such as catalysts will usually suffer strong ion beam modification effects during sputtering. This has been documented for some years [77,78] and has also been discussed in standard guides for specimen handling and depth profiling published by ISO [20,21] and ASTM [79]. Reference [77] provides a database of more than a hundred entries consisting of binary and ternary metal compounds containing O, N, F, S, and Cl, in which the surface composition is altered by sputtering.

XPS, however, can in principle provide nondestructive chemical in-depth analysis for a range of depths, limited to a few nanometers, by using a variation of the information depth in a method related to the EAL (cf. Section 16.2) for element-specific photoelectrons in a certain phase. The information depth can be varied either by changing stepwise the angle of photoemission (angle-resolved XPS

[ARXPS]) or by changing the energy of the exciting x-ray photons (see Chapter 7). The extraction of chemical depth profile information from careful ARXPS experiments involves a complex mathematical procedure with latent ambiguity [80], particularly since the method is strictly speaking applicable only to very flat surfaces (see Chapter 8). It may be relatively straightforward for catalyst surfaces with well-defined morphologies, as, for instance, with stratified layers, but it will fail in most cases for real multiphase systems, e.g., supported catalysts. The extraction of chemical depth profile information from variable x-ray energy XPS measurements avoids the changes in geometry required in ARXPS, but needs a synchrotron radiation source. The method has been applied successfully to powdered zeolite catalysts [81,82] consisting of Na-type faujasite zeolite (NaY) and H-type faujasite zeolite (HY). The HY zeolite was prepared from the NaY zeolite through a triple ion exchange followed by calcination. Al- and Na-rich overlayers at the external surfaces of NaY zeolite particles, and a gradual decrease in the Al: Si ratio from the external surface to the bulk for HY zeolite particles, were found. Another application was that of the differentiation of surface and bulk oxygen species, and their relation to the catalytic activity, of $La_{0.8}Sr_{0.2}CoO_3$ perovskite-type oxidation catalysts [83].

16.10 QUANTITATIVE SURFACE ANALYSIS OF CATALYSTS: COMPOSITION, DISPERSION, AND COVERAGE

In most cases, it is XPS that has been used to acquire quantitative data concerning catalyst surfaces. The quantitative data of interest are elemental concentration ratios; dispersion of the catalytically active component, or promoter, in the case of supported catalysts; and the coverage of the support by the promoter. For a given loading and support surface area, dispersion and coverage are not independent parameters. Dispersion is defined as the ratio of surface promoter atoms to the total number of promoter atoms and is correlated with the catalyst activity.

There are two principal ways in which surface concentration data may be obtained from the evaluation of XPS intensities. One relies on a first-principle description of the photoelectron emission from a solid surface and the other on empirically determined elemental sensitivity factors. Either approach can be used in its simplest form to estimate elemental surface concentration ratios from XPS intensity ratios for a catalyst surface, provided the sample is homogeneous and isotropic, i.e., all elements are uniformly distributed in the surface layer sampled by XPS. However, a heterogeneous catalyst is in fact just that—heterogeneous; it can be both multiphase and of a complex structure. Operative terms here are porosity, inner and outer surface, texture, segregation, etc. Nevertheless, the simple procedures have been used extensively in catalyst characterization studies for straightforward interpretations of XPS intensity ratios although, in many cases, it was not established that the samples met the requirements of the procedure. Papers lacking this validation are, from both scientific and quality assurance points of view, substandard. For the limiting case of high-surface area ($>200\,m^2/g$), high-dispersion, supported catalysts, it has been shown [84] that XPS provides "bulk characteristic" rather than "surface characteristic" promoter: support element concentration ratios, because the average depth of information in the method results in the signal intensity consisting of an integral over the inner pore system.

Taking into account dispersion and coverage for the important class of high area supported catalysts, appropriate quantitative models have been developed that are based on the idea that such samples can be modeled as stacks of support material sheets each covered by promoter particles (Figure 16.11 or Ref. [84]). The surface area of the support and the density of the support material define the corresponding sheet thickness d. For example, a typical porous silica is characterized by $d = 2.6$ nm. When considering this stratified layer model it turns out that the simple expression

$$(I_p/I_s)_{\text{monodisperse promoter}} = (p/s)\,(\sigma_p/\sigma_s) \tag{16.1}$$

can be derived for the case of monodisperse (i.e., single atomic distribution) promoter particles. Here "p" and "s" refer to promoter and support, respectively, so that I_p/I_s is the measured intensity ratio of the chosen promoter and support elements; p/s is their atomic ratio; and σ_p/σ_s is the ratio of their

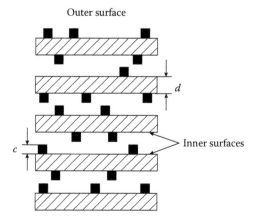

FIGURE 16.11 Stratified layer model for supported catalysts developed for the quantitative interpretation of XPS intensity data. The support is modeled as a stack of sheets of thickness d. This thickness is determined by the density ρ_s and the surface area A_s of the support by $d = 2/\rho_s A_s$. The promoter is represented by single-sized cubes with dimension c. Within the model the surface coverage Θ can be calculated for a given promoter weight fraction x and density ρ_p by $\Theta = x/(1 - x)A_s\rho_p c$. (Based on Kerkhof, F.P.J.M. and Moulijn, J.A., *J. Phys. Chem.* 83, 1612, 1979.)

photoemission cross-sections. Equation 16.1 is a good approximation if the difference in kinetic energies of the respective photoelectrons is small [84]. Crystallite growth, or sintering phenomena of the promoter on the support, can also be taken into account with the assumption of single-sized cubic promoter crystallites. In that case I_p/I_s depends also on the crystallite dimension c (Figure 16.11). The relation between the promoter: support XPS intensity ratios characteristic of the coarsely dispersed promoter particles and the monodisperse ones is given by

$$(I_p/I_s)_{\text{coarsely dispersed promoter}}/(I_p/I_s)_{\text{monodisperse promoter}} = (1-e^{-\alpha})/\alpha \qquad (16.2)$$

where α is a dimensionless crystallite size defined by $\alpha = c/\lambda$. Hence it follows that the mean crystallite size c of the promoter can be determined from the respective XPS intensity ratios and Equation 16.2 when λ, the inelastic mean free path-length of the photoelectrons originating from the promoter and traveling through the promoter, is known [84]. Once the mean crystallite size is known the mean dispersion of the promoter can be estimated.

The above stratified layer model [84] has been developed [85] further by considering random orientations of the stacks, which should be closer to the real situation. The authors also analyzed the dependence of I_p/I_s on the promoter particle shape. The result was that, for genuinely randomly oriented samples, the XPS signal from the promoter phase, present as convex particles of equal size, is determined by the surface: volume ratio of the promotor compound. This property is proportional to dispersion. Thus, in the extended model the experimental XPS I_p/I_s ratio is a direct measure of the dispersion. I_p/I_s can be converted into dimensions characteristic of some of the likely geometries of particles on the support (e.g., layers with a uniform thickness d, spheres with diameter $3d$, or half-spheres with radius $2.25d$ [85]), together with an appropriate coverage. However, the following parameters must be known exactly: elemental concentrations and densities of promoter and support, photoemission cross-sections characteristic of the respective promoter and support core levels, inelastic mean-free path lengths of the respective photoelectrons, analyzer transmission, and specific area of support. Once again it should be emphasized that application of the stratified layer model [84], as well as that of the randomly oriented layer model [85], requires single-sized and uniformly distributed promoter particles. Because in many practical situations that is not the case, Kaliaguine et al. [86] have developed another important extension accounting for two populations of crystallites on the outermost support sheet, characterized by different crystallite sizes and coverages. Relevant practical situations described by this model are, for instance, repartition of supported material at the outer surface, and promoter segregation from the pores to the outer surface. Its application requires that information on the crystallite size and coverage for both populations is obtained by independent methods such as electron microscopy or XRD.

Finally it must be said that extreme care must be exercised in the interpretation of XPS intensity ratios recorded from catalysts. The accuracy of the determination of dispersion and coverage data in this way suffers from the general inhomogeneity of practical samples; from the effects of surface roughness arising from the fact that most samples are investigated in the form of powder; and from problems in estimating parameters such as inelastic mean free paths, material densities, etc. However, when the standard techniques for catalyst characterization, e.g., selective chemisorption, TEM, and XRD line profile analysis, are unable to estimate dispersion data or crystallite sizes, it can be useful to try XPS as an alternative means of trouble-shooting. This might involve specialized extensions [87,88] of the models presented in Refs. [84,85] or the development of entirely new approaches to fit individual situations. Again, only detailed descriptions of the data reduction procedures used to reach this goal will meet the requirements of best practice in XPS summarized in Ref. [16].

Another, more direct way of estimating the coverage of a catalyst support by promoter species should be mentioned. This approach relies on the unique surface specificity of ISS. In Ref. [89] coverage data obtained by ISS from a titania support (surface area $\approx 50\,m^2/g$) with different WO_3 loadings have been reported (Figure 16.12). The coverage of the support by monolayer-like W species was monitored

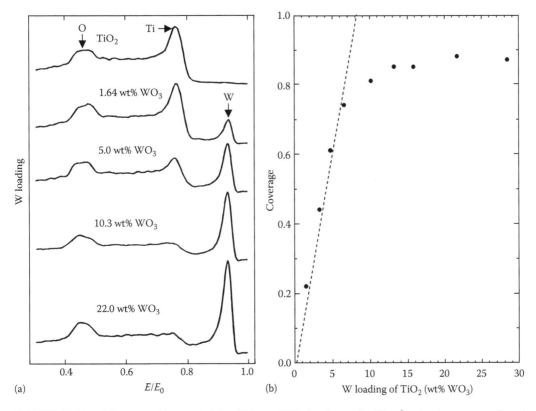

FIGURE 16.12 ISS spectra (a) recorded for different WO_3 loadings of a 50-m^2/g titania support, plotted together with the spectrum of the unloaded support. Surface coverage data determined by application of Equation 16.3 are shown in (b). The dotted lines represent theoretical coverage obtained by assuming monolayer-like dispersion of the W phase and 1:1 stoichiometry of the reaction between WO_3 and surface hydroxyls on the support. (Redrawn from Fiedor, J.N., Houalla, M., Proctor, A., and Hercules, D.M., *Surf. Interface Anal.*, 23, 234, 1995. With permission.)

by following the changes in the Ti ISS intensity (Figure 16.12a), characteristic of the support, normalized to the O ISS peak intensity. The surface coverage Θ was estimated from

$$\Theta = 1 - (Ti/O)_{cat} / (Ti/O)_{sup} \tag{16.3}$$

where Ti/O are the ISS peak intensity ratios of Ti and O measured without (subscript "sup"), and with (subscript "cat"), WO_3 loading.

Details concerning the assumptions to be made in surface coverage measurement by ISS are discussed in Ref. [90]. The main result is shown in Figure 16.12b, in which surface coverages estimated by Equation 16.3 are plotted. Up to loadings of 8–10 wt% WO_3 a monolayer-like dispersion of W species was observed. The leveling off of the plot of coverage for higher loadings indicates multilayer formation. The advantage of direct ISS analysis of the support coverage is that it also can be applied to systems where a chemisorption approach suffers from the severe problem that a molecule would have to be found that adsorbs, in this example, exclusively on tungsten oxide and not on the TiO_2 support.

16.11 STATE OF THE ART AND OUTLOOK

The development of surface analytical instrumentation has been very extensive during the last 10 years. Technical solutions for overcoming the problems of charging of insulators have been successfully developed for XPS, ToF-SIMS, and AES spectrometers. As a result, improved "chemical resolution" in studies of oxide and other nonconductive catalyst samples including supported metal catalysts has been achieved. This improvement is being exploited currently in the identification of chemical species relevant to heterogeneously catalyzed reactions, with increased sensitivity. With regard to lateral resolution, surface characterization by ToF-SIMS and AES can now be performed at resolutions in the nanometer range <100 nm. When XPS and XAS are used in conjunction with synchrotron radiation facilities, surface chemical information on the nanometer-scale may also be provided. Such progress now enables more detailed studies of phase formation and aging to be carried out for practical catalysts.

Catalyst models built on realistic supports and prepared as thin films have made big steps forward, and based on this progress the situation at the surface of a real working catalyst may be simulated much more precisely. Additionally, by avoiding charging problems, the full performance capability of state-of-the-art surface spectroscopies may be exploited successfully in the identification of catalytically active species.

The other drawback in the application of UHV-based surface spectroscopies to catalysis, which has always made studies of active catalyst surfaces difficult, if not impossible, is the "pressure gap," but this is being eliminated a step at a time. The millibar level has been reached in recent *in situ* XPS and XAS studies of over reaction model catalysts, and the challenge of the future will be to close the gap still further. The operational availability of synchrotron radiation is providing the major support to this important advance. In the future the development of intense pulsed x-rays provided by a free-electron laser will offer new options for surface chemical analysis. One of these options is that of time-resolved observations of chemical reactions over catalysts, where such reactions may proceed on time scales of a millionth of a second or less.

Last but not least, the progress in standardization of surface analytical techniques under ISO TC 201, reached by international experts during the last 15 years, must be mentioned here. It provides effective support for the improvement of the overall quality of catalyst characterization using surface chemical analysis, by describing validated methods of analysis and data reduction.

REFERENCES

1. B. Delmon, Surface analysis and the complexity of real catalysts, *Surf. Interface Anal.* 9, 195 (1986).
2. D. Ogletree, H. Bluhm, G. Lebedev, C. Fadley, Z. Hussain, and M. Salmeron, A differentially pumped electrostatic lens system for photoemission studies in the millibar range, *Rev. Sci. Inst.* 73, 3872 (2002).

3. H. Bluhm, M. Hävecker, A. Knop-Gericke, E. Kleimenov, R. Schlogl, D. Teschner, V. Bukhtiyarov, D. Ogletree, and M. Salmeron, Methanol oxidation on a copper catalyst investigated using in situ X-ray photoelectron spectroscopy, *J. Phys. Chem. B* 108, 14340 (2004).

4. A. Knop-Gericke, M. Hävecker, Th. Schedel-Niedrig, and R. Schlögl, High-pressure low-energy XAS: A new tool for probing reacting surfaces of heterogeneous catalysts, *Top. Catal.* 10, 187 (2000).

5. J. W. Niemantsverdriet, *Spectroscopy in Catalysis*, VCH Verlagsgesellschaft mbH, Weinheim, New York, 1993, Chapter 3.

6. R. Austermann, D. Denley, D. Hart, P. Himelfarb, R. Irwin, M. Narayana, R. Szentirmay, S. Tang, and R. Yeates, Catalyst characterization, *Anal. Chem.* 59, 68R (1987).

7. ISO 18115:2001, Surface chemical analysis—Vocabulary ISO, Geneva.

8. G. Margaritondo, C. Coluzza, and R. Sanjinés, *Photoemission: From Hertz and Einstein to Spectromicroscopy, Photoemission: From the Past to the Future* (C. Coluzza, R. Sanjinés, and G. Margaritondo, Eds.), Ecole Polytechnique Fédérale de Lausanne, Lausanne, Switzerland, 1992, pp. 44–95.

9. J. Kawai, *Encyclopedia of Analytical Chemistry* (R. A. Meyers, Ed.), Wiley, Chichester, 2000, pp. 13288–13315; F. de Groot, High-resolution x-ray emission and x-ray absorption spectroscopy, *Chem. Rev.* 101, 1779 (2001).

10. D. Harrison Jr., Sputtering models—A synoptic view, *Rad. Effects* 70, 1 (1983).

11. H. J. Borg and J. W. Niemantsverdriet, *Applications of Secondary Ion Mass Spectrometry in Catalysis and Surface Chemistry, Catalysis: A Specialist Periodical Report*, Vol. 11 (J. J. Spivey and A. K. Agarwal, Eds.), The Royal Society of Chemistry, Cambridge, 1994, pp. 1–50.

12. M. J. P. Hopstaken, R. Linke, W. J. H. van Gennip, and J. W. Niemantsverdriet, *Applications in Catalysis, ToF-SIMS: Surface Analysis by Mass Spectrometry* (J. C. Vickerman and D. Briggs, Eds.), Surface Spectra and IM Publications, Manchester and Chichester, 2001, pp. 697–725.

13. B. Horrell and D. Cocke, Application of ion-scattering spectroscopy to catalyst characterization, *Catal. Rev. Sci. Eng.* 29, 447 (1987); H. Niehus, W. Heiland, and E. Taglauer, Low-energy ion scattering at surfaces, *Surf. Sci. Rep.* 17, 213 (1993).

14. J. Klein, Ch. P. Li, D. Hercules, and J. Black, J. C. Klein, Ch. P. Li, D. M. Hercules, and J. F. Black, Decomposition of copper compounds in X-ray photoelectron spectrometers, *Appl. Spectrosc.* 38, 729 (1984).

15. G. Beamson and D. Briggs, *High Resolution XPS of Organic Polymers: The Scienta ESCA 300 Database*, John Wiley & Sons, Chichester, 1992, Chapter 12.

16. Standard Practice for Reporting Data in Auger Electron Spectroscopy and X-Ray Photoelectron Spectroscopy (E996-94), *Annual Book of ASTM Standards*, Vol. 03.06, ASTM International, West Conshohocken, PA, 2003, p. 796.

17. J. Grimblot, L. Gengembre, and A. D'Huysser, XPS: A routine technique to characterize surface of practical catalysts? *J. Electron Spectr. Relat. Phenom.* 52, 485 (1990).

18. B. Peplinski, W. E. S. Unger, and I. Grohmann, Characterization of Cu–Zn–Al oxide catalysts in the precipitaed, calcined and reduced state by means of XPS with the help of a finger-print data base, *Appl. Surf. Sci.* 62, 115 (1992).

19. H. Shimada, N. Matsubayashi, M. Imamura, T. Sato, and A. Nishijima, Determination of external surface composition of zeolite particles by synchrotron radiation XPS, *Catal. Lett.* 39, 125 (1996).

20. ISO 18116:2005, Surface chemical analysis—Guidelines for preparation and mounting of specimens for analysis.

21. ISO/FDIS 18117, Surface chemical analysis—Handling of specimens prior to analysis, international standard under development.

22. G. A. Somorjai, *Chemistry in Two Dimensions—Surfaces*, Cornell University Press, Ithaca, NY, 1991.

23. G. Hoflund, D. Asbury, and R. Gilbert, A characterization study of platinum-tin oxide films supported on Al_2O_3, *Thin Solid Films* 129, 139 (1985); H. Borg, L. van den Oetelaar, and J. Niemantsverdriet, Preparation of a rhodium catalyst from rhodium trichloride on a flat, conducting alumina support studied with static secondary ion mass spectrometry and monochromatic X-ray photoelectron spectroscopy, *Catal. Lett.* 17, 81 (1993).

24. L. Coulier, J. A. R. van Veen, and J. Niemantsverdriet, TiO_2-supported Mo model catalysts: Ti as promoter for thiophene HDS? *Catal. Lett.* 79, 149 (2002).

25. J. Niemantsverdriet, A. Engelen, A. de Jong, W. Wieldraaijer, and G. Kramer, Realistic surface science models of industrial catalysts, *Appl. Surf. Sci.* 144/145, 366 (1999).

26. H.-J. Freund, M. Bäumer, J. Libuda, T. Risse, G. Rupprechter, and S. Shaikhutdinov, Preparation and characterization of model catalysts: From ultrahigh vacuum to in situ conditions at the atomic dimension, *J. Catal.* 216, 223 (2003).

27. O. Shekhah, W. Ranke, and R. Schlögl, Styrene synthesis: In situ characterization and reactivity studies of unpromoted and potassium-promoted iron oxide model catalysts, *J. Catal.* 225, 56 (2004).

28. P. Thüne, J. Loos, A. de Jong, P. Lemstra, and J. Niemantsverdriet, Planar model system for olefin polymerization: The Phillips CrO$_x$/SiO$_2$ catalyst, *Top. Catal.* 13, 67 (2000).

29. E. Ruckenstein and M. Malhotra, Splitting of platinum crystallites supported on thin, nonporous alumina films, *J. Catal.* 41, 303 (1976); A. Jiménez-González and D. Schmeisser, Electron spectroscopic studies of Mo- and Pt-modified γ-Al$_2$O$_3$ model catalysts, *J. Catal.* 130, 332 (1991); B. Frederick, G. Apai, and T. Rhodin, *Surf. Sci.* 250, 59 (1991).

30. C. Coluzza, J. Almeida, T. dell'Orto, F. Gozzo, H. Berger, D. Bouvet, M. Dutoit, S. Contarini, and G. Magaritondo, Spatially localized energy shifts in the photoemission spectroscopy of insulators, *J. Appl. Phys.* 76, 3710 (1994).

31. U. Gelius, B. Wannberg, Y. Nakayama, and P. Baltzer, ESCA studies of polymers, and insulators using monochromatic x-ray excitation, in: *Photoemission: From the Past to the Future* (C. Coluzza, R. Sanjinés, and G. Margaritondo, Eds.), Ecole Polytechnique Fédérale de Lausanne, Lausanne, Switzerland, 1992, pp. 143–211.

32. ISO 19318:2004, Surface chemical analysis—x-ray photoelectron spectroscopy—reporting of methods used for charge control and charge correction. ISO Geneva.

33. A. R. Walker, European Patent No. EP 0 243 060 B1.

34. S. C. Page, European Patent No. EP 0 458 498 B1.

35. Physical Electronics PHI 5000 VersaProbe, Product brochure, Chanhassen, MN; Physical Electronics PHI Quantera, Product brochure, Chanhassen, MN, 2004.

36. E. Ünveren, E. Kemnitz, S. Hutton, A. Lippitz, and W. Unger, Analysis of highly resolved x-ray photoelectron Cr 2p spectra obtained with a Cr$_2$O$_3$ powder sample prepared with adhesive tape, *Surf. Interface Anal.* 36, 92 (2004).

37. NIST X-ray Photoelectron Spectroscopy Database, NIST Standard Reference Database 20, Version 3.4 (Web Version), Data compiled and evaluated by Charles D. Wagner, Alexander V. Naumkin, Anna Kraut-Vass, Juanita W. Allison, Cedric J. Powell, and John R. Rumble Jr., http://srdata.nist.gov/xps/index.htm

38. ISO 15472:2001, Surface chemical analysis—x-ray photoelectron spectrometers—calibration of energy scales.

39. P. Swift, Adventitious carbon—The panacea for energy referencing? *Surf. Interface Anal.* 4, 47 (1982); S. Kohiki and K. Oki, Problems of adventitious carbon as an energy reference, *J. Electron Spectr. Rel. Phenom.* 33, 375 (1984).

40. Th. Gross, A. Lippitz, W. E. S. Unger, A. Lehnert, and G. Schmid, Static charge correction in XPS on Cu–Zn–Al oxide catalyst precursors with the help of deposited Pd particles as a reference material, *Appl. Surf. Sci.* 78, 345 (1994).

41. G. Moretti, XPS Studies of characterized Cu/Al$_2$O$_3$, Zn/Al$_2$O$_3$ and CuZn/Al$_2$O$_3$, catalysts, *Surf. Interface Anal.* 17, 745 (1991).

42. Y. Okamoto, K. Fukino, T. Imanaka, and S. Teranishi, Surface characterization of copper(II) oxide-zinc oxide methanol-synthesis catalysts by x-ray photoelectron spectroscopy. 2. Reduced catalysts, *J. Phys. Chem.* 87, 3747 (1983); G. Moretti, M. Lo Jacono, G. Fierro, and G. Minelli, Surface characterization of CuO–ZnO–Al$_2$O$_3$ methanol-synthesis catalysts by XPS, *Surf. Interface Anal.* 9, 246 (1986); G. Moretti, G. Fierro, M. Lo Jacono, and P. Porta, Characterization of CuO–ZnO catalysts by X-ray photoelectron spectroscopy: Precursors, calcined and reduced samples, *Surf. Interface Anal.* 14, 325 (1989).

43. Th. Gross, K. Richter, H. Sonntag, and W. E. S. Unger, A method for depositing well defined metal particles onto a solid sample suitable for static charge referencing in X-ray photoelectron spectroscopy, *J. Electron Spectrosc. Relat. Phenom.* 48, 7 (1989).

44. Th. Gross, M. Ramm, H. Sonntag, W. E. S. Unger, H. M. Weijers, and E. H. Adem, An XPS analysis of different SiO$_2$ modifications employing a C 1s as well as an Au 4f$_{7/2}$ static charge reference, *Surf. Interface Anal.* 18, 59 (1992).

45. W. Unger, Th. Gross, O. Böse, A. Lippitz, Th. Fritz, and U. Gelius, VAMAS TWA2 Project A2: Evaluation of static charge stabilization and determination methods in XPS on non-conducting samples. Report on an inter-laboratory comparison, *Surf. Interface Anal.* 29, 535 (2000).

46. C. Wagner and A. Joshi, The auger parameter, its utility and advantages: A review, *J. Electron Spectrosc. Relat. Phenom.* 47, 283 (1988).

47. C. Wagner, D. Passoja, H. Hillery, T. Kinisky, H. Six, W. Jansen, and J. Taylor, *J. Vac. Sci. Technol.* 21, 933 (1982).

48. A. Hess, E. Kemnitz, A. Lippitz, W. E. S. Unger, and D.-H. Menz, ESCA, XRD, and IR characterization of aluminum oxide, hydroxyfluoride, and fluoride surfaces in correlation with their catalytic activity in heterogeneous halogen exchange reactions, *J. Catal.* 148, 270 (1994).

49. C. S. Fadley, in: *Electron Spectroscopy: Theory, Techniques and Applications*, Vol. 2 (C. R. Brundle and A. D. Baker, Eds.), Academic Press, New York, 1978, pp. 1–156; D. Sarma, P. V. Kamath and C. N. R. Rao, *Chem. Phys.* 73, 71 (1983); B. Veal and A. Paulikas, *Phys. Rev. B: Condensed Matter* 31, 5399 (1985).

50. E. Ünveren, E. Kemnitz, S. Hutton, A. Lippitz, and W. E. S. Unger, Analysis of highly resolved x-ray photoelectron Cr 2p spectra obtained with a Cr_2O_3 powder sample prepared with adhesive tape, *Surf. Interface Anal.* 36, 92 (2004).

51. S. Kohiki, Photoemission from small Pd clusters on Al_2O_3 and SiO_2 substrates, *Appl. Surf. Sci.* 25, 81 (1986); S. Kohiki and S. Ikeda, Photoemission from small palladium clusters supported on various substrates, *Phys. Rev. B* 34, 3786 (1986).

52. I. Jirka, An ESCA study of copper clusters on carbon, *Surf. Sci.* 232, 307 (1990).

53. L. Nosova, M. Stenin, Y. Nogin, and Y. Ryndin, EXAFS and XPS studies of the influence of metal particle size, nature of support and H_2 and CO adsorption on the structure and electronic properties of palladium, *Appl. Surf. Sci.* 55, 43 (1992).

54. G. Moretti and P. Porta, New advancements in the theory of the Auger parameter: Applications to the characterization of small metallic particles, *Surf. Interface Anal.* 20, 675 (1993).

55. A. Sartre, M. Phaner, L. Porte, and G. Sauvion, STM and ESCA studies of palladium particles deposited on a HOPG surface, *Appl. Surf. Sci.* 70/71, 402 (1993).

56. S. DiCenzo, S. Berry, and E. Hartford, Photoelectron spectroscopy of single-size Au clusters collected on a substrate, *Phys. Rev. B* 38, 8465 (1988).

57. W. Grünert, U. Sauerlandt, R. Schlögl, and H. Karge, XPS investigations of lanthanum in faujasite-type zeolites, *J. Phys. Chem.* 97, 1413 (1993).

58. E. Shpiro, G. Antoshin, and Kh. Minachev, in: *Catalysis on Zeolites* (D. Kalló and K. M. Minachev, Eds.), Academiai Kiadó, Budapest, 1988, pp. 43–93 and references cited therein.

59. B. Sexton, T. Smith, and J. Sanders, Characterization of copper-exchanged Na-A, X and Y zeolites with X-ray photoelectron spectroscopy and transmission electron microscopy, *J. Electron Spectrosc. Relat. Phenom.* 35, 27 (1985).

60. D. Briggs, A. Brown, and J. C. Vickerman, *Handbook of Static Secondary Ion Mass Spectrometry (SIMS)*, John Wiley & Sons, Chichester, 1989, Library of Spectra, pp. 17–155. The Static SIMS Library (release 2.0 of the CD-ROM version), SurfaceSpectra, 1999, Manchester, UK.

61. Standard Practice for Reporting Mass Spectral Data in Secondary Ion Mass Spectrometry (SIMS) (E1504-92), *Annual Book of ASTM Standards*, Vol. 03.06, ASTM International, West Conshohocken, PA, 2003, p. 829.

62. E. Ünveren, E. Kemnitz, U. Oran, and W. E. S. Unger, Static TOF-SIMS surface analysis of a CCl_2F_2 activated chromia catalyst used for a Cl/F exchange reaction, *J. Phys. Chem. B* 108, 154, (2004).

63. W. E. S. Unger, G. Lietz, H. Lieske, and J. Völter, SIMS and FAB-MS surface studies of Pt–Sn/Al_2O_3 and Pt_3Pb model catalysts, *Appl. Surf. Sci.* 45, 29 (1990).

64. C. Defossé and P. Canesson, Potentiality of photoelectron spectroscopy in the characterization of surface acidity: Photoelectron and infrared spectroscopic comparative study of pyridine adsorption on NH_4–Y zeolite activated at various temperatures, *J. Chem. Soc. Faraday Trans.* 172, 2565 (1976).

65. C. Defossé, P. Canesson, P. Rouxhet, and B. Delmon, Surface characterization of silica-aluminas by photoelectron spectroscopy, *J. Catal.* 51, 269 (1978).

66. R. Borade, A. Sayari, A. Adnot, and S. Kaliaguine, Characterization of acidity in ZSM-5 zeolites: An x-ray photoelectron and IR spectroscopy study, *J. Phys. Chem.* 94, 5989 (1990).

67. M. Huang, A. Adnot, and S. Kaliaguine, Characterization of basicity in alkaline cation faujasite zeolites—An XPS study using pyrrole as a probe molecule, *J. Catal.* 137, 322 (1992).

68. J. F. Watts and M. M. Chehimi, XPS investigations of acid–base interactions in adhesion. Part 5. The determination of acid–base properties of polymer surfaces: Present status and future prospects, *Int. J. Adhes. Adhesives* 15, 91 (1995); N. Shahidzadeh-Ahmadi, M. M. Chehimi, F. Arefi-Khonsari, N. Foulon-Belkacemi, J. Amouroux, and M. Delamar, A physicochemical study of oxygen plasma-modified polypropylene, *Colloids Surf. A Physicochem. Eng. Aspects* 105, 277 (1995).

69. A. Chilkoti and B. Rattner, Chemical derivatization methods for enhancing the analytical capabilities of x-ray photoelectron spectroscopy and static secondary ion mass spectrometry, in: *Surface Characterization of Advanced Polymers* (L. Sabbatini and P. Zambonin, Eds.), VCH Verlagsgesellschaft, Weinheim, 1993, Chapter 6.

70. S. Ono, O. Böse, W. Unger, Y. Takeichi, and S. Hirano, Characterization of lithium niobate thin films derived from aqueous solution, *J. Am. Ceram. Soc.* 81, 1749 (1998).

71. T. Cheung, X-ray photoemission of oxidized Sn–Pt and Pb–Pt composites, *Surf. Sci.* 177, L887 (1986); T. Cheung, X-ray photoemission studies of Pt–Sn and Pt–Pb bimetallic systems, *Surf. Sci.* 177, 493 (1986).

72. S. Y. Lai and J. C. Vickerman, Carbon monoxide hydrogenation over silica-supported ruthenium–copper bimetallic catalysts, *J. Catal.* 90, 337 (1984).

73. J. H. Sinfelt, G. H. Via, G. Meitzner, and F. W. Lytle, Structure of bimetallic catalysts, in: *ACS Symposium Series No. 288, Catalyst Characterization Science* (M. L. Deviney and J. L. Gland, Eds.), American Chemical Society, Washington, DC, 1985, Chapter 22; L. R. Sharpe, W. E. Heinemann, and R. C. Elder, EXAFS spectroelectrochemistry, *Chem. Rev.* 90, 705 (1990).

74. Pt-Re/Al$_2$O$_3$: J. H. Onuferko, D. R. Short, and M. J. Kelley, X-ray absorption spectroscopy and x-ray photoelectron spectroscopy of Pt–Re: Al$_2$O$_3$ catalysts: A comparison, *Appl. Surf. Sci.* 19, 227 (1984); G. Meitzner, G. H. Via, F. W. Lytle, and J. H. Sinfelt, Structure of bimetallic clusters—Extended X-ray absorption fine-structure (EXAFS) of Pt–Re and Pd–Re clusters, *J. Phys. Chem.* 87, 6354 (1987); D. Bazin, H. Dexpert, and P. Lagarde, Characterization of heterogeneous catalysts: The EXAFS Tool, in: *Topics in Current Chemistry*, Vol. 145, Springer-Verlag, Heidelberg, 1988, pp. 69–80; Pt–Sn/Al$_2$O$_3$ and Pt–Sn/SiO$_2$; B. A. Sexton, A. E. Hughes, and K. Foger, An X-ray photoelectron spectroscopy and reaction study of Pt–Sn catalysts, *J. Catal.* 88, 466 (1984); G. Meitzner, G. H. Via, F. W. Lytle, S. C. Fung, and J. H. Sinfelt, Extended x-ray absorption fine structure (EXAFS) studies of platinum–tin catalysts, *J. Phys. Chem.* 92, 2925 (1988).

75. J. C. Klein and D. M. Hercules, Surface analysis of Raney nickel alloys, *Anal. Chem.* 53, 754 (1981).

76. S. Kasztelan, J. Grimblot, and J. Bonelle, Study of the atomic distribution in NiO–MoO$_3$–gamma-Al$_2$O$_3$ catalysts by ion-scattering spectroscopy (ISS) influence of the experimental conditions, *J. Chim. Phys.* 80, 793 (1983).

77. G. Betz and G. K. Wehner, Sputtering of multicomponent meterials, in: *Sputtering by Particle Bombardment—Physics and Applications* (R. Behrisch, Ed.), Springer-Verlag, Berlin, 1981, Chapter 1.

78. R. Kelly, Phase changes in insulators produced by particle bombardment, *Nucl. Instr. Methods* 182/183, 351 (1981); R. Kelly, in: *Ion Bombardment Modification of Surfaces: Fundamentals and Applications* (O. Auciello and R. Kelly, Eds.), Elsevier Science, Amsterdam, 1984, p. 79.

79. Standard Guide for Depth Profiling in Auger Electron Spectroscopy (E1127-91), *Annual Book of ASTM Standards*, Vol. 03.06, ASTM International, West Conshohocken, PA, 2003, p. 812.

80. P. J. Cumpson, Angle-resolved XPS and AES: Depth-resolution limits and a general comparison of properties of depth-profile reconstruction methods, *J. Electron Spectrosc. Rel. Phenom.* 73 (1995) 25.

81. H. Shimada, N. Matsubayashi, M. Imamura, T. Sato, and A. Nishijima, Determination of external surface composition of zeolite particles by synchrotron radiation XPS, *Catal. Lett.* 39, 125 (1996).

82. H. Shimada, K. Sato, N. Matsubayashi, M. Imamura, T. Saito, and K. Furuya, XPS depth profiling of powdered materials, *Appl. Surf. Sci.* 145, 21 (1999).

83. M. Imamura, N. Matsubayashi, and H. Shimada, Catalytically active oxygen species in La$_{1-x}$Sr$_x$CoO$_{3-\delta}$ studied by XPS and XAFS spectroscopy, *J. Phys. Chem. B* 104(31), 7348 (2000).

84. F. P. J. M. Kerkhof and J. A. Moulijn, Quantitative analysis of XPS intensities for supported catalysts, *J. Phys. Chem.* 83, 1612 (1979).

85. H. P. C. E. Kuipers, H. C. E. van Leuven, and W. M. Visser, The characterization of heterogeneous catalysts by XPS based on geometrical probability. 1: Monometallic catalysts, *Surf. Interface Anal.* 8, 235 (1986).

86. S. Kaliaguine, A. Adnot, and G. Lemay, A model for the quantitative analysis of ESCA intensity ratios for supported catalysts with partial surface segregation, *J. Phys. Chem.* 91, 2886 (1987).

87. I. Grohmann and Th. Gross, Quantitative analysis of X-ray photoelectron spectroscopy intensities for dealuminated zeolites and supported catalysts where the promoter-forming elements also occur in the support, *J. Electron Spectrosc. Relat. Phenom.* 53, 99 (1990).

88. M. A. Stranick, M. Houalla, and D. M. Hercules, Determination of the distribution of species in supported metal catalysts by X-ray photoelectron spectroscopy, *J. Catal.* 103, 151 (1987).

89. J. N Fiedor, M. Houalla, A. Proctor, and D. M. Hercules, Monitoring the surface coverage of W/TiO$_2$ catalysts by ion scattering spectroscopy, *Surf. Interface Anal.* 23, 234 (1995).

90. M. A. Eberhardt, M. Houalla, and D. M. Hercules, Ion scattering and electron spectroscopic study of the surface coverage of V/Al$_2$O$_3$ catalysts, *Surf. Interface Anal.* 20, 766 (1993).

17 Surface Analysis of Biomaterials

Marek Jasieniak, Daniel Graham, Peter Kingshott,
Lara J. Gamble, and Hans J. Griesser

CONTENTS

17.1   INTRODUCTION

The need for new and improved materials for human health care and other biotechnological purposes has driven a vast amount of research. As the average life span increases in modern societies, the need for implants to replace or supplement natural body tissues has grown markedly. Knee and hip implants are now routine surgical procedures, and many other devices and procedures save lives or restore function and comfort. Perhaps less spectacular but in numbers by far the most frequent biomedical devices, contact lenses are used by millions of wearers to compensate for impaired vision.

In addition to human health care, materials science influences many other biotechnology fields, such as drug research, veterinary medicine, agriculture, horticulture, aquaculture, and food science. As an example, plant genomics relies on the use of microarrays, and in veterinary medicine the requirement exists for better diagnostic tools.

In all these applications, the common scientific denominator is the encounter by a synthetic material with a biological environment, and it is the chemical composition and properties (such as surface roughness, phase separation, etc.) of the surface of the material that determine the interfacial forces affecting approaching proteins and other biomolecules, as well as cells and tissue, thus governing the ensuing biological responses. As a result, it is essential for interpretation of the observed biological responses that the surface of the biomaterial under investigation is characterized appropriately and in sufficient detail to ensure reliable interpretation. Castner and Ratner [1] have emphasized the importance of detailed surface analysis of biomaterials for the interpretation of bio-interfacial interactions such as protein adsorption and cell colonization; unfortunately, the biomaterials literature contains numerous examples of insufficient, and at times manifestly incorrect, surface characterization and ensuing interpretation of bio-responses.

Often, new materials have been used in bio-applications by adaptation from related fields, for example, the use of polyurethanes (PUs) as materials for implants, derived from the commercial polymer Lycra used for clothing. The use of existing polymeric and metallic materials for biomedical implants and other biotechnological applications offers great advantages in terms of cost, development times, and regulatory approval. A popular approach to the combination of the advantages of existing polymers with the requirements of biomaterials applications is to use surface modification or thin-film coating strategies that maintain the desirable bulk properties of the materials while at the same time conferring biocompatibility. (The word "biocompatibility" should be used with great caution despite its frequent use, because bio-interfacial requirements vary with the application, and no single surface will be universally biocompatible for pacemakers, contact lenses, hip implants, and other devices). An illustrative case concerns silicone hydrogel contact lenses sold for continuous wear for 30 days and nights [2] (i.e., with no need for removal at night or cleaning); they consist of a silicone hydrogel material that provides the right flexibility, tear strength, transparency, etc., but is too hydrophobic to be used by itself, and so the lens also contains an ultrathin (~20 nm) hydrophilic plasma-coated layer that confers appropriate interfacial properties, enabling low fouling and wear comfort.

Any attempts to modify biomaterials surfaces must be accompanied by rigorous surface analyses, again in order to understand the rationale for the observed biological responses. In this chapter, methods appropriate to the surface analysis of biomaterials will be reviewed. Practical advice will be given on the capabilities, limitations, and pitfalls inherent in the usage of the techniques. Experience shows that it is often necessary to employ several techniques of analysis; as biomaterials surfaces are often complex in composition and analytical requirements, a careful choice of methods is essential in order to be able to compensate for the limitations of each method and to provide complementary information from several techniques, which often gives a more detailed picture.

Surface analytical methods are useful not only for the analysis of the surfaces of materials intended for biological applications, but also for the study of the consequences of the exposure of synthetic materials to biological media. A key early event in the sequence of biological responses is typically the adsorption of a layer of proteins onto a bare surface. Many proteins are highly surface active and their structure is indeed designed for them to function at interfaces. Thus, they accumulate rapidly at the surface of a solid material. The study of protein layers adsorbed at a solid surface is, however, extremely challenging. Typical questions are as follows: what proteins adsorb from a complex mixture such as blood, how much, and in what conformation, and how does the adsorbed initial protein monolayer determine the subsequent biological events? Surface analysis techniques have made valuable contributions to the study of these questions, though biological methods are also essential tools for gaining information on the multiple aspects of such questions.

The main techniques to be discussed for the characterization of biomaterials surfaces are X-ray photoelectron spectroscopy (XPS) and time-of-flight secondary ion mass spectrometry (ToF-SIMS), since they probe the composition of biomaterials over depths comparable to the range that governs interfacial interactions. Other techniques, such as infrared (IR) spectroscopy, have been used, but their information depths are much greater, leading to the danger that they may give information that is irrelevant or even misleading to the interpretation of observed biological responses.

17.2 XPS ANALYSIS OF BIOMATERIALS SURFACES

Full descriptions of the XPS technique have already been given in Chapters 3, 8, and 10. In this section, we concentrate on how XPS can be used to advantage in the study of the surfaces of biomaterials.

17.2.1 Applicability of XPS

From the biomaterials point of view, the two most important properties of XPS are its surface-specificity, and its ability to detect all elements (except H and He) on a surface and to provide quantitative information on their concentrations. The information depth of only 2–5 nm (depending on the angle of emission between the surface and the detector) is comparable with the thickness of material that determines the interfacial reactions defining biological responses such as protein adsorption. Quantification is straightforward if a sample is homogeneous within the information depth, since the intensity of the photoelectron signal from an element is then proportional to the concentration of that element, but fully homogeneous samples of all types, not just biomaterials, are rare. However, using the approach described by Tougaard in Chapter 8, even inhomogeneous samples can provide good quantification. Such inhomogeneity can, for example, be found in some polymer surfaces, where the surface layers may not be of the same composition as material deeper inside the polymer, due to preferential segregation driven by interfacial energy. In contact with air, some polymers tend to expose the more hydrophilic elements of their chains at the interface. XPS has also been used extensively to detect and quantify surface segregation in metal alloys, which in the context of biomaterials can occur in orthopedic implants, for example, Cr in CoCr.

If it is suspected that the biomaterial interaction layer is indeed inhomogeneous, then either the above-mentioned Tougaard approach can be tried, or alternatively angle-resolved XPS (ARXPS). On no account should so-called depth profiling be used, since the incident ion beam will cause substantial damage to the layer, leading to erroneous interpretation. ARXPS is based on the fact that electrons of a particular kinetic energy in a solid have an inelastic mean free path (IMFP) associated with that energy. The consequence is that if the kinetic energy is fixed, then as the sample is rotated so that the takeoff angle, α, into the analyzer is progressively increased, so the information depth is progressively decreased. However, the technique is suitable only for very flat surfaces. An alternative based on the same principle is to maintain the angle fixed, but to vary the kinetic energy by varying the incident exciting x-ray photon energy, in a synchrotron. The relationship between takeoff angle and information depth is given by equation:

$$I = I_{\rm o}\, \exp\left(\frac{-d}{\lambda \cos \alpha}\right) \qquad\qquad (17.1)$$

where

    d is the information depth
    $I_{\rm o}$ is the initial intensity at depth $d = 0$
    I is the detectable intensity at depth d
    λ is the IMFP

One of the recent developments in XPS relevant to biomaterials is the imaging capability [3]. The spatial distributions of selected individual elements can now be mapped with resolutions of a few micrometers. This is not as good as the spatial resolution achievable with AES and ToF-SIMS, but then AES should not be used on polymeric biomaterials in particular because of the extensive damage caused by the electron beam. ToF-SIMS imaging is discussed in Section 17.3.

A physical process peculiar to some polymeric biomaterials is that of their mobility at or near room temperature, in the form of diffusion and reptation of polymer chains. The modification of a hydrophilic polymer surface can lead to a movement of some polymer chains from the surface into the near-surface regions, driven by interfacial energy and translational entropy. The surface composition will therefore show a time-dependent change, which can be monitored easily by XPS, provided that there is a suitable marker signal that can be related to the change. In the case of surface-attached amine groups, the burial inside mobile polymers can be monitored by the decrease in the N 1s signal [4,5].

A further valuable application of XPS to biomaterials is in the study of the problem of surface contaminants on synthetic materials. Such contaminants might include processing aids, extrusion lubricants, surface-active additives in polymers, adsorbed adventitious hydrocarbon molecules, silicones, and others. XPS is well suited to their detection, even at coverages well below a monolayer. In some cases, for example, adventitious hydrocarbons on polymers, the spectral differences between the contaminant and the material are so subtle that spectra must be recorded at the highest possible resolution, involving the use of monochromatic radiation.

With polymeric biomaterials such as PUs, the possibility of sample degradation during analysis must be considered. In the case of a nonmonochromated source, the anode, although water-cooled, is in close proximity to the specimen, and for many polymeric biomaterials there is the danger of significant degradation by heat and Bremsstrahlung. Sample degradation during nonmonochromatic XPS analysis has been documented extensively in many textbooks and reviews, but most instruments in operation nowadays employ a monochromatic source. With an x-ray monochromator, the source of heating is displaced from the sample position (and of course the Bremsstrahlung is eliminated); however, it should not be assumed that sample degradation during monochromatized analysis can be neglected. While sample damage is usually small and often negligible within the finite signal-to-noise ratios of analyses, with more radiation-sensitive polymers such as those containing carbon–halogen bonds there can still occur substantial effects within the usual analysis conditions and times. One such example is the significant degradation observed in the course of monochromatic XPS analysis of antibacterial coatings containing brominated furanones; a signal assignable to bromine ion increased in intensity over time periods of the order of 6–10 min, showing that a significant percentage of C–Br bonds had suffered radiation damage [6].

17.2.2 Freeze Hydration XPS

One of the limitations of XPS is the need to perform analysis on a surface dehydrated as a result of exposure to UHV, which may alter the surface composition. Some polymers, such as PUs, are known to alter their surface composition according to the contacting medium; in contact with air, the more hydrophobic phase of a PU will be enriched at the surface, whereas in water the more hydrophilic component is surface-enriched. This adaptation of the polymer surface to its environment to minimize the interfacial energy has the unfortunate consequence that the surface composition recorded by XPS analysis may not be the same as that existing in the aqueous biological environment; hence, the interpretation of XPS data acquired to study observed biological responses may be questionable. Analysis can be performed on frozen hydrated samples to avoid this [6–8], although it is a complex experiment that requires specialized equipment and skills.

Some biocompatible materials that contain significant amounts of loosely bound water, such as hydrogels (e.g., bulk poly(2-hydroxyethyl methacrylate) [poly(HEMA)], a commonly used component of contact lenses), also present difficulties for UHV analysis because the water held in them will desorb under vacuum, effectively creating a virtual leak in the UHV system. These materials have such an affinity for water that even drying on a laboratory bench for days is not sufficient to remove enough water to achieve the vacuum required for XPS analysis. In addition, even if it appears that an adequate vacuum has been achieved, once the x-ray source and charge compensation electron gun impinge on the surface, further (and sometimes unexpected) outgassing may occur. A cold stage experiment (i.e., maintaining the sample at a temperature that will freeze water in vacuum) is one possible approach for analyzing such samples. However, the time required to freeze, pump, and analyze one sample greatly limits the number of samples that can be analyzed in a day, and the cost of individual sample analysis (taking into account significant increase in time and labor for analysis, as well as liquid nitrogen consumption) is increased. In addition, a layer of ice commonly develops on such samples, and steps need to be taken to evaporate all but the loosely bound water, otherwise the analysis will show primarily oxygen from ice. More detail about cold stage analysis can be found in a review article on biomedical surface analysis by Castner and Ratner [1].

For a simple quantitative analysis of samples such as hydrogels, a more cost-effective approach may be used to lyophilize (i.e., freeze-dry) the sample prior to analysis. This freeze-drying approach uses a separate vacuum system designed to remove water from biological samples, and is most effective in removing most of the water from hydrogels prior to their introduction to a UHV chamber. One risk during this process is the introduction of surface contaminants such as hydrocarbon or PDMS (poly(dimethylsiloxane)) from pump oil or vacuum greases in lyophilizers used to prepare samples for surface analysis; this can be prevented by the use of liquid nitrogen traps. Care must also be taken not to expose the lyophilized hygroscopic samples to air for long periods of time or water will be re-adsorbed. Even using the lyophilizing approach, it may be necessary to restrict number of samples in order to achieve the appropriate analysis pressures.

17.2.3 Contamination Problems

One of the most challenging aspects of the XPS analysis of biomaterials surfaces is ensuring cleanliness of the surface and the avoidance of possible misinterpretations due to unrecognized contaminants. The problem is trivial scientifically but extremely important in practice. Many materials become easily contaminated, and it is very difficult to avoid almost instantaneous adsorption of environmental contaminants onto high-energy surfaces such as metal oxides. The analyst needs to gain sufficient experience to recognize the unavoidable residual amounts of hydrocarbon contamination that accumulate, even with careful handling, between sample cleaning and insertion into the XPS instrument. Hydrocarbon contamination adsorbs onto sample surfaces very rapidly from the laboratory air and can potentially also occur from hydrocarbon material in the vacuum chamber itself (pump oil, etc.), though in well-designed and well-tended chambers this is not a serious issue. Many new systems have oil-free pumps and ion pumps or turbo pumps in the analysis chambers, which greatly minimize this contamination issue. Older chambers can use cold traps to limit contamination from hydrocarbon vapors in the pumps. However, many of the biomaterial surfaces are prepared outside the vacuum system and are unavoidably exposed to contaminants in the air (in the laboratory and during transfer from the preparation laboratory to the analysis laboratory). Common causes can be grease used to seal desiccators, contaminated solvents, or even exhaust fumes in buildings with bad ventilations systems.

If possible, sample cleaning should be carried out in an inert atmosphere and the sample should then be introduced into the instrument through an airlock in the same atmosphere. Typically, the C content can be between 5 and 10 at.%, despite best protocols, on surfaces such as silicon wafers. And while the contaminant is usually designated as hydrocarbon, often there may be some O associated with it, indicating partially oxidized hydrocarbons, such as fatty acids, fatty acid alcohols, or fatty acid esters. This is seen readily in the high energy tail to the C 1s peak at ~285 eV, associated with hydrocarbons. The challenge then is for the analyst to be able to distinguish between the signal from this hydrocarbon and that from processes performed on parallel samples, such as surface modification or protein adsorption. The interpretation of the chemical changes either following surface modification procedures, or as a result of thin film coatings, can be jeopardized if hydrocarbon contamination is not properly considered.

Usually easier to detect, but equally difficult to eliminate, are organosilicone contaminants. Unlike hydrocarbons, organosilicones can migrate along surfaces and spread through a laboratory by both transfer and migration. They can arise from many sources, such as plastic bags and the caps of vials used for storing samples, being present as plasticizers in many such articles, as well as in Latex gloves and other articles used in laboratories. Since silicones are surface active, they become enriched on surfaces, are readily transferred to any contacting surface, and can spread onto laboratory benches, tweezers used for sample handling, analysis stubs, and the like. It is good practice to store tweezers in an organic solvent that removes any organosilicone that may have been picked up during sample handling, and to change the solvent frequently. It takes time and experience to identify such sources of contamination and to take appropriate countermeasures. In some laboratories, silicone-free areas have been established, in which no material or article containing organosilicones is permitted. Samples should be stored only in glass or additive-free polystyrene containers. Unfortunately, a number of important biomaterials themselves contain organosilicones, and when investigating those, extra care must be taken. With such measures and vigilance, contamination by silicones can be eliminated. The issue is even more pressing for ToF-SIMS analyses, where even smaller amounts of silicones are readily observed.

Unfortunately, there are many studies reported in the biomaterials literature where inexperienced analysts have not recognized the problem. A monolayer or so of organosilicone surface contamination may affect markedly the bio-interfacial properties of a biomaterial, and cast doubt on interpretations that are based on the surface chemistry of a presumed uncontaminated material.

17.2.4 SPECTRAL INTERPRETATION IN BIOMATERIALS

The analysis of C 1s XPS signals from polymeric biomaterials can be quite challenging, and the literature contains many examples of erroneous interpretations. It is essential that agreed binding energy values are used for fitting components to the observed C 1s signal. Unconstrained fitting (i.e., letting the software find the best-fit binding energy values) can lead to what may look like attractive fits but which actually have little physical meaning. It is advisable to constrain the fitting by selecting binding energy values for those components that are expected to be present, and then to use judgment and caution when considering whether the resultant fit has indeed any physical meaning. Often it is useful to consider alternative possibilities for the surface chemistry, to introduce constraints into the fitting routine, and then to check the quality of the resultant fit. It also helps to consider how many components can actually be justified. Perhaps it is opportune to recognize, and state, that the results of fitting are often not clear-cut and that there may be more than one physically acceptable set of components. The introduction of additional peaks may well improve the quality of the fit, but the question must always be asked as to whether such addition has physical meaning. Gengenbach et al. [9] have investigated the problem using the C 1s signal from a chemically diverse polymer surface, and concluded that the use of residual least squares can help to decide whether the addition of a further peak is statistically meaningful or not, given limited signal-to-noise ratios in

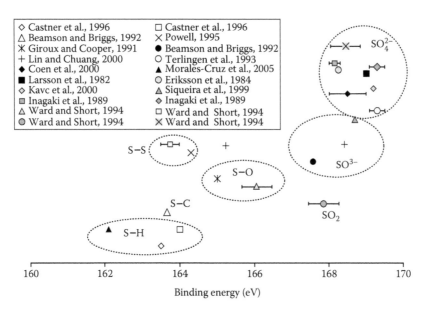

FIGURE 17.1   Binding energies for S 2p in organo-sulfur compounds, with S in various oxidation states. The binding energies were re-normalized to C–C at 285.0 eV. (From Siow, K.S., PhD thesis, University of South Australia, Mawson Lakes, South Australia, Australia, 2007.)

spectra, but the issue of physical meaning of components remains a matter of judgment coupled with careful assessment of what chemistries might be expected on the sample surface. It is essential to be aware of the limitations in the reliability and information content of component fits. At times, a complementary technique such as ToF-SIMS may help to decide whether all the fitted C 1s components are actually physically significant.

A further problem can arise with regard to the choice of reference binding energies of photoelectron peak components, for curve fitting. The binding energies of components chosen in the N 1s signal have been controversial at times, and the literature contains various binding energies for imines, imidazoles, and azides, but positions of 399.3 eV for amine N and 400.0 eV for amide N have been well documented [9,10]. Likewise, surveying the literature on binding energies of the 2p peaks for S and P, Siow [11] found that reported values for various oxidation states of S and P (sulfonate/sulfate/sulfite and phosphate/phosphonate/phosphite, respectively) varied significantly (Figures 17.1 and 17.2). As a result, Siow used well-defined compounds such as chondroitin sulfate to collect his own set of reference binding energies for various S- and P-containing groups, measured on the same instrument, in order to interpret the oxidation states of the two elements on plasma-prepared surfaces.

17.3   TOF-SIMS ANALYSIS OF BIOMATERIALS SURFACES

Static ToF-SIMS has become an effective method for the molecular surface characterization of polymeric and biological materials. The combination of extreme surface sensitivity (~1 nm information depth), high information content (numerous peaks), extremely low detection limits for many elements, molecular specificity, and sub-micrometer lateral resolution provides surface information unavailable by other techniques. As for XPS, ToF-SIMS has been applied to the characterization of biomaterials prior to biological exposure and then again after contact with biological media. In particular, ToF-SIMS is well suited to the study of the adsorption of biomolecules onto materials surfaces since, for example, adsorbed proteins can be detected at amounts as low as 0.1 ng/cm^2 [12,13]. There are several excellent reviews describing the application of ToF-SIMS to biomaterials surfaces [14–18].

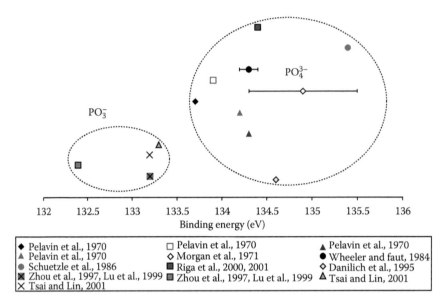

◆ Pelavin et al., 1970	□ Pelavin et al., 1970	▲ Pelavin et al., 1970
▲ Pelavin et al., 1970	◇ Morgan et al., 1971	● Wheeler and faut, 1984
● Schuetzle et al., 1986	■ Riga et al., 2000, 2001	◇ Danilich et al., 1995
⊠ Zhou et al., 1997, Lu et al., 1999	■ Zhou et al., 1997, Lu et al., 1999	△ Tsai and Lin, 2001
✕ Tsai and Lin, 2001		

FIGURE 17.2 Binding energies for P 2p in organo-phosphorus compounds, with P in various oxidation states. The binding energies were re-normalized to C–C at 285.0 eV. (From Siow, K.S., PhD thesis, University of South Australia, Mawson Lakes, South Australia, Australia, 2007.)

17.3.1 Methodology

In a ToF-SIMS experiment, ions from a primary source impinge on a surface, which in turn leads to the creation of secondary ions as well as neutrals. The secondary ions are electrostatically extracted into a mass spectrometer and their mass-to-charge ratios, m/q, are determined by measuring the time it takes for the ions to be ejected from the surface and arrive at the detector. For the analysis of sputtered neutrals, post-ionization is applied using a pulsed laser. The ejected neutrals then become accessible for mass analysis, and since their number is much greater than that of the ions, a significant increase in instrumental sensitivity is achieved [19].

By its very nature, SIMS is destructive. However, by using an ion beam of sufficiently low current, it is possible to limit beam damage and thereby derive data from a virtually intact surface. This version, called static SIMS (SSIMS), thus operates under conditions where the sample surface is not significantly altered during the analysis (hence static). In practice, meeting the requirements of the static conditions regime means that <0.1% of the surface atomic sites are struck by the primary ion beam during the time of measurement [20].

The main analytical limitations of this technique originate from the difficulties in quantifying the concentrations of surface species due to matrix effects. However, as long as a surface species M is in exactly the same chemical environment, the secondary ion emission is proportional to the coverage of the surface by the component M [21]. Accordingly, for chemically similar materials, it is possible to perform semiquantitative analysis.

17.3.1.1 Primary Ion Sources

To date, most ToF-SIMS experiments have been performed using a Ga liquid metal ion gun. It is now clear that this monatomic source delivers poor information on biological surfaces, since it produces too low a yield of higher mass secondary molecular ions, which contain much more structural information than the lower mass ions. The sensitivity and the yield decrease dramatically with increasing molecular weight (MW) and with the complexity of the analyte [22]. Thus, in order to generate a signal of sufficient intensity, long acquisition times may be required, frequently far beyond the static limit. As a result, the surface structure can be substantially damaged during the analysis, when spectra would no longer be representative of the original material.

The problem is that Ga projectiles penetrate easily into the surface, depositing most of their energy at considerable depths, with only a small portion of the energy available at the surface itself [23]. This energy transfer mechanism results in only a small yield of secondary ions.

The development of sources of polyatomic cluster ions such as SF_5, Au_n, Bi_n, and C_n has been a major instrumental development in the area of molecular ToF-SIMS surface analysis [24,25]. In order to evaluate their potential for molecular surface analysis, a series of monatomic (Ga, Cs, Au, and Bi) as well as polyatomic (SF_6, Au_n, Bi_n, and C_{60}) ion sources were applied to model organic compounds [26]. It was shown that cluster ion beams increased the secondary yield by more than three orders of magnitude compared to that from the use of Ga. While C_{60} primary ions offer the highest sensitivity and the lowest fragmentation, the large spot size of the beam has until now hampered the analysis of small (sub-micrometers) areas, but recent advances in ion beam technology have overcome that problem. The recently designed Ionoptika C_{60} primary ion column, with a spot size as small as 300 nm [27], has made possible the application of such a cluster ion source to the imaging of biological objects (lipids, proteins, cells, and tissues). There is growing evidence that the C_{60} ion beam can also be used in depth profiling of organic materials [28–30], avoiding the carbonization problem that occurs when applying LMIG sources to organic samples.

Cluster ions interact strongly with a surface, a consequence of the larger collision cross-section compared to monatomic projectiles. The result is a higher efficiency of energy transfer, that is, greater ionization of surface species. In addition, the damage accumulation rate is lower with cluster ions than with monatomic ions. The reason is that each atom in a cluster carries only a fraction of the initial polyatomic cluster energy. For example, when a 15 kV voltage is used to accelerate an Au_4 cluster, then the average energy released from each gold atom on impact is 15/4 = 3.75 keV. In the case of C_{60}^+, instead of a particle at 15 keV hitting the surface, there may be up to 60 collisions each of average energy ~250 eV. The use of C_{60}^+ results in smaller damage cross-sections. The analysis of tissues by ToF-SIMS imaging with cluster ions can cover a mass range extending to about 1000 Da [31]. In recent, unpublished work, identifiable peaks have been observed, with good signal-to-noise, from covalently immobilized protein layers up to $m/z = 2500$, which, for proteins such as lysozyme, represents significant fractions of the molecule.

17.3.1.2 Charge Compensation

A common method for minimizing the positive charging of samples is that of the provision of leakage paths by use of a conducting grid [32], combined with a balancing negative charge from a pulsed electron beam. Conducting grids in contact with the sample surface can provide a leakage path, which prevents charge buildup when the analyzed area is centered within a grid square. In ToF-SIMS, however, the sample is flooded with low energy electrons (20–30 eV) between each primary ion pulse. It is crucial that no electrons are emitted when the high positive voltage is applied to the sample, otherwise sputtering of the surface with high-energy electrons would result. In some cases, control over the positive charge buildup can be achieved by increasing the raster area.

17.3.1.3 Sample Handling

ToF-SIMS samples must be compatible with the UHV environment ($<10^{-7}$ Pa). As for XPS analysis, mobile polymer surfaces, hydrated materials, polymers containing dissolved gases (usually air), and volatile components can interfere with analysis and interpretation. Fortunately, most ToF-SIMS instruments are equipped with liquid-nitrogen-based cold stages, which allow analysis of samples that contain volatile components, including water of hydration, at room temperature.

In the analysis of highly hydrated specimens (e.g., hydrogels), however, avoiding artifacts related to sample preparation is still difficult. Study of the surface chemistry of hydrogel polymers has attracted great attention, as this class of polymers is widely used in a variety of biomedical technologies. Low-temperature sample handling methods have been developed and applied for instance to the surface analysis of hydrogel contact lenses [33]. In specific cases, qualitative analysis of changes upon hydration have shown the emergence of components of lower surface energy at the surface and the detection of contaminants of low surface energy, such as PDMS.

The presence of collagen molecules at the surface of a collagen-modified poloxamine hydrogel has been studied by a deep-freezing approach [34]. Positive and negative mass spectra of the frozen samples confirmed the surface collagen, while images using selected molecular ions showed that the collagen was uniformly distributed. However, due to the hydrated character of the hydrogel matrices, a thin ice layer was present on the surface of the frozen sample ($-120°C$), resulting in a series of peaks in the positive spectrum characteristic of water clusters, with the formula $[H(H_2O)n]^+$. Application of gentle heating caused significant sublimation of water and the disappearance of its pattern near $-70°C$. Thus, exposure of the matrix to thermal treatment allowed proper characterization of the system.

A combination of ToF-SIMS and in-vacuum cryo-sectioning has been developed for the *in situ* preparation and analysis of frozen nonhydrated samples [35]. Water adsorption from the residual vacuum onto the cold surface poses a problem as it interferes with the surface-specific ToF-SIMS analyses. In order to eliminate that preparative artifact, a model structure with a defined composition was subjected to various thermal treatments. It was established that there was only a small temperature interval near $-110°C$ in which the adsorbed water was removed.

17.3.1.4 Contaminants

ToF-SIMS can provide very high sensitivity for the analysis of the outermost surface layers, with the information depth for molecular ions <1 nm [36], but this advantage also exacerbates the problems arising from surface contamination. The spectra of secondary ions originating from the contamination will be superimposed on the spectrum of the sample under investigation [37], and, as for XPS, this can lead to ambiguity of interpretation, particularly with hydrocarbon contaminants. The most common contaminant observed in ToF-SIMS spectra is PDMS; polymer additives have also been detected [38,67]. Removal of siloxane contaminants in an ultrasonic bath of hexane is effective, but it can be applied only to metals, ceramics, and some polymeric materials; it cannot be applied to many polymers or to chemically or biologically modified surfaces, such as the examples of plasma functionalized polypropylene, RGD-modified polymeric surfaces, etc., described below. If such samples are found to be contaminated, then they should be discarded. Other common contaminants include sodium and potassium, and their presence can affect the accuracy of an analysis. Fortunately, they are water-soluble and, in most cases, can be eliminated by washing with deionized water.

17.3.1.5 Peak Overlap

Advances in mass spectrometry that optimize ion yield may complicate peak resolution. Count rate and mass resolution are often compromised [39]. There are only a few examples of the application of peak fitting routines to static ToF-SIMS spectra [40,41]. The development and application of peak routines would increase the analytical reliability of mass spectra by making the identification of integration limits and peak positions more straightforward.

17.3.1.6 Topographical Effects

Interpretation of ToF-SIMS images from biological specimens can be problematic because the images are influenced not only by the sample chemistry but also by topographical features [42].

Processing of the images to obtain clear contrast between chemically distinct regions and topographical effects, and to identify chemical species, can be a formidable challenge, particularly when working with organic and biological molecules that have similar features [43]. Application of multivariate analysis methods (see Section 17.4) has been found to be extremely valuable in resolving the topographical problem [44,45].

17.3.1.7 Examples of Application

The applications, information content, and limitations of ToF-SIMS are best illustrated by specific examples. Since ToF-SIMS analysis requires a high level of experience and consideration of the specifics of samples, it is less straightforward than for XPS to make general recommendations for procedures.

Due to its very high sensitivity, ToF-SIMS is well suited to the study of the uniformity of thin coatings. Even minute coating defects can lead to observable substrate signals. An example is given in Figure 17.3, which shows small regions of the ToF-SIMS spectra recorded from a commercial glass slide (used in bio-assays) before (Figure 17.3a), and after (Figure 17.3b), coating with a plasma polymer from n-heptylamine (Jasieniak and Griesser, unpublished). The original glass surface contained a small amount of hydrocarbon contamination and some Al. The plasma polymer coating contains amine groups on a hydrocarbon backbone. The after spectrum (Figure 17.3b) reveals the

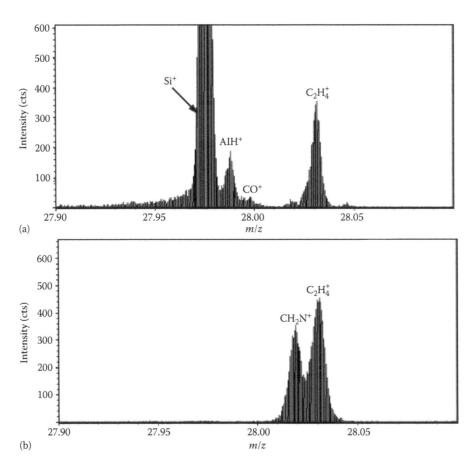

FIGURE 17.3 Positive SSIMS spectra of glass before (a) and after (b) heptylamine plasma deposition. (From Jasieniak, M. and Griesser, H.J., unpublished data.)

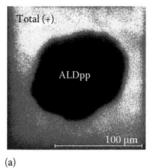
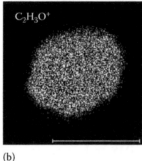

(a)　　　　　　　　　　　　　(b)

FIGURE 17.4 ToF-SIMS images of patterned FEP with propionaldehyde plasma polymer: (a) total positive ion image and (b) image of CH_2CHO^+, a main polymer structural fragment. (From Nguyen, P.C.T., PhD thesis, University of South Australia, Mawson Lakes, South Australia, Australia, 2006.)

N-containing radical characteristic of the amine, but complete absence of a signal from Si. The plasma polymerization process is therefore believed to have formed highly uniform coatings, and indeed this analysis showed that there was complete coverage even at a thickness of ~18 nm.

The second example illustrates the imaging capability of ToF-SIMS. Plasma polymerization of propanal through a stainless steel mask containing circular holes was used to deposit a microarray of dots, with aldehyde surface functionality, on a fluorinated ethylene propylene (FEP) substrate [46]. The micro-patterned surface was imaged by ToF-SIMS (Figure 17.4); the maps revealed that the process did indeed fabricate the desired functional microarray with good spatial definition (the laser drilled holes were somewhat irregularly circular, as shown by optical microscopy, and this is reproduced in the SIMS image), and that the modified areas were aldehyde-functionalized. After coupling oligonucleotides to the surface aldehyde groups, mass signals assignable to the oligonucleotides were recorded (not shown), located only on the microdots, confirming the suitability of the approach for the fabrication of ssDNA microarrays.

While planar surfaces are easier to analyze, ToF-SIMS can also be used for the analysis of three-dimensional (3D) samples. The chemical and morphological characteristics of 3D porous poly(lactic-co-glycolic acid) (PLGA) scaffolds were investigated using ToF-SIMS (Figure 17.5) (Jasieniak and Griesser, unpublished). The major peak in the positive mass spectrum at $m/z = 56$ amu is assignable to the $C_3H_4O^+$ fragment, which derives from the PLGA structure.

A central theme in biomaterials research is the design and fabrication of bioactive surfaces. As opposed to nonbioactive materials such as PUs and titanium, bioactive biomaterials consist of synthetic materials whose surfaces are coated with layers of biologically active molecules, which are intended to signal to the biological environment that a desirable biological response can ensue. One example is the covalent attachment (grafting) onto materials surfaces of proteins that support cell attachment and proliferation; this is thought to be important for tissue engineering. Covalently grafted proteins can often be detected by XPS (unless the underlying biomaterial is a polymer with a similar elemental composition), but when it is required to graft small synthetic oligopeptides that replicate an active region of a larger protein, XPS analysis becomes restricted by its sensitivity limits. ToF-SIMS, on the other hand, can detect extremely small amounts of surface-grafted biomolecules, as shown by the example (Jasieniak and Griesser, unpublished) of the grafting of the oligopeptide GRGDSP onto polypropylene via covalent interfacial immobilization. Polypropylene was surface functionalized with aldehyde and epoxy groups and by the plasma polymerization of allyl glycidyl ether, and the GRGDSP oligopeptide was then anchored by covalent reaction (Figure 17.6). The process of immobilization was monitored by ToF-SIMS.

The ToF-SIMS spectra looked superficially similar and were processed using principal components analysis (PCA) (see next section) to bring out the spectral differences. The scores plot

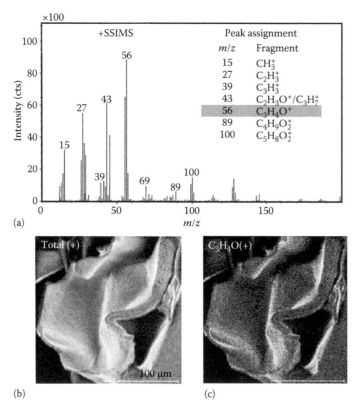

(a)

(b)   (c)

FIGURE 17.5 ToF-SIMS characteristics of a PLGA scaffold: (a) positive mass spectrum, (b) total positive ion image, and (c) image of $C_3H_4O^+$, PLGA structural fragment. (From Jasieniak, M. and Griesser, H.J., unpublished data; sample prepared by Croll, T., University of Melbourne, Victoria, Australia.)

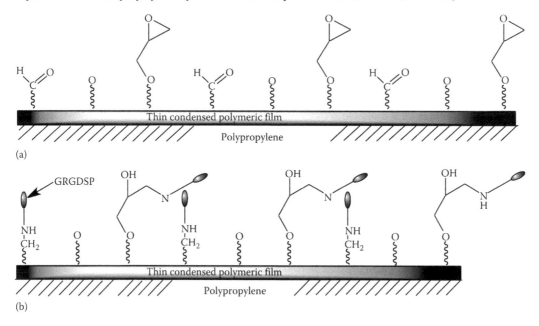

(a)

(b)

FIGURE 17.6 Schematic diagram of polypropylene (a) modified with aldehyde and epoxide functionalities, and (b) grafted with covalently immobilized GRGDSP. (From Jasieniak, M. and Griesser, H.J., unpublished data.)

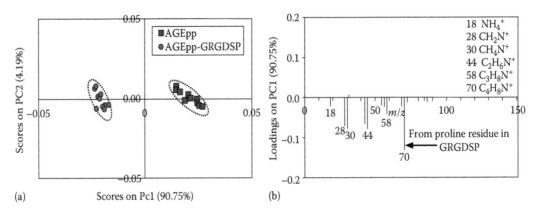

FIGURE 17.7 PCA analysis of ToF-SIMS spectra recorded before and after GRGDSP immobilization onto epoxy plasma modified polypropylene: (a) scores plot and (b) loadings plot on PC1. (From Jasieniak, M. and Griesser, H.J., unpublished data.)

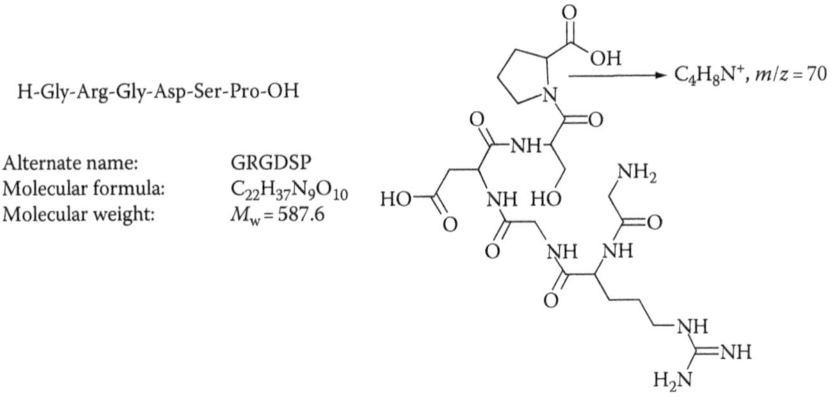

H-Gly-Arg-Gly-Asp-Ser-Pro-OH

Alternate name: GRGDSP
Molecular formula: $C_{22}H_{37}N_9O_{10}$
Molecular weight: $M_w = 587.6$

$C_4H_8N^+$, $m/z = 70$

FIGURE 17.8 Origin of the mass signal at $m/z = 70$ amu present in the mass spectrum of AGEpp-GRGDSP.

(Figure 17.7a) showed clear differences between the plasma-coated layer and the subsequent GRGDSP grafting. Such differences could, however, also arise from the presence of surface contaminants on one of the samples; it was important to check that the differences could be attributed to chemically meaningful peaks. Figure 17.7b shows that the negative loading peaks were all from nitrogen-containing fragments such as immonium ions from the amino acids, or smaller fragments, that is, consistent with grafted GRGDSP. It is often instructive to consider the peaks with high loadings as they provide the most reliable spectral information. For example, the high loading peak at 70 amu reflects the presence of a $C_4H_8N^+$ fragment, which is the fingerprint immonium ion originating from the amino acid proline, part of GRGDSP (Figure 17.8).

17.4 PCA OF ToF-SIMS SPECTRA

Due to the complexities inherent in ToF-SIMS spectra, it is often challenging to digest and interpret even a small number of spectra. Such spectra typically contain hundreds of peaks, and a comparison of their relative intensities can quickly become daunting. For example, if a set of spectra each contained only 100 peaks, creating cross plots comparing pairs of each of the peaks would require the generation of a huge number of plots. This is complicated further by the fact that the intensities of many of the peaks in a given spectrum are correlated, since the molecular species giving rise to them come from the same types of surface chemical structures. ToF-SIMS images add another layer of complexity, since for a 256 by 256 pixel image there would have to be 65,536 spectra to process. Fortunately, there are tools available to the ToF-SIMS analyst to aid in data interpretation, including spectral libraries, peak databases, MW and molecular fragment calculators, and multivariate analysis methods.

Recently, the use of multivariate analysis methods has become widespread in the ToF-SIMS community. Of the many multivariate analysis methods available, PCA is by far the most common [47–55]. PCA can be a powerful analysis aid for analyzing ToF-SIMS spectra, but as with any tool it is necessary to understand when to use it and how to apply it. When used properly, PCA can

- Reduce large data sets into more manageable variables.
- Help find trends within data sets that are not readily apparent from other data analysis methods.
- Find differences between samples when the chemistry is very similar (e.g., proteins/peptides).
- Identify peaks of interest for unknowns.
- Show trends in surface contaminants.
- Allow use of all the data.
- Eliminate user bias.
- Simplify interpretation of complex data sets.

PCA or other multivariate methods are not always necessary; for instance, where the goal of a ToF-SIMS experiment might be the verification of the presence or absence of known representative peaks for a given compound. That could be achieved simply by looking at the spectra. Also, if there were only one or two spectra per sample, and only a few different sample types, using PCA would not be sensible. PCA looks for directions of maximum variance between samples. For the differences to be significant, enough data need to be recorded to achieve statistically significant results. This is particularly important when dealing with samples that can have large spot-to-spot variability in relative ion yields due to compositional, height, or matrix effect variations. Finally, it is often not helpful to use multivariate analysis methods such as PCA on data sets where there is no real control of sample variables. If a data set has been recorded in which a batch of samples were prepared in several different ways with several different materials, PCA would probably be able to separate the samples, but would find it very difficult to determine what the differences mean or where they originated. The best data sets to use with PCA are those in which the samples were prepared in a controlled experiment, where care was taken to assure that only one process variable was changed across the sample set.

17.4.1 Description of the PCA Method

Detailed explanations of PCA have been given in the literature [56–58]. Briefly, it is a multivariate analysis method that determines the greatest directions of variance within a data set. This is performed through the singular value decomposition of the variance–covariance matrix from a data set, which produces the characteristic roots (eigenvalues) and characteristic vectors (eigenvectors) of the variance–covariance matrix. The result of this process is a set of new uncorrelated variables called principal components (PCs), which are linear combinations of the original variables (ToF-SIMS peak intensities). As each PC is subtracted successively, in decreasing order of magnitude, from the data matrix, the next largest PC describes the next greatest direction of variance. This is important to note since it means that the data described by PC2 comprise the data matrix after PC1 has been subtracted, and not the original data matrix itself. The output from PCA consists of three matrices: the scores, the loadings, and the residuals. Scores describe the relationship (spread) between the samples as evidenced by the new PC axes. Mathematically, scores are the projection of the sample data onto a PC axis. Loadings describe how the original variables relate to the PC axes. Mathematically, the loadings are the direction cosines between the original variables and the PC axes. The residual matrix contains the remaining variance not described in the scores and loadings, and represents random noise in the data.

The scores and loadings must be used together to understand the data set. Figure 17.9 shows the PC1 scores and loadings for a hypothetical set of SIMS data. In this figure, the PC1 scores are plotted against the sample number. It can be seen that the samples are separated into three groups along the PC1 axis. Group 1 samples have negative scores, group 2 samples are grouped around the origin, and group 3 samples have positive scores. The loadings plot can be used to determine the main variables (peak intensities) responsible for the separation seen between the samples, and shows that peaks (variables) 1, 3, and 4 have positive loadings, while peaks 2 and 5 have negative loadings on PC1. Peaks with positive loadings correspond to samples with positive scores, that is, they will have higher relative intensities in the spectra of samples with positive scores on the same PC axis. Peaks with negative loadings correspond to samples with negative scores, and will therefore have higher relative intensities in the spectra of samples with negative loadings. Figure 17.10 illustrates this for peaks 3 and 5; peak 3 is seen to have the highest relative intensity for group 3 samples, while peak 5 is seen to have the highest relative intensity for the group 1 samples.

Subsequent PCs can be interpreted in a similar way, though it should be remembered that the data used for each PC comprise the data matrix minus the previous PC. This means that it is not possible to go back to the original data directly to see raw data trends seen in the PCs after PC1.

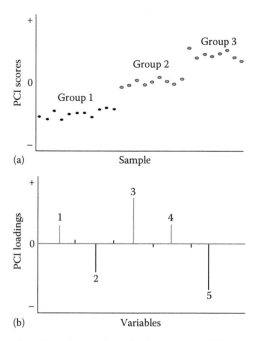

FIGURE 17.9 PC1 scores and loadings from a hypothetical data set. PC1 separates the samples into three groups as seen in the scores plot (a). The PC1 loadings (b) show that peaks 1, 3, and 4 have positive loadings, corresponding to samples with positive scores (group 3). Peaks 2 and 5 have negative loadings and correspond to samples with negative scores (group 1).

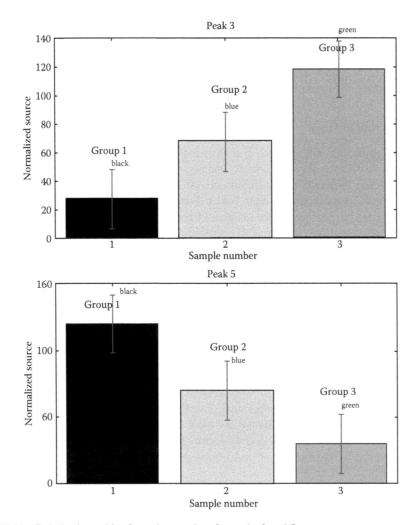

FIGURE 17.10 Relative intensities from the raw data for peaks 3 and 5.

17.4.2 Sequence of Steps for Applying PCA

Set out below is the series of steps that should be taken when planning and carrying out PCA of the results of ToF-SIMS analysis. Much of what follows applies equally well to other surface analytical techniques.

17.4.2.1 Planning

Obviously, before any samples are prepared or analyses made, it is necessary to decide on the aims of the analysis in terms of the information sought. In addition, if PCA is to be applied to the resultant data, there are further considerations. PCA will find the maximal differences between any samples in a data set, but if the analysis is not carefully planned then the interpretation will be difficult, if not impossible. It is thus essential to minimize the number of variables between samples in the set. Ideally, there should be only one variable per sample set, so that data interpretation is at its simplest.

Other considerations include the decision on how many duplicate samples should be analyzed, in order to provide adequate statistics, and the likely homogeneity of the sample surfaces. Homogeneous samples require fewer spectra recorded than do inhomogeneous samples. Self-assembled monolayer samples on gold tend to be very reproducible spot-to-spot, resulting in tight grouping in PCA. On the other hand, spectra taken on samples with adsorbed proteins tend to vary significantly spot-to-spot, resulting in significant scatter within groups in PCA. As a general rule, it is suggested that three to five spectra each be acquired on duplicate samples for homogeneous samples, and five to seven spectra each on three to five samples for nonhomogeneous samples.

17.4.2.2   Collection of Data

With statistical significance still in mind, it is important to collect an adequate number of spectra from each sample. In practice, this means more than two spectra per sample, otherwise there will not be enough confidence to decide if any differences observed are significant. Within reason, it is always better to collect more spectra than fewer. Furthermore, as mentioned above, duplicate samples should always be provided so that variations from sample to sample can be checked, thereby further improving the statistical confidence. It goes without saying that during collection of data the instrumental parameters should be maintained as constant as possible, otherwise uncertainties can creep into the analysis.

17.4.2.3   Calibration of Spectra

Consistency is again essential here. In the analysis of any one set of samples, the same set of peaks should always be used for mass calibration. The internal consistency of the calibration should be checked by spectral overlay, as demonstrated in Figure 17.11. Since calibration errors can diverge with increasing m/q, small errors in the low mass region can become large errors in the high mass region. If possible, a peak above $m/q = 100$ should be used, so that the centroid accuracy in the high mass range is improved. It is especially important to check if spectral calibration using auto-calibration routines, often included in packaged software, are used.

In Figure 17.11, the left-hand side shows the result of the initial calibration attempt, when all the spectra were calibrated to the same set of peaks using the same methodology. Significant variation

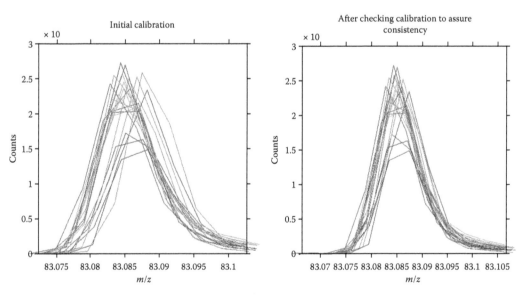

FIGURE 17.11   Checking for calibration consistency.

can be seen. On the right, the same spectra are seen after checking and adjusting the initial calibration. There is now closer coincidence of the spectra once consistency was ensured from one spectrum to another.

17.4.2.4   Selection of Peaks

One of the reasons for using PCA is the elimination of user bias, that is, the selection by the analyst of what are regarded as key peaks. On the other hand, the inclusion of all peaks can introduce undesirable variables into the analysis. It is better to start with a set containing more peaks than at first thought necessary, since it is easier subsequently to reduce the data set rather than to return to the original spectra to acquire more. Time will also be saved since there would then be no need to recreate the data matrix. As well, inclusion of all the major peaks from the outset may reveal unexpected trends, perhaps with regard to contaminants or sample chemistry. As an example, it was found [50] that in the spectra from a series of dodecanethiol SAMs, a series of low mass peaks, often ignored because of their ubiquity, actually contained valuable information about the ordering of the monolayers.

However, despite the above, it may often be appropriate to select a subset of the peaks from a set of spectra. Such a selection will depend on the nature of the sample set, and must always be based on logical decisions and valid assumptions. Rejection of peaks because they do not seem important is not a good reason. For example, if all peaks in the spectra were to be used in the PCA of adsorbed proteins, differences in the substrate peaks could overwhelm those from the proteins. In Ref. [59], it was found that if a subset of only amino-acid-related peaks were to be chosen, separation of the different proteins was much clearer.

A variety of methods of peak selection has been used in PCA. Some typical criteria include the selection of all peaks above a given intensity (e.g., above 1000 c/s) and selection of all peaks with intensities greater that the neighboring background (e.g., all peaks above three times the background in the particular region). Selection can also be performed by binning, that is, by combining or grouping spectra into a bin covering a small range of the spectral region. However, although PCA time is thereby shortened, it should be realized that the high mass resolution of ToF-SIMS is being wasted to a certain extent, since peaks from different species will be combined in a single bin. Should PCA throw up spectral differences, then it would be necessary to go back and check each bin for the peaks responsible.

After peak selection for PCA, peak areas need to be measured in each spectrum, a process that requires some care. Integration limits should be chosen so that the measured area refers only to the selected peak, with no overlap with adjacent peaks, particularly important in spectral regions where several peaks may overlap. Thus, the limits should be placed as close as sensibly possible to a peak, with overall emphasis again on consistency of choice.

An obvious requirement is that for a given set of samples, the same chosen set of peaks should be used in all spectra, irrespective of whether or not they appear in the spectra from some of the samples. It is good practice, when selecting peaks and setting the integration limits, to overlay several representative spectra from each sample to ensure that all relevant peaks have been selected and that the integration limits are consistent.

The software packages of most manufacturers contain automatic routines for integration, leading to peak area measurement, but it is advisable to turn off any automated calibration functions, since they could alter the calibration previously performed.

17.4.2.5   Normalization of Data

Data normalization in multivariate analysis has been reviewed in Ref. [60]. Such normalization is carried out to eliminate variations in the data that may not be related to the actual chemical differences of interest. These variations might include fluctuations in the primary ion current or the detector efficiency. Many different normalization methods have been used in ToF-SIMS data [48–51,53–55,59,61–74], sometimes to the total intensity, or to the sum of selected peaks, or to the most intense peak in the spectrum. When most peaks in a spectrum are selected, it is probably best to normalize to the total intensity, but when only a subset of peaks has been chosen, then normalization to the sum of the intensities of those peaks is a logical choice. If normalization is based on a single peak, care must be taken that bias is not being introduced into the results.

17.4.2.6   Preprocessing of Data

Although normalization, as described above, is one form of data preprocessing, it is usual to apply other preprocessing steps as well, before PCA. Such procedures might include mean centering, auto-scaling, log scaling, mean scaling, and taking the square root of the data matrix. Different centering and scaling methods for multivariate analysis have been reviewed in Ref. [75]. The procedures all make assumptions about the data [76], and it is necessary to be aware of what these assumptions are, and if they are valid for a particular data set.

As a minimum procedure, mean centering is advised for PCA of ToF-SIMS data. It consists of subtracting the mean of a given variable from the variables across each spectrum, so that the values of that variable within the data set will be spread across a common mean of zero. Such centering makes it easier to check that differences between samples are indeed due to sample variations.

17.4.2.7   Interpretation after PCA

Once the spectral data have undergone PCA, they should be examined closely to make sure the results are sensible. The PCA scores and loadings plots must be scrutinized and checked against the raw data, and if any discrepancies are obvious, it may be necessary to review the original assumptions made during the preprocessing steps. Alternative preprocessing methods might be needed in order to understand better the results or to clarify observed trends. As mentioned above, it is vital to understand the assumptions, so that different approaches can be tried based on logical reasons.

In summary, PCA can be a powerful analytical method when applied properly to ToF-SIMS data. Careful planning of experiments is necessary so that the effectiveness of the analysis is maximized. Experiments in which only one process variable is allowed to change are ideal for PCA, since then trends seen in the PCA results can be related to the process variable of interest, thus simplifying interpretation.

As with any multivariate method, PCA requires preprocessing of the data to put them in a form suitable for the analysis. Users should understand the assumptions being made when using a given preprocessing method, so they can interpret the results better. Data preprocessing can require some trial and error, but can soon be learnt by using a systematic approach to the analysis trends in the analysis method when dealing with different types of ToF-SIMS data. It should be remembered that PCA is simply a tool to aid the ToF-SIMS analyst.

17.5   SURFACE TOPOGRAPHY CHARACTERIZATION OF BIOMATERIALS SURFACES

The surface topography of biomaterials and biomedical implants also affects the biological responses [77,78]. Thus, it is important to characterize this topography and, when performing surface modification or coating steps, to probe for any alterations in surface topography that may cause bio-interfacial interactions to be superimposed on concurrent changes to the surface chemical composition. Atomic force microscopy (AFM) is often ideally suited to this task. The application of AFM topography measurements to biomaterials and biomedical devices is no different from that of the AFM characterization of other types of samples, and the reader is referred to Chapter 5. The surface topographies of some biomaterials, such as hydroxyapatite-coated orthopedic implants and some tissue engineering scaffolds, on the other hand, are too rough for study by AFM, and scanning electron microscopy (SEM) is commonly used for the topographical characterization of such samples, using well-established routine experimental protocols.

Determination of the surface topography, even if it does not change between the samples under comparison, is helpful because then surface roughness can be factored into XPS analyses of, for example, thin film coatings [79]. Often, nonideal topographies are encountered when analyzing biomedical devices by XPS. For curved or rough substrates, instead of a single takeoff angle, there will be a distribution of photoelectron takeoff angles present in the sampled area. The incorporation of slope frequency histograms of the surface topography features, obtained by AFM or scanning tunnelling microscopy (STM), into the XPS analysis of overlayer thicknesses [79], is particularly useful for determining the thickness of thin coatings (<10 nm) on biomedical devices such as contact lenses. Approximate approaches and limitations are also discussed in Ref. [79]; for moderate surface topographies, it is possible to use approximations such as the average slope to determine accurate overlayer thicknesses.

17.6 ANALYSIS OF ADSORBED BIOMOLECULE LAYERS

Surface analysis techniques are of relevance not only for the characterization of biomaterials prior to biological testing, but also for help in elucidating the consequences of interactions between biomaterials and biological systems. Biological media contain thousands of different proteins, many of which are surface active and can play a role in host responses to implants and the triggering of the failure of devices. For example, adsorbed fibrinogen (Fg) promotes platelet adhesion, and bacteria can colonize adsorbed protein layers. Many of the studies to date have focused on the adsorption of individual proteins or model systems, with mixtures of a few of these proteins competing in surface adsorption. In addition, there have been numerous experiments in which one protein component has been labeled (e.g., with a radioisotope such as ^{125}I) and added to serum or plasma to provide quantitative adsorption data from a complex solution. Such studies include the pioneering work of Vroman who defined the role that Fg plays in blood-contacting surfaces [80]. However, on account of the complex nature of biological media, an understanding of the precise mechanisms of protein–surface interactions that dictate biological responses is still a long way off. In human serum, over 3000 proteins have so far been identified using classical proteomics approaches [81] (the "proteomics" terminology describes activities concerning proteins equivalent to those that are described by genomics in the case of genes), and an estimated 50,000 molecular forms of the plasma proteins might be present at any given time [82]. However, the task can be simplified by considering only the proteins that make up about 99% of all those in plasma [83], that is, albumin, IgGs, transferrin, Fg, IgAs, α-2-macroglobulin, IgMs, α-1-antitrysin, C3 complement, lipoprotein A, α-1-acid-glycoprotein, apolipoprotein B, apolipoprotein A-1, factor H, ceruloplasmin, C4 complement, complement factor B, prealbumin, C9 complement, C1q complement, and C8 complement. Andrade and Hlady defined the "big twelve" involved in biocompatibility as those proteins in plasma with a concentration of >1 mg/mL [84].

Despite the numerous publications on protein adsorption, there have been very few studies that correlate such adsorption *in vitro* with what actually takes place *in vivo*, and in particular with the identification of which proteins, other than those already observed, adsorb *in vivo* to biomaterials surfaces. Perhaps the failure to achieve better biocompatibility *in vivo* is related to the 1% of low abundance proteins that are poorly understood, or to different molecular forms of proteins, existing *in vivo*, that might provide different functional analogues compared to those investigated *in vitro*. Unidentified proteins may adsorb to biomaterials from complex media, and work in tandem with proteins of known function to accelerate host responses. One report has shown that it is possible to identify, using matrix-assisted laser desorption/ionization ToF-MS (MALDI-ToF-MS) (see Section 17.6.3.1), some proteins that adsorb to contact lenses when worn by patients, and also other proteins adsorb that could not be identified [85]. Other issues are the detection limits of the various techniques used to determine the amount of adsorbed protein, and the question as to what is the minimal surface density of protein adsorption that can trigger an adverse response. For example, a critical surface concentration for Fg as low as $5 \, ng/cm^2$ can trigger platelet adhesion [86].

Surface analytical techniques assist in elucidating biomolecule adsorption processes onto biomaterials, and can often detect adsorbed biomolecules at very low concentrations, complementing biochemical methods such as ELISA (enzyme-linked immunosorbent assay) [87] for the study of interactions between biomaterials surfaces and biological environments.

17.6.1 METHODS OF STUDYING PROTEIN–SURFACE INTERACTIONS

Over the years, many techniques have been developed and applied to the investigation of protein adsorption onto biomaterials, resulting in improvements in the understanding of the physicochemical mechanisms responsible for protein–surface interactions (Table 17.1). However, a sufficient understanding of protein adsorption for the rational design of improved biomaterials remains a challenge. In addition to developing new methods of assessing protein adsorption, a multi-technique analysis approach is necessary, as individual methods on their own are not capable of providing a complete picture of the process. The methods should be able to provide different types of information about the adsorption process, including the amount of adsorbed protein, the rate of adsorption, the conformation or orientation of the adsorbed protein layer, and the extent of competitive adsorption from complex biological media. Detection of time-dependent changes in the structure of the protein layers and identification of unknown adsorbates that exist in real biological media are highly desirable. The aim of this section is not to provide an exhaustive account of all techniques available, or to describe each of the techniques in great detail, but to give a description of some of the more common methods of assessing protein adsorption and a broad overview of some of the emerging techniques.

TABLE 17.1
Techniques Used to Study Protein Adsorption onto Biomaterials Surfaces

Technique	Quantitative Information	Structure/ Conformation	Kinetics	Competition/More than One Protein
Ellipsometry	Yes	Yes	Yes	Yes
Neutron reflectrometry	Yes	Yes	Yes	Yes
SPR	Yes	Potentially	Yes	potentially
ATR-FTIR	Yes	Yes	Yes	Yes
OWLS	Yes	Yes	Yes	Potentially
IRAS	Yes	Yes	Yes	Yes
TIRF	Yes	Yes	Yes	Yes
AFM (SPM)	No	Yes	Potentially	Potentially
XPS	Semi	Some	*Ex situ*	No
Surface-MALDI	Potentially	Potentially	Potentially	Yes
ToF-SIMS	Potentially	Yes	Potentially	Yes
QCM	Yes	Yes	Yes	Potentially
Radiolabeling	Yes	No	Yes	Yes
ELISA	Qualitative	Potentially	Yes	Yes
Electrophoresis	Qualitative	Potentially	No	Yes
SFA	Potentially	Yes	Yes	Potentially
CD	No	Yes	No	No

Note: IRAS, IR reflection-absorption spectroscopy; TIRF, total internal reflection fluorescence; and SFA, surface force apparatus.

When performing protein adsorption studies, the key questions that need to be asked before deciding on which is the best techniques to use include

1. What is the material surface to be investigated?
2. How much protein needs to be detected?
3. Is it a fundamental study of single protein adsorption, or is a complex protein solution like serum to be used?
4. Is kinetic information required?
5. Are physical properties of the protein layer (e.g., viscoelasticity) required?
6. Is label-free detection of the adsorbed proteins needed?
7. Is knowledge of the activity of the protein layer (e.g., its capacity to attach cells) wanted?

The answers will determine the choice of techniques to be used. The techniques differ quite considerably in the methods of detection and the ease of use. In many cases, the methods are simply extensions of existing surface analytical tools that have been applied to other materials, and very few have been developed specifically for the study of protein adsorption. Associated with each technique is the added uncertainty of some of the assumptions that are made to generate the data, thus quite often cross-correlations between methods are necessary for accurate determinations. Instead of covering all the techniques in detail, a summary will be provided of selected methods, including XPS, ToF-SIMS, quartz crystal microbalance (QCM), radiolabeling, and surface-MALDI-MS. The combination of these techniques provides the following information: (1) a quick estimate of the extent of protein adsorption (XPS), (2) detection at very high sensitivity of minute surface-adsorbed quantities (ToF-SIMS), (3) the kinetics and surface properties of adsorbed protein layers (QCM-D), (4) accurate quantitative determination of protein adsorption (radiolabeling), and (5) detection and potential identification of proteins adsorbed from complex solutions, for example, plasma (surface-MALDI).

17.6.2 XPS Characterization of Protein Adsorption

One of the strengths of XPS is that it can be quantitative. While XPS cannot be used to distinguish between various proteins, since all tend to have very similar amounts of carbon, oxygen, and nitrogen atomic species, it can be used to determine (either qualitatively or quantitatively) the total amount of protein adsorbed onto surfaces. Some amino acids do contain sulfur, but most proteins have those acids in such small quantities that the amount of sulfur is near or below the detection limit of the technique; thus, quantification of protein adsorption using sulfur is not practical for most applications.

In general, nitrogen is a good indicator of protein adsorption because there is a significant amount of nitrogen in proteins (typically ~15 at.% for proteins that are not heavily glycosylated). For proteins adsorbed onto substrates that do not contain significant amounts of nitrogen, the N 1s atomic percentage can be used for quantification of protein adsorption [12,70,88–90]. XPS was used to detect the adsorption of Fg onto a variety of surfaces in a study on the detection limits of XPS and ToF-SIMS by Wagner et al. [12]. Calibration of the XPS signal against the amount of protein on mica, PTFE, and two nitrogen-containing plasma polymers was achieved by radiolabeling of the protein. For substrates such as mica and PTFE that do not contain nitrogen, following the increase in the N atomic percentage was a convenient indication of protein adsorption on the surface. Detection of protein concentrations down to 10–25 ng/cm² on mica (corresponding to N ≈ 0.5–1 at.%) was reported. It should be noted that absolute quantification of proteins is difficult without some calibration step such as the radiolabeling of controls prior to analysis. However, relative amounts can be easily determined.

On substrates that contain significant amounts of nitrogen, such as the allylamine and heptylamine plasma polymers, also discussed by Wagner et al. [12], an analysis of the concentration of nitrogen present as the protein exposure is increased is no longer a straightforward means of following protein adsorption amounts. If both the protein and the substrate have similar atomic percentages of nitrogen, there may not be significant differences before and after the protein adsorption. For substrates such as the amine polymers, it is possible to follow a distinct signal for the protein in the high-resolution C 1s spectra. As was shown in Ref. [12], the amide carbon (N-C=O, at 288.2 eV) gave rise to a distinct peak not found from the amine plasma polymer controls, which contained amine and hydrocarbon species (typically C attached to amine gives rise to a component at ~286.4 eV when referenced to hydrocarbon C 1s at 285.0 eV). The amide peak could be followed in order to determine increasing amounts of protein adsorption where the N percentage was not changing significantly, but the detection limit was significantly worse (200 ng/cm^2 for the allylamine substrate).

Another example of difficult protein quantification using the N 1s signal was reported by Martin et al. [90], in which the protein osteopontin (OPN) was immobilized on poly(HEMA) using 1,1-carbonyldiimidazol (CDI) chemistry. Poly(HEMA) does not contain nitrogen, but the CDI linker does. While the surface with CDI had noticeably lower nitrogen than after the subsequently adsorbed OPN, it would have been difficult to quantify the amount of OPN on the surface using simply the atomic percentage of N. While the N 1s signal due to OPN increases with increasing OPN concentration, that from the CDI is attenuated by the OPN overlayer. At smaller surface concentrations of OPN, the N 1s signal from the CDI was still apparent. The authors found that in this particular case they were able to use the N 1s signal at high resolution to distinguish between the amide and amine nitrogen peaks from the protein (at ~400 eV) and the imidazol ring nitrogen of the CDI (at 401.4 eV).

An alternative approach is to use the attenuation of a signal from a unique, or marker, element present in the substrate, the attenuation arising from the presence of the protein overlayer. Thus, as for the analysis of thin film coatings in other applications, one can use an overlayer model and the same equation as is used in ARXPS (Equation 17.1).

Another method is to use an element in the overlayer to compare the observed atomic percentage or intensity with that of an infinitely thick overlayer, which serves as a theoretical reference value I_0:

$$\frac{I}{I_0} = 1 - \exp\left(\frac{-d}{\lambda \cos \alpha}\right) \tag{17.2}$$

Assuming a mean free path of 3 nm, typical for many organic materials, either of Equations 17.1 or 17.2 can then be used to obtain an estimate for the thickness of the protein layer, and, assuming coverage to be uniform and the density of the protein layer to be between 1.2 and 1.4 g/cm^3, the adsorbed protein amount in ng/cm^2 can also be derived [91], but for more accurate measurements, calibration using radiolabeling or another technique is indicated.

One important detail that should always be considered with any quantitative analysis of this type is the standard deviation across the surface. An example of this variability is found in the study by Wagner et al. [12] described above. A high degree of variability in the data across the surfaces was reported (using an analysis spot size of the order of 800 μm^2), most notably on the PTFE surface, which led to a degree of uncertainty in the quantitative analytical results. The authors hypothesized that surface roughness, as well as the patchy protein adsorption seen on fluorocarbon surfaces, were likely causes for the high variability in the data. In that study, two spots across four samples were analyzed. Analyzing duplicate samples with multiple analysis spots is recommended, when possible, for a more accurate determination of the amount of adsorbed protein.

17.6.2.1 XPS Analysis of Microarray Chemistry: A DNA Example

Another common adsorbed biomolecule is DNA, as seen in DNA microarrays, and again the quantitative nature of XPS can be used to determine their surface concentrations [92–94]. A recent study has used careful comparison of ^{32}P-radiolabeling to calibrate the P and N signals from XPS against the amount of thiol-coupled DNA probe on a gold surface, as well as the subsequent addition of the target DNA after hybridization [94]. XPS systems are now available that can map atomic species with spatial resolution down to about 10 μm. The XPS imaging capability allows easier identification of microarray dots that would otherwise be difficult to find. This imaging capability, along with the progressive decrease in analyzed area over the years, can allow analysis of a single microarray dot (~100 μm in diameter). An example of this type of analysis is given in Ref. [8], in which DNA microarrays on commercial polymer substrates were analyzed by XPS and ToF-SIMS. The dots were identified in the imaging mode, allowing for exact registration of the analyzed area (~55 μm) with a microarray dot. Quantification of the DNA before and after hybridization was used to calculate the efficiency of hybridization over a range of conditions, without the potential interference and possible inconsistencies of fluorescence labels. Since the underlying polymer substrate contained nitrogen, the phosphorus signal from the DNA backbone was used to quantify unambiguously the DNA.

When analyzing species such as proteins, it should be remembered that XPS is an UHV technique and that, in general, proteins (as well as cells and tissue samples) require a hydrated environment to maintain their structural conformation. In a hydrated environment, the hydrophobic amino acids tend to be found on the inside of the protein, while the hydrophilic amino acids are found on the outside. When proteins are introduced into a vacuum environment, many have been found to rearrange (denature) so that the hydrophobic amino acids are presented to the vacuum interface. Various techniques can be used to maintain protein conformation in vacuum, such as glutaraldehyde fixation [55], trehalose coating [55], or cold stage analysis (discussed in Section 17.2.2). However, all these techniques involve adding chemical species (i.e., glutaraldehyde and trehalose), or potentially leave ice on the surface (in cold stage analysis), which will affect the overall atomic composition results calculated from XPS. Fortunately, protein denaturation does not seem to affect the relative quantification of proteins by XPS under UHV. While the structure may rearrange, the relative amounts of C, N, and O can still be analyzed.

17.6.3 ToF-SIMS Analysis of Adsorbed Protein Layers

Static ToF-SIMS is a very useful technique for the characterization of adsorbed protein layers due to its chemical specificity and surface sensitivity. With its basis in mass spectrometry, ToF-SIMS yields information about the molecular structure of the analyzed surface layers. It has been used in many studies for the characterization of adsorbed protein films (e.g., Refs. [13,53–55,59,63,68–73]), and can provide a fingerprint mass spectrum for each protein, based on detecting the parent and fragment ions of each of the constituent amino acids [52]. Due to the high mass resolution of ToF detection, signals from the protein can readily be separated from those of the substrate. In cases of spectral similarity or when probing for extremely low amounts of adsorbed proteins [13], PCA analysis is invaluable.

At the most basic level, ToF-SIMS can be used to look for the presence of adsorbed protein molecules, by comparing spectra recorded on biomaterials surfaces before and after exposure to protein solution. Due to its extremely high sensitivity for many elements, the technique can identify proteins present in ultralow amounts not detectable by other techniques [13]. Using the imaging mode of ToF-SIMS, the lateral homogeneity of the distribution of protein molecules on surfaces can also be investigated. It has also been used to identify which proteins are present in an adsorbed layer containing several different proteins. In this case, since most proteins have the same amino acid fragments, the relative intensities of the fragments must be measured for protein identification [59]. Using spectra from single layers of a protein as a reference, it is possible to determine the relative surface composition of simple mixed protein layers (e.g., binary layers) [69–72], but because of the similarity of protein spectra it is not currently possible to quantify proteins present in complex adsorbate layers (e.g., deposited from blood plasma). Gaining information such as the difference between two adsorbed protein films can be difficult due to the large number of peaks in the ToF-SIMS spectra. In cases where there is an absence of unique peaks for different samples, PCA lends itself to the analysis of spectral differences [53,63,73].

An illustrative example of the utility of the ToF-SIMS analysis of protein adsorption is that of spectra recorded from poly(*N*-isopropyl-acrylamide) (pNIPAM) graft coatings before and after contact with a solution of either bovine serum albumin or lysozyme (Cole, Jasieniak, and Griesser, unpublished). The XPS spectrum of pNIPAM is similar to those of many proteins, and it is therefore difficult to look for protein adsorption on this particular polymer by XPS, with adequate sensitivity. The polymer pNIPAM is known to possess a lower critical solution temperature of 32°C: below this temperature the polymer exists in a well-hydrated state, but above it the polymer is in a much denser (collapsed) conformation, having lost part of its water of hydration. ToF-SIMS was used to investigate whether in the well-hydrated state the pNIPAM graft coating would resist full protein adsorption, and also whether above the transition temperature there might be higher amounts of protein present. The spectra are not shown as they look superficially very similar for all samples analyzed. Upon application of PCA analysis, it was found that the spectra recorded after exposure of pNIPAM graft coatings in the hydrated state at 20°C to the two protein solutions did not differ statistically from the reference spectrum of the same graft coating that had not been exposed to proteins (Figure 17.12). Thus, it could be concluded that within the sensitivity of ToF-SIMS analysis, this graft polymer coating resisted completely the adsorption of both proteins.

When pNIPAM coated samples were exposed to solutions of the proteins at 37°C, on the other hand, PCA analysis yielded clearly defined statistical differences between exposed and unexposed samples (Figure 17.12) for both proteins. Another feature of interest was that for 2 out of the 10 spectra recorded on the sample exposed to albumin, the scores differed considerably from those of the other 8, indicating that there was significantly more protein present at those 2 analysis spot locations. This is a good illustration of the possible lateral variations in adsorbed protein layers. Such variations might arise, for example, from coating defects caused by dust particles on the substrate surface (the coatings were not prepared in a clean room environment), or from the clumping of proteins. For BSA (bovine serum albumin) adsorbed onto a steel stub in order to obtain reference albumin spectra (black points in Figure 17.12), however, the data scores fell within a tight cluster, indicating that there was no inhomogeneous clumping of the protein on that surface. This reinforces the need for multiple analyses at different locations on a given sample, so that lateral inhomogeneities in coatings and protein adsorption can be identified and taken into account in interpretation. The lysozyme adsorption data were more uniform (Figure 17.12), with the spread in the data scores

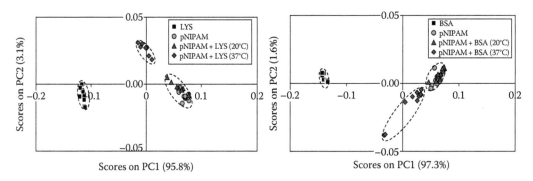

FIGURE 17.12　PCA analysis of ToF-SIMS spectra recorded from pNIPAM graft coatings as-prepared, and after exposure to albumin or lysozyme solutions at 20°C and 37°C, and on the proteins adsorbed onto a SIMS sample stub. The peaks from the stainless steel substrate were removed from the spectra of the proteins on stubs in order to obtain protein reference spectra. (From Cole, M., Jasieniak, M., and Griesser, H.J., unpublished data.)

originating in the signal-to-noise ratios of the peaks, indicating that for this sample (and the reference sample) the coating and the adsorbed protein layer were laterally uniform, and hence suggesting that the deviations in the albumin data were caused by coating defects.

As for XPS, the question must be considered as to how the vacuum environment might affect adsorbed proteins. Preservation of the adsorbed protein structure by glutaraldehyde or trehalose fixation has been reported [55]; cryo-analysis is also possible, but the issues discussed above for XPS are equally relevant to SIMS analyses.

17.6.3.1 MALDI-ToF-MS for the Detection of Biomolecules on Surfaces (Surface-MALDI-MS)

The development of MALDI-ToF-MS has been the platform for now well-established MS methods in the field of proteomics. In MALDI-MS, the sample is irradiated with a pulsed laser. Mixing the analyte with a highly UV-absorbing matrix results in fast evaporation upon adsorption of laser light, and volatilized molecules that are ionized in the process are extracted as in SIMS and analyzed by a mass spectrometer; ToF detection has become the norm. The matrix crystals that embed the analyte create efficient volatilization and improved ionization efficiencies, and enable detection of higher MW species [96]. In contrast to SIMS, little or no fragmentation of the analyte molecules occurs. On the one hand, this enables determination of the molecular mass, but on the other hand, the chemical information content inherent in mass spectra is not available in the standard method, although various variants of MALDI-MS have been devised for various specific purposes.

MALDI-ToF-MS is a method for the accurate determination of the molecular mass of macromolecules such as proteins [97], with a mass range up to 550,000 Da [98], an accuracy of 10^{-3} to 10^{-4} [99], and very low detection limits (i.e., attomoles/nanograms) [100]. In the conventional version, the analyte is dissolved in a solution together with photoactive molecules (e.g., *trans*-3,5-dimethoxy-4-hydroxy cinnamic acid/sinapinic acid). This mixture is applied to a sample stub and the solvent evaporated to yield photomatrix crystals with encapsulated protein molecules. The crystals are irradiated with a pulsed UV laser ($\lambda = 337$ nm, 3 ns pulse width) for ablation. Analyte molecules are ionized in the expanding plume [101].

More recently, the method has been adapted to the analysis of surface-located analyte molecules, including detection of adsorbed proteins on biomaterials surfaces such as contact lenses [102]. In this variation, standard MALDI instrumentation and sample preparation are used, except that the matrix solution is applied to the material surface after the adsorption of the analyte/protein molecules. In this way, it becomes possible to look for the presence and identity, via their molecular masses, of several proteins coexisting in adsorbed layers, a task that is beyond other surface analysis techniques.

In the surface-MALDI-MS technique for the analysis of adsorbed protein layers, the matrix solution is applied directly to the protein-coated biomaterial surface, after it has been thoroughly rinsed to remove loosely adherent protein molecules. As the solvent evaporates, the adsorbed proteins become soluble and are encapsulated in the growing matrix crystals (Figure 17.13). The proteins are then ionized and detected in the same manner as in conventional MALDI.

Figure 17.14 shows a surface-MALDI-MS spectrum recorded from a soft contact lens after it had been worn for 10 min by a human subject [102]. A number of different protein species are detected as indicated by the unique ions in the spectrum. These include human tear lysozyme (peaks A and B) and several other unidentified low MW protein species (peaks C, D, and E). The latter may be either unique proteins or proteolytic fragments of other tear proteins, and are not detectable by other methods.

Surface-MALDI-MS is a highly sensitive, complementary method for studying the composition of adsorbed multicomponent protein layers. In addition to detecting the presence of known proteins, the method can also be used for the detection of unknown species, whose biological activity could be involved in biological responses to biomaterials and devices. An example of this is given in

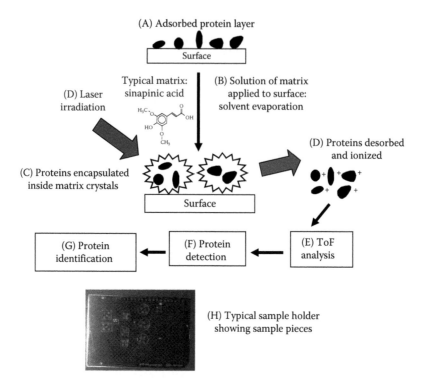

FIGURE 17.13   Schematic representation of the surface-MALDI-MS experimental approach.

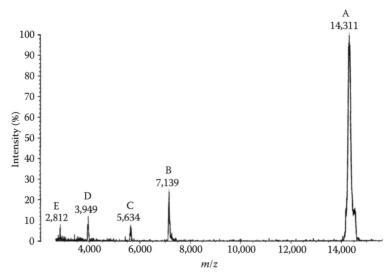

FIGURE 17.14   Surface-MALDI-MS spectrum recorded from a contact lens surface. A and B peaks correspond to human tear lysozyme. C, D, and E are other proteins or tear protein fragments. (From Kingshott, P., St. John, H.A.W., Chatelier, R.C., and Griesser, H.J., *J. Biomed. Mater. Res.*, 49, 36, 2000. With permission.)

Figure 17.15, which shows a surface-MALDI-MS spectrum of a PET surface coated with a monolayer of Fg, and exposed to a solution of 10% fetal bovine serum [103]. The surface preparation was aimed at mimicking the conditions of macrophage adhesion to Fg-exposed surfaces in which the cells are suspended in serum. The results show that even after optimum packing of Fg on the surface,

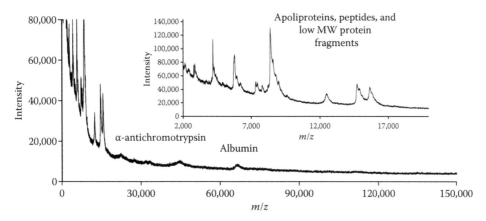

FIGURE 17.15 Surface-MALDI-MS spectrum recorded from a Fg-coated PET surface. (From Ademovic, Z., Holst, B., Kahn, R.A., Jørring, I., Brevig, T., Wei, J., Winter-Jensen, B., and Kingshott, P., *J. Mater. Sci.: Mater. Med.*, 17, 203, 2006. With permission.)

low MW proteins from serum can either attach to the Fg or adsorb in the gaps between Fg molecules. Parallel radiolabeling experiments showed that no Fg was displaced during the experiment. This study highlights the fact that studying protein adsorption from complex solutions may require further development of the more sensitive and specific tools, such as surface-MALDI-MS, in order to unravel the detailed mechanisms of bio-interfacial interactions and responses by complex biological systems.

Other biomaterial surfaces to which surface-MALDI has been applied to analyze the composition of protein adlayers include those of polysaccharides [104–106], plasma polymers surfaces [103,107,108], nitinol wire used in stents [109], and hemodialysis membranes [110].

There are several as yet unresolved limitations of surface-MALDI-MS. First, the adsorbed proteins need to be eluted from the surface by the matrix solution, otherwise they will not become encapsulated inside the crystals, which is a prerequisite for ionization. Thus, the more strongly adsorbed the proteins are on the biomaterial surface, the less likely they will be able to be eluted. Typically, low MW species are easily eluted; however, large, structurally unstable proteins may stay adsorbed. For this reason, the matrix solution is usually applied immediately after the adsorption experiment to minimize any time-dependent strengthening of surface binding. Second, due to suppression effects, the detectability of proteins can be affected. When too much protein or too many species are present, phase separation can occur during matrix crystal formation, resulting in poor distributions of protein inside the crystals and thus in loss of ionization efficiency and in probable loss of signal. The third limitation is instrumental: the degradation of signal at high MWs (>100 kDa) due to reduced signal transmission through the instrument. It is possible to detect very high MW species (e.g., Fg at 340 kDa) under ideal conditions (data not shown), but this is not achieved routinely. One possible method of overcoming this limitation could be the direct digestion of adsorbed proteins, followed by detection of the peptides unique to each adsorbed protein. Despite these limitations, surface-MALDI-MS provides unique information concerning protein adlayers not accessible by other techniques, and with high sensitivity, in particular in demonstrating differences in protein profiles as a function of materials surface chemistry.

17.6.4 QUARTZ CRYSTAL MICROBALANCE (QCM)

A surface analysis technique designed specifically for the characterization of adsorption processes onto thin layer surfaces, the QCM or thickness-shear mode resonator, is based on an AT-cut

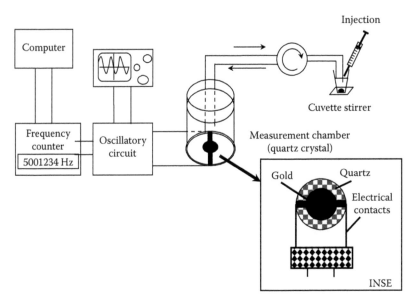

FIGURE 17.16 Schematic diagram of the QCM setup. (Adapted from Steinem, C., Janshoff, A., Wegener, J., Ulrich, W.P., Willenbrink, W., Sieber, M., and Galla, H.J., *Biosens. Bioelectron.*, 12, 787, 1997. With permission.)

piezoelectric quartz crystal with metal electrodes on both sides, as indicated in Figure 17.16 (inset) [111,112]. By applying an RF voltage across the electrodes a resonance is setup from mechanical oscillations near the fundamental resonance frequency (typically 5 or 10 MHz) of the crystal [111]. Material deposited or adsorbed on the surface of the crystal causes a reduction in the fundamental frequency, f_0, according to [112]

$$\Delta f = \frac{-2f_0^2}{A\sqrt{\rho\mu}}\Delta m \tag{17.3}$$

where
   Δf is the change in resonance frequency (Hz)
   Δm is the change in mass (ng)
   ρ is the density of the quartz crystal (kg/m^3)
   μ is the shear modulus of the quartz crystal (kg/m/s^2)
   f_0 is the fundamental resonance frequency of the crystal (MHz)
   A is the electrode area (m^2)

Commercial systems are designed to measure mass changes up to about 100 μg reliably and rapidly, with a detection limit of the order of 1 ng [111]. Equation 17.3 assumes that added mass is simply an extension of the crystal mass, and the model holds well for air measurements, but cannot be applied to the accurate determination of the amount of adsorbed protein in a liquid. This is due to interference or nonlinear changes to the resonance frequency for viscoelastic layers such as adsorbed proteins, caused by slip, internal dissipation, and an entrapped water layer [113]. However, decreases in the resonance frequency can be used elegantly to follow the kinetics of protein adsorption [113,114], while changes in the dissipation factor, D, can be used to infer structural information about adsorbed proteins.

   An example of the information obtainable by QCM is given in Figure 17.17, in which on the left is shown the QCM resonance frequency changes for the adsorption of horse spleen ferritin onto gold from a series of solutions with various concentrations. An immediate decrease in frequency

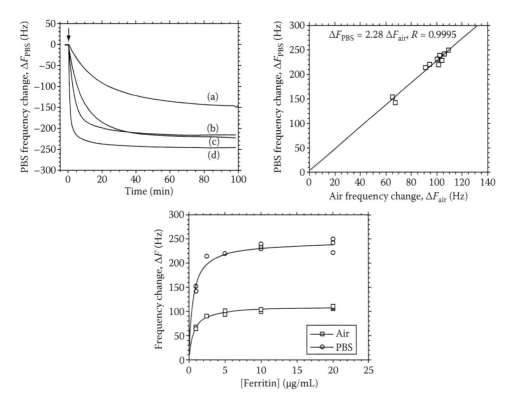

FIGURE 17.17 QCM characterization of the adsorption of ferritin onto gold [120]. (From Caruso, F., Neil Forlong, D., and Kingshott, P., *J. Coll. Interface Sci.*, 186, 129, 1997. With permission.)

occurs for all concentrations with the maximum ΔF ranging from 145 ± 5 Hz for adsorption at $1.0\,\mu g/mL$ to 230 ± 15 Hz at concentrations above $2.5\,\mu g/mL$, where saturation coverage is reached. The right-hand plot shows that the adsorption of ferritin onto gold can be fitted to Langmuir-type behavior; however, larger increases in ΔF are observed for the measurements in phosphate buffered saline due to viscoelastic dampening caused by large proteins. The third plot relates the frequency changes in buffer versus those in air; a calibration factor can be derived from the linearity of the plot and used to estimate the surface coverage and thickness of the ferritin layer. This calibration factor takes into consideration that in PBS the mass uptake is a combination of the protein mass and the mass of coupled water. In all cases less than a monolayer adsorbs on the gold surface ($6.4 \pm 0.3\,mg/m^2$ for observed saturation coverage versus $9.2\,mg/m^2$ calculated from the dimensions of ferritin), and the results can be correlated with other techniques such as SPR (surface plasmon resonance), AFM, and XPS [113].

More detailed analysis of the QCM data for protein adsorption in aqueous buffer solution can provide structural information about protein adlayers, from concurrent measurements of the change in resonance frequency, f, and the so-called dissipation factor, D [111–114]. The dissipation factor is a measure of the extent of viscoelastic dampening induced by the structural mobility of the protein molecules coupled to the QCM crystal, which is related to the interfacial protein conformation. The experiment is performed by periodically switching on and off the driving voltage of the piezoelectric crystal oscillator and recording the voltage decay, which is then fitted to an exponentially damped sinusoidal function to provide both f and D [112]. Figure 17.18 shows QCM data for the adsorption of two structurally similar forms of hemoglobin (met-Hb and HbCO) at their isoelectric points, onto a hydrophobic C18 self-assembled monolayer on a 5 MHz gold-coated quartz crystal [114].

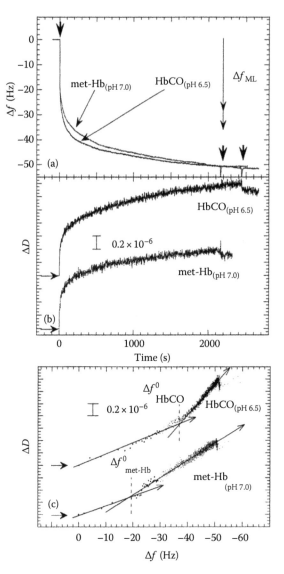

FIGURE 17.18 QCM analysis of the adsorption of two structurally similar forms of hemoglobin at their isoelectric points onto a C18 self-assembled monolayer. (From Höök, F., Rodahl, M., Kasemo, B., and Brzezinski, P., *Proc. Natl. Acad. Sci. USA*, 95, 12271, 1998. With permission.)

Figure 17.18a and b shows the simultaneous decrease in f and increase in D for both proteins, and it is evident that very similar adsorption kinetics apply. The increase in D indicates that more energy is dissipated as protein adsorption takes place. When the changes in D (ΔD) are plotted against changes in f (Δf) (cf. the D–f plot in Figure 17.18c), more subtle information can be extracted about the adsorption process of the two proteins. The different slopes in the D–f plots indicate that at least two different kinetic processes occur during adsorption, and the slightly different forms of Hb show more significant differences in the dissipation shift for the same frequency change. There are two phases in the adsorption process for both proteins. The slope of the initial phase is similar for both proteins, but the onset of the second phase is delayed for HbCO, despite the same final Δf value (Figure 17.18a). That is, the break occurs at 20 Hz for met-Hb compared to 38 Hz for HbCO, and is attributed to the lower structural stability or more rigid structure of the met-Hb form caused by Hb dissociation into subunits and by denaturation. The slope of the second adsorption phase for

HbCO is steeper by a factor of 1.3 than that for met-Hb. Additionally, the Δf at the end of the initial stage is indicative of a monolayer of protein, whereas the final Δf corresponds to a bilayer of protein. The higher apparent dissipation for HbCO from the D–f plots, as compared to met-Hb, is caused by a more flexible, hydrated second layer of coupled protein in its native state. Finally, conversion of met-Hb to HbCO after adsorption results in similar structural perturbations to the layer, in comparison with that seen during the adsorption of HbCO alone [114].

The main disadvantage of the QCM technique is the difficulty in determining the exact mass of the adsorbed protein, since the mass detected is a sum of the molecular mass of the protein and the associated water. Complementary techniques based on optical detection, such as SPR, or optical waveguide light spectroscopy (OWLS), which are not sensitive to associated water, are necessary to obtain direct adsorbed masses. However, the QCM technique is very powerful for acquiring both kinetic data and insights into the film structure, viscoelastic properties, and protein conformation changes. Analysis of the data with viscoelastic models can also be performed.

17.6.5 RADIOLABELING

Radiolabeling is not a surface analysis technique *per se* but is presented here briefly as it is regarded as the most reliable method of obtaining quantitative values of adsorbed protein coverage on biomaterials surfaces, with sensitivities down to a few ng/cm^2 being readily achieved [116]. The most common radiochemical method employs Na ^{125}I labeling, using either the so-called iodine monochloride (ICI) [117] or the chloramin T methods [118]. However, there are a number of factors that need to be taken into account during a measurement. First, any free ^{125}I not consumed during labeling must be removed by either dialysis or gel filtration or by combinations of the two. This is to ensure that the level of protein is not overestimated. Second, the labeling procedure can itself affect profoundly the protein stability and influence the measurement. It has been shown by MALDI-MS and chromatography that fragile proteins such as IgG need soft labeling methods such as the ICI method, whereas more stable proteins such as human serum albumin (HSA) can tolerate the harsher chloramine-T method [119]. Finally, it has to be assumed that the labeling does not influence the adsorption behavior, since labeled and unlabeled proteins are usually mixed, and the final quantitative measurement relies on both adsorbing equally. Despite these uncertainties, the method can also be used for measuring the amount of a specific protein that adsorbs from complex solutions such as plasma, when a known concentration of labeled protein is added to the mixture.

REFERENCES

1. D. G. Castner and B. D. Ratner, *Surf. Sci.*, **500**, 28 (2002).
2. http://www.cibavision.com.au/lenses/night_and_day.shtml.
3. C. Y. Lee, G. M. Harbers, D. W. Grainger, L. J. Gamble, and D. G. Castner, *J. Am. Chem. Soc.*, **129**, 9429 (2007).
4. X. Xie, T. R. Gengenbach, and H. J. Griesser, *J. Adhesion Sci. Technol.*, **6**, 1411 (1992).
5. T. R. Gengenbach, X. Xie, R. C. Chatelier, and H. J. Griesser, *J. Adhesion Sci. Technol.*, **8**, 305 (1994).
6. S. A. Al-Bataineh, L. G. Britcher, and H. J. Griesser, *Surf. Interface Anal.*, **38**, 1512 (2006).
7. B. D. Ratner, T. A. Thomas, D. Shuttleworth, and T. A. Horbett, *J. Coll. Interface Sci.*, **83**, 630 (1981).
8. K. Lewis and B. Ratner, *J. Coll. Interface Sci.*, **159**, 77 (1993).
9. T. R. Gengenbach, R. C. Chatelier, and H. J. Griesser, *Surf. Interface Anal.*, **24**, 271 (1996).
10. T. R. Gengenbach, R. C. Chatelier, and H. J. Griesser, *Surf. Interface Anal.*, **24**, 611 (1996).
11. K. S. Siow, PhD thesis, University of South Australia, Mawson Lakes, South Australia, Australia (2007).
12. M. S. Wagner, S. L. McArthur, M. Shen, T. A. Horbett, and D. G. Castner, *J. Biomater. Sci., Polym. Ed.*, **13**, 407 (2002).
13. P. Kingshott, S. McArthur, H. Thissen, D. G. Castner, and H. J. Griesser, *Biomaterials*, **23**, 4775 (2002).

14. D. Leonard and H. J. Mathieu, *J. Anal. Chem.*, **365**, 3 (1999).
15. H. J. Mathieu, Y. Chevolot, L. Ruiz-Taylor, and D. Leonard, *Adv. Polym. Sci.*, **162**, 1 (2003).
16. H. J. Mathieu, *Surf. Interface Anal.*, **32**, 3 (2001).
17. A. M. Belu, D. G. Graham, and D. G. Castner, *Biomaterials*, **24**, 3635 (2003).
18. R. Michel and D. G. Castner, *Surf. Interface Anal.*, **38**, 1386 (2006).
19. http://www.ion-tof.com/technique-timeofflight-IONTOF-TOF-SIMS-TIME-OF-FLIGHT-SURFACE-ANALYSIS.htm.
20. D. Briggs, *Surface Analysis of Polymers by XPS and Static SIMS*, Cambridge University Press, Cambridge, 1998.
21. A. Benninghoven, *Surf. Sci.*, **299/300**, 246 (1994).
22. C. W. Diehnelt, M. J. Van Stiponk, and E. A. Schweikert, *Int. J. Mass Spectrom.*, **207**, 111 (2001).
23. Z. Postawa, B. Czerwinski, N. Winograd, and B. J. Garrison, *J. Phys. Chem. B*, **109**, 11973 (2005).
24. E. A. Jones, J. S. Fletcher, C. E. Thompson, D. A. Jackson, N. P. Lockyer, and J. C. Vickerman, *Appl. Surf. Sci.*, **252**, 6844 (2006).
25. E. A. Jones, N. P. Lockyer, and J. C. Vickerman, *Int. J. Mass Spectrom.*, **260**, 146 (2007).
26. F. Kollmer, *Appl. Surf. Sci.*, **231–232**, 153 (2004).
27. http://www.ionoptika.com/.
28. A. G. Shard, P. J. Brewer, F. M. Green, and I. S. Gilmore, *Surf. Interface Anal.*, **39**, 294 (2007).
29. J. S. Fletcher, X. A. Conlan, E. A. Jones, G. Biddulph, N. P. Lockyer, and J. C. Vickerman, *Anal. Chem.*, **78**, 1827 (2006).
30. A. G. Sostarecz, C. M. McQuaw, A. Wucher, and N. Winograd, *Anal. Chem.*, **76**, 6651 (2004).
31. A. Brunelle, D. Touboul, and O. Laprévote, *J. Mass Spectrom.*, **40**, 985 (2005).
32. H. W. Werner and A. E. Morgan, *J. Appl. Phys.*, **47**, 1232 (1976).
33. D. J. Hook, P. L. Valint Jr., L. Chen, and J. A. Gardella Jr., *Appl. Surf. Sci.*, **252**, 6679 (2006).
34. A. Sosnik, R. N. S. Sodhi, P. M. Brodersen, and M. V. Sefton, *Biomaterials*, **27**, 2340 (2006).
35. J. Möller, A. Beumer, D. Lipinsky, and H. F. Arlinghaus, *Appl. Surf. Sci.*, **252**, 6709 (2006).
36. J. C. Vickerman and D. Briggs, *TOF-SIMS Surface Analysis by Mass Spectrometry*, 1st edn., IM Publications, UK and Surface Spectra Limited, Chapter 5 (2001).
37. U. Oran, E. Ünveren, T. Wirth, and W. E. S. Unger, *Appl. Surf. Sci.*, **227**, 318 (2004).
38. J. B. Cliffa, D. J. Gasparb, P. J. Bottomleya, and D. D. Myrolda, *Appl. Surf. Sci.*, **231–232**, 912 (2004).
39. A. J. Fahey and S. Messenger, *Int. J. Mass Spectrom.*, **208**, 227 (2001).
40. T. Stephan, *Plant Space Sci.*, **49**, 859 (2001).
41. S. Rangarajan and B. J. Tyler, *J. Vac. Sci. Technol. A: Vac. Surf. Films*, **24**, 1730 (2006).
42. B. J. Tyler, *Appl. Surf. Sci.*, **252**, 6875 (2006).
43. A. Willse and B. Tyler, *Anal. Chem.*, **74**, 6314 (2002).
44. B. J. Tyler, G. Rayala, and D. G. Castner, *Biomaterials*, **28**, 2412 (2007).
45. P. C. T. Nguyen, PhD thesis, University of South Australia Mawson Labes, South Australia, Australia (2006).
46. M. C. Biesinger, P.-Y. Paepegaey, N. S. McIntyre, R. R. Harbottle, and N. Petersen, *Anal. Chem.*, **74**, 5711 (2002).
47. X. V. Eynde and P. Bertrand, *Surf. Interface Anal.*, **25**, 878 (1997).
48. D. J. Graham, D. D. Price, and B. D. Ratner, *Langmuir*, **18**, 1518 (2002).
49. D. J. Graham and B. D. Ratner, *Langmuir*, **18**, 5861 (2002).
50. S. L. McArthur, M. S. Wagner, P. G. Hartley, K. M. McLean, H. J. Griesser, and D. G. Castner, *Surf. Interface Anal.*, **33**, 924 (2002).
51. S. J. Pachuta, *Appl. Surf. Sci.*, **231–232**, 217 (2004).
52. M. S. Wagner and D. G. Castner, *Langmuir*, **17**, 4649 (2001).
53. N. Xia, C. J. May, S. L. McArthur, and D. G. Castner, *Langmuir*, **18**, 4090 (2002).
54. N. Xia and D. G. Castner, *J. Biomed. Mater. Res. Pt. A*, **67A**, 179 (2003).
55. S. Wold, K. Esbensen, and P. Geladi, *Chemometrics Intelligent Lab. Syst.*, **2**, 37 (1987).
56. J. E. Jackson, *J. Qual. Technol.*, **12**, 201 (1980).
57. J. E. Jackson, *A Users's Guide to Principal Components*, John Wiley & Sons Inc., New York (1991).
58. J.-B. Lhoest, M. S. Wagner, C. D. Tidwell, and D. G. Castner, *J. Biomed. Mater. Res.*, **57**, 432 (2001).
59. S. N. Deming, J. A. Palasota, and J. M. Nocerino, *J. Chemometrics*, **7**, 393 (1993).
60. H. B. Lu, C. T. Campbell, D. J. Graham, and B. D. Ratner, *Anal. Chem.*, **72**, 2886 (2000).
61. M. Shen, M. S. Wagner, D. G. Castner, B. D. Ratner, and T. A. Horbett, *Langmuir*, **19**, 1692 (2003).
62. V. H. Perez-Luna, T. A. Horbett, and B. D. Ratner, *J. Biomed. Mater. Res.*, **28**, 1111 (1994).
63. A. Chilkoti, B. D. Ratner, and D. Briggs, *Anal. Chem.*, **65**, 1736 (1993).

64. A. Chilkoti, A. E. Schmierer, V. H. Perez-Luna, and B. D. Ratner, *Anal. Chem.*, **67**, 2883 (1995).
65. G. Coullerez, S. Lundmark, E. Malmstrom, A. Hult, and H. J. Mathieu, *Surf. Interface Anal.*, **35**, 693 (2003).
66. N. Medard, C. Poleunis, X. Vanden-Eynde, and P. Bertrand, *Surf. Interface Anal.*, **34**, 565 (2002).
67. S. Ferrari and B. D. Ratner, *Surf. Interface Anal.*, **29**, 837 (2000).
68. M. S. Wagner, T. A. Horbett, and D. G. Castner, *Langmuir*, **19**, 1708 (2003).
69. M. S. Wagner, T. A. Horbett, and D. G. Castner, *Biomaterials*, **24**, 1897 (2003).
70. M. S. Wagner, M. Shen, T. A. Horbett, and D. G. Castner, *Appl. Surf. Sci.*, **203–204**, 704 (2003).
71. M. S. Wagner, M. Shen, T. A. Horbett, and D. G. Castner, *J. Biomed. Mater. Res.*, **64A**, 1 (2003).
72. M. S. Wagner, B. J. Tyler, and D. G. Castner, *Anal. Chem.*, **74**, 1824 (2002).
73. S. L. McArthur, M. W. Halter, V. Vogel, and D. G. Castner, *Langmuir*, **19**, 8316 (2003).
74. R. Bro and A. K. Smilde, *J. Chemometrics*, **17**, 16 (2003).
75. M. S. Wagner, D. J. Graham, M. D. Ratner, and D. G. Castner, *Surf. Sci.*, **570**, 78 (2004).
76. A. Curtis and C. Wilkinson, *Biomaterials*, **18**, 1573 (1997).
77. H. G. Craighead, C. D. James, and A. M. P. Turner, *Curr. Opin. Solid State Mater. Sci.*, **5**, 177 (2001).
78. L. Vroman and A. L. Adams, *J. Biomed. Mater. Res.*, **3**, 43 (1969).
79. G. S. Omenn, *Proteomics*, **5**, 3223 (2005).
80. N. L. Anderson and N. G. Anderson, *Mol. Cell Proteom.*, **1**, 845 (2002).
81. http://www.hupo.org/.
82. J. D. Andrade and V. Hlady, Plasma protein adsorption: The big twelve, *Ann. N. Y. Acad. Sci.*, **516**, 158 (1987).
83. P. Kingshott, H. A. W. St. John, R. C. Chatelier, and H. J. Griesser, *J. Biomed. Mater. Res.*, **49**, 36 (2000).
84. W. B. Tsai, J. M. Grunkemeier, and T. A. Horbett, *J. Biomed. Mater. Res.*, **44**, 2243 (2000).
85. J. M. van Emon (Ed.), *Immunoassay and Other Bioanalytical Techniques*, CRC Press (2007).
86. T. A. Barber, S. L. Golledge, D. G. Castner, and K. E. Healy, *J. Biomed. Mater. Res. A*, **64**, 38 (2003).
87. B. L. Beckstead, D. M. Santosa, and C. M. Giachelli, *J. Biomed. Mater. Res. A*, **79**, 94 (2006).
88. S. M. Martin, R. Ganapathy, T. K. Kim, D. Leach-Scampavia, C. M. Giachelli, and B. D. Ratner, *J. Biomed. Mater. Res. A*, **67A**, 334 (2003).
89. H. Fitzpatrick, P. F. Luckham, S. Eriksen, and K. Hammond, *J. Coll. Interface Sci.*, **1**, 149 (1992).
90. C. J. May, H. E. Canavan, and D. G. Castner, *Anal. Chem.*, **76**, 1114 (2004).
91. D. Y. Petrovykh, H. Kimura-Suda, L. J. Whitman, and M. J. Tarlov, *J. Am. Chem. Soc.*, **125**, 5219 (2003).
92. C. Y. Lee, P. Gong, G. M. Harbers, D. W. Grainger, D. G. Castner, and L. J. Gamble, *Anal. Chem.*, **78**, 3316 (2006).
93. D. S. Mantus, B. D. Ratner, B. A. Carlson, and J. F. Moulder, *Anal. Chem.*, **65**, 1431 (1993).
94. D. Briggs, *Comprehensive Polymer Science*, C. Booth and C. Price (Eds.), Vol. 1: *Polymer Characterization*, Pergamon Press, Oxford, pp. 543–599 (1989).
95. R. C. Chatelier, H. A. W. St. John, T. R. Gengenbach, P. Kingshott, and H. J. Griesser, *Surf. Interface Anal.*, **25**, 741 (1997).
96. M. Karas and F. Hillenkamp, *Anal. Chem.*, **60**, 2299 (1988).
97. R. Kaufmann, P. Chaurand, D. Kirsch, and B. Spengler, *Rapid Commun. Mass Spectrom.*, **10**, 1199 (1996).
98. T. W. D. Chan, A. W. Cilburn, and P. J. Derrick, *Org. Mass Spectrom.*, **27**, 53 (1992).
99. A. Overberg, M. Karas, and F. Hillenkamp, *Mass Spectrom.*, **5**, 128 (1991).
100. S. Jespersen, W. M. A. Niessen, U. R. Tjaden, J. Vandergreef, E. Litborn, U. Lindberg, and J. Roeraade, *Rapid Commun. Mass Spectrom.*, **8**, 581–584 (1994).
101. R. Zenobi and R. Knochenmuss, *Mass Spectrom. Rev.*, **17**, 337 (1998).
102. P. Kingshott, H. A. W. St. John, R. C. Chatelier, and H. J. Griesser, *J. Biomed. Mater. Res.*, **49**, 36 (2000).
103. Z. Ademovic, B. Holst, R. A. Kahn, I. Jørring, T. Brevig, J. Wei, B. Winter-Jensen, and P. Kingshott, *J. Mater. Sci.: Mater. Med.*, **17**, 203 (2006).
104. P. Kingshott, H. A. W. St. John, R. C. Chatelier, and H. J. Griesser, *Polym. Sci. Eng.*, **76**, 81 (1997).
105. K. M. McLean, S. L. McArthur, R. C. Chatelier, P. Kingshott, and H. J. Griesser, *Coll. Surf. B: Biointerfaces*, **17**, 23 (2000).
106. P. Kingshott, H. A. W. St. John, and H. J. Griesser, *Anal. Biochem.*, **273**, 156 (1999).
107. P. Kingshott, PhD thesis, University of New South Wales, Sydney, Australia (1998).
108. P. Kingshott, H. A. W. St. John, R. C. Chatelier, F. Caruso, and H. J. Griesser, *Polym. Prepr.*, **38**, 1008 (1997).

109. B. Clarke, P. Kingshott, Y. Rochev, X. Hou, A. Gorelov, and W. M. Carroll, *Acta Biomater.*, **3**, 103 (2007).
110. I. Ishikawa, Y. Chikazawa, K. Sato, M. Nakagawa, H. Imamura, S. Hayama, H. Yamaya, M. Asaka, N. Tomosugi, H. Yokoyama, and K. Matsumoto, *Am. J. Neph.*, **26**, 372 (2006).
111. F. Höök, M. Rodahl, P. Brzezinski, and B. Kasemo, *Langmuir*, **14**, 729 (1998).
112. M. Rodahl, F. Höök, P. Brzezinski, and B. Kasemo, *Rev. Sci. Instrum.*, **66**, 3924 (1995).
113. F. Höök, M. Rodahl, P. Brzezinski, and B. Kasemo, *J. Coll. Interface Sci.*, **208**, 63 (1998).
114. F. Höök, M. Rodahl, B. Kasemo, and P. Brzezinski, *Proc. Natl. Acad. Sci. USA*, **95**, 12271 (1998).
115. C. Steinem, A. Janshoff, J. Wegener, W. P. Ulrich, W. Willenbrink, M. Sieber, and H. J. Galla, *Biosens. Bioelectron.*, **12**, 787 (1997).
116. W.-B. Tsai, J. Grunkemeier, and T. A. Horbett, *J. Biomed. Mater. Res.*, **44**, 130 (1999).
117. T. A. Horbett, Techniques in protein adsorption studies, *Techniques in Biocompatibility Testing*, D. F. Williams (Ed.), CRC Press, Boca Raton, FL, pp.183–214 (1986).
118. M. Balcells, D. Klee, M. Fabry, and H. Höcker, *J. Coll. Interface Sci.*, **220**, 198 (1999).
119. M. Holmberg, K. B. Stibius, L. Nielsen, S. Ndoni, N. B. Larsen, P. Kingshott, and X. L. Hou, *Anal. Biochem.*, **363**, 120 (2007).
120. F. Caruso, D. Neil Forlong, and P. Kingshott, Characterisation of fernitin adsorption onto gold, *J. Coll. Interface Sci.*, **186**, 129–140 (1997).

18 Adhesion Science and Technology

John F. Watts

CONTENTS

18.1 INTRODUCTION

It is now four decades since the surface analysis methods x-ray photoelectron spectroscopy (XPS), Auger electron spectroscopy (AES), and secondary ion mass spectrometry (SIMS) became commercially available, and in that time they have moved from their original home in the laboratory of the surface chemist and surface physicist to become methods of applied surface analysis. The wide range of disciplines within which these methods have been successfully applied can be readily appreciated by the diverse contents of this book. Some areas, such as corrosion, catalysis, and polymers, have featured in the literature since the early 1970s while other disciplines have come to surface analysis rather more recently.

Adhesion research has made use of surface analysis since the mid-1970s, but by the end of that decade there were, perhaps, only two or three research groups worldwide that had a strong presence in both adhesion and surface analysis. By the end of the following decade, the situation had changed significantly and surface analysis was then seen to feature prominently in adhesion conferences and journals. This change was partly a result of the wider availability of surface analysis equipment, and also reflected a shift in emphasis of the adhesion community. It was no longer acceptable merely to report performance data for the plethora of systems available, but there was now a requirement to consider interfacial chemistry, of both the adhesion process and the locus of failure, in rather more detail, and to relate relevant parameters back to the performance of an adhesive joint or organic coating.

The adhesion phenomena that are considered in this chapter are essentially those that occur between organic polymers and metal oxides, and their description builds on previous reviews [1,2]. Before any investigation of adhesion can commence, a thorough appreciation of surface analysis in the fields of polymers and oxidation is an advantage. As adhesion loss is often associated with environmental degradation of the metal substrate, some experience in this field can also be valuable. The description of adhesion as a multidisciplinary, multifaceted subject also applies when considering the use of surface analysis in an adhesion investigation.

The role of surface analysis in adhesion research can be divided into three general areas: assessment of substrate surface characteristics prior to bonding, identification of the exact locus of failure of an adhesive joint or organic coating following some kind of mechanical or chemical perturbation or during a durability test, and analysis of interface chemistry with a view to gaining a better understanding of the adhesion process itself. From a chronological point of view, the above sequence has been essentially the same as that in which the use of surface analysis has developed in adhesion, and is the approach that will be adopted in the main body of this chapter. In addition, the nature of interfacial interactions in adhesion and the contributions that surface analysis can make in this area are considered, together with suggestions for the best course of action for the investigations of the very complex formulations that are commonplace in adhesives and coatings.

The goal of almost all investigations of adhesion is to produce a better bond. This may be stronger, more durable, tougher or cheaper; there are many criteria that may be applied. The contribution of surface analysis investigations will be that of providing a better understanding of the system, thus enabling an improved prescription of substrate pretreatment, formulation chemistry, systems chemistry, etc., in order to achieve the elusive better bond. Throughout this chapter such considerations are the focus of discussion and the relevance of the surface analytical data to the overall systems approach is reviewed continually.

The last decades have seen several innovations and refinements of equipment that have had an impact on the application of surface analysis methods, particularly XPS and SIMS, to adhesion studies. In XPS, the use of x-ray monochromators, delivering a line-width of better than 0.4 eV, has now become routine, with attendant improvements in spectral resolution (particularly important for resolution of the C 1s spectrum of organic materials) and in turnkey charge compensation methodologies. XPS data can now be acquired routinely from sub-100 μm regions and the information can be processed to give maps of concentration, overlayer thickness, etc. The ultimate resolution in this mode of operation is of the order of 10 μm. Parallel-detection imaging XPS (iXPS) can reach an ultimate image resolution of 1 μm. Static SIMS (SSIMS, see Chapter 4) has seen the time-of-flight (ToF) mass analyzer become the analyzer of choice, and the quadrupole mass analyzer has now been consigned to a supporting role on multi-technique or DSIMS systems. Resolution and transmission of ToF analyzers have improved dramatically, and a spatial resolution of ~100 nm can be achieved with a liquid-metal-ion source in certain modes of operation. The biggest advance in ToF-SIMS recently has been the use of cluster sources, and those currently in favor are Bi_x^+ for high resolution work and C_{60}^+ for biological and organic studies (see Chapter 4 for more details). The major advantage of the use of a cluster-beam source in SIMS is the improved yield at high masses, which is particularly important for polymer analysis, together with the apparent lack of damage when used on certain polymeric systems and at relatively high fluencies. The use of a C_{60}^+ source enables sputter depth-profiling of organic materials, and recent work has indicated that this capability will add an important new dimension to the use of surface analysis in adhesion studies.

18.2 CHARACTERISTICS OF THE SOLID SUBSTRATE

Because the forces responsible for adhesion act over very short distances (~0.5 nm), it is self-evident that any extraneous material between polymer and substrate may have a deleterious effect on the level of adhesion achieved. Such material may be organic or inorganic and the occurrence of both, and the manner in which adhesion may be affected, is well documented.

18.2.1 ORGANIC CONTAMINATION

One of the earliest examples of the problem-solving capability of surface analysis is from the American automobile industry. In the early 1970s, the industry was faced with impending federal legislation regarding improved corrosion resistance (against perforation) of automobile bodies. There was a body of evidence that poor performance in an accelerated corrosion test (the salt spray test) was related to the concentration of surface carbon on the steel sheet stock. An investigation by Hospadaruk et al. [3] provided a direct correlation between surface carbon, as measured by AES and a combustion method, and the time taken to arrive at failure in the salt spray test. Thus, AES together with XPS rapidly became accepted methods for the assessment of surface contamination. In a later investigation, Iezzi and Leidheiser used scanning Auger microscopy (SAM) to provide further details of the deleterious effect of the adventitious hydrocarbon layer by describing the manner in which it interfered with the formation of a uniform conversion coating [4]. An example of the extent to which handling can change the surface chemistry of a metal surface is indicated in Figure 18.1. An XPS survey spectrum of a clean aluminum foil is compared there with the spectrum recorded after handling contamination by a female operative. The level of carbon is seen to increase but, in addition, prominent Si 2p and Si 2s peaks are observed, associated with the use of a hand cream containing silicone oils (the precise chemical nature of the silicon was established unambiguously by ToF-SIMS).

Prior to painting, sheet steel is invariably treated with an inorganic conversion coating based on zinc phosphate, which produces a characteristic acicular surface morphology resistant to corrosion and providing enhanced coatings adhesion. The presence of patches of oily deposit was shown to interfere with the phosphate growth mechanism, resulting in areas devoid of the phosphate deposit; these areas subsequently showed inferior adhesion and corrosion resistance behavior [4].

The desirability of low levels of surface hydrocarbon is now well established, and XPS and AES have both provided means by which the concentration, type, and thickness of contamination can be readily evaluated. Nevertheless, a great deal of work is still undertaken in this area with the aim of establishing the reasons for premature failure of an adhesive joint or coating. The organic contamination is often more complex than just the presence of an oily or greasy nonspecific hydrocarbon would imply. An example is the presence, at the bonding surface, of release agents based on fluorocarbons or poly(dimethyl siloxane). Compounds of this type are found on backing papers of adhesive tapes, preimpregnated composite stick (prepreg), and preformed composite structures, and it is not unknown for them to remain on, or be transferred to, the substrate prior to bonding, leading to low strength in the subsequent bond. Analysis of the substrate failure surface by XPS or SIMS can lead to rapid and unequivocal identification of the contaminant. A good example of this problem is that of the adhesive bonding of glass and carbon-fiber-reinforced plastic components, and XPS has been able to identify the source of poor bond performance to be associated with residues from such backing, or release, papers [5,6]. The bonding surfaces may be protected by a release paper, whose purpose is to protect the surface from handling damage, but unfortunately removal of the paper often leaves behind a residue of the release agent, which is then responsible for poor bond properties. The solution to this dilemma is an additional treatment, following removal of the release paper, in order to ensure a clean surface devoid of fluorine- or silicon-containing products [5,6].

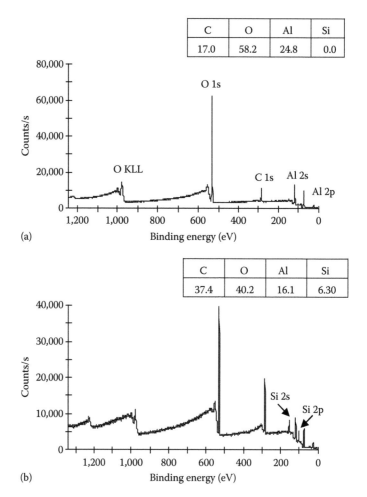

C	O	Al	Si
17.0	58.2	24.8	0.0

C	O	Al	Si
37.4	40.2	16.1	6.30

FIGURE 18.1   XPS survey spectra from aluminum: (a) as received and (b) following handling contamination. The boxes above the spectra indicate the respective surface compositions, in at.%, of each element. (Courtesy of M.-L. Abel and S. Doughty.)

18.2.2 OXIDE FILMS AT METAL SURFACES

All metals, with the exceptions of the very noble ones, form oxide films on their surfaces on exposure to the atmosphere, and the compositions of these surface oxides are of critical importance to the adhesion scientist. In some cases, it is possible to draw parallels between the results of exposure to reactive gases in UHV clean conditions and of ambient oxidation; this is certainly true for pure iron exposed to water vapor in UHV [7], which results in the formation of an FeOOH surface layer. Mild steel responds in the same manner, when cleaned by an abrasion process and passivated by air exposure. The early literature on the composition of metal surfaces following atmospheric exposure has been reviewed by Castle [8], and his review provides not only a sound understanding of the hierarchy of layers that form on metal surfaces, but is also an important starting point for those wishing to understand the adhesion of organic systems to such substrates.

The situation with alloys is often rather different since the oxide film may be formed by the oxidation of some, rather than all, of the constituent elements. For example, the surface oxide on stainless steel is a mixed Fe/Cr spinel enriched in chromium relative to the alloy composition, while that on cupronickel coinage metal is NiO. The composition of the surface phase is usually a function of equilibrium thermodynamics and process parameters such as oxygen partial pressure.

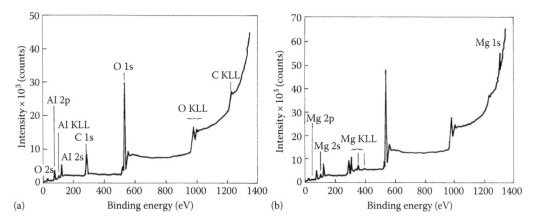

FIGURE 18.2 XPS survey spectra of aluminum alloys following chromic acid anodizing: (a) commercial purity aluminum that showed extremely poor durability and (b) alloy BS L157 (Al–5Cu–1Mg–1Si–1Mn), which exhibited extremely good durability despite the high concentration of magnesium in the surface analysis. (From Poole, P. and Watts, J.F., *Int. J. Adhes. Adhes.*, 5, 33, 1985. With permission.)

There are well-documented cases in which a change in process parameters has led to the formation of a different surface phase and hence to an adverse effect on adhesion. Such a case is that of the annealing of some Al–Mg alloys; should the oxygen partial pressure in the closed-coil annealing process become too low, then the surface phase will consist of MgO rather than Al_2O_3. The resulting adhesion loss occurs because MgO is a brittle friable oxide that spalls readily from the parent metal, in marked contrast to Al_2O_3 which is a tenacious, strongly adherent oxide. Extension of the interpretation of such observations to other systems must, however, be treated with caution. The presence of a magnesium rich surface phase associated with poor levels of adhesion does not necessarily mean that surface magnesium is always a predictor or diagnostic of poor adhesion. In the surface bonding of the above-mentioned Al–Mg alloys, etched alloy surfaces gave the expected poor performance and led to a cautionary warning [9]. Subsequent, more comprehensive investigations [10], however, indicated that if the Mg was chemically incorporated into a mixed oxide film (rather than as a surface-segregated layer of MgO), as occurs in the thick anodized film on Al–Mg alloys, then good bond performance could be achieved. This is illustrated graphically in Figure 18.2, where the XPS spectrum (Figure 18.2a) of commercial purity aluminum with poor durability is compared with that (Figure 18.2b) from an anodized Mg-containing Al alloy that, although having a substantial concentration of Mg in the spectrum, had excellent durability [10]. The commercial treatments used prior to the adhesive bonding of aluminum for structural purposes involve the electrochemical thickening of the oxide (i.e., anodizing) in acid media. This can lead to oxide films with thicknesses of around 1 μm, for which AES has been widely used in a depth-profiling mode to monitor thickness and composition. An example shown in Figure 18.3 illustrates Auger depth profiling through the interface region of an adhesive joint prepared on a very thin metallic aluminum substrate [11].

In general, it is now regarded as a requirement of any investigation of adhesion that the adherend surfaces are fully characterized by surface analysis before any adhesion tests are undertaken. This does, of course, beg the question concerning the inevitable hydrocarbon contamination. Such contamination will always be present in the spectrum recorded prior to bonding, but, in the case of a thin adventitious layer (<10 nm, say), it will be displaced on application of the polymer phase by specific interactions between the substrate and the adhesive or coating, sufficient to overcome the weak dispersion forces between the contaminant and the metal oxide. Documented evidence of failure occurring in a preexistent contaminant layer, referred to as a weak boundary layer by the adhesion community, is rare.

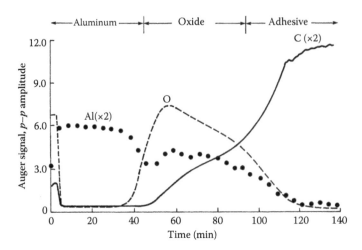

FIGURE 18.3 Sputter depth profile, by AES, through the interface region of an adhesive joint on a phosphoric acid anodized ultrathin aluminum substrate. (From Solomon, J.S., Hanlin, D., and McDevitt, N.T., in *Adhesion and Adsorption of Polymers Part A*, Plenum Publishing Corp., New York, 1980, 103–122. With permission.)

18.2.3 CARBON FIBER COMPOSITE MATERIALS

In order to achieve strong interfacial bonds between the fibers and the resin matrices of composites, carbon fibers must be treated in either an electrochemical or a gas phase process. XPS has been used widely to study the surface chemical modifications brought about by such treatments, and the early literature relating to these studies was reviewed by Castle and Watts in 1988 [12]. In essence, the understanding of the surface properties has been reached by detailed peak fitting of the C 1s spectra from carbon fibers following electrochemical treatment, and by the use of a labeling method known as chemical derivatization. Work relating to peak fitting of the C 1s spectra of carbon fibers is discussed at length by Sherwood in Chapter 14. In chemical derivatization, a small molecule or ion is used to label a specific chemical group that is thought to be present on the material surface but cannot be detected uniquely by XPS. In the case of carbon fibers, Ba^+, Ag^+, and Mg^{2+} have all been used to good effect to label acidic sites (–COOH) that are present at very low concentrations at the fiber surface. A variation of the chemical derivatization methodology is the determination, by XPS, of adsorption isotherms, which were found [13] to be of the Langmuir type, allowing the monolayer coverage of the labelant (Ag^+ or Mg^{2+} ions), and hence the surface acidity, to be estimated. The source of nitrogen in the XPS spectra of carbon fibers has long been a topic of debate and provides a useful illustration of the chemical specificity obtainable with XPS. On going from untreated to treated (in NH_4CO_3 electrolyte) fibers, the binding energy of the N 1s peak shifts to a slightly higher value [14], reflecting a change in the nitrogen chemistry from that in the residue of the poly(acrylo nitrile) precursor to that in the ammonium ions adsorbed from the electrolyte. ToF-SIMS has also been applied to the analysis of carbon fiber surfaces and shown to provide information complementary to that from XPS studies [15,16], while inverse gas chromatography has been found to be a useful complementary method in conjunction with XPS and ToF-SIMS for the study of carbon fiber acidity [17,18]. The term "inverse" indicates that the material of interest consists of the stationary phase in the gas chromatographic column, rather than of the volatile probe injected into the column. The experiment itself is very straightforward and can be carried out readily using standard gas chromatography (GC) equipment. A column (of stainless steel or polytetra fluoroethylene (PTFE), typically of diameter 3 mm), packed with the material of interest, is placed between the injector and the detector of the GC. The analytical measurement consists of the injection through the column of volatile probes with known properties carried on an inert gas (usually nitrogen). The retention time for each probe is related to the surface properties of the stationary phase. The usual practice is to carry out a series of injections for an homologous series of nonpolar probes such as the *n*-alkanes, and then to compare the results with data acquired using probes of known acido-basic properties (such as chloroform, tetrahydrofuran, and diethyl ether). Practical and thermodynamic aspects of the technique are dealt with in Ref. [18].

The thorny question of the level of electrochemically induced topography that cannot be distinguished by scanning electron microscopy (SEM) now seems to have been resolved by scanning probe microscopy scanning tunneling microscopy (STM) and atomic force microscopy (AFM). Recent studies have shown conclusively the development of edge features and exposed basal planes following electrochemical treatment [19]. In addition, ToF-SIMS has been shown to be invaluable for the study of adsorption of organic matrices or matrix components on fiber surfaces [20].

Although chemical analysis of the substrate is now a straightforward matter, research has also concentrated on the extension of such analyses to other properties such as surface polarity and acido-basic characteristics. These are extremely important and point the way to a predictive approach to bond integrity. Such measurements can then be compared with interface analyses obtained from either thin layers of the organic phase deposited on the substrate or sectioning of a real system. All aspects of such methods are discussed in further detail in Section 18.4.

18.3   FAILURE ANALYSIS: IDENTIFICATION OF THE LOCUS OF FAILURE

In the use of surface analysis for the failure analysis of fracture surfaces, there have been some spectacular successes over the years, and it could be argued that it was the success in this area that made such a strong impression on the adhesion community, leading to the adoption of surface analysis so enthusiastically. XPS, in particular, can be said to provide a chemical overview of the failure in a manner analogous to the overview of morphology provided by SEM fractography. Nowadays, identification of the locus of failure of adhesive joints and organic coatings is a routine undertaking carried out with great success in many laboratories around the world. ToF-SIMS, with its molecular specificity, has played an important role in the identification of organic residues at failure interfaces, and, rather than concentrating on a thin (<3 nm) layer of adhesive or coating on a metal substrate, it is now possible to relate failure to individual components in a formulation. Failure analysis can occasionally lead to improvements in the design of an organic system. In this section, several types of failure analysis investigation are reviewed, and in order to group them in a logical order, they are categorized according to substrate type.

18.3.1   Adhesion to Brass

One of the first published investigations into practical adhesion phenomena, in which surface analysis techniques were used, was that into the adhesion of rubber to brass by van Ooij in the mid-1970s. Subsequently, this particular research was extended and refined, and publications by the same author on the topic cover more than two decades. The adhesion of vulcanized rubber to brass is of critical importance in the tire industry since the steel wires used for tire reinforcement are brass-plated to ensure good bonding to the rubber of the tire. The need to brass plate the wires had been known for many years, but it was not until van Ooij's benchmark publications of 1977 that the chemistry involved in this particular adhesion process became fully understood.

In order to be able to apply XPS in particular to the study of adhesion, van Ooij [21] used thin brass strips 1 cm wide rather than attempting to study the very fine wires themselves. When investigating the 70/30 brass alloy used in the tire industry, he found [22] that its surface consisted of a thin film of Cu(I) oxide on top of a thicker layer of copper and zinc oxide. When the alloy is heated, the zinc oxide grows in preference to the copper oxide, and in extreme cases leads to the solid state reduction of the surface Cu_2O phase by the metallic zinc. The mechanism of adhesion of rubber to the brass substrate is expected to result in the formation of both copper and zinc sulfides, by reaction between the elemental sulfur, which yields an active sulfide ion at the oxidized brass surface, and

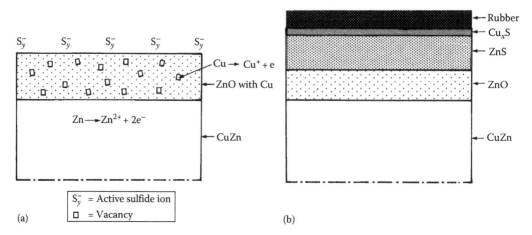

FIGURE 18.4 Schematic representation of rubber-to-brass adhesion, based on XPS analyses: (a) situation prior to vulcanization, indicating the formation of the active sulfide ion from elemental sulfur at the brass surface and (b) formation of Cu_xS following vulcanization. (From van Ooij, W.J., *Surf. Tech.*, 6, 1, 1977. With permission.)

both the metallic copper, which is subsequently reduced to a Cu(I) ion, and zinc ions, which are present in the inorganic film on the brass. In fact, it is the Cu_xS phase that is responsible for the extent of adhesion between the two phases; if the concentration of Cu_xS is optimized at the interface, it can lead to improved adhesion as a result of the catalytic effect of the sulfide on vulcanization, but if in too high a concentration, the sulfide will lead to embrittlement of the mixed copper and zinc sulfide film. The situation that is thought to exist at the rubber/brass interface is illustrated schematically in Figure 18.4. The consequences of this unequivocal link between alloy surface condition and adhesion performance were of great technological importance, since tire manufacturers were then able to predict with good accuracy the expected adhesion that would be achieved between their own particular rubber formulation and brass-plated steel wires from any particular source. The potential existed, as in the steel cleanliness example, to use surface analysis in a quality assurance role.

In subsequent work [23–25], van Ooij developed the use of model compounds (in the case of natural rubber and squalene) to simulate the physical and chemical interactions at the interface of the brass/natural rubber system. Cuts in a tire carcass may lead to exposure of the steel/brass interface, and the galvanic effects that give rise to dezincification may then become important [26]. The use of duplex Zn–Ni on Zn–Co alloys as replacements for brass has been proposed in order to circumvent the problems associated with the corrosion of brass [27]. The electrodeposited duplex layers were optimized in terms of the process parameters and adhesion levels achievable with natural rubber, and were found to be comparable in performance to the traditional steel/brass system. The advantages of such a system are that the corrosion problems are much reduced [28]. The important conclusion arising from the work of van Ooij is that by arriving at a complete understanding of a traditional system, it is possible to predict exactly the developments needed for an improved system. Van Ooij also pioneered the use of SSIMS in such adhesion studies and published a comprehensive treatise on the SSIMS spectra of a series of synthetic rubbers [29].

In the application of protective organic coatings to brasses and bronzes, there appear even now to have been relatively few studies that have made use of surface analysis. One such study is that of Castle et al. [30] into the localized protection of aluminum-brass (Cu-76, Zn-22, Al-2) power station condenser tubes by a commercial acrylic lacquer. By careful examination, using XPS, of the failure surfaces produced by mechanical testing following exposure to hot saline solution for several days, it was possible to identify the failure mechanism as that associated with electrochemical activity within a developing crevice. In the study, by XPS or AES, of metal surfaces that are the result of electrochemical activity, it is often informative to report the ratio of characteristic ions from the aqueous exposure medium as a means of deducing the prior electrochemical history of the electrode. The excess of chloride over sodium ions in the surface analysis was indicative of the development of an anodic crevice, which is in marked contrast to the cathodic delamination case histories so often seen in coatings failure and is quoted in the next section. The analyses on both sides of the failure path showed the same excess that became more marked with increasing coating thickness. At longer exposure times, the anodic crevice conditions, which were solely a result of electrochemical activity within the crevice formed between the aluminum-brass substrate and the detached acrylic coating, broke down, probably as a result of mass transport through the thickness of the polymer film. This example illustrates clearly the importance of taking into account corrosion phenomena when possible adhesion failure mechanisms are being considered. In this case, the mineral layer present on the brass, following power station operation, was protective [31] and acted as a cathode to the brass anodic under-film region that had been cleaned prior to the application of the acrylic lacquer. This situation is not usually encountered; the more common situation being that of the exposed metal acting as a local anode to the coated metal that becomes cathodic.

It is perhaps worthwhile at this point to review briefly the electrochemical processes that are associated with metallic corrosion. The reaction responsible for the degradation of the metal (Me) occurs at the anode, and for this reason the process is often described as anodic dissolution:

$$Me \rightarrow Me^{x+} + xe^- \tag{18.1}$$

The reaction leads to the production of an Me^{x+} ion and x electrons each with a single negative charge. As electrical neutrality must be maintained, the electrons are consumed in the cathodic reaction that occurs on an adjacent portion of the electrode termed the cathode. In aqueous electrochemistry, the reaction involves water and oxygen as well as the electrons:

$$2H_2O + O_2 + 4e^- \rightarrow 4OH^- \tag{18.2}$$

This reaction is referred to as the cathodic reduction of water (and oxygen) and leads to an increase in pH (alkalinity) in the environs of the cathode. In the aluminum-brass example, coatings failure occurred in the region of the anodic surface, while in the examples in the next section, detachment of coatings from the cathodic surface is described. The latter process is known as cathodic delamination or disbondment.

18.3.2 Adhesion of Organic Coatings to Steel

Because mild steel is such a technologically important substrate, there have been many studies carried out on the adhesion to it of organic systems, predominantly in the form of coatings. The main difficulty that arises is that any exposure trial in an aqueous environment will lead to the production of copious amounts of corrosion product (i.e., rust) that can make any subsequent investigation of failure mechanism extremely difficult. Various authors have sought to circumvent this problem in different ways, including the separation of the anodic and cathodic sites by the use of a sacrificial anode such as zinc, or an impressed current method (frequently used in the cathodic protection of underground pipelines), or the use of aqueous environments containing low partial pressures of oxygen, which, in turn, maintain the level of anodic activity (rusting) at a modest level.

Knowledge accumulated from the investigation of failures associated with low carbon steels (often referred to by the generic name of mild steel) has allowed the interfacial chemistry of the failure process to be well understood. Implicit in the mechanism of adhesion loss is the influence of corrosion, already mentioned, and the recognition that the electrochemical process of rusting will have a marked effect on durability of the coating or the adhesive joint. Thus, in a test in which the joint or coating is exposed to an aqueous or humid environment, it is important to consider the effect of such exposure on both substrate and polymer. In the case of exposure at the free corrosion potential (the potential that freely rusting steel will assume, also known as the rest potential), there is the possibility that post-failure corrosion will cloud the analytical issues; care must be taken not to confuse such post-failure activity with that responsible for failure in the first place.

The first recognition of the need to study degradation in both phases was in an investigation into the adhesion of polybutadiene to steel exposed to a saline solution, either at rest potential or with an impressed cathodic potential [32,33]. It was shown by XPS that two characteristic types of failure occurred. At the exposed metal, cathodically generated alkali brought about interfacial failure and subsequent blistering, which caused, for the specimen at rest potential, an extension of the anodic (rusting) site. In the case of cathodic polarization (to simulate cathodic protection), the surface remained very clean with a low level of carbon as shown by XPS, Figure 18.5a. Outside this region, downward diffusion of solvated cations (i.e., cations surrounded by a tightly bound sheath of water molecules), through the coating, led to a failure process in which a thin (~4 nm) layer of polymer remained adhering to the substrate, Figure 18.5b. The latter process has also been invoked by other authors for polybutadiene [34] and other coatings [35,36]. The failure mechanism was at the time ascribed to alkaline hydrolysis of the near-interface polymer, and terms such as "active saponification" were often used. Some years later a more plausible explanation was offered [37], on the basis that the presence of alkali cations was not a necessary condition for such failure, and that it was the presence of water within the polymer that was responsible for failure. The strands of the argument can be brought together as follows: the damage to the polymer in this type of failure is reversible and shares all the characteristics of the phenomenon known in the paint industry as wet adhesion (in which tests are carried out in warm water). The explanation assumes that the locus of failure characterized by the XPS spectrum of Figure 18.5b is the result of the downward diffusion of water molecules through the thickness of the paint film; the molecules accumulate near the oxide/polymer interface and lead to swelling and plasticization of the polymer. Such damage is reversible. The role of the solvated cations is to control the rate at which water molecules reach the proximity of the interface. It has been postulated that the thermodynamic activity of water molecules within the polymer phase may be all-important in determining the kinetics of the process [38]. What little information there is on the activity of water in the solution to which coatings have been exposed does indeed support this hypothesis [39].

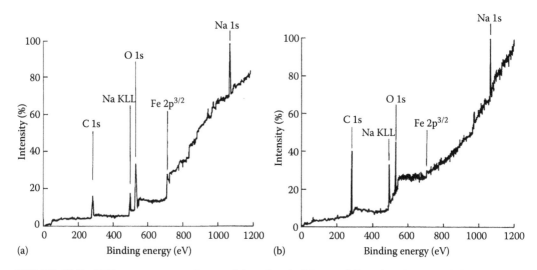

FIGURE 18.5 XPS survey spectra from polybutadiene/mild steel failure interfaces: (a) blistered region representative of cathodic delamination and lateral diffusion of hydroxyl ions and (b) wet adhesion failure resulting from the downward diffusion of water molecules. (From Watts, J.F. and Castle, J.E., *J. Mater. Sci.*, 18, 2987, 1983. With permission.)

All the investigations referred to above have used the approach of exposing samples to the chosen environment within the laboratory, peeling back the polymeric phase with tweezers or by a similar method, and then analyzing the interfacial metal and polymer surfaces. For true interfacial failure, this means that the metal oxide surface, which has a very high surface free energy, will inevitably become covered with adventitious hydrocarbon contamination, as this leads to a reduction in the surface free energy and is the thermodynamically favored process (see, e.g., the spectrum of Figure 18.5a). The challenge for the analyst is then the attribution of the carbon signal in the spectrum to either, or both, contamination or coating/adhesive residues. This is no easy task, but may be tackled by using fine structure within the C 1s spectrum, angle-resolved XPS, quantitative surface analysis, and molecular definition of the organic residue by ToF-SIMS. The alternative is to separate the polymer from the metal adherent within the spectrometer using either custom-made or commercial devices [40,41].

Although the assay of carbon can provide an indication as to whether the failure is truly interfacial or is cohesive within the polymer phase, much additional information can be revealed concerning polymer degradation chemistry by peak-fitting the C 1s spectrum. In the case of epoxy coatings, the formation of carboxylate species brought about by alkaline hydrolysis, as well as carbonate residues, has been identified by Dickie and colleagues [34–36] at interfacial failure surfaces. The carbonate is ascribed to the degradation of the urea and urethane components of the paint, although other authors have attributed a similar observation to the presence of dissolved carbon dioxide in the electrolyte [36]. Using angle-resolved XPS, Watts et al. were able to determine the orientation of the polymeric fragments remaining on the steel surface, and then to relate their observations to chain scission within the polymer [42], which was shown to occur adjacent to the residual epoxy groups. The latter groups always remain in the outermost region of the thin (2–3 nm) overlayer remaining on the metal substrate. On the interfacial polymer side of the failure they were below the surface.

The native oxide present on mild steel is often chemically modified prior to painting by application of a conversion coating intended to improve corrosion resistance. Such treatments are often used in the automobile and consumer goods industries, in which those based on zinc phosphate and chromate are commonplace. Dickie et al. [34] have studied the failure of coated prephosphated steel and have commented on the chemical degradation that appears to be a precursor to adhesion loss. It is well known that the under-film alkalinity associated with cathodic delamination can bring about phosphate dissolution [43]. In the study of a simple phosphoric acid wash treatment, Watts [44] noted that a surface iron phosphate phase was deposited to which adhesion was enhanced, in comparison with adhesion to the bare steel substrate. For chromate coatings, the interfacial chemistry can be complex. Figure 18.6 illustrates fine structure within the Cr 2p spectrum from a chromate coating applied to steel in the as-coated condition and following cathodic delamination [45]. The decrease in hexavalent chromium is a result of the reduction reaction occurring at the cathode surface, which in turn leads to loss of adhesion. In this case, the cathodic exposure conditions have, unusually, brought about degradation of the inorganic phase rather than that of the interfacial bonding.

The adhesive bonding of mild steel adherents has received scant attention apart from a very thorough investigation by Kinloch et al. [46] in the early days of the application of surface analysis to adhesion. The reasons for such lack of attention are associated mainly with the gross corrosion of the substrate that occurs on environmental exposure, particularly in liquid water environments, leading to back deposition of rust on all exposed surfaces, making the interpretation of analysis very uncertain. There are ways in which this problem can be circumvented by using unrealistically short exposure times, or using aqueous environments with low partial pressures of oxygen. Attempts have been made to develop inorganic thin-film pretreatments for mild steel, and although such treatments show promise they are far from being commercially useful at the present time [47–49].

In the characterization of failure interfaces on steel and other substrates, XPS has been the most useful analytical technique. While the strengths of XPS in such applications are well known and have been extensively reviewed [1,2], there is of course one major shortcoming with standard XPS,

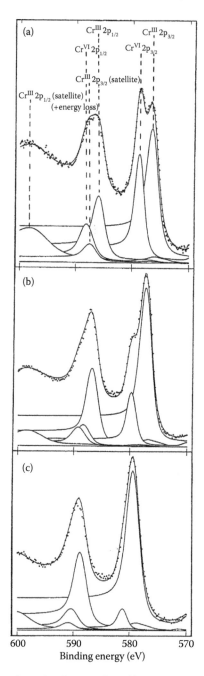

FIGURE 18.6 Cr 2p XPS spectra from the chromated steel/epoxy system: (a) metal surface following chromate treatment and (b) interfacial metal surface following cathodic delamination. Note the reduction in the Cr^{VI} component when compared with the as-received substrate prior to coating. The underside of the polymer coating (c) shows a very low concentration of Cr^{VI}. (From Murase, M. and Watts, J.F., *J. Mater. Chem.*, 8, 1007, 1998. With permission.)

and that is the lack of spatial resolution. All the investigations described above have employed XPS in its standard (area integrating) form in which several square millimeters contribute to the signal. This integration has its advantages, not least because there are then no problems concerning the provision of a representative analysis, but nowadays XPS of small regions is common place either as small area XPS (SAX, generally taken to mean XPS point analysis from a region 10–50 μm in diameter) or as iXPS (chemical maps at an image resolution of a few micrometers or better, see Chapter 3). The various ways in which SAX and iXPS can be achieved have recently been reviewed [50]. Examples of the application of these reduced area methods of XPS to adhesion science appeared fairly rapidly after their introduction [51–54], and they are now taken as a routine option in such an application.

The adhesive bonding of mild steel to a commercial rubber-toughened epoxy adhesive has been investigated using lap shear geometry; specimens were exposed to water containing a low partial pressure of oxygen (to ensure that the level of rusting remained low). Following the aqueous exposure the joints were pulled in a mechanical testing machine and the fracture surfaces examined by iXPS, SAX, and ToF-SIMS [51]. The failure surfaces consisted, visually, of a small central region of cohesively failed adhesive surrounded by regions of apparently interfacial failure. In the spectrometer used in this work, the output in the dispersive plane of the analyzer is essentially a small portion of the electron kinetic energy spectrum, while in the nondispersive plane, the output axis represents distance in a linear direction across the specimen. A position-sensitive detector provides a form of output known as an $E–X$ plot (representing energy and distance at the detector). The $E–X$ images from the work described above [51] indicated clearly that the oxide was thicker at the edge of the joint, but in the central region both Fe^{metal} and Fe^{oxide} contributions were apparent in the Fe 2p image, suggesting that the oxide was of about the same thickness as the air-formed film present when the joints were assembled. The SAX analyses showed a very low concentration of carbon at the interfacial metal surfaces, indicating that the failure mechanism, in this case, was interfacial in nature. The Na^+ assay showed that failure had been produced by cathodic delamination. A two-stage process was observed, the faster process giving rise to a zero-volume debond, in which interfacial bonds are broken but there is no separation of polymer phase from the metal substrate, followed by a slower separation and crack opening stage, which led to the development of a wide crevice in which mass transport could easily take place.

SAX was used in conjunction with ToF-SIMS to record elemental line scans across the metal and adhesive failure surfaces. Examples of these line scans from the metal surface are shown in Figure 18.7 and need a brief explanation. The island of epoxy adhesive remaining in the central region of the fracture surface is readily identified in the C 1s XPS line scan of Figure 18.7a, the carbon concentration of 30–35 at.% being entirely consistent with an interfacial failure. The O 1s and Fe 2p data of Figure 18.7b and c, respectively, delineate the iron oxide on the exposed metal (i.e., where the O 1s intensity is higher than that of the adhesive, ~45% as against 30%); the Fe 2p signal increases gradually as the crevice tip (adjacent to the intact adhesive) is approached and as the contribution from the Fe^{metal} signal rises. The N 1s excursion, Figure 18.7d, at the crevice tip results from enhanced adsorption of nitrogenous species on the cathodic surface. The reasons for this are unclear. The presence of a sharp rise (Figure 18.7e) of Na^+ concentration (even though the specimens were exposed in ultrapure water) shows that cathodic conditions prevailed at the tip of the disbondment crevice formed between the detached adhesive and the metallic substrate, and in Figure 18.7f can be seen a corresponding rise in OH^- concentration in the same region. These observations identify explicitly the under-film alkalinity, generated by cathodic activity, that is responsible for adhesion loss. The chloride ion signal (Figure 18.7g) reaches a maximum slightly away from the crevice tip. The aromatic ion $C_6H_5^+$ is tracked in Figure 18.7h and, as expected, identifies the exposed epoxy adhesive. It is interesting to note that in the profiles in Figure 18.7 each peak appears at slightly different points within the failure region, demonstrating the role that the electrode potential within the crevice plays in the interfacial chemistry. The manner in which the maxima of the various species occur as a function of distance within the crevice is analogous to that in conventional chromatography, where the distance travelled by the various fronts in, for example, thin layer chromatography can be related to the chemistry of a particular species. It can be envisaged that the XPS and ToF-SIMS data of Figure 18.7 describe the deposition of chemical species at local conditions of pH and electrode potential within the crevice. Using this approach, it should be feasible to map potential at a crevice tip by the appropriate ionic additions to the aqueous solution to which the tip is exposed.

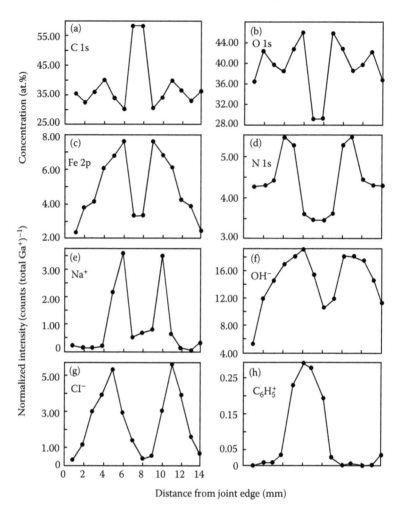

FIGURE 18.7 Line scans from a series of point analyses of the metal substrate of a failed adhesive joint by SAX and ToF-SIMS: (a) C 1s, (b) O 1s, (c) Fe 2p, (d) N 1s, (e) $^{23}Na^+$, (f) $^{17}OH^-$, (g) $^{35}Cl^-$, and (h) $^{77}C_6H_5^+$. (From Davis, S.J. and Watts, J.F., *J. Mater. Chem.*, 6, 479, 1996. With permission.)

The above example emphasizes further the fact that in order to understand the interfacial reactions responsible for adhesion loss, the electrochemical response of the substrate must be understood. Such an understanding becomes even more essential when studying stainless steels where the polymer must interact with the air-formed passive film. In order to achieve the interaction, many, mainly acidic, pretreatments have been developed on an empirical basis. Examples of the type of investigation required can be found in the work of Gettings and Kinloch, who investigated both the surface characteristics [55] and the nature of the failure surfaces [56] in some detail.

18.3.3 ZINC SURFACES

Although the adhesion of polymers to zinc itself is of relatively little interest, much work has been carried out on adhesion phenomena associated with zinc-coated steel such as hot-dipped galvanized steel (HDGS), or electro-galvanized steel (EGS). Such substrates are notoriously difficult to paint and are therefore often used in conjunction with a conversion coating of the type discussed in the previous section.

The interaction of a poly(vinyl chloride) (PVC) formulation with HDGS has been studied by Dickie et al. [57,58]. They showed that the PVC resin underwent extensive dehydrochlorination at the metal surface, an effect which could be reduced somewhat by the addition of a resin modifier such as dicyandiamide. Their conclusions were based on quantitative XPS analyses and careful binding energy determinations of the Cl $2p_{3/2}$ core level. The concentration of chlorine at the interfacial failure surface was about half that within the bulk of the adhesive, and about 5 at.% of this element within the interfacial region was present as an ionic compound, amine hydroxychloride (Cl $2p_{3/2}$ = 197.7 eV compared with 199.7 eV for the adhesive component). Scant attention was paid to the zinc chemistry, which is a pity since zinc is particularly amenable to study by way of the Zn Auger Parameter (the sum of the Zn $2p_{3/2}$ photoelectron and Zn $L_{2,3}M_{4,5}M_{4,5}$ Auger peak energies) and has been shown to be useful in the study of pretreatment processes [59]. The Auger Parameter concept is particularly useful in the study of samples subject to electrostatic charging (such as inorganic pretreatment layers), since the increase in binding energy of the photoelectron peak is exactly offset by a reduction in the kinetic energy of the Auger peak. The chemical information available by this method is limited to Auger transitions that involve three core-like electrons, but in such cases (F, Mg, Al, Si, Cu, Zn, Ge, etc.), the chemical information available may be more reliable than the use of the photoelectron binding energy alone. For example, the Zn Auger Parameter values for ZnO and $ZnCl_2$ are 2009.8 eV and 2009.2 eV, respectively, using Al K_α radiation.

The adhesion of organic systems to zinc was also studied by van Ooij et al. [53], who extended the coating variables to include pretreatment of the metal and the degree of cure of the organic system. Their work is important in that, perhaps for the first time, it indicated that the degree of cure of a fully formulated system might be detectable by careful consideration of the ToF-SIMS spectrum. They also reiterated the idea, already reported anecdotally by both Hammond et al. [36] and Watts and Castle [60], that certain of the coating constituents can be adsorbed preferentially at the metal oxide/polymer interface. With the increasing sophistication of analytical tools, this idea received ever greater support. Using an XPS elemental mapping method, Haack et al. [52] have shown convincingly, by determining the lateral distribution of the elemental markers for cathodic and anodic reactions, that both anodic and cathodic areas of activity can be observed at the failure surface of adhesively bonded EGS, and they associated failure within the anodic regions to the poor cohesive strength of the zinc corrosion products. Indeed, the readiness with which Zn forms the so-called white rust seems to suggest that the words of caution regarding mild steel studies are just as relevant to zinc.

In an investigation making use of SAX [61] and ToF-SIMS [62], to study the locus of failure of adhesively bonded HDGS, the advantage of spatially resolved surface chemical information in understanding the failure process has been demonstrated clearly. It was shown that once an adhesive joint was immersed in water there is the possibility that discrete electrochemical cells might develop, which would then exacerbate failure, with adhesion being lost very rapidly in areas of cathodic activity. At a micro-cathode in one of these cells, cations from the solution (e.g., Na^+ or Mg^{2+}), even if present at very low concentrations, will adsorb on the substrate and decorate the cathodic site. The greater detection limits of ToF-SIMS allow even the trace quantities contained in ultrapure water to be used analytically in this manner.

Failure as a result of aqueous exposure is sometimes associated with a characteristic locus of failure rich in minor components of the adhesive or coating formulation. These segregated layers will generally form hydrophilic interfaces with the main body of the polymer, and rapid failure is observed as a result of water diffusion along them. An interesting example of this phenomenon is the segregation of the photo-initiator package in a radcure coating on HDGS [63]. Subsequent reduction of the amount of the initiator in the formulation produced a system with the required performance.

18.3.4 ALUMINUM ALLOYS

As might be expected, it is in the aerospace industry that so much research into the pretreatment and adhesive bonding of aluminum and its alloys has been carried out. The early work was reviewed by

Venables [64], who showed conclusively that for the successful adhesive bonding of aluminum adherents, a complex pretreatment schedule must be followed that provided a characteristic microporous morphology. The introductory article by Davis [65] on surface analysis includes also many examples concerning adhesive bonding. The various methods of pretreatment have been reviewed by Olefjord and Kozma [66] and Clearfield et al. [67]. Some of the initial studies of the failure of adhesively bonded aluminum were performed on failed Boeing wedge test specimens, in which failure was ascribed to hydration of the anodic oxide layer [64]. This provided a weakened region within which failure could occur. The suggestion led to the investigation of a number of hydration inhibitors that were thought to be able to arrest the chemical degradation and thus improve joint durability. The assumption that such a hydration mechanism is the only cause of failure in aluminum adhesive joints is questionable, although electrochemical studies have shown unambiguously that sub-adhesive hydration can, and does, occur [68], the sticking point being the extent to which the anodic oxide is converted into a voluminous hydration product with low cohesive strength. An important observation by Watts et al. was that in cyclic exposure trials of T-peel specimens, the damage caused by immersion in water was partially reversible in the drying stage of the cycle [69]. This indicated a phenomenon similar to the wet adhesion process described in Section 18.3.2, and ruled out gross hydration of the aluminum oxide phase, which would have involved a large increase in volume. The latter hydration, if it occurs, would not be reversible, and the original bond integrity would not be retrievable, contrary to observation. Analysis of the failure surfaces by XPS showed that failure had occurred close to, but not at, the oxide/polymer interface [41]. It was assumed that a thin layer of adhesive remained on top of the porous oxide and also filled the outer porous structure. As water and other aggressive species will diffuse in from the edge of an adhesive joint, failure will proceed from the edge toward the center in the manner illustrated for adhesively bonded steel in Figure 18.7.

Care must be taken when selecting a sample for analysis, to ensure that it is the corrosion- or exposure-induced failure surface that is chosen, and that the central island of adhesive is not inadvertently included in the analysis. This does not pose a problem in the case of microanalysis methods such as AES or ToF-SIMS, but can be a very real concern in the case of XPS when used in its standard, large area, mode. The situation is, however, simplified to a certain extent as the central region of adhesive that has failed in a cohesive manner can usually be identified visually (e.g., a rubber-toughened epoxy may be black, which contrasts well with the metallic substrate). Figure 18.8a illustrates schematically the appearance of a failed T-peel test piece from the above investigation [41]; apparent interfacial failure at the specimen edge can be identified visually or with a low-power optical microscope, as one interfacial surface will be that of the uncoated metal and the other will consist of the entire thickness of the adhesive. SAX spectra were recorded from both the edges of the joint and from the central region, and are shown in Figure 18.8b through d. The nitrogen peak (at a binding energy of ~400 eV) is diagnostic of the adhesive, which can also be recognized by the intense C 1s line. The two spectra recorded from the edge of the specimen also show Al 2p and 2s peaks, which, when taken with the C 1s line, indicate that, rather than being interfacial in nature, the failure is actually within the adhesive, and there is a thin layer of polymer remaining on the substrate side of the failed joint. Clearly, substrate hydration has not occurred, since one of the characteristics of a voluminous corrosion product, as far as surface analysis is concerned, is a low concentration of adventitious carbon. In situ XPS studies of miniature peel specimens (using Ar^+ profiling) have indicated a very deep transition zone between the metal oxide of the substrate and the polymeric adhesive. This observation has confirmed directly the existence of an extended mechanical interphase, where gross penetration of the adhesive into the porous pretreatment layer occurs, which had been hypothesized on the basis of morphological observations by SEM and transmission electron microscopy (TEM) [70]. Studies making use of parallel electron energy-loss spectroscopy (EELS) in a TEM have provided important analytical information regarding the complete penetration of adhesive in some anodizing pretreatments [71,72]. It is likely that the use of spectroscopic EELS in a scanning transmission electron microscopy (STEM) will go some way toward unraveling the complex changes that occur in the anodic oxide layer on exposure to water, which is an important mechanistic precursor to the failure of the adhesive bond itself.

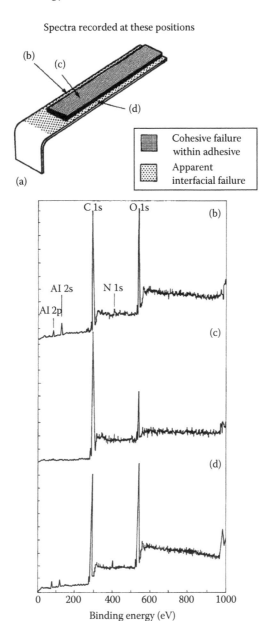

FIGURE 18.8 T-peel failure of adhesively bonded aluminum exposed to water at 50°C for 10 weeks: (a) schematic of joint. SAX (250 μm) spectra recorded from the edge regions (b) and (d), and from the central cohesive region (c). (From Watts, J.F., Blunden, R.A., and Hall, T.J., *Surf. Interface Anal.*, 16, 227, 1990. With permission.)

iXPS can provide important evidence, on the micrometer scale, of the manner in which failure has occurred, and Figure 18.9 shows XPS images from the failure surface of adhesively bonded aluminum tested in a wedge cleavage test geometry.* The failure path changes from one metal/adhesive interface to the other, and the Si 2p images show that an organosilane adhesion promoter, added to the adhesive formulation, has segregated to one interface but not the other.

---

* The wedge cleavage test (also known as the Boeing wedge test) is a simple form of a double cantilever beam geometry as described by ASTM D3702, with separation achieved by a wedge rather than a mechanical testing machine. This facilitates the testing of many samples at the same time.

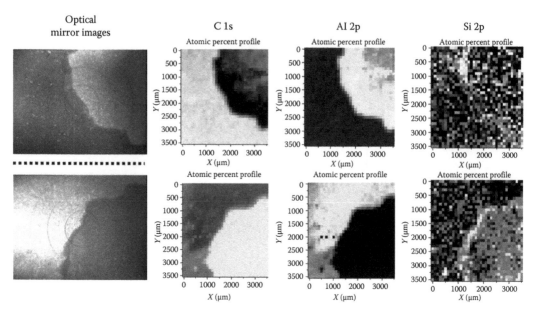

FIGURE 18.9 (See color insert following page 396.) SAX images (field of view 3.5 mm) recorded from the failure surfaces of adhesively bonded aluminum using an epoxy resin to which had been added 1% of an organosilane. The optical images (LHS) show failure surfaces of the joint opened out as a book along the dotted line. Adhesive and metal regions are identified by the intense C 1s and Al 2p images, respectively, in the middle. The Si 2p images (RHS) show how the silane has segregated to one failure surface (toward the RHS of these images), but not the other.

18.3.5 Composite Materials

Although the study of carbon fibers can be traced back to the early days of XPS, and much is now known about the surface chemistry and morphology induced by electrolytic treatments, it was only with the advent of high spatial resolution SIMS (and to a lesser extent XPS) that composite fracture surfaces started to receive the attention of the surface analysis community. This is perhaps not surprising as the failure surfaces are complex, consisting of fibers of 6 μm diameter in a polymeric matrix. Attempts have been made to employ SAM to establish failure mechanisms [73,74], but it is now generally accepted that ToF-SIMS is the technique of choice for the analysis of composite fracture surfaces and the investigation of the interfacial chemistry of adhesion of such systems. Initial work with SIMS using quadrupole mass spectrometers was impressive and showed that the untreated fiber and matrix failed interfacially, while, on the other hand, an electrochemically oxidized fiber within a resin matrix failed predominantly cohesively [75]. In these early studies, the fiber was characterized by CN^- (present from the poly(acrylonitrile) precursor) and the resin by Cl^- from the residual epichlorohydrin. Both ion signals were strong enough to provide good quality maps, but subsequent work showed that CN^- was also often present in the matrix (e.g., in the case of an amine-cured epoxy) so that this approach was clearly not universally applicable. Later studies used ToF-SIMS [76,77], which provides an important route for the identification of major fragments of the resin at the fracture surface, in turn enabling the components in the resin formulation responsible for fiber/matrix adhesion to be identified. ToF-SIMS analyses carried out on double cantilever beam test pieces of a carbon fiber composite fabricated from prepreg material (fiber tows preimpregnated with partially cured resin that allows rapid and straightforward composite manufacture) showed that the fracture surfaces were decorated with poly(dimethyl siloxane) that had diffused from the outer surfaces (protected by a peel film) into the bulk during cure [78]. The thorny question of the *ex situ* fracture of composite specimens and of their subsequent transfer through air to the spectrometer has been addressed in the author's laboratory. The use of an *in situ* fracture stage has been shown to improve spectral quality and to make diagnostic peaks more readily observable, as opposed to a plethora of ubiquitous C_xH_y peaks [79]. Recent work has seen the development of a fully instrumented computer-controlled *in situ* stage providing mechanical data in the form of a load extension curve prior to controlled fracture [80]. In that paper, it was established that even trace amounts of PDMS, present in a laboratory atmosphere, may adsorb on a fracture surface and lead to confusion in the interpretation of high-resolution SIMS spectra. Such problems are easily avoided by *in situ* testing, and the approach is particularly important if a silicon-based reinforcement is being studied.

18.3.6 CERAMICS

Most works associated with the adhesion of polymeric materials to ceramic substrates have been carried out in relation to the microelectronics industry. Buchwalter et al. [81–83] have shown the importance of the substrate surface condition in the level of adhesion achieved by a polyimide film to SiO_2, Al_2O_3, and MgO. Their work revealed the importance of specific interactions of the acid–base type in adhesion, and formed part of the burgeoning body of data in this area. In essence, the approach maintains that the forces of adhesion can be treated quantitatively, at a microscopic level, by regarding such interactions as those between Lewis or Brönsted acids and bases. The basic tenor of this concept, of which the Lewis approach is the more widely applicable, is that one component in the adhesion couple can be regarded as the Lewis acid (electron acceptor) and the other as a Lewis base (electron donor), the specific interaction responsible for adhesion being the formation of a Lewis acid–base adduct or complex. The work of Buchwalter et al. [81–83] showed clearly how the extent of donor–acceptor forces at the polymer/ceramic interface could influence the locus of failure. The forces required to bring about failure are related to the types of polyimide precursors employed. For example, the peel strengths of polyimides on SiO_2 and Al_2O_3 are higher than on an MgO substrate, and the locus of failure is cohesive within the polymer. On MgO substrates, the peel strength is higher for an ester-derived polyimide than it is for an acid-derived variety. The failure is of a mixed-mode type in the case of the acid polyimide, and is ascribable to carboxylate salt formation and the degradation of MgO by the formation of strong acid–base interactions between the basic MgO and the polyamic acid precursor. With the more neutral ester polyimide, the locus of failure was within a weak boundary layer of the polyimide itself [83]. The question of the acid–base characteristics of metal oxides and polymers is of crucial importance in adhesion and are considered again later.

Oxide substrates, in general, and alumina, in particular, are not good electrical conductors, and may give rise to a phenomenon known as vertical differential charging (VDC) in the resultant XPS spectrum. Such a phenomenon may be observed in the XPS analysis of insulating materials, and results from a change in the electrostatic potential of the specimen as a function of the depth over which the analysis takes place (around 5 nm). This means that the resultant spectrum is factually the convolution of a series of spectra, which are shifted a small amount, relative to each other, on the energy scale. In practice, the result is a spectrum, recorded by the spectrometer, that is broader than anticipated for the particular spectrometer conditions employed (i.e., pass energy, analyser slits, and x-ray source), and shows very little, if any, fine structure, as this will have been destroyed by the convolution effect. In adhesion studies of polymer layers on ceramics, particular care must be taken to ensure that the presence of VDC does not lead to an erroneous assignment of the failure characteristics.

The potential difficulty of VDC is well illustrated by the work of Taylor and Watts as part of an investigation of the adhesion of a photocured resin to quartz, alumina, and silicon [84]. The spectra of Figure 18.10 were taken from the oxide failure surfaces of an adhesive joint that had failed following aqueous exposure [84]. Figure 18.10a shows the C 1s spectrum recorded using achromatic Al K_α radiation from a standard twin-anode assembly; the spectrum is rather broad and featureless, and there is a temptation to ascribe failure to a weak boundary layer of preexisting hydrocarbon contamination. When the spectrum was recorded again using monochromatic Al K_α radiation,

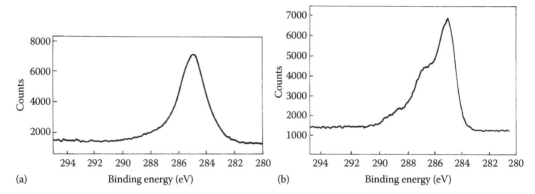

FIGURE 18.10 C 1s spectra from a failure surface of adhesively bonded alumina recorded with (a) achromatic Al K_α and (b) monochromatic Al K_α radiation. (From Watts, J.F. and Taylor, A.M., *J. Adhes.*, 46, 161, 1994. With permission.)

Figure 18.10b, with proper charge control (i.e., an electron flood gun), the C 1s spectrum then revealed much fine structure that could be related directly to polymer composition. The adhesive employed was an aromatic methacrylate material with an aliphatic reactive diluent added at a level of about 10%. The absence of a $\pi \rightarrow \pi^*$ shake-up satellite in the spectrum of Figure 18.10b indicates that the locus of failure was associated with a segregated layer of the reactive diluent. This conclusion was confirmed subsequently by quantitative angular-resolved XPS [85], which enabled the overlayer thickness to be estimated at about 1–2 nm. Reformulation of the resin removed the segregated monolayer of reactive diluent and changed the nature of failure to a cohesive mode located within the aromatic adhesive. The reformulation also improved the durability of the joint [86]. The results of this work are important in that they provide a good illustration of VDC, and also emphasize the problems associated with the examination of thin polymeric films on insulators. They also show how the conclusions from surface analysis can be fed back into the formulation process to enable a weak boundary layer to be removed, thus providing a product with superior properties.

18.3.7 SUMMARY

In the analysis of surfaces produced by adhesion failures at interfaces, XPS, supported by ToF-SIMS, can provide a great deal of information relating to the characteristics of the locus of failure. XPS can identify and quantify thin polymer overlayers and deduce their orientation, and it can monitor degradation in both polymer and metal oxide, and the diffusion of active species within the disbondment crevice. The latter can in turn be related to electrode potential. Failure as a result of the ingress of water is a common occurrence, and such a failure is characterized by a thin (1–2 nm) overlayer of polymer remaining on the substrate. Failure as a result of electrochemical activity such as cathodic disbondment is much closer to the classical concept of interfacial separation. ToF-SIMS is able to identify thin overlayers with chemical resolution far superior to that of XPS, and, as all polymer coatings and adhesives are invariably multicomponent systems, the identification of which of the individual constituents of the formulation have segregated to the polymer/metal interface and the correlation of such data with durability and other performance data assume critical importance. Many of the components of such systems will yield similar XPS spectra, so the additional provision of ToF-SIMS analysis enables their identification in an unambiguous manner. The definition of the exact locus of failure will, however, remain the preserve of XPS as a result of the ease with which quantitative analyses can be obtained, together with the determination of overlayer thicknesses. As both XPS and ToF-SIMS are available in spatially resolved modes (10 μm for SAX and sub-micrometer for ToF-SIMS), heterogeneities at the locus of failure can be resolved quite readily, be they the lateral distribution of a minor component by ToF-SIMS or a contour map of overlayer thickness by XPS.

Investigation of failure surfaces is often the starting point in an adhesion study, and has resulted in many successful surface analysis investigations. There is, however, much additional information that is required to be able to interpret the surface analysis data successfully: a knowledge of substrate characteristics prior to coating or bonding, the manner in which the adherent behaves electrochemically, the formulation of the polymeric phase, etc. are all important parameters. The forensic analysis of failure surfaces, undertaken with the aim of determining the cause of failure, can by itself sometimes lead to a fuller understanding of interfacial characteristics responsible for bonding, but, in general, this is not so. It is better to start with a knowledge of bonding characteristics and use that as a basis for understanding the exact mechanism of failure, although such a route is experimentally difficult. The analysis of buried interfaces between polymers and metal oxides is very challenging, and in the following section, some of the ways in which this can be achieved are reviewed.

18.4 PROBING THE BURIED INTERFACE

Adhesion is a consequence of the chemical or physical interaction between two surfaces, one of which is a solid and the other a liquid, temporarily more mobile. As a consequence of the way in which adhesion is achieved practically, the interface or interphase region, where the bonds responsible for adhesion occur, is buried below many micrometers, or even millimeters, of solid substrate and solidified (generally cross-linked) polymer. The dimensions of the interphase region are likely to be of the order of nanometers at the most (unless an extended mechanical interphase is present of the type found in the adhesive bonding of anodized aluminum alloys), so that direct examination of interphase chemistry is best considered as an exercise in the analysis of a deeply buried interface. This situation is encountered frequently by those working in microelectronics, and in corrosion and oxidation research. Removal of material has usually been accomplished with an energetic ion beam, either as part of a dynamic SIMS analysis or in conjunction with AES, or (less often) XPS, in order to produce by sputtering a profile of composition with depth (Chapter 10). Such an approach is, however, not appropriate in adhesion studies as the organic material will undergo gross degradation, and eventually graphitization, in the ion beam, although, as noted earlier, C_{60} (and perhaps other cluster beams operated at low energy) offers promising routes to the depth profiling of polymers, but such opportunities have yet to be fully explored in adhesion studies. Thus, the study of interphase chemistry has led to the development of some rather ingenious ways in which the sample, often a real adhesive joint or a coated substrate, and containing a buried interface, can be prepared for surface analysis [87]. In the following section, the more successful approaches that have been used are described briefly.

Mechanical, chemical, or electrochemical perturbations can sometimes provide useful results as described in Section 18.3, but rely on the interphase being the weakest link, which is not always so. Exposure of the metal substrate by causing the polymer to swell with N-methyl pyrrolidone has been proposed, and works well for some coatings [88]; indeed, the length of time taken to displace the polymer from the substrate by swelling has been suggested as a test for estimating the strength of adhesion of the coating/substrate system [89]. There is of course a danger that air exposure will modify the interphase chemistry, particularly if reactive species are exposed, but such exposure can be avoided by the use of an argon glove box attached to the spectrometer [90].

The opposite approach is removal of the metal from the adhesive joint, as used extensively by Watts and Castle [33] for the investigation of interfacial chemistry. In the case of ferrous alloys, the metal substrate can be dissolved in a methanolic iodine or bromine solution, leaving the oxide film supported on the polymer or adhesive. The supported film can then be mounted for analysis and ion sputtering in order to obtain a depth profile through the oxide toward the oxide/polymer interface. Notwithstanding the danger of ion-beam-induced reduction, the authors were able to show convincingly the formation of an Fe(II)-containing carboxylate interphase between polybutadiene and mild steel [33]. Identification of the role of the polymer as an agent in the reduction of Fe(III) oxide was not in itself surprising as the polymer was known to cure by an oxidative process, and the reduction effect was later confirmed by Mössbauer spectroscopy [91]. For aluminum alloys, a similar stripping approach can be adopted using sodium hydroxide solution [92].

Mechanical sectioning at cryo-temperatures in order to prevent polymer spreading and smearing can also be useful for interface analysis. The well-known ball-cratering technique used with AES for gross depth profiling has been developed with a cold stage to enable polymer films to be sectioned successfully. However, electron-excited Auger analysis can yield chemical information for only a few elements, so that the natural extension of the method was the development of a taper-section stage for use with SAX [87]. The design of such a stage was based on the use of an auxiliary high vacuum (HV) chamber attached to the preparation chamber of the spectrometer. Taper-sections of ~1° incline angle could be cut on the cooled specimen using a conventional tungsten carbide end mill. The specimen could then be transferred directly to the analysis position, where the angle of taper then became equivalent to a depth of about 1 μm if a 100 μm SAX facility were used. This has proved invaluable for the analysis of chemistry in polymer [90] and inorganic systems [59].

A more sophisticated variation on the taper-sectioning process is the use of a microtome to cut sections at very low angles. This method has been developed in the author's laboratory and is termed ultralow angle microtomy (ULAM) [93–95]. In this approach, a large area (histological) microtome is employed, and angled blocks are used to set the interface of interest at a very low angle (0.03° to 2°, depending on the depth resolution requirements) to the plane of the microtome blade. In this way, it is possible to investigate interfaces between layers of organic coatings in any SAX system, an x-ray spot size of 15 μm yielding a depth resolution of 13 nm with the smallest ULAM taper angle. Figure 18.11 shows a depth profile obtained in this way from a polyvinylidene fluoride (PVdF) topcoat on a polyurethane primer [94]. ToF-SIMS analysis identified the segregation of a polyacrylic additive as a 50 nm layer at the interface between the two coatings [95].

An alternative route to buried interface analysis, but one that moves away from the real situation, is the use of thin-film model substrates. A methodology has been described [11] in which a thin film of aluminum (0.5–1 μm) was deposited on a secondary adherent and then anodized and bonded in the usual fashion. The thin film was then separated from its backing and the sandwich specimen of

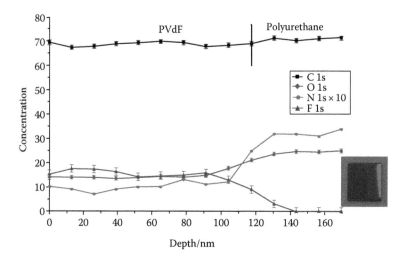

FIGURE 18.11 Depth profile obtained from an XPS line scan across a sample cut from a PVdF topcoat applied to a polyurethane layer, using ULAM. The ULAM angle was 0.03°, and the XPS measurements were carried out using a spot size of 15 μm, providing a theoretical depth resolution of 13 nm. The sample is shown at the lower right, where three distinct regions can be identified, viz., on the LHS the outer surface of the PVdF, in the middle the sectioned PVdF layer, and on the extreme RHS the exposed polyurethane primer. (From Hinder, S.J., Lowe, C., Maxted, J.T., and Watts, J.F., *J. Mater. Sci.*, 40, 285, 2005. With permission.)

Al/Al$_2$O$_3$/adhesive was presented for analysis by AES or XPS with the metal side uppermost. The depth profile shown in Figure 18.3 is from this type of specimen.

Another modeling approach involves the deposition of an ultrathin film onto a solid substrate. If the film is thin enough, XPS or ToF-SIMS can then be used to look through the film to observe interfacial chemistry directly. Such an approach was first used in a study of the process of polymer metallization for Cr, Fe, Ni [96,97], Ti [98], and Al [99], and is now applied to the study of thin polymer films on metallic substrates [100]. The best results have been achieved using very dilute polymer solutions (0.01–0.02 w/w%), polished substrates, and very high resolution (monochromatic Al K$_\alpha$) XPS [101]. The information that can be gleaned from such studies is impressive but relies on careful peak-fitting of the C 1s spectra, or indeed high resolution ToF-SIMS spectra, as are discussed in the next section.

18.5 ORGANOSILANE ADHESION PROMOTERS

The use of organofunctional silanes as adhesion promoters is well documented [102], and much is now known about their mode of operation [103]. Surface analysis has played, and continues to play, an important role in the understanding of the action of these complex materials, which have been studied on glass, and on steel and other metals. There is now general agreement that the enhancement of adhesion properties is the result of covalent bonding between the silane molecule and hydrated metal oxide, which is consistent with the theoretical models of the last four decades. However, one of the assumptions of these models is the existence of a uniform, toothbrush-like monolayer of molecules at the oxide surface. Such molecular organization has been observed in some cases but does not occur with all molecules of this type showing adhesion enhancing properties.

Two important investigations dealing with silane organization on iron surfaces were published some 30 years ago. Bailey and Castle demonstrated how XPS could be employed to monitor adsorption from the liquid phase, by constructing adsorption isotherms of surface concentration (as determined by XPS) as functions of solution concentration [104]. The methodology has also been used to assess the adsorption of Ag$^+$ and Mg^{2+} ions on carbon fiber surfaces, the surface concentration of acidic sites [13], and the adsorption of macromolecules onto the surface of an intrinsically conducting polymer, using ToF-SIMS instead of XPS [105]. In the earlier work, Bailey and Castle studied the adsorption of two ethoxysilanes on iron and recorded the data reproduced in Figure 18.12. There

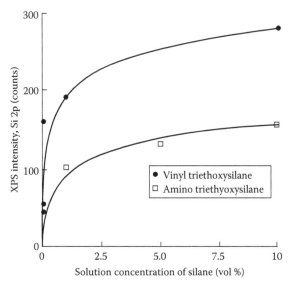

FIGURE 18.12 XPS adsorption isotherms for methanolic solutions of vinyl and amino triethoxysilanes on iron. (From Bailey, R. and Castle, J.E., *J. Mater. Sci.*, 12, 2049, 1977. With permission.)

was enhanced adsorption of the vinyl triethoxysilane relative to the amino-propyl-triethoxysilane molecule. The isotherms were of the Temkin type, indicating chemisorption, which for the vinyl silane can be ascribed to a specific interaction between the silicon end of the hydrolyzed silane molecule and the hydrated iron oxide surface. The amino silane molecule showed a lower coverage, thought to be the result of the propensity of either end of the molecule (silane or amino) to interact with the metal surface, so that the molecular orientation was less clear; it would have been anticipated that the molecule would lie parallel to the metal surface, unlike the vinyl silane that would adopt a bristle-like conformation. As can be seen in Section 18.7, the situation was resolved to a certain extent two decades later with the aid of computer chemistry.

The other important observation from the same era was made by Gettings and Kinloch using SSIMS [106]. When studying the fracture surfaces of adhesive joints fabricated using various silane-based primers, they were able to draw a direct correlation between good joint durability and the presence of an $m/z = 100^+$ ion in the SIMS spectrum at the failure surface. The ion was attributed to an Fe–O–Si covalent bond and was present only for the γ-glycidoxy-propyl-trimethoxysilane (GPS); for the first time, improved durability could be ascribed to primary bond formation. Actually, the assignment of the $m/z = 100^+$ peak in the (quadrupole) SIMS spectrum could not be certain since the spectrum was recorded at unit mass resolution, but the question was resolved, to a large extent, by the ToF-SIMS investigation of Davis and Watts [107] in which the organization of GPS molecules on iron was studied. The layer deposited from 2% methanolic solution was shown to be poorly ordered and of 1.7 nm thickness. In the SIMS spectrum, there was initially no peak at $m/z = 100^+$, but on gentle ion etching, a well-defined peak at that mass appeared. Although alternative assignments for the peak exist (e.g., $FeOC_2^+$ or SiO_3^+), they can be eliminated by consideration of the relative intensities of related ions [107].

In a series of papers, using ToF-SIMS, Abel et al. have cataloged the manner in which specific interactions between GPS, aluminum, and a structural adhesive provided enhanced hydrodynamic stability of adhesive bonds. In their work, a 1% aqueous solution of GPS was applied to grit-blasted aluminum and cured at 93°C. High resolution ToF-SIMS indicated, unambiguously, the formation of a covalent bond between the silanol groups and the aluminum surface, identified by an Al–O–Si$^+$ fragment, as shown in Figure 18.13 [108]. Having established this important interaction, the thin-film approach was then used to construct adsorption isotherms of various components, with the nature of interfacial bonding being identified by the application of high-resolution ToF-SIMS to specimens from the plateau region of the isotherm. Molecules studied were all related to epoxy adhesives, specifically an epoxy analogue molecule (diethanolamine) [109]; to major components of a commercial structural adhesive, that is, the curing agent toluene diisocyanate [110]; and to the epoxy component of the diglycidyl ether of bisphenol A [111], as well as to the fully formulated adhesive itself [112]. The outcome of this extended investigation consisted of a complete picture of the manner in which this epoxy-terminated organosilane adhesion promoter is able to improve the performance of an adhesive joint. Initially, the silane interacts to form a covalent bond with the AlOOH present on the aluminum surface, then the terminal epoxy group on the adhesion promoter is able to form a covalent bond with the curing agent, which, in turn, is able to bond with a dicycidyl ether of bisphenol A (DGEBA) molecule, and finally the usual epoxy cross-linking chemistry takes over. The end result is that covalent bonding dominates from the inorganic film on the metal substrate through to the bulk of the adhesive, as indicated in the scheme below:

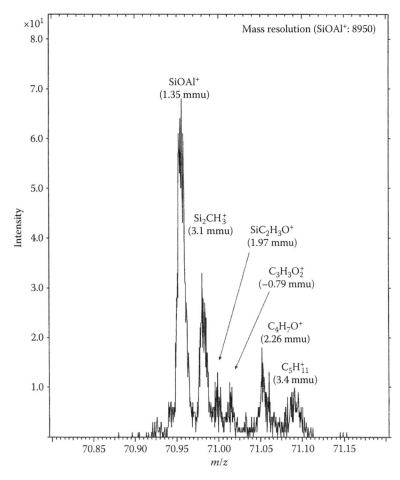

FIGURE 18.13   High-resolution ToF-SIMS spectrum in the $m/z = 71$ region, recorded from aluminum treated with GPS. The intense component labeled AlOSi$^+$ is diagnostic of covalent bond formation between an aluminum substrate and the silanols of the hydrolyzed silane. (From Abel, M.-L., Fletcher, I.W., Digby, R.P., and Watts, J.F., *Surf. Interface Anal.*, 29, 115, 2000.)

As covalent bonds are very resistant to displacement by water, the macroscopic properties of the joint are improved in the presence of water. The work indicates the way in which the provision of adsorption isotherms, as well as establishing the general type of adsorption, is also able to provide, in the case of chemisorption, an indication of the solution concentration that should be used to deposit a layer of monolayer, or at least saturation, coverage on the metallic substrate. The latter parameter is equivalent to the plateau region of the adsorption isotherm and varies from a few percent in the case of a small molecule to <<0.1 w/w% for macromolecules [113]. While both XPS and ToF-SIMS are important for establishing the extent of adsorption, ToF-SIMS is preferred in the case of competing organic molecules as it is able to distinguish the identity of the adsorbate with more certainty, particularly at low coverages. In the identification of specific interactions, ToF-SIMS is probably essential in obtaining the complete picture of the exact nature of the interface. The use of XPS alone in this regard should be eschewed, and it is fair to say that both are required in investigations of this type.

Although iron and aluminum have received most attention, organosilane deposition and organization has also been studied on many other metal surfaces, as reviewed by Abel et al. [114], including stainless steels [115], EGS [116], aluminum, titanium, [117,118], and copper [119]. The work of Boerio and coworkers is particularly significant in that they have combined XPS with vibrational spectroscopies, such as reflection adsorption infrared spectroscopy (RAIRS) and surface-enhanced Raman spectroscopy (SERS), to provide enhanced chemical information. SERS is applicable only to highly Raman-active compounds, so that in the field of adhesion the only study possible has been that of the interaction of silver with organic materials [120,121]. Silanes are used widely in the size that is applied to glass fibers immediately after manufacture, and the interaction was studied by Yates and West using an Ag L_α source to record the Si 1s/Si KLL Auger Parameter [122], a method also used by Cave et al. to study the interaction of octadecyl-trichlorosilane with aluminum [123]. The measurement of this Auger Parameter is particularly useful as, in principle, it allows the polarizability of the oxygen atom in Si–O compounds to be determined, which, in turn, can be related to specific bond formation in a manner that could not be achieved merely by recording the Si 2p photoelectron spectrum. The most extensive investigation of the interaction of organosilanes with glass fibers has been carried out by Jones et al. [124–126].

18.6   ACID–BASE INTERACTIONS IN ADHESION

18.6.1   Evaluation of Acid–Base Interactions in Adhesion

For many years now, it has been clear that, in describing the nature of interfacial interactions responsible for adhesion, it is incorrect merely to describe a polar contribution to the surface free energy and then arrive at an interface free energy term using a geometric mean approach [127]. A more rigorous description of such forces can be provided by adopting the donor–acceptor approach pioneered by Fowkes [128]. This is now widely accepted by the adhesion community, with many workers seeking to clarify and quantify the role that acid–base interactions have to play in adhesion phenomena. Progress up to the early 1990s was summarized in a Festschrift in honor of Fowkes, which provided a comprehensive overview of the achievements to that time [129]. Work in this field at that time set the scene for the use of surface analysis techniques for the estimation of the acid–base properties of metal oxides and polymers, and represents now yet one more methodology in the surface analyst's armory that can be used to good effect by the adhesion community. The advantages of using surface-specific techniques are immediately obvious to those working in applied surface science; conventional methods such as flow microcalorimetry, Fourier transform infra-red spectroscopy (FT-IRS), and the like rely generally on high surface area solid oxides, which means that either the film must be removed from the metal and ground to a powder, or the polymer taken into solution. Such methods are perfectly acceptable in the study of homogeneous materials but cannot be used to measure concentration gradients, which might occur in the case of the surface treatment of a metal substrate, or the occurrence of segregation effects in coatings or adhesives. In addition, the important contribution of inverse gas chromatography (see Section 18.2.3) must be mentioned, which provides a complementary surface-specific method for the determination of acid–base properties of metals and oxides [130,131].

It is appropriate at this stage to review the quantification of acid–base forces based on the exothermic enthalpy of acid–base interaction ($-\Delta H_{AB}$). The acid–base contribution to the work of adhesion, W_{AB}, was defined by Fowkes [111] as

$$W_{AB} = fN\left(-\Delta H_{AB}\right) \tag{18.3}$$

where
   N is the number of acid–base pairs available for bonding at the interface
   f is an entropy factor, specific to a particular system, which allows the conversion of enthalpy to free energy

Although originally taken as unity, f is now known to have values as low as 0.15.

The exothermic enthalpy of acid–base interaction can be predicted using Drago's equation [132]:

$$-\Delta H_{AB} = E_A E_B + C_A C_B \tag{18.4}$$

in which each Lewis acid or base is characterized by two parameters E and C, where E is related to the susceptibility of a species to undergo an electrostatic interaction, and C to the susceptibility to take part in a covalent bond. The subscripts A and B refer to acidic and basic species, respectively. Equation 18.4 can be used to predict ΔH_{AB} for a polymer adsorbed on an inorganic substrate provided that the E and C parameters for these materials are known.

The enthalpy of formation of a Lewis acid–base adduct can also be estimated using Gutmann's [133] donor and acceptor numbers, DN and AN, where DN characterizes the basicity and AN the acidity, of Lewis species. The enthalpy is given (very approximately) by the simple relationship:

$$-\Delta H_{AB} = \frac{DN \bullet AN}{100} \tag{18.5}$$

The accuracy of the Drago and Gutmann methods has been appraised critically by Jensen [134].

18.6.2 XPS CHEMICAL SHIFT AND ACID–BASE INTERACTIONS

It was pointed out by Chehimi [135], on the basis of XPS analyses by Burger and Fluck [136], that there is an excellent correlation between XPS chemical shifts and $-\Delta H_{AB}$, for a series of quick frozen solutions of $SbCl_5$ with a series of Lewis bases in dichloroethane. As Drago's E and C parameters were available for all but one of the Lewis bases and for $SbCl_5$, the thermochemical term could be evaluated using Equation 18.4. If these parametric data were plotted against the Sb $3d_{5/2}$ binding energy, referenced to the Cl 2p binding energy, the linear relationship shown in Figure 18.14 was obtained. In the original work [136], a relative XPS binding energy value (Sb $3d_{5/2}$–Cl 2p) of 322.75 eV was found for the trimethylphosphate:$SbCl_5$ adduct in dichloroethane. The values of Drago's parameters for the base were not available, but by interpolation of the data of Figure 18.14, the enthalpy of formation of this complex was estimated by Chehimi as -89.9 kJ mol^{-1}. It is this type of relationship that has been used in the determination of the acid–base properties of polymers in the solid state.

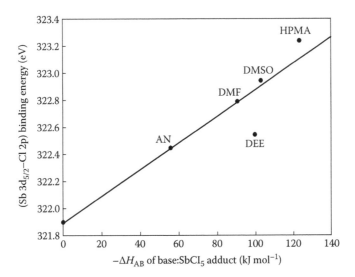

FIGURE 18.14 Correlation of XPS data with $-\Delta H_{AB}$ for $SbCl_5$:Lewis base complexes in quickly frozen solutions of dichloromethane. Key to data points: AN, acetonitrile; DEE, diethylether; DMF, dimethylformamide; DMSO, dimethyl sulfoxide; HPMA, hexamethylphosphoramide. (From Chehimi, M.M., *J. Mater. Sci. Lett.*, 10, 908, 1991. With permission.)

18.6.3   QUANTITATIVE ACID–BASE CHARACTERISTICS OF THE POLYMER

The experimental procedures have been described in detail in references [137,138]. In brief, FT-IRS was used to determine $-\Delta H_{AB}$ of polymer:solvent couples in very dilute solutions ($\sim$0.02 mol dm^{-3}) using the method of Fowkes et al. [139]. XPS measurements were made on thin films of the candidate polymers after exposing them to the organic vapor for a normalized vapor pressure time of 4 atm-min at 20°C. Such an exposure ensured that the volume of polymer probed by XPS (i.e., approximately the outer 5 nm) was saturated with organic vapor, and that the chemical shift in the XPS line characteristic of the organic vapor was a measure of the extent of any acid–base interactions. The method is illustrated in Figure 18.5 for a series of polymers exposed to trichloromethane, and it can be seen that there is indeed a linear correlation between the thermochemical term measured by FT-IRS and the XPS Cl 2p binding energy of the organic vapor used to interrogate the solid polymer. In order to use this method for unknowns, it is necessary first to establish a calibration curve of the type shown in Figure 18.15, for a particular solvent using a series of homopolymers, and then the magnitude of $-\Delta H_{AB}$ of the interaction between that solvent and a more complex polymeric system, such as a fully formulated coating or adhesive, can be interpolated in the manner illustrated for trimethylphosphate from the data of Figure 18.14.

Clearly, $-\Delta H_{AB}$, as determined in the manner described above for a polymer with a particular solvent, is not an intrinsic property of the polymer. To obtain an intrinsic ΔH_{AB}, it is necessary to record data for the candidate polymer with two solvents of known acid–base characteristics. The values of Drago's E and C parameters for the polymer can then be found by evaluation of Equation 18.4 for the two acid–base adducts (the polymer with solvent 1 [E_{A1} and C_{A1}] and with solvent 2 [E_{A2} and C_{A2}]), which provide the thermochemical terms $-\Delta H_{AB1}$ and $-\Delta H_{AB2}$, respectively. The most convenient method of obtaining the solution is the graphical method of Fowkes [128], illustrated in Figure 18.16.

With orthogonal axes representing the E and C parameters, and by rearranging Equation 18.2, it is possible to define two lines that have slopes of C_{A1}/E_{A1} and C_{A2}/E_{A2}, and intercepts with the ordinate of $-\Delta H_{AB1}/E_{A1}$ and $-\Delta H_{AB2}/E_{A2}$, for the respective interactions of the polymer with solvent 1 and solvent 2. The intercepts of the two lines provide numerical values of Drago's parameters for the polymer under consideration. For the hypothetical data of Figure 18.16, the polymer characteristics are $E_B = 0.30$ and $C_B = 4.35$ (kcal mol^{-1})$^{0.5}$. In this manner, it is possible to estimate Drago's parameters for systems that are not amenable to the more straightforward FT-IRS method. Virtually, all commercial products fall into this category, since segregation or depletion of minor components are

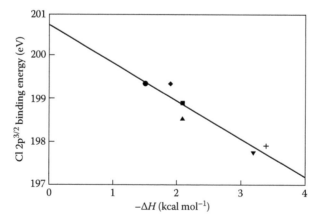

FIGURE 18.15   Correlation of Cl 2p$_{3/2}$ binding energy with $-\Delta H_{AB}$ for trichloromethane absorbed in a series of polymers. Key to data points: +, aromatic moisture cured urethane; •, poly(vinyl acetate); ▲, poly(methyl methacrylate); ▼, poly(ethylene oxide); ♦, poly(n-butyl methacrylate); ■, poly(cyclohexyl methacrylate). (From Chehimi, M.M., Watts, J.F., Castle, J.E., and Jenkins, S.N., *J. Mater. Chem.*, 2, 209, 1992. With permission.)

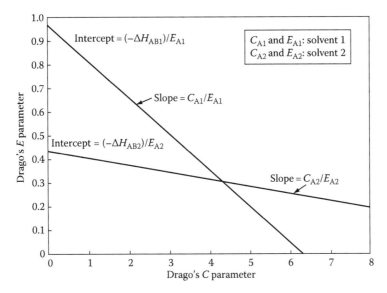

FIGURE 18.16 Graphical method for the estimation of the E and C parameters of a polymer using two solvents of known E and C values, and a knowledge of the exothermic acid–base interaction between the polymer and solvent 1 ($-\Delta H_{AB1}$) and solvent 2 ($-\Delta H_{AB2}$). (From Fowkes, F.W., *J. Adhes. Sci. Tech.*, 1, 2, 1987. With permission.)

known to occur at both the inorganic/organic interface and the free surface. Such phenomena make the use of a bulk solution method such as FT-IRS uncertain for the elucidation of acid–base properties.

Other XPS approaches have been used to identify and rank such acid–base properties of polymers. Watts and Chehimi sought to develop a solid-state titration method in which candidate polymers were cast onto soda-lime glass, the extent of uptake of the sodium ion, a very mobile Lewis acid, being indicative of the basicity of the polymer [140]. The same authors employed angular-resolved XPS to study the orientation of the carbonyl group of poly(methyl methacrylate), which is basic, when the polymer was cast onto acidic or basic substrates [141]. A pronounced orientation of the functional group on the acidic substrate was found. The study was extended using monochromatic XPS and the thin-film approach of Leadley and Watts, described in Section 18.4; subtle changes in the C 1s spectrum were observed, as shown in Figure 18.17 [142], and ascribed to the nature of the interaction between polymethyl methacrylate (PMMA) and metal oxide surfaces. Schematics of the type of interactions proposed, on the basis of the high-resolution XPS data, are shown in Figure 18.18 [142].

The thin-film approach has been shown to be very informative when dilute solutions of commercial polymer formulations are applied to metallic substrates, as the segregation of minor components can be observed readily [45,63]. The adsorption of amine molecules, as analogues of an epoxy adhesive, was studied by both XPS and SIMS [143–146], and the complementary nature of the two techniques was demonstrated once again. The existence of a strong donor–acceptor interaction with anodized aluminum was postulated in the later work, and the integrity of the adsorption process appeared to be related to the existence and density of Brönsted sites on the Al_2O_3 surface [145].

The work described above established XPS firmly as a flexible method for the evaluation of the acid–base properties of homopolymers, but the technologically more important advantage of XPS is its ability to analyze thin modified or segregated layers not amenable to the traditional forms of analysis [147]. The molecular probe technique was used to good effect by Shahidzadeh and coworkers in the study of the plasma treatment of poly(propylene) film [148–150]. Their work identified the need to select basic probe molecules for the assessment of acidic surfaces and dimethyl sulfoxide was shown to be a good choice, although the photoelectron cross-section for sulfur is low, as it is for

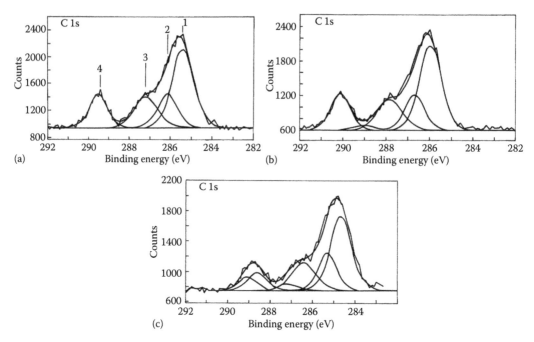

FIGURE 18.17 C 1s XPS spectra of thin films (~2 nm) of PMMA on oxidized substrates of increasing basic character: (a) silicon (acidic), (b) aluminum (weakly basic), and (c) nickel (strongly basic). The numbered components in (a) refer to the schematic of Figure 18.18. Additional components in (b) and (c) are related to the specific interactions shown in Figure 18.18. (From Leadley, S.R. and Watts, J.F., *J. Adhes.*, 60, 175, 1997. With permission.)

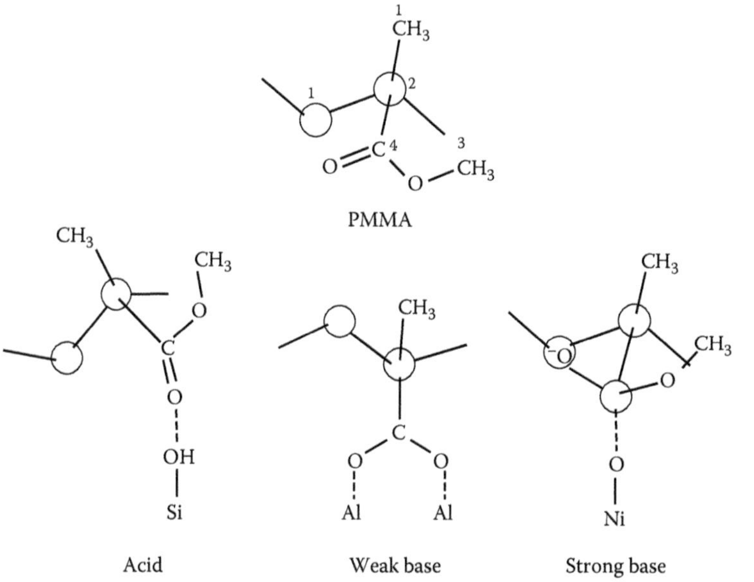

FIGURE 18.18 Schematic representations of the types of bonding thought to give rise to the spectra of Figure 18.17. The acid surface is oxidized silicon (cf. Figure 18.17a), the weak base oxidized aluminum (cf. Figure 18.17b), and the strong base oxidized nickel (cf. Figure 18.17c). (From Leadley, S.R. and Watts, J.F., *J. Adhes.*, 60, 175, 1997. With permission.)

the Cl 2p core level used for trichloromethane, which puts an effective limit on the detectability of the molecular probe, that is, the number (but not the strength) of the acid–base pairs detectable.

18.6.4   ACID–BASE PROPERTIES OF INORGANIC SURFACES

In this area, the chemical specificity of XPS again provides a powerful route to the determination of the acid–base properties of surfaces. Indeed, there appears to be a direct linear correlation between certain XPS data and the isoelectric point of the solid surface (IEPS), as pointed out by Delamar [151]. By summing the separation between the energy of the photoelectron peak from oxidized metal and that of the accepted value from the metal, with the separation of the energy of the O 1s peak from that of an arbitrarily defined binding energy, and then plotting the summation against IEPS, a linear relationship was found. The Brönsted characteristics of metal surfaces were studied by Watts and Gibson [152], who extended the cation exchange method of Simmons and Beard [153] to include both anion and cation exchanges. The method is based on the fact that, in the presence of an aqueous medium, the hydroxyl groups on a hydrated metal surface may act as either acids or bases to produce cationic (MOH_2^+) or anionic (MOH^-) complexes, where M is the metal under consideration [152,153]. The concentration of each of these species can then be determined by using a cationic (K^+) or anionic (PO_4^{2-}) exchange process. By carrying out the exchanges at a series of solution pH values, the acid equilibrium constants for the formation of the metal complexes can be determined. The IEPS is the value of solution pH at which the surface concentration of cationic complex ions (MOH_2^+) is equal to the concentration of anionic species (MOH^-), and it can be determined conveniently as the mean of the two constants determined by the ion exchanges procedures described above. The method has been criticized by Delamar [154], who suggested possible improvements to obtain data of greater reliability. Notwithstanding the shortcomings, the method has been used to assess the acid–base properties of oxidized titanium and iron surfaces under both idealized [153] and real situations [152]. In a detailed investigation, Kurbatov et al. [155,156] showed how slight changes in the hydroxylated surface layer present on iron surfaces could be effected by surface preparation protocols.

18.6.5   CONCLUDING REMARKS

Application of the concepts of acceptor–donor interaction has been investigated fairly thoroughly in adhesion science. The foundations have been laid that will allow XPS and ToF-SIMS to play important roles in such investigations, and they are now important weapons in the armory of the surface analyst who carries out research in the adhesion arena. Although information concerning the acid–base properties of polymer and inorganic surfaces is generally difficult to obtain, it is clear that the goal of achieving a predictive approach to adhesion and to the hydrolytic stability of adhesive joints and coated substrates is now much closer than hitherto. The opportunity to reverse-engineer interface chemistry, be it by the design of specific substrate pretreatments or by the incorporation of chosen functional groups in the polymer formulation, is now at hand. Indeed, surface analysis can contribute at all stages of our understanding of both the adhesion and the failure process. It is possible to predict the nature of interfacial chemistry in the manner described above, examine it directly using the methodologies of Section 18.4, and then examine the locus of failure in some detail as described in Section 18.3. It is possible to say that the time is at hand when a cradle-to-grave approach to adhesion studies can be adopted.

18.7   COMPUTER CHEMISTRY AND MOLECULAR MODELING

Although not strictly part of a chapter dealing with experimental surface analysis techniques for solving adhesion and related problems, discussion of the use of computer techniques for the modeling of adhesion phenomena is included, as it is an important theoretical approach, which is often used in conjunction with surface analysis. Although computer chemistry methods based on quantum mechanics, for example, self-consistent field (SCF) molecular orbital calculations, have been available for many years, they have not been used by the adhesion community, and it was the development of molecular simulation packages running on workstations in the 1990s that provided the impetus in the applied fields of adhesion and catalysis. The difference between the two approaches is well known but is worth repeating at this juncture. In the SCF approach, *ab initio* calculations are made and all electron–electron interactions, as well as overlap integrals, are calculated. In order to simplify the computational process, an approximation is often made in which only the valence electrons are considered, that is, those electrons actually taking part in chemical interactions. A popular SCF method that employs this valence shell approximation is the complete-neglect-of-differential-overlap approach, in which certain other electron–electron interactions are also neglected. The use of these methods in adhesion has been reviewed by Cain [157], and an elegant example of the application of SCF calculations to adhesion is the study of acrylate and methacrylate esters on aluminum oxide by Holubka et al. [158]. The mechanism of the interaction of dicyandiamide (a widely used epoxy curing agent) has also been reported [159].

Molecular simulation methods, on the other hand, use semiempirical quantum chemical calculations and do not estimate the probability of chemical bond formation. They use essentially molecular dynamics, based on the integration of Newton's laws of motion, over many (several million) very short (femtosecond) time increments. The molecules under study are represented by spheres of known mass interconnected elastically, and are defined by a force field that contains data relevant to the molecular structure such as atomic masses, bond lengths, bond angles, etc. Other factors that affect molecular motion such as temperature and secondary bonds must also be included. The first step is optimization of the geometry of the molecular structures by calculating the total energy as described by the appropriate energy expression. Molecular motion, for instance of a molecule relative to a solid surface, can then be calculated for all the atomic masses present by imparting an initial velocity to them and evaluating the response after many time increments. Further information is available in standard texts on computational chemistry such as Ref. [160].

Molecular dynamics have been used by Sennett et al. to study the adsorption of epoxy analogues and of silanes on alumina and hematite surfaces [161,162], while Kinloch and Hobbs have investigated the effect of chain length on adsorption characteristics and related this to bond performance [163]. Taylor et al. have shown that the failure of a photocured resin, described in Section 18.3.6 [84–86], in which 1–2 nm of aliphatic reactive diluent remained on the inorganic substrate, was consistent with a monolayer of such material at the interface. The orientation of the molecule was 30° to 40° to the solid surface, which indicated a monolayer coverage of the same dimensions as that determined by angular-resolved XPS [85,86]. The limitations of molecular dynamics, so far as the application to adhesion is concerned, are associated with establishing the initial location of the molecule on the solid substrate, but can be overcome by using an adsorption routine in the manner described by Davis and Watts in a study of silane adsorption [107]. Using a commercial molecular dynamics suite of programs,* they studied the initial approach of the molecule to the inorganic surface to determine the local energy minimum, which was then used as input data for molecular dynamics. The trajectory file associated with simulation was interrogated, and the structure with the lowest total energy frame extracted. In this manner, it was possible to study the adsorption characteristics of silane molecules on an inorganic substrate without the uncertainty regarding the initial interaction site. The adsorption of three organosilanes was investigated, and the lowest energy frame for the three molecules on a periodic FeOOH lattice is shown in Figure 18.19 [49]. Two of these studies repeated the work of Bailey and Castle [104], and it is gratifying that the data for vinyl triethoxy silane (Figure 18.19a) were in complete accord with those from XPS. The result for amino propyl triethoxysilane (APS) was particularly informative in that the XPS data showed no preference

---

* MSI Cerius Sorption software.

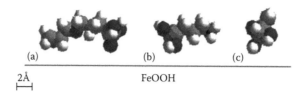

(a)   (b)   (c)

2Å   FeOOH

FIGURE 18.19 (See color insert following page 396.) Molecular dynamics simulations of the positional arrangement of fully hydrolyzed organosilane molecules on an FeOOH substrate, obtained using the Cerius^2 Sorption package: (a) GPS, in which the silane head is at the right-hand end of the molecule, (b) γ-APS, in which the silane head is at the left-hand end of the molecule, and (c) vinyl triethoxysilane, in which the silane head of the molecule is adjacent to the FeOOH surface. The scale bar is 0.2 nm. The diagram was produced using the original (color) computer graphics files obtained by Davis. (From Tatoulian, M., Arefi-Khonsari, F., Shahidzadeh-Ahmadi, N., and Amoroux, J., *Int. J. Adhes. Adhes.*, 15, 177, 1995.)

for adsorption via either functional group, and this conclusion was confirmed and clarified by molecular dynamics, which indicated that either the amino or the silane ends of the molecule may approach the substrate, resulting in a parallel orientation for the molecule on the surface (Figure 18.19b). For glycidoxy-propyl-trimethoxysilane, the situation is known to be more complicated, since XPS and ToF-SIMS data showed that the overlayer was some 1.7 nm thick, implying a complex overlayer. The molecular dynamics simulation of Figure 18.19a shows that either end group may interact with the substrate, as for APS, but that the layer is far from parallel. These examples indicate clearly that although the molecular simulation routines have an important role to play in adhesion science, they must not be relied upon in isolation to provide unambiguous results. As is often the case, calculations must be supported by experimental data. The attraction of the molecular simulation approach may lie more in its use in screening candidate materials for a comprehensive research program.

18.8   PROSPECTS FOR THE FUTURE

In this chapter, the development of the application of surface analytical techniques to adhesion research has been described. Nowadays, the use of XPS and ToF-SIMS for the forensic analysis of interfacial failure surfaces is relatively straightforward, and may provide direct information regarding the actual mechanism of adhesion. There has always been a tacit assumption that adhesion phenomena show very little lateral variation, which is probably an oversimplification. Adhesion scientists can learn something from corrosion investigations in this connection; SAM has been shown to be a powerful tool in understanding the breakdown of passivity around inclusions [164], and if used intelligently may well help in understanding adhesion mechanisms too. For example, the interactions of organic systems with heterogeneities in the microstructure may provide preferential sites for the initiation of failure. As both XPS and ToF-SIMS can now be performed with good spatial resolution, the ability to assess the uniformity of a failure surface is achievable. This is particularly important in the case of composites with fiber and other reinforcements, but it also allows the identification of localized effects such as electrochemistry on the micrometer scale.

ToF-SIMS has had a profound impact on the manner in which interface analysis is carried out, and enables the identification of specific components in a commercial formulation (not readily achievable by XPS unless there is a well-defined chemical shift or elemental marker unique to each component). The time for the use of model compounds in adhesion research has probably now passed, and collaborative research between the adhesion scientist, who is also a surface analyst, with industrial chemists responsible for the formulation of coatings, adhesives, and sealants must be the way forward for the further understanding of adhesion phenomena and the development of new products. Needless to say, this is a philosophy the author embraces enthusiastically.

XPS, ToF-SIMS, and AES are now mature analytical techniques with a body of understanding and standard operating procedures (from the likes of TC201) [165] in the open literature. In addition, and within a similar timescale, these methods have been developed in individual ways to allow for their use specifically in adhesion science investigations; in effect the two have matured together. Given the ready availability of surface analysis in universities and other analytical laboratories worldwide, it seems that the two activities are inextricably entwined for some time to come.

ACKNOWLEDGMENTS

It is a pleasure to thank those involved in adhesion research at the University of Surrey over the last two decades. In particular, the contributions of Professor Jim Castle and Dr. Marie-Laure Abel have been pivotal in many of the ideas developed in this chapter.

REFERENCES

1. J F Watts, The use of XPS for the analysis of organic coating systems, in *Organic Coatings 1*, Eds. A Wilson, H Prosser, and J W Nicholson, Applied Science Publishers, London, United Kingdom, pp. 137–187, 1987.
2. J F Watts, The interfacial chemistry of adhesion: Novel routes to the holy grail? in *Adhesion—Current Research & Applications*, Ed. W Possart, Wiley-VCH Verlag GmbH, Germany, pp. 1–16, 2005.
3. V Hospadaruk, J Huff, R W Zurilla, and H T Greenwood, SAE Technical Paper Series 780186 (1978).
4. R A Iezzi and H Leidheiser, Jr., *Corrosion*, **37**, 28 (1981).
5. B M Parker and R Waghorne, *Composites*, **13**, 280 (1982).
6. A J Kinloch, G K A Kadokian, and J F Watts, *Phil Trans Roy Soc*, **A338**, 83 (1992).
7. M W Roberts and P R Wood, *J Electron Spec*, **11**, 431 (1977).
8. J E Castle, in *Corrosion Control by Coatings*, Ed. H Leidheiser, Jr., Science Press, Princeton, NJ, pp. 435–454, 1979.
9. A J Kinloch and N R Smart, *J Adhesion*, **12**, 23 (1981).
10. P Poole and J F Watts, *Int J Adhes Adhes*, **5**, 33 (1985).
11. J S Solomon, D Hanlin, and N T McDevitt, Adhesive-adherend bond joint characterization by Auger electron spectroscopy and x-ray photoelectron spectroscopy, in *Adhesion and Adsorption of Polymers Part A*, Ed. L -H Lee, Plenum Publishing Corp., New York, pp. 103–122, 1980.
12. J E Castle and J F Watts, The study of interfaces in composite materials by surface analytical techniques, in *Interfaces in Polymer, Ceramic and Metal Matrix Composites*, Ed. H Ishida, Elsevier Science, pp. 57–71, 1988.
13. C A Baillie, J F Watts, and J E Castle, *J Mater Chem*, **2**, 939 (1992).
14. M R Alexander and F R Jones, *Surf Interface Anal*, **22**, 230 (1994).
15. M R Alexander and F R Jones, *Carbon*, **33**, 569 (1995).
16. M R Alexander and F R Jones, *Carbon*, **34**, 1093 (1996).
17. P E Vickers, J F Watts, C Perruchot, and M M Chehimi, *Carbon*, **38**, 675 (2000).
18. B Lindsay, M -L Abel, and J F Watts, *Carbon*, **45**, 2433 (2007).
19. P A Zhdan, D Grey, and J E Castle, *Surf Interface Anal*, **22**, 290 (1994).
20. P E Vickers, M E Turner, M -L Abel, and J F Watts, *Composites*, **29A**, 1291 (1998).
21. W J van Ooij, *Surf Sci*, **68**, 1 (1977).
22. W J van Ooij, *Surf Tech*, **6**, 1 (1977).
23. W J van Ooij, *Rubber Chem Tech*, **52**, 605 (1979).
24. W J van Ooij, *Rubber Chem Tech*, **57**, 421 (1984).
25. W J van Ooij and M E F Biemond, *Rubber Chem Tech*, **57**, 686 (1984).
26. J Giridhar, W J van Ooij, and J H Ahn, *Kautsch Gummi Kunstst*, **44**, 348 (1991).
27. J Giridhar and W J van Ooij, *Surf Coat Tech*, **52**, 17 (1992).
28. J Giridhar and W J van Ooij, *Surf Coat Tech*, **53**, 35 (1992).
29. W J van Ooij and M Nahmias, *Rubber Chem Tech*, **62**, 656 (1989).
30. J E Castle, Z B Luklinska, and M S Parvizi, *J Mater Sci*, **19**, 3217 (1984).
31. J E Castle, D C Epler, and D Peplow, *Corr Sci*, **16**, 145 (1976).

32. J E Castle and J F Watts, Cathodic disbondment of well characterized steel/coating interfaces, in *Corrosion Control by Organic Coatings*, Ed. H Leidheiser, NACE, Houston, TX, pp. 78–86, 1981.
33. J F Watts and J E Castle, *J Mater Sci*, **18**, 2987 (1983).
34. R A Dickie, J S Hammond, and J W Holubka, *Ind Eng Chem Prod Res Dev*, **20**, 339 (1981).
35. J W Holubka, J S Hammond, J E deVries, and R A Dickie, *J Coat Tech*, **52**, 63 (1980).
36. J S Hammond, J W Holubka, J E deVries, and R A Dickie, *Corr Sci*, **21**, 239 (1981).
37. J E Castle, J F Watts, P J Mills, and S A Heinrich, The effect of solution composition on the interfacial chemistry of cathodic disbondment, in *Corrosion Protection by Organic Coatings*, Eds. M W Kendig and H Leidheiser, Jr., The Electrochemical Society, Pennington, NJ, pp. 68–83, 1987.
38. J E Castle, *Organic Coatings: 53rd International Meeting of Physical Chemistry*, Ed. P -C Lacaze, AIP Conference Proceedings 354, AIP Press, Woodbury, New York, pp. 432–447 (1995).
39. X H Jin, K C Tsay, A Elbasir, and J D Scantlebury, The adhesion and bisbanding of chlorinated rubber on mild steel, in *Corrosion Protection by Organic Coatings*, Eds. M W Kendig and H Leidheiser, Jr., The Electrochemical Society, Pennington, NJ, pp. 37–47, 1987.
40. R Cayless and D L Perry, *J Adhes*, **26**, 113 (1988).
41. J F Watts, R A Blunden, and T J Hall, *Surf Interface Anal*, **16**, 227 (1990).
42. J F Watts, J E Castle, and S J Ludlam, *J Mater Sci*, **21**, 2965 (1986).
43. T R Roberts, J Kolts, and J H Steele, SAE Technical Paper Series 800443 (1980).
44. J F Watts, *J Mater Sci*, **19**, 3459 (1984).
45. M Murase and J F Watts, *J Mater Chem*, **8**, 1007 (1998).
46. M Gettings, F S Baker, and A J Kinloch, *J Appl Polym Sci*, **21**, 2375 (1977).
47. R A Cayless and L B Hazell, European Patent 0331 284 A1 (1989).
48. S J Davis, J F Watts, and L B Hazell, *Surf Interface Anal*, **21**, 460 (1994).
49. S J Davis, PhD thesis, The Deposition and Organisation of Inorganic and Organic Adhesion Promoters on Iron Surfaces, University of Surrey, 1995.
50. J F Watts and J Wolstenholme, *An Introduction to Surface Analysis by XPS and AES*, John Wiley & Sons Ltd., Chichester, United Kingdom, 2003.
51. S J Davis and J F Watts, *J Mater Chem*, **6**, 479 (1996).
52. L P Haack, M A Bolt, S L Kaberline, J E deVries, and R A Dickie, *Surf Interface Anal*, **20**, 115 (1993).
53. W J van Ooij, A Sabata, and A D Appelhans, *Surf Interface Anal*, **17**, 403 (1991).
54. A M Taylor, J F Watts, H Duncan, and I W Fletcher, *J Adhes*, **46**, 145 (1994).
55. M Gettings and A J Kinloch, *Surf Interface Anal*, **1**, 165 (1979).
56. M Gettings and A J Kinloch, *Surf Interface Anal*, **1**, 189 (1979).
57. J E deVries, J W Holubka, and R A Dickie, *J Adhes Sci Tech*, **3**, 189 (1989).
58. J E deVries, L P Haack, J W Holubka, and R A Dickie, *J Adhes Sci Tech*, **3**, 203 (1989).
59. Y Yoshikawa and J F Watts, *Surf Interface Anal*, **20**, 379 (1993).
60. J F Watts and J E Castle, *J Mater Sci*, **19**, 2259 (1984).
61. M F Fitzpatrick, J S G Ling, and J F Watts, *Surf Interface Anal*, **29**, 131 (2000).
62. M F Fitzpatrick and J F Watts, *Surf Interface Anal*, **27**, 705 (1999).
63. S R Leadley, J F Watts, A Rodriguez, and C Lowe, *Int J Adhes Adhes*, **18**, 193 (1998).
64. J D Venables, *J Mater Sci*, **19**, 2431 (1984).
65. G D Davis, Characterization of surfaces, in *Adhesive Bonding*, Ed. L -H Lee, Plenum Press, New York, pp. 139–173, 1991.
66. I Olefjord and L Kozma, *Mater Sci Tech*, **3**, 860 (1987).
67. H M Clearfield, D K McNamara, and G D Davis, Adherend surface preparation for structural adhesive bonding, in *Adhesive Bonding*, Ed. L -H Lee, Plenum Press, New York, pp. 203–237, 1991.
68. G D Davis, P L Whisnant, and J D Venables, *J Adhes Sci Tech*, **9**, 433 (1995).
69. J F Watts, J E Castle, and T J Hall, *J Mater Sci Lett*, **7**, 176 (1988).
70. J P Sargent, *Int J Adhes Adhes*, **14**, 21 (1994).
71. A J Kinloch, M S G Little, and J F Watts, *Acter Mater*, **48**, 4543 (2000).
72. J F Watts, D Bland, A J Kinloch, and V Stolojan, *Surf Interface Anal*, in press.
73. C Cazeneuve, J E Castle, and J F Watts, *Interfacial Phenomena in Composite Materials '89*, Ed. F R Jones, Butterworth and Co. Ltd., pp. 88–96, 1989.
74. C Cazeneuve, J F Watts, and J E Castle, *J Mater Sci*, **25**, 1902 (1990).
75. F R Jones, P Denison, A Brown, P Humphrey, and J Harvey, *J Mater Sci*, **23**, 2153 (1988).
76. M J Hearn and D Briggs, *Surf Interface Anal*, **17**, 421 (1991).
77. S Yumitori, D Wang, and F R Jones, *Composites*, **25**, 698 (1994).
78. P E Vickers, L Boniface, A C Prickett, and J F Watts, *Composites*, **A31**, 559 (2000).
79. A C Prickett, P A Smith, and J F Watts, *Surf Interface Anal*, **31**, 11 (2001).

80. A R Wood, N Benedetto, N Hooker, E Scullion, P A Smith, and J F Watts, *Surf Interface Anal*, submitted.
81. L P Buchwalter, *J Adhes Sci Tech*, **1**, 341 (1987).
82. L P Buchwalter, *J Adhes Sci Tech*, **4**, 697 (1990).
83. T S Oh, L P Buchwalter, and J Kim, *J Adhes Sci Tech*, **4**, 303 (1990).
84. J F Watts and A M Taylor, *J Adhes*, **46**, 161 (1994).
85. A M Taylor, J F Watts, J Bromley-Barratt, and G Beamson, *Surf Interface Anal*, **21**, 697 (1994).
86. A M Taylor, C H McLean, M Charlton, and J F Watts, *Surf Interface Anal*, **23**, 342 (1995).
87. J E Castle and J F Watts, *Adv Mater J*, **1**, 16 (1990).
88. W J van Ooij, T H Tisser, and M E F Biedmond, *Surf Interface Anal*, **6**, 197 (1984).
89. W J van Ooij, R A Edwards, A Sabata, and J Zappia, *J Adhes Sci Tech*, **7**, 897 (1993).
90. A N MacInnes, PhD thesis, The Interlayer Formed between Iron and Acrylic Latex, University of Surrey, 1990.
91. H Leidheiser, S Music, and G W Simmons, *Nature*, **297**, 667 (1982).
92. J Liu, PhD thesis, The Adhesion Characteristics of a Laminated Aluminium/Polyester System, University of Surrey, 1993.
93 S J Hinder, C Lowe, J T Maxted, and J F Watts, *J Mater Sci*, **40**, 285 (2005).
94. S J Hinder and J F Watts, *Surf Interface Anal*, **36**, 1032 (2004).
95. S J Hinder, C Lowe, J T Maxted, and J F Watts, *Surf Interface Anal*, **36**, 1575 (2004).
96. J M Burkstrand, *Appl Phys Lett*, **33**, 387 (1978).
97. J M Burkstrand, *J Vac Sci Technol*, **20**, 440 (1980).
98. F S Ohuchi and S C Freilich, *J Vac Sci Technol*, **A4**, 1039 (1986).
99. P Stoyanov, S Akhter, and J M White, *Surf Interface Anal*, **15**, 509 (1990).
100. M M Chehimi and J F Watts, *J Adhes Sci Tech*, **6**, 377 (1992).
101. S R Leadley and J F Watts, *J Electron Spec*, submitted.
102. E P Plueddeman, *Silane Coupling Agents*, Plenum Press, New York, 1992.
103. K L Mittal, *Silane and Other Coupling Agents*, VSP, Zeist, the Netherlands, 1992.
104. R Bailey and J E Castle, *J Mater Sci*, **12**, 2049 (1977).
105. M-L Abel, M M Chehimi, A M Brown, S R Leadley, and J F Watts, *J Mater Chem*, **5**, 845 (1995).
106. M Gettings and A J Kinloch, *J Mater Sci*, **12**, 2511 (1977).
107. S J Davis and J F Watts, *Int J Adhes Adhes*, **16**, 5 (1996).
108. M-L Abel, I W Fletcher, R P Digby, and J F Watts, *Surf Interface Anal*, **29**, 115 (2000).
109. M-L Abel, A Rattana, and J F Watts, *Langmuir*, **16**, 6510 (2000).
110. A Rattana, M-L Abel, and J F Watts, *Int J Adhes Adhes*, **26**, 28 (2006).
111. A Rattana, M-L Abel, and J F Watts, *J Adhes*, **81**, 963 (2005).
112. A Rattana, J D Hermes, M-L Abel, and J F Watts, *Int J Adhes Adhes*, **22**, 205 (2002).
113. J F Watts, S R Leadley, J E Castle, and C J Blomfield, *Langmuir*, **16**, 2292 (2000).
114. M-L Abel, J F Watts, and R Digby, *J Adhes*, **80**, 291 (2004).
115. W J van Ooij and A Sabata, in *Polymer/Inorganic Interfaces*, Eds. R L Opila, F J Boerio, and A W Czandera, Materials Research Society, Pittsburgh, PA, pp. 155–160, 1993.
116. S G Hong and F J Boerio, *Surf Interface Anal*, **21**, 650 (1994).
117. F J Boerio, C A Gosselin, R G Dillingham, and H W Liu, *J Adhes*, **13**, 159 (1981).
118. D J Ondrus and F J Boerio, *J Colloid Interface Sci*, **124**, 349 (1988).
119. R Chen and F J Boerio, *J Adhes Sci Tech*, **4**, 453 (1990).
120. J T Young and F J Boerio, *Surf Interface Anal*, **20**, 341 (1993).
121. Y M Tsai, F J Boerio, W J van Ooij, D K Kim, and T Rau, *Surf Interface Anal*, **23**, 261 (1995).
122. K Yates and R H West, *Surf Interface Anal*, **5**, 113 (1983).
123. N G Cave, A J Kinloch, S C Mugford, and J F Watts, *Surface Interface Anal*, **17**, 120 (1991).
124. D Wang, F R Jones, and P Denison, *J Mater Sci*, **27**, 36 (1992).
125. D Wang and F R Jones, *Surf Interface Anal*, **20**, 457 (1993).
126. D Wang and F R Jones, *Composites*, **50**, 215 (1994).
127. A Zettlemoyer, in *Interface Conversion for Polymer Coatings*, Eds. P Weiss and C D Cheever, Elsevier, New York, pp. 208–237, 1968.
128. F W Fowkes, *J Adhes Sci Tech*, **1**, 2 (1987).
129. K L Mittal and H R Anderson (Eds.), *Acid–Base Interactions: Relevance to Adhesion Science and Technology*, VSP, Utrecht, the Netherlands, 1991.
130. M -L Abel and J F Watts, Inverse gas chromatography, in *Handbook of Adhesion*, 2nd edition, Ed. D E Packham, John Wiley & Sons Ltd., Chichester, United Kingdom, pp. 252–254, 2005.

131. M -L Abel and J F Watts, Inverse gas chromatography and acid–base interactions, in *Handbook of Adhesion*, 2nd edition, Ed. D E Packham, John Wiley & Sons Ltd, Chichester, United Kingdom, pp. 255–257, 2005.
132. R S Drago, G C Vogel, and T E Needham, *J Am Chem Soc*, **93**, 6014 (1971).
133. V Gutmann, *The Donor–Acceptor Approach to Molecular Interactions*, Plenum Press, New York, 1978.
134. W B Jensen, *J Adhes Sci Tech*, **5**, 1 (1991).
135. M M Chehimi, *J Mater Sci Lett*, **10,** 908 (1991).
136. K Burger and E Fluck, *Inorg Nucl Chem Lett*, **10**, 171 (1974).
137. M M Chehimi, J F Watts, J E Castle, and S N Jenkins, *J Mater Chem*, **2**, 209 (1992).
138. M M Chehimi, J F Watts, W K Eldred, K Fraoua, and M Simon, *J Mater Chem*, **4**, 305 (1994).
139. F M Fowkes, D O Tischler, J A Wolfe, L A Lannigan, C M Ademu-John, and M J Halliwell, *J Polym Sci, Polym Chem Edn*, **22**, 547 (1984).
140. M M Chehimi and J F Watts, *J Adhes*, **41**, 81 (1993).
141. M M Chehimi and J F Watts, *J Electron Spec Relat Phenom*, **63**, 393 (1993).
142. S R Leadley and J F Watts, *J Adhes*, **60**, 175 (1997).
143. S Affrossman, N M D Brown, R A Pethrick, V K Sharma, and R J Rurner, *Appl Surf Sci*, **16**, 469 (1983).
144. S Affrossman and S M MacDonald, *Langmuir*, **10**, 2257 (1994).
145. C Fauquet, P Dubot, L Minel, M-G Barthes-Labtousse, M Rei Vilar, and M Villante, *Appl Surf Sci*, **81**, 435 (1994).
146. S Affrossman and S M MacDonald, *Langmuir*, **12**, 2090 (1996).
147. M M Chehimi, Acid–base properties of polymers in the solid state, in *Handbook of Advanced Materials Testing*, Eds. N P Cheremisinoff and P N Cheremisinoff, Marcel Dekker, New York, Chapter 33, 1996.
148. N Shahidzadeh, PhD thesis, Corrélation entre les propriétés acido-basiques des film de polypropyléne traits par plasma hors équilibre et les mécanismes dádhésion, Université Pierre et Marie Curie Paris VI, 1996.
149. M Tatoulian, F Arefi-Khonsari, N Shahidzadeh-Ahmadi, and J Amoroux, *Int J Adhes Adhes*, **15**, 177 (1995).
150. N Shahidzadeh-Ahmadi, M M Chehimi, F Arefi-Khonsari, J Amoroux, and M Delamar, *Plasmas Polym*, **1**, 27 (1996).
151. M Delamar, *J Electron Spec*, **53**, c11 (1990).
152. J F Watts and E M Gibson, *Int J Adhes Adhes*, **11**, 105 (1991).
153. G W Simmons and B C Beard, *J Phys Chem*, **91**, 1143 (1987).
154. M Delamar, *J Electron Spec*, **67**, R1 (1994).
155. G Kurbatov, E Darque-Ceretti, and M Aucouturier, *Surf Interface Anal*, **18**, 811 (1992).
156. G Kurbatov, E Darque-Ceretti, and M Aucouturier, *Surf Interface Anal*, **20**, 402 (1993).
157. S R Cain, *J Adhes Sci Tech*, **4**, 333 (1990).
158. J W Holubka, R A Dickie, and J C Cassatta, *J Adhes Sci Tech*, **6**, 243 (1992).
159. J W Holubka and J C Ball, *J Adhes Sci Tech*, **4**, 443 (1990).
160. K B Lipkowitz and D B Boyd (Eds.), *Reviews in Computational Chemistry*, VCH Publishers Inc., New York, 1990.
161. M S Sennett, W X Zukas, and S E Wentworth, *Comp Polym Sci*, **2**, 124 (1992).
162. M S Sennett, S E Wentworth, and A J Kinloch, *J Adhes*, **54**, 23 (1995).
163. P M Hobbs and A J Kinloch, *J Adhes*, **66**, 1 (1998).
164. J E Castle, L Sun, and H Yan, *Corrosion Sci*, **36**, 1093 (1994).

19 Electron Spectroscopy in Corrosion Science

James E. Castle

CONTENTS

19.1 INTRODUCTION

There are two branches of the topic generally known as "corrosion." Firstly there is "failure analysis," i.e., the forensic studies that reveal the possible causes of a failure, and secondly there is "corrosion science," i.e., the study of metals with the ultimate purpose of developing alloys with a better corrosion performance under given conditions. The two fields are strongly interrelated since the forensic study provides an understanding of the conditions that new alloys must tolerate, and the scientific study provides the understanding that relates particular symptoms of corrosion to a probable mechanism of failure.

In both these fields, it is necessary to think carefully of the relevance of using a surface-sensitive technique, such as electron spectroscopy. In most cases, the users of metallic alloys are not concerned with corrosion thicknesses measured in micrometers—generally it is only when wastage reaches thicknesses of tens of micrometers, that concern arises, so what help is information on thickness scales of nanometers? Firstly, it should be recalled that the oxide thickness that keeps bright metals bright is only some 2 nm—perfect for characterization by electron spectroscopy; secondly, that inhibitors, or alloying components, that act to prevent the corrosion rate from becoming high enough to cause damage, may do so at submonolayer coverage, again excellent for electron spectroscopic analysis.

In studying corrosion it is above all else the interaction between a surface (usually of a metal) and its environment that is being studied. The first level of interaction is that of adsorption of a monolayer of oxidant on the clean metal surface, but in normal ambient environments this takes place so quickly as to be immaterial to the kinetics of corrosion. Indeed the first 2–3 nm of reaction product form in fractions of a second so that, unless the reaction can be slowed down by use of low-pressure, near-vacuum environments, it also is too fast to study by electron spectroscopy. However, once these layers are formed then the rate of change at ambient temperatures becomes very slow, and this so-called passivated surface is suitable for study by both x-ray photoelectron spectroscopy (XPS) and Auger electron spectroscopy (AES). In this chapter, the importance of such studies to the understanding of the environmental interactions of metals and alloys will be discussed. Most importantly, the focus will be on the manner in which the studies should be undertaken so as to preserve relevant surface chemistry and to obtain this information in a cost-effective manner.

A passive layer degenerates into what would be recognized as corrosion by most engineers, in two discrete ways. Firstly, in a gaseous atmosphere, there will be the influence of heat. A rise in temperature permits diffusion of ions by way of point defects and other imperfections in the crystalline lattice of the film, leading to a steady increase in thickness by one or other of a number of well-understood "rate laws" [1,2]. Rapid corrosion may then be associated with the flaking or "spalling" of this thick corrosion product, and XPS has an important role in discovering the nature of the weak interfaces that permit this damage. Secondly, in an aqueous environment, there will be dissolution of the passive layer in whole or in part, in both cases accompanied by continuous reformation of corrosion product, leading to loss of metal—i.e., aqueous corrosion. The path of this form of corrosion is often determined by the nature of the alloying elements present in the film. Often, one element or the other will be enriched in the thin surface layer and thus lead to repassivation of the metal. XPS has had a commanding role in exploring this mechanism of corrosion protection. Finally, since attack in an aqueous medium is electrochemical in nature, there is always the possibility that the anodic and cathodic areas can become separated in a manner that focuses attack on the anodic area. This then leads to the destructive form of corrosion known as pitting. AES has proved to be a powerful analytical tool in the study of pitting corrosion—relating the location of the pits to the underlying metallurgical composition, and then revealing the distribution around the pit of both environmental and alloying elements, each of which play a part in the prognosis for its future development.

As has been made clear in Chapters 3, 8, and 10, XPS is a powerful technique for the analysis of relatively large and uniform areas of a sample surface: it has the power to obtain chemical state information for most elements, and the depth resolution to provide separate analyses of both the surface down to a depth of 1–2 nm, and the metal lying immediately below the surface layer [3]. This is the essential requirement for the study of a passivated surface. AES has an excellent spatial resolution, but for many elements has rather poor resolution of chemical state. However, knowledge of the elements present is often all that is necessary for analysis of the pit chemistry and for separation of anodic and cathodic areas by virtue of their surface composition. Although they are powerful tools, neither XPS nor AES are low cost. Thus they will normally be available as a service provided at a location that is remote from the corrosion laboratory. Cost-effective use requires that the questions asked of the techniques be carefully framed, and that the samples to be examined are handled and transported in a manner that preserves their surface chemistry. This chapter is directed toward the problems for surface analysis in the context of both failure analysis and corrosion science. It is, therefore, intended to provide a guide to practical use for these purposes rather than a discourse on the research findings arising from the use of electron spectroscopy. Some emphasis will be placed on the information available in the so-called survey scan, which in many cases will provide all the information necessary to make progress in an investigation of corrosion.

19.2 INITIAL INSPECTION

Initial inspection can add a great deal to a successful outcome from electron spectroscopy. It is also important to have well-framed and appropriate questions to ask of the analysis, which can often be established by discussion of the history of the sample or test piece, and by provision of some indication of the manner in which the results will be used. Prior examination with optical and scanning electron microscopy (SEM) and, for thick films, the results of energy dispersive x-ray (EDX) analysis will assist in the selection of the most fruitful regions of the sample for mechanistic interpretation. In considering what might be achieved by XPS it should be kept in mind that corrosion is a continuing process and some indication of how it might develop in future is of real importance to the client. A single analysis, no matter how detailed, is only an instantaneous snapshot of the surface. It is much more useful if the sample to be analyzed is one of a set in which specimens of the same material are each subjected to a particular corrosion process, with one of the process parameters being changed incrementally between specimens. In corrosion science, that parameter would usefully be the exposure time. In a gaseous environment, it might be the temperature of exposure that is varied, or, for electrochemically corroded samples, the current flow and potential. In the latter case, other variable parameters would be the pH of the electrolyte, or the concentrations of inhibitors and of aggressive ions such as chloride. From the above discussion, it will be seen that in an ideal case the design of an experiment will include XPS as a matter of course, providing one of the response variables for assessment of the factors to be included in that design. It should be noted that an XPS spectrum will often reveal elements that have been overlooked as potential factors in the experiment—elements such as sodium and calcium from glassware might be found at significant levels; buffering agents such as borates are frequently seen on the surface; and very high levels of organic adsorbents can arise if glassware has been cleaned with aggressive detergents. An initial discussion should try to establish the elements likely to be present in order that they can be assayed and discounted. In the case of samples exposed to gas-phase oxidation, it is possible that minor elements in the metal will be strongly concentrated at interfaces. A guide to the likely position of different elements in the hierarchy of oxide layers can be obtained from an Ellingham diagram, which shows the free energy of oxidation of pure metals as a function of temperature [4,5]. As a first approximation, it is likely that the oxides of alloying elements will be stacked in the order found in the Ellingham diagram, with those of the least negative-free energy of oxidation being found furthest from the underlying metal.

Some examples of the above suggestions will be given and discussed in more detail in subsequent sections. However, such a degree of planning is unlikely to be possible in forensic analysis, where single samples are normally produced for examination. Where possible, samples of the surface taken from areas of differing flow rates, different surface orientation, i.e., downward or upward facing, and differing oxygen potential or heat transfer rate, should be sought. A meaningful comparison might then be possible, throwing light on a possible accelerating factor in the corrosion mechanism.

As a result of this initial discussion the samples and type of investigation can be categorized. Table 19.1 serves as a reminder of the main points that need to be considered.

TABLE 19.1
Initial Assessment of Corrosion Samples

History	Appearance	Analysis	Elements Sought
Source and type: Laboratory or field test piece			
Aqueous	Metallic	XPS: film thickness; composition; stoichiometry; OH ions; enrichment factors	Inhibitors, environment, alloying
	Opaque adherent	AES or small area XPS (SAXPS), taper section (ball cratering)	Segregation of valence states or alloying elements
	Opaque, nonadherent	Surface, and both sides of failure interface	Possible segregation to weak interface layer
	Nonuniform	XPS multiarea	Transport to insoluble deposit
	Pitted	AES of pit and periphery	Include Cl⁻ and S⁻
Gaseous	Metallic	Areas of light tarnish?	Alloying elements, sulfur?
	Opaque adherent	Ball cratering	Gradient in chemistry
	Opaque, nonadherent	Identify spalling interface	Alloying element plus S or V
Source and type: Failed material from machinery, pipelines, or architectural cladding			
Aqueous	Unstable or discolored patina	Attempt removal of film for surface and interface analysis or surface analysis plus depth profile	Alloying elements and ions known to be corrosive, sulfate and chloride. Relate findings to Pourbaix diagram
	Corrosion–erosion	Examine eroded area and surroundings	Hydroxide instead of oxide?
	Loose friable oxide	Seek source of deposit and identify possible gradients driving transport	Sequence of valence states
	Visible pitting	Seek evidence for microelectrochemical activity	Peripheral deposits of alkali, alkaline earth hydroxides
	Corrosion-assisted stress or fatigue cracks	Establish whether cracked surface is cathode or crack tip is anode	Chloride at crack tip?
Gaseous	Thin tarnish film	XPS and angle-resolved XPS	Are oxides in sequence expected from free energy considerations?
	Heavy surface oxide	Examine spalled interfaces for segregation	Minor alloying elements and sulfur

19.3 METALLIC SURFACE

It will sometimes happen that the surface presented for examination has a metallic appearance. As indicated in Table 19.1, this can arise with samples from either aqueous or gaseous environments, but is more likely to occur in samples from a test environment rather than in those, such as failed components, presented for forensic examination.

Before considering the manner in which corrosion processes might yield a bright metallic surface, it is useful to consider the changes likely to arise within the surface layers of a metal on exposure to the atmosphere. There are normally several distinct layers on a technological surface, even after exposure under relatively clean conditions (Figure 19.1) [6]. The innermost layer, adjacent to the metal, is likely to be an oxide that, in the case of an alloy exposed at ambient temperature, will be a mixture of all the alloying components, often in near proportion to their concentrations in the alloy. The surface of this oxide may be converted to a hydroxide and to this will be attracted both a layer of adsorbed water and the polar ends of organic molecules

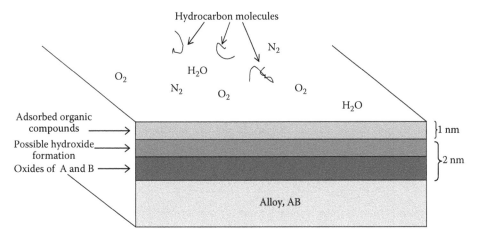

FIGURE 19.1   The hierarchy of layers on a metal surface. (From Castle, J.E., *J. Adhes.*, 84, 368, 2008.)

such as surfactants, personal care products, and fabric conditioners, all of which are to be found in a normal unfiltered environment. XPS will register signals from each of these layers, and also from the metal immediately underlying them, to a total depth of about 5 nm. The outermost layers will consist of weakly adsorbed molecules, such as nonpolar organic molecules and permanent gases, none of which is likely to survive the vacuum environment of a spectrometer as an adsorbed layer.

The influence of the strongly adsorbed layers can be seen in the shape and characteristics of an XPS survey scan. Figure 19.2 shows XPS spectra from the surface of an Fe/17Cr/2Mo alloy in three states, rather typical of the condition in which they might be offered for more detailed analysis. The spectrum (a) in Figure 19.2 is from a surface that had been metallographically polished in air using dry alumina (1 μm grade) on a velvet cloth. Such a sample would give the analysis to be expected from a clean sample after air exposure, with the concentrations of alloying elements in the oxide film being similar to those of the metal itself, as predicted in 1949 by Cabrera and Mott [7]. The C 1s peak is prominent in this spectrum but corresponds to a layer thickness of ca. 0.3 nm, that is, about one monolayer. Spectrum (b) is from the same surface and shows the serious influence of poor handling—residues from a finger placed on the surface are sufficient to obscure some of the detail available in the spectrum (and apparently, to remove some of the Fe from the surface layer). By contrast, spectrum (c) appears to show the metallic peaks in great detail. This is the result of a brief, 30 s, ion etch applied *in situ* after the specimen had been loaded into the spectrometer. This is a procedure sometimes used by operators of instruments who are less interested in the problems of corrosion than in having spectra typical of pristine samples. Ion etching does irreparable damage to the chemical state information available in XPS spectra, and should not be used except in well-understood situations (see Chapter 10), or for obtaining elemental depth profiles, in which it is recognized that the chemical states observed may be of lower valence than those formed during the corrosion process.

While those in an XPS service laboratory understand well that a thumb print will damage the surface, this is not always understood by those sending samples for analysis. The increase in C 1s intensity shown in Figure 19.2b is accompanied by a reduction in the signal from the metallic elements, and the inelastic electron backgrounds have undergone corresponding increases. There is really no way to recover such a situation—ion etching restores a clean-looking spectrum but will change the surface chemistry; chromium is often converted to carbide by etching, and oxides are reduced to lower valence states or even to the metal. To avoid such problems it is useful if the analyst gives the client some simple prior guidance as to the correct means of transport of samples to the

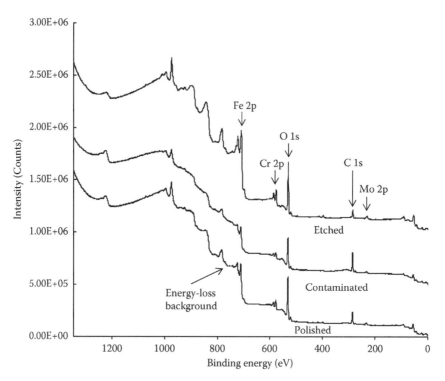

FIGURE 19.2   XPS spectra from a metallurgically prepared surface of an Fe/Cr alloy after different treatments. (a) Handled correctly; (b) after contact with thumb; and (c) after a brief (30 s) ion etch.

service laboratory. It is recommended that samples be wrapped gently in clean aluminum foil or lint-free paper, such as lens tissue, before being placed in a plastic bag. Alternatively, if practicable, glass tubes are excellent. The aim is to reduce any contact with organic vapors, such as plasticizers from soft polymers sometimes used in the manufacture of containers.

A survey scan offers the first indication, not only of the state of the sample, but of significant changes in the surface composition. In the lower part of Figure 19.3 is the survey spectrum recorded after the surface of a 316 stainless steel had been electrochemically passivated in a sulfate solution of slightly acidic pH. In this it can be seen immediately that the Cr signal is more prominent than that of the Fe signal in the air-formed film of Figure 19.2a. Comparison of Figures 19.2a and 19.3 reveals one of the most important differences between the two great subdivisions of corrosion science, namely gaseous and aqueous corrosion. Oxidation in the gas phase leaves behind all the oxidation products for subsequent investigation. By contrast, aqueous corrosion proceeds in part by dissolution of material so that in most cases, but not all, the corrosion product is available. Cr appears enriched in the surface layers giving rise to the spectrum of Figure 19.3, because Fe compounds have been dissolved selectively from the surface. Changes such as these are important to the corrosion scientist in trying to understand the mechanism of protection offered by a particular alloy, or trying to predict the future performance of a metallic component.

Not only are the Cr 2p peaks more prominent in Figure 19.3, but the spectrum background in the region of the Fe peaks is very different from that under the Cr, by contrast with the spectrum of the ion-etched specimen in Figure 19.2c. There the backgrounds under the Fe and Cr peaks are quite similar. Such a change in background can be used by electron spectroscopists to establish that the surface has been depleted in Fe relative to the bulk alloy: they will recognize that the rising

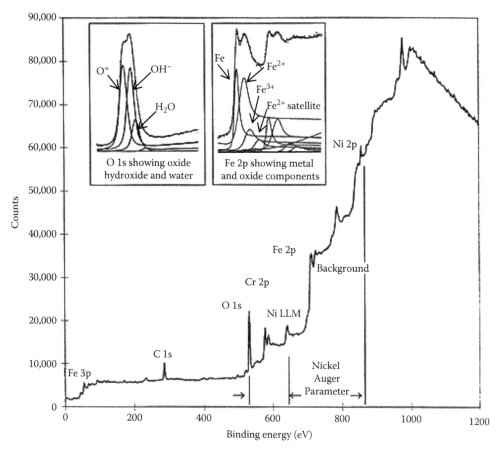

FIGURE 19.3 (Lower part) An XPS survey spectrum from the surface of Type 316 stainless steel after electrochemical passivation. Below the spectrum the spectral features used to calculate an Auger Parameter, in this case that of nickel, are indicated. (Insets) Detailed spectra of the O 1s and Fe 2p regions, deconvoluted to reveal contributions from various chemical states.

background on the high binding energy side of the Fe peaks in Figure 19.3 is caused by Fe photoelectrons, originating in the metal substrate, losing energy as they straggle to the surface through an overlayer that is relatively low in Fe content [8–10].

A survey scan is thus much more than a simple register of the peaks to be studied further in detailed high-resolution or "narrow" scans: the use of backgrounds on the high binding energy sides of peaks adds a great deal to the understanding of the near-surface structure of otherwise unknown samples. By way of example, Figure 19.4 shows the survey scans of three high-alloy steel test pieces sent for examination. Each had been immersed in an acid bath at 90°C for different times and potentials as part of a program to discover a treatment for corrosion protection [11]. They are typical of the sets of samples sent by corrosion scientists for XPS analysis, with very little guidance as to how the surface is composed and structured. Notice how the background beneath the Mo (BE 412 eV) peak from sample 1 is similar to that under the Cr peak (BE 575 eV) of sample 2, but neither is at all similar to the Fe background that slopes steeply away from the peak. This shows immediately that sample 1 had Mo-rich surface layers while samples 2 and 3 had Cr-rich surfaces. In addition, sample 3 has a much thicker surface layer than that on sample 2, as indicated by the greater slope of the Fe background. Such information, gleaned rapidly from a survey scan is of great help in deciding how best to continue with the examination of previously unknown samples.

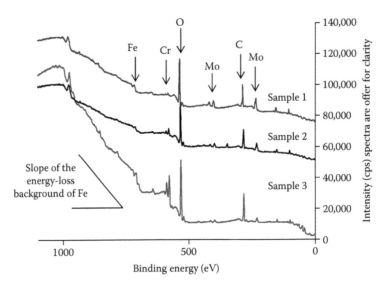

FIGURE 19.4 Use of the survey scan for qualitative assessment of surface structure in a set of steel samples treated for different times and at different potentials. In sample 1, both Fe and Cr seem to be depleted, and the surface species is Mo-rich as indicated by the decreasing background associated with the Mo spectral structure. Sample 2 is similar to normally passivated stainless steel (cf. Figure 19.3). Sample 3 is enriched in Cr, and carries a thicker film, as evidenced by the increased slope of the background associated with Fe. (From Turnbull, A., Ryan, M., Willetts, A., and Zhou, S., *Corros. Sci.*, 45, 1051, 2003.)

19.4   AQUEOUS CORROSION

19.4.1   ELECTROCHEMICAL ASPECTS OF CORROSION

The study of aqueous corrosion is inseparable from electrochemistry and to understand how to use XPS to its full potential a few concepts need to be set down. The fundamental step of all corrosion processes is oxidation, i.e.,

$$M = M^{n+} + ne^-$$

(19.1)

Typically, in acidic electrolytes the metal ions (cations) enter solution leading to wastage of the metal. This process requires the continual removal of electrons which would otherwise cause the electropotential of the metal to move in a negative sense until the point is reached at which the positive ions can no longer form. In the presence of dissolved oxygen the electrons are removed by the cathodic reaction

$$O_2 + 2H_2O + 4e^- = 4OH^-$$

(19.2)

Or, in acidic solutions

$$2e^- + 2H^+ = H_2$$

(19.3)

In electrochemical studies of corrosion it is so arranged that these two reactions occur at the surfaces of well-separated electrodes joined by a conductor. XPS can then be used to provide an analysis of these surfaces and of the films that moderate the rate of reaction.

It can be seen from Equations 19.1 through 19.3, that, since they balance the interactions of both ions and electrons, their equilibria will be influenced by the concentrations of the metal and hydrogen

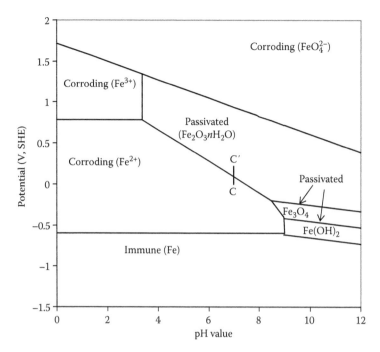

FIGURE 19.5 The Pourbaix diagram for iron. The line C'–C represents the position of the series of analyses of the cupronickel alloy whose stability diagram is shown in Figure 19.6.

ions in the electrolyte, by the pH value, and by the distribution of electrons in the metals, that is, the electropotential. The universal method of depicting these equilibria was created by Pourbaix in his eponymous "Pourbaix diagrams," in which the concentrations of metal ions are plotted on a map of pH vs. potential [12].

An example is given in Figure 19.5. Pourbaix suggested that if a metal surface yielded ions at a concentration of $<10^{-6}$ mol L^{-1} then, for many purposes, it could be considered as noncorroding. The lines joining concentrations at this equilibrium value establish the boundaries between significant fields on the plot. To one side of the boundary the metal is noncorroding—on the other side it will corrode. In fact, Pourbaix identified three possible states: immune (i.e., cathodically inhibited), passivated, and corroding. In the passive state, the equilibrium condition of Equation 19.1 should give rise to corrosion but the rate of corrosion is limited kinetically by formation of a thin insoluble layer.

The extent of reaction at any time is determined by the total charge, Σne^-, which has passed through the electrode and the knowledge that 96,500 C (i.e., 1 Fd) will cause the deposition or solution of 1 mol of a univalent ion. Taking Cu dissolving as Cu^+ ions as an example, the loss of 3 nm (the XPS sampling depth) requires a current to flow through the surface at a current density of 100 μA cm^{-2} for only 40 s (4 mA s). The power of XPS is that it can determine the compounds and ions involved in the corrosion process before there is any visible sign of reaction, and before the surface is swamped by secondary corrosion products. This, and its ability to recognize valence states, provides it with an unique ability to identify the passivating films within the zones of the Pourbaix diagram. Knowledge of the thermodynamically stable phase indicates to the analyst what to expect; any departure from that expectation then points to a possible reason for corrosion. The best determinations of such departures can be made by exposing the metal at a controlled pH and electropotential to a given charge transfer, say 0.1–0.5 A s. An example is given in Figure 19.6 for the copper alloy 706 (Cu/9.8%Ni/1.7%Fe) immersed in 3.4% NaCl solution [13]. Here the aim was to discover the possible reasons for the known action of Fe in improving the performance of Cu/Ni alloys in

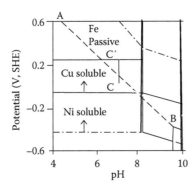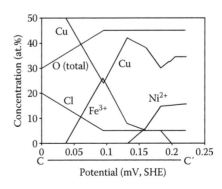

FIGURE 19.6 Left hand: Fragment of a Pourbaix diagram for the alloy CuNiFe. Right hand: Surface composition as a function of potential, measured by XPS along the line CC′ in the left-hand diagram. Note the increase in iron within the passive region (potential ca. 0.1 mV) for iron metal, and the high chloride value at lower potentials. (From Kato, C., Ateya, B.G., Castle, J.E., and Pickering, H.W., *J. Electrochem. Soc.*, 127, 1897, 1980.)

seawater. Pourbaix diagrams for alloys are not normally available, but a first step was to overlay the diagrams of the pure metal components of the alloy and a fragment of such an overlay diagram in given in the left-hand side of Figure 19.6. Seven XPS analyses were made at intervals along the line C–C′. The results are given in the right-hand side of Figure 19.6 in which the concentrations of the detected species, in at.%, are plotted against an extended potential scale. Within this region of the combined Pourbaix diagrams both nickel and copper form soluble species and therefore corrode. Iron, however, is predicted to have a stable passivating film (Figure 19.5) for potentials greater than those represented by the line A–B in Figure 19.6. The XPS analyses showed that, as predicted by the diagram, the concentration of iron reached a maximum within the region in which pure iron is passive. Thus, even the very small concentration of iron in the alloy was sufficient to give a passivating film of iron hydroxy oxide. The analyses also showed up some unexpected results. Close to point C the surface film had a composition close to Cu_2Cl_2, while near point C′ the iron oxide on the surface was replaced by a form of nickel oxide.

It is the passivation of the surface by such surface films that enables the use of metals in water and aqueous solutions. XPS has played a major role in identifying these surface films. Excellent examples of the use of XPS in the field of passivation can be found throughout the literature of the past 30 years [14–17], and, since passivation is especially important, it will be discussed separately, illustrating how XPS enables the analyst to identify all the key components of the passivating film. The stability of passive films, that has made possible the easy transfer of test pieces from the passivating medium to the high vacuum of the spectrometer, does not apply to the other two fields of the Pourbaix diagram, and the use and role of XPS in active corrosion and in the "immune state," more correctly called "cathodically inhibited state," will be considered briefly.

When working with test pieces that have been cathodically inhibited, the investigator is faced with the problem that the surface chemistry is likely to change rapidly on withdrawal from the electrolyte, both because of the sudden loss of the controlling electropotential on breaking the circuit, and because of possible exposure to an oxidizing atmosphere. Elaborate transfer mechanisms [18] have been devised by researchers in the field of electrochemical corrosion to minimize these changes, and Figure 19.7 shows an *in-situ* cell used in the laboratory of Castle et al. [19]. Despite these problems, conventional XPS has been very useful in the study of the behavior of alloys that have been held at a potential in the cathodic region. For example, it can happen that the more active components of the alloy will dissolve progressively, a process known as dealloying [20]. XPS can detect the early stages in this by following the changes in composition of the metal phase. One unusual

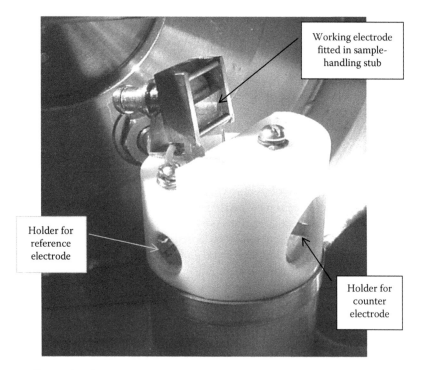

FIGURE 19.7 Electrochemical cell built into the entry lock of an electron spectrometer. (From Castle, J.E., Guascito, M.R., Salvi, A.M., and Decker, F., *Surf. Interface Anal.*, 34, 619, 2002.)

example of dealloying, connected with the development of dental alloys, has shown how angle-resolved XPS can be used to measure the depth of dealloying [21].

There is another telltale sign to look for in establishing that a metallic surface has been in a benign situation with regard to the local electrochemistry, and that is the presence of ions such as magnesium (from seawater) or calcium (from freshwater) on the surface. These arise from the fact that, just as the formation of metal ions is minimized at cathodic potentials, that of an excess of OH⁻ ions from the equilibria, described by Equations 19.2 and 19.3, is stimulated. The establishment of a high pH value in the hydrodynamic boundary layer gives conditions for precipitation of poorly soluble hydroxides such as $Mg(OH)_2$ or $Ca(OH)_2$ [22]. The presence of such ions is detected easily by XPS and should always be reported, together with evidence of OH⁻ ions deduced from the oxygen spectrum or from assessment of the stoichiometry. The corrosion scientist can then consider what might be driving the surface to the cathodic potential, possibly an anodic potential created within a nearby crevice or by contact with a dissimilar metal.

In completing this brief overview of the importance of electrochemistry in understanding the metal surface exposed to an electrolyte, a surface carrying a visible corrosion layer should be considered, since this is the type of surface most likely to be presented to the analytical service. On a Pourbaix diagram such a surface would be categorized as "actively corroding." Separation of the anodic and cathodic reactions may be achievable in an electrochemical cell but on a metallic surface immersed in a corrosive medium they both occur on the same surface. XPS will then sample the consequences of both types of activity and the analysis thus represents a mean value. A difficulty arises, since cations diffusing away from the corroding surface are likely to encounter the OH⁻ ions created at the cathodic areas and give rise to a precipitate of oxide and hydroxide [23], or a more complex patina on the surface. As the patina increases in thickness a pH gradient can develop, leading to segregation of layers of different composition [24]. The history of the corrosion is thus encapsulated and with care can be stripped away for analysis.

19.4.2 Thin Passivating Films

The phenomenon of passivation is a very special topic in the field of corrosion since it enables the use of metals in circumstances which would otherwise be impossible. As was seen in the above brief review of electrochemical corrosion, passivation is represented by an isolated zone within a pH vs. potential diagram. In practice, when a fresh metallic surface is exposed under passivating conditions the electrochemical current density is initially high, indicating rapid dissolution but, within a few seconds, falls to a very low value (ca. $1\,\mu A\ cm^{-2}$), corresponding to a passivated surface. The use of ellipsometry in the 1930s showed that the passivated surface was covered with a thin film but it was not until the use of XPS in the 1970s that the composition and thickness of the film on important metals such as stainless steels were established [13,14].

XPS characterization of passive films poses rather unique problems within the general field of thin-film analysis. In the microelectronics industry thin films are grown from known reagents on well-prepared substrates, whereas passive films are generated by seemingly random processes of dissolution and reprecipitation. Typical questions posed to the analyst by the corrosion science industry center on "discovering the reason for the good performance of a passivating film on a given alloy in a particular environment in such a way that its use can be extended into a more challenging environment or used to develop a new range of alloys." A survey of the literature shows that the most sought-after information is that listed in Table 19.2. This represents a good guide for a satisfactory report, following XPS characterization [25]. Each of these topics will be considered briefly in turn.

19.4.2.1 Thickness of a Contamination Layer

Surface contamination is always likely to occur when dealing with passive films. It probably arises, in the main, as the test piece passes through the meniscus layer when being withdrawn from the electrolyte. As a result, the quantitative analysis is likely to return values of several tens of at.% for the carbon concentration. If reported in this unmodified form to the client, such an analysis is often misleading. Equally, if the carbon is simply discounted and removed entirely from the analysis, important information is lost since the thickness of the contamination influences the quality and intensity of the spectra from both oxide and underlying metal. Thus, the best option is to estimate the thickness of the contamination layer and then to remove the contribution of carbon from the quantitative analysis. A good estimate can be obtained using the relationship:

$$d_{carbon} = -\lambda_c \cos\theta \ln(1 - n_c) \tag{19.4}$$

TABLE 19.2
Items to Consider When Characterizing Passive Films

1	Thickness of any contamination layer
2	Mean composition of the probed depth
3	Enrichment of the surface relative to the alloy
4	Thickness of the oxide
5	Valence state of all oxidized species
6	Stoichiometric analysis to assess hydroxyl content
7	Concentration of inhibiting (PO_4^{3-}) or damaging ions (Cl^-)
8	Location of the above ions

---

where

d_{carbon} is the thickness of the contamination layer, based on the organic components of the C 1s peak

λ_c is the effective attenuation length (EAL) of electrons at the energy of the C 1s peak

θ is the electron take-off angle

n_c is the atom fraction of carbon in the analysis (i.e., at.%/100) [25,26]

Using this relationship and an EAL value of 3.3 nm for organic material, it is found that a layer containing 40 at.% carbon is equivalent to a thickness of 1.20 nm.

19.4.2.2 Mean Composition of the Analyzed Depth

The contamination layer will attenuate XPS peaks of different energies to differing extents. Thus a fully corrected analysis requires that the area of each peak should be reduced by the factor $\exp(-d_{carbon}/\lambda \cos \theta)$, where λ is the EAL for transport through organic carbon at the energy of the appropriate peak. Smith [26] showed that values of the EAL (nm) for organic overlayers could be fitted to a regression analysis given by the curve, $0.0016(KE)^{0.7608}$, where KE is the kinetic energy of the chosen peak (eV). This has been found to be satisfactory for correction of the analysis. Following this, the analysis should be renormalized to 100 at.%, omitting carbon. The effect of this correction will be rather small if the peaks concerned are close in binding energy, such as O 1s, Cr 2p, and Fe 2p in the spectrum of stainless steel, but much more significant in the case of Cl 2p, S 2p, and Cu 2p, in the spectrum of corroded copper. Where possible the metallic and nonmetallic components of a peak should be separated and listed as separate entries in the quantification, likewise for the O^{2-} and OH^- components of the O 1s peak.

19.4.2.3 Enrichment of the Surface Relative to the Alloy

The passive layer, which is usually of the order of 2 nm in thickness, represents a large fraction of the acquired signal but in many cases the signal from the underlying metal phase is just visible. Continuing with the passivation of stainless steel as an example, the detailed or narrow scans, as exemplified in the inset figures of Figure 19.3, of the principal peaks for the major elements of the sample, should be examined. For the metallic elements the valence states corresponding to the metal and the oxide can be recognized. This crucial information enables separation of the spectrum originating from the underlying metal from that of the oxide itself.

Having available the separate analyses of the metal and the nonmetal components of the alloy elements, either the enrichment or the depletion in the oxide of a given element can be assessed by comparing the cation fraction (e.g., $Z^{m+}/(Z^{m+} + X^{n+} + Y^{p+} + \cdots)$) for a given element, with the atomic fraction (e.g., $Z/(Z + X + Y + \cdots)$) of the same element in the underlying alloy. The ratio of these fractions is known as the enrichment factor [27].

The use of cation fractions in this manner was first reported in 1975 when Asami and Hashimoto published a series of studies that became a hallmark of their work with various coauthors. Studying a series of FeCr alloys by means of XPS, they compared the cation ratio $Cr^{III}/(Fe^{II} + Fe^{III} + Cr^{III})$ for the ions in the surface "oxide" layer with the composition of the metal [28]. For the surface passivated electrochemically there was a rapid rise in this ratio at a bulk composition of ca. 18 at.% Cr (see Figure 19.8a). This is the concentration of chromium at which passivation becomes possible and at which similar steels are usually referred to as "stainless." The cation and atomic ratios for the metal immediately below the oxide, i.e., that visible in the XPS spectrum, remained the same as that of the alloy throughout the whole range of compositions. No surface enrichment was found for surfaces that had not been passivated electrochemically.

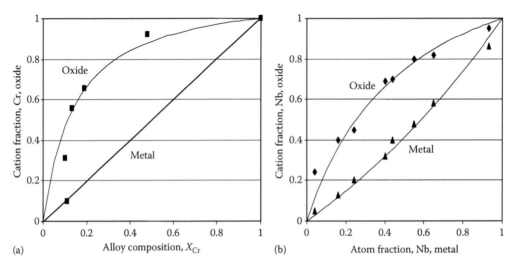

FIGURE 19.8   Results of XPS analyses of surface enrichment in alloys. (a) Enrichment of Cr in the passive oxide film formed on FeCr alloys. (From Asami, K., Hashimoto, K., and Shimodaira, S., *Corros. Sci.*, 18, 151, 1978.) (b) Enrichment of Nb in the surface oxide on NbMn alloys. In each case, the compositions of the oxides lie on a smooth curve, while those of the underlying alloys are close to those of the bulk compositions. (From Castle, J.E. and Asami, K., *Surf. Interface Anal.*, 36, 220, 2004. With permission.)

Determination of the enriched species in the oxides formed on alloys, using XPS, gives an important clue as to the method by which corrosion resistance is conferred in media of widely different characters. Such strong enrichment of an alloying element in a thin film can be a hallmark of passivation. A summary of the enrichment for many alloys has been set out by Castle and Asami [29]. They also showed that, for many alloys showing an enriched oxide composition, the cation fraction lies on a smooth curve across the whole range of alloy compositions as exemplified by the MnNb alloys in Figure 19.8b. In many cases there was little change in the metal composition underlying the oxide, indicating that enrichment occurs by selective dissolution.

19.4.2.4   Thickness of the "Oxide"

A measurement of film thickness (without undertaking depth profiling using an ion gun) is possible only if a peak component can be ascribed to the metallic form for at least one of the alloying elements. This is likely to be possible for passive films. The most straightforward measurement assumes that the film is homogenous and of uniform thickness; this is not always so for the products of corrosion but it does provide a number representing an average value. As set out in most texts, the method applies to pure substrates, e.g., silicon or aluminum, rather than to alloys [30,31], the intensity of the signal from the oxide being related to film thickness using the standard form of the Beer–Lambert equation,

$$I_{ox,z}^{d} = I_{ox,z}^{\infty} \ \exp(-d/\lambda_{z} \cos\theta) \tag{19.5}$$

where
$I_{ox,z}^{d}$ is the observed intensity from the element z in oxidized form
$I_{ox,z}^{\infty}$ is the intensity from the same element in a thick (>10 nm) sample of the same material
EAL [32] for the electron energy corresponding to the chosen peak is given by λ_{z}
θ is the electron take-off angle relative to the sample normal
d is the required thickness of the surface film

The value of d can also be obtained from the attenuation of the substrate signal, using the unoxidized component of the peak obtained for the element z. Thus

$$I_{me,z}^{d} = I_{me,z}^{\infty}\,(1 - \exp(-d/\lambda_{z}\cos\theta)) \tag{19.6}$$

Notice that the value of the EAL for the unoxidized and oxidized forms of λ_{z} will not differ greatly, and thus it is convenient to take a ratio of Equations 19.5 and 19.6, so eliminating instrumental constants and x-ray flux from the measurement, i.e.,

$$I_{me,z}^{d}/I_{ox,z}^{d} = I_{me,z}^{\infty}/I_{ox,z}^{\infty}\,(\exp(d/\lambda_{z}\cos\theta - 1)) \tag{19.7}$$

The ratio $I_{me,z}^{\infty}/I_{ox,z}^{\infty}$ is often given the symbol R^{∞} and can be determined for pure substances as a constant for a particular element/oxide combination. Equation 19.7 has been shown to give outstanding precision in determination of film thickness on silicon, which has been adopted as an international standard [31]. In fact, R^{∞} is not very different from unity for most elements of interest to corrosion scientists. The principal problem in using the method is that corrosion studies concern mainly alloys, in which the proportions of the alloying elements in the oxide will differ from those in the alloy. In that case a correction term must be introduced to compensate for the enrichment or depletion of an element in the oxide film. Usually the analyst will already have values for such a term. It is the enrichment factor represented by the ratio of the cation fraction to the atomic fraction, i.e., for an alloy, XYZ;

$$F_{z} = \{Z^{m+}/(Z^{m+} + X^{n+} + Y^{p+})\}/\{Z/(Z + X + Y)\} \tag{19.8}$$

Equation 19.7 thus becomes

$$I_{me,z}^{d}/I_{ox,z}^{d} = I_{me,z}^{\infty}/I_{ox,z}^{\infty}\,(\exp(d/\lambda_{z}\cos\theta) - 1) \tag{19.9}$$

Or, rearranging and setting $I_{me,z}^{d}/I_{ox,z}^{d} = R^{d}$ and taking R^{∞} as unity we obtain

$$d = \lambda_{z}\cos\theta\ln(F_{z}R_{z}^{d} + 1) \tag{19.10}$$

This provides a simple method of quoting a mean thickness that will be of use to the corrosion scientist. A more detailed description of other methods used to estimate thickness of overlayers has been given elsewhere, and the reader is also referred to the work of Tougaard for a more specialized discussion of the use of relevant information in the energy-loss background [33] (see also Chapter 8).

19.4.2.5 Valence States of All Oxidized Species

In some cases, especially with the transition elements, there is the possibility of multiple valence states. The problem then is not only that of separation of the metallic and oxide states in XPS spectra, but also of the determination of the concentrations of the different species within the oxide phase. Figure 19.3 includes a peak for the Fe(II) state in the peak-fit for Fe, a distinct possibility since the positions for the 2p peaks of Fe(III) and Fe(II) are well known, and the prominent peak for metallic Fe is present as well to act as an internal reference. However, such a reference peak is not always available, and it can then be difficult to differentiate between peaks corresponding to different valence states. Secondary features such as satellites can prove particularly useful. Figure 19.9 shows the presence of satellites in the iron oxides and how, albeit qualitatively, it is possible to use them to distinguish between the iron valence states. The satellites associated with Cr(III) are also useful features in the Cr_2O_3 spectrum [34]. The characterization of valence state has been particularly important in understanding the formation of passive films on pure Fe, as in the work of

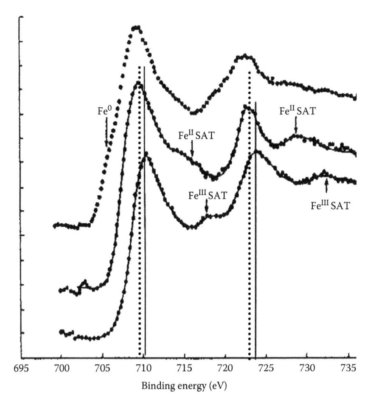

FIGURE 19.9 Detailed XPS spectra from iron oxides grown *in situ* within the spectrometer. Lower spectrum, Fe_2O_3; center, FeO; upper, mixed oxide similar to Fe_3O_4. The solid and dotted vertical lines indicate the binding energies (y) of the Fe(III) and Fe(II) components, respectively, showing the rather small shift between them. Thus the satellite structure is important in helping to identify the two states. The uppermost spectrum contains both Fe(III) and Fe(II) species, and the sharp valley between the principal peaks is generated by the overlapping satellites. The binding energies (y) of the Fe metal peak is also indicated.

Strehblow [35]. There the variation in film composition as a function of electropotential was the important feature. It may be necessary to extend the acquisition window beyond that conventionally used in data systems to include the satellite peaks. Ferrous (i.e., Fe(II)) ions are easily oxidized to the ferric state in air under ambient conditions. However, one of the important findings from early work using XPS was that the ferrous state appears to be stabilized by its incorporation with Cr(III) into the passive film formed on stainless steels [36,37]. Such incorporation enables transfer between electrolyte and spectrometer without any problems occurring from air exposure. Although problem-free transfer has been confirmed by comparison with similar analyses carried out on samples transferred under protective atmospheres, this stabilization is unusual. The possible oxidation of ions to higher valence states is always a problem in analysis of samples removed from electrolytes under nonpassivating conditions.

19.4.2.6 Stoichiometric Analysis to Assess Hydroxyl Content

The hydroxide ion is an important constituent of many films, protective and nonprotective, formed on metal surfaces whenever water is present in the environment. Its presence is seen in the oxide product most likely to form on iron, where the doubling of the O 1s XPS peak is a signature not only of "rust" FeOOH, but also of FeOOH as a stable film on brightly polished iron held in a relatively dry atmosphere. The OH$^-$ ion is also a key element in the stable green patina, $Cu(OH)_2CuCO_3$, formed on copper when used architecturally, e.g., as roofing material, but there the doubling of the oxygen peak is not clearly resolvable. In other instances, such as on stainless steel, the hydroxide content might be greater than expected, so that the dominant component of the O 1s peak is misrepresented as oxide rather than hydroxide. For all these reasons it is important to examine the stoichiometry of the overall analysis using the valence states considered most likely to be present, from the curve fitting that has been applied. If the cation and anion charges are not in electrical balance then there is a good probability that hydroxyl ions are present since hydrogen content cannot be monitored by XPS. A perfect balance will not be possible, since it will be influenced by the presence of concentration gradients in the film and by the presence of water and organic forms of oxygen in the contamination layer, but if there is a large deficit of positive charge when taking O^{2-} to be the anion, then it is probable that a much better match will be obtained when OH^- is included [38]. By assuming a perfect balance in stoichiometry an approximate figure for the concentration of compounds in the form of hydroxides can be obtained. That can then be compared with the concentration actually calculated by peak fitting of the oxygen peak, and perhaps corrected further by allowing for the presence of bound water in the corrosion product.

19.4.2.7 Concentration and Location of Inhibiting or Damaging Ions

The location, within the passive layer, of ions such as PO_4^{3-}, often a component of inhibitor formulations, or Cl^-, often associated with premature film breakdown, will almost certainly require angle-resolved XPS, using spectra gathered at two or more angles, or, alternatively, depth profiling using ion beam etching [39]. Either option will require additional instrumental resources and would normally be undertaken after initial discussion with the client. In looking for the location of ions there need not be too much concern about damage to the chemical state of the host lattice or of the ion species being sought—the observation of a peak in the depth distribution of the relevant atom is all that is needed. Auger depth profiling is used for a number of purposes in corrosion studies and is the most frequent reason for the application of AES in a corrosion investigation. Predominantly, it is used to measure the thickness of a thin (submicrometer) layer, usually by determining the extent of the oxygen profile. Often this is performed by selecting an "end point" in the oxygen profile, for example, the thickness at which the oxygen intensity drops to a value midway between the maximum and the minimum values. The "Round Robin" reported by Marcus and Olefjord [40] revealed the good reproducibility of this measurement, made at a number of collaborating laboratories. AES depth profiling is very useful in finding the position, relative to the oxygen profile, of any adverse anions such as chloride and sulfide, or of any film-enhancing ions such as phosphate or borate.

19.4.3 SURFACE FILMS FORMED UNDER NONPASSIVATING CONDITIONS

Metals undergoing corrosion normally form thin adherent layers, several micrometers in thickness, and often with a friable or porous surface. These layers limit the rate of attack but over a period of time the underlying wastage of metal can be quite considerable. This is the type of corrosion that is most frequently encountered in forensic studies and for which XPS provides one of the tools of investigation. Corrosion requires material transport across the layer and thus a soluble species is implicated. It also requires a driving force and this often takes the form of a gradient in oxygen potential. Both the nature of the soluble species and the oxygen gradient can be recognized by XPS analysis. In this type of investigation it is necessary to analyze the loose surface material and also to be able to expose the inner parts of the film, though not necessarily the interface itself, for analysis.

In the surface layers it is often possible to recognize states of higher valency that point to the presence of an oxygen-rich environment. Commonly found examples are cupric ions, Cu(II), on copper-based alloys, and ferric ions, Fe(III), on steels and iron alloys. Steels in particular are susceptible to attack and rusting in the presence of ambient air and, while giving good service in sealed, oxygen-free, water systems, will corrode rapidly if the system is open to air for prolonged periods. A surface layer of FeOOH is an indicator of exposure to wet and oxygenated conditions. The inset diagrams of Figure 19.3 show the XPS hallmark associated with a rusty surface. In fact much the same spectrum can be found on brightly polished mild steel as soon as it has been exposed to the ambient atmosphere; thus visible rusting is not in itself a necessary indicator. Corrosion of copper alloys is often first recognized by the presence of a dark tarnish or colored patina on the surface. Figure 19.10 shows portions of XPS spectra recorded from two areas on a single coupon of Cu/9.8Ni/1.7Fe cupronickel that had been corroded in NaCl solution. A dark tarnish located near one side of the coupon gave the lower spectrum; the rest of the sample appeared bright, and was covered with a protective film containing Cu(I), possibly as a chloride, as revealed in the upper spectrum. Cu and Ni ions in their divalent states have very typical satellites associated with the 2p peak that can be seen clearly in the lower spectrum of Figure 19.10 [41].

For copper to form corrosion products in an aqueous solution either oxygen or a positive electropotential needs to be present. Cu ions entering solution as the cuprous (Cu^+) ion become oxidized and precipitated as (typically) a basic salt, $Cu(OH)_2CuX_2$. Except in the special case of the green patina formed on architectural copper, $Cu(OH)_2CuCO_3$, the tarnish films are not protective and their

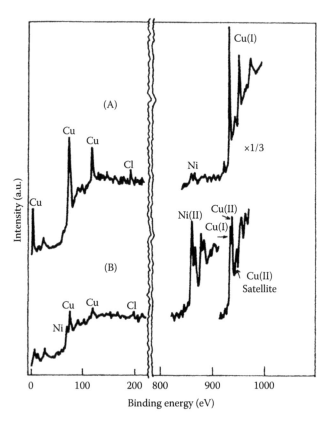

FIGURE 19.10 Selected regions of XPS survey spectra from two areas on a Cu/9.8Ni/1.7Fe alloy corroded in NaCl solution. Lower spectrum: area covered with dark tarnish, or patina. Upper spectrum: bright area typical of other areas of the sample. Note how satellites on the Ni and Cu peaks identify the divalent states, even within a typical survey scan. (From Kato, C., Ateya, B.G., Castle, J.E., and Pickering, H.W., *J. Electrochem. Soc.*, 127, 1899, 1980. With permission.)

presence is a sign of continuing corrosion. Frequently the tarnish starts in one area of a sample and spreads across the surface. Thus even a small test coupon will have areas where Cu is in the cupric state and adjacent areas where it is in the cuprous form. XPS analyses of both areas would be needed in order to provide a true indication of the nature of the corrosion process. This is an example of the transport of corrosion products in soluble form through the electrolyte, and in flow conditions this will often lead to deposition occurring in a remote part of the circuit, so that all that is recognized in the XPS analysis is the composition of the dissolving surface. An extreme example of this can be found in large oil-filled transformers, in which the diffusion of copper from copper foils interleaved with paper insulation results in deposits on the insulation [42]. The paper is wound, bandage-style, around a central conductor and copper foil is interleaved with the windings at intervals of about 25 layers. Analysis of each successive layer of paper then reveals the Cu concentration profiles characteristic of diffusion, as shown in Figure 19.11. Here the excellent repeatability of the analysis should be noted, on both surfaces, and over a range of concentrations which reached a maximum of only 1.1 at.%. The profiles also give an indication of the manner in which Cu ions can migrate through any overlayer, including thick deposits of corrosion products.

Although the normal types of corrosion product cannot be separated like a stack of paper, the principle, that corrosion products contain the evidence of the transport process involved, is important. Generally, when presented with a sample carrying a loose surface deposit, it is preferable to remove it and, in so doing, expose the inner surface for analysis. This can be carried out by careful scraping but in many cases the use of a double-sided adhesive tape works well. The inner layers will often reveal much of the corrosion story. The analyst should look for evidence of corrosion-specific ions, such as chloride or sulfate; there may also be a segregated layer of one of the alloying elements, nickel compounds in the case of copper–nickel alloys, or even aluminum compounds in the unique case of aluminum–brass in seawater [43]. Having analyzed the separated layers, a brief ion-etch and further analysis of each surface will often establish whether the failure has occurred close to an interface, or within the body of a discrete layer (see Figure 19.12). The light ion etch allows the direction of the local concentration gradients to be revealed without incurring the high level of chemical damage likely to arise from a full-depth ion-etch profile [41]. The profiles show that the inner surface of the fracture face had high concentrations of Cu and Cl ions, while the outer surface

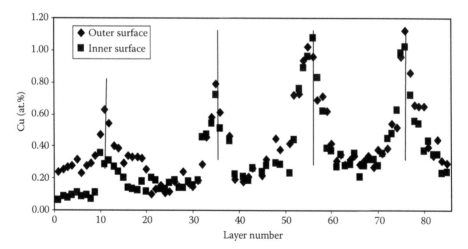

FIGURE 19.11 Analysis by XPS of the concentrations of Cu on each side of successive paper windings in an oil-filled transformer. The windings are interleaved with Cu foil at intervals of about 25 layers. The vertical lines indicate the positions of the foil layers. It can be seen that Cu appears on the paper windings between the foil layers, indicating diffusion from the foils. (From Whitfield, T.B., Castle, J.E., Saracco, C., and Ali, M., *Surf. Interface Anal.*, 34, 176, 2002. With permission.)

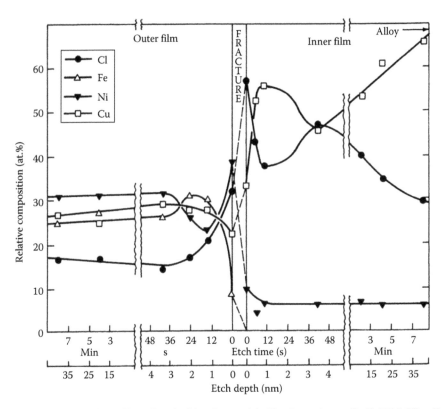

FIGURE 19.12 XPS depth profiles of each side of a tarnish film formed on a Cu/9.4Ni/1.7Fe alloy after corrosion in NaCl solution. The tarnish film was removed from the alloy with adhesive tape. (From Kato, C., Ateya, B.G., Castle, J.E., and Pickering, H.W., *J. Electrochem. Soc.*, 127, 1899, 1980. With permission.)

had a low level of Cl, but relatively high levels of the alloying elements. This type of analysis is vital to the understanding of some corrosion mechanisms.

19.4.4 CORROSION INHIBITION AND PROTECTION

Metals used in sealed systems are often protected against corrosion by use of inhibitors (e.g., benzotriazole for copper alloys [44], or phosphates and organic compounds for steels [45]). It is to be expected that appropriate traces of the inhibitors on the metal surfaces would be found, but if not then it must be suspected that the intended inhibitor has either become exhausted or was not used at the correct concentration in the first place. In quantifying this type of analysis it must be remembered that many inhibitors act at the monolayer level of adsorption, while the published relative sensitivity factors apply only to analysis integrated over a depth related to the EAL. A correction, to give an approximate value for the concentration of the inhibitor film on the same basis as other elements, can be made by multiplying the intensity by the value of λ expressed in atom layers of the principal component present in the surface film.

In flow-through systems, or in open waters, metals may be protected by application of a cathodic potential, which may be achieved by the use of either an applied voltage, or by attachment of a sacrificial anode such as magnesium. When successfully applied the surface is likely to have an excess, relative to anions such as chloride or sulfate, of the alkaline or alkaline earth elements, such as sodium, magnesium, or calcium. It is useful, in reporting an analysis, to draw attention to any obvious imbalance in the concentrations of ions derived from the medium in which the corrosion occurred. Just as cathodic surfaces have an excess of cations, so also anodic surfaces are likely to have an excess of anions.

19.4.5 CHEMICAL STATE

XPS analysis is particularly valuable because it can provide information about the valence state of most of the key elements (remembering that certain elements, notably Cu, require the use of an Auger peak in the spectrum in addition to the photoelectron peaks, to identify the valence states). The chemical states of compounds found in corrosion films will often give an important clue as to the mechanism of attack. Evidence for the presence of gradients in oxygen potential may be found in the sequence of ions found in a depth distribution through the corrosion product. Compounds of a metal in different valence states can often have quite different solubilities, which will have a strong impact on corrosion processes. Knowing which chemical species is present on a surface enables the corrosion specialist to relate the condition of the metal to a possible location within the Pourbaix diagram, a great help in telling whether exposure conditions are safe. Thus every means should be used to identify valence states wherever possible.

The Wagner chemical state diagram [46] is very useful, particularly when XPS is used to study a sequence of exposure trials [47]. It takes the form of a plot of the KE of a prominent Auger peak against the BE of a prominent photoelectron peak, for various compounds of an element, where the KEs and BEs are measured in the same spectrum. Diagonal lines can then be drawn across the plot, each representing a constant value of the Auger Parameter, i.e., (KE + BE). Figure 19.13 shows data for Cu plotted on such a diagram. XPS was used [46] to characterize brass surfaces exposed to very pure water at 95°C with oxygen concentrations in the parts per billion to parts per million range, in an attempt to solve a technological problem. The problem was that of copper entering the

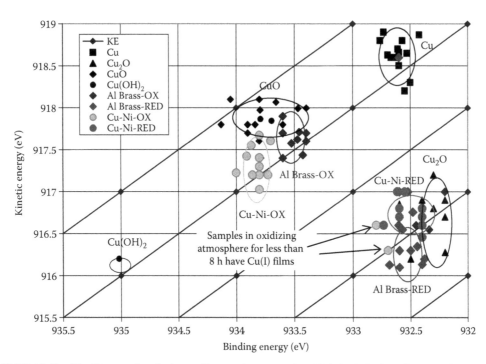

FIGURE 19.13 The Wagner chemical state diagram for copper, in which the KE of the Cu LMM Auger peak is plotted against the BE of the Cu 2p photoelectron peak; the diagonal lines each correspond to constant Auger parameter, that is (KE + BE). Clusters of experimental points recorded from corroded brass and cupronickel alloys appear in the CuO region if the exposure leading to corrosion was in oxidizing conditions, or in the Cu_2O region if it was in reducing conditions. However, the important points are those in the Cu_2O region originating from exposures to high oxygen levels for short times. (From Dooley, B.L., Castle, J.E., and Zhdan, P.A., *Power Plant Chem.*, 7, 561, 2005. With permission.)

water/steam system of a turbogenerator as a result of the solubility of a copper compound. Most of the points in the figure cluster around either the cupric state, for oxygen greater than a critical level, or cuprous, below that level. The exceptions are two points (arrowed) in the cuprous region found after exposure at high oxygen levels for short exposure times. The research showed that there was an induction period before Cu(II) precipitated on the surface. The corrosion significance is that Cu(I) is a nonequilibrium state in a high oxygen potential and it was this that gave the enhanced solubility leading to the industrial problem. The satellite peaks of Fe and Cr have already been cited for evidence of valence state, but those of the divalent states of Co, Ni, and Cu also have well-known uses [48].

19.4.6 AQUEOUS CORROSION—CONCLUDING SUMMARY

In the introduction to this chapter it was pointed out that the analyst would generally have to deal with samples originating either from a forensic investigation into a failure, or from a research project designed to improve corrosion resistance. It is useful to review the section on aqueous corrosion with this in mind. In doing so it should be remembered that failure by pitting is a special case of corrosion which is discussed in a later section. What is considered here is general corrosion leading to wastage of a large area of metal.

In the forensic situation the problem is to aid the investigator (e.g., a corrosion engineer) in discovering how the corroded surface came to be in its state of failure. The design operating conditions will be known, and from them there should be some information about the surface layers that were expected to provide corrosion resistance. Problems can arise when a departure from that design occurs, and the surface chemistry can often reveal the form of departure. In ideal cases the conditions and the properties of the surface film will be related to the Pourbaix diagram. An obvious question is whether the surface film is that which was expected. For example, Fe_3O_4 is often considered to be the protective film on steel in high-temperature water so that, if XPS reveals Fe_2O_3 or FeOOH, it might be concluded that the oxygen level in the aqueous environment was too high. A gradient of oxygen potential through the thickness of the film will perhaps reveal the driving force for corrosion—this gradient will be indicated by a lower valence state in the inner layers relative to that found on the surface. Material removed by corrosion might be deposited uniformly as an outer layer, but is likely to have been deposited irregularly, or even in remote locations, in flow systems. In all these cases transport through the aqueous phase will have played an important role, and XPS should be used to identify the soluble species. Guidance can then be offered on the presence of elements, on valence states, or on compounds known to have high solubilities. Investigation of thick films will require exposure of the inner layers, best achieved by mechanical means and not by ion etching, since the latter is almost certain to change the valence states present. Material can be pulled from the surface using adhesive tape or might be removable by careful scraping with a blade.

In a research project there is much more scope for examination of samples as a function of one of the operational variables, and by this means to discover the underlying reasons why a given film has protective properties. Much space has already been given to the study of protective films but one further example might be useful. In it an Fe/Cr/Mo alloy was studied in order to understand better the role of molybdenum in stainless steels [49]. Figure 19.14 shows the XPS results in the form of the enrichment factors for Mo and Cr as a function of a series of potentials covering the active and passive regions of the alloy. Within the passive range the expected threefold enrichment of Cr is found, but there is no particular contribution from Mo. However, at the limit of the active range, immediately before the onset of passivity, it is Mo that is strongly enriched. Thus, in the context of research into the repassivation of steels that have been damaged by, e.g., scratching or wear, it is found that it is Mo that is likely to be the metal on which an oxide film first regrows. This finding could easily have been missed if analyses had been confined to the fully developed passive surface.

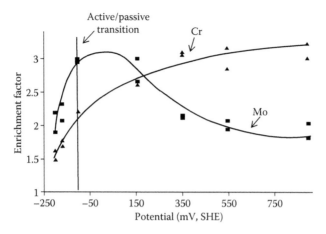

FIGURE 19.14 Enrichment factors for Mo and Cr in an Fe/Cr/Mo alloy, at a series of potentials straddling the active-passive transition, calculated from XPS analysis. (From Castle, J.E., Yang, X.F., Qiu, J.H., and Zhdan, P.A., *Modifications of Passive Films*, Institute of Materials, London, 1994.)

19.5 GASEOUS OXIDATION

19.5.1 BASIC PRINCIPLES

Most alloys remain bright and shiny, even when exposed to dry air for long periods at ambient temperatures. The surfaces are protected, as has been explained already, by about 2 nm of oxide in which the concentrations of the alloying elements are similar to those of the alloy itself. In the case of oxidation in air under ambient conditions, the XPS spectra from chromium and iron in stainless steel look very similar, especially with regard to the backgrounds on which the peaks are superimposed. This observation indicates that the concentrations of the elements in the air-formed oxide are the same as those in the steel—there is no concentration gradient—confirming a prediction that was made in their seminal paper on the early stages of metal oxidation by Cabrera and Mott in 1949 [7]. On heating, the situation begins to change and alloying elements become segregated into layers within the film. For example, after oxidation at 100°C, the metallic ratio in the surface films formed on Cu/Ni alloys began to differ from that of the alloy, Cu becoming enriched in the films on the copper-rich alloys, and Ni enriched on the nickel-rich alloys, as shown in Figure 19.15. After oxidation at 300°C the innermost layer on the alloys consisted of NiO while the outer layer was mainly Cu_2O with some CuO at the outer extremes of the film [50]. The formation of segregated layers on alloys is the norm, and gives rise to a gradual slowing in the rate of oxidation as the ability of the reactants, metal, and oxygen, to interact, is impeded by the oxide. In fact the rate of oxidation is controlled by the diffusion of interstitial ions or vacancies through the crystal structures present in the oxide film, or along any interfaces or boundaries traversing the film. Oxidation is initially fast and so films grow rapidly beyond the point at which the entire film can be "seen" by XPS. Thus XPS is most useful when employed either in examining the very early stages of oxidation, or the interfaces within the oxide. Interfaces can be deleterious, giving easy interlaminar failure and flaking or "spalling" of the oxide layer, or they can be beneficial, acting as a barrier to diffusion.

The most profitable and widespread use of XPS in the field of research into gas-phase oxidation has been associated with the discovery of methods encouraging the formation of beneficial layers on the surfaces of alloys. Such formation could be produced by minor alloying additions [51], ion implantation [52], or particular forms of mechanical working [53]. In many cases, the changes that are induced influence the early stages of oxidation, i.e., when the thin films are amenable to analysis by XPS. Sometimes the changes influence the grain structure or texture of the oxide film and the combination of atomic force microscopy (AFM) and XPS is then particularly effective.

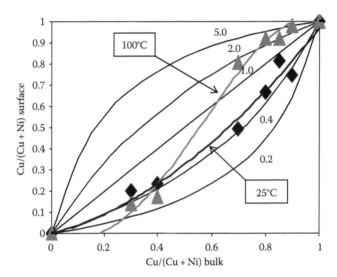

FIGURE 19.15 Oxidation of CuNi alloys in air, studied by XPS. A grid of curves is shown, in which each curve is characterized by a ranking number, i.e., a rank of 1.0 corresponds to a surface composition that is the same as the bulk alloy. Values >1 indicate Cu enrichment and values <1 indicate Ni enrichment (cf. also Figure 19.8, for Cr and Nb). The data for several Cu/Ni alloys show that Ni is enriched in the oxide for all compositions at room temperature but at 100°C copper oxides dominate for compositions >60%. (From Castle, J.E. and Nasserian-Riabi, M., *Corros. Sci.*, 15, 537, 1975.)

The importance of heat treatment *in vacuo* has been described by Brooker et al. [54,55], using AES spectra acquired at temperature as a sample of an 80%Fe20%Cr alloy was heated on a hot stage. The results of their AES measurements are shown in Figure 19.16. At temperatures up to 360°C, the surface oxide contained both iron and chromium ions. As the temperature was increased further, chromium oxide replaced iron completely as a result of a solid-state redox reaction. There were other significant changes; by 640°C the carbon concentration at the surface had decreased to zero as a result of reaction with the chromium oxide and, as the oxide was destroyed, so sulfur segregated

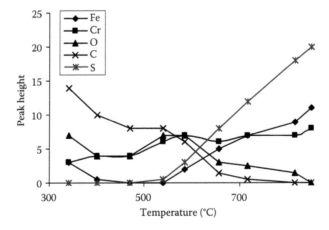

FIGURE 19.16 Changes in the surface composition of a steel heated on the hot-stage of a scanning Auger microscope, as determined by the peaks heights in AES spectra. Sulfur segregates to the surface on heating above 550°C, replacing oxygen and carbon that decrease to zero. (From Brooker, A.D., Castle, J.E., Cohen, J.M., and Waldron, M.B., *Metal Technol.*, 2, 66, 1984; Brooker, A.D., A study by Auger electron and x-ray spectroscopies of vacuum brazing, PhD thesis, University of Surrey, Surrey, U.K., 1986.)

to the surface, becoming very apparent as the temperature reached 700°C. Moreover, as soon as the chromium oxide was lost then iron reappeared in the surface film, giving a final composition that reflected the composition of the bulk alloy. Recently, Greeff et al. [56,57] have examined the competition between sulfur segregation and oxygen adsorption as a function of temperature and oxygen exposure. Their work illustrates how the thermally driven changes in the surfaces of metals will often involve the replacement of an oxide by another that is thermodynamically more stable. Manganese oxide [58] and silicon oxide are two that frequently form the most stable surfaces on steels. The message for the corrosion scientist from this type of study is that vacuum-annealing is not necessarily the best way to start an experiment.

One advantage for those studying high-temperature oxidation of metals is that once quenched to room temperature, the films are stable and unlikely to change. The distribution of elements within the films, and their thickness, can then be studied by AES depth profiling. Hakiki et al. [59] have used depth profiles in the form of the Cr/Fe ratio to show that iron oxide segregates as an outer layer on 304 stainless steel and that the thickness of this layer increases steadily with oxidation temperature. Just how reproducible this method is for measuring the thickness of an oxide was demonstrated in a study by Greyling et al. [60] of the oxidation in air of a W720 maraging steel over the temperature range 300°C–600°C. They used AES depth profiling to measure oxide thicknesses, extending up to 500 nm, after oxidation at 10 chosen temperatures; the measurements at each temperature fell on smooth curves with remarkably little scatter [60]. Their papers [59,60] provide good examples of quantitative depth profiles.

19.5.2 Oxide Spalling

In contrast to the ample literature on the early stages of metal oxidation there is much less dealing with XPS analysis in forensic studies of failure by metal oxidation. Principally this lack is because the oxide formed is likely to be in the thickness range of millimeters rather than nanometers. However, EDX analysis of polished cross-sections enables the distribution of elements to be established with ease. The study of oxide spalling is one area in which XPS can be valuable. Oxide spalling results from the creation of stress in surface layers with differing coefficients of thermal expansion, under the impact of thermal cycling during the normal operation of chemical and power plant. It is damaging as a corrosion process, leading to accelerated metal loss; damaging to the plant, because accumulations of spalled oxide can impede flow; and of concern in the nuclear industry, where it can lead to the transfer of radioactive material to nonactive parts of the circuit.

Spalling reveals weak interfaces or layers of easy shear failure within the oxide structure. Such interfaces result from the accumulation of either voids, caused by mismatch of the diffusion processes taking place in the layers of the oxide, or poorly bonded atoms such as sulfur. Examples may be found in the work of Grabke [61] that again shows differences in the segregation of sulfur and other elements to the interfaces revealed by spalling. Research on methods that might provide increased resistance to spalling has led to the introduction of alloying elements that provide "pegs" in the form of oxide intrusions into the grain boundaries of the metal phase. The study of oxides within the grain boundary is a rare example of the use of polished cross-sections for SAM analysis, since the good spatial resolution, both laterally and in-depth, of that technique, allows improved visibility of the oxide intrusions, relative to that obtainable by EDX.

Examination of the two surfaces exposed at a spalling interface can show if failure is within a single layer, when the same oxide composition will be found on each side; between layers, when a differing oxide composition will be found; or is at the location of a weak boundary layer, in which case a gratuitous element may have accumulated, perhaps at the monolayer level. The easiest opportunity to examine an interface arises when loose oxide can be removed as flakes from the surface by gentle prising or by use of adhesive tape. Other more extreme methods have involved glueing a rigid support to the outer surface and shattering the oxide with a blow, by deformation in a vice, or by thermal shock by immersion in liquid nitrogen. The possibility that the glue might have penetrated the oxide must be kept in mind so that elements derived from the glue can be discounted in the analysis. Variations on these methods include the deposition of metallic films by electroplating conducting oxides, such as Fe_3O_4, or ion-plating at an elevated temperature, whereupon on cooling the differential expansion pulls the interface apart along its weakest line, usually that between the oxide and the metal The latter method has been used to good effect in the study of silicon-containing layers at the oxide/metal interface on certain high alloy steels [62].

Notwithstanding the use of such methods, the removal of material in flakes large enough for XPS cannot always be achieved. In this case the preparation of tapered sections by ball cratering and their examination by AES is useful. Flis et al. [63] have given a good example of the use of point analysis to demonstrate the accumulation of carbon beneath the nitride layer on a plasma-nitrided steel. The series of point analyses were made across the crater obtained by the use of a ball-cratering device.

19.6 LOCALIZED CORROSION

19.6.1 PITTING CORROSION

Researchers in the field of corrosion are likely to choose XPS for studies in which the surface is expected to be uniform in composition over large areas, and in which the main need is information on surface composition and on the valence state of the elements present. By contrast AES will be chosen when it is necessary to examine areas of highly localized corrosion, such as cracks and pits.

Although AES allows the possibility of excellent microscopy, the use of an electron beam as the excitation source brings with it all the problems of electrostatic charging, as found in SEM. Unlike in SEM, however, the charging problem cannot be overcome by coating the surface, which would simply mask the chemical information in the surface layer. AES is thus overwhelmingly a technique for the study of metallic surfaces and the very thin layers of oxides and minerals that form on them.

It is likely that the corrosion scientist intending to use AES will already have some understanding of the sample surface, gained by the use of SEM and of EDX analysis. An idea of the contrast between the information provided by EDX and that arising from use of AES can be derived from Figure 19.17 [64]. Figure 19.17a shows the SEM image of a typical inclusion group in a sample of a stainless steel with a composition similar to that of AISI316. The steel had been exposed to a solution of 0.5 M NaCl, 0.5 M Na_2SO_4, and 0.08% H_2O_2 for 10 s in order to initiate corrosion at the inclusion [65]. The inclusion was formed from MnS that had nucleated on the side of a central oxide particle, as revealed clearly by the distribution of manganese in the x-ray image (Figure 19.17d). The oxide particle contained some manganese but the most intense signal comes from the MnS. Figure 19.17e shows that no Cu was present in the inclusion but some was distributed throughout the steel, albeit at a very low level. Information of this nature is commonplace in the corrosion literature, and the corrosion scientist might be forgiven for believing that that was all that could be learnt about the specimen, but the particular value of AES in corrosion studies is revealed in the scanning Auger microscopy (SAM) images of the same region (Figure 19.17b and c). Those SAM images show that MnS had been replaced in the surface layers by CuS, which formed a protective layer, preventing further hydrolysis of the MnS. The surface film of CuS was found only on the MnS phase, and the Mn present in the oxide inclusion remained unaltered. The SAM images were obtained simultaneously with the x-ray images, using the same electron beam (15 kV) to excite both, and what can be seen here is a comparison of surface information with that from the bulk. An SEM fitted with both x-ray and Auger electron detection becomes a microprobe analyzer operating in two depth zones, the micrometer and the nanometer, and doing so simultaneously. The ability of AES to reveal surface chemical detail that eludes x-ray mapping is what makes the technique so valuable to the corrosion scientist [65].

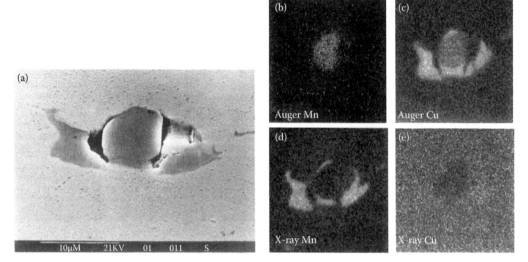

FIGURE 19.17 Pitting attack at an inclusion in a stainless steel. (a) SEM image of corrosion around an inclusion group on the surface. (b) and (c) SAM, and (d) and (e), EDX images, respectively, of the area around the inclusion. The SAM images show how CuS forms on the MnS inclusion. (Micrographs courtesy of M.A. Baker; From Baker, M.A., University of Surrey, Surrey, U.K., private communication.)

19.6.2 IMAGE ACQUISITION

The acquisition of a set of images always requires choices to be made in order to make the most efficient use of the available instrument time. The choices to be made will involve: spatial resolution, i.e., the number of pixels in the image for a given magnification; energy resolution, in particular whether chemical state information is to be resolved; and the signal/noise ratio in the image, determined largely by the dwell time at each pixel. An image of resolution 128 × 128 pixels will contain more than 16,000 analyses per peak and, since the peak intensity must be defined by the intensity at two or more points in the spectrum, the result is 32,000 analyses per element. A minimum description of the surface state in, for example, a study of chloride attack on a manganese sulfide inclusion in a 316 stainless steel, would require analysis of nine elements; Fe, Cr, Ni, Mo, O, C, Mn, S, and Cl. The requirement is thus for some 300,000 separate analyses. In order to obtain a reasonable signal/noise ratio in the image a minimum acquisition time of 10 ms per analysis would be required, which, allowing for the computer overhead incurred in moving the beam, setting the analysis energy and analyzer parameters, and recording the data, adds up, in the present state of the technology, to a minimum total acquisition time of 2 h. Since the intensity relating to a given element is defined by only two sets of counts, peak top and peak bottom, it contains no chemical state information. The acquisition times may seem surprisingly long compared with the time taken for mapping using EDS on a typical SEM. However, EDS is a parallel analysis method, i.e., all energies are acquired simultaneously, whereas an electron spectrometer is a serial analyzer, and the volume analyzed in SAM is so much smaller then in x-ray analysis.

The inherently long acquisition times have important consequences for the use of SAM in corrosion science. An engineer studying the structure of solid-state devices will know exactly where to direct attention in order to confirm the integrity of a manufactured structure, but the corrosion scientist has no initial certainty about where to examine a surface in order to understand what might have led to its degradation. To make efficient use of the expensive instrumentation the work to be carried out needs to be well planned. One area of forensic investigation which is often cost-effective is that of pitting corrosion.

The nucleation of a pit at the site of an inclusion, as shown in Figure 19.17, is a frequent occurrence. The local chemistry depends on the nature of the inclusion and EDX is valuable in identifying

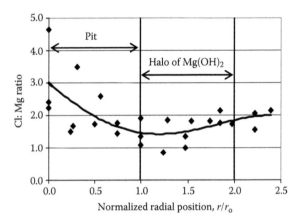

FIGURE 19.18   The Cl/Mg ratio derived from AES line scans across three small pits. The center of the pit has excess Cl and the periphery excess Mg. (From Castle, J.E., *Surf. Interface Anal.*, 9, 345, 1986.)

that; however, SAM is of exceptional value in helping to decide whether or not the pit is electrochemically active [67]. An active pit will have been anodic in the central region and cathodic in the surrounding material. As was discussed in the brief review of electrochemistry, metal enters solution as positive ions at the anode while electrons travel to the adjacent cathodic region and are consumed by reaction with dissolved oxygen to produce OH^- ions. Charge neutrality in the pit is balanced by the accumulation of anions, such as Cl^-, that can be detected by AES in the SAM image: the higher concentration of OH^- ions in the surrounding region is likewise balanced by excess cations, such as Na^+ or Mg^{++}, which will also show up in the SAM map. Once the pit becomes inactive or passivated the potential necessary to maintain these excess concentrations is no longer present and these characteristic signatures of activity fade rapidly. An example of such pit chemistry is given in Figure 19.18. Line scans formed by AES point analysis across three small pits (ca. >3 μm in diameter) formed on Type 316 stainless steel in $MgCl_2$ solution were plotted in terms of the Cl/Mg ratio. The merged data from all three pits are given in Figure 19.18, plotted against a normalized distance from the center of the pit. At the center the Cl concentration is greatest. Around the periphery of the pit the ratio reaches a minimum value and it is in this, cathodic region, that a halo of $Mg(OH)_2$ in often seen [68].

It has to be said that the complete separation of anodic and cathodic sites on a micrometer scale is mainly a laboratory exercise, and is difficult to observe in a forensic investigation, but chloride accumulations within pits can be observed, as can halos of insoluble magnesium hydroxide on pits formed in seawater. Each of these have their origin in the electropotentials associated with pitting attack and are a strong pointer to active pits. In some cases pits are covered with a cap of corrosion product formed as the cation effluent from the pit meets the surrounding OH-rich medium [69]. These should be removed before SAM analysis in order that the pit composition can be revealed.

When carrying out corrosion research, and when pits as obvious as that seen in Figure 19.17 are absent, a systematic approach is needed to identify sites and areas of electrochemical activity on the surface. As has already been discussed, immersion of a metal in an electrolyte frequently leads to electrochemical activity and to the establishment of anodic and cathodic sites on the surface. The surface chemistry, i.e., the ions adsorbed and the material deposited as a result of corrosion, along with the rapid changes caused by the differences in the electropotential, can be studied by SAM. Ions or material found on a particular area of the surface must be associated with either (a) current flow through, or (b) a particular electropotential on, that area of the surface. The current flow or potential difference then implies electrochemical activity and hence the sites that were active at the time of removal from the electrolyte can be established. The circumstances which can lead to ions being segregated within the surface layer have been discussed in a review by Castle [65]. As an example, consider the change in the pH within the boundary layer as a result of the electrochemical current density at any local cathodes on the surface. The change in concentration of OH⁻ ions may lead to the solubility product of some metal hydroxides being exceeded, leading to the formation of a deposit, which then decorates the cathodic areas. This is such a useful device for the investigation of surface electropotentials, and the associated current flows, that it is often useful to add marker ions, such as Mg^{2+}, to the test electrolyte. Magnesium ions will, at a critical concentration, cause the precipitation of $Mg(OH)_2$, which is readily detectable by AES in the SAM image [23]. Having bracketed, with SAM, the magnesium concentration and pH value at which a deposit appears, the corresponding current density at which the cathode was operating can be determined readily [66–68].

At an anode, deposition of a salt depends on the concentration of metal ions and thus on the anodic current flowing through the exposed surface. This will give rise, for example, to the chloride deposits that are readily observable by AES and have been shown in SAM maps by Daud [67], Castle [68], and Baker and Castle [69,70]. Chlorides, however, are mostly soluble and therefore likely to be found in occluded cells hidden from the limited depthwise view of the electron spectrometer. Moreover, the metal ions within the pit cannot be probed by varying the concentration of chloride ions in the electrolyte; the link with the concentration in the pit is too indirect. Complementary studies may instead be made using effects that depend on local potential rather than on current density. In this case the deposition of a marker on the surface can be achieved by exploiting a redox reaction. Typical of such reactions is the reduction of Cu ions to Cu metal at the cathode. Marker deposition can be achieved using any ions with an insoluble product produced as a result of a redox reaction within the required potential range. Obviously, the ions chosen should not be the same as any forming the substrate composition, nor should they stimulate corrosion.

The presence of potential differences on the surface can lead to species transport recognizable in an Auger map. Comparison of the microscale chemistry determined by SAM with that found on electrodes held at known potentials enabled Castle and Ke to estimate the potential differences set up at the site of an incipient pit [71]. Figure 19.19 from their work illustrates how sulfur generated

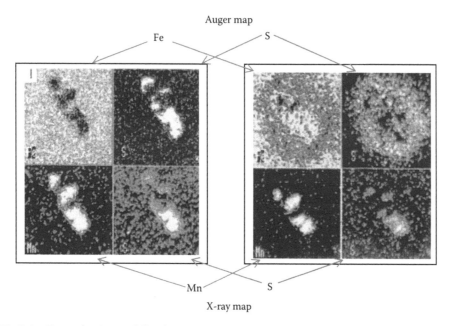

FIGURE 19.19 (See color insert following page 396.) Study by SAM and EDX of corrosion at an inclusion, showing the separation of anodic and cathodic sites. The left-hand maps show the surface before exposure, and the right-hand ones after exposure to a corrosive environment for 35 min. Changes in surface chemistry can be recognized in the Auger maps but not in the x-ray maps. (From Castle, J.E. and Ke, R., *Corros. Sci.*, 30, 409, 1990.)

by hydrolytic attack on a sulfide inclusion is captured by the cathodic annulus surrounding the anode associated with the inclusion.

19.6.3 SPECIAL TYPES OF LOCALIZED CORROSION

On occasions an analyst will be asked to assist in studies of rather specialized forms of localized corrosion. A few notes are given below on three of the more usual of these forms, but in general the particular analysis approach used will need to be worked out by discussion with the originator of the investigation.

19.6.3.1 Stress Corrosion Cracking

Stress corrosion, as the name suggests, arises from the conjoint influences of tensile stress and a corrosive environment. Usually the environment is not so corrosive as to lead to general corrosion and the metal surface may appear passivated over the greater part. Typically, all the corrosive activity is associated with a crack tip, with the major part of the cracked interface becoming rapidly repassivated. It differs from embrittlement cracking in that the ions causing the damage are derived from the environment in which the metal is immersed and not from segregation within the metal itself. The volume of solution in the active region is miniscule and disperses rapidly once the aggravating condition is removed by relaxing the applied stress. Auger analysis of the crack tip is usually undertaken by prepreparing test pieces to fit a fracture stage in an SAM, and then using the stage to open the crack fully in an ultra-high vacuum (UHV) environment. Comparison of the film on the repassivated surface with that within the active region may lead to an understanding of the local mechanism, but being able to maintain the characteristics of the active tip until it is fixed in place by exposure in UHV is difficult. Freezing the test piece in liquid nitrogen immediately on withdrawal from the test environment can help.

19.6.3.2 Filiform Corrosion

Filiform attack is recognizable from the network of threadlike tunnels produced just beneath the surface of a paint or an oxide film. The head of the attack is the anode of an electrochemical system in which the tail behind the head is the cathode. The problem for the surface analyst is that of exposing the head and tail for analysis before the potential that hold the key ions in place has dissipated.

19.6.3.3 Breakaway Oxidation

Localized corrosion in a surrounding gas phase has many colloquially descriptive names but a generic title would be breakaway corrosion. The characteristic feature is a runaway rate of attack that is locally much greater than the norm predicted by the rate laws. An example is that of the behavior of low-grade steels in the CO_2 of a first-generation nuclear power plant, in which the slow accumulation of carbon close to the oxide/metal interface eventually interferes with the formation of a reasonably protective oxide. Again there is a need for strong interaction between analyst and investigator for the effective use of SAM in mechanistic studies.

19.7 CONCLUDING COMMENTS

The key message that this chapter has for the electron spectroscopist is that the interpretation based on his or her analysis can be much enhanced by a basic understanding of the factors that fit the most likely surface chemistry. In the aqueous phase these will be, for a given alloy, the pH and electropotential plus the presence or otherwise of specific ions such as chloride. The Pourbaix diagram provides a useful guide to what might be expected in a surface film. In the gas phase the order in which oxides appear in the layers exposed by the depth profile will be determined by their free energy of oxidation. The Ellingham diagram provides a guide to this. Often it will be departures from the expected that provide the investigator with the mechanistic clues to the corrosion mechanism. The best outcome usually arises from close cooperation between analyst and corrosion scientist; corrosion is an ongoing, active, process and the discontinuous snapshots provided by XPS or AES do not allow the whole story to be revealed. Samples connected by a sequential series of electropotentials, exposure time, pH range, or other sets of consistent factors are likely to give a successful outcome. However chance observations can often be useful—magnesium hydroxide deposits that show the location of cathodic areas on a metal surface are a case in point. In this context it is useful to have as full a history of the sample or test piece as possible.

REFERENCES

1. T. B. Grimley, Continuous oxide films, in *Corrosion, Metal/Environment Reactions*, Vol. 1, 3rd edition (Eds. L. L. Shreir, R. A. Jarman, and G. T. Burstein), Butterworth, Heineman, Oxford, Ch. 1.9, pp. 1.254–1.267 (1994).
2. J. E. Castle, Discontinuous oxide films, in *Corrosion, Metal/Environment Reactions*, Vol. 1, 3rd edition (Eds. L. L. Shreir, R. A. Jarman, and G. T. Burstein), Butterworth, Heineman, Oxford, Ch. 1.10, pp. 1.268–1.292 (1994).
3. B. Elsener and A. Rossi, *Electrochim. Acta*, **37**, 2269 (1992).
4. F. D. Richardson and J. H. E. Jeffes, *J. Iron Steel Inst.*, **160**, 261 (1948).
5. http://www.engr.sjsu.edu/ellingham/ellingham_tool_p1.php.
6. J. E. Castle, *J. Adhes.*, **84**, 368 (2008).
7. N. Cabrera and N. F. Mott, *Rep. Prog. Phys.*, **12**, 163 (1948–1949).
8. S. Tougaard, *Appl. Surf. Sci.*, **100/101**, 1 (1996).
9. J. E. Castle and A. M. Salvi, *J. Vac. Sci. Technol.*, **A19**, 1170 (2001).
10. J. E. Castle, H. Chapman-Kpodo, A. Proctor, and A. M. Salvi, *J. Electron Spectrosc. Relat. Phenom.*, **106**, 65 (1999).
11. A. Turnbull, M. Ryan, A. Willetts, and S. Zhou, *Corros. Sci.*, **45**, 1051 (2003).
12. M. Pourbaix, *Atlas of Electrochemical Equilibria in Aqueous Solutions*, Pergamon, New York (1966).
13. C. Kato, B. G. Ateya, J. E. Castle, and H. W. Pickering, *J. Electrochem. Soc.*, **127**, 1897 (1980).
14. I. Olefjord and L. Wegrelius, *Corros. Sci.*, **31**, 89 (1990).
15. C. R. Clayton and Y. C. Lu, *J. Electrochem. Soc.*, **133**, 2465 (1986).
16. S. Mischler, H. J. Mathieu, and D. Landolt, *Surf. Interface Anal.*, **11**, 182 (1988).
17. P. Marcus and I. Olefjord, *Corros. Sci.*, **28**, 589 (1988).
18. S. Haquipt, C. Calinski, H. W. Hoppe, H. D. Speckmann, and H. H. Stehblow, *Surf. Interface Anal.*, **9**, 357 (1986).
19. J. E. Castle, M. R. Guascito, A. M. Salvi, and F. Decker, *Surf. Interface Anal.*, **34**, 619 (2002).
20. http://corrosion.ksc.nasa.gov/dealloying.htm.
21. G. Hultqvist and H. Hero, *Corros. Sci.*, **24**, 789 (1984).
22. J. E. Castle and R. T. Tremaine, *Surf. Interface Anal.*, **1**, 49 (1979).
23. A. M. Beccaria and P. Traverso, in *ECASIA 97* (Eds. I. Olefjord. L. Nyborg, and D. Briggs), Wiley, Chichester, p. 285 (1997).
24. D. C. Epler and J. E. Castle, *Corrosion*, **35**, 451 (1979).
25. J. E. Castle, *J. Vac. Sci. Technol. A*, **25**, 1 (2007).
26. C. Smith, *J. Elec. Spectrosc. Relat. Phenom.*, **148**, 21 (2005).
27. J. E. Castle and J. H. Qiu, *Corros, Sci.*, **29**, 591 (1989).
28. K. Asami, K. Hashimoto, and S. Shimodaira, *Corros. Sci.*, **18**, 151 (1978).
29. J. E. Castle and K. Asami, *Surf. Interface Anal.*, **36**, 220 (2004).
30. I. Olefjord, P. Marcus, H. J. Mathieu, and S. Hofmann, in *ECASIA 95* (Eds. H. J. Mathieu, B. Reihl, and D. Briggs), Wiley, Chichester, p. 188 (1985).
31. M. P. Seah, *Surf. Interface Anal.*, **37**, 300 (2005).
32. C. J. Powell and A. Jablonski, NIST Electron Effective-Attenuation-Length Database, NIST Standard Reference Database 82, http://nist.gov/srd/surface.htm.
33. S. Tougaard, *Surf. Interface Anal.*, **26**, 249 (1998).
34. E. Unveren, E. Kemnitz, S. Hutton, A. Lippitz, and W. E. S. Unger, *Surf. Interface Anal.*, **36**, 92 (2004).

35. H.-H. Strehblow, *Surf. Interface Anal.*, **12**, 363 (1988).
36. K. Asami, K. Hashimoto, and S. Shimodaira, *Corros. Sci.*, **16**, 387 (1976).
37. J. H. Gerretsen, J. H. W. de Wit, and J. C. Rivière, *Corros. Sci.*, **31**, 545 (1990).
38. J. E. Castle and C. R. Clayton, *Corros. Sci.*, **17**, 7 (1977).
39. A. Schneider, D. Kuron, S. Hofmann, and R. Kircheim, *Corros. Sci.*, **31**, 191 (1990).
40. P. Marcus and I. Olefjord, *Corros. Sci.*, **28**, 589 (1988).
41. C. Kato, B. G. Ateya, J. E. Castle, and H. W. Pickering, *J Electrochem. Soc.*, **127**, 1890 (1980).
42. T. B. Whitfield, J. E. Castle, C. Saracco, and M. Ali, *Surf. Interface Anal.*, **34**, 176 (2002).
43. J. E. Castle, D. C. Epler, and D. B. Peplow, *Corros. Sci.*, **16**, 145 (1976).
44. T. Kosec, I. Miloševand, and B. Pihlar, *Appl. Surf. Sci.*, **253**, 8863 (2007).
45. C.-O. A. Olsson, P. Agarwal, M. Frey, and D. Landolt, *Corros. Sci.*, **42**, 197 (2000).
46. C. D. Wagner and A. Joshi, *J. Electron Spectrosc. Relat. Phenom.*, **47**, 283 (1988).
47. B. L. Dooley, J. E. Castle, and P. A. Zhdan, *Power Plant Chem.*, **7**, 561 (2005).
48. J. E. Castle, *Nature Physical Sci.*, **234**, 93 (1971).
49. J. E. Castle, X. F. Yang, J. H. Qiu, and P. A. Zhdan, in *Modifications of Passive Films* (Eds. P. Marcus, B. Baroux, and M. Keddam), Institute of Materials, London, pp. 70–75 (1994).
50. J. E. Castle and M. Nasserian-Riabi, *Corros. Sci.*, **15**, 537 (1975).
51. C. Maffiotte, M. Navas, M. L. Costano, and A. M. Lancha, *Surf. Interface Anal.*, **30**, 161 (2000).
52. A. Gutierrez, M. F. Lopez, F. J. Perez Trujillo, M. P. Hierro, and F. Pedraza, *Surf. Interface Anal.*, **30**, 130 (2000).
53. E. M. Muller-Lorenz, H. J. Grabbke, B. Eltester, and M. Lucas, in *ECASIA 97* (Eds. I. Olefjord. L. Nyborg, and D. Briggs), Wiley, Chichester, p. 341 (1997).
54. A. D. Brooker, J. E. Castle, J. M. Cohen, and M. B. Waldron, *Metal Technol.*, **2**, 66 (1984).
55. A. D. Brooker, A study by auger electron and x-ray spectroscopies of vacuum brazing, PhD thesis, The Library University of Surrey, Surrey, U.K., (1986).
56. A. P. Greeff, C. W. Louw, J. J. Terblans, and H. C. Swart, *Corros. Sci.*, **42**, 991 (2000).
57. A. P. Greeff, C. W. Louw, and H. C. Swart, *Corros. Sci.*, **42**, 1725 (2000).
58. J. E. Castle and M. J. Durbin, *Carbon*, **13**, 23 (1975).
59. N. E. Hakiki, M. F. Montemor, M. G. S. Ferreira, and M. da Cunha Bela, *Corros. Sci.*, **42**, 687 (2000).
60. C. J. Greyling, I. A. Kotze, and P. E. Viljoen, *Surf. Interface Anal.*, **16**, 293 (1990).
61. H. J. Grabke, *Surf. Interface Anal.*, **30**, 112 (2000).
62. J. P. Coad, G. Tappin, and J. C. Rivière, *Surf. Sci.*, **117**, 629 (1982).
63. J. Flis, J. Mankowski, and T. Zakroczmski, *Corros. Sci.*, **42**, 313 (2000).
64. M. A. Baker, University of Surrey, private communication.
65. J. E. Castle, *Proc. 12th Int. Corr. Congress*, **5B**, NACE International, 3982 (1993).
66. J. E. Castle, L. Sun, and H. Yan, *Corr. Sci.*, **21**, 1093 (1994).
67. R. Daud, PhD thesis, University of Surrey (1985).
68. J. E. Castle, *Surf. Interface Anal.*, **9**, 345 (1986).
69. M. A. Baker and J. E. Castle, *Corros. Sci.*, **34**, 667 (1993).
70. M. A. Baker and J. E. Castle, *Corros. Sci.*, **33**, 1295 (1992).
71. J. E. Castle and R. Ke, *Corros. Sci.*, **30**, 409 (1990).

Index